# GEOLOGICAL ENGINEERING

Cover illustration by Luis I. González de Vallejo

# GEOLOGICAL ENGINEERING

**Luis I. González de Vallejo**
*Universidad Complutense de Madrid*

**Mercedes Ferrer**
*Instituto Geológico y Minero de España*

with a Foreword by M.H. de Freitas
*Imperial College, London*

CRC Press
Taylor & Francis Group
Boca Raton London New York Leiden

CRC Press is an imprint of the
Taylor & Francis Group, an **informa** business

A BALKEMA BOOK

*CRC Press/Balkema is an imprint of the Taylor & Francis Group, an informa business*

© 2011 Taylor & Francis Group, London, UK

Typeset by Vikatan Publishing Solutions (P) Ltd, Chennai, India

Printed and bound in Poland by Poligrafia Janusz Nowak, Poznán

Authorized translation from the Spanish language edition, entitled INGENIERÍA GEOLÓGICA by GONZÁLEZ DE VALLEJO, LUIS, published by Pearson Educación, S.A. Copyright © Pearson Educación, S.A., 2002.

English translation by Bill Newton, Pauline Moran and Valerie Stacey from Gabinete Lingüístico of the Fundación General de la Universidad Complutense de Madrid.

English technical review by M.H. de Freitas, Imperial College, London.

All rights reserved. No part of this publication or the information contained herein may be reproduced, stored in a retrieval system, or transmitted in any form or by any means, electronic, mechanical, by photocopying, recording or otherwise, without prior permission in writing from the publisher. Innovations reported here may not be used without the approval of the authors.

Although all care is taken to ensure integrity and the quality of this publication and the information herein, no responsibility is assumed by the publishers nor the author for any damage to the property or persons as a result of operation or use of this publication and/or the information contained herein.

Published by: CRC Press/Balkema
P.O. Box 447, 2300 AK Leiden, The Netherlands
e-mail: Pub.NL@taylorandfrancis.com
www.crcpress.com – www.taylorandfrancis.co.uk – www.balkema.nl

*Library of Congress Cataloging-in-Publication Data*

*Applied for*

ISBN: 978-0-415-41352-7 (Hbk)

# BRIEF CONTENTS

| | |
|---|---|
| ABOUT THE AUTHORS | xv |
| CONTRIBUTORS | xvii |
| FOREWORD | xix |
| PREFACE | xxi |

## PART I – FUNDAMENTALS

| | |
|---|---|
| 1 INTRODUCTION TO GEOLOGICAL ENGINEERING | 3 |
| 2 SOIL MECHANICS AND ENGINEERING GEOLOGY OF SEDIMENTS | 19 |
| 3 ROCK MECHANICS | 109 |
| 4 HYDROGEOLOGY | 223 |

## PART II – METHODS

| | |
|---|---|
| 5 SITE INVESTIGATION | 263 |
| 6 ROCK MASS DESCRIPTION AND CHARACTERISATION | 327 |
| 7 ENGINEERING GEOLOGICAL MAPPING | 351 |

## PART III – APPLICATIONS

| | |
|---|---|
| 8 FOUNDATIONS | 369 |
| 9 SLOPES | 401 |
| 10 TUNNELS | 451 |
| 11 DAMS AND RESERVOIRS | 501 |
| 12 EARTH STRUCTURES | 535 |

## PART IV – GEOLOGICAL HAZARDS

| | |
|---|---|
| 13 LANDSLIDES AND OTHER MASS MOVEMENTS | 555 |
| 14 SEISMIC HAZARD | 595 |
| 15 PREVENTION OF GEOLOGICAL HAZARDS | 625 |

| | | |
|---|---|---|
| **APPENDIX A** | **CHARTS FOR CIRCULAR AND WEDGE FAILURE ANALYSIS** | **643** |
| **APPENDIX B** | **PRESSURE UNITS CONVERSION CHART** | **653** |
| **APPENDIX C** | **SYMBOLS AND ACRONYMS** | **657** |
| **APPENDIX D** | **LIST OF BOXES** | **663** |
| **APPENDIX E** | **PERMISSIONS TO REPRODUCE FIGURES AND TABLES** | **665** |
| **INDEX** | | **671** |

# CONTENTS

ABOUT THE AUTHORS — xv

CONTRIBUTORS — xvii

FOREWORD — xix

PREFACE — xxi

## PART I – FUNDAMENTALS

**1 INTRODUCTION TO GEOLOGICAL ENGINEERING** — 3

1.1 DEFINITION AND IMPORTANCE OF GEOLOGICAL ENGINEERING — 4

1.2 THE GEOLOGICAL ENVIRONMENT AND ITS RELATION WITH ENGINEERING — 6

1.3 GEOLOGICAL FACTORS AND GEOTECHNICAL PROBLEMS — 8

1.4 METHODS AND APPLICATIONS IN GEOLOGICAL ENGINEERING — 15

1.5 INFORMATION SOURCES IN ENGINEERING GEOLOGY — 16

1.6 HOW THIS BOOK IS STRUCTURED — 16

RECOMMENDED READING — 17

REFERENCES — 17

**2 SOIL MECHANICS AND ENGINEERING GEOLOGY OF SEDIMENTS** — 19

2.1 INTRODUCTION — 20
 The nature of soils — 20
 Soils in geotechnical engineering — 20

2.2 SOIL DESCRIPTION AND CLASSIFICATION. PHASE RELATIONSHIPS — 23
 Types of soils — 23
 Particle size distribution — 23
 Plasticity — 24
 Phase relationships — 26

2.3 FLOW OF WATER THROUGH SOILS — 28
 Total head. Bernoulli's Theorem — 29
 Hydrostatic conditions — 29
 Ground water flow — 30
  Basic concepts. Head loss and permeability — 30
  Hydraulic head and hydraulic gradient — 31
  Darcy's law — 31
 Steady flow in an isotropic medium — 33
 Anisotropic soil conditions — 36
 Permeability and water flow in stratified soils — 38

2.4 EFFECTIVE STRESS — 40
 Soil phases and soil structure — 40
 Saturated soils. The principle of effective stress — 41
 Seepage forces and piping — 44
 Loading saturated soils — 50
  The concept of consolidation — 50
  Concepts of loading with and without drainage — 51
  Undrained loading in saturated soils — 52

2.5 CONSOLIDATION AND COMPRESSIBILITY — 56
 Normally consolidated and over-consolidated soils — 56
 Horizontal stresses in the ground — 62
 Influence of complementary factors on soil behaviour — 63
 The oedometer test — 65

2.6 SHEAR STRENGTH OF SOILS — 71
 Failure criterion — 71
 The direct shear test — 72
 Behaviour of soils subjected to shear stress — 76
  Granular soils — 76
  Clay soils — 78
 The triaxial test — 79
  The test apparatus — 79

|     |     | Types of test | 81 |
| --- | --- | --- | --- |
|     |     | The uniaxial compression test | 85 |
| 2.7 | INFLUENCE OF MINERALOGY AND FABRIC ON THE GEOTECHNICAL PROPERTIES OF SOILS | | 85 |
|     | Clay minerals in engineering geology | | 86 |
|     | Physico-chemical properties | | 88 |
|     | Geotechnical properties and mineralogical composition | | 89 |
|     | Microfabric of clayey soils | | 89 |
|     | Geotechnical properties and microfabric | | 93 |
|     | Summary | | 94 |
| 2.8 | ENGINEERING GEOLOGICAL CHARACTERISTICS OF SEDIMENTS | | 94 |
|     | Colluvial deposits | | 95 |
|     | Alluvial deposits | | 95 |
|     | Lacustrine deposits | | 95 |
|     | Coastal deposits | | 95 |
|     | Glacial deposits | | 96 |
|     | Deserts and arid climate deposits | | 97 |
|     | Evaporitic deposits | | 98 |
|     | Tropical soils | | 98 |
|     | Volcanic soils | | 99 |
| 2.9 | PROBLEMATIC SOILS | | 100 |
|     | Swelling and shrinking clays | | 101 |
|     | Dispersive soils | | 103 |
|     | Saline and aggressive soils | | 104 |
|     | Collapsible soils | | 104 |
|     | The action of ice and permafrost | | 106 |
|     | Soft sensitive soils | | 106 |
|     | Soils susceptible to liquefaction | | 106 |
| RECOMMENDED READING | | | 107 |
| REFERENCES | | | 107 |
| **3 ROCK MECHANICS** | | | **109** |
| 3.1 | INTRODUCTION | | 110 |
|     | Definition, objectives and scope | | 110 |
|     | Rock and soil | | 112 |
|     | Rock masses | | 113 |
| 3.2 | PHYSICAL AND MECHANICAL PROPERTIES OF ROCKS | | 116 |
|     | Rock characteristics | | 116 |
|     | Physical properties of intact rock | | 118 |
|     | Rock classification for geotechnical purposes | | 122 |
|     | Rock mass classification | | 124 |
|     | Weathering of rock | | 125 |

|     |     | Weathering processes | 125 |
| --- | --- | --- | --- |
|     |     | Weathering of intact rock | 126 |
|     |     | Weathering of rock masses | 127 |
|     | Groundwater | | 129 |
|     |     | Permeability and water flow | 129 |
|     |     | Effects of water on the properties of rock masses | 129 |
| 3.3 | STRESS AND STRAIN IN ROCKS | | 131 |
|     | Force and stress | | 131 |
|     | Stress on a plane | | 132 |
|     | Stress in three dimensions | | 138 |
|     | Strength and failure | | 139 |
|     |     | Basic concepts | 139 |
|     |     | Failure mechanisms | 140 |
|     | Stress-strain relationships in rock | | 141 |
|     | Strength criteria | | 144 |
| 3.4 | STRENGTH AND DEFORMABILITY OF INTACT ROCK | | 147 |
|     | Strength and strength parameters | | 147 |
|     |     | Effects of anisotropy and pore pressure on strength | 147 |
|     | Failure criteria | | 149 |
|     |     | Mohr-Coulomb criterion | 149 |
|     |     | Hoek-Brown's criterion | 150 |
|     | Deformability | | 150 |
|     | Strength and deformability laboratory tests | | 154 |
|     |     | Uniaxial compression test | 154 |
|     |     | Triaxial compression test | 159 |
|     |     | Tensile strength tests | 162 |
|     |     | Sonic velocity | 164 |
|     |     | Limitations of laboratory tests | 164 |
| 3.5 | DISCONTINUITIES | | 165 |
|     | Influence on rock mass behaviour | | 165 |
|     | Types of discontinuities | | 166 |
|     | Characteristics of discontinuities | | 168 |
|     | Shear strength of discontinuity planes | | 170 |
|     |     | Barton and Choubey criterion | 172 |
|     |     | Discontinuities with infilling | 175 |
|     |     | Direct shear strength laboratory test | 175 |
|     | Permeability and water pressure | | 177 |
| 3.6 | STRENGTH AND DEFORMABILITY OF ROCK MASSES | | 179 |
|     | Rock mass strength | | 179 |
|     |     | Failure criteria for isotropic rock masses | 181 |
|     |     | Failure criteria for anisotropic rock masses | 186 |

|  |  |  |
|---|---|---|
| Summary | 187 |  |
| Rock mass deformability | 187 |  |
| *In situ* deformability tests | 188 |  |
| Geophysical methods | 188 |  |
| Empirical correlations | 189 |  |
| Permeability and water pressure | 193 |  |
| Scale effect | 195 |  |

3.7 *IN SITU* STRESS ... 201
    Origin and types of *in situ* stress ... 201
    Geological and morphological factors which influence the state of stress ... 203
    Methods for measuring *in situ* stress ... 205
        Measuring the direction of stresses by geological methods ... 206
        Estimating stress magnitude from empirical relationships ... 207
        Instrumental methods for measuring orientation and magnitude of stress ... 207

3.8 ROCK MASS CLASSIFICATIONS ... 215
    RMR Classification ... 216
    Geomechanical classifications in practice ... 216

RECOMMENDED READING ... 220

REFERENCES ... 221

# 4 HYDROGEOLOGY ... 223

4.1 HYDROGEOLOGICAL BEHAVIOUR OF SOILS AND ROCKS ... 224
    Types of aquifers and their behaviour ... 224
    Piezometric level ... 227
    Water movement in aquifers ... 228

4.2 HYDROGEOLOGICAL PARAMETERS ... 230
    Porosity ... 230
    Storage coefficient ... 231
    Permeability ... 232
    Transmissivity ... 233

4.3 FLOW. DARCY'S LAW AND FUNDAMENTAL FLOW EQUATIONS IN POROUS MEDIA ... 233
    Darcy's law ... 233
    Darcy's velocity and real velocity ... 234
    Generalization of Darcy's law ... 235
    Continuity equation for steady flow ... 236
    Laplace equation ... 236
    Poisson's equation ... 237
    Flow equation in transitory regime ... 237

4.4 EVALUATION METHODS FOR HYDROGEOLOGICAL PARAMETERS ... 238
    Pumping tests ... 238
    Injection tests ... 248
    Tracer tests ... 249

4.5 SOLUTION METHODS ... 251
    Analytical methods ... 251
    Flow nets ... 252
    Numerical methods ... 253

4.6 CHEMICAL PROPERTIES OF WATER ... 255
    Chemical quality of groundwater ... 255
    Physical-chemical processes. Water-aquifer interaction ... 256
    Contamination of groundwater ... 257
    Anthropogenic activities ... 257
    Mechanisms of ground water contamination ... 258

RECOMMENDED READING AND REFERENCES ... 259

# PART II – METHODS

# 5 SITE INVESTIGATION ... 263

5.1 PLANNING AND DESIGN ... 264
    Aims and importance ... 264
    Planning site investigations ... 264

5.2 PRELIMINARY INVESTIGATIONS ... 268
    Desk-based study ... 268
    Aerial photo and remote sensing interpretation ... 269
        Aerial photo interpretation ... 269
        Remote sensing ... 270
    The walk-over survey ... 273
    Preliminary site investigation report ... 275

5.3 ENGINEERING GEOPHYSICS ... 275
    Surface geophysics ... 276
        Electrical methods ... 276
        Seismic methods ... 277
        Electromagnetic methods ... 282
        Gravity methods ... 285
        Magnetic methods ... 285
    Borehole geophysics ... 286
        Geophysical logging ... 286
        Seismic logging inside boreholes ... 287
        Seismic tomography ... 288

5.4 BOREHOLES, TRIAL PITS, TRENCHES AND SAMPLING ... 289
    Borehole drilling ... 289
        Rotary drilling ... 289
        Auger drilling ... 291
        Percussion drilling ... 292

|  |  | Special boreholes | 293 |
|---|---|---|---|
|  |  | Number and depth of boreholes | 293 |
|  |  | Borehole data presentation | 293 |
|  |  | Trial excavations | 293 |
|  |  | Geotechnical sampling | 294 |
|  |  | Borehole logging | 297 |
|  | 5.5 | *IN SITU* TESTS | 301 |
|  |  | Standard penetration test (SPT) | 301 |
|  |  | Probing penetrometers | 302 |
|  |  | Cone penetration test (CPT) | 303 |
|  |  | Field vane test | 305 |
|  |  | Schmidt hammer test | 305 |
|  |  | Point load test | 306 |
|  |  | Shear strength test on discontinuities | 308 |
|  |  | Tilt test | 310 |
|  |  | Pressuremeter test | 311 |
|  |  | Plate loading test on soils | 311 |
|  |  | Dilatometer test | 312 |
|  |  | Plate loading test on rock | 313 |
|  |  | Flat jack test | 313 |
|  |  | Seismic methods | 316 |
|  |  | Measuring *in situ* stress | 316 |
|  |  | Permeability tests | 316 |
|  |  | Permeability tests on soils | 316 |
|  |  | Permeability tests on rock | 317 |
|  | 5.6 | GEOTECHNICAL INSTRUMENTATION | 319 |
|  |  | Displacement measurements | 319 |
|  |  | Pore pressure and water level measurements | 322 |
|  |  | Stress measurements | 324 |
|  | RECOMMENDED READING | | 325 |
|  | REFERENCES | | 325 |
| 6 | ROCK MASS DESCRIPTION AND CHARACTERISATION | | 327 |
|  | 6.1 | METHODOLOGY | 328 |
|  | 6.2 | DESCRIPTION AND ZONING | 331 |
|  | 6.3 | INTACT ROCK CHARACTERISATION | 331 |
|  |  | Identification | 332 |
|  |  | Weathering | 332 |
|  |  | Strength | 332 |
|  | 6.4 | DESCRIPTION OF DISCONTINUITIES | 335 |
|  |  | Orientation | 335 |
|  |  | Spacing | 336 |
|  |  | Persistence | 337 |
|  |  | Roughness | 338 |
|  |  | Strength of discontinuity wall | 340 |
|  |  | Aperture | 341 |
|  |  | Filling | 342 |
|  |  | Seepage | 343 |
|  | 6.5 | ROCK MASS PARAMETERS | 343 |
|  |  | Number and orientation of discontinuity sets | 344 |
|  |  | Block size and fracture degree | 344 |
|  |  | Degree of weathering | 347 |
|  | 6.6 | ROCK MASS CLASSIFICATION AND CHARACTERISATION | 349 |
|  | RECOMMENDED READING | | 349 |
|  | REFERENCES | | 350 |
| 7 | ENGINEERING GEOLOGICAL MAPPING | | 351 |
|  | 7.1 | DEFINITION | 352 |
|  | 7.2 | TYPES OF MAPS | 352 |
|  |  | Classification | 352 |
|  |  | Content of engineering geological maps | 354 |
|  |  | Classification and geotechnical properties of soils and rocks | 354 |
|  |  | Hydrogeological conditions | 357 |
|  |  | Geomorphological conditions | 357 |
|  |  | Geodynamic processes | 357 |
|  | 7.3 | MAPPING METHODS | 358 |
|  |  | Geotechnical zoning | 358 |
|  |  | Representing data | 358 |
|  |  | Computer aided mapping | 360 |
|  |  | Geotechnical cross-sections | 360 |
|  | 7.4 | DATA COLLECTION | 360 |
|  | 7.5 | APPLICATIONS | 361 |
|  |  | Land and urban planning | 361 |
|  |  | Engineering | 361 |
|  | RECOMMENDED READING | | 365 |
|  | REFERENCES | | 365 |

# PART III – APPLICATIONS

| 8 | FOUNDATIONS | | 369 |
|---|---|---|---|
|  | 8.1 | INTRODUCTION | 370 |
|  |  | Basic design criteria | 370 |
|  |  | Stages in foundation design | 371 |
|  | 8.2 | SHALLOW FOUNDATIONS | 371 |
|  |  | Types of shallow foundations | 371 |
|  |  | Ultimate bearing capacity | 372 |
|  |  | Basic definitions | 372 |
|  |  | Calculating the ultimate bearing capacity | 373 |

|  | Ultimate bearing capacity in undrained conditions | 374 |
|  | Ultimate bearing capacity in drained conditions | 375 |
|  | Factor of safety. Safe bearing capacity | 375 |
|  | Distribution of pressures under shallow foundations | 376 |
|  | Stress distribution under loaded areas | 378 |
|  |    Fundamentals. Criteria for use | 378 |
|  |    Point load on an elastic half-space | 379 |
|  |    Vertical stresses under the corner of a uniformly loaded rectangle | 379 |
|  |    Stresses under a uniformly loaded circular area | 380 |
|  | Settlement in soils | 382 |
|  |    General considerations | 382 |
|  |    Immediate and consolidation settlement | 382 |
|  |    Immediate and primary consolidation settlements in saturated clays | 383 |
|  |    Settlements in granular soils | 384 |
|  |    Settlements in stiff clays | 384 |
| 8.3 | DEEP FOUNDATIONS | 385 |
|  | Types of pile | 386 |
|  | Single piles | 387 |
|  | Ultimate load capacity of a pile | 389 |
|  | Pile groups | 391 |
|  | Negative friction on piles | 391 |
|  | Laterally loaded piles | 392 |
| 8.4 | FOUNDATIONS ON ROCK | 392 |
| 8.5 | FOUNDATIONS IN COMPLEX GEOLOGICAL CONDITIONS | 394 |
|  | Expansive soils | 394 |
|  | Collapsible soils | 396 |
|  | Karstic cavities | 396 |
|  | Volcanic cavities | 396 |
|  | Soft and organic soils | 397 |
|  | Anthropogenic fills | 397 |
| 8.6 | SITE INVESTIGATION | 398 |
|  | Stages in site investigations | 398 |
| RECOMMENDED READING | | 400 |
| REFERENCES | | 400 |

# 9 SLOPES — 401

| 9.1 | INTRODUCTION | 402 |
| 9.2 | SITE INVESTIGATIONS | 403 |
| 9.3 | FACTORS INFLUENCING SLOPE STABILITY | 404 |
|  | Stratigraphy and lithology | 404 |
|  | Geological structure and discontinuities | 404 |
|  | Hydrogeological conditions | 405 |
|  | Geomechanical properties of soil and rock masses | 408 |
|  | *In situ* stresses | 408 |
|  | Other factors | 409 |
| 9.4 | TYPES OF SLOPE FAILURE | 410 |
|  | Soil slopes | 410 |
|  | Rock slopes | 411 |
|  |    Plane failure | 411 |
|  |    Wedge failure | 412 |
|  |    Toppling | 413 |
|  |    Buckling | 414 |
|  |    Non-planar failure | 414 |
| 9.5 | STABILITY ANALYSIS | 415 |
|  | Introduction | 415 |
|  | Limit equilibrium methods | 415 |
|  |    Soil slopes | 417 |
|  |    Rock slopes | 426 |
|  | Stress-strain methods | 432 |
|  | Geomechanical slope classification | 433 |
|  |    Slope mass rating (SMR) | 433 |
| 9.6 | STABILIZATION MEASURES | 434 |
|  | Introduction | 434 |
|  | Stabilization methods | 435 |
|  |    Modifying the geometry | 435 |
|  |    Drainage methods | 436 |
|  |    Resistant structural elements | 439 |
|  |    Walls and retaining elements | 440 |
|  |    Surface protection measures | 441 |
| 9.7 | MONITORING AND CONTROL | 443 |
| 9.8 | SLOPE EXCAVATION | 445 |
|  | Rippability criteria | 447 |
| RECOMMENDED READING | | 449 |
| REFERENCES | | 449 |

# 10 TUNNELS — 451

| 10.1 | INTRODUCTION | 452 |
| 10.2 | SITE INVESTIGATION | 453 |
| 10.3 | INFLUENCE OF GEOLOGICAL CONDITIONS | 454 |
|  | Geological structure | 457 |
|  | Discontinuities | 458 |
|  | Intact rock strength | 459 |
|  | Hydrogeological conditions | 460 |
|  | *In situ* stress | 461 |
|  |    Methods of analysis | 462 |
|  |    Effects of high stress on tunnelling | 464 |

| | | |
|---|---|---|
| 10.4 | GEOMECHANICAL DESIGN PARAMETERS | 464 |
| | Geological and geomechanical data | 464 |
| | Strength and deformability | 465 |
| | Magnitude and direction of *in situ* stress | 466 |
| |    Estimation of *K* from the TSI index | 466 |
| |    Sheorey's method | 471 |
| | Water inflow and pressure | 471 |
| 10.5 | ROCK MASS CLASSIFICATIONS FOR TUNNELLING | 472 |
| | Q System | 472 |
| | SRC rock mass classification | 476 |
| | Suggested criteria for the application of rock mass classifications | 480 |
| 10.6 | TUNNEL SUPPORT DESIGN USING ROCK MASS CLASSIFICATIONS | 480 |
| | Tunnel support based on RMR classification | 481 |
| | Tunnel support based on the Q index | 483 |
| 10.7 | EXCAVABILITY | 483 |
| 10.8 | TUNNEL EXCAVATION AND SUPPORT METHODS IN ROCK | 484 |
| | Excavation methods | 487 |
| | Stages of excavation | 489 |
| | Support systems | 489 |
| | Ground improvement | 491 |
| | The New Austrian Tunnelling Method | 491 |
| | Portals | 492 |
| 10.9 | TUNNEL EXCAVATION AND SUPPORT METHODS IN SOIL | 493 |
| | Non-mechanical excavation methods | 493 |
| | Semi-mechanical excavation methods | 493 |
| | Tunnel excavation with tunnel boring machines | 494 |
| 10.10 | GEOLOGICAL ENGINEERING DURING TUNNEL CONSTRUCTION | 495 |
| | RECOMMENDED READING | 499 |
| | REFERENCES | 499 |

## 11 DAMS AND RESERVOIRS — 501

| | | |
|---|---|---|
| 11.1 | INTRODUCTION | 502 |
| 11.2 | TYPES OF DAMS AND AUXILIARY STRUCTURES | 503 |
| | Types of dams | 503 |
| |    Embankment dams | 504 |
| |    Concrete dams | 504 |
| | Auxiliary structures | 506 |
| 11.3 | SITE INVESTIGATION | 507 |
| | Planning site investigation | 507 |
| |    Preliminary and feasibility studies | 508 |
| |    Selecting the type of dam | 508 |
| |    Design | 508 |
| |    Construction | 508 |
| |    Operation | 509 |
| | Site investigation methods | 509 |
| 11.4 | ENGINEERING GEOLOGICAL CRITERIA FOR DAM SELECTION | 513 |
| | General criteria | 513 |
| | Foundation conditions | 513 |
| | Availability of materials | 514 |
| | Siting of auxiliary structures | 514 |
| | Conditions for embankment dams | 515 |
| | Conditions for concrete dams | 515 |
| | Environmental considerations | 515 |
| 11.5 | GEOLOGICAL MATERIALS FOR DAM CONSTRUCTION | 516 |
| | Site investigations for dam materials | 516 |
| | Types of materials | 516 |
| |    Cores | 516 |
| |    Rockfills and ripraps | 517 |
| |    Filters and drains | 517 |
| |    Aggregates | 517 |
| 11.6 | RESERVOIR WATER TIGHTNESS | 518 |
| 11.7 | PERMEABILITY OF DAM FOUNDATIONS | 519 |
| | Uplift pressures | 519 |
| | Erosion | 519 |
| | Leakage control | 521 |
| 11.8 | RESERVOIR SLOPE STABILITY | 521 |
| 11.9 | ENGINEERING GEOLOGICAL CONDITIONS FOR DAM FOUNDATIONS | 523 |
| | General conditions | 523 |
| | Loads on dam foundations | 523 |
| | Dam foundation failure mechanisms | 524 |
| | Stress distributions in dam foundations | 527 |
| | Foundation improvement measurements | 528 |
| | Dam foundation problems and possible remedial measures | 529 |
| 11.10 | SEISMIC ACTIONS AND INDUCED SEISMICITY | 532 |
| | RECOMMENDED READING | 533 |
| | REFERENCES | 533 |

## 12 EARTH STRUCTURES — 535

| | | |
|---|---|---|
| 12.1 | INTRODUCTION | 536 |

| 12.2 | DESIGN METHODOLOGY | 537 |
| 12.3 | MATERIALS | 540 |
| | Earthfill embankments | 540 |
| | Rockfill embankments | 541 |
| | Coarse rockfill | 545 |
| 12.4 | IMPLEMENTATION AND CONTROL | 545 |
| 12.5 | EMBANKMENTS ON SOFT SOILS | 548 |
| 12.6 | EMBANKMENTS ON SLOPES | 550 |
| | REFERENCES AND RECOMMENDED READING | 551 |

# PART IV – GEOLOGICAL HAZARDS

## 13 LANDSLIDES AND OTHER MASS MOVEMENTS — 555

| 13.1 | INTRODUCTION | 556 |
| 13.2 | SLOPE MOVEMENTS | 556 |
| | Types of slope movements | 557 |
| | Landslides | 557 |
| | Flows | 560 |
| | Rock falls | 561 |
| | Rock avalanches | 562 |
| | Lateral displacements | 562 |
| | Causes of slope movements | 563 |
| | Rainfall and climatic conditions | 565 |
| | Changes in water level | 567 |
| | Erosion | 567 |
| | Earthquakes | 568 |
| | Volcanism | 569 |
| | Human actions | 569 |
| 13.3 | INVESTIGATION OF LANDSLIDES | 570 |
| | General field surveys | 570 |
| | Analysis of the processes | 574 |
| | Detailed investigations | 576 |
| | Stability analysis | 580 |
| | Monitoring | 581 |
| | Alarm systems | 582 |
| 13.4 | CORRECTIVE MEASURES | 582 |
| | Stabilisation and protection against rock falls | 583 |
| 13.5 | COLLAPSE AND SUBSIDENCE | 585 |
| | Types of movements and their causes | 585 |
| | Collapse | 586 |
| | Subsidence | 587 |
| | Investigation of the processes | 587 |
| | Corrective measures | 589 |
| 13.6 | PREVENTION OF RISKS FROM MASS MOVEMENTS | 589 |
| | Susceptibility and hazard maps | 591 |
| | Slope movement maps | 591 |
| | Collapse and subsidence maps | 592 |
| | RECOMMENDED READING | 593 |
| | REFERENCES | 593 |

## 14 SEISMIC HAZARD — 595

| 14.1 | INTRODUCTION | 596 |
| 14.2 | FAULTS AND EARTHQUAKES | 596 |
| | Faults as the source of earthquakes | 596 |
| | Stick-slip regimes and the seismic cycle | 597 |
| | The seismic fault model | 598 |
| | Slip rates and recurrence periods | 599 |
| | Geological recording of fault activity | 600 |
| | The study of seismic faults | 600 |
| 14.3 | SEISMICITY STUDIES | 604 |
| 14.4 | SEISMIC HAZARD ANALYSIS | 606 |
| | Deterministic method | 606 |
| | Probabilistic methods | 608 |
| 14.5 | SEISMIC SITE RESPONSE | 609 |
| | Design earthquake | 610 |
| | Seismic parameters of ground motion | 610 |
| | Modification of ground motion by local conditions | 611 |
| 14.6 | GROUND EFFECTS INDUCED BY EARTHQUAKES | 613 |
| | Liquefaction potential | 613 |
| | Landslides induced by earthquakes | 615 |
| | Fault rupture | 616 |
| 14.7 | APPLICATIONS TO GEOLOGICAL ENGINEERING | 617 |
| | Seismic hazard studies applied to site assessment | 617 |
| | Seismic microzonation | 617 |
| | Seismic vulnerability assessment | 619 |
| | RECOMMENDED READING | 622 |
| | REFERENCES | 622 |

## 15 PREVENTION OF GEOLOGICAL HAZARDS — 625

| 15.1 | GEOLOGICAL HAZARDS | 626 |
| 15.2 | HAZARD, RISK AND VULNERABILITY | 627 |
| 15.3 | SAFETY CRITERIA IN GEOLOGICAL ENGINEERING | 631 |
| 15.4 | PREVENTION AND MITIGATION OF GEOLOGICAL HAZARDS | 638 |

| | | | |
|---|---|---|---|
| 15.5 HAZARD AND RISK MAPS | 639 | **APPENDIX C** SYMBOLS AND ACRONYMS | 657 |
| **RECOMMENDED READING** | 641 | **APPENDIX D** LIST OF BOXES | 663 |
| **REFERENCES** | 642 | | |
| **APPENDIX A** CHARTS FOR CIRCULAR AND WEDGE FAILURE ANALYSIS | 643 | **APPENDIX E** PERMISSIONS TO REPRODUCE FIGURES AND TABLES | 665 |
| **APPENDIX B** PRESSURE UNITS CONVERSION CHART | 653 | **INDEX** | **671** |

# ABOUT THE AUTHORS

## Luis I. González de Vallejo

Professor of Geological Engineering at the Complutense University of Madrid, where he is also Director of the MSc Courses in Geological Engineering. He has dedicated his professional career in geological engineering to consulting, research and teaching, and he has conducted a large number of geological and geotechnical investigations for the design and construction of tunnels, dams and foundations in Spain and Central and South America, including landslide and earthquake hazard analysis, large excavations and site assessment for nuclear power plants and radioactive waste disposal.

He has written over 120 papers in journals and proceedings as well as five books. Associate Editor of Soils and Rocks and Member of the Editorial Board of several scientific journals, he has been invited to present the 2nd Mariano Ruiz Vazquez Lecture, at the Academy of Engineering of Mexico, in 2007, and the XXVII Manuel Rocha Lecture, at the Portuguese Geotechnical Society, in 2010.

## Mercedes Ferrer

Senior Research Officer at the Geological Survey of Spain and Associate Lecturer on Rock Mechanics at the Complutense University of Madrid, where she graduated in Geological Sciences and obtained her doctorate for her research on the deformability and failure mechanisms of soft rocks. She has carried out research projects on geological hazards in Spain, Italy and Central and South America, particularly on landslides and geo-hazard mapping for urban planning, mitigation and prevention purposes.

She has written over 100 papers and research reports on geological hazards, landslides and slope stability. At present she is leading a research project on the causes and failure mechanisms of the mega-rockslides of volcanic islands flanks.

# CONTRIBUTORS

**Luis Ortuño,**
    Uriel y Asociados and Technical University of Madrid
    Chapter 2 (Sections 2.3 to 2.6); Chapter 8 (Section 8.2 and 8.3).

**Carlos Oteo,**
    Professor of Geotechnical Engineering
    Chapter 12; Chapter 2 (Section 2.2 and 2.9); Chapter 8 (Section 8.5); Chapter 10 (Section 10.9).

**Alfredo Iglesias,**
    Geological Survey of Spain
    Chapter 4.

**Ricardo Oliveira,**
    COBA and New University of Lisbon
    Chapter 11 (Section 11.3).

**Andres Carbó,**
    Complutense University of Madrid
    Chapter 5 (Section 5.3).

**Alfonso Muñoz,**
    Complutense University of Madrid
    Chapter 5 (Section 5.3).

**Ramón Capote,**
    Complutense University of Madrid
    Chapter 14 (Section 14.2).

**Meaza Tsige,**
    Complutense University of Madrid
    Chapter 2 (Section 2.7).

**Claudio Olalla,**
    Technical University of Madrid
    Chapter 3 (Failure Mechanisms, in Section 3.3).

**Julián García-Mayordomo,**
    Geological Survey of Spain
    Chapter 14 (Section 14.3).

**Carmen Antón-Pacheco,**
    Geological Survey of Spain
    Chapter 5 (Remote Sensing, in Section 5.2).

# FOREWORD

Geological Engineering is concerned with identifying and understanding the geological controls on engineering properties. These controls can range in scale and character from the properties of the materials that make up a volume of ground to the distribution of those materials, including the fluids they contain, and their boundaries within a volume of ground. These controls arise from geological history and can be seen, touched and studied using measurements made in-situ, representative samples and appropriate experiments.

In addition to these there are those controls that cannot be seen and touched, and consequently are much more difficult to study; they arise mainly from changes in stress in response to engineering the ground. In most cases this results in changes in porosity, moisture content, quality of pore fluid and quantity of flow. Here the stresses produced by geological history and gravity interfere with those of construction history; the changes produced vary in both space and time.

Many of these changes are coupled – they do not occur in isolation. For example, shear displacement of a rough joint surface can cause dilation of the joint, which increases its permeability and permits pore pressures within it to decrease and effective stress within it, and hence the frictional resistance of its surface contacts, to increase. These coupled responses are complex interactions that are at the heart of the ground's response to engineering and to predict that response requires geology, materials science, mechanics and engineering each to be applied with appropriate knowledge of the other three subjects.

Where can a student begin with such a task? The answer to that is in many parts but for geologists these subjects begin to make sense when studied in the field. However, this knowledge has to be applied – how is that to be learnt? Here the geologist must cross disciplines and learn the language of other subjects. The reverse is necessary for those who do not start as geologists; they have to apply their materials science, mechanics and engineering to the materials, structures and processes of geology.

Crossing disciplines is the most difficult challenge to overcome in geotechnical engineering. Reading learned papers and case histories, attending lectures and completing coursework exercises are all essential, as is learning from professionals in geotechnical engineering. But underpinning all these endeavours are reliable and well written textbooks as these form the bedrock of student learning. Professor Luis González de Vallejo and Dr. Mercedes Ferrer have now produced an account of many of the topics described above that is accessible to English speaking readers from both science and engineering. The many illustrations enable this text to provide a unique means of communication to its readers, unmatched by other texts in the field, because they impart a sense of the scale and variability that have to be addressed when working in the subject. This is a contribution to geotechnical engineering that will support and improve the teaching and application of Engineering Geology for years to come.

*Michael H. de Freitas*
*Reader Emeritus in Engineering Geology*
*Imperial College, London*

# PREFACE

This book is the outcome of the professional and teaching experience of the authors since 1980, especially their participation in the MSc courses in Geological Engineering and undergraduate courses in Engineering Geology and Rock Mechanics at the Complutense University, Madrid. The need for a text book which brings together the main topics dealt with in these courses led to the writing of this book. First published in Spanish in 2002 and then in Italian in 2005, both editions have been extremely well received by academics and students, as well as by geo-engineering practitioners. This English edition updates the earlier versions and includes many useful suggestions made by colleagues and readers.

Engineering Geology has evolved significantly in recent years, from what was primarily a scientific discipline, into Geological Engineering, a mainly technical discipline embracing geological and engineering education, as can be seen from the course content at many universities. Independently of the changes in these disciplines, the authors have attempted to highlight the importance of the core values, especially an understanding of geology as the fundamental basis for Engineering Geology and Geological Engineering, providing engineering geologists and geological engineers with the knowledge they require of geological and engineering sciences and their applications.

Geological Engineering, as a discipline applied to engineering and the environment, is highly significant in socio-economic terms, and ranges from geotechnical investigations for building foundations to large-scale public works and infrastructure, providing appropriate solutions for the geological and environmental site characteristics. The role of Geological Engineering is essential for optimizing investment and planning construction projects adequately, since ground behaviour problems are an important uncertainty factor and therefore a risk. Another important application of Geological Engineering is to reduce the enormous social impact of the damage caused by natural catastrophes. Geological hazards can be mitigated to a considerable extent if prevention and control measures are adopted and here Geological Engineering is of fundamental importance. This book addresses these areas in four sections: Fundamentals, Methods, Applications and Geological Hazards.

## Acknowledgements

The authors are most grateful to Prof. Michael de Freitas, *maestro* and an inspiration for generations of engineering geologists worldwide, who revised the manuscript and contributed substantially with his suggestions to improving the content of this book.

They also are very grateful to Dr. Janjaap Blom, Senior Publisher at CRC Press/Balkema, for his invaluable support and personal co-operation with the authors. Special thanks to Richard Gundel, Production Editor at CRC Press/Balkema for his professional guidance during the production process of the book. The authors thank Bill Newton, Coordinator of the Gabinete Lingüístico of the Fundación General de la Universidad Complutense, Madrid, Pauline Moran and Valerie Stacey for their excellent work on the English translation of the text.

The teaching and learning experience shared with teachers and students of the MSc courses in Geological Engineering at the Complutense University of Madrid, has been a permanent stimulus for the authors to write this book. We would also like to thank Dr. Juan Miguel Insúa, José Ángel Rodríguez Franco, Prof. Mike Rosenbaum, Beatriz Blanco and Dra. Teresa Hijazo for their generous help in the preparation of the manuscripts. The authors are grateful for the valuable comments and suggestions by Dr. Nick Barton and Dr. Josep Gili.

*Luis I. González de Vallejo*
*Mercedes Ferrer*
October 2010

# PART I

# FUNDAMENTALS

# 1

# INTRODUCTION TO GEOLOGICAL ENGINEERING

1. Definition and importance of geological engineering
2. The geological environment and its relation with engineering
3. Geological factors and geotechnical problems
4. Methods and applications in geological engineering
5. Information sources in engineering geology
6. How this book is structured

# 1.1 Definition and importance of geological engineering

**Geological engineering** is the application of geological and engineering sciences to design and construction in civil, mining and petroleum engineering, and to the environment. The aim of this discipline is to ensure that the geological factors which affect engineering activities are considered and adequately interpreted, as well as to mitigate the consequences of geological and environmental hazards.

Although there are differences between **geological engineering** and **engineering geology**, in this book both terms are considered to be equivalent (*Box 1.1*).

Engineering geology emerged with the development of large-scale civil engineering projects and urban growth, and by the mid 20th century had become established as a separate, specialized branch of the geological sciences. While engineering development made rapid progress in the last century, it was the catastrophic failure of several large engineering works that pointed out the need of geological investigations applied to engineering. Among these events were the failure of dams for geological reasons and their grave consequences, including the loss of hundreds of human lives, as in the dam failures in San Francisco (California, 1928), at Vajont (Italy, 1963) and at Malpasset (France, 1959), landsliding during the building of the Panama Canal in the early decades of the 20th century and the collapse of slopes on the Swedish railways in 1912.

The development of other related sciences, such as **soil mechanics** and **rock mechanics,** was the basis for modern **geotechnical engineering,** where engineering geology provides solutions to construction problems from a geological point of view (*Figure 1.1*). Geotechnical

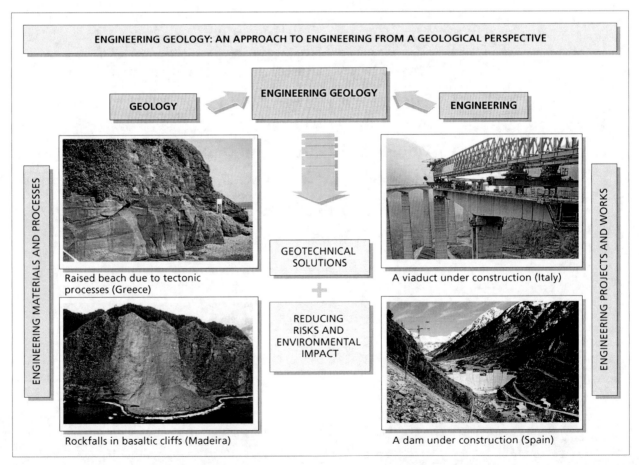

*Figure 1.1*  Engineering geology, geology and engineering.

## Box 1.1

### Geological engineering: education and professional practice

Education in geological engineering is based on a sound knowledge of geological and engineering sciences, the mechanical behaviour of soils and rocks, and their response to changes imposed by engineering works. Site and ground investigation methods to analyse and model geo-materials and geological processes form an essential part of this discipline.

Engineering geologists and geological engineers have a scientific and technical education, and a training applicable to the solution of the geological and environmental problems which affect engineering, and therefore they should be able to answer the following questions:

1. Where to site a civil engineering facility or industrial plant so that it will be geologically secure and economically feasible.
2. How to select the alignment for communication or transportation infrastructure to ensure favourable geological conditions.
3. How to assess that building foundations are geologically and geotechnically safe and economically feasible.
4. How to excavate a slope that is both stable and economically feasible.
5. How to excavate a tunnel or underground facility so that it is stable.
6. How to locate geological materials for dams, embankments and road construction.
7. The remedial measures and ground treatments needed to improve ground conditions and control instability, seepages, settlements, and collapse.
8. The geological and geotechnical conditions required to store urban, toxic and radioactive wastes.
9. How to prevent or mitigate geological hazards.
10. What geologic and geotechnical criteria must be taken into account in land use and urban planning, and to mitigate environmental impact.

### Applied geology, engineering geology and geological engineering

— *Applied geology* or *geology for engineers* is the geology used in engineering practice. This is the branch of geology which deals with its application to the needs of civil engineering. It does not necessarily imply the use of engineering geological methods for the study and solution of geological problems in engineering.
— *Engineering geology* and *geological engineering* are different from applied geology in that in addition to geological knowledge, education and training is required in the problems of the ground for engineering works, site investigation methods and the classification and behaviour of soils and rocks in relation to civil engineering; this field also includes practical knowledge of soil mechanics, rock mechanics and hydrogeology (Fookes, 1997).
— *Engineering geology* and *geological engineering* are equivalent disciplines, although in some countries there is a difference depending on whether the university where these courses are offered is oriented more towards a geological training (engineering geology) or towards engineering (geological engineering). An engineering geologist can be defined as a specialist geologist (scientist) in contrast with geological (or geotechnical) engineers who are trained as engineers with additional geological knowledge (Turner, 2008).

---

engineering integrates ground engineering techniques applied to foundations, reinforcement, support, ground improvement and excavation, and the disciplines of soil mechanics, rock mechanics and engineering geology mentioned above. Recently, the term **geo-engineering** has been coined to describe the field that deals with all aspects of engineering geology, rock mechanics, soil mechanics and geotechnical engineering (Bock et al., 2004).

At the beginning of the 21st century, one of the top priority areas for engineering geology is sustainable development. The inevitable confrontation between consequences of progress and geological processes, the uncontrollable sprawl of modern cities into geologically adverse areas and the damage caused by natural hazards can easily threaten the environment's fragile balance.

Nowadays, the need for geological studies of the ground before initiating large-scale works is fully recognized, and such studies are an obligatory part of engineering practice. This requirement also applies to works on a smaller scale, but often with a more direct impact people's daily lives,

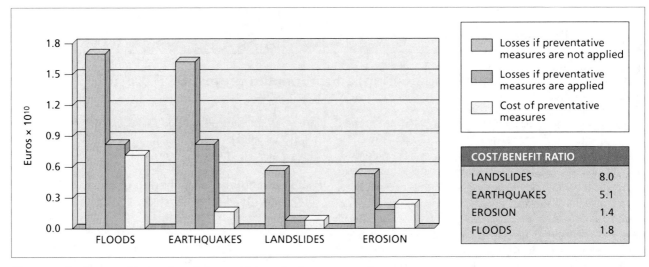

30 year projection, considering the maximum risk hypothesis.
Cost/benefit ratio: losses from geological hazards less losses if preventative measures are applied, divided by the cost of the preventative measures.

Figure 1.2   Economic losses from geological hazards in Spain (IGME, 1987).

such as home and building construction, where geotechnical surveys are also needed.

The importance of engineering geology is particularly important in two main fields of activity. The first is engineering projects and related works where the ground constitutes the foundation, excavation, storage or construction material. Included in this field are the main types of infrastructure projects: buildings, hydraulic or maritime works, industrial plants, mining installations, power stations, etc. The role of engineering geology in these projects is fundamental to ensuring safety and economic viability. The second field is the prevention, mitigation and control of geological hazards and risks, and the management of environmental impact of public works and industrial, mining or urban activities.

Both of these fields are of great importance to a country's gross national product as they are directly related to the infrastructure, construction, mining and building sectors. However, the impacts of geo-environmental hazards on society and the environment can be incalculable if no preventive or control measures are taken (Figure 1.2).

## 1.2 The geological environment and its relation with engineering

The geological environment is in continuous evolution through processes affecting rock and soil materials and the natural environment as a whole. Anthropogenic environments, such as cities, infrastructures or public works frequently intrude on regions which are geologically unstable, modifying the geological processes or sometimes even triggering them. The search for harmonious solutions between the geological and the anthropogenic environments requires an understanding of the factors which set them apart, in order to avoid erroneous interpretations. The most important differentiating factors are:

— Geological and engineering scale.
— Geological and anthropological time.
— Geological and engineering language.

The study of geology begins with a spatial view of Earth's physical phenomena, on a range of **scales** from the cosmic to the microscopic. **Time** is measured in millions of years. In engineering, spatial and time scales are adjusted to the reach of human activities. Most geological processes, such as orogenesis or lithogenesis, take place over millions of years and shape such diverse phenomena as the properties and characteristics of materials and the occurrence of seismic or volcanic processes. Man as a species appeared in the Quaternary period, some 2 million years ago, quite recent compared with the 4,600 million years of the life of the planet Earth. However, human activity can dramatically affect specific natural processes such as erosion, sedimentation, and even climate. Whether natural processes can be speeded up or modified is one of the fundamental questions to consider in engineering geology. Many of the geotechnical properties of geological materials, such as permeability, alterability, strength or deformability, and processes such as dissolution, subsidence or expansivity, may be substantially modified by human action.

## Box 1.2

### El Berrinche landslide, Tegucigalpa (Honduras)

This landslide occurred on October 30, 1998, in the aftermath of Hurricane Mitch. The hurricane devastated Central America, causing more than 25,000 deaths and incalculable economic losses. Its effects were aggravated by intense deforestation and urban encroachment on unstable hillsides. Landslides on some of the shanty-covered, overpopulated slopes surrounding the city of Tegucigalpa wreaked tremendous damage. Hundreds of households lost members of their family and experienced permanent economic setback in what was perhaps the costliest natural catastrophe in the history of a country, with major social losses and economical damages, in terms of homes destroyed and people affected by the landslides.

The El Berrinche landslide destroyed the neighbourhood of same name and partially affected others. It caused the blockage of the Choluteca River, which diverted its course into inhabited areas, flooding the lower parts of the city and causing many deaths. A river of mud swept along huge amounts of vegetation, carrying with it vehicles and parts of houses, and reaching a height of several meters above the street level, damaging the city infrastructure.

In Tegucigalpa, these areas were known to be within a risk zone, and some maps had even been prepared to that effect. In 1958, during a previous event, a large number of houses were destroyed on the slopes located in front of the El Berrinche hillside.

The intense rainfall that Hurricane Mitch released onto Tegucigalpa became a true test of the evaluation of the ground behaviour and its susceptibility to landslides. Clearly different behaviour was noticed between areas as a function of the type of geological material present, with lithology controlling the relative stability of slopes. In fact, the largest landslides took place in slopes formed in mudstone and siltstone with intercalations of greywake and clayey sandstone (Valle de Angeles Group), all highly degradable and weathered, while in the other lithologic group outcroppings in the zone, and formed by massive tuffs (Padre Miguel volcanoclastic Group) only isolated rock falls occurred.

A view of the landslide affecting part of the city of Tegucigalpa.

Comparing geological and human time is fundamental to appreciate the possible consequences of geological factors and hazards. Most projects are expected to have a service life of 50–100 years; it is, however, accepted practice to demand geological and environmental safety guarantees of periods of 500–1000 years in areas which can be affected by flooding, earthquakes, etc. There are cases where geological stability must be ensured for even longer periods, such as in the storage of radioactive waste, where periods of more than 10,000 years are envisioned.

On a human scale, many geological processes and most large-scale natural hazards have a very low probability of occurrence. Engineering planning and design must take into account the great variability in the frequency with which geological processes occur, from almost instantaneous processes, like earthquakes, to very slow ones, such as dissolution and erosion.

**Mapping scales** as a means of spatial representation are another differential aspect to be considered. In geology, the scales are adjusted to the dimensions of the phenomena or the geological units, formations and structures which need to be represented. Most geological maps use scales of between 1/1,000,000 and 1/50,000, whereas in engineering the most frequent scales are between 1/10,000 and 1/500. Regional geological maps allow factors to be identified which, although not within the specific project area, may be necessary in order to appreciate regional geological aspects or the presence of hazards whose scope may affect the zone under survey. Small-scale geological maps are the norm in geotechnical, lithological and thematic cartography, where discontinuities, hydrogeological data, materials, etc. are represented on the same scale as the project documents.

Another problem which often arises when integrating geological data into engineering projects is the lack of **communication** between these two fields. Independently of the geological or engineering terminology itself, there tend to be differences in approach and in the evaluation of results, depending on the point of view from which the problem is being addressed. Engineering deals with materials whose properties vary within narrow margins, do not change substantially over time, and can be tested in laboratory, such as concrete, steel, etc. In geology, however, the majority of materials are anisotropic and heterogeneous, they have extremely variable properties and undergo alterations and changes over time.

In an engineering project the data must be quantifiable and allow modelling. In geology, numerical quantification is not always easy, and simplification of a wide range of variation in properties to figures that fall within narrow margins can be difficult and, at times, it is impossible to achieve numerical precision that satisfies project requirements. While in engineering very precise knowledge of construction materials is usually available, geological and geotechnical information is generally based on a limited number of surveys. As a result, there is an uncertainty factor present in geotechnical studies which affects most projects. An understanding of these differences and the use of a common language appropriate to the aims of a project is fundamental to the practice of engineering geology. Engineering geology has methods at its disposal to quantify or express geological data in a way which allows them to be integrated into numerical modelling and into decision-making processes during planning and construction.

**Statistics** is an important tool for the analysis of very variable or even random data. The study of certain phenomena with insufficiently known periodicity can be approached from probability analysis with acceptable results, as is the case in specific geological hazards. The quantification of a set of engineering geological properties for construction applications is possible through systems of **geomechanical classifications** of rock masses. The **factor of safety** concept, normally used in engineering to express the degree of stability of the work underway, is also integrated into geological engineering practice. By including these and other procedures, with relation above all to knowledge of the geological medium and its interaction with construction activities, geological factors affecting safety and engineering issues can be defined, evaluated and integrated.

## 1.3 Geological factors and geotechnical problems

Given the diversity of the geological environment and the complexity of its processes, engineering solutions must be found for those geological factors that may create problems for project execution.

The most important problems are related to geological hazards which may affect the safety or viability of a project. Of a secondary but still crucial importance are all the geological factors which affect the technical or economic aspects of the project. These factors and their influence on geotechnical problems are shown in *Tables 1.1* to *1.4*.

*Tables 1.1* and *1.2* show the possible influence of lithology and geological structure on the geotechnical behaviour of rock and soil materials. *Tables 1.3* and *1.4* show how water and materials are affected by different geological processes, causing geotechnical problems. To sum up, the following conclusions are reached:

— Geological factors are the cause of most geotechnical problems.
— Water is one of the factors with the highest incidence affecting the geotechnical behaviour of materials.

## Table 1.1 INFLUENCE OF LITHOLOGY ON THE GEOTECHNICAL BEHAVIOUR OF THE GROUND

| Lithology | Characteristic factors | Geotechnical problems |
|---|---|---|
| Hard rocks | — Hard and abrasive minerals | — Abrasivity (Photo A)<br>— Excavation difficulties |
| Soft rocks | — Medium to low strength<br>— Alterable minerals | — Slope failures (Photo B)<br>— Deformability in tunnels<br>— Change of properties over time |
| Hard soils | — Medium to high strength | — Problems in foundations with expansive clays and collapsible soils |
| Soft soils | — Low to very low strength | — Settlements of foundations (Photo C)<br>— Slopes failures |
| Organic and biogenic soils | — High compressibility<br>— Metastable structures | — Subsidence (Photo D) and collapse |

Photo A — Granite with quartz, plagioclase and micas.

Photo B — Slope failures in open cast mines, southern Spain.

Photo C — Leaning tower of Pisa.

Photo D — Settlement of the basilica Nª Sª de Guadalupe, Mexico City, built on soft lacustrine soils affected by subsidence.

## Table 1.2 GEOLOGICAL STRUCTURES AND GEOTECHNICAL PROBLEMS

| Geological structures | Characteristic factors | Geotechnical problems |
|---|---|---|
| Faults and fractures (Photo A) | — Very continuous surfaces; variable thickness | Failures, instabilities, seepages and alterations |
| Bedding planes (Photo B) | — Medium-highly persistent surfaces; little separation | Failures, instabilities and seepages |
| Discontinuities (Photo B) | — Small-medium persistence; closed or open | Failures, instabilities, seepages and weathering |
| Folds (Photo C) | — Surfaces with high continuity or persistence | Instabilities and seepages |
| Foliation, schistosity (Photo D) | — Surfaces with low continuity; closed features | Anisotropic behaviour dependent on the orientation |

Photo A   Normal fault.

Photo B   Strata and joints.

Photo C   Folds in quartzite.

Photo D   Folded schist.

## Table 1.3 EFFECTS OF WATER RELATED GEOLOGICAL PROCESSES

| Water-related geological processes | Effects on materials | Geotechnical problems |
| --- | --- | --- |
| Dissolution (Photo A) | — Loss of material in soluble rocks and soils<br>— Karstification | — Cavities<br>— Subsidence<br>— Collapse |
| Erosion – piping (Photo B) | — Loss of material, sheetwash<br>— Piping, internal erosion<br>— Gully erosion | — Subsidence<br>— Collapse<br>— Settlement<br>— Piping and undermining<br>— Silting |
| Chemical reactions (Photo C) | — Changes in chemical composition | — Attacks on cement, aggregates, metals and rocks |
| Weathering (Photo D) | — Changes in physical and chemical properties | — Loss of strength<br>— Increased deformability and permeability |

Photo A — Gypsum karst, southeast Spain.

Photo B — Erosion and gullies in pyroclastic deposits, Guatemala.

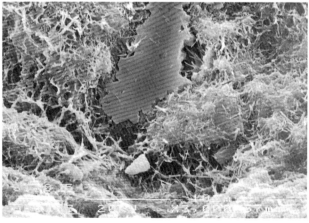

Photo C — Concrete attacked by sulphates: formation of ettringite in the form of very fine fibres and carbonate crystals.

Photo D — Weathering in Tertiary materials, central Spain.

## Table 1.4 INFLUENCE OF GEOLOGICAL PROCESSES ON ENGINEERING AND THE ENVIRONMENT

| Geological processes | Effects on the physical environment | Geo-environmental problems and actions |
|---|---|---|
| Seismicity (Photo A) | – Earthquakes, tsunamis<br>– Ground movements and failures, landslides, liquefaction | – Damage to population and infrastructure<br>– Anti-seismic design<br>– Preventive measures<br>– Emergency plans |
| Volcanism (Photo B) | – Volcanic eruptions<br>– Changes in relief<br>– Tsunamis and earthquakes<br>– Collapse and large scale slope movements | – Damage to population and infrastructure<br>– Monitoring systems<br>– Preventive measures<br>– Evacuation plans |
| Uplift and subsidence (Photo C) | – Long term morphological changes<br>– Long term changes in coastal dynamics and sea levels | – Monitoring and control measures |
| Erosion-sedimentation (Photo D) | – Medium term morphological changes<br>– Short term hydrological changes<br>– Silting | – Increased risk of flooding and landslides<br>– Protection measures for river beds and coasts |

*(continued)*

Photo A — Building destroyed in the Mexico earthquake, 1985.

Photo B — Lava flow in the Teneguía eruption, Canary Islands, 1971.

Photo C — Palacio de Bellas Artes, Mexico City, affected by the subsidence of the Mexico valley soils.

Photo D — Silting of riverbed to above road level, requiring excavation to an artificial channel, northwest Argentina.

| Table 1.4 | INFLUENCE OF GEOLOGICAL PROCESSES ON ENGINEERING AND THE ENVIRONMENT (CONT.) | |
|---|---|---|
| **Geological processes** | **Effects on the physical environment** | **Geo-environmental problems 4 and actions** |
| Slope movements (Photo E) | — Landslides, rock falls, subsidence<br>— Short and medium term morphological changes, diversion of river beds | — Damage to populations and infrastructures<br>— Impounding of river beds<br>— Stabilization, control and preventive measures |
| Changes in water table (Photo F) | — Changes in aquifers<br>— Changes in soil properties<br>— Drying out and waterlogging<br>— Subsidence and instability of slopes | — Problems in foundations<br>— Effect on crops and irrigation<br>— Drainage measures |
| Tectonic processes | — Natural stress<br>— Seismicity<br>— Instabilities | — Rock bursts in mines and deep tunnels<br>— Long term deformations in underground works<br>— Design measures in tunnels and mines |
| Geochemical processes | — High temperatures<br>— Thermal anomalies<br>— Presence of gases | — Hazards from gas explosion<br>— Difficulties during tunnel construction |

Photo E     Damage to motorways caused by landslides, southern Spain.

Photo F     Subsidence along active faults caused by water extraction from wells, Celaya, Mexico.

## Box 1.3

### The failure of Aznalcóllar Dam: An example of underestimation of the geological and geotechnical conditions with serious environmental consequences

The tailing dam of Aznalcóllar (southern Spain), owned by the mining company Boliden-Apirsa, was 28 m high when it failed on April 25, 1998. The safety conditions of the dam had been checked three years earlier and both the owners and those responsible for the design confirmed that it fulfilled all construction and safety requirements. This conclusion was reiterated just 5 days before the disaster.

The failure of the dam released a 4.5 $Hm^3$ of liquid mine waste into the river Agrio, and from there into the Guadiamar, tributary of the Guadalquivir. The surrounding land was flooded, contaminated with acid water containing heavy metals, affecting all the surrounding ecosystems in the area, including Doñana National Park.

The dam was founded on a Miocene overconsolidated high-plasticity clay formation, known as blue marl, which contains frequent shear surfaces with slickensides.

These blue marls have been extensively studied and the problems they cause were well known, especially in the stability of slopes of roads and railways of southern Spain. Their strength can be very low when they come into contact with water and when high pore pressures are generated along shear surfaces. According to the expert reports, the failure of the dam was due to a failure in the clay substratum, causing the foundation of the dam to slide forward (see Box 11.3, Chapter 11).

After the event, it became clear that the geological and geotechnical factors which caused the dam failure were not adequately taken into account, and that the monitoring systems were not operative, both fundamental aspects in geological engineering.

The Aznalcóllar dam after failure.

— Geological processes may modify the behaviour of materials, affecting the physical medium and causing geotechnical problems.

Thus it is evident that geotechnical problems can often require expensive solutions to be adopted. Depending on the scope of the problems, projects could be modified or sites relocated. For example, foundations might have to be laid more deeply because of the insufficient bearing capacity of the ground at depths nearer the surface. In contrast, favourable geotechnical conditions provide greater security for the work site and also mean work can go ahead uninterrupted, which has a significant influence on the cost and delivery schedule for the completed work.

In general terms, the **conditions a site must meet** to be geologically and geotechnically suitable are as follows:

— The absence of active geological processes which present unacceptable risks for the project.
— Adequate bearing capacity of the ground for the structural foundations.
— Materials with strength enough to be stable in surface or underground excavations.
— Watertight geological formations for storing water and solid or liquid wastes.
— Availability of materials for the construction of earth works.
— Easy extraction of materials for excavation.

The relationship between the geological factors and geotechnical problems, and the difference between favourable and unfavourable geotechnical conditions, make clear that the starting point for any geotechnical site investigation must be geological knowledge. **Interpreting geology from the perspective of engineering geology allows the behaviour of ground to be defined and predicted.** The potential for advances in geotechnical engineering that can be provided by geology is extensively described by de Freitas (2009).

## 1.4 Methods and applications in geological engineering

Geological engineering is based on geology and on the mechanical behaviour of soils and rocks. It requires a knowledge of site investigation techniques, mechanical, instrumental and geophysical, as well as a knowledge of methods for ground analysis and modelling. Methodology used in geological engineering studies follows the sequence described in *Table 1.5*, in general terms.

To develop the methodological sequence three types of models must be defined (*Figure 1.3*):

**Table 1.5  METHODOLOGICAL PROCEDURES FOR GEOLOGICAL ENGINEERING AND ENGINEERING DESIGN**

1. Identification of geological materials and geological processes. Analysis of geomorphological, structural, lithological and groundwater conditions.
2. Site and ground investigation.
3. Defining the spatial distribution of materials, structures and discontinuities.
4. Defining the hydrogeological, *in situ* stress and environmental conditions.
5. Characterization of geomechanical, hydrogeological and chemical properties.
6. Characterization of the geological materials to be used in the construction.
7. Selection of design parameters of the ground to be used in stability analyses for excavations, earth structures, foundations, etc.
8. Modelling of ground behaviour under construction and operational conditions.
9. Assessment of ground treatments to control seepages, settlements, instability, etc. and to improve ground conditions.
10. Analysis of geological hazards and environmental impact on engineering design.
11. Geological and geotechnical monitoring and control during construction and operational service.

— The geological model.
— The geomechanical model.
— The ground behaviour model.

The **geological model** represents the spatial distribution of materials, tectonic structures, geomorphologic and hydrogeological data, and other characteristics of the site and its area of influence. The **geomechanical** model gives a geotechnical and hydrogeological description of the materials and their geomechanical classification. The **ground behaviour model** describes the response of the ground during and after construction.

This methodology constitutes the basis for the following applications of geological engineering to civil and mining engineering and to the environment:

— Transport infrastructures.
— Hydraulic and maritime works.
— Urban, industrial and service buildings.
— Power stations.
— Mining and quarrying.
— Storage for urban, industrial and radioactive waste.
— Regional and urban planning.
— Civil defence and emergency planning.

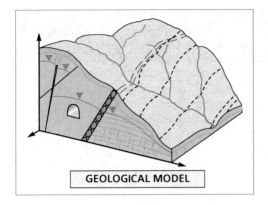
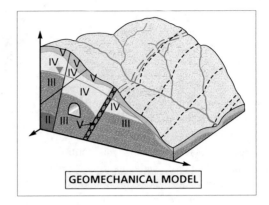
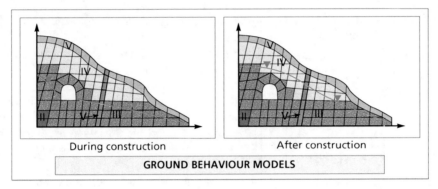

*Figure 1.3*   Examples of modelling in geological engineering.

## 1.5 Information sources in engineering geology

The main periodical publications in engineering geology/geological engineering are published by national and international associations, which regularly hold congresses and symposia, as well as publishing reviews or bulletins. The most important associations include:

— International Association for Engineering Geology and the Environment (IAEG)
— Association of Environmental and Engineering Geologists (AEG)
— International Society for Rock Mechanics (ISRM)
— International Society for Soil Mechanics and Geotechnical Engineering (ISSMGE)

Periodical publications include:

— Engineering Geology (Elsevier)
— Environmental & Engineering Geoscience Journal (GSA and AEG)
— Quarterly Journal of Engineering Geology and Hydrogeology (The Geological Society of London)
— Bulletin of Engineering Geology and the Environment, IAEG (Springer)
— Géotechnique (T. Telford)
— Journal of Geotechnical and Geoenvironmental Engineering, ASCE.
— International Journal of Rock Mechanics and Mining Sciences, ISRM (Elsevier)
— Canadian Geotechnical Journal (NRC Research Press)
— International Journal of Geomechanics, ASCE.
— Soils and Rocks, ABMS and SPG.
— Rock Mechanics and Rock Engineering (Springer)

## 1.6 How this book is structured

This book provides an introduction to geological engineering and engineering geology, their fundamentals and basic concepts, methodologies and main applications. The study of geological engineering requires a sound knowledge of geology. Emphasis has been made throughout the text to point out how the geology is closely related to engineering problems, as this is one of the principal objectives of engineering geology. Examples are given to illustrate these issues. However, this book does **not include basic descriptions of geological concepts.**

This book has 15 chapters, divided into four parts. **Part I** deals with the fundamentals of geological engineering.

Special attention is paid to basic concepts of soil and rock mechanics as well as hydrogeology (Chapters 1 to 4). **Part II** deals with site investigations methods with a description of the different procedures for identifying the properties and geomechanical characteristics of materials (Chapters 5 and 6; geotechnical mapping is included in Chapter 7). **Part III** describes the different applications of geological engineering, focussing on the most important fields of application to foundations, slopes, tunnels, dams and earth structures (Chapters 8 to 12). **Part IV** deals with geological hazards most relevant to geological engineering, focussing on the prevention, mitigation and control. Chapter 13 deals with landslides and other mass movements, Chapter 14 with earthquakes and Chapter 15 with prevention and mitigation of geological hazards.

# Recommended reading

## General and background to engineering geology

Blyth, F.G.H. and de Freitas, M.H. (1984). A geology for engineers. 7th ed. Arnold, London.
Culshaw, M.G. (2005). From concept towards reality: developing the attributed 3D geological model of the shallow subsurface. Ql. Jl. of Eng. Geol. and Hydrogeol., 38, 231–284.
de Freitas, M.H. (2009). Geology; its principles, practice and potential for Geotechnics. The 9th Glossop Lecture, Geological Society of London. Ql. Jl. of Eng. Geol. and Hydrogeol., 42, 397–441.
Fookes, P.G. (1997). Geology for engineers: the geological model, prediction and performance. Ql. Jl. of Eng. Geol. and Hydrogeol., 30, 293–424.
Fookes, P.G., Baynes, F.J. and Hutchinson, J.N. (2000). Total geological history. A model approach to the anticipation, observation and understanding of site conditions. Geo2000. Int. Conf. on Geotechnical & Geological Engineering, Melbourne. Vol. 1: Invited Papers. Technomic Publishing Co., 370–460.
Johnson, R.B. and DeGraff, J.V. (1988). Principles of engineering geology. J. Wiley & Sons. New York.
Legget, R.F. and Karrow, P.F. (1983). Handbook of geology in civil engineering. McGraw-Hill, New York.
Parriaux, A. (2009). Geology: basics for engineers. CRC Press/Balkema. The Netherlands.
Price, D.G. (2009). Engineering geology. Principles and practice. Edited and compiled by M. H. de Freitas. Springer.
Rahn, P.H. (1996). Engineering geology: an environmental approach. 2nd ed. Prentice Hall.
Terzaghi, K. (1960). From theory to practice in soil mechanics. John Wiley and Sons, New York.

## Education and training in engineering geology

Bock, H. et al. (2004). The Joint European Working Group of the ISSMGE, ISRM and IAEG for the definition of professional tasks, responsibilities and co-operation in ground engineering. Engineering Geology for Infrastructures Planning in Europe: A European Perspective. Hack, Azzam and Charlier, Eds. Lecture Notes in Earth Sci. Springer, Berlin, 104:1–8.
de Freitas, M.H. (1994). Keynote Lecture: Teaching and training in engineering geology: professional practice and registration. Proc. 7th Cong. of the Int. Assoc. of Engineering Geology, Lisbon. Oliveira, Rodrigues, Coelho & Cunha Eds. Balkema. Vol. 6, pp. LVII–LXXV.
Knill, J. (2003). Core values: the first Hans-Cloos lecture. (2003) Bull. of Eng. Geol. and the Environment, 62 (1), 1–34. Springer.
Oliveira, R. (2008). Geo-engineering education and training. The past and the future. Proc. 1st Int. Cong. on Education and Training in Geo-Engineering Sciences. Manoliu & Radulescu Eds., CRC Press, Taylor & Francis Group, London, 79–86.
Rengers, N. and Bock, H. (2008). Competency-oriented curricula development in Geo-engineering with particular reference to engineering geology. Proc. 1st Int. Cong. on Education and Training in Geo-Engineering Sciences. Manoliu & Radulescu Eds., CRC Press, Taylor & Francis Group, London, 101–110.
Turner, A.K. (2008). Education and professional recognition of engineering geologists and geological engineers in Canada and the United States. Proc. 1st Int. Cong. on Education and Training in Geo-Engineering Sciences. Manoliu & Radulescu Eds., CRC Press, Taylor & Francis Group, London, 111–118.
Turner, A.K. (2010). Defining competencies for geo-engineering: implications for education and training. Proc. 11th IAEG Congress, Auckland, New Zealand. Balkema, The Netherlands (in press).

# References

Bock, H. et al. (2004). The Joint European Working Group of the ISSMGE, ISRM and IAEG for the definition of professional tasks, responsibilities and co-operation in ground engineering. Engineering Geology for Infrastructures Planning in Europe: A European Perspective. Hack, Azzam and Charlier, Eds. Lecture Notes in Earth Sci. Springer, Berlin, 104:1–8.
de Freitas, M.H. (2009). Geology; its principles, practice and potential for Geotechnics. The 9th Glossop Lecture,

Geological Society of London. Ql. Jl. of Eng. Geol. and Hydrogeol., 42, 397–441.

Fookes, P.G. (1997). Geology for engineers: the geological model, prediction and performance. Ql. Jl. of Eng. Geol. and Hydrogeol., 30, 293–424.

IGME (1987). Impacto económico y social de los riesgos geológicos en España. Instituto Geológico y Minero de España (Geological Survey of Spain), Madrid.

Turner, A.K. (2008). Education and professional recognition of engineering geologists and geological engineers in Canada and the United States. Proc. 1st Int. Cong. on Education and Training in Geo-Engineering Sciences. Manoliu & Radulescu Eds., CRC Press, Taylor & Francis Group, London, 111–118.

# 2

# SOIL MECHANICS AND ENGINEERING GEOLOGY OF SEDIMENTS

1. Introduction
2. Soil description and classification. Phase relationships
3. Flow of water through soils
4. Effective stress
5. Consolidation and compressibility
6. Shear strength of soils
7. Influence of mineralogy and fabric on the geotechnical properties of soils
8. Engineering geological characteristics of sediments
9. Problematic soils

## 2.1 Introduction

## The nature of soils

Soils are formed when pre-existing rock masses, the "parent" rock, is broken down and disintegrated by environmental processes. There are three kinds of processes:

— **Physical:** temperature changes and the action of water, wind or glaciers break rocks down. Changes in temperature cause expansion and contraction at different rates in different minerals, producing internal changes and fissuring. Water can break rock down in several ways: (i) by eroding fragments; (ii) by freezing, directly producing internal stresses due to the increased volume of ice compared with water; (iii) by alternating wet/dry cycles over time. These physical actions break down the parent rock into smaller fragments which are then separated by active agents, such as water, wind or gravity, and transported to other sites where they are eroded further. This means that physical activity creates rock particles which further disintegrate to form soil.
— **Chemical:** chemical phenomena affect rock in different ways: (i) hydration: e.g. transformation of anhydrite or sulphate hemi-hydrate into gypsum or sulphate di-hydrate; (ii) dissolution of salts, such as sulphates, in water; (iii) oxidation of iron-bearing minerals by environmental agents; (iv) cementation, caused by water containing carbonates or other minerals, previously dissolved from other rock. This means that the effect of chemical activity may be disintegration and/or cementation.
— **Biological action:** bacterial activity decomposes organic matter and acts as a catalyzing agent that can affect the reactions of inorganic constituents and mix organic matter with other particles derived from physical and chemical actions.

All these processes result in the breakdown (alteration or weathering) and transformation of rock, and create a **weathering profile** (*Figure 2.1*). In this profile, the parent rock is farthest from the surface and in the lowest position; the soil is at the top. **Residual soils** have developed *in situ* and remain in their original location. **Transported soil** e.g. colluvial or alluvial deposits, have been moved from the location of their formation.

Different processes involved in soil formation are shown in *Figure 2.2*.

## Soils in geotechnical engineering

Anthropogenic activity such as excavations or ground levelling, can modify the existing environmental conditions and

| LITHOLOGICAL COLUMN | LOVE (1951) LITTLE (1961) | VARGAS (1951) | SOWERS (1954, 1963) | CHANDLER (1969) | | GEOL. SOC. ENG. GROUP (1970) | DEERE & PATTON (1971) | |
|---|---|---|---|---|---|---|---|---|
| | IGNEOUS ROCKS | IGNEOUS, BASALTS & SANDSTONES | IGNEOUS & METAMORPHIC | MARLS & SILTSTONES | | IGNEOUS ROCKS | IGNEOUS & METAMORPHIC | |
| | VI SOIL | RESIDUAL SOIL | UPPER ZONE | V COMPLETELY WEATHERED | PARTIALLY WEATHERED | VI RESIDUAL SOIL | HORIZON IA | RESIDUAL SOIL |
| | | | | IV | | | HORIZON IB | |
| | V COMPLETELY WEATHERED | YOUNG RESIDUAL SOIL | INTERMEDIATE ZONE | | | V COMPLETELY WEATHERED | HORIZON IC (SAPROLITE) | |
| | IV HIGHLY WEATHERED | | | III | | IV HIGHLY WEATHERED | IA TRANSITION WITH WEATHERED ROCK (SAPROLITE) | TRANSITION ZONE |
| | III MODERATELY WEATHERED | LAYERS OF DISINTEGRATED ROCK | PARTIALLY WEATHERED ZONE | | | III MODERATELY WEATHERED | | |
| | | | | II | | | | |
| | II SLIGHTLY WEATHERED | | | | | II SLIGHTLY WEATHERED | IB PARTIALLY WEATHERED | |
| | | | | | | IB ALMOST UNWEATHERED | | |
| | I FRESH ROCK | FRESH ROCK | FRESH ROCK | I FRESH ROCK | | IA FRESH ROCK | FRESH ROCK | |

*Figure 2.1* The weathering profile.

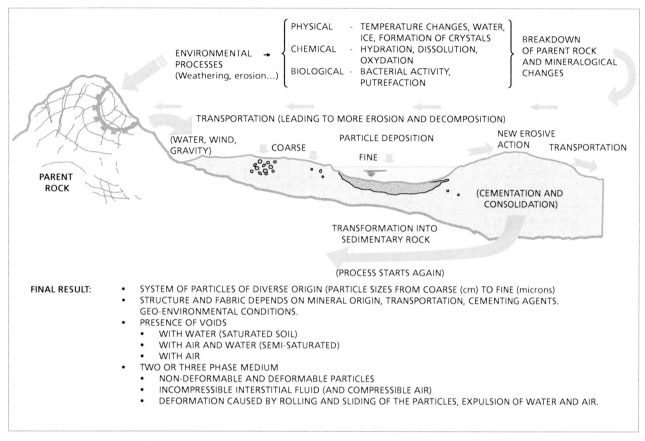

Figure 2.2   Origins of soils from rocks.

how the soil reacts will depend on many factors including the soil composition and geology, human activity and how far the engineering project is adapted to the natural environment.

The ground response is therefore complex and depends both on the existing materials in the area and on the actions and forces applied to the soils. How a rock mass responds will depend on its strength, weathering, and discontinuities. Soils, formed by loose materials, will have a different response. Figure 2.3 shows the following characteristic properties of soils:

— Soils are formed of small individual particles (ranging from microns to a few centimetres) which, for practical purposes, can be considered **non-deformable**.
— Between these uncemented or slightly cemented particles are voids with a total volume that can approach and sometimes exceed the same order of magnitude as the volume occupied by the particles (from half to several times greater).
— A soil is either a two-phase or three-phase system (solid, liquid ± gas).
— The voids, pores and interstices may be full of water, as in **saturated soils**, or contain air and water, as in **semi-saturated soils**. The degree of saturation conditions the response of the material as a whole. Under normal temperature and pressure conditions, water is considered to be incompressible.

The chemical composition of the soil differs depending on the original parent rock and the changes produced by particle weathering, deposition and cementation. Its components may include highly deformable organic matter, as well as silica, salts and carbonates, which help to cement the particles.

From a geological engineering point of view, **soil** is defined as an aggregate of uncemented or weakly cemented materials, with weak points of contact, which can be separated by low energy mechanical means, or by agitating in water.

Soil response to actions derived from engineering activities results in an interactive displacement of rotating and slipping particles and so this response depends on the:

— Proportions of the various solid materials in a unit volume of reference soil.
— Particle size and distribution.
— Ratio of the total volume to the volume of solids; the higher the ratio, the more deformable the soil will be.
— Average size of voids.

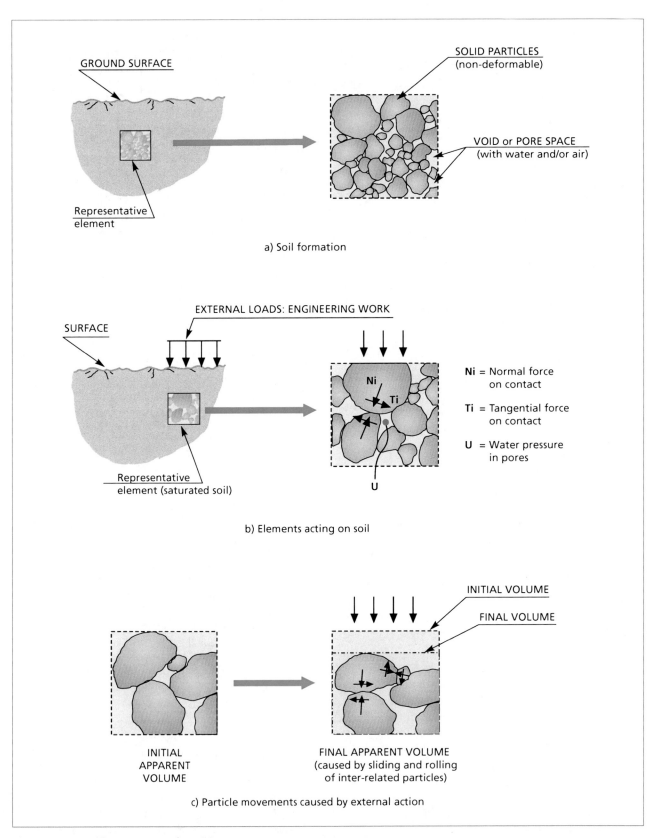

Figure 2.3   Soil as a system of particles.

The complexity of soil behaviour means that the following problems must be considered:

— Problems of **deformability**, produced by loads and external actions, causing normal and shear stresses on inter-particle contact. High deformability can lead to failure, where the change in apparent volume resulting from a change in external loading mobilises **strength**, a property determined by deformability rather than by particle failure.
— Problems of **flow**. Water flow in the soil conditions its response to applied load because load-induced deformations develop over the time the soil takes to expel or absorb water. This **consolidation** process is needed to stabilize the changes induced within a soil by external actions.

## 2.2 Soil description and classification. Phase relationships

## Types of soils

Soil is a complex material, very variable both in chemical composition and particle size. Studying soils requires language and terminology that can be easily understood by specialists in different countries and in different fields. The particles which make up soils are therefore classified into four main groups, depending on their size:

— **Gravels:** grain size between 8–10 cm and 2 mm with grains visible to the naked eye. Low water retention because of surface inactivity and large inter-particle spaces.
— **Sands:** particle size between 2 and 0.060 mm, still recognizable with the naked eye. They do not form continuous aggregates when mixed with water and readily separate from it.
— **Silts:** particle size from 0.060 to 0.002 mm. Some classifications give a lower value of 0.005 mm, but for practical purposes there is hardly any difference between the two. Better water retention than larger sized particles. A silt and water paste placed on the palm of the hand readily exudes water when tapped.
— **Clays:** formed from smaller particles than silt (0.002 mm) with gel-sized particles resulting from chemical changes. They are formed mainly from silicate minerals composed of chains of tetrahedral and octahedral elements (with the silicate ion at the centre of regular structures) and are joined by weak covalent bonds that allow water molecules to enter the chains. This may sometimes cause volume increases which return to their former state as the water evaporates. The resulting structure has a high water retention capacity with small inter-particle spaces and a large absorbent particle surface area. As a result clays are generally problematic materials, requiring a long time for consolidation or expulsion of water under loads (see Sections 2.7 to 2.9).

These particle sizes can be found naturally occurring in various proportions, the percentages of which are defined by laboratory analyses and dictate a description of the soil.

## Particle size distribution

Particle size analysis is carried out to determine the percentage by weight of particles within the different size ranges. For particles larger than 0.075 mm the sieving method is used. The sedimentation method is used for particles of 0.075 mm or less, using a hydrometer. In the first method a sample of soil is dried and the particles are disaggregated, then sifted by shaking through a series of sieves. These normally decrease in size in geometric progression, with a scale factor of 2. The material remaining in each sieve is weighed and the percentage of material $C_j$ that passes through a sieve with diameter $D_j$ can be determined from the given initial weight of the sample:

$$C_j = \frac{\sum_{i=j+1}^{n+1} W_i}{W} \times 100 \qquad W = \sum_{i=1}^{n+1} W_i$$

where $W$ is the total dry weight of the sample, and $W_i$ is the weight retained by the sieve with diameter $D_j$. $W_{n+1}$ is the weight that passes through the finest sieve used and is retained by the solid base on which the column of sieves sits.

This data can be used to show the soil **particle size distribution** as a curve, plotting $C_j$ against log $D_j$. Figure 2.4 shows the corresponding to: 1) sand with gravels, 2) fine dune sand, 3) silty sand, 4) silt, 5) silty clay.

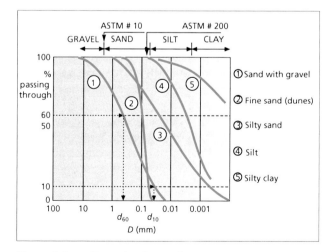

Figure 2.4   Example of a grading curve.

A clearer definition of the soil particle size distribution curve is obtained by using two coefficients:

- **Coefficient of uniformity,** $C_u$, is the ratio between the diameter of the sieve where 60% of the material passes through, and of the sieve where 10% passes through (Figure 2.4). If $C_u$ is less than 5, the soil has a uniform grain size; if $C_u$ is between 5 and 20 it is slightly uniform, and if $C_u > 20$ it is a well-graded soil. The more uniform the soil particle size distribution curve, the more uniform the void size will be. This makes the soil less dense and more liable to erosion.
- The **fine particle content** is the percentage of soil that passes through ASTM sieve N° 200 (0.075 mm). This percentage indicates the proportion of clay and silt contained in the soil, and is related to potential water retention. The greater the content of fine particles, the more difficult it will be to expel water.
- **Coarse-grained soils** have more than 50% of the particles larger than 0.075 mm (retained by ASTM sieve N° 200), and **fine-grained soils** have more than 50% equal to or smaller than 0.075 mm (passes through sieve N° 200).

## Plasticity

Measuring the soil particle size distribution is the first step in soil classification, but in some soils this is not so clear, e.g. in mixtures of silts, clays and sand. For this reason, agricultural indexes are used which define the soil consistency based on its water content. **Consistency** is the ratio of the water mass in the soil to the solid particle mass. The **water content** is the weight of the water in the soil divided by the weight of dry soil. The weight of the water is calculated from the difference between the weight of the soil sample before and after it is oven-dried for the time required for the water to evaporate

**Atterberg** defined three limits: (i) **shrinkage limit** or transition between solid and semi-solid state; (ii) **plastic limit**, PL, separating semi-solid from plastic state; and (iii) **liquid limit**, LL, separating plastic from semi-liquid state. The last two of these limits, the most commonly used in practice, are determined from the soil fraction that passes through ASTM sieve N° 40 ~~(0.01 mm)~~. 0.425 mm

The **plastic limit** is determined by kneading dry soil with a little water to form small balls and then rolling them out with the palm of the hand on a smooth surface to a diameter of about 3 mm and a length of about 25–30 mm. If at this stage the rolls crack into pieces of about 6 mm long, their water content corresponds to that of the plastic limit (found by oven-drying various rolls in similar conditions). If they do not crack, the rolls are reshaped into a ball and rolled in the hand, until they lose water content and re-rolled, this being repeated until they start cracking.

There are two main methods for the determination of the **liquid limit**: the percussion method (originally proposed by Casagrande) and the fall cone method (originally proposed by the Geotechnical Commission of the Swedish State Railways). The liquid limit can be obtained using the percussion method by kneading dry soil (previously disaggregated) with sufficient water to make a suspension with the consistency of yoghurt, and putting this into the **Casagrande cup** mould (Figure 2.5). Using a grooving tool, a groove about 2 mm wide at its lowest point is then cut across the centre of the mass. The mould is placed on

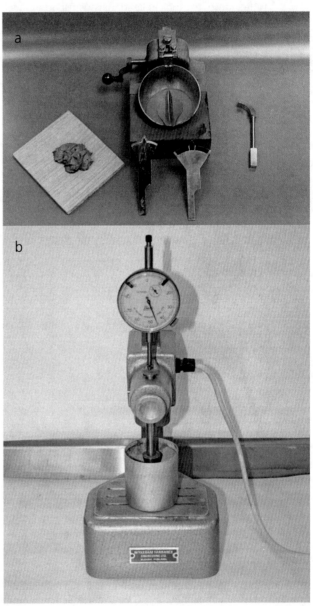

Figure 2.5  a) Casagrande device to determine the liquid limit, showing the moulded clay on the cup with the groove already cut. Three types of grooving tool are also shown. b) Fall cone apparatus.

a base and subjected to regular blows, which are counted. The liquid limit is the water content of the sample when the groove closes along a distance of some 12 mm after 25 blows. As it is difficult to achieve this, the water content is determined by interpolation from several samples, in which 12 mm closure has to be achieved with more or less 25 blows. At least three tests for the same soil should be made at varying moisture content.

The **fall cone test** apparatus consists of a metal cone of a certain weight and a certain apex angle suspended vertically with its apex just touching a horizontally levelled sample of soil (Figure 2.5). The cone is then released and allowed to penetrate the soil under its own weight, and the depth of penetration is measured.

Depending on the countries, the weight and the apex angle of the cone differ. According to the original definition (Swedish Standard SS 02 71 20) the liquid limit is the water content of a remoulded clayey soil at which the depth of penetration of the cone with a mass of 60 g and an apex angle of 60° is 10 mm. According to the British method (BS 1377) the liquid limit is the moisture content at which a cone of apex angle of 30° and weigh of 80 g penetrates 20 mm in 5 s when allowed to drop into the soil. To calculate the liquid limit, three or more tests at varying moisture contents of soil must be conducted, and the corresponding cone penetration depths, $d$, measured. If a semilogarithmic graph is plotted with moisture content, $w$, versus cone penetration depth, $d$, the moisture content corresponding to $d = 20$ mm is the liquid limit.

The depth of penetration is related with the undrained shear strength (see Section 2.6) of clayey soils (Hansbo, 1957):

$$S_u = k \frac{W}{d^2}$$

where $S_u$ is the undrained shear strength, $W$ is the weight of the cone, $d$ is the depth of penetration and $k$ is a constant depending on the apex angle and the degree of remoulding of the soil (for remoulded clays, $k = 0.8$–$0.85$ for an apex angle of 30° and $k = 0.27$–$0.29$ for an apex angle of 60°; for undisturbed clays and an apex angle of 30°, $k = 1.0$). These coefficients indicate that the undrained shear strength at the liquid limit is 1.59–1.7 kPa for the 60/60 fall cone and 1.57–1.67 kPa for the 30/80 fall cone. Ranges for this value between 1.6 and 2.3 have been suggested by different authors.

The fall cone method can also be used to obtain the plastic limit. In this case, the weigh of the cone is 240 g; the moisture content corresponding to a cone penetration depth of 20 mm will be the plastic limit.

Once the $LL$ and $PL$ have been found, a point representing each soil sample can be obtained from the **Casagrande Plasticity Chart** (Figure 2.6), which shows the ratio between the liquid limit, $LL$, and the **plasticity index** $PI$ ($PI = LL - PL$ represents the moisture content interval for passing from a semi-solid to a semi-liquid state). From a series of practical studies, Casagrande defined soils with $LL > 50$ as having high plasticity, i.e. they absorb a large amount of water and may experience considerable plastic deformation. Below this value, soils are considered to have low plasticity. He also defined an **A-line** (Figure 2.6) parallel to the direction in which samples are generally distributed when plotted on this chart.

The A-line and the low and high plasticity criteria are used to define various zones in the Casagrande Chart, shown in Figure 2.6. According to studies by Casagrande, silty soils, with an appreciable organic content, have a lower moisture content when passing from a semi-solid to a semi-liquid state, and are situated below the A-line, while clays are found above it. As a result, various types of soil can be defined: clays with low plasticity (CL), clays with high plasticity (CH), silts and organic soils with low plasticity (ML-OL), and silts and organic soils with high plasticity. In practice, the point corresponding to the known $LL$ and $PI$ values is recorded, so giving

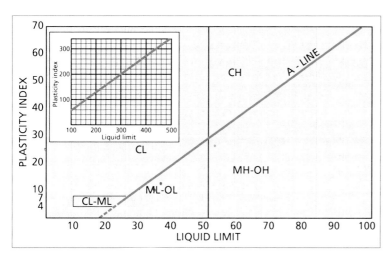

Figure 2.6    Casagrande plasticity chart.

| Major divisions (1) | Subdivisions (2) | USCS symbol (3) | Typical names (4) | Laboratory classification criteria (5) | |
|---|---|---|---|---|---|
| Coarse-grained soils (More than 50% retained on No. 200 sieve) | Gravels (More than 50% of coarse fraction retained on No. 4 sieve) | GW | Well-graded gravels or gravel-sand mixtures, little or no fines | Less than 5% fines* | $C_u \geq 4$ and $1 \leq C_c \leq 3$ |
| | | GP | Poorly graded gravels or gravelly sands, little or no fines | Less than 5% fines* | $C_u < 4$ and/or $1 > C_c > 3$ |
| | | GM | Silty gravels, gravel-sand-silt mixtures | More than 12% fines* | Minus No. 40 soil plots below the A-line |
| | | GC | Clayey gravels, gravel-sand-clay mixtures | More than 12% fines* | Minus No. 40 soil plots on or above the A-line |
| | Sands (50% or more of coarse fraction passes No. 4 sieve) | SW | Well-graded sands or gravelly sands, little or no fines | Less than 5% fines* | $C_u \geq 6$ and $1 \leq C_c \leq 3$ |
| | | SP | Poorly graded sands or gravelly sands, little or no fines | Less than 5% fines* | $C_u < 6$ and/or $1 > C_c > 3$ |
| | | SM | Silty sands, sand-silt mixtures | More than 12% fines* | Minus no. 40 soil plots below the A-line |
| | | SC | Clayey sands, sand-clay mixtures | More than 12% fines* | Minus No. 40 soil plots on or above the A-line |
| Fine-grained soils (50% or more passes the No. 200 sieve) | Silts and clays (liquid limit less than 50) | ML | Inorganic silts, rock flour, silts of low plasticity | Inorganic soil | PI < 4 or plots below A-line** |
| | | CL | Inorganic clays of low plasticity, gravelly clays, sandy clays, etc. | Inorganic soil | PI > 7 and plots on or above A-line** |
| | | OL | Organic silts and organic clays of low plasticity | Organic soil | LL (oven dried)/LL (not dried) < 0.75 |
| | Silts and clays (liquid limit 50 or more) | MH | Inorganic silts, micaceous silts, silts of high plasticity | Inorganic soil | Plots below A-line |
| | | CH | Inorganic highly plastic clays, fat clays, silty clays, etc. | Inorganic soil | Plots on or above A-line |
| | | OH | Organic silts and highly plastic organic clays | Organic soil | LL (oven dried)/LL (not dried) < 0.75 |
| Peat | Highly organic | PT | Peat and other highly organic soils | Primarily organic matter, dark in colour, and organic odor | |

$C_u$ (coefficient of uniformity) = $D_{60}/D_{10}$; $C_c$ (coefficient of curvature) = $(D_{30})^2/(D_{10} \times D_{60})$.
* "Fines" are those soil particles that pass the No. 200 sieve. For gravels and sands with between 5 and 12% fines, use of dual symbols is required (i.e., GW-GM, GW-GC, GP-GM, or GP-GC).
** If $4 \leq PI \leq 7$ and PI plots above A-line, then dual symbols (i.e., CL-ML) are required.

Figure 2.7   Unified Soil Classification System (USCS).

a classification that completes soil identification and shows the predominance of either the clay or silt fraction.

The **activity** is defined as the ratio of the plasticity index to the percentage of clay size particles. This ratio shows the degree of plasticity of the clay size fraction.

Casagrande completed this system of identification with particle size distribution data and developed the widely-used **Unified Soil Classification System** shown in Figure 2.7

In practice, the content of some chemical components is also determined in order to complete soil identification: organic material, (to determine the compressible part of the soil particles), sulphate content (to determine potential dissolution or attacks on concrete), and carbonate content (as a possible cementing agent). The mineralogical content of the clay fraction is determined when the soil shows high plasticity or other features of unfavourable geotechnical behaviour (e.g. swelling).

## Phase relationships

To define a soil's original state, several factors have to be determined including: (i) the relative concentration of solids, (ii) the relative volume of voids, and (iii) the relative water content in an elemental volume of soil. A simplified physical model, equivalent to the sample volume, is usually used for this, as shown in Figure 2.8. The model assumes that the whole volume of loose particles will be concentrated as a mass of solid, leaving the rest of the volume occupied by voids.

Two initial indexes defining soil state are porosity and void ratio. The **porosity** ($n$) is the ratio of the volume of the voids to the total volume of the soil. The **void ratio**, ($e$) is the ratio of the volume of the voids to the volume of solids. From the simplified model shown in Figure 2.8 the following expressions can be obtained:

$$n = \frac{e}{1+e} \qquad e = \frac{n}{1-n}$$

Porosity $n$ is usually used for rocks and the void ratio $e$ for soils. The void ratio normally varies between 0.30 and 1.30, although in loose soils with organic material it can reach values of 3 or more. The greater the void volume, the higher the void ratio will be, implying a weaker or softer soil with greater deformability.

# SOIL MECHANICS AND ENGINEERING GEOLOGY OF SEDIMENTS

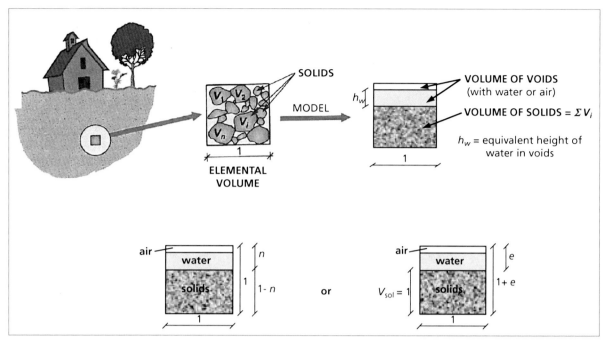

Figure 2.8    Simplified model equivalent to a representative soil sample.

Various parameters are used to estimate the relative concentration of solids and water:

— **Bulk density**, $\rho$, is the ratio of total mass to total volume (g/cm³ or kg/m³):

$$\rho = \frac{M_s + M_w}{V}$$

— **Dry density**, $\rho_{dry}$, is the ratio of solids to total volume:

$$\rho_{dry} = \frac{M_s}{V}$$

— **Specific gravity of solid soil particles**, $G$, is the ratio of the solid weight to the weight of an equal volume of water:

$$G = \frac{W_s}{V_s\, \gamma_w}$$

it usually ranges from 26 to 28 kN/m³.

— **Unit weight of solids**, $\gamma_s$, is the ratio of the weight of the solid soil particles to the volume of solids:

$$\gamma_s = \frac{W_s}{V_s}$$

— **Unit weight or specific weight**, $\gamma$, is the ratio of total weight to total volume:

$$\gamma = \frac{W_s + W_w}{V}$$

but depends on saturation.

— **Bulk unit weight**, $\gamma_{ap}$, is the ratio of the real weight of the soil sample (solid particles + natural water content at time of weighting) to total volume:

$$\gamma_{ap} = \frac{W}{V}$$

it normally varies from 15 to 21 kN/m³.

— **Dry unit weight**, $\gamma_d$, is the ratio of the solid weight (dry weight) to the total volume:

$$\gamma_d = \frac{W_s}{V}$$

it usually ranges from 13 to 19 kN/m³.

— **Saturated unit weight**, $\gamma_{sat}$, is the ratio of the solid weight plus the weight of water in the voids (assuming that the soil is saturated) to the total volume:

$$\gamma_{sat} = \frac{W_{sat}}{V}$$

it varies from 16 to 21 kN/m³.

— **Unit weight of water**, $\gamma_w$, is the ratio of the weight of water to the volume of water ($\gamma_w = 9.81$ kN/m³).

— **Water content**, $w$, is the ratio of the weight of water in the sample to the weight of solids:

$$w = \frac{W_w}{W_s}$$

It normally varies from 5–8% in granular soils, and from 60–70% in clayey soils, although in some organic and marshy deposits, such as peat, it can reach values of 300–400%.

— **Degree of saturation,** $S_r$, is the ratio of the volume of water contained in the sample to the volume of voids:

$$S_r = \frac{V_w}{V_{voids}}(\%)$$

it varies from 0 to 100%.

From *Figure 2.8* the following expressions can be deduced:

$$\gamma_d = \frac{G}{1+e} \qquad \gamma_{sat} = \frac{G+e\gamma_w}{1+e}$$

$$\gamma_{ap} = \gamma_d(1+w) \qquad S_r = \frac{W}{W_{sat}}$$

Defining the natural water content is equivalent to identifying the natural consistency of the soil so it is usual to compare the natural water content with the liquid and plastic limits. This can be done as shown in *Figure 2.9*, by plotting the depth, the natural water content and the water content corresponding to the liquid and plastic limits. This diagram gives an idea of the consistency and whether the samples represent different soils. The **liquidity index, LI** or $I_L$, is defined by the ratio of the natural water content of the soil sample minus water content at the plastic limit to the index of plasticity. It normally varies between 0 and 1, but may be negative in very dry soils:

$$LI = \frac{w - PL}{LL - PL}$$

In sands, where water retention and plasticity are very low or absent, a comparison of this type is not usually

| Table 2.1 | PROPERTIES OF COARSE GRAIN SOILS | | | |
|---|---|---|---|---|
| Coarse-grain soils | Relative density $Dr$ (%) | Dry unit weight $\gamma_d$ (kN/m³) | Water content $w$ (%) | Void ratio $e$ |
| Very loose | 0–15 | <14.0 | >16 | >0.9 |
| Loose | 15–35 | 14.0–16.0 | 12–16 | 0.65–0.9 |
| Medium dense | 35–65 | 16.0–17.5 | 8–12 | 0.55–0.65 |
| Dense | 65–85 | 17.5–18.5 | 6–8 | 0.4–0.55 |
| Very dense | 85–100 | >18.5 | <6 | <0.4 |

| Table 2.2 | PROPERTIES OF FINE GRAIN SOILS | | | |
|---|---|---|---|---|
| Fine soils | Liquidity index, $I_L$ | Dry unit weight $\gamma_d$ (kN/m³) | Water content $W$ (%) | Void ratio $e$ |
| Very soft | 1.00–0.80 | <14.0 | >55 | >1.30 |
| Soft | 0.80–0.65 | 14.0–16.0 | 40–55 | 1.0–1.3 |
| Firm | 0.65–0.40 | 16.0–17.0 | 25–40 | 0.7–1.0 |
| Stiff | 0.40–0.25 | 17.0–18.0 | 15–25 | 0.5–0.7 |
| Hard | <0.25 | >18.0 | <15 | <0.5 |

made. Comparison of water content is made, however, with the concentration of solids. For this purpose, the following aspects are considered: *a*) maximum dry unit weight (maximum possible content of solids in a given volume), $\gamma_{max}$, corresponding to a minimum void ratio, $e_{min}$, and *b*) minimum dry density (lowest possible content of solids in a given volume), $\gamma_{min}$, corresponding to a maximum void ratio, $e_{max}$. These two values can be determined in the laboratory. The **relative density** ($D_r$) or **density index** ($I_D$) is generally used to describe the degree of compaction of coarse-grained soils. It can be obtained from the following expression:

$$D_r = \frac{e_{max} - e}{e_{max} - e_{min}} = \frac{\gamma_{d\,max}}{\gamma_d} \cdot \frac{\gamma_d - \gamma_{d\,min}}{\gamma_{max} - \gamma_{d\,min}}$$

where $\gamma_d$ is the apparent dry unit weight of the granular soil in the sample, and $e$ its void ratio. From $D_r$ the relative compactness of the sample can be evaluated according to *Table 2.1* which gives the properties of coarse grained sandy soils. *Table 2.2* gives the properties of fine-grained soils, such as silts and clays.

## 2.3 Flow of water through soils

This section describes the basic concepts of water flow through porous media and through soils. General concepts of Hydrogeology are described in Chapter 4.

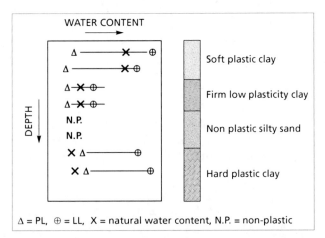

△ = PL, ⊕ = LL, X = natural water content, N.P. = non-plastic

*Figure 2.9* Situation of a real soil between consistency limits.

# Total head. Bernoulli's Theorem

For problems involving water seeping through soils under hydraulic gradient, the expression of energy at a certain point of the fluid can be defined by Bernoulli's Theorem:

$$H = z + \frac{u}{\gamma_w} + \frac{v^2}{2g}$$

where $H$ is the **total hydraulic head**, with the three elements:

— $z$: the **elevation head** above a chosen reference plane.
— $u/\gamma_w$: the **pressure head**, where $u$ is the pore water pressure at the point in question and $\gamma_w$ the unit weight of the water (9.81 kN/m$^3$).
— $v^2/2g$: the **velocity head,** where $v$ is the discharge velocity at the point under consideration and $g$ the gravitational acceleration.

The first two represent potential energy in terms of position, and the third term corresponds to kinetic energy; they are all expressed in units of length.

In the ideal case of a perfect incompressible fluid subject to a steady flow regime, Bernoulli showed that the total hydraulic head remains constant (Figure 2.10). Therefore, between any two points of the fluid in motion, the total energy produced by the head $H$ is maintained. This energy is simply transferred from some components to others (elevation, pressure or velocity head):

$$z_A + \frac{u_A}{\gamma_w} + \frac{v_A^2}{2g} = z_B + \frac{u_B}{\gamma_w} + \frac{v_B^2}{2g}$$

However, flow of water in the ground encounters obstacles, and any obstacle to the flow between two points also produces a head loss $\Delta H$. Also, **for there to be any flow there must be a difference in the total hydraulic head, i.e. an hydraulic gradient, for the water to flow from the higher head ($H_A$) towards the lower head ($H_B$)**. The difference $\Delta H = H_A - H_B$ represents the energy used to overcome the resistance along the path of flow, i.e. the amount of energy used in the process of flowing from A to B.

# Hydrostatic conditions

A very common specific example of where the total head remains constant is when the water is static (i.e. under hydrostatic conditions), or the pore water is static, with the hydrostatic pressure depending on the depth below the water table. Although the viscosity of the water is not zero and it is not a perfect fluid, there is no movement, so there is no reason to take obstacles into consideration. In addition, as the discharge velocity is zero, Bernoulli's Theorem is reduced to a binomial:

$$h = z + \frac{u}{\gamma_w}$$

where $h$ is the **piezometric level or head, relative to a defined horizontal datum.**

This equation, with the condition that $h$ remains constant throughout the liquid mass, allows the **water pressure** to be calculated at any point in the fluid. A simple example is a watertight container filled with water, such as a swimming pool (Figure 2.11). First, an arbitrary reference plane (i.e. a datum), $z = 0$ is chosen. Two points are selected in the liquid mass, one on the surface (A) and the other at an intermediate depth (B) for which the water pressure is to be calculated. From Bernoulli's equation it is known that $h_A = h_B$, where the piezometric head at A is:

$$h_A = z_A + \frac{u_A}{\gamma_w} = z_A$$

since it is on the surface of the water and its pressure is the same as atmospheric pressure (taken to be 0).

Now taking point B:

$$h_B = z_B + \frac{u_B}{\gamma_w} = h_A = z_A$$

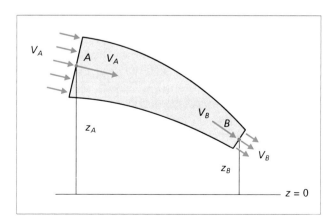

Figure 2.10   Bernoulli's theorem.

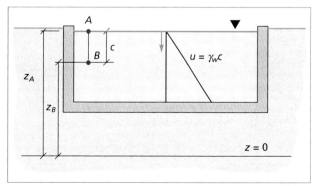

Figure 2.11   Hydrostatic pressure calculation.

> ## Box 2.1
>
>
>
> ### Using open pipe piezometers
>
> The concepts above can be applied directly when pore water pressure at any point in the ground (*B*) is required.
>
> An open pipe piezometer is inserted into the ground to the desired depth. After a time water will rise to level *A* to balance its pressure at *B*.
>
> Conditions inside the pipe are hydrostatic (with no loss of head) so piezometric values at *A* and *B* are equal. Therefore, water pressure at *B* will be:
>
> $$u_B = \gamma_w (z_A - z_B) = \gamma_w c$$
>
> and
>
> $$c = u_B / \gamma_w$$
>
> so that **the height of the water measured in an open pipe piezometer at any point in the ground is equal to the water pressure at that point divided by the unit weight of water.** Likewise, the pore water pressure in the ground at the base of an open pipe piezometer is equal to the height of the column of water in the piezometer multiplied by the unit weight of water.
>
>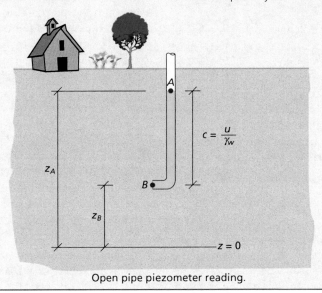
>
> Open pipe piezometer reading.

therefore:

$$u_B = \gamma_w (z_A - z_B) = \gamma_w c$$

This equation can be expressed as follows: the hydrostatic pressure at any point of a fluid at rest and situated at a given depth below the free surface of the fluid is equal to the unit weight of the fluid multiplied by that depth.

It therefore follows that in hydrostatic conditions the water pressure increases linearly with depth (*Figure 2.11*).

# Ground water flow

## Basic concepts. Head loss and permeability

Soil is a mass of particles with voids or pores that interconnect in such a way that water can flow through it. The flow path can be very tortuous because the water has to get round the many obstacles formed by the soil particles (*Figure 2.12*), and so there is loss of hydraulic head during the process.

The flow rate is a function of the soil grain size distribution. The particles of granular soils, such as sand, are of considerable size, and so are the pores between them. Head loss is small because it is easy for water to flow through them. However, in fine soils, such as clay, where particle size is measured in microns, the pores are also extremely small. In this case, it is much more difficult for water to flow and head loss will be considerable.

If the **coefficient of permeability**, *k*, of a soil is defined as a parameter influencing the velocity at which water flows through its fabric, it will be seen that permeability will depend on:

— Grain size and the distribution of grain sizes (and therefore pores) within the soil; the smaller the soil particles, the lower the value of *k*.

# SOIL MECHANICS AND ENGINEERING GEOLOGY OF SEDIMENTS

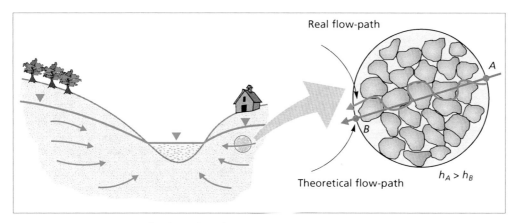

Figure 2.12   Seepage paths through soils.

- Soil density, taking into account that even with the same grain size distribution, the denser the ground is the closer the particles will be, thus the lower the void volume and value of $k$.
- The shape and orientation of the particles. If sedimentation conditions produce preferred orientations, permeability may vary considerably, depending on the direction of flow.

The coefficient of permeability was first expressed by Darcy in 1856. It is measured in units of velocity (m/s, m/day or cm/s) and is perhaps the hydraulic parameter which varies most depending on soil type. Some typical values are included in Table 2.3.

## Hydraulic head and hydraulic gradient

One characteristic of ground water flow is that it moves very slowly. A high velocity value would be around 0.6 m/min, giving an extremely small **velocity head** ($v^2/2g$) of only $5 \times 10^{-6}$ m. This value is negligible in comparison with the terms $z$ and $u/\gamma_w$, and is in fact much lower than the precision needed to measure the elevation head ($z$) at any point (Lambe and Whitman, 1979). In practice, therefore, the expression of hydraulic head in soils can be reduced to the elevation and pressure heads, i.e. the head measured in the piezometer, and often called the piezometric head:

$$h = z + \frac{u}{\gamma_w}$$

Water flows through soils from a point ($A$) with higher piezometric head ($h_A$) to another point ($B$) with lower head ($h_A > h_B$). Taking into consideration that the head loss $\Delta h = h_A - h_B$ takes place along length $L$, which is the distance separating the two selected points along a flow line, **the hydraulic gradient can be defined as the loss of piezometric head per unit length:**

Table 2.3   GUIDELINE VALUES FOR SOIL PERMEABILITY

| Type of soil | $k$ (cm/s) |
| --- | --- |
| Poorly-graded gravel (GP) | ≥1 |
| Uniform gravel (GP) | 0.2–1 |
| Well graded gravel (GW) | 0.05–0.3 |
| Uniform sand (SP) | $5 \times 10^{-3}$–0.2 |
| Well graded sand (SW) | $10^{-3}$–0.1 |
| Silty sand (SM) | $10^{-3}$–$5 \times 10^{-3}$ |
| Clayey sand (SC) | $10^{-4}$–$10^{-3}$ |
| Low plasticity silt (ML) | $5 \times 10^{-5}$–$10^{-4}$ |
| Low plasticity clay (CL) | $10^{-5}$–$10^{-8}$ |

(Powers, 1992).

$$i = \frac{\Delta h}{L}$$

## Darcy's law

There are two types of water flow: laminar and turbulent. The flow regime is considered to be laminar when there is no interference between the paths taken by the particles of water (flow lines). Otherwise, flow is turbulent. For the purposes of geotechnical engineering ground water flow regime is usually considered to be laminar, except in cases of highly permeable soils, or where the flow is through karst or large fractures. For laminar flow **Darcy's law** can be applied and the discharge velocity can be assumed to be proportional to the hydraulic gradient:

$$v = k \frac{\Delta h}{L} = ki$$

where $k$ is the permeability of the medium (expressed by the permeability coefficient), and $v$ the average discharge

## Box 2.2

### Calculating pore water pressure. Worked example

The stratigraphic column below the horizontal ground level of a wide valley is formed by 3 m of coarse gravels over a 12 m clay deposit. Beneath the clay is a substratum of highly permeable fractured sandstone. The water table in the gravel layer is 0.6 m below ground surface. In the sandstone substratum, however, conditions are artesian, with a piezometric height of 6 m above ground level.

Because the gravel layer is very permeable, conditions within it are assumed to be hydrostatic. Determine the following data for the clay layer:

a) Pore water pressure distribution.
b) Hydraulic gradient.
c) Pore water pressure at an intermediate point $P$, lying at a depth of 6 m below the surface ($\gamma_w = 9.81$ kN/m$^3$).

### Solution:

a) Water pressures at the base of the clay layer will coincide with those at the top of the sandstones. To calculate these, reference points $C$ and $O$ in the Figure are taken. Point $O$ is the height the water would reach in an open pipe piezometer situated at $C$. To simplify, the reference plane ($z = 0$) is assumed to be at the same level as $C$.

As already described, once water pressure equilibrium is reached, the regime in the piezometric pipe will be hydrostatic and therefore:

$$h_O = h_C$$
$$h_O = z_O + u_O/\gamma_w = z_O = 21 \text{ m} \Rightarrow h_C = 21 \text{ m}$$
$$u_C = \gamma_w (z_O - z_C) = 9.81 \times 21 = 206.01 \text{ kPa}$$

Points $A$ and $B$ are used to obtain the water pressure at the top of the clay layer. As conditions in the gravels are hydrostatic, this will again give:

$$h_A = h_B$$
$$h_A = z_A + u_A/\gamma_w = z_A = 14.4 \text{ m} \Rightarrow h_B = 14.4 \text{ m}$$

therefore

$$u_B = \gamma_w (z_A - z_B) = 9.81 \times 2.4 = 23.54 \text{ kPa}$$

As can be seen, $h_C > h_B$, so there will be upward flow.

b) $i = \Delta h/L = (h_C - h_B)/L_{CB} = (21 - 14.4)/12 = 0.55$

c) Point $P$ lies 9 m above $C$ and the pressure at that level is given directly from the line joining 23.54 kPa to 206.01 kPa. Bearing in mind that there is a 0.55 m loss of water pressure for each upward metre of flow:

$$h_P = h_C - (0.55 \times 9) = 16.05 \text{ m}$$

and

$$h_P = 16.05 = z_P + u_P/\gamma_w = 9 + u_P/\gamma_w$$
$$\Rightarrow u_P = 69.16 \text{ kPa}$$

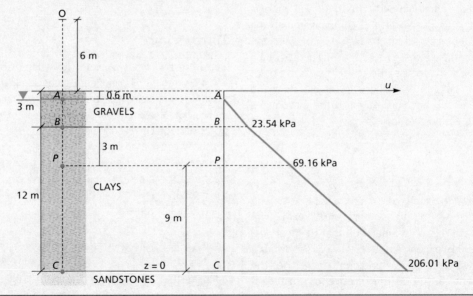

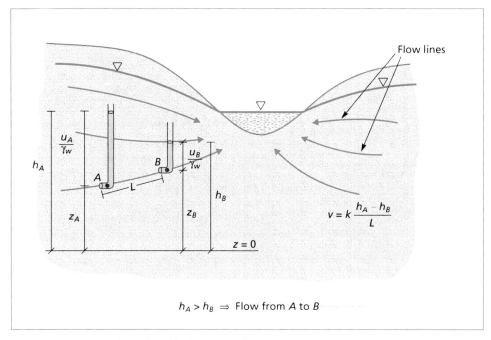

*Figure 2.13* Loss of head and hydraulic gradient.

velocity through a macroscopic cross section of soil, i.e. the apparent discharge velocity along the theoretical flowlines in *Figure 2.13*.

## Steady flow in an isotropic medium

As already mentioned, the coefficient of permeability may depend on the direction of flow. This coefficient can be considered as a tensor in a three-dimensional space, so the generalised form of Darcy's Law can be expressed as (*Figure 2.14*):

$$v_x = -k_x \frac{\partial h}{\partial x}; \quad v_y = -k_y \frac{\partial h}{\partial y}; \quad v_z = -k_z \frac{\partial h}{\partial z}$$

where

— $v_x, v_y, v_z$ are flow velocity components, on the $x$, $y$ and $z$ axes.
— $k_x, k_y, k_z$ are coefficients of permeability along the three main orthogonal directions.
— $-(\partial h/\partial x)$, $-(\partial h/\partial y)$, $-(\partial h/\partial z)$ are the hydraulic gradients on the three selected axes (note the $(-)$ sign; this is needed in the mathematical formulation, as discharge velocity is in the direction of decreasing head $h$).

If it is assumed that:

— Water is incompressible.

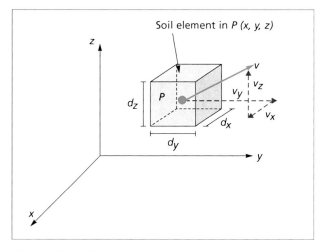

*Figure 2.14* Seepage velocity vector.

— $v$ and $u$ are exclusive functions of position ($x$, $y$ and $z$) where $u = \gamma_w h$ (*Figure 2.13*).
— Soil density is constant and the soil is saturated.

then the continuity equation in three dimensions (mass conservation) can be expressed in mathematical form. This states that in a steady flow regime, the difference between the volume of water entering a soil element per unit of time and the volume of water leaving must be zero (e.g. there are no springs or sinkholes present in the element). Therefore:

$$\frac{\partial v_x}{\partial x} + \frac{\partial v_y}{\partial y} + \frac{\partial v_z}{\partial z} = 0$$

Taking into account Darcy's Law generalized to three dimensions, this can now be written:

$$\frac{\partial v_x}{\partial x} = -k_x \frac{\partial^2 h}{\partial x^2}; \quad \frac{\partial v_y}{\partial y} = -k_y \frac{\partial^2 h}{\partial y^2};$$

$$\frac{\partial v_z}{\partial z} = -k_z \frac{\partial^2 h}{\partial z^2}$$

then:

$$k_x \frac{\partial^2 h}{\partial x^2} + k_y \frac{\partial^2 h}{\partial y^2} + k_z \frac{\partial^2 h}{\partial z^2} = 0$$

Finally, if the medium is isotropic ($k_x = k_y = k_z$):

$$\frac{\partial^2 h}{\partial x^2} + \frac{\partial^2 h}{\partial y^2} + \frac{\partial^2 h}{\partial z^2} = 0; \quad \text{written } \nabla^2 h = 0$$

This is **Laplace's equation**, which is applied to many problems of flow, such as the transmission of heat and electricity or, as in this case, water through a porous medium. A characteristic of this equation, which is often difficult to resolve analytically, is that it can be resolved in graphic form. This is done by plotting two series of curves that are orthogonal to each other and fulfil certain conditions (Figure 2.15).

One series represents **equipotential lines**, along which the piezometric head is constant. The other represents **flow lines**, which are perpendicular to the equipotential lines and tangential to the flow velocity vector at each point, (i.e. there is no flow perpendicular to them).

The procedure to follow is shown in the following simple, two-dimensional example (Figure 2.16): A watertight diaphragm wall is embedded halfway down into a permeable alluvial layer. Underneath there is a substratum with permeability 10 times lower than the alluvial layer (which means that in comparison this layer can be considered impermeable, with flow only through the upper layer.) The diaphragm wall protrudes above the ground surface and is used to dam up water to a certain height, so that the difference in height of the water surface on either side of it is $\Delta h$.

The graphic solution is obtained by the trial-and-error sketching of the flow net in the following steps:

1. The geometrical conditions are plotted to scale.
2. The boundary flowlines and equipotentials are drawn as follows:

   — Line CD is an equipotential; and all its points have the same piezometric head as point A because there are no total head losses in the retained mass of water between A and C; i.e. it is hydrostatic (Figure 2.11).
   — Line FG is an equipotential with the piezometric head of point B for the same reasons.
   — Line HI is an impervious boundary. As there is no flow through it, velocity is tangential to it and so it constitutes a flow line.
   — Line DEF is also an impervious boundary and constitutes another flow line.

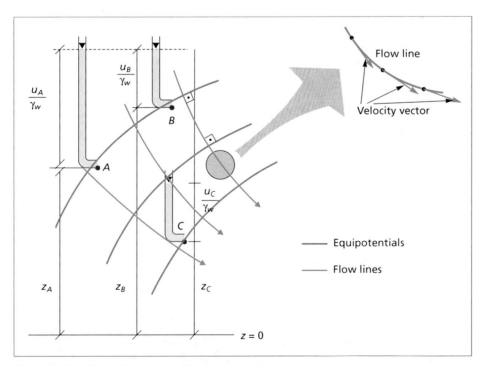

Figure 2.15   Graphic solution of Laplace's equation.

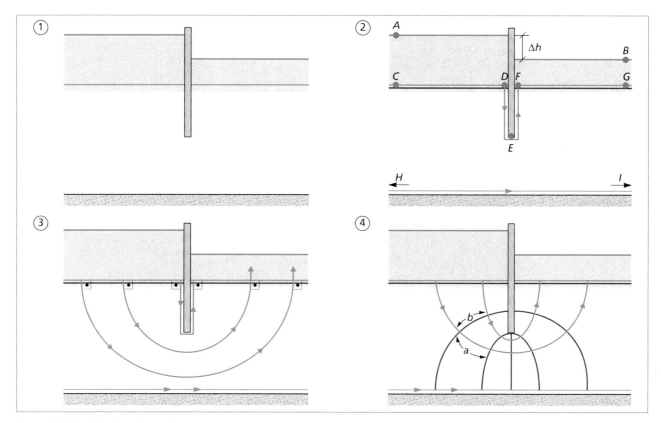

*Figure 2.16* Constructing a flow net.

3. A few additional flow lines are then plotted, perpendicular to the known boundary equipotentials.
4. The equipotential lines necessary to obtain curvilinear squares are then plotted so that both sets of curves are perpendicular to each other.
5. The result is checked and corrected if necessary. Corrections are usually needed to obtain better squares and orthogonality. This procedure is helped by checking if the diagonals of the curvilinear squares are also orthogonal, or if a circle can be drawn inside them.

Once an acceptable flow net has been plotted, it can be assumed that Laplace's equation for the field of flow has been satisfactorily solved. The main features of a flow net plotted this way are:

— **The total head loss ($\Delta h$) is distributed equally between the equipotentials.**
— **A flow channel is the "streamtube" between two adjacent flow lines.**
— **All flow channels carry the same volume of water per unit time, i.e. the same discharge.**

*Figure 2.17* shows the flow net obtained and some features of how to use it. If $N_f$ is the number of **flow channels** plotted, then in this case $N_f = 3$. On the other hand, the total head loss $\Delta h$ is distributed in $N_d = 6$ successive drops in potential. Given that the head loss between successive equipotentials is always the same, $dh = \Delta h/N_d$ will be lost between each.

If any element on the grid is now selected (element X in *Figure 2.17*), the total volume of water flowing across it will be:

$$q_x = k \frac{dh}{a_x} b_x = k \frac{\Delta h / N_d}{a_x} b_x$$

and given that the flow net plotted is squared ($b_x = a_x$):

$$q_x = k \frac{\Delta h}{N_d}$$

If the flow channels are taken to carry the same discharge, the total volume of water flow per unit of time will be:

$$Q = k \, \Delta h \frac{N_f}{N_d}$$

To calculate the pore water pressure at any point P, identify the equipotential line on which it is located; this gives its piezometric head which equals elevation head + pressure head. Subtracting the elevation head, $z$, gives the pressure

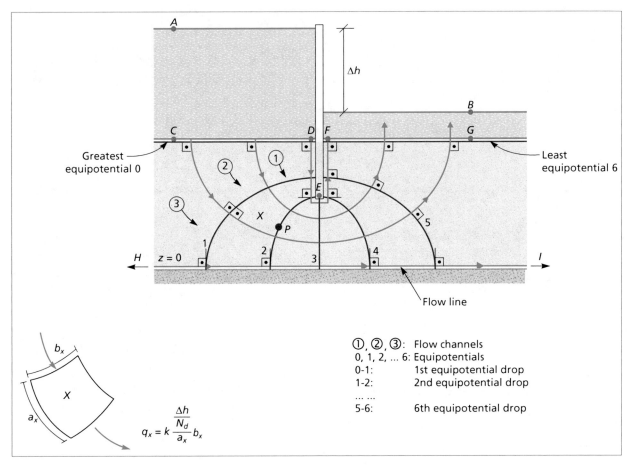

*Figure 2.17* Example of a flow net.

head, $h_p$. Thus, if point P in *Figure 2.17* is on the third equipotential line, two drops in piezometric head have occurred:

$$h_P = h_A - 2\frac{\Delta h}{N_d} = z_P + \frac{u_P}{\gamma_w}$$

then

$$u_P = \gamma_w \left( h_A - 2\frac{\Delta h}{N_d} - z_P \right)$$

If point P does not coincide with any of the equipotential lines plotted, the grid can be made denser in the area of P until an equipotential is made to pass through it.

## Anisotropic soil conditions

As already shown, if the soil is anisotropic the continuity equation is a function of the permeability. The two-dimensional case gives:

$$k_x \frac{\partial^2 h}{\partial x^2} + k_z \frac{\partial^2 h}{\partial z^2} = 0$$

This expression can be solved by making one of the following changes in coordinates:

$$X' = x; \quad Z' = z\sqrt{\frac{k_x}{k_z}}$$

or, alternatively:

$$X' = x\sqrt{\frac{k_z}{k_x}}; \quad Z' = z$$

since with the new coordinates, it is reduced in both cases to:

$$\frac{\partial^2 h}{\partial X'^2} + \frac{\partial^2 h}{\partial Z'^2} = 0$$

which is again Laplace's equation, and can be resolved graphically.

As a result, the flow net can be plotted by changing the scale of the space in which it is drawn; i.e. by transforming it. The flow net is then redrawn as if the medium were isotropic, and finally the transform is reversed to reveal the real flow net (*Box 2.3*).

## Box 2.3

### Flow net in an anisotropic medium. Worked example

The attached figure shows an example of *Figure 2.16* in a situation where the horizontal permeability is 9 times greater than the vertical. The steps to be followed to draw the flow net are:

1. Draw the geometry of the problem to scale (horizontal scale = vertical scale) marking points of interest (e.g. *P*, for calculating pore water pressures).
2. Select the axis to change and draw the geometry in the transformed space. In the case study, the simplest change is the one where the vertical axis does not change; this means the thickness of the permeable medium, the height of the water and height of the diaphragm wall remain the same (i.e. the original drawing is the same except for horizontal distances such as *P*).
3. Draw the flow net following the illustration in *Figures 2.16* and *2.17* as if the ground was isotropic.
4. Undo the change of axial dimension to obtain the flow net in real space where horizontal scale = vertical scale. Now the net will no longer fulfil the conditions of Laplace's equation with regard to orthogonality between equipotentials and flowlines.

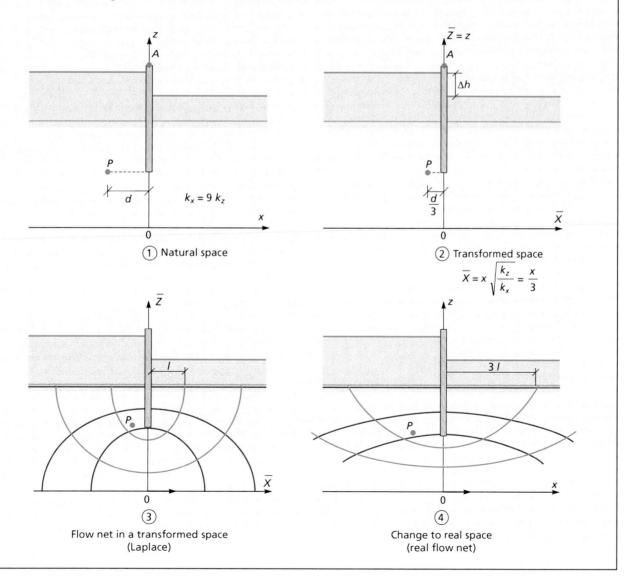

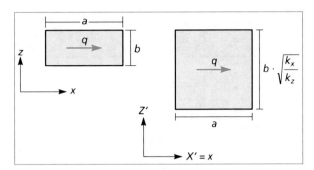

*Figure 2.18*  Space transforms required in an anisotropic medium.

*Figure 2.18* shows in diagram form a hypothetical element of ground, parallel to its coordinate axes, in both the real and the transformed space.

The discharge passing through the element will be the same, in both the real and the transformed space, and the same will occur with the head loss (*dh*) between the two limit equipotentials of the element. Therefore, real space will give:

$$q_x = k_x \frac{dh}{a} b$$

and transformed space:

$$q_x = \bar{k} \frac{dh}{a} b \sqrt{\frac{k_x}{k_z}}$$

where $\bar{k}$ would be the equivalent permeability of the transformed space.

Equalizing both expressions gives:

$$k_x = \bar{k} \sqrt{\frac{k_x}{k_z}}$$

and finally

$$\bar{k} = \sqrt{k_x k_z}$$

Therefore, the total volume of the water flow per unit time would be:

$$Q = \bar{k} \Delta h \frac{N_f}{N_d} = \sqrt{k_x k_z} \, \Delta h \frac{N_f}{N_d}$$

an expression which is valid for both spaces.

# Permeability and water flow in stratified soils

Soil deposits frequently consist of successive or alternating layers with different characteristics. This is typical of alluvial sediments, where it is very common to find alternating sub-horizontal arrangements of materials with widely differing grain sizes, and therefore different permeability. In these cases it is often useful to define the "equivalent permeability", representing the water flow through the strata as a whole.

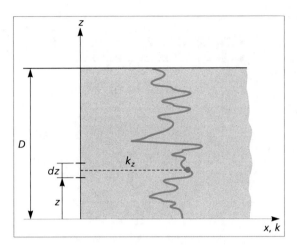

*Figure 2.19*  Permeabilty and steady flow through stratified medium.

## Vertical flow

*Figure 2.19* shows the theoretical case of a soil deposit of thickness (*D*), where permeability varies continuously with depth.

Considering vertical flow conditions through this stratified medium, it is easy to understand that if the volume of water flowing vertically through any horizontal section of the deposit, per unit time, is constant i.e. if flow was steady its discharge will also be constant.

If $k_v$ and $i_v$ are the permeability and the equivalent gradient respectively for the whole stratum (considering it as if it were a uniform layer), the previous observation gives:

$$v_z = k_z i_z = k_v i_v \implies i_z = \frac{k_v i_v}{k_z}$$

where:

— $k_z$ is the real vertical permeability of the soil at generic elevation *z*.
— $i_z$ is the real vertical hydraulic gradient at generic elevation *z*.

The piezometric head loss throughout the whole thickness *D* of the soil deposit will be:

$$\Delta h = i_v D = \int_0^D i_z dz \implies i_v D = \int_0^D \frac{k_v i_v}{k_z} dz = k_v i_v \int_0^D \frac{dz}{k_z}$$

and isolating $k_v$:

$$k_v = \frac{D}{\int_0^D \frac{dz}{k_z}}$$

## Box 2.4

### Permeability calculation. Worked example

Find the equivalent vertical and horizontal permeability of stratified ground composed of two layers of silty sands of thickness $L_1$ and permeability $k_1$, with a gravel layer of thickness $L_2$ and permeability $k_2$ sandwiched between them (*Figure A*). Direct application of the equation obtained will give:

$$k_v = \frac{L_1 + L_2 + L_1}{\frac{L_1}{k_1} + \frac{L_2}{k_2} + \frac{L_1}{k_1}}$$

$$k_h = \frac{1}{L_1 + L_2 + L_1}\left[k_1 L_1 + k_2 L_2 + k_1 L_1\right]$$

*Figure B* shows the hypothetical ground profile for an embankment dam site consisting of 20 m of silty sand alluvium over an impermeable stratum. Once the embankment is built, significant seepage through the alluvium is detected. Further investigation reveals the presence of a thin continuous layer of highly pervious gravels 0.10 m thick which went undetected in the preliminary site investigation. To determine the equivalent horizontal permeability of the stratified deposit and compare it with the one assumed in the project, the $k_h$ equation is applied directly in the conditions shown in the figure. This gives:

$$k_h = \frac{1}{20}\left[9.9 \cdot 10^{-5} + 0.1 \cdot 0.01 + 10 \cdot 10^{-5}\right]$$

$$\approx 6 \cdot 10^{-5} \text{ m/s}$$

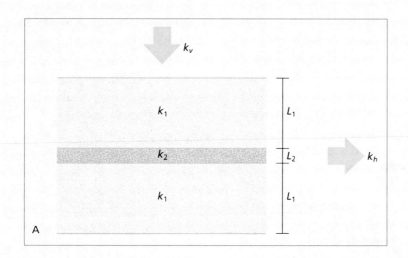

A

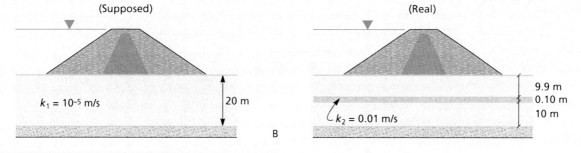

B

Therefore, for a ground profile formed by $n$ strata with thickness $L_i$ and permeability $k_i$, the equivalent vertical permeability will be:

$$k_v = \frac{\sum_{i=1}^{n} L_i}{\sum_{i=1}^{n} \frac{L_i}{k_i}}$$

As can be seen, the equivalent (real) permeability is 6 times greater than supposed. Bearing in mind that discharge is proportional to permeability, the infiltration recorded will be 6 times greater than what was initially expected.

### Horizontal flow

In this case, the hydraulic gradient has to be the same through the vertical section of the whole soil profile considered. So, if $k_h$ and $i_h$ are the equivalent permeability and the equivalent gradient respectively for horizontal flow, and flow is steady, the discharge through the whole soil mass will be:

$$Q_h = k_h i_h D = \int_0^D k_z i_h dz$$

and isolating $k_h$:

$$k_h = \frac{1}{D}\int_0^D k_z dz$$

This means that for a ground profile formed by $n$ strata with thickness $L_i$ and permeability $k_i$, the equivalent horizontal permeability will be:

$$k_h = \frac{\sum_{i=1}^{n} k_i \cdot L_i}{\sum_{i=1}^{n} L_i}$$

## 2.4 Effective stress

### Soil phases and soil structure

Soil is a material made up of particles with voids or pores between them. It normally has three distinct phases (*Figure 2.20*):

— Solid: **particles**.
— Liquid: **water**, which totally or partly fills the pores.
— Gas: **air**, which totally or partly occupies the pores.

This multiphase character is the main cause of difficulty for understanding soil behaviour in relation to external forces, because its response depends on the complex interaction between these different phases.

Observation of natural sedimentation processes in soil on a microscopic scale shows that the grains, when loaded from

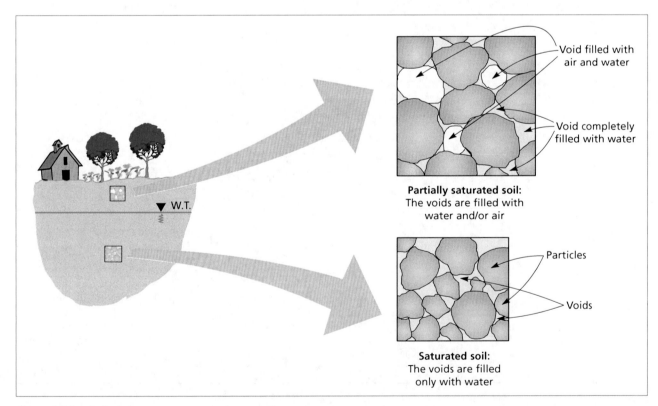

*Figure 2.20* Soil phases.

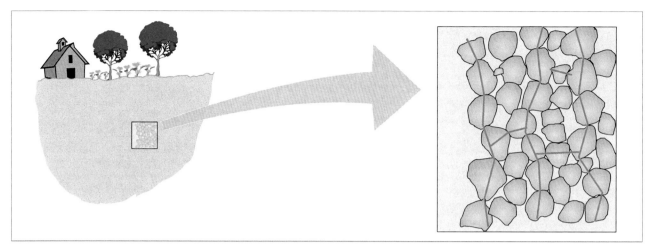

*Figure 2.21* Soil structure and operation of stress transmission "chains".

above, tend to organise themselves into mainly sub-vertical "chains". Further observation of how these forces from above are transmitted (basically, those due to gravity, i.e. the weight of the overlying soil) shows that transmission is precisely through grain-to-grain contact in these chains, and that particles outside them hardly receive or transmit any pressure (*Figure 2.21*).

If new forces from new geological processes or building work are added to existing soil, the soil fabric will tend towards a new state, characterized by a new **structure**.

Assuming that soil particles and water are undeformable, as is reasonable for practical purposes and normal construction activities, the new soil structure will be the result of the arrangement of particles that have slipped and rolled past each other. Examples of this are:

— **Soil compression** (reduction of volume) which basically consists of a reduction in the pore volume, i.e. a rearrangement of particles to form a denser structure with smaller spaces between them. If the soil is saturated, the reduction in the volume of the spaces must be associated with the expulsion of the same volume of pore water.
— **Soil swelling** (increase of volume) consists of an increase in the pore volume, i.e., a rearrangement of particles into a more open structure, with larger spaces between them. Again, in the case of fully saturated soil, the increase in the volume of the spaces will be associated with the absorption of an equal volume of water captured from the surrounding saturated soil.

Some basic characteristics of soil behaviour can be summarized from the descriptions above:

— Geological processes lead to a specific arrangement of particles, and a specific structure, characterized by a series of preferred orientations, both from geometry (spatial distribution) and from stress transmission (stress distribution).
— The presence of preferred orientations gives the soil a marked **anisotropic** character, i.e. its response to external forces (strength and deformability) will depend on the direction of the forces applied.
— Modifying the stress state may cause rearrangement of particles and new preferred directions. The new structure will depend both on applied forces (magnitude and direction) and on the original situation (initial structure). As a result, soil response (strength and deformability) will be a function of its **stress history**.

## Saturated soils. The principle of effective stress

As can be seen from the section above, a "microscopic" study of soil behaviour is complex, because of its structure and pressure transmission mechanisms. Clays and other fine soils are even more complicated because their particles are so small that the forces of gravity become less relevant then physical-chemical factors and so this microscopic approach is generally only used for research.

Classic soil mechanics has tended to study soil behaviour from a macroscopic perspective, as if it were a continuous medium. Even when simplifying it like this, different soil phases need to be considered to analyse their interaction and establish a basic theoretical framework, such as the one for saturated soils suggested by Terzaghi:

"The stress at any point on a plane through a soil mass can be calculated from the total principal stress, $\sigma_1$, $\sigma_2$, $\sigma_3$, acting on that point. If the soil pores are full of water under

## Box 2.5

### Shear stress and Terzaghi's principle

Terzaghi's principle refers only to the principal normal stress, and, by extension, to normal stress on any other orientation of the axes.

It is worth looking at what happens to shear stress. To do this, consider the state of stress of a saturated soil element (for simplicity, this is assumed to have horizontal strain in only two main directions). Its total main stresses $\sigma_1$ and $\sigma_3$ will be the result of stresses produced throughout its geological history plus those added by the load of the building constructed on the surface. If these stresses are known, the corresponding Mohr circle can be represented (shown in red in the attached figure). This will define the total state of stress of the soil element. Maximum shear stress will be given by the radius of the circle:

$$\tau_{max} = \frac{\sigma_1 - \sigma_3}{2}$$

The **pore water pressure**, $u$, referred to in the principle, is the pressure that would be registered by a piezometer situated at the same depth as the element. In the figure, the water conditions are defined by the water table, and are therefore hydrostatic. As a result, the height of the column of water inside the piezometer ($u/\gamma_w$) will reach the water table. Applying Terzaghi's principle, the principal effective stresses will be:

$$\sigma'_1 = \sigma_1 - u; \quad \sigma'_3 = \sigma_3 - u$$

With this data, a new Mohr circle can be plotted (shown in blue in the figure). As shown, it is identical to that of total stresses, but it is displaced on the x-axis at magnitude $u$ from pore pressure. Thus shear stresses are the same for a given $(\sigma_1 - \sigma_3)$ regardless of whether total stresses or effective stresses are considered.

The above result may also be demonstrated analytically using the same principle, since:

$$\tau'_{max} = \frac{\sigma'_1 - \sigma'_3}{2}$$
$$= \frac{\left[(\sigma_1 - u) - (\sigma_3 - u)\right]}{2} = \frac{(\sigma_1 - u - \sigma_3 + u)}{2} =$$
$$= \frac{\sigma_1 - \sigma_3}{2} = \tau_{max}$$

although perhaps the simplest explanation is the well-known fact that water cannot support shear stresses, so that those already existing in saturated soil must be completely carried by the solid soil skeleton.

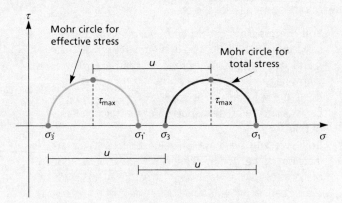

pressure $u$, the total principal stress will be composed of two parts. One part, $u$, called neutral pressure or **pore pressure**, acts on water and solid particles in all directions and with equal intensity. The differences $\sigma'_1 = \sigma_1 - u$, $\sigma'_2 = \sigma_2 - u$, $\sigma'_3 = \sigma_3 - u$ represent an excess of pressure on the neutral pressure $u$ and act exclusively on the solid phase of the soil. These fractions of total principal stress are known as **effective stress**.

Any measurable effect due to a change of stress, such as compression, distortion or modification of the shear strength of a soil is brought about exclusively by changes in effective stress".

As a main corollary, if there is no change in either volume or shape (i.e. no distortion) in a saturated soil, then effective stress are unchanged. This means that total and

## Box 2.6

### Stress in a homogeneous soil layer. Worked example

The stratigraphic profile in the attached *Figure (a)* is formed by a deposit of fine sands of thickness 10 m and saturated unit weight of 21 kN/m³. The water table is at ground surface level and the water conditions are hydrostatic (without flow). Plot total vertical stresses, pore pressures and effective vertical stress distribution.

### Solution:

**a) Total vertical stress**

Where there is a horizontal ground surface similar to that shown in *Figure (a)*, it is usually assumed that the vertical and horizontal directions correspond to the principal stress directions (see the soil element shown in *Figure (b)*).

Total vertical stress on a horizontal section of soil lying at a certain depth $z$ can be defined as the weight ($W$) of the column of soil that lies above that section, divided by its area ($S$). So, if a point in the ground ($P$) is assumed, like the one in *Figure (b)*, the total vertical stress will be:

$$\sigma_v = W/S$$

where $W$ is the sum of the weight of all the materials in the column (solid soil particles and pore water). To clarify this concept, four points, $A$, $B$, $C$ and $D$, situated at different depths, are shown in *Figure (c)*. Total vertical stresses are calculated as follows:

*Point A*: as this is at the surface, it is at atmospheric pressure and is therefore used as the reference pressure:

$$\sigma_v^A = 0$$

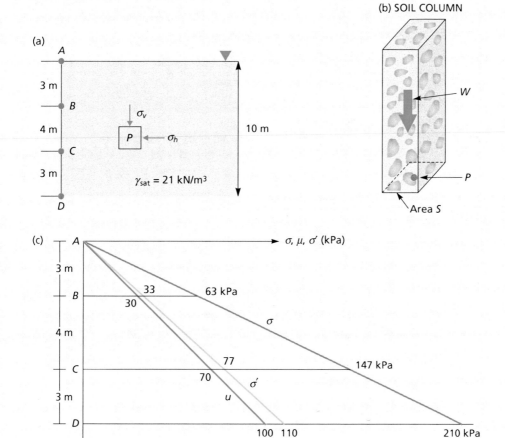

*Point B*: at a depth of $z = 3$ m. The soil that lies over it is saturated and has a unit weight $\gamma_{sat} = 21$ kN/m³ (unit weight includes particle and pore water weights). Thus, assuming a horizontal area $S = 1$ m² for the soil column:

$$\sigma_v^B = \frac{W}{S} = \frac{\gamma_{sat} \cdot z_B \cdot 1 \cdot 1}{1} = \gamma_{sat} \cdot z_B = 63 \text{ kN/m}^2$$

To summarize, **total vertical stress at a point in the soil at depth $z$ is equal to the unit weight of the ground lying above it, multiplied by the depth $z$.**

*Point C*: this point is at a depth of 7 m in the same saturated layer. Its total vertical stress will therefore be:

$$\sigma_v^C = \gamma_{sat} \cdot z_C = 21 \cdot 7 = 147 \text{ kN/m}^2$$

Total vertical stress at $C$ can also be expressed as the stress at point $B$ above, plus the stress generated by the weight of the soil column between $B$ and $C$, i.e.:

$$\sigma_v^C = \sigma_v^B + \gamma_{sat} \cdot (z_C - z_B) = 63 + 21 \cdot 4 = 147 \text{ kN/m}^2$$

This way of calculating stress can be directly applied when several strata with layers of different densities are involved.

*Point D*: using the previous concepts:

$$\sigma_v^D = \sigma_v^C + \gamma_{sat} \cdot (z_D - z_C) = 147 + 21 \cdot 3 = 210 \text{ kN/m}^2$$

b) **Pore pressures**

As the water conditions are hydrostatic, pore pressure at depth $z$ below the water table is given by the unit weight of the water multiplied by this depth. Taking $\gamma_w = 10$ kN/m³ will give:

$u_A = 0$ (at the surface of the water table at atmospheric pressure)

$u_B = \gamma_w \cdot 3 = 10 \cdot 3 = 30$ kN/m²

$u_C = \gamma_w \cdot 7 = 10 \cdot 7 = 70$ kN/m²

$u_D = \gamma_w \cdot 10 = 10 \cdot 10 = 100$ kN/m²

c) **Effective stresses**

Finally, applying Terzaghi's principle will give:

$$\sigma_v'^A = \sigma_v^A - u_A = 0$$
$$\sigma_v'^B = \sigma_v^B - u_B = 63 - 30 = 33 \text{ kN/m}^2$$
$$\sigma_v'^C = \sigma_v^C - u_C = 147 - 70 = 77 \text{ kN/m}^2$$
$$\sigma_v'^D = \sigma_v^D - u_D = 210 - 100 = 110 \text{ kN/m}^2$$

The corresponding stress distributions are drawn in the previous figure.

pore water stresses could be modified by the same amount, without the solid soil skeleton undergoing any change:

$$\sigma'_{initial} = \sigma_{initial} - u_{initial}$$

In a closed system where a change in applied pressure results in a change of equal magnitude to the pore pressure $\Delta\sigma = \Delta u = K$ (where $K$ is a constant). Under these circumstances a change to the initial applied stress ($\Delta\sigma$) produces a final effective stress of:

$$\sigma'_{final} = \sigma_{initial} + \Delta\sigma - (u_{initial} + \Delta u) =$$
$$\sigma_{initial} + K - u_{initial} - K = \sigma_{initial} - u_{initial} = \sigma'_{initial}$$

i.e. under these conditions a change in applied pressure results in no change in effective stress.

## Seepage forces and piping

Water flowing through soil exerts a frictional force on it. As already mentioned, for flow to occur there must be a difference in piezometric head, so that water flows from points of higher total head $h_A$ to those of lower head $h_B$. The difference $\Delta h = h_A - h_B$ represents the energy used to overcome the resistance the soil fabric creates to the flow of water through it.

This means that if the forces which resist flow are less than the forces which accompany flow, soil particles may be dragged along by the water, a phenomenon which may cause serious problems in a geotechnical context such as internal erosion (*Figure 2.22*).

Forces resistant to erosion depend on the soil cohesion, its particle size distribution and density. The soils most susceptible to erosion by water are fine, uniform, loose sands. The erosive force of the water in turn depends on its hydraulic gradient ($i = \Delta h/l$).

As shown in *Figure 2.22*, this phenomenon is generally localized and is usually due to heterogeneous soil behaviour (when in both a natural and a compacted state), and the presence of fissures or other factors that can concentrate flow.

If the flow is concentrated enough and there is sufficient hydraulic gradient near the outflow surface, the soil particles at the surface may be dragged away. This will lead to an increase in hydraulic gradient and thus erosive forces because the

## Box 2.7

## Stress in stratified soils. Worked example

The stratigraphic column under the horizontal surface of a wide valley is composed of 3 m of coarse gravels lying on 12 m of clay deposits. Below the clays is a layer of highly permeable fractured sandstone. The hydrogeological conditions are hydrostatic, with the water table situated at 0.60 m below ground surface. The bulk unit weights of the different soil strata are:

— Gravels (above the water table): $\gamma_g^1 = 16.8$ kN/m³
— Saturated gravels (below water table): $\gamma_g^2 = 20.8$ kN/m³
— Clay (saturated): $\gamma_c = 21.6$ kN/m³

Draw total vertical stress, pore pressure and effective vertical stress distributions in the soil layers (taking $\gamma_w = 10$ kN/m³).

### Solution:

S, A, B and C are taken as the calculation reference points. As can be seen, a change takes place at these points, due either to the presence of the water table or the stratigraphy.

a) **Total vertical stress**

Point S: this point is located at the surface, so:

$$\sigma_v^S = 0$$

Point A:

$$\sigma_v^A = \gamma_g^1 \cdot z_A = 16.8 \cdot 0.6 = 10.08 \text{ kPa}$$

Point B:

$$\sigma_v^B = \sigma_v^A + \gamma_g^2 \cdot (z_B - z_A) = 10.08 + 20.8 \cdot 2.4 = 60 \text{ kPa}$$

Point C:

$$\sigma_v^C = \sigma_v^B + \gamma_c \cdot (z_C - z_B) = 60 + 21.6 \cdot 12 = 319.2 \text{ kPa}$$

b) **Pore pressures**

$u_S = 0$ (at atmospheric pressure)

$u_A = 0$ (surface of water table at atmospheric pressure)

$u_B = \gamma_w \cdot (z_B - z_A) = 10 \cdot 2.4 = 24$ kPa

$u_C = \gamma_w \cdot (z_C - z_A) = 10 \cdot 14.4 = 144$ kPa

c) **Effective vertical stresses**

$\sigma_v'^S = \sigma_v^S - u_S = 0$

$\sigma_v'^A = \sigma_v^A - u_A = 10.08 - 0 = 10.08$ kPa

$\sigma_v'^B = \sigma_v^B - u_B = 60 - 24 = 36$ kPa

$\sigma_v'^C = \sigma_v^C - u_C = 319.2 - 144 = 175.2$ kPa

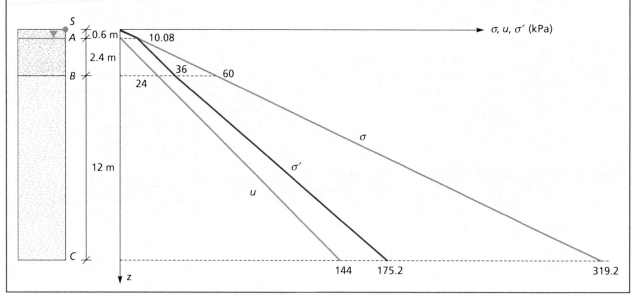

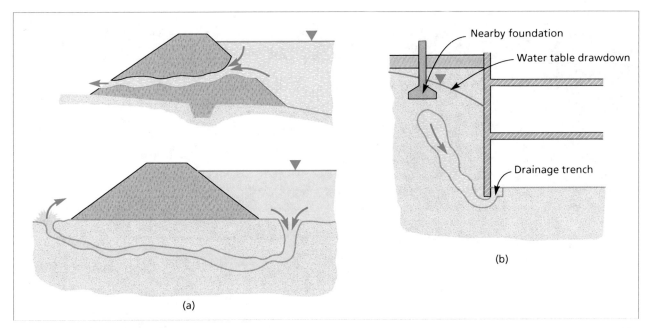

Figure 2.22  a) Piping phenomena in an embankment dam caused by internal erosion. b) Piping in an excavation caused by internal erosion.

approximate difference in piezometric head ($\Delta h$) is maintained, but the seepage path ($L$) is shortened due to loss of soil. This may cause progressive internal erosion in the soil, which in extreme cases can lead to the failure of an engineering structure.

Laboratory tests can be carried out to establish how susceptible soil is to internal erosion (see Section 2.9), and Chapter 12.

The frictional drag created as water flows pass and around soil particles is known as a seepage force because it exists only when seepage (i.e. groundwater flow) exists.

A simple method for establishing equilibrium conditions in relation to seepage forces is shown in *Figure 2.23*. Three possible situations in a constant head permeameter are illustrated. A sample of sand with height ($L$) is confined between wire meshes in the container. Above the soil a depth of water is kept at a constant height at all times (point $D$). Below it is a tube connecting the permeameter to an adjacent container; this is always full of water (up to point $A$) and can be moved up and down as required. Three open piezometers ($P1$, $P2$ and $P3$) lead out from inside the soil sample. If the base of the soil sample is taken as a reference plane ($z = 0$), a continuous measure of piezometric heads is simple, the only requirement being a calibrated rule, as shown in *Figure 2.23*. The water level reading on the rule is equal to the piezometric head $h$ ($h = z + u/\gamma_w$).

In *Figure 2.23* a) the height of the water in the adjacent container ($A$) is made to coincide with the water level in the permeameter ($D$). The conditions are hydrostatic, i.e., without any flow:

$$h_A = h_B = h_C = h_D = h_{P1} = h_{P2} = h_{P3}$$

This can be confirmed easily just by observing that the water levels in the permeameter, the adjacent container and the piezometers are at the same height. The pore pressures at each end of the soil mass are simply calculated as follows:

$$h_A = z_A + \frac{u_A}{\gamma_w} = z_A =$$

$$= L + \Delta L \begin{cases} h_A = h_C = z_C + \dfrac{u_C}{\gamma_w} = L + \dfrac{u_C}{\gamma_w} \Rightarrow u_C = \Delta L \cdot \gamma_w \\ h_A = h_B = z_B + \dfrac{u_B}{\gamma_w} = 0 + \dfrac{u_B}{\gamma_w} \Rightarrow u_B = (L + \Delta L) \cdot \gamma_w \end{cases}$$

The total vertical pressures are as follows:

Point C: $\sigma_{vC} = \Delta L \gamma_w$

Point B: $\sigma_{vB} = \Delta L \gamma_w + L \gamma_{sat}$

where $\gamma_{sat}$ = saturated unit weight of the soil in the permeameter.

Thus the effective vertical stresses:

$$\sigma'_{vC} = \sigma_{vC} - u_C = \Delta L \gamma_w - \Delta L \gamma_w = 0$$

$$\sigma'_{vB} = \sigma_{vB} - u_B = L(\gamma_{sat} - \gamma_w)$$

*Figure 2.23* b) shows a situation in which the water level in the adjacent container is at height $\Delta h$ above the free surface of the permeameter, giving a difference in piezometric head through the sample. Assuming there is no head loss where there is no soil (paths $AB$ and $CD$), gives:

Point B: $h_B = z_B + \dfrac{u_B}{\gamma_w} = h_A = z_A = L + \Delta L + \Delta h$

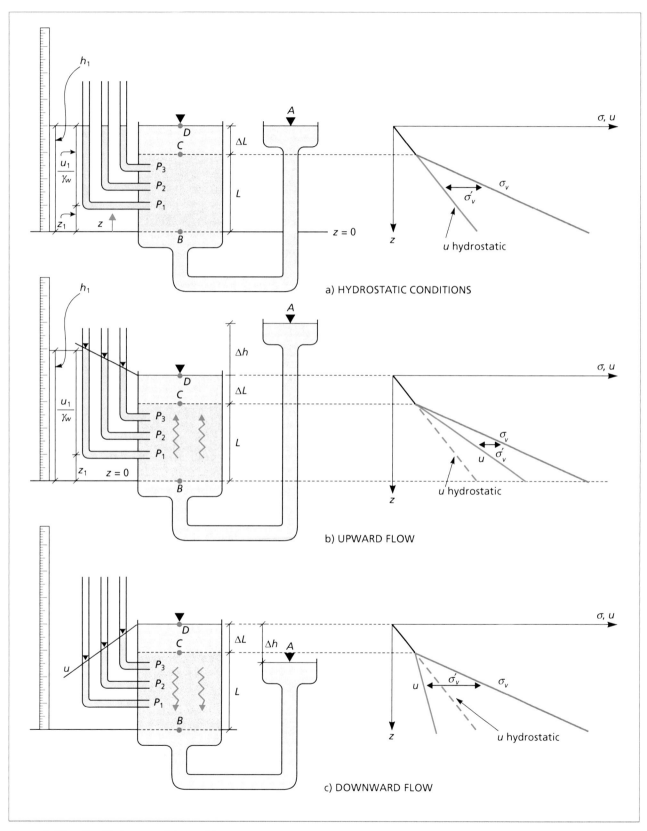

*Figure 2.23* Equilibrium conditions with seepage.

Point C: $h_C = z_C + \dfrac{u_C}{\gamma_w} = h_D = z_D = L + \Delta L$

Here the difference in piezometric head between the base and the top of the soil sample is equal to $\Delta h$, the flow in the soil mass is in an upward direction ($h_B > h_C$), and the resulting hydraulic gradient is: $i = \Delta h/L$.

Another way of checking seepage conditions is by direct observation of the piezometers. *Figure 2.23* b) shows clearly that the water level in piezometer $P_1$ is higher than in $P_2$, and this in turn is greater than in $P_3$, showing that $h_{P1} > h_{P2} > h_{P3}$ and that there is upward flow. The gradient can also be determined directly simply by reading the water height in each piezometer on the calibrated rule (i.e the real piezometric heads) and dividing these by flow paths between the piezometers, which are the differences in geometric height:

$$i = \dfrac{h_{P1} - h_{P2}}{z_2 - z_1} = \dfrac{h_{P2} - h_{P3}}{z_3 - z_2} = \dfrac{h_{P1} - h_{P3}}{z_3 - z_1}$$

For pore pressures at each end of the soil mass, this will give (remembering that $z_B = z_0 = 0$ = datum):

$$h_B = h_A = L + \Delta L + \Delta H = z_B + \dfrac{u_B}{\gamma_w} = 0 + \dfrac{u_B}{\gamma_w} \Rightarrow$$
$$\Rightarrow u_B = (L + \Delta L + \Delta h) \cdot \gamma_w$$
$$h_C = h_D = L + \Delta L = z_C + \dfrac{u_C}{\gamma_w} = L + \dfrac{u_C}{\gamma_w} \Rightarrow$$
$$\Rightarrow u_C = \Delta L \cdot \gamma_w$$

As can be deduced from the above equations and the piezometers in *Figure 2.23*, where there is upward flow, pore pressures in the soil mass are greater than in hydrostatic conditions. However the total vertical stress has not changed (the height of saturated soil at each point and the depth of water CD are unchanged from the hydrostatic case), effective vertical stress will be less. At point B this will give:

$$\sigma'_{vB} = \sigma_{vB} - u_B = (\Delta L \cdot \gamma_w + L \cdot \gamma_{sat}) - (L + \Delta L + \Delta h) \cdot \gamma_w$$
$$\sigma'_{vB} = L \cdot (\gamma_{sat} - \gamma_w) - \Delta h \cdot \gamma_w$$

The above expression suggests that if the difference in head $\Delta h$ is increased enough, effective vertical stress in the soil may be reduced to zero, a situation in the field of upward flow known as **piping**. In these conditions, soil without cohesion loses its shear strength completely and starts to behave like a fluid. Quicksands are the classic example of this.

The above expression can be formulated as a function of the hydraulic gradient $i = \Delta h/L$:

$$\sigma'_{vB} = L \cdot (\gamma_{sat} - \gamma_w) - i \cdot L \cdot \gamma_w = L \cdot (\gamma_{sat} - \gamma_w - i \cdot \gamma_w)$$

so that piping would occur at a specific hydraulic gradient $i_C$, or **critical hydraulic gradient**:

$$\gamma_{sat} - \gamma_w - i_C \cdot \gamma_w = 0 \;\Rightarrow\; i_C = (\gamma_{sat} - \gamma_w)/\gamma_w$$

Bearing in mind that a common order of magnitude for the specific saturated weight of a soil is $\gamma_{sat} = 20$ kN/m$^3$ and that the specific weight of water is in the region of $\gamma_w = 10$ kN/m$^3$, the critical gradient is usually found at around $i_C = 1$.

Due to the wide variety of ground conditions that can occur, the examples shown in *Figure 2.22* are just individual cases that are useful to illustrate examples of internal erosion and piping in general. When dealing with real problems involving water flow, it is essential to ensure an adequate factor of safety as a safeguard against internal erosion and piping phenomena.

*Figure 2.23* c) shows a third seepage alternative in which the adjacent container is at a lower level than the permeameter. Here the difference in the resulting piezometric head $\Delta h$ is the opposite of the previous example. Pore pressures at the two ends will be:

Point B: $h_B = z_B + \dfrac{u_B}{\gamma_w} = h_A = z_A = L + \Delta L - \Delta h$

Point C: $h_C = z_C + \dfrac{u_C}{\gamma_w} = h_D = z_D = L + \Delta L$

The difference in piezometric head is equal to $\Delta h$ but on this occasion the seepage in the soil mass is downwards ($h_C > h_B$) with hydraulic gradient $i = \Delta h/L$.

By observing the piezometers, it can be seen that the water level in piezometer $P3$ is higher than in $P2$, and this in turn is higher than $P1$, indicating that $h_{P3} > h_{P2} > h_{P1}$ and that the seepage is downward. As in the previous case, the gradient can also be determined directly from the piezometers.

For the pore pressures at each end of the soil mass, this will give (remembering that $z_B = z_0 = 0$ = datum):

$$h_B = h_A = L + \Delta L - \Delta h = z_B + \dfrac{u_B}{\gamma_w} = 0 + \dfrac{u_B}{\gamma_w} \Rightarrow$$
$$\Rightarrow u_B = (L + \Delta L - \Delta h) \cdot \gamma_w$$
$$h_C = h_D = L + \Delta L = z_C + \dfrac{u_C}{\gamma_w} = L + \dfrac{u_C}{\gamma_w} \Rightarrow$$
$$\Rightarrow u_C = \Delta L \cdot \gamma_w$$

Pore pressures in the soil mass are lower than in hydrostatic conditions so effective vertical stress will therefore have increased.

## Box 2.8

### Piping conditions. Worked example

The stratigraphic column under the horizontal surface of a wide valley is composed of 3 m of coarse gravels lying on 12 m of clay deposits. Below the clays is a layer of highly permeable fractured sandstone. The water table in the gravel layer lies 0.6 m below ground surface. In contrast, in the sandstone layer the water is under artesian conditions, with a piezometric height of 6 m above the surface of the ground. The apparent unit weights of the different soil strata are:

— Gravels (above the water table): $\gamma_g^1 = 16.8$ kN/m³
— Saturated gravels (below the water table): $\gamma_g^2 = 20.8$ kN/m³
— Clay (saturated): $\gamma_c = 21.6$ kN/m³

A large dry excavation has been projected in the valley, for which the water level has to be drawdown at the base of the excavation. Determine the depth at which conditions for piping would be reached if:

a) Artesian conditions are maintained in the sandstones.
b) Drainage wells are bored to lower the piezometric height in the sandstones by 6 m (unit weight of water $\gamma_w = 9.81$ kN/m³).

**Solution:**

a) The artesian conditions in the sandstone layer mean that if a piezometer is installed at e.g. point C, the water would rise up to 6 m above the surface of the valley; i.e.:

$$\frac{u_C}{\gamma_w} = 21 \text{m} \Rightarrow u_C = 21 \cdot 9.81 = 206.01 \text{ kPa}$$

Total vertical stress at C is:

$$\sigma_{vC} = 21.6 \cdot z$$

Assuming the ground the ground the ground the ground has no strength but only weight, piping will occur when:

$$\sigma'_{vC} = \sigma_{vC} - u_C = 0 \Rightarrow \sigma_{vC} = u_C$$

so that equalling the two previous expressions will give:

$$z = \frac{206.01}{21.6} = 9.54 \text{ m} \Rightarrow d = 15 - 9.54 = 5.46 \text{ m}$$

b) Using the same operation as in the previous example:

$$\frac{u_C}{\gamma_w} = 15 \text{m} \Rightarrow u_C = 15 \cdot 9.81 = 147.15 \text{ kPa}$$

$$\sigma_{vC} = 21.6 \cdot z$$

$$z = \frac{147.15}{21.6} = 6.81 \text{m} \Rightarrow d = 15 - 6.81 = 8.19 \text{ m}$$

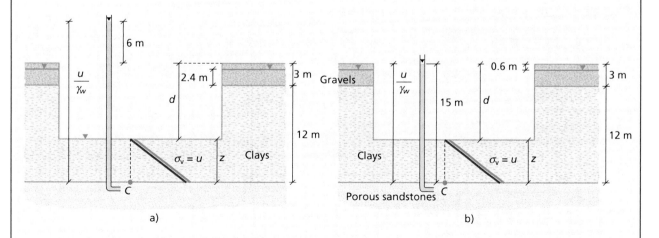

a)           b)

At point B this will give:

$$\sigma'_{vB} = \sigma_{vB} - u_B = (\Delta L \cdot \gamma_w + L \cdot \gamma_{sat}) - (L + \Delta L - \Delta h) \cdot \gamma_w$$

$$\sigma'_{vB} = L \cdot (\gamma_{sat} - \gamma_w) + \Delta h \cdot \gamma_w$$

## Loading saturated soils

### The concept of consolidation

When loads are applied to the soil, immediate changes take place in the total stress acting on it ($\Delta\sigma$). In the case of saturated soil, Terzaghi's principle establishes that such increments in total stress may lead to increased effective stress and/or pore pressures, but always in such a way that they agree with the fundamental equation of the principle, that is:

— Before the load is applied the total stress on an element of saturated soil is:

$$\sigma_0 = \sigma'_0 + u_0$$

— After the load ($\Delta\sigma$) is applied:

$$\sigma_0 + \Delta\sigma = (\sigma'_0 + \Delta\sigma') + (u_0 + \Delta u)$$

— Therefore:

$$\Delta\sigma = \Delta\sigma' + \Delta u$$

For practical purposes the effect of a finite load is considered to be limited to a certain area of influence in its near surroundings (Figure 2.24). As a result, only this area of soil will undergo changes in stress and measurable increases in pore pressure ($\Delta u$). In the rest of the soil, the initial equilibrium conditions ($\sigma_0$, $u_0$) will be essentially unaltered.

As described in Section 2.3, the difference in pore pressure (and piezometric head $h$) will cause a flow of water from inside the area of influence (higher $h$) to the outside (lower $h$). The process will obviously be temporary, because while the flow occurs the excess pore pressures originating inside the area of influence will gradually decrease. In fact, the flow will stop when these excess pore pressures are reduced to zero and equilibrium pore pressures are reached, again in accordance with the boundary hydrogeological conditions ($u = u_0$; $\Delta u = 0$). These concepts can be formulated according to Terzaghi's principle as follows:

— Immediately after the application of the load:

$$\Delta\sigma = \Delta\sigma'_{initial} + \Delta u_{initial}$$

— After a certain time (t):

$$\Delta\sigma = \Delta\sigma'_t + \Delta u_t$$

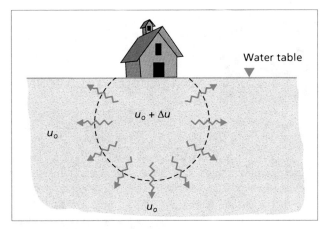

Figure 2.24  Excess pore pressure induced by ground loading (Lancellotta, 1995).

where:

$$\Delta u_t < \Delta u_{initial} \Rightarrow \Delta\sigma'_t > \Delta\sigma'_{initial}$$

— When equilibrium is finally attained:

$$\Delta\sigma = \Delta\sigma'_{final} + \Delta u_{final}$$
$$\Delta u_{final} = 0$$
$$\Delta\sigma = \Delta\sigma'_{final}$$

To sum up, the phases that take place when a saturated soil is loaded are:

1. Loading leads to an immediate increase in total stress ($\Delta\sigma$) in a defined area of influence.
2. According to Terzaghi's principle, $\Delta\sigma$ instantly divides into an increment in effective stress $\Delta\sigma'_{initial}$ and an increment in pore pressure $\Delta u_{initial}$, usually called excess pore pressure.
3. The development of $\Delta u_{initial}$ produces a difference in piezometric head between the soil situated within the area of influence and the rest of the soil, causing flow.
4. As the flow progresses, the excess pore pressures $\Delta u_{initial}$ inside the area of influence gradually decrease, and effective stress increases by the same amount, in accordance with Terzaghi's principle.
5. When equilibrium is finally reached and excess pore pressures disappear ($\Delta u = 0$), the entire increment in total stress applied at origin will have been wholly transformed into effective stress.

This process of dissipation of excess pore pressure generated by the application of a load to the ground is known as **consolidation**. Consolidation can be also defined as the gradual reduction in volume of a fully saturated soil due to the lowering of its pore pressure.

As with all problems involving seepage, the greater or lesser ease of flow and the corresponding dissipation of excess pore pressures will depend on the ground permeability. This means that in highly permeable granular soils there will be very rapid flow and dissipation will take place almost simultaneously when the load is applied (rapid consolidation). Conversely, in clays with very low permeability the flow will be slow, and dissipation will be spread over a considerable period of time (slow consolidation).

## Concepts of loading with and without drainage

Two basic concepts are derived from the mechanisms described above: **undrained loading conditions** and **drained loading conditions**.

The example in *Figure 2.24* showed that if the layer of saturated soil is composed of soils with low permeability, the transitory flow induced by excess pore pressure may last over a long time period; i.e. the less permeable the soil is, the slower the flow will be, and the more time excess pore pressure will take to dissipate and reach the final equilibrium defined by the hydrogeological boundary conditions.

In fact, with very low permeability soils, such as clays, it is reasonable to assume that hardly any appreciable flow occurs after instantaneous loading. As a result, there will be hardly any dissipation of the excess pore pressures after the load is applied. This is usually called "undrained" loading, as the water with excess pore pressure has not had time to "drain" out of the area of influence. As a complementary concept, remembering that in saturated soil any change in volume must be linked to variations in the volume of the voids within it, through the expulsion or absorption of water, it is obvious that in conditions of undrained loading there is no change in soil volume.

Undrained loading is a relative concept because whether the flow (or drainage) is easier or not after the application of a load and the consequent dissipation of excess pore pressure will depend on a series of other factors, shown in *Figure 2.25*:

— Ground permeability.
— Speed of loading.
— Proximity of highly permeable soils or drainage layers.

For example, when carrying out stability analysis for embankments to be built at a normal rate on a layer of saturated clays with low permeability, usual practice assumes undrained loading conditions (generally the most unfavourable

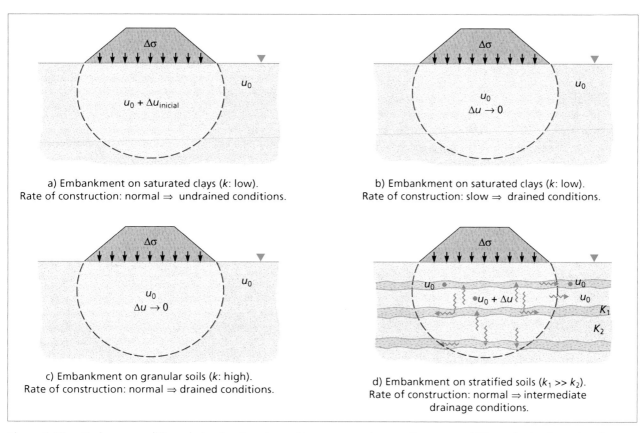

Figure 2.25  Drainage conditions during loading depending on ground properties, permeability and rate of construction.

hypothesis). Evidently, loading cannot be said to be "instantaneous" as embankment construction involves the placing and compaction of a considerable number of soil layers, a process that can take several days or weeks. However, if the permeability of the ground is low or very low, a normal construction process may be quick enough to prevent significant drainage taking place in the area of influence of the embankment, and it is therefore reasonable to assume **undrained conditions**. In other words, in situations with low permeability soils, a normal construction rate may be considered "fast" or "immediate" in geotechnical terms, although not in terms of real time.

In contrast, if the embankment were to be built on the same soil but slowly enough to allow all the excess pore pressure generated at each stage to dissipate gradually, then in spite of the impervious soil, the loading process would be slow enough to consider **drained conditions** (which are more favourable for stability). In fact, this would involve a process were loading is applied incrementally in small "instantaneous" increases or steps, leaving enough time between each one for excess pore pressure dissipation (i.e. consolidation) to occur. In this way, pore overpressures at any point in time would be limited to those associated with each small load increment or step, instead of those produced by the full height of the embankment.

Continuing with the same example, if the embankment were to be built on a very permeable soil, e.g. a medium to coarse grained sand, dissipation of excess pore pressures and the production of flow to reach equilibrium would take place very quickly, almost simultaneously with loading. Therefore, for practical purposes, the increments in total applied stress can be considered as immediately transformed into increments in effective stress. In these circumstances, even if loading takes place "quickly", **drained conditions** exist.

Finally, the presence of drainage layers in the area of excess pore pressure speeds up the dissipation process considerably as it facilitates the flow of water. This may occur in a stratified soil where low permeability clay layers alternate with higher permeability layers. In this case, the loading conditions may even be considered to be drained, depending on the proximity of the permeable horizons and the speed of construction.

At any given moment, the real situation will always lie somewhere in between truly drained and undrained conditions as these represent the extremes of the transitory dissipation process of excess pore pressures. As will be seen later, soil shear strength also depends on drainage conditions. This is evident from the second part of Terzaghi's principle, which states that "any measurable effect due to a change in stress, such as compression, distortion or the modification of shear strength of a soil, **is due exclusively to changes in effective stresses.**" Given that effective stresses vary throughout the transitory process of dissipation, the shear strength of the soil will also vary. In practice, therefore, it is essential to determine the different drainage conditions applicable to each particular problem.

## Undrained loading in saturated soils

At this point, it is clearly important to know how $\Delta\sigma'$ and $\Delta u$ are distributed throughout the transitory process, starting when a load is applied. The previous section mentioned that in soils with low permeability the moment "immediately" after the application of the increment in total stress is of special interest as it may be considered as similar to the undrained condition.

Reproducing these conditions in the laboratory is relatively simple as it is enough to carry out tests which prevent water flowing into or out of the soil sample tested. Alternatively, "quick" tests can be carried out where the speed of loading guarantees that there is practically no drainage. The main difficulty here stems from the fact that the increments in effective stress and pore pressures originated by increments in total stresses depend on the direction of loading.

In order to clarify this concept, *Figure 2.26* shows the most common types of laboratory loading assuming drained conditions. *Table 2.4* is a summary of the initial distribution of stresses when drainage is prevented.

If a soil is saturated and its fabric far more compressible than the water in its pores, a change in all-round pressure on the fabric $\Delta\sigma$ will result in a change of equal magnitude in the pressure of its pore water $\Delta u$, i.e. $\Delta u = \Delta\sigma$. The ratio $\Delta u/\Delta\sigma$ is called the pore pressure parameter $B$ and in saturated soils this is usually close to 1.0.

The simplest case is isotropic loading, in which the soil is subjected to equal increments of total stresses in three principal directions. In the absence of drainage, if the soil is saturated ($B = 1$), all the increment in total stress will be transmitted to the water in the pores, so effective stresses will not vary:

$$\Delta\sigma_1 = \Delta\sigma_2 = \Delta\sigma_3 = \Delta\sigma = \Delta u \Rightarrow$$

$$\Rightarrow \begin{cases} \Delta\sigma'_1 = \Delta\sigma - \Delta u \\ \Delta\sigma'_2 = \Delta\sigma - \Delta u \\ \Delta\sigma'_3 = \Delta\sigma - \Delta u \end{cases} \Rightarrow \begin{cases} \Delta\sigma'_1 = 0 \\ \Delta\sigma'_2 = 0 \\ \Delta\sigma'_3 = 0 \end{cases}$$

Thus, in accordance with Terzaghi's principle, the soil will not undergo any noticeable changes in either volume or shape in spite of loading; it will not be distorted and its shear strength will not be modified. If drainage is then permitted (by opening a valve in the test apparatus) the process of dissipation of excess pore pressures, i.e. consolidation, will begin until equilibrium is finally reached, expressed by:

$$\Rightarrow \begin{cases} \Delta\sigma'_1 = \Delta\sigma \\ \Delta\sigma'_2 = \Delta\sigma \, , \quad \Delta u = 0 \\ \Delta\sigma'_3 = \Delta\sigma \end{cases}$$

# SOIL MECHANICS AND ENGINEERING GEOLOGY OF SEDIMENTS

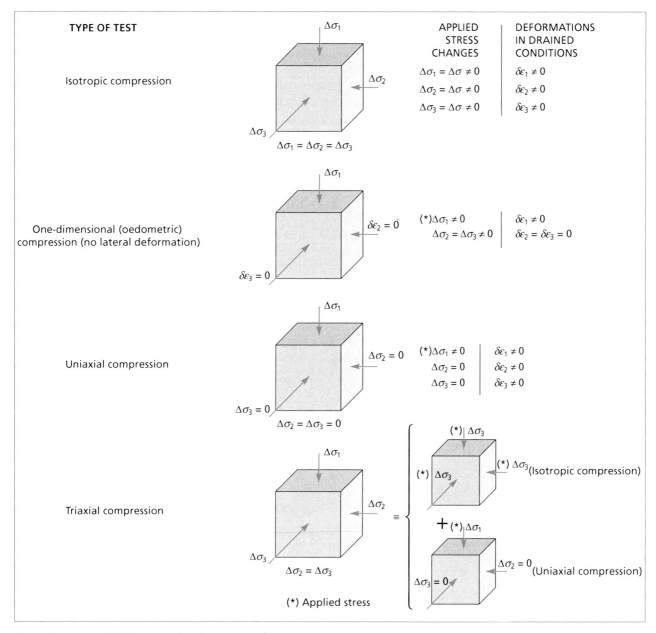

Figure 2.26  Typical laboratory loading systems for isotropic ground.

| Table 2.4 | STRESS DISTRIBUTION IN UNDRAINED CONDITIONS IN COMMON LOADING SYSTEMS | |
|---|---|---|
| **Type of load** | **Stress relationships** | **Notes** |
| Isotropic compression | $\Delta u = \Delta \sigma \Rightarrow \Delta \sigma' = 0$ | In general $\Delta u = B \cdot \Delta \sigma$<br>For saturated soils $B = 1$ |
| One-dimensional compression | $\Delta u = \Delta \sigma_1 \Rightarrow \Delta \sigma'_1 = 0$ | |
| Uniaxial compression | $\Delta u = A \cdot \Delta \sigma_1$ | For soft soils $A > 0.5$<br>For stiff soils $A < 0.5$ |
| Triaxial compression | $\Delta u = \Delta \sigma_3 + A \cdot (\Delta \sigma_1 - \Delta \sigma_3)$ | In general $\Delta u = B [\Delta \sigma_3 + A \cdot (\Delta \sigma_1 - \Delta \sigma_3)]$<br>For saturated soils $B = 1$ |

# Box 2.9

## Stress distribution. Worked example

The ground beneath a large lake consists of a 50 m thick deposit of clays with a rock substratum. The bed of the lake is flat and 20 m deep. The action of geological processes generates clays in suspension which in a very short time form a 2 m thick layer of sediments that completely covers the bottom of the lake.

Assuming the free water surface remains unchanged, determine the total vertical stress, pore pressure and effective vertical stress distributions:

a)  In the original situation.
b)  Immediately after the deposition of 2 m of clay, assuming this is deposited instantaneously.
c)  Final conditions once equilibrium is reached and the pore overpressures have been dissipated.

(Assume that the saturated unit weight of the clays is constant and equal to $\gamma_{sat} = 20$ kN/m³, that the unit weight of water is $\gamma_w = 10$ kN/m³, and that the rock substratum is impervious for practical purposes. The surface of the water in the lake is taken as the origin of the depth axis, $z$.)

## Solution:

### a) Original conditions

Given that the clay deposit is uniform (with constant unit weight), points A and B in the figure can be selected to obtain the stress distribution.

### Total vertical stress

*Point A*: at the bottom of the lake, so the only material above it is the 20 m high water column. If $z_w$ is the depth of the lake:

$$\sigma_v^A = \gamma_w \cdot z_w = 10 \text{ kN/m}^3 \cdot 20 \text{ m} = 200 \text{ kPa}$$

*Point B*: at the bottom of the clay deposit, so its total vertical stress will be that of point A, plus the stress corresponding to the weight of the column of saturated clay between A and B (remember that the saturated unit weight already includes the weight of the water that completely fills the soil pores):

$$\sigma_v^B = \sigma_v^A + \gamma_{sat} \cdot (z_B - z_A) = 200 + 20 \cdot 50 = 1{,}200 \text{ kPa}$$

### Pore pressures

*Point A*: as conditions are hydrostatic, the water pressure is given by the product of the unit weight of water and the depth of the point, measured from the surface of the water in the lake (the water table):

$$u_A = \gamma_w \cdot z_w = 10 \cdot 20 = 200 \text{ kPa}$$

*Point B*:

$$u_B = \gamma_w \cdot z_B = 10 \cdot 70 = 700 \text{ kPa}$$

### Effective vertical stress

*Point A*:

$$\sigma_v'^A = \sigma_v^A - u_A = 0 \text{ kPa}$$

*Point B*:

$$\sigma_v'^B = \sigma_v^B - u_B = 1{,}200 - 700 = 500 \text{ kPa}$$

(Note: These stresses do not depend on the height of the free water surface above the bottom of the lake, and whatever the depth of the lake, the resulting effective stresses are the same as those that would exist if the water level was at the top of the clay).

### b) Immediately after "instantaneous" sedimentation of an additional 2 m of clay

As the area of the lake is very large, it is reasonable to assume for practical purposes that the sediment is of infinite lateral extent. This means that any vertical section would be a plane of symmetry (as there is no difference between the vertical sections). Thus when an extensive (infinite) load is placed on the soil, strain can only be vertical, corresponding to **one-dimensional compression**, with zero lateral strain.

As seen above, immediately after loading, if the soil is not permeable, there would not have been time for drainage to take place. For one-dimensional loading without drainage, the increment change in total vertical stress (in this case an increase) is transformed into an equal increase in pore water pressure, and the effective stresses thus do not vary.

### Total vertical stress

*Point A*: after sedimentation, 18 m of water and 2 m of saturated clays are loaded on point A, therefore:

$$\sigma_v^A = 10 \cdot 18 + 20 \cdot 2 = 220 \text{ kPa}$$

(the increment in total vertical stress is $\Delta\sigma_v = 20$ kPa).

*Point B*:

$$\sigma_v^B = \sigma_v^A + \gamma_{sat} \cdot (z_B - z_A) = 220 + 20 \cdot 50 = 1{,}220 \text{ kPa}$$

### Effective stresses

As these stresses do not vary immediately on loading, they are the same as those in the original situation:

$$\sigma_v'^A = 0 \text{ kPa} \qquad \sigma_v'^B = 500 \text{ kPa}$$

### Pore pressures

These are given by Terzaghi's principle:

*Point A*:

$$u_A = \sigma_v^A - \sigma_v'^A = 220 - 0 = 220 \text{ kPa}$$

*Point B*:

$$u_B = \sigma_v^B - \sigma_v'^B = 1{,}220 - 500 = 720 \text{ kPa}$$

reflecting the increase in pressure of $\Delta\sigma_v = \Delta u = 20$ kPa compared with the initial situation.

### c) Final situation

The increase in pore water pressure noted above will cause an upward flow of water through the whole clay deposit because the rock substratum is impervious and the only drainage boundary is the ground surface (the lake bed). As the soil consolidates and excess pore pressure is reduced, effective stresses will increase according to Terzaghi's principle. The final hydrogeological equilibrium conditions will be the same as those in the original situation, i.e. hydrostatic pressures defined by the water level in the lake.

### Total vertical stresses

These correspond to the stress increment and are therefore are the same as those in section (b) above:

$$\sigma_v^A = 220 \text{ kPa} \qquad \sigma_v^B = 1{,}220 \text{ kPa}$$

### Pore pressures

Once the excess pore water pressures have been dissipated, pore pressures will be determined by the final conditions (c) that will be the same as those in the original situation:

$$u_A = 200 \text{ kPa} \qquad u_B = 700 \text{ kPa}$$

### Effective vertical stress

By applying Terzaghi's principle it can be proved that the increment in total stress will have been completely transformed into increased effective stress:

$$\sigma_v'^A = \sigma_v^A - u_A = 220 - 200 = 20 \text{ kPa}$$
$$\sigma_v'^B = \sigma_v^B - u_B = 1{,}220 - 700 = 520 \text{ kPa}$$

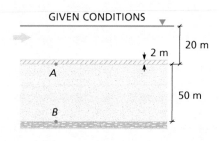

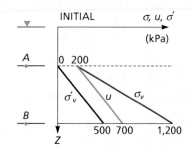

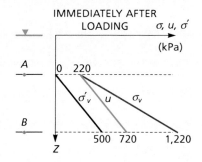

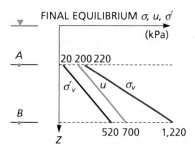

The sample will change its volume but not its shape. Consolidation occurs in one-dimensional or "oedometric" compression tests, in which lateral deformation of the soil sample is prevented while vertical stresses are applied. If drainage is impeded the applied increment in vertical total stress ($\Delta\sigma$) is fully transmitted to the pore water, and so effective stresses do not vary. If drainage is then permitted a process of dissipation (consolidation) will take place, and when equilibrium is reached, will give:

$$\Delta u = 0$$
$$\Delta\sigma'_1 = \Delta\sigma_1$$

Here the sample changes both its volume and its shape (it shortens in the direction of $\Delta\sigma_1$).

The two cases above are clearly exceptional, because the excess pore pressure to be dissipated is generated with little associated particle movement. However, in triaxial loading conditions when $\sigma_1 \neq \sigma_3$ an additional change in pore pressure can be generated by the movement of particles as they shear past each other, even in the absence of drainage. In such cases, Skempton (1954) proposed the following expression for the excess pore pressure in saturated soils, shown in Table 2.4:

$$\Delta u = \Delta\sigma_3 + A \cdot (\Delta\sigma_1 - \Delta\sigma_3)$$

where $A$ is the pore pressure parameter depending on the type of soil, and varies throughout the loading process.

If increments in total stress due to loading are known, and resulting excess pore pressure can be measured, then increments in effective stress can be calculated by applying Terzaghi's principle. Other loading systems can produce other stress distributions.

## 2.5 Consolidation and compressibility

### Normally consolidated and over-consolidated soils

#### Processes of consolidation

The structure and stress-strain characteristics of a soil depend on its geological history. Figure 2.27 represents the simple case of a laterally extensive deposit of sediment in a watery environment, e.g. marine or lacustrine clays and silts, over a period of time.

If the sediment surface is horizontal and covers a wide area (infinite for practical purposes), any vertical section through the sediment can be considered as a plane of symmetry because other vertical sections will be no different. Consequently, there are no tangential stresses in the vertical planes (and therefore in the horizontal planes also), the principal stress directions being vertical and horizontal. In addition, the weight of any newly deposited sediment will only produce vertical deformation of the existing deposit. These conditions are called **zero lateral deformation** or **one-dimensional deformation**.

Figure 2.27a shows how vertical effective stress can be calculated if the unit weight of the soil and pore pressure conditions (in this case hydrostatic) are known for any soil element at a particular moment (1) in its geological history, e.g. point A, lying at depth $z$ below a sediment surface under water whose level is constant as shown:

$$\sigma^1_{v(A)} = \gamma_w \cdot h_w + \gamma_{sat} \cdot z_1$$
$$u_{(A)} = \gamma_w \cdot (h_w + z_1)$$
$$\sigma'^1_{v(A)} = \sigma^1_{v(A)} - u_{(A)} = (\gamma_{sat} - \gamma_w) \cdot z_1$$

where $\gamma_{sat}$ is the saturated unit weight of the sediment.

At this moment (1), element A will have a certain void ratio ($e_1$). Representing its state in ($\sigma'_v$, $e$) space will give point 1 in Figure 2.27b.

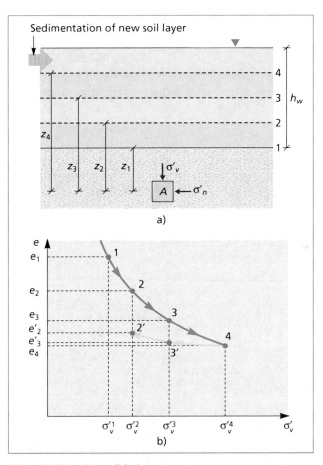

Figure 2.27   Consolidation processes.

If the process of sedimentation continues, with time a new thickness of soil will be deposited and the surface of the deposit will be raised to position 2 in *Figure 2.27a*. This causes an increment increase in vertical and horizontal stresses in the element under consideration. With respect to vertical stress, once the deposit is consolidated and the resulting excess of pore pressure are dissipated, the new effective vertical stress at A will be:

$$\sigma'^2_{v(A)} = (\gamma_{sat} - \gamma_w) \cdot z_2$$

The increment increase in effective stress applied in this way $[\Delta\sigma'_v = (\gamma_{sat} - \gamma_w) \cdot (z_2 - z_1)]$ will produce compression of the soil and therefore a reduction in its void ratio, so that the new state will be represented by point 2 in *Figure 2.27b*.

As sedimentation continues, new increases in effective vertical stresses and further reductions in void ratio will occur. Joining the representative points for each instant of this process will give a curve (1–2–3–4) similar to that shown in *Figure 2.27b*. This curve, known as the **virgin compression line**, represents the history of the element during the sedimentation or loading process. It also represents all the soil elements at each moment of the sedimentation process. Thus, points 1, 2, 3 and 4 will show the evolution of state $(\sigma'_v, e)$ of an element (A) as it is buried under ever greater depths at a given moment in the history of the soil deposit.

*Figure 2.28* shows an interesting aspect of soil behaviour. It reproduces the virgin compression line with special emphasis on the state of two elements, A and B, situated at different depths at a given instant. If an increment in effective stress $(\Delta\sigma'_v)$ is applied to the whole deposit, it is simple to show what the new states of those elements will be on the curve: points A' and B'. Notice that the reduction in void ratio of element A (the compression it has undergone) is greater than that of B. In short, **the greater the initial level of stress, the stiffer (less deformable) the soil will be**. This behaviour can be understood by observing that the void ratio of B was less than that of A, showing that its structure was denser.

## Unloading processes

In *Figure 2.27*, suppose that sedimentation stops when state 4 is reached and an erosion process is initiated due to a change in the geo-environmental conditions. Just as the addition of a new soil layer entailed an increase in effective stress and compression (i.e. reduction in void ratio), the removal of sediment or soil layers will involve unloading, and result in expansion of the deposit soil (i.e. increase in void ratio).

*Figure 2.27* shows that when unloading occurs the $(\sigma'_v, e)$ path followed does not retrace the virgin compression line but forms a new and flatter curve (4–3'–2').

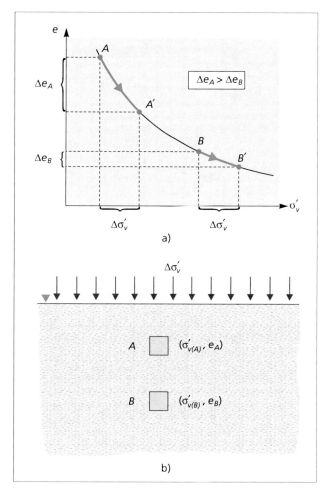

*Figure 2.28* Increase in ground stiffness with stress level.

This shows that sediment "remembers" its past history. When a sediment is unloaded to a value $\sigma_{vx}$, its void ratio (e) is not the same as when it was loaded for the first time to $\sigma_{vx}$. This means that for the same state of stress (e.g. $\sigma'^3_v$), the void ratio ($e_3$) during the original loading process is greater than the void ratio ($e'_3$) during unloading, i.e. when **effective vertical stresses are equal, soil in the process of unloading shows a denser structure (i.e., stronger and less deformable)**.

From the above description, two fundamental concepts can be stated in relation to the state of soil and its predictable behaviour:

— When the sedimentation process is at 1, the effective vertical stress in the element is $\sigma'^1_v$. This is also the maximum effective vertical stress borne by the soil element up to that moment. The same is true of states 2, 3 and 4. The soil has not been subjected to greater effective vertical stresses in any of these states than those at the time of observation. In these conditions

## Box 2.10

### Vertical and volumetric strain in one-dimensional conditions

The figure below shows a soil element of area $S_0$ with an initial height $H_0$. If it is subjected to an increase in effective vertical stress, at the same time as lateral strain is prevented, a reduction of the initial height $\Delta H$ due to compression (*) can be observed.

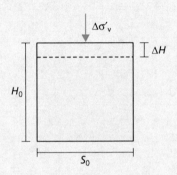

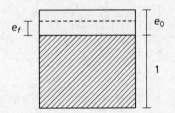

(*) Note: compressions are considered positive.

Under these conditions, the increment in vertical strain will be the same as the increment in volumetric strain:

$$\delta \varepsilon_v = \frac{\Delta H}{H_0} = \frac{\Delta H S_0}{H_0 S_0} = \frac{\Delta V}{V_0} = \delta v$$

In practice, stress history is usually represented on axes ($\sigma'$, e), so volumetric strain has to be expressed as a function of the void ratio.

The figure also shows a characteristic prism of unit volume of solid material before and after specific volumetric compression. From this figure it is simple to obtain:

— Original volume of soil

$$V_0 = 1 + e_0$$

— Final volume of soil

$$V_f = 1 + e_f$$

— Volumetric strain:

$$\delta v = -\frac{\Delta V}{V_0} = -\frac{V_f - V_0}{V_0} =$$
$$= -\frac{(1 + e_f) - (1 + e_0)}{1 + e_0} = \frac{e_0 - e_f}{1 + e_0}$$

---

the soil is said to be **normally consolidated**. The virgin compression line therefore represents the history, or ($\sigma_v$, e) states, of normally consolidated soils.

— Conversely, at points 3' and 2' the effective vertical stress of the element at any of these moments is lower than the maximum stress borne throughout its entire geological history (i.e. point 4). Thus, at the moment represented by point 3' the effective vertical stress is $\sigma_v'^3$, but the maximum for that element is $\sigma_v'^4$. The same occurs at the instant represented by 2'. In cases such as these, where the soil has undergone effective vertical stresses greater than those it bears at the moment of observation, the soil is said to be **overconsolidated**.

For a quantitative definition of **overconsolidation**, two fundamental parameters are used:

— **Preconsolidation pressure** ($\sigma_p'$) which is the maximum effective vertical stress of the soil element throughout its stress history.

— The **overconsolidation ratio** (OCR), which is the ratio between the maximum effective vertical stress in the past (the preconsolidation pressure) and the actual effective vertical stress:

$$OCR = \frac{\sigma_{v\ max}'}{\sigma_{v\ actual}'}$$

Thus, the different moments in the consolidation history selected in *Figure 2.27* give:

Moment (1): $OCR(1) = \dfrac{\sigma_v'^1}{\sigma_v'^1} = 1$ (NC)

## Box 2.11

### Calculating the degree of overconsolidation. Worked example

In a normally consolidated clay deposit the water table is at the surface. An erosion process lowers the ground surface by 3 m. Assuming that the water table coincides at all times with the ground surface, find the degree of overconsolidation induced by the erosion process.

(For clays take $\gamma_{sat} = 21$ kN/m³, and for pore water $\gamma_w = 9.81$ kN/m³).

### Solution:

As the water table is always at the surface, the total stresses, pore pressures, and effective vertical stresses can be represented by the following expressions:

$$\sigma_v = \gamma_{sat} \cdot z$$
$$u = \gamma_w \cdot z$$
$$\sigma'_v = (\gamma_{sat} - \gamma_w) \cdot z$$

where $z$ is the depth measured from the surface at any given time.

The attached table shows effective vertical stresses for different depths before and after erosion, as well as the degree of overconsolidation. The required OCR-depth ratio after erosion is shown in the figure.

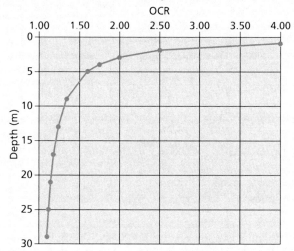

| Initial depth (m) | Final depth (m) | $\sigma'_{v\,initial}$ (kPa) | $\sigma'_{v\,final}$ (kPa) | OCR |
|---|---|---|---|---|
| 4 | 1 | 44.76 | 11.19 | 4.00 |
| 5 | 2 | 55.95 | 22.38 | 2.50 |
| 6 | 3 | 67.14 | 33.57 | 2.00 |
| 7 | 4 | 78.33 | 44.76 | 1.75 |
| 8 | 5 | 89.52 | 55.95 | 1.60 |
| 12 | 9 | 134.28 | 100.71 | 1.33 |
| 16 | 13 | 179.04 | 145.47 | 1.23 |
| 20 | 17 | 223.8 | 190.23 | 1.18 |
| 24 | 21 | 268.56 | 234.99 | 1.14 |
| 28 | 25 | 313.32 | 279.75 | 1.12 |
| 32 | 29 | 358.08 | 324.51 | 1.10 |

Moment (2): $OCR(2) = \dfrac{\sigma'^2_v}{\sigma'^2_v} = 1$ (NC)

Moment (3): $OCR(3) = \dfrac{\sigma'^3_v}{\sigma'^3_v} = 1$ (NC)

Moment (4): $OCR(4) = \dfrac{\sigma'^4_v}{\sigma'^4_v} = 1$ (NC)

Moment (3'): $OCR(3') = \dfrac{\sigma'^4_v}{\sigma'^3_v} > 1;\ \sigma'_p = \sigma'^4_v$

Moment (2'): $OCR(2') = \dfrac{\sigma'^4_v}{\sigma'^2_v} > OCR(3') > 1;\ \sigma'_p = \sigma'^4_v$

As can be seen, the overconsolidation ratio is equal to 1 for normally consolidated (NC) states and greater than 1 for overconsolidated states.

### Reloading processes

*Figure 2.29* shows the states already analysed and includes the effects of an additional change in geological history. Once state 2' is reached, erosion (unloading) ceases and sedimentation (reloading or recompression) begins again. It can also be seen here that this does follow the former path along the unloading curve (2'–3'–4) but takes another new path (2'–3''–4''), although often very close to it.

In fact, if the unloading was not very significant, then the unloading and reloading paths will practically coincide. This has interesting implications which will be dealt with later.

It can also be seen from *Figure 2.29* that once reloading reaches the maximum historical stress $\sigma_v'^4$ (preconsolidation stress), the later states are increasingly closer to the virgin compression line, and end up lying on its continuation (points 5 and 6). This suggests that the reloading process somehow progressively erases the soil "memory" so that it finally "forgets" it ever underwent a cycle of unloading and reloading. In fact, points 5 ($\sigma_v'^5$, $e_5$) and 6 ($\sigma_v'^6$, $e_6$) of the history described would be exactly the same if the soil had only undergone virgin compression 1–2–3–4–5–6, with no intermediate unloading. These points (from just beyond 4'') again correspond to normally consolidated states.

### Deformability of normally consolidated and overconsolidated soils

Supposing that the geological history of a soil element is given by the path shown in *Figure 2.30*, at the moment of observation the effective vertical stress in the element ($\sigma_v'^2$) is known from the position of the ground surface and the water table. Then it is useful to calculate the unit settlement ($\delta\varepsilon_v$) which will result from an increment of increased effective stress $\Delta\sigma_v' = \Delta\sigma_v'^4 - \Delta\sigma_v'^2$, similar to that caused by a specific construction project for normally and overconsolidated states.

*Figure 2.30* shows that if the soil is normally consolidated (point 2), the reduction in void ratio will be $\Delta e^{NC} = e_2 - e_4$. However, if the soil is overconsolidated (point 2'), the reduction in void ratio will be considerably less, $\Delta e^{OC} = e_{2'} - e_4$, and so will the vertical deformation (settlement). In other words, if **conditions are equal, the deformability of overconsolidated soil is considerably less than that of normally consolidated soil,** which underlines the importance of evaluating this aspect correctly in practice.

Some of the procedures used, based on laboratory tests, are described later. It should be pointed out, however, that appropriate engineering geological investigation is essential to obtain a reliable, though only qualitative, estimation of the situation.

Another interesting deduction that can be made from *Figure 2.30* is that deformations produced in an unloading-reloading curve are recoverable (elastic). E.g. starting at point 2', a complete reloading-unloading cycle (2'–4–2') can take place and go back to the same void ratio, which shows no irrecoverable (plastic) deformation has occurred. However, when the virgin compression line is followed to any extent (in normally consolidated states), irrecoverable plastic deformations do occur. Thus, starting from point 2 and applying the same load cycle (starting at $\sigma_v'^2$, increasing the stress to $\sigma_v'^4$, and then unloading again to $\sigma_v'^2$), the path taken by successive states of the soil element in space ($\sigma_v'$, $e$) will now

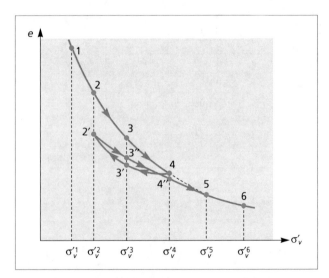

Figure 2.29  Void ratio-effective stress relationship. Recompression curve.

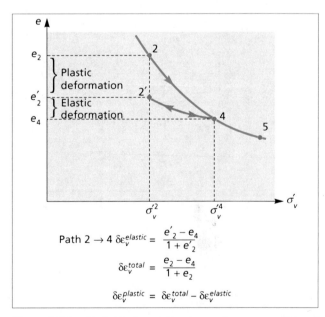

Figure 2.30  Deformation behaviour of overconsolidated and normally consolidated soils.

be as shown as a series of dots (2–4–2'). As can be seen here, in this case the loading-unloading cycle does not return to the same void ratio (the same soil volume). The irrecoverable (plastic) deformation will be the difference $\Delta e = e_2 - e'_2$, and the recoverable (elastic) part will be $\Delta e = e'_2 - e_4$, corresponding to the path along the unloading-reloading curve. To sum up, **overconsolidated soils will behave approximately as elastic materials, while normally consolidated soils will always show elastic and plastic deformation.**

### Plotting the stress history on a semilogarithmic scale

For most soils the above curves can be converted approximately into straight lines using a logarithmic scale for axis $\sigma'_v$. Figure 2.31 shows a diagram of the stress history as described in previous sections with the new axes.

If the **compression index** $c_c$ is defined as the slope of the virgin compression line, and the **swelling index** $c_s$ is defined as the slope of the unloading-reloading curve, it is straightforward to calculate variations in void ratio (and therefore vertical unit deformation) for any increment in effective vertical stress. So, to find the void ratio variation when moving from state 2 to state 3 on the virgin compression line:

$$e_2 - e_3 = c_c \log \frac{\sigma'^3_v}{\sigma'^2_v}$$

or for the path of a reloading between 2' and 3':

$$e_{2'} - e_{3'} = c_s \log \frac{\sigma'^3_v}{\sigma'^2_v}$$

Therefore, if starting from a known state ($\sigma'^0_v, e_0$), and applying an increase in effective vertical stress $\Delta\sigma'_v$, the final void ratio ($e$) will be given by the expression:

$$e_0 - e = c_c \log \frac{\sigma'^0_v + \Delta\sigma'_v}{\sigma'^0_v}$$

for normal consolidated states, or by:

$$e_0 - e = c_s \log \frac{\sigma'^0_v + \Delta\sigma'_v}{\sigma'^0_v}$$

for overconsolidated states.

Although details are given later on how the compression and swelling indexes can be determined in the laboratory, there are certain empirical correlations that

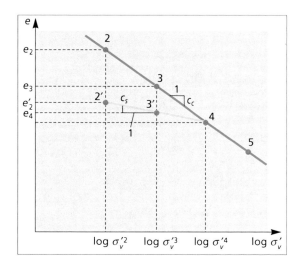

Figure 2.31   One-dimensional loading processes presented on a semi-logarithmic scale.

allow an estimation of the order of magnitude of the compression index (the swelling index is usually less critical because it is generally between 1/5 and 1/10 of the compression index).

The hypothesis of semilogarithmic linearity was put forward by Skempton (1970), Figure 2.32. It shows various series of points ($\log \sigma'_v, e$) representing states of a large number of normally consolidated argillaceous samples at different depths. The void ratios were determined from borehole samples, and vertical stresses from a mean density characteristic of each sediment and the depths of the samples. As can be seen, the sample depths range from a few decimetres below the sea bed to some 3,000 m below, which more than covers the usual range of stresses affecting engineering works. A complementary series of approximate dividing lines is shown, representing the order of magnitude of the liquid limits ($W_L$ in the figure) in the soils tested. Two main conclusions can be drawn from Figure 2.32:

— The points ($\log \sigma'_v, e$) representing **the virgin compression lines for each clay can be reasonably adjusted with straight lines.** The areas of greatest dispersion in relation to the linearity hypothesis seem to correspond to the shallowest samples, which may be due to errors in estimating the void ratio in the laboratory.
— The inclination of the virgin compression line (the compression index) increases as the liquid limit of the soil increases. Given that the greater the slope, the more compressible the soil will be (there is greater variation in the void ratio for the same increase in effective vertical stress), it can be concluded that, **if other circumstances are equal, the more plastic the soil is, the greater its compressibility will be.**

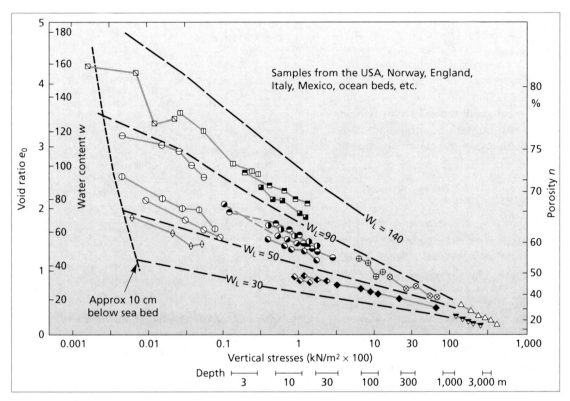

*Figure 2.32* Sedimentation compression curves for normally consolidated argillaceous sediments (modified from Skempton, 1970).

## Horizontal stresses in the ground

Section 2.4 shows how total and effective vertical stresses can be calculated from the apparent unit weights of the different strata present, and from the hydrogeological conditions of the surrounding area. However, the calculation of horizontal stresses pose a special problem because, like the void ratio, they depend very directly on the stress history of the soil.

In one-dimensional conditions (zero lateral deformation), effective horizontal stress is proportional to vertical stress, and their ration is called the **coefficient of earth pressure at rest** ($K_0$):

$$\sigma'_h = K_0 \cdot \sigma'_v$$

In normally consolidated soils $K_0$ is constant and can be estimated empirically using a simplification of Jáky's expression:

$$K_0^{NC} = 1 - \sin \phi'$$

where $\phi'$ is the angle of internal friction of the soil.

Considering the usual range of $\phi'$ in soils, the coefficient of earth pressure at rest $K_0^{NC}$ is always lower than 1 and generally around 0.5. As a result, normally consolidated soil will have effective horizontal stresses that are a fraction of vertical ones.

More generally, and for all types of states, including overconsolidated, an approximation of $K_0$ can be obtained using the empirical expression (Mayne and Kulhawy, 1982):

$$K_0 = (1 - \sin \phi') \cdot \left[ \left( \frac{OCR}{OCR_{max}^{(1-\sin\phi')}} \right) + \frac{3}{4} \cdot \left( 1 - \frac{OCR}{OCR_{max}} \right) \right]$$

OCR and $OCR_{max}$ are used to determine $K_0$ in overconsolidated states. OCR is the degree of overconsolidation ratio at the moment of observation, while $OCR_{max}$ is the maximum degree of overconsolidation ratio experienced by the soil in an unloading-reloading curve; i.e. the ratio between effective vertical preconsolidation pressure ($\sigma'_p$) and minimum effective vertical stress within the curve. This is:

$$OCR = \frac{\sigma'_{v\,max}}{\sigma'_{v\,actual}} = \frac{\sigma'_p}{\sigma'_{v\,actual}}$$

$$OCR_{max} = \frac{\sigma'_{v\,max}}{\sigma'_{v\,min}} = \frac{\sigma'_p}{\sigma'_{v\,min}}$$

## Box 2.12

### Calculating settlement. Worked example

A large landfill is planned on 10 m of normally consolidated clays with an underlying rock stratum. The water table is at ground surface. Clay samples extracted from the stratum at an intermediate point at a depth of 5 m provide the following soil properties: saturated unit weight, $\gamma_{sat} = 20$ kN/m$^3$, void ratio, $e_0 = 0.8$, and compression index, $c_c = 0.15$. Determine the settlement of the clay layer if the increase in vertical stress due to the landfill load is $\Delta\sigma = 80$ kPa.

### Solution:

The hypothesis of uniform loading over an infinite lateral extent, allows one-dimensional conditions to be assumed. The figure shows the effective stress distribution corresponding to the initial and final situations (unit weight of water is $\gamma_w = 10$ kN/m$^3$).

Taking that the increment in effective stress is constant throughout the thickness of the clay layer, the midpoint can be taken as representative of the whole stratum:

— Final void ratio:

$$e_0 - e = c_c \log \frac{\sigma_v'^0 + \Delta\sigma_v'}{\sigma_v'^0}$$

$$\Rightarrow 0.8 - e = 0.15 \cdot \log \frac{50 + 80}{50} \Rightarrow e \approx 0.74$$

— Vertical unit strain:

$$\delta\varepsilon_v = \frac{\Delta H}{H_0} = \frac{e_0 - e_f}{1 + e_0} = \frac{0.8 - 0.74}{1 + 0.8} = 0.033$$

— Total settlement (clay thickness $H_0 = 1{,}000$ cm):

$$0.033 = \frac{\Delta H}{1000} \Rightarrow \Delta H = 33 \text{ cm}$$

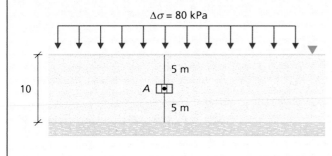
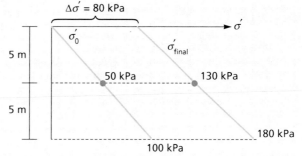

As an example, *Figure 2.29* shows that in an unloading curve OCR = OCR$_{max}$, given that at each moment minimum stress coincides with actual stress. However, in a reloading process, OCR$_{max}$ is greater than OCR. State 3", for example, would give:

$$OCR = \frac{\sigma_v'^4}{\sigma_v'^3}; \quad OCR_{max} = \frac{\sigma_v'^4}{\sigma_v'^2}$$

Finally, for normally consolidated states OCR = OCR$_{max}$ = 1, so the expression of $K_0$ can be reduced to that of Jáky, which has previously been referred to.

## Influence of complementary factors on soil behaviour

Previous sections have analysed a simple example of sedimentation and erosion but there are other factors that influence soil behaviour. Changes in hydrogeological conditions (such as the height of the water table) produce stress changes that can be considered as consolidation or overconsolidation processes. The analysis of stresses associated with complex geological processes e.g. tectonic forces, unloading and desiccation is more complex.

There are other phenomena not directly associated with stress changes but which also have a direct influence

## Box 2.13

### Calculating the coefficient of earth pressure at rest ($K_0$) and horizontal stress. Worked example

The table below shows the history of effective vertical stresses in an element of soil. If the soil has an angle of internal friction of $\phi' = 28°$, determine the evolution of the coefficient of earth pressure at rest ($K_0$) and the effective horizontal stresses as the vertical effective stress on the element is increased, then decreased, then increased again, according to the following pattern of loading, unloading and reloading.

$\sigma'_v$ (kPa) = 1, 2, 3, 4, 5, 6, 7, 8, 9, 10, 9, 8, 7, 6, 5, 4, 3, 2, 1, 2, 3, 4, 5, 6, 7, 8, 9, 10, 11, 12

| $\sigma'_v$ | OCR | OCR$_{max}$ | $K_0$ | $\sigma'_h$ | Observations |
|---|---|---|---|---|---|
| 1.00 | 1.00 | 1.00 | 0.53 | 0.53 | Virgin compression (NC) |
| 2.00 | 1.00 | 1.00 | 0.53 | 1.06 | |
| 3.00 | 1.00 | 1.00 | 0.53 | 1.59 | |
| 4.00 | 1.00 | 1.00 | 0.53 | 2.12 | |
| 5.00 | 1.00 | 1.00 | 0.53 | 2.65 | |
| 6.00 | 1.00 | 1.00 | 0.53 | 3.18 | |
| 7.00 | 1.00 | 1.00 | 0.53 | 3.71 | |
| 8.00 | 1.00 | 1.00 | 0.53 | 4.24 | |
| 9.00 | 1.00 | 1.00 | 0.53 | 4.77 | |
| 10.00 | 1.00 | 1.00 | 0.53 | 5.31 | |
| 9.00 | 1.11 | 1.11 | 0.56 | 5.02 | Unloading (OC) |
| 8.00 | 1.25 | 1.25 | 0.59 | 4.71 | |
| 7.00 | 1.43 | 1.43 | 0.63 | 4.39 | |
| 6.00 | 1.67 | 1.67 | 0.67 | 4.05 | |
| 5.00 | 2.00 | 2.00 | 0.73 | 3.67 | |
| 4.00 | 2.50 | 2.50 | 0.82 | 3.26 | |
| 3.00 | 3.33 | 3.33 | 0.93 | 2.80 | |
| 2.00 | 5.00 | 5.00 | 1.13 | 2.26 | |
| 1.00 | 10.00 | 10.00 | 1.56 | 1.56 | |
| 2.00 | 5.00 | 10.00 | 0.98 | 1.96 | Re-loading (OC) |
| 3.00 | 3.33 | 10.00 | 0.79 | 2.36 | |
| 4.00 | 2.50 | 10.00 | 0.69 | 2.76 | |
| 5.00 | 2.00 | 10.00 | 0.63 | 3.16 | |
| 6.00 | 1.67 | 10.00 | 0.59 | 3.55 | |
| 7.00 | 1.43 | 10.00 | 0.56 | 3.95 | |
| 8.00 | 1.25 | 10.00 | 0.54 | 4.35 | |
| 9.00 | 1.11 | 10.00 | 0.53 | 4.75 | |
| 10.00 | 1.00 | 1.00 | 0.53 | 5.31 | Virgin compression (NC) |
| 11.00 | 1.00 | 1.00 | 0.53 | 5.84 | |
| 12.00 | 1.00 | 1.00 | 0.53 | 6.37 | |

**Solution:**

Applying Mayne and Kulhawy's (1982) expression, the results shown in the table are obtained.

The figure below shows the evolution of effective stress on axes ($\sigma'_h$, $\sigma'_v$). It can be seen that in normally consolidated conditions $K_0$ is constant and equal to $K_0 = 1 - \sin \phi' = 0.53$; so that the "stress path" is linear.

Once unloading begins, $K_0$ gradually increases. This means that for the same vertical stress, the effective horizontal stress is greater than it was under normally consolidated conditions. The example also shows that for a degree of overconsolidation of 4 or higher, effective horizontal stresses may be even higher than the vertical stresses.

When maximum unloading is completed and reloading begins $K_0$ gradually decreases, running along a stress path that is slightly separated from the unloading path, between it and the path of the normally consolidated states.

Finally, on reaching the virgin compression line once more, the coefficient of earth pressure at rest returns to that of normally consolidated soils, and the stress path rejoins the original curve defined by that state.

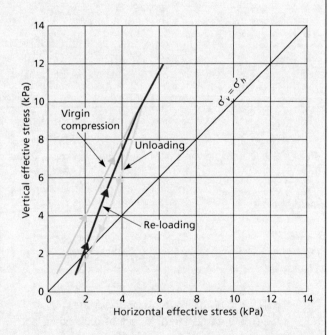

on soil behaviour. These include reworking of original fabric whilst soft by burrowing animals (bio-turbation), chemical cementation, hardening or overconsolidation through creep under constant loading, dissolution of bonding agents, etc.

In situations other than those shown in the previous section, the coefficient of earth pressure at rest cannot be estimated using the given empirical expressions. Instead, it has to be determined on site, using, for example, pressuremeters or hydraulic fracturing tests, although all these methods have some limitations (see Chapter 5).

## The oedometer test

### Description of the test

Although the loading conditions for a foundation do not normally induce a state of zero lateral (one-dimensional) deformation, this test model is widely used, with some modifications, to estimate settlements produced by embankments, footings, rafts, and similar foundations, especially on fine saturated soils such as silts and clays.

In the laboratory, the **oedometer test** is used to study one-dimensional compressibility of soils. This is carried out in an apparatus called an **oedometer**.

The oedometer consists of a metal ring enclosing a soil specimen, usually taken from an undisturbed soil sample (Figures 2.33 and 2.34). Porous stones are placed at the top and bottom of the sample to allow water drainage. The metal ring and its sample are then placed in a cell filled with water, to maintain complete saturation at all times.

A loading cap is then placed on the upper porous stone and vertical load applied to its centre with the compression of the specimen that results being measured by a dial gauge. The load is increased in stages, with each successive load usually double the previous one. At each stage, vertical compression of the soil sample is measured.

The rigidity of the metal ring containing the sample prevents any lateral deformation so compression is one-dimensional. In these conditions, as has already been shown, when a new load increment is placed on a saturated soil with low permeability, all the incremental increases in total vertical stress, $\Delta\sigma_v$, are instantly transmitted to the pore pressure ($\Delta u = \Delta\sigma_v$), and effective stress therefore does not vary ($\Delta\sigma' = 0$). Then, as the excess pore pressure created by this loading gradually dissipates due to drainage through the porous stones, effective stress increases and the soil compresses (consolidates). In an oedometric test each increment in load has to be maintained for long enough to ensure that the consolidation process has been completed. This is generally achieved (although not always) with load intervals of about 24 hours between the application of successive loads.

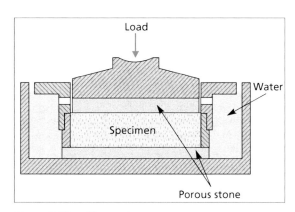

Figure 2.33   The oedometer.

Figure 2.34   Oedometer bench.

The position of the soil specimen in an oedometric cell is shown in the diagram in Figure 2.33. For practical purposes, the specimen represents a layer of soil between two permeable layers (the porous stones), with a very extensive load applied to it (in one-dimensional conditions). Under these conditions the oedometer test could be used to reproduce the conditions described above.

Figure 2.35 shows the pore pressure distribution ($u_0$) before the load application. It is hydrostatic and is determined by the water level in the cell. Assuming that the ground being analysed has low permeability, the application of a load increment $\Delta\sigma_v$ will immediately cause an increment in pore pressure of equal magnitude: $\Delta u_i = \Delta\sigma_v$.

The top and bottom ends of the soil specimen will be the first to drain and relieve their excess pore pressures and this will happen quickly as they are nearest the porous stones. Further away from the porous stones, inside the soil sample, the flow path to the free draining ends is longer and initial pore overpressures will take longer to dissipate. The centre of the soil specimen in Figure 2.35 is the furthest away from the drainage limits so it will take the longest time to consolidate. Therefore, at any time ($t$) after loading, the excess pore pressure present will vary between one point and another depending on the distance to the drainage limits. *Figure 2.35c* shows some pore pressure distributions at different times after loading.

*Figure 2.36a* shows the change of excess pore pressure for a time $t_1$ after a load increment $\Delta\sigma_v$. Observation of any point $P$ lying at a depth $z$ in the soil specimen, will show that at that time excess pore pressure is $\Delta u^{(t)}$, and the resultant dissipation has induced an increment in effective vertical stress $\Delta\sigma_v'^{(t)}$. It is evident from Terzaghi's principle that the sum of both has to be equal to the increment in initial total vertical stress:

$$\Delta\sigma_v = \Delta\sigma_v'^{(t)} + \Delta u^{(t)}$$

If $\sigma_{v0}'$ is the effective stress and $u_0$ the pore pressure present at $P$ before the load stage is applied, then:

— At the moment of loading ($\Delta\sigma_v$):

$$\sigma_v'^{(i)} = \sigma_{v0}'$$
$$u_i = u_0 + \Delta\sigma_v$$

— At any time $t$:

$$\sigma_v'^{(t)} = \sigma_{v0}' + \Delta\sigma_v'^{(t)}$$
$$u^{(t)} = u_0 + \Delta u^{(t)}$$

— When consolidation has finished:

$$\sigma_v'^{(final)} = \sigma_{v0}' + \Delta\sigma_v a$$
$$u^{(final)} = u_0$$

In any case, as what is usually referred to is "excess pore pressure" above the equilibrium pressure, or increments in effective stress, only these increments are normally represented graphically (*Figure 2.36b*).

### Plotting the results

The usual practice in oedometric testing is to carry out a series of loading steps followed by one or two unloading steps. These are usually plotted on a graph showing vertical unit deformation ($\varepsilon_v\%$) or void ratios on the y-axis and effective vertical stresses of each loading step on the x-axis. Since what is really measured is the vertical compression ($\Delta H$) of a soil specimen with an initial thickness $H_0$, to determine the void ratio after each loading step the following ratios are used:

$$\frac{\Delta H}{H_0} = \frac{e_0 - e}{1 + e_0} \Rightarrow e = e_0 - \frac{\Delta H}{H_0}(1 + e_0)$$

where $e_0$ is the initial void ratio of the specimen.

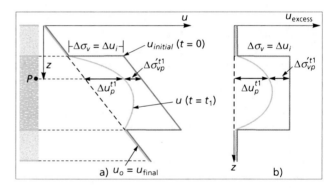

Figure 2.36  Pore pressure dissipation and increase in effective stress.

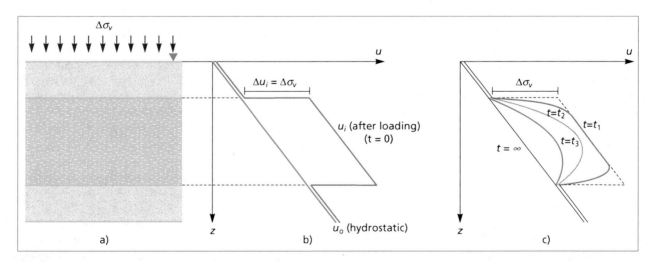

Figure 2.35  Pore pressure paths in an oedometer.

*Figure 2.37* shows the oedometric curves for a test in which an intermediate unloading-reloading cycle has been carried out, Lancellota (1995). The clay sample was extracted from a depth of 13.20 m, and had a void ratio of $e_0 = 1.50$. The first curve is plotted on a natural scale and the second on a semi-logarithmic one.

The curves shown are analogous to the curves previously described in this section. In fact, the main differences between the field and the laboratory curve for a soil are usually due to disturbance of the sample during its extraction and handling. To transform the laboratory curve into the field curve, which represents *in situ* ground behaviour, a series of graphical corrections are made.

### Obtaining the field compressibility curve in normally consolidated soils

The graphical procedure to follow is shown in *Figure 2.38*. In theory, if neither soil water content nor initial void ratio ($e_0$) vary with extraction of the sample, the laboratory curve should pass through point A, which represents the *in situ* state of the soil at the depth the sample was taken. Also, as already described, for normally consolidated soil, the successive points $(e, \sigma'_v)$ have to be in an approximately straight line, which represents the virgin compression line on a semi-logarithmic scale. However, the disturbance produced in the sample by extraction makes the laboratory curve appear as shown in *Figure 2.38*. As can be seen, the slope on the first curved portion slowly increases as effective vertical stress approaches the original *in situ* state $(\sigma'_{v0})$ but it does not reach point A. For larger effective stress, a straight portion is reached, though with a gentler slope than the field virgin compression line. Schmertmann (1955) observed that the straight sections of the laboratory compression curves for samples with different degrees of disturbance intersected the field virgin compression curve at approximately $0.42e_0$. Therefore, it is accepted that the laboratory curve will coincide with the field curve of the soil for this void ratio value. As a result, if a line is drawn from point $0.42e_0$ in the laboratory curve to point A at $e_0$, representing the initial *in situ* state, this line gives the estimated field or *in situ* virgin compression curve of the soil.

### Obtaining the field curve in overconsolidated soils. Preconsolidation pressure calculation

*Figure 2.39* shows the procedure, also proposed by Schmertmann (1955), for reconstructing the field compressibility curve from the laboratory curve and determining the preconsolidation pressure in an overconsolidated soil sample. At least one unloading-reloading cycle is needed during the test, which is carried out in the following steps:

— From point A, representing the initial *in situ* state, a line is plotted parallel to the unloading-reloading curve (u – r).

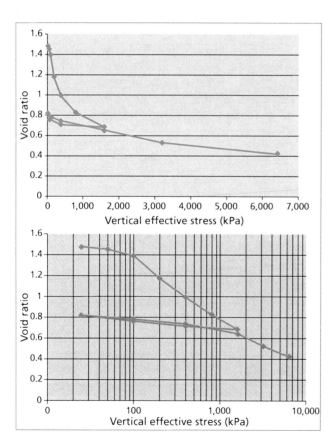

*Figure 2.37* Plotting results from an oedometer test (Lancellotta, 1995).

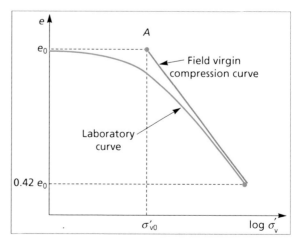

*Figure 2.38* Obtaining the field virgin compression curve of a normally consolidated soil from an oedometer test.

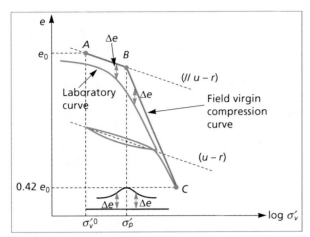

Figure 2.39  Schmertmann's procedure for constructing the field compression curve of a preconsolidated soil.

— A value is assumed for the preconsolidation pressure ($\sigma'_p$) and point B is obtained; the appropriateness of this assumed value is checked in step 4.
— B is joined to point C of the laboratory curve where it reaches $0.42\, e_0$, to obtain the field virgin compression curve.
— Differences in the void ratio ($\Delta e$) between the laboratory curve and the postulated field curve are plotted. If the assumed preconsolidation pressure is correct, the values of $\Delta e$ will be symmetrical with respect to $\sigma'_p$. If not, another preconsolidation pressure is estimated and the process is repeated. As already mentioned, it is very useful to have access to geological evidence allowing the degree of overconsolidation of the soil to be established, even if only qualitatively.

## Soil compressibility parameters

### Compression and expansion indices

Once the field curve of the ground is known, its compression and the expansion indices can be obtained, by determining the inclination of the corresponding unloading-reloading and virgin compression lines. This is done simply by selecting two points on each of the curves and applying the expressions shown previously.

### The coefficient of volume compressibility and the oedometric modulus

The representation of the oedometric curve in space $\sigma'_v, e$ can be easily transformed to axes $\sigma'_v, \varepsilon_v$, which is useful as it allows ground deformations to be visualized directly. Figure 2.40 can then be assumed to show the virgin compression line of the normally consolidated soil in Figures 2.27 or 2.29 on a real scale, with axes $\left(\sigma'_v, \varepsilon_v\right)$.

To relate the increments in strain and increments in effective stresses in one-dimensional loading conditions, two parameters are usually used:

— The **oedometric modulus**, $E_m$, defined by the expression:

$$E_m = \frac{\Delta \sigma'_v}{\Delta \varepsilon_v} \quad (N/m^2)$$

corresponds to the definition of a deformation modulus and coincides with the inverse of the slope of the virgin compression line so that when the stress level increases the slope diminishes and $E_m$ increases, demonstrating that the soil becomes stiffer with increasing stress.

— The **coefficient of volume compressibility**, $m_v$, defined as the volume change per unit increase in effective stress ($m^2/MN$), or the inverse of the oedometric modulus:

$$m_v = \Delta \varepsilon_v / \Delta \sigma'_v \quad (m^2/N)$$

As can be deduced from the definition above, the oedometric modulus varies continuously along the virgin compression line, increasing at the same rate as the increase in effective vertical stress. It is really the inverse of the tangent of the curve at each point, so its correct mathematical expression is:

$$E_m = \frac{d\sigma'_v}{d\varepsilon_v} \quad (N/m^2)$$

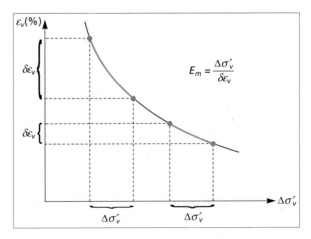

Figure 2.40  Oedometric curve represented in terms of vertical deformation and definition of the oedometric modulus.

# Box 2.14

## Calculating consolidation time. Worked example

A very large embankment is built over a normally consolidated clay layer of thickness $H$ m. The clay is below the water table. Assuming that the loading conditions are one-dimensional, find the time taken to reach half of the total settlement under the following conditions:

a) If there is only one drainage boundary of sand at the top of the clay layer.
b) If there are two drainage boundaries of sand, one at the top and one at the base of the clay.
c) If there are three permeable levels of sand, one at the top and one at the base of the clay, and a third, thin layer, halfway through the clay.

### Solution:

The attached figure shows the three drainage situations, together with the drainage paths in each case.

Assuming that in case c) the intermediate drainage level is thin enough to have no influence on the thickness $H$ of the clay, the total settlement ($S_\infty$) will be the same for all three hypotheses. Halfway through the settlement process, the degree of consolidation for all three cases will be:

$$U = \frac{S_t}{S_\infty} = \frac{0.5 \cdot S_\infty}{S_\infty} = 0.5 \Rightarrow U(\%) = 50\%$$

Applying the $U - T_v$ relationship included in Table 2.5, $T_v = 0.196$, again in all three cases. Therefore, the expression for the time factor $T_v$, in each one of the hypotheses will be:

Case a) $\quad 0.196 = \dfrac{c_v \cdot t_a}{H^2} \Rightarrow t_a = \dfrac{0.196 \cdot H^2}{c_v}$

Case b) $\quad 0.196 = \dfrac{c_v \cdot t_b}{(H/2)^2} \Rightarrow t_b = \dfrac{0.196 \cdot H^2}{4\,c_v}$

Case c) $\quad 0.196 = \dfrac{c_v \cdot t_c}{(H/4)^2} \Rightarrow t_c = \dfrac{0.196 \cdot H^2}{16\,c_v}$

As can be seen, the time needed to reach a certain degree of consolidation is proportional to the square of the drainage path. This means that in hypothesis c) settlement will be reached in a quarter of the time of case b), and in a sixteenth of the time of case a). Obviously this time ratio is valid for any degree of consolidation.

This example highlights the importance of an engineering geological ground profile description to detect the presence of interbedded pervious layers.

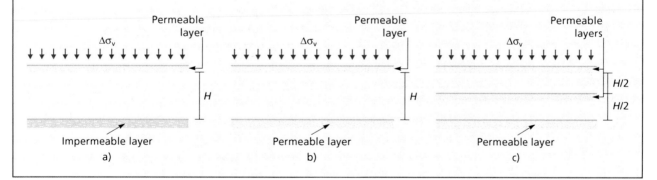

Similarly, the correct expression of the coefficient of volume compressibility is:

$$m_v = \frac{d\varepsilon_v}{d\sigma'_v} \quad (m^2/N)$$

To use both $E_m$ and $m_v$ in practice, the virgin compression line is usually discretized into rectilinear segments (load stages) small enough for a constant oedometric or coefficient of volume compressibility to be assumed in each of the straight sections.

### Estimating consolidation times

As described earlier, in a saturated soil with low permeability the increase in effective stress and associated settlement after loading is not immediate but requires a certain time

# Box 2.15

## Settlement-time curves. Worked example

Settlement of 30 mm has occurred in a layer of normally consolidated clay underlying a building foundation in the 300 days since the building load became operative. The clay layer is bounded at the top and bottom by permeable strata. According to laboratory data, the settlement corresponds to a degree of consolidation $U = 25\%$ of the clay layer. Draw the settlement-time curve for a period of 10 years (assuming that the lateral extent of the foundation area compared with the thickness of the clay is enough for conditions to be considered one-dimensional or oedometric).

### Solution:

From the data given the total oedometric consolidation settlement can be calculated:

$$U = \frac{S_t}{S_\infty} \Rightarrow S_\infty = \frac{S_t}{U} = \frac{30}{0.25} = 120 \text{ mm}$$

From Table 25 it can also be deduced that the time factor $T_v$ for $U = 25\%$ is $T_v = 0.0491$. Recalling the expression for the time factor:

$$T_v = \frac{c_v \cdot t}{H^2}$$

and substituting the known data will give:

$$0.0491 = \frac{c_v \cdot 300}{H^2} \Rightarrow$$

$$\Rightarrow \frac{c_v}{H^2} = 0.0491/300 = 1{,}636 \cdot 10^{-4} \text{ days}^{-1}$$

Note that $c_v/H^2$ is a constant, since $c_v$ is the coefficient of consolidation and $H$ the drainage path (half the initial thickness of the clay in this case as it drains at both ends). In these conditions, the corresponding settlement can be determined for any degree of consolidation $U$ and, from the table $U - T_v$, the associated time factor and time needed to reach the degree of consolidation can be selected according to the following expression:

$$U \to \begin{cases} S_t = U \cdot S_\infty \\ T_v \to t = \dfrac{T_v}{c_v/H^2} = \dfrac{T_v}{1{,}636 \cdot 10^{-4}} \text{ days} \end{cases}$$

The following table and figure show the results and the settlement-time curve required.

| U% | $T_v$ | t (days) | t (years) | S (mm) |
|---|---|---|---|---|
| 5 | 0.0017 | 10.39 | 0.03 | 6 |
| 10 | 0.0077 | 47.05 | 0.13 | 12 |
| 15 | 0.0177 | 108.15 | 0.30 | 18 |
| 20 | 0.0314 | 191.82 | 0.53 | 24 |
| 25 | 0.0491 | 300.00 | 0.82 | 30 |
| 30 | 0.0707 | 431.98 | 1.18 | 36 |
| 35 | 0.0962 | 587.78 | 1.61 | 42 |
| 40 | 0.126 | 769.86 | 2.11 | 48 |
| 45 | 0.159 | 971.49 | 2.66 | 54 |
| 50 | 0.196 | 1197.56 | 3.28 | 60 |
| 55 | 0.238 | 1454.18 | 3.98 | 66 |
| 60 | 0.286 | 1747.45 | 4.79 | 72 |
| 65 | 0.342 | 2089.61 | 5.72 | 78 |
| 70 | 0.403 | 2462.32 | 6.75 | 84 |
| 75 | 0.477 | 2914.46 | 7.98 | 90 |
| 80 | 0.567 | 3464.36 | 9.49 | 96 |
| 85 | 0.684 | 4179.23 | 11.45 | 102 |
| 90 | 0.848 | 5181.26 | 14.20 | 108 |
| 90 | 1.129 | 6898.17 | 18.90 | 114 |

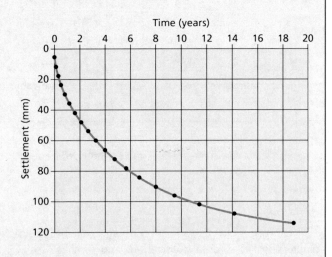

to be complete. This effect can easily be studied in the oedometer.

The higher or lower settlement velocity of a soil depends on the **coefficient of consolidation**, $c_v$, expressed by:

$$c_v = \frac{k_v \cdot E_m}{\gamma_w} \ (m^2/s)$$

where

- $k_v$ is the coefficient of vertical permeability of the soil in the oedometer (drainage takes place vertically, towards the porous stones).
- $E_m$ is the oedometric modulus of the soil, or the inverse of $m_v$.
- $\gamma_w$ is the unit weight of water.

It has already been seen that $E_m$ increases during the consolidation process as effective vertical stress increases. Conversely, $k_v$ decreases (the soil becomes less permeable the more compressed it is). For load stages that are not too big, the product $k_v \cdot E_m$ remains approximately constant, so that $c_v$ can also be assumed to be constant.

The water overpressure is maximum in the first few moments after a load is applied and therefore the flow of water and the speed of settlement i.e. speed of consolidation are relatively fast. Then, as the excess water pressure goes down, the flow rate and settlement velocity slow down.

The evolution of this process can be seen in *Figure 2.41* which represents the typical appearance of the settlement-time ratio on a real scale, with positive downward settlement. In the oedometric test, each time a load stage is applied the settlement produced can be measured at regular intervals and the settlement-time evolution drawn using **Casagrande's method**, a graphical construction using a logarithmic scale on the time axis. From the resulting **consolidation curve**, the consolidation coefficient can be deduced for the load stage applied.

The **degree of consolidation**, $U$, of a layer of soil at a certain time ($t$) after loading is the ratio between the settlement produced up to that moment ($S_t$), and the total settlement that will be produced when all excess pore pressure has dissipated at an infinite time ($S_\infty$), i.e. when the

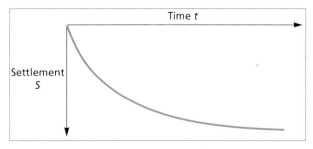

*Figure 2.41* Settlement-time ratio after applying total vertical stress increment.

| Table 2.5 | RELATIONSHIP BETWEEN TIME FACTOR AND DEGREE OF CONSOLIDATION | | |
|---|---|---|---|
| U (%) | $T_v$ | U (%) | $T_v$ |
| 0 | 0 | 50 | 0.196 |
| 5 | 0.0017 | 55 | 0.238 |
| 10 | 0.0077 | 60 | 0.286 |
| 15 | 0.0177 | 65 | 0.342 |
| 20 | 0.0314 | 70 | 0.403 |
| 25 | 0.0491 | 75 | 0.477 |
| 30 | 0.0707 | 80 | 0.567 |
| 35 | 0.0962 | 85 | 0.684 |
| 40 | 0.126 | 90 | 0.848 |
| 45 | 0.159 | 95 | 1.129 |

whole increase in total stress has been transformed into an increase in effective stress:

$$U = \frac{S_t}{S_\infty}$$

The **time factor** $T_v$ I is a dimensionless number defined by the following ratio:

$$T_v = \frac{c_v \cdot t}{H^2}$$

where:

- $t$ is the time passed since the application of the new load.
- $H$ is the drainage path, defined as the longest route water will have to take in the layer of soil to reach the permeable boundary; in the oedometer this will be half the thickness of the specimen taking into account that the porous stones for drainage are above and below the specimen.

Terzaghi and Fröhlich demonstrated that the time factor and the degree of consolidation are interrelated, as shown in *Table 2.5* (with $\Delta\sigma_v = \Delta u_0$ constant throughout the thickness of the soil layer). For most practical cases the times needed for different degrees of consolidation can be estimated from this Table.

## 2.6 Shear strength of soils

### Failure criterion

The shear strength of a soil cannot be considered as a single constant parameter as it depends on such factors as the nature of the soil, its structure, bonds and degree of deformation, and particularly on its state of stress and fluid pressure in its pores (water or air and water).

The best known failure criterion for soils is that of Mohr Coulomb, which relates the normal effective stresses and tangential stresses acting on any plane of the soil at the time of failure. Using this criterion, shear strength for a saturated soil can be expressed by:

$$\tau = c' + (\sigma_n - u) \tan \phi'$$

where:

$\tau$ = shear strength of the soil along a specific plane
$\sigma_n$ = total normal stress acting on the same plane
$u$ = pore water pressure
$c'$ and $\phi'$ = shear strength parameters related to effective stress: cohesion and angle of shearing resistance.

The above equation represents a straight line in space $(\sigma', \tau)$ that is called the **failure envelope of the soil** (*Figure 2.42*). For every normal effective stress value on a plane that crosses an element of the soil, the line of this envelope gives the maximum tangential stress mobilized along the plane.

Some relevant aspects can be deduced from *Figure 2.42*:

— Effective cohesion is the ordinate at the origin of the failure envelope. It therefore represents the maximum tangential stress that can be mobilized on any plane when normal effective stress on that plane is zero.
— The maximum tangential stress mobilized on a plane becomes greater as the normal effective stress acting on that plane increases. In other words, the higher its level of effective stress, the more resistance the soil can generate.
— The line of resistance described acts as an "envelope", separating two possible states, that below it, where the soil is "stable" i.e. not failed, from impossible ones above it, because the soil under those conditions would fail and so could never reach this state. Thus:

- Point (1) shown in *Figure 2.42* represents a state of failure.

- Point (2) represents a combination $(\sigma', \tau)$ that has a certain factor of safety, since for any particular normal effective stress tangential stress is lower than the maximum that can be mobilized.
- Point (3) represents an impossible state in that it lies above the failure envelope, which means that it has exceeded the maximum combination $(\sigma', \tau)$ of the failure criterion and is therefore not compatible with the strength of the soil.

Taking into consideration the basic concepts of the stress tensor and the Mohr circle, conditions for failure on a given plane can be related to stresses acting along other planes. *Figure 2.43* shows three Mohr circles in a space $(\sigma', \tau)$ which basically represent three stress states of a soil element. If the soil shear strength parameters $(c', \phi')$ give the failure envelope shown, then it can be observed that:

— The state of stress represented by circle (a) has a safety margin as it does not reach the soil failure envelope. It is therefore a safe possible state (the soil has not failed).
— The state of stress represented by circle (b) indicates a failure situation. Point (O) represents the combination $(\sigma'_f, \tau_f)$ on a plane crossing the soil element, where conditions for the failure criterion described are reached.
— The state of stress shown by circle (c) is impossible because there would be planes of orientation crossing the soil element on which failure conditions $(\sigma', \tau)$ would be exceeded (i.e. all planes with by points on the circle lying above the failure envelope).

It can be deduced from the above analysis that when failure conditions are reached in a soil element, the Mohr circle that represents its state of stress will be tangential to the failure envelope. Also, the tangential point will represent the particular plane along which these failure conditions are reached.

## The direct shear test

### Test description

The direct shear test apparatus is shown diagrammatically in *Figure 2.44*. It consists of a rigid steel box, usually square and divided into two halves, into which the soil specimen is placed. Above it is a loading plate on which a vertical load (N) can be applied. This is all put into a larger steel container which can be filled with water to carry out the test in saturated conditions. To facilitate drainage, porous stones can be placed above and below the specimen. The shear stress in the soil is induced by displacing the lower part of the shear box horizontally while movement of the upper part is completely prevented.

To carry out a full test on a particular soil, three identical samples of the same material are tested under three

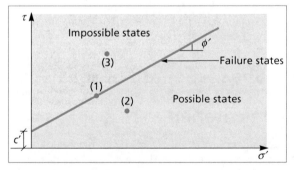

Figure 2.42   Failure criterion.

## Box 2.16

### Calculating shear and principal stresses. Worked example

The intrinsic strength parameters of a soil are $c' = 0$, $\phi' = 30°$. Assuming that in an element of this soil failure has been reached on a plane at 45° to the horizontal, with a value $\sigma'_f = 10$ kPa, find:

- The shear stress at failure $\tau_f$.
- The orientation and magnitude of the principal stress in the soil element.

### Solution:

The solution to the problem is shown graphically in the adjoining figure. The steps to follow are:

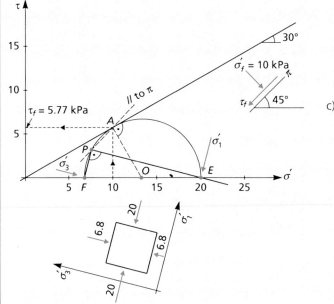

a) A vertical line is plotted from $\sigma' = 10$ kPa on the x-axis until it intersects the failure envelope. From the point obtained (A) it can be deduced that shear stress at failure is $\tau_f = 5.77$ kPa. (The same value could have been reached by applying the failure envelope equation for the given value of the normal effective stress).

b) Point (A) represents the stresses on the failure plane of the soil element considered. As these are failure stresses, the Mohr circle will be tangent to the failure envelope, (A) being the point of tangency. Therefore, by drawing a perpendicular to the failure envelope from (A) to the x-axis, the centre (O) of the required circle is obtained, allowing the circle with centre (O) and radius (OA) to be drawn. The intersection of the circle with the x-axis gives the values of the principal stresses in the soil. Measuring directly from the graph gives: $\sigma'_1 \approx 20$ kPa and $\sigma'_3 \approx 6.8$ kPa.

c) Since (A) represents the stress in the soil on a plane at 45° to the horizontal, a line is drawn from (A) 45° to the horizontal, (i.e. parallel to this plane), and where this intersects the Mohr circle again, the pole (P) is obtained. From (P) the lines (PE) and (PF) are drawn, to obtain the orientations of the major and minor principal planes, respectively (the directions of the principal stresses will be perpendicular to these planes). The stress state of the soil element in a cartesian system formed by the principal axes is shown in the diagram below the figure.

different vertical loads ($N_1$, $N_2$, $N_3$), or, what amounts to the same thing, under three different normal stresses; it is enough to divide each load (N) by the cross section area of the sample ($S_0$) to obtain the acting normal stress.

In each of the individual tests, the following measurements are taken at the same time as the lower part of the box is made to move horizontally at a constant velocity:

— The force ($F_H$) necessary to prevent movement of the upper half of the box. Dividing this force by the section area of the sample ($S_0$) gives the tangential stress ($\tau$) acting at each instant on the shear plane.

— Vertical displacement of the sample. Bearing in mind that the walls of the shear box are rigid, like those of the oedometer, measurement of vertical strain ($\delta\varepsilon_v$) will give the change in volume of the specimen ($\delta v$) directly, since:

$$\delta\varepsilon_v = \frac{\Delta H}{H_0} = \frac{\Delta H \cdot S_0}{H_0 \cdot S_0} = \frac{\Delta V}{V_0} = \delta v$$

where:

$H_0$ is the initial height of the specimen (4.2 cm in normal shear boxes).

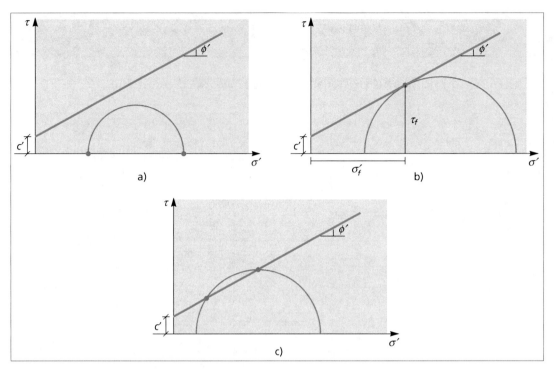

*Figure 2.43*  The soil failure envelope and the Mohr circle. Possible (a, b) and impossible (c) states.

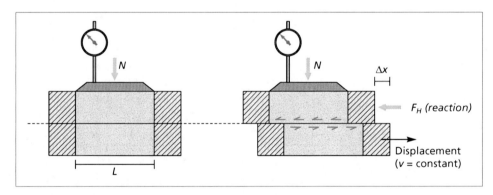

*Figure 2.44*  Direct shear test apparatus.

$S_0$ is the section of the specimen (usually 36 cm² for boxes with sides $L = 6$ cm).
$V_0 = H_0 \cdot S_0$ is the original volume of the soil.
$\Delta H$ is the vertical displacement (positive if it is shortening).
$\Delta V$ is the variation in soil volume (positive if it is compression).

The test procedure is usually carried out in the following stages:

a) A total vertical stress $\sigma_{n1} = \dfrac{N_1}{L^2}$ is applied.
b) If the test is carried out with a saturated sample and the container flooded, the sample is usually allowed to consolidate until the excess pore pressure generated has dissipated. This phase is similar to one of the stages in the oedometric test so a settlement-time curve can be plotted and checked when consolidation has finished, as this is the moment it can be assumed that the total stress applied has been completely transformed into effective stress.
c) The lower part of the shear box is given a constant horizontal velocity which is transmitted to the upper box through the sample; the force ($F_{H1}$) so transmitted is measured by the reaction it generates to itself, at set time intervals, so that the tangential stress at each instant is:

$$\tau = \frac{F_{H1}}{L^2}$$

In this phase, if the shear velocity is low enough to allow dissipation of the excess pore pressure generated by the application of tangential stresses, the test can be considered as drained. In these conditions the results will be expressed directly as effective stress ($\sigma_n = \sigma'_n$). Therefore, given that in the direct shear apparatus drainage cannot be controlled or pore pressure measured with time, it is important to select a low enough velocity of shearing, which obviously depends on the type and permeability of the soil being tested.

d) Vertical displacement of the specimen is measured at the same time intervals to obtain changes in soil volume with time.

The same procedure is carried out with two more identical soil specimens, but these specimens are subjected to greater normal stresses ($\sigma_{n2}$) and ($\sigma_{n3}$).

In relation to the shear strength mobilization, *Figure 2.45* shows the qualitative results of a completely drained test such as the one described. The x-axis shows the horizontal displacement ($\Delta x$) between the upper and lower boxes of the shear box and the y-axis shows tangential stress ($\tau$) measured for each value of this displacement. The following points of interest can be seen in *Figure 2.45*:

— Whatever the normal effective stress applied, the tangential stress that is mobilized gradually increases with the progressive displacement of the shear box until it reaches a maximum ($\tau_f$).
— The greater the initial normal effective stress, the greater the maximum tangential stress reached.
— The greater the initial normal effective stress, the more pronounced the initial slope of the tangential stress-horizontal displacement curve will be, indicating that soil stiffens with the level of stress.

*Figure 2.45b* shows the maximum tangential stresses of the previous curves together with initial normal effective stress. It can be seen that the representative points ($\sigma'_n$, $\tau_f$) of the three tests can be approximately joined with a straight line. This is the failure envelope from which the parameters ($c'$, $\phi'$) can easily be obtained.

## Advantages and disadvantages of the direct shear test

The test described above has advantages and disadvantages. The main advantages are:

— It is quick and inexpensive.
— It uses basic principles.
— The preparation of samples is simple.
— With larger shear boxes coarse-grained material can be tested.
— The same principles can be used, with some modifications, to determine the shear strength of discontinuities in rock, concrete-soil contact, etc.
— It can be used to measure strength operating after large displacements, especially in clays, so revealing their "residual" strength.

Some of its disadvantages are:

— A failure surface is a requirement.
— Stress distribution at the shear surface is not uniform.
— In general, pore pressures cannot be measured, so the only way of controlling drainage is by varying the horizontal displacement velocity.

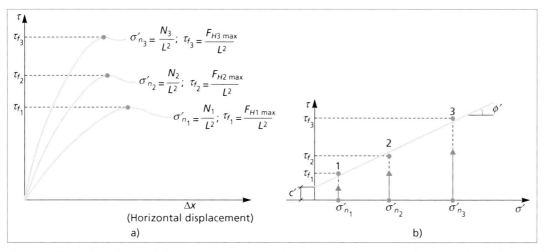

Figure 2.45  Obtaining the failure envelope and the shear strength parameters from a series of direct drained shear tests.

# Behaviour of soils subjected to shear stress

In spite of its limitations, the simplicity of the direct shear test helps to establish concepts about particular models of soil behaviour, which can then be applied to other types of forces acting on it. Models for the two main soil groups—granular and clayey soils—are described in the following sections.

## Granular soils

Suppose that three samples of the same sand are tested with three different densities. For the sake of simplicity, it is assumed that all three samples are tested under drained conditions, so that total and effective stress will always coincide.

Sample 1 is very loose sand, with a high void ratio ($e_1$). Sample 2, of the same sand, is denser than sample 1, with a lower void volume and a lower void ratio ($e_2$). Sample 3 is very dense, with a lower void volume than the other two and therefore the lowest void ratio ($e_3$) of the three samples. Each of the three samples is placed in a direct shear box and the same normal effective stress is applied to all three:

$$\sigma'_{n1} = \frac{N_1}{L^2}$$

They are then subjected to shear as described above.

Figure 2.46 shows the qualitative results obtained from these tests over time, based on the different parameters involved, where:

— $\tau_f$: shear stress.
— $\Delta x$: horizontal displacement between the upper and lower parts of the shear box.
— $\tau_f$: mobilized tangential stress.
— $\Delta V$: change in volume.
— $e$: void ratio.
— $\sigma'_n$: effective vertical stress applied.

The void ratio can be obtained for each stage of the test by applying the following expression:

$$\frac{\Delta H}{H_0} = \frac{e_0 - e}{1 + e_0} \Rightarrow e = e_0 - \frac{\Delta H}{H_0}(1 + e_0)$$

where:

$H_0$ is the initial height of the sample.
$e_0$ is the initial void ratio of each sample ($e_1$, $e_2$ and $e_3$ respectively in this case).
$\Delta H$ is the measured vertical displacement (positive if it is compression).

A detailed observation of the graphs shown in Figure 2.46 leads to the following points of interest:

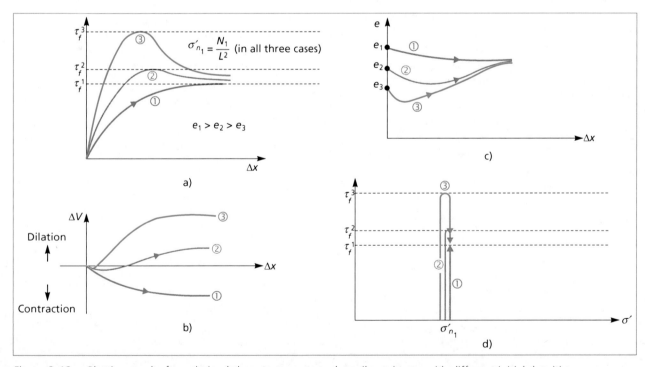

Figure 2.46   Plotting results from drained shear tests on granular soil specimens with different initial densities.

- **Low density sample (1):**
  - Mobilized tangential stress increases with horizontal displacement ($\Delta x$) until it reaches a maximum value ($\tau_f^1$). From this moment on, it remains constant even if horizontal displacement continues (Figure 2.46a).
  - From the beginning of the test, settlement of the loading plate is observed, i.e. the volume of the sample is reduced when subjected to shear stresses: it contracts. A certain degree of horizontal displacement is reached after which no further appreciable changes in volume are observed (Figure 2.46b).

- **Medium density sample (2):**
  - As in sample 1, mobilized tangential stress increases with horizontal displacement ($\Delta x$). In this case the slope of the curve ($\tau, \Delta x$) is greater, and it also reaches a maximum tangential stress (peak strength = $\tau_f^2$) which is clearly greater than ($\tau_f^1$). From this it can be seen that sample 2, which is denser than sample 1, is also stiffer and more resistant. However, if horizontal displacement continues, mobilized tangential stress decreases until it finally converges with ($\tau_f^1$) (Figure 2.46a).
  - At the beginning of the test, settlement of the loading plate occurs, i.e. the sample decreases in volume (contraction). However, when a certain point is reached, an increase in volume (dilation) may occur. Finally, as in the case above, if there is enough displacement a state is reached beyond which no further appreciable changes in volume are observed (Figure 2.46b).
  - Figure 2.46c shows the changes in volume just described in terms of the void ratio, but it also shows an interesting feature of the behaviour of granular soils. When the state described occurs in which appreciable changes in volume no longer occur and the tangential stresses of samples 1 and 2 converge, the void ratios of both also converge.

- **High density sample (3):**
  - The last test shows a greater slope to the curve ($\tau, \Delta x$), together with maximum mobilized tangential stress. The peak strength ($\tau_f^3$) observed is much greater than the maximum tangential stresses reached in the other two tests. In fact, the most dense sample shows stiffer behaviour and turns out to be considerably stronger. In any case, just as in the tests carried out on less dense samples, with sufficient horizontal displacement mobilized tangential stress decreases until it finally converges with ($\tau_f^1$) (Figure 2.46a).
  - At the beginning of the test the loading plate may go down slightly, perhaps due to readjustments in the shear apparatus, but very soon afterwards uplift displacements begin to be recorded. This shows that the behaviour of the dense sample is clearly dilative, tending to increase in volume when subjected to shear. As in the cases above, with enough displacement a state is reached beyond which there are no further appreciable changes in volume (Figure 2.46b).
  - The dense sample also tends to converge towards a single void ratio and reach a state where further displacement does not result in more changes in volume or modifications in the tangential stress, which remains at approximately equal to ($\tau_f^1$) (Figure 2.46c).

These three shear tests, whose stress paths can be shown in ($\sigma_n', \tau$) space (Figure 2.46d) can be repeated with identical samples to the previous ones, but now subjecting them to greater normal effective stresses. Figure 2.47 shows a diagram of the three resulting failure envelopes, showing how the friction angle (at peak) depends directly on the initial density of the soil.

As has been seen, the relationship between the initial density or compactness of a particular granular soil and its strength is very marked, so much so that in practice approximate correlations are usually available between compactness (determined from in situ tests like the SPT; see Chapter 5), and the angle of shearing resistance, as shown in Table 2.6.

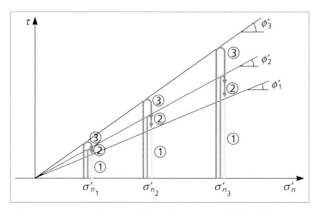

Figure 2.47  Failure envelopes in terms of initial density.

| Table 2.6 | RELATIONSHIP BETWEEN SPT VALUES, ANGLE OF FRICTION AND RELATIVE DENSITY IN COARSE GRAIN SOILS | | |
|---|---|---|---|
| N (SPT) | Relative density | % | $\phi$ (°) |
| 0–4 | Very loose | 0–15 | <28 |
| 4–10 | Loose | 16–35 | 28–30 |
| 10–30 | Moderately dense | 36–65 | 30–36 |
| 30–50 | Dense | 66–85 | 36–41 |
| >50 | Very dense | 86–100 | >41 |

Although compactness is a prime factor, whether a granular soil is more or less strong also depends on other factors, especially **the shape of the particles, and the grain size distribution.** The influence of these three factors on the strength is easy to see. It is self-evident that the shape of rounded particles makes it easier for them to slip and roll past each other than for irregular angled grains, so a soil made up of irregular particles will have greater resistance to shear. The grain size distribution in a uniform soil means that most particles are of similar size, so the maximum size of the voids will depend directly on particle size. In contrast, a well-graded soil has particles of many different sizes, so medium-sized grains can fill the voids between coarser grains and finer grains the voids between medium grains. This means that the grains can be more tightly packed and so, logically, a well-graded soil can have a stronger, denser structure than a uniform soil and it will be more difficult for larger particles to slip and roll past each other than smaller particles.

Table 2.7 shows some approximate values of the angle of friction expected for different degrees of compactness and SPT.

## Clay soils

For simplicity, the tests described below are assumed to have been carried out in drained conditions, i.e. allowing any excess pore pressure caused by increments in both tangential and normal stresses during the test to dissipate completely. It is also assumed that the soil tested is a reconstituted clay, i.e. made in the laboratory from a suspension. This working hypothesis simplifies and idealizes the formation of a clay deposit by ignoring such complementary effects of increased strength from ageing, cementation and similar diagenetic processes.

### Low plasticity clays

Figure 2.48 shows the one-dimensional consolidation process in low plasticity clay, reconstituted in the laboratory from an aqueous suspension. As described in Section 2.5, points (1), (2) and (3) correspond to normally-consolidated states, while points (4) and (5) correspond to overconsolidated states under a preconsolidation pressure equal to state (3). Once each of the five previous states is reached (when consolidation in each of them is completed), the soil is subjected to drained direct shear tests. Figure 2.48b shows the stress paths ($\sigma'_n$, $\tau$) of the five tests, while Figure 2.48c shows the corresponding curves ($\tau$, $\Delta x$). To make it clearer, only the curves for the tests on samples (2), (3) and (4) are shown in this last figure. Finally, it is assumed that a vertical displacement gauge is available to determine changes in volume during shear.

The following behaviour models can be deduced from the results obtained:

- **Normally consolidated samples (1, 2 and 3):**
  — Mobilized tangential stress increases with horizontal displacement ($\Delta x$) until maximum peak value $\tau_{max}$ is reached. This peak is hardly noticeable as $\tau$ descends very rapidly to a value of $\tau_{NC} \approx \tau_{max}$, which remains constant even if horizontal displacement continues. If an unloading-reloading cycle is carried out, approximately the same levels of tangential stress as before will be reached.
  — The failure envelope is defined by an angle of internal friction ($\phi'_{NC}$) and zero effective cohesion ($c' = 0$). In the absence of other effects, e.g. bonds or other cementation, the strength of a normally consolidated clay with low plasticity is usually entirely described by friction; i.e. cohesion is not apparent.
  — Soil volume tends to reduce (i.e. become contractive) during shear, although as in the case of tangential stress, it also reaches a certain magnitude of horizontal displacement beyond which no further appreciable changes in volume are observed.

- **Overconsolidated samples (4 and 5):**
  — The inclinations of the curves ($\tau$, $\Delta x$) are greater than in normally consolidated curves, and mobilize their maximum tangential stress with smaller deformations, i.e. they are stiffer.
  — The maximum tangential stresses reached are clearly greater than those of normally consolidated soil tested under the same initial normal stresses. In fact, the stress paths on the plane ($\sigma'$, $\tau$) exceed the failure envelope of normally consolidated states and reach a peak strength above that of the envelope defined by $c' = 0$, $\phi'_{NC}$.
  — The failure envelope of consolidated states is defined by an apparent cohesion and angle of shearing resistance ($c'$, $\phi'_{OC}$).

| Table 2.7 | RELATIONSHIP BETWEEN ANGLE OF FRICTION AND RELATIVE DENSITY IN GRANULAR SOILS | | |
|---|---|---|---|
| | Angle of friction (degrees) | | |
| Type of soil | Loose | Moderately dense | Dense |
| Non-plastic silts | 26 to 30 | 28 to 32 | 30 to 34 |
| Fine to medium uniform sands | 26 to 30 | 30 to 34 | 32 to 36 |
| Well-graded sands | 30 to 34 | 34 to 40 | 38 to 46 |
| Mixture of sands and gravels | 32 to 36 | 36 to 42 | 40 to 48 |

SOIL MECHANICS AND ENGINEERING GEOLOGY OF SEDIMENTS

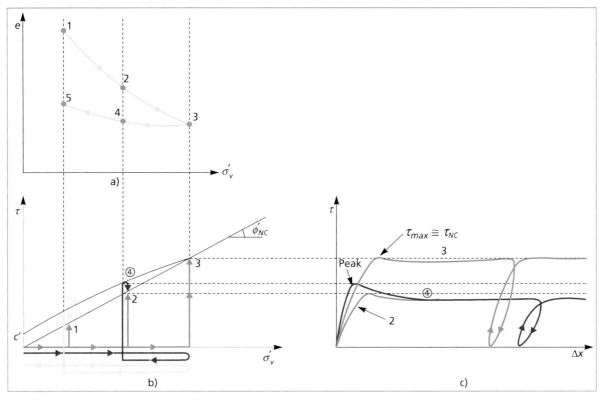

Figure 2.48  Drained shear test curves on low plasticity clay specimens (modified from Burland, 1988).

— As the deformations continue, once peak value is reached tangential stresses are reduced and tend towards those of a normally consolidated soil with the same stress levels.
— Samples with a low OCR may contract to some extent, but as overconsolidation increases they become dilative.
— If there is enough deformation, this will produce a state in which deformations may continue without changes in tangential stress or in soil volume.

### High plasticity clays

Figure 2.49 shows the same testing procedure as Figure 2.48 but applied in this case to a high plasticity clay.

Comparing Figures 2.48 and 2.49 shows that the main difference between the clays is how they behave after maximum tangential stress is reached. In the soils with a high clay content, there may be a very marked reduction of mobilized strength as the deformations accumulate, producing the development of a resistance envelope clearly below the one given by $c' = 0$, $\phi'_{NC}$. This is **residual strength**, defined by the residual shear strength parameters ($c'_r = 0$, $\phi'_r < \phi'_{NC}$).

The mechanism that explains the decrease in strength down to the level of residual strength is related to the platelet shape of the particles that make up clay minerals. As the level of shear strain increases, the particles are gradually reoriented and end up parallel to each other, a weaker arrangement than the original one created by sedimentation. Particle reorientation is usually concentrated in a thin band where failure is triggered (Lupini et al., 1981).

That the strength of high plasticity soils may be reduced obviously has important implications for engineering projects if, for example, these are carried out on slopes where sliding has already occurred and where the level of strain may have caused near residual conditions.

The direct shear test can be used for studying residual strength in the laboratory. To reach the required level of deformation the procedure consists of carrying out various complete tests with the shear box, moving it back once the maximum allowed horizontal displacement permitted has been reached, and repeating the test as many times as needed. It was to avoid this that the ring-shear apparatus was developed, permitting unlimited shear displacement.

## The triaxial test

### The test apparatus

The **triaxial test** is the most widely used laboratory test for studying the shear strength of soils. Although it has some

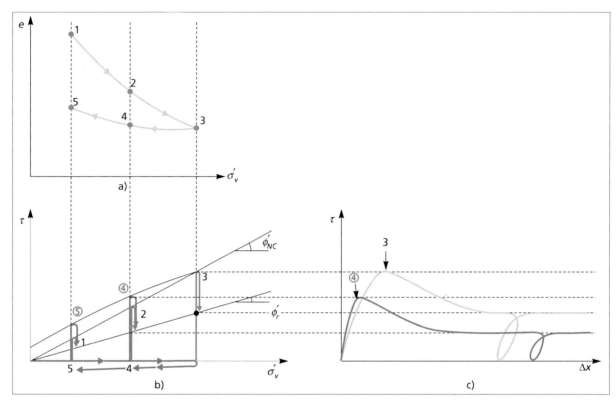

*Figure 2.49* Drained shear test curves on high plasticity clay specimens (modified from Burland, 1988).

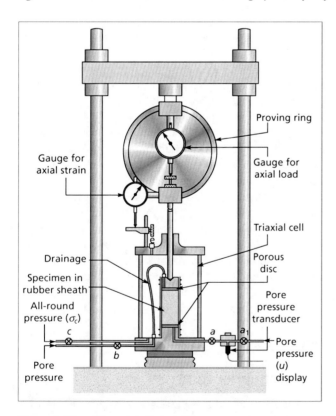

*Figure 2.50* Diagram of triaxial apparatus.

limitations, it is extremely versatile and gives reliable and varied information on soil strength under different conditions, which can be controlled as required.

The triaxial test (*Figures 2.50* and *2.51*) requires the preparation of a solid cylindrical specimen with a height which is double its diameter, surrounded by a rubber membrane. If the intention is to allow drainage and control pore pressures inside the specimen, porous stones are placed both at the base and at the top of the cylinder.

The cylindrical specimen is then placed on a central pedestal of a cylindrical cell, which is filled with water that can be pressurized via a pipe and valve c (*Figure 2.50*). Taking into account that the fluid pressure is exerted with the same intensity in all directions (i.e. all-round), total isotropic pressure $\sigma_1 = \sigma_2 = \sigma_3 = \sigma_c$ can be applied to the specimen within the cell.

A plastic tube or drainage line is connected to the upper part of the cylindrical specimen through the porous stones. This allows the pore pressure ($u$) of the water filling the soil voids, and which is measured at $a$, to be controlled (via pipe and valve $b$); this also controls the water into and out of the cylinder. This means that if the soil is saturated, the reduction or increase in its void volume must be associated with the outflow or inflow of the same volume of water. The system for measuring this is connected to valve b) and allows changes in soil volume during tests with drainage to be measured at all times.

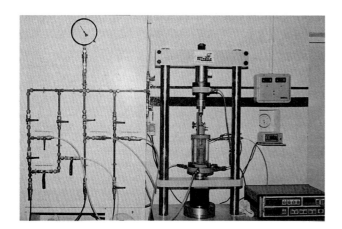

*Figure 2.51* Triaxial test equipment.

Finally, on the pedestal below the cylinder there is a third pipe directly connected to the specimen through the lower layer of porous stones. This pipe is connected to a pore pressure transducer (pipe and valve *a*) for continuous measurement of pore pressure in the specimen.

The loading system described only allows the application and control of isotropic stresses, both total and effective. To create shear stresses a deformation mode is imposed to introduce tangential stress, so creating shear stress.

In the case of the triaxial test, the whole cell is subjected to a controlled upward movement at constant velocity. To counteract this movement, there is a very rigid piston which sits on the upper part of the solid cylindrical specimen and projects the cell to bear on a dynamometer (in this case a loading ring) which measures the vertical load transmitted to it from the base of the cell through the sample ($\Delta\sigma_1$). When the sample can carry no more increments of load it fails, and $\sigma_{1\,failure}$ is thus defined under the all-round pressures operating. In the triaxial test, therefore, total principal stresses, pore water pressures and effective stress can be controlled:

$$\sigma_1 = \sigma_c + \Delta\sigma_1; \quad \sigma'_1 = \sigma_1 - u$$
$$\sigma_2 = \sigma_3 = \sigma_c; \quad \sigma'_2 = \sigma'_3 = \sigma_c - u$$

As can be seen from the above expressions:

— The loading system applied is not completely general but has axial symmetry ($\sigma_2 = \sigma_3$).
— Maximum tangential stress at each instant during the test is given by (see construction of the Mohr circle, *Box 2.16*):

$$\tau = \frac{\sigma_1 - \sigma_3}{2}$$

Finally, axial shortening in the solid cylindrical specimen can be measured continuously, using a gauge situated between the cell and the loading ram so enabling shear stress to be plotted as a function of axial displacement and axial strain.

## Types of test

Although the versatility of the triaxial apparatus allows many different tests to be carried out, the three most characteristic types are:

— Consolidated—Drained (CD)
— Consolidated—Undrained (CU)
— Unconsolidated—Undrained (UU)

In each of these, there are two different phases for loading the sample and taking it to failure:

— First the application of a specific isotropic all-round pressure ($\sigma_c$), and its associated and equal increase in pore pressure ($\Delta u = \Delta\sigma_c$) which may or may not allowed to drain (i.e. $\Delta u \to 0$)
— Then the introduction of tangential stresses through upward movement of the cell and the resulting increase in the main vertical stress up to failure.

In all cases, as in the direct shear test, three identical soil specimens are made to fail by being subjected to increasing isotropic stresses ($\sigma_{c1}$, $\sigma_{c2}$, $\sigma_{c3}$) in the first phase.

## Consolidated and drained triaxial test (CD test)

*Figure 2.52* shows the two basic phases of the test. First, all-round pressure ($\sigma_c$) is applied to the saturated sample and the pore pressure ($u_0$) desired is obtained by allowing the sample to drain freely. According to the concepts described in Section 2.4, increments in stress applied instantaneously will lead to an initial increase in pore pressure (measured at a) and the resulting effective stresses will be as in Terzaghi's principle. If drainage is allowed (via *b*), the resulting excess pore pressure will slowly dissipate, depending on the permeability of the soil, until complete consolidation is reached. At that point, pore pressure will return to either its original value ($u_0$ in this case) before $\sigma_c$ was applied, or to some other value as required. The volume of water discharged from the sample (which equals the change in volume of the sample) is controlled by valve *b*, which remains open during the process of pore pressure dissipation; from this it follows that the pressure at *b* must also be controlled to equal that at *a*, ($u_0$) in this case. Effective stresses acting on the soil will then give:

$$\sigma'_1 = \sigma_c - u_0$$
$$\sigma'_2 = \sigma_c - u_0$$
$$\sigma'_3 = \sigma_c - u_0$$

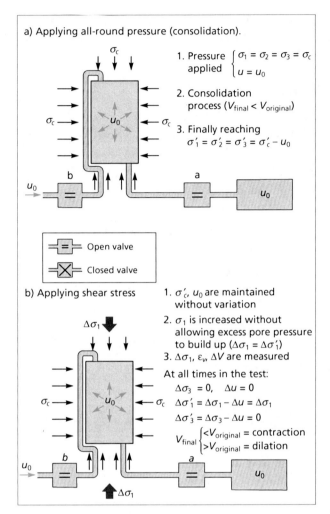

Figure 2.52   CD triaxial test stages.

The corresponding reduction in volume produced by the isotropic increment of effective stresses can be measured in the drainage control system and so, starting with the saturated sample, the volume of water expelled will be equal to the decrease in volume of the sample.

Once consolidation is complete the shear phase can begin. Because this can also cause pore pressure to change, the all-round pressure and pore pressure from the previous phase are kept constant by having valves *a* and *b* open, and the cell is moved upwards by the loading ram. As the test is carried out with drainage, the speed of the test is selected that is slow enough to guarantee the resulting excess pore pressure dissipates continuously. This can be controlled by reading the pore pressure transducer, which should give a reading of around $u_0$ at all times. Throughout the process, the increase in vertical stress $\Delta\sigma_1 = \Delta\sigma'_1$ is measured, along with the variation in specimen volume $\Delta V$ and the resulting axial shortening produced, $\varepsilon_v$. The difference ($\sigma_1 - \sigma_3 = \Delta\sigma_1$) is called the **deviator** stress and it represents double the maximum tangential stress at each instant of the test.

In CD triaxial tests, the variations throughout the test in the deviator stress and change of volume in the cylinder specimen against $\varepsilon_v$ are usually plotted as already described for direct shear test.

In a normal test programme, failure in three specimens of soil, prepared in the same way, is induced by applying to each a different isotropic effective consolidation stress in the first phase (for example, ($\sigma_c - u$ = 100, 200 and 300 kPa, respectively). In each test, failure is reached with a different vertical stress $\sigma_{1f} = \sigma'_{1f}$. The higher the initial effective all-round pressure, the greater the vertical stress. So, three Mohr circles in effective stresses can be plotted simply on a diagram $\sigma'$, $\tau$ (Figure 2.53), given that:

— The minor principal stress is equal to effective consolidation pressure in the first phase ($\sigma_c - u$).
— The diameter of the Mohr circle is the deviator stress at failure $(\sigma_1 - \sigma_3)_f = \Delta\sigma_{1f}$.

In the direct shear test the points representing failure of each sample were plotted and defined a straight line, and in this case something similar occurs: the circles have an approximately common tangent. The failure envelope in effective stresses is obtained by plotting the common tangent to the three circles, and from this the shear strength parameters of the soil can be deduced ($c'$, $\phi'$).

### Consolidated and undrained triaxial test with pore water pressure measurement (CU test)

Figure 2.54 shows the basic phases of this test. The first is consolidation under isotropic effective stress and is identical to the first stage of the CD test. Once consolidation has taken place, valve *b*, used for drainage and for adjusting pore pressure, is closed and the shear phase begins by moving the cell upwards at the same time as vertical displacement of the specimen pushes it against the loading ram.

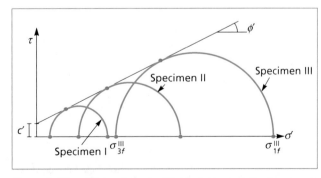

Figure 2.53   Mohr circles in CD triaxial tests (effective stress).

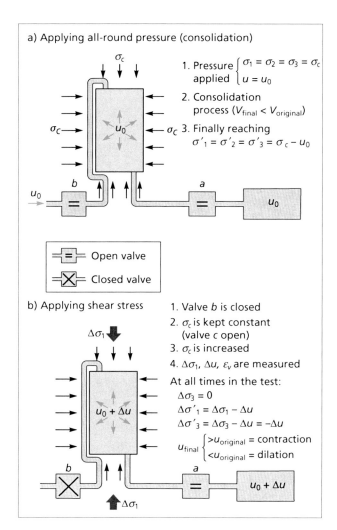

*Figure 2.54* CU triaxial test stages.

Throughout the process, measurements are taken of the increment generated in total vertical stress ($\Delta\sigma_1$), the pore pressure variation in the specimen ($\Delta u$), measured via valve *a*, and the axial shortening ($\varepsilon_v$) produced. In CU tests, the variation in the deviator stress and the changes in specimen volume against $\varepsilon_v$ are usually plotted during the test.

When failure is reached an increment in total vertical stress ($\Delta\sigma_1)_f$ will have occurred. As pore pressure at that particular instant is known ($u_f = u_0 + \Delta u_f$), effective stress at the time of failure can easily be determined, and the corresponding Mohr circle plotted:

| State | Total stress | Pore pressure | Effective stress |
|---|---|---|---|
| Initial | $\sigma_1 = \sigma_3 = \sigma_c$ | $u_0$ | $\sigma'_1 = \sigma'_3 = \sigma_c - u_0$ |
| Failure | $\sigma_{1f} = \sigma_c + \Delta\sigma_{1f}$<br>$\sigma_{3f} = \sigma_c$ | $u_0 + \Delta u_f$ | $\sigma'_{1f} = (\sigma_c + \Delta\sigma_{1f}) -$<br>$- (u_0 + \Delta u_f)$<br>$\sigma'_{3f} = \sigma_c - (u_0 + \Delta u_f)$ |

As in the previous case, a normal test programme requires failure to be induced in three specimens prepared in the same way by applying the same method of incremental isotropic effective consolidation stress, so that effective parameters for soil strength ($c'$, $\phi'$) can be determined from plotting the tangent common to the three resulting Mohr circles on effective stresses.

### Some points of interest on the absence of drainage and pore water pressure response

As described, in the shear phase of the CU test the drainage valve remains closed. As a result, if the specimen is saturated, as is usual in this test, water cannot get in or out of the cylinder specimen, and the volume will remain constant throughout the entire shear phase ($\Delta V = 0$).

Under these conditions, if the test soil is contractive, i.e. it tends to reduce its volume when subjected to shear, this will be reflected in an increase in pore pressure ($\Delta u > 0$) during the test. The explanation of this phenomenon is obvious as to reduce soil volume, enough water must be released (leading to a transitory rise in pore pressure) to provoke the discharge of the volume of water needed for contraction. Therefore, if release of pore water is prevented, the increment in pore pressure increase generated will not be able to dissipate and will continue to accumulate and increase with a progressive build-up of shear stress.

Conversely, if the test soil is dilative, i.e. if it tends to increase in volume when subjected to shear, this will be reflected in a reduction in pore pressure ($\Delta u < 0$) during the test. Once again, the explanation of this phenomenon is straightforward: it is the opposite effect of the mechanism described for contractive soils.

Keeping in mind the concepts described in relation to stresses induced in saturated soils by loading processes without drainage, pore overpressure for the traxial test in saturated soil is given by:

$$\Delta u = \Delta\sigma_3 + A(\Delta\sigma_1 - \Delta\sigma_3)$$

where parameter *A* depends on the type of soil. Taking into account that in the shear phase total all-round pressure remains constant ($\Delta\sigma_3 = 0$), the above expression is reduced to:

$$\Delta u = A(\Delta\sigma_1)$$

and at the instant of shear will be:

$$\Delta u_f = A_f(\Delta\sigma_{1f})$$

where $\Delta\sigma_{1f}$ is positive; as a result, the "sign" of $\Delta u_f$ depends exclusively on $A_f$.

Table 2.8 gives the orders of magnitude of parameter A at failure for some soils, and also the tendency for volume changes of these same soils in tests with drainage. This table also shows the direct relationship between the dilative or contractive character of a particular soil, as well as the pore pressure response when drainage is prevented.

## Unconsolidated Undrained triaxial test (UU test)

The particular feature of this type of triaxial test is that the valve for drainage and applying pore pressure (b) is always closed. In the first phase, only an isotropic cell pressure ($\sigma_1 = \sigma_3 = \sigma_c$) is applied to prevent drainage. If the specimen is saturated, and as there is no drainage, all the total isotropic stress in the chamber is transmitted to the interstitial water and there is therefore no variation in effective stresses in the soil. Even if three different chamber stresses are applied to three identical specimens, initial effective stresses will be the same in all three samples. This is why when the shear phase is carried out, also without drainage, the deviator stress at failure $\Delta\sigma_{1f}$ is always the same. In this phase, the increase in total vertical stress $\Delta\sigma_1$ and axial strain $\varepsilon_v$ are measured.

Figure 2.55 shows the three Mohr circles at failure of the three specimens tested. They are expressed in total stresses (the only ones measured) and have the same diameter i.e. the same deviator stress at failure; in fact, if pore pressure at the moment of failure in each test is ignored, only one circle for effective stress would be obtained (shown dotted).

This would be the same for the three samples, and would be tangential to the failure envelope defined by the effective parameters ($c'$, $\phi'$) of the test soil.

As can be seen, the circles in total stress have a horizontal line as a common tangent. The point where this line intersects the ordinate axis is the **undrained shear strength**, ($S_u$) coinciding with the radius of the circles, for both total and effective stresses. $S_u$ therefore represents the maximum

| Table 2.8 | RANGE OF VALUES OF THE PORE PRESSURE COEFFICIENT A AT FAILURE | |
|---|---|---|
| Soil type | $\Delta V$ in drained triaxial tests | $A_f$ in undrained triaxial tests |
| Sensitive clay | High contraction | +0.75 to +1.5 |
| Normally consolidated clay | Contraction | +0.50 to +1.0 |
| Compacted sandy clay | Slight contraction | +0.25 to +0.75 |
| Slightly overconsolidated clay | Slight or no contraction | +0.00 to +0.5 |
| Compacted clayey gravel | Dilation/ contraction | −0.25 to +0.25 |
| Highly overconsolidated clay | Dilation | −0.50 to 0.0 |

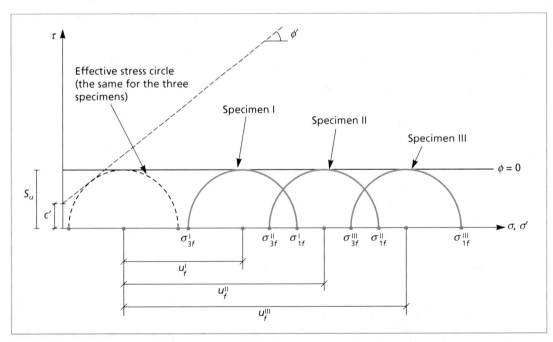

Figure 2.55   Mohr circles obtained from UU triaxial test (total stress).

tangential stress that can be mobilized by the soil, brought to shear failure in undrained conditions from its initial effective stress state.

This test is useful because it is quick and simple. It obviously does not allow effective shear strength parameters ($c'$, $\phi'$) to be determined as not even pore pressure is measured during the test. Nevertheless, it gives the maximum tangential stress available in the soil for an initial state of effective stress. If it is assumed that the samples extracted are representative of the soil *in situ*, and their original conditions were not altered during extraction, using this test allows an approximate determination of the maximum shear stress available in relation to loading processes where undrained conditions are assumed.

The undrained shear strength can be also determined by the fall cone test described in Section 2.2.

## The uniaxial compression test

This test is an special version of the triaxial test where axial stress is applied but without all-round pressure. A cylindrical specimen is placed in a press and failure is induced by compression without any lateral confinement, i.e. where $\sigma_3 = 0$. It can only be carried out in predominantly cohesive soils because otherwise, as there is no lateral confinement, the specimen could disintegrate spontaneously. Although the specimen is in direct contact with the air (without a rubber membrane), it can be assumed from how quickly failure is reached and from the imperviousness of the soils tested with this procedure, that no there is no dissipation of pore pressure inside the sample.

*Figure 2.56* shows the Mohr circle in total stresses obtained in this test. The least total stress $\sigma_3$ is zero, and the uniaxial compression strength (usually called $q_u$) is the deviator stress ($\sigma_1 - \sigma_3 = \sigma_1 = q_u$). The radius of the Mohr circle will be the undrained shear strength, $S_u$. The increment in total vertical stress $\Delta\sigma_1$ and axial strain $\varepsilon_v$ are measured, and the corresponding stress-strain curve is usually plotted. Clays can be classified according to their uniaxial compressive strength as shown in *Table 2.9*.

*Table 2.9* **RELATIONSHIP BETWEEN UNDRAINED SHEAR STRENGTH AND CONSISTENCY OF CLAYEY SOILS**

| Consistency | Undrained shear strength (kN/m$^2$) |
|---|---|
| Very soft | <20 |
| Soft | 20–40 |
| Firm | 40–75 |
| Stiff | 75–150 |
| Very stiff | 150–300 |
| Hard | >300 |

## 2.7 Influence of mineralogy and fabric on the geotechnical properties of soils

Soils are formed from solid particles, fluids, gases and voids, and can be grouped into two classes according to particle size, as seen in Section 2.2:

— **Coarse-grained or granular soils** (with a predominant grain size larger than 0.075 mm), formed mainly of quartz, feldspar and calcite, and, less frequently, of sulphates, salts and volcanic glasses.
— **Fine-grained or fine soils** (with more than 50% of the grain size equal to or smaller than 0.075 mm), formed mainly from silts and clay minerals, such as kaolinites, illites, smectites, and organic material.

The two classes of soil are differentiated by particle size analysis. The behaviour of coarse granular soil particles is usually stable and resistant (*Figure 2.57*), while fine soils can form finely layered and laminar structures of variable behaviour and can be geotechnically unstable. Granular soils are not plastic and their strength depends basically on the angle of internal friction, which is conditioned by the shape, size and degree of solid particle packing. Such soils are considered to be **frictional soils**. Fine soils are plastic, with strength depending both on internal friction between the solid particles and on the cohesive forces present in them. For this reason they are also known as **cohesive soils.**

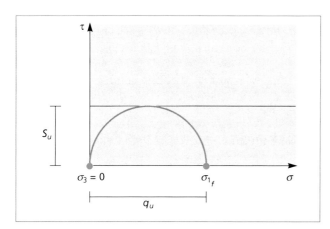

*Figure 2.56* Mohr circle obtained from uniaxial compression test (total stress).

Figure 2.57   Structure of granular soils seen under optical microscope.

The geological factors listed in *Table 2.10*, help to determine other soil properties such as density, porosity, compressibility, and changes in volume.

# Clay minerals in engineering geology

Clay minerals are layer silicates belonging to the larger class of sheet silicates called phyllosilicates, characterized by their layered structure. There is a wide range of clay minerals with very different physical and chemical properties, although as a result of their layered structure, flattened morphologies and perfect separation between the layers are common to most of them.

The tiny crystals of these minerals are less than 2 μm in size. They are the most common minerals on the earth's surface and form part of fine-grained soils and sedimentary rocks.

Structurally they have two basic units bonded together by common oxygens, one formed from tetrahedrons connected to the three oxygens of the basal vertices, with a thickness of 3Å, and the other from octahedrons bonded together by common edges, with a thickness of 4Å. The centre of the tetrahedrons is occupied by $Si^{4+}$, which is frequently replaced by $Al^{3+}$ and at times by $Fe^{3+}$. The centre of the octahedrons is normally occupied by $Al^{3+}$, $Mg^{2+}$ and $Fe^{2+}$, and at times by $Fe^{3+}$, $Li^+$ and other transition elements. In order to maintain electrical neutrality, $Al^{3+}$ should occupy two thirds of the octahedral positions, whilst $Mg^{2+}$ occupies all of them. Minerals are differentiated according to whether they contain aluminium (dioctahedral minerals) or magnesium (trioctahedral).

There are different types of clay minerals, depending on the number of layers in their structure. These can be further differentiated into sub-groups according to the degree of ordering and the type of isomorphic substitution.

| Table 2.10 | GEOLOGICAL FACTORS CONDITIONING THE GEOTECHNICAL PROPERTIES OF SOILS |
|---|---|
| **Geological factors** | **Main characteristics** |
| Type of soil | — Residual soils and parent rock<br>— Transported soils and depositional conditions |
| Geo-environmental conditions | — Particle size distribution and porosity<br>— Water content<br>— Ground water geochemistry<br>— Confining pressure and temperature |
| Mineral composition | — Percentage of clay minerals<br>— Structural composition<br>— Specific surface area, electrical charge and ionic exchange capacity<br>— Pore water composition |
| Soil fabric and post-sedimentary processes | — Soil structure and microfabric<br>— Discontinuities<br>— Weathering<br>— Diagenetic changes<br>— Consolidation and sediment loading |

## Type 1.1

**The kaolin group.** The composition of this group is $Al_4Si_4O_{10}(OH)_8$, with a tetrahedral layer occupied by $Si^{4+}$ and an octahedral layer occupied by $Al^{3+}$ with a thickness of 7Å. They are therefore dioctahedral minerals without any isomorphic substitutions, although some types can be differentiated depending on the degree of disorder in the stacking of layers (*Figures 2.58* and *2.59*). When **kaolinite** is well ordered it forms pseudo-hexagonal columns. **Halloysite** belongs to this group; it shows a high degree of disorder, with one variety of 7Å and another of 10Å. The latter type includes a layer of water of 2.9Å between two tetrahedral-octahedral layers of 7Å. The water layer is irreversibly lost at 60°C, reducing spacing to 7Å. Halloysite frequently shows tubular morphologies, and in other cases irregular or globular shapes (*Figure 2.60*).

## Type 2.1

This has a structure formed by two tetrahedral layers with an octahedral layer intercalated to form a "sandwich" with a basal spacing of 9.5Å.

**The illite group.** Illites have a basal spacing of 10Å (*Figures 2.58* and *2.61*) with a layer charge between 0.9 and 0.7. They have many similarities with micas, especially muscovite. The fact that composition is very varied casts doubt on whether illites exist as a mineral in sedimentary rocks,

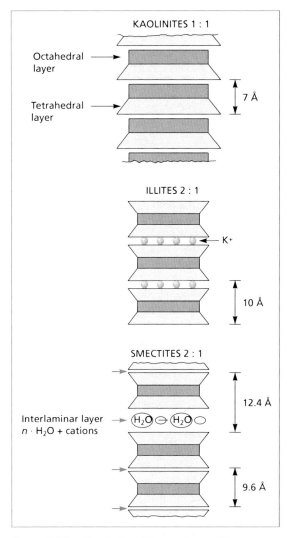

*Figure 2.58* Kaolinites, illites and smectites.

reference being made instead to illitic minerals. However, the name remains unchanged because of its interlayering with smectites. A general simplified formula would be:

$$K_{0.74} (Al_{1.56} Mg_{0.28} Fe_{0.22}) (Si_{3.4} Al_{0.6}) O_{10} (OH)_2$$

Illite particles commonly have a compact but planar morphology, although fibrous illites of diagenetic origin have also been described in sandstones.

**The smectite group.** The composition of smectites is $Al_1 Si_4 O_{10} (OH)_8$, with two tetrahedral layers occupied by $Si^{4+}$ and one octahedral layer by $Al^{3+}$. They are characterized by a layer charge of between 0.6 and 0.3, and a presence of weakly hydrated cations, which facilitates the penetration of water molecules.

Amongst the dioctahedral smectites the most frequent mineral is **montmorillonite**. The layer charge is octahedral, as can be deduced from the ideal structural formula:

*Figure 2.60* Halloysite seen under electron microscope (x 205.200).

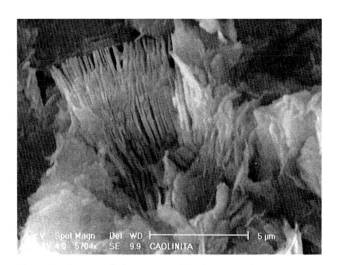

*Figure 2.59* Kaolinite seen under electron microscope.

*Figure 2.61* Illite seen under electron microscope.

$Na_{0.33}(Al_{1.67}Mg_{0.33})Si_4O_{10}(OH)_2$. A prominent feature of smectites is the property of incorporating hydration water from 0 to 100% humidity. The interlayer cations are surrounded by water molecules, which increases basal spacing. Anhydrous Na-smectite has a basal spacing of 9.6Å, which becomes ≈12.4Å, 15.2Å and 18Å when 1, 2 or 3 water molecules respectively are incorporated. During the expansion of smectite the interlayer cation may be replaced by another cation (*Figures 2.58* and *2.62*).

## Identification of clay minerals

The methods most commonly used are X-ray diffraction, differential thermal analysis and electronic microscopy.

# Physico-chemical properties

The physicochemical properties of clay minerals are related to the exchange processes in the interlayer area and the size of crystals and aggregates in the clay particles. The processes of adsorption and cation exchange are the causes of hydration and swelling of the basal spacing. They depend on the cation exchange capacity, which is expressed in centimoles of charge (+) per kilogram (cmolc/kg) or milli-equivalents per 100 grams of soil (meq/100 g).

The **size** of the clay particles is very small, with a range between tens of Å and a few μm. This produces a high specific surface area where electrostatic interactions take place, depending on the pH, the exchanged cations and the salinity of the medium.

The **specific surface area** (area of surface per unit mass) is expressed in m²/g. A distinction is made between an external surface area, where interactions related to surface charges and broken particle edges occur, and an internal surface area, where interlayer exchanges take place.

*Figure 2.62*  Smectite seen under electron microscope.

The **electric charge** present in clay minerals is one of the most significant properties influencing the interaction between clays. The origin of this charge is due to three factors related to the structure and formation of the minerals:

— The charge defect on clay surfaces caused by isomorphic substitution.
— Absorption of anions or cations due to imperfections on the crystal surface, especially on the edges of clays.
— Ionization of the surface, principally in hydroxyls (Al-OH and/or Mg-OH) which act like reversible electrodes of $H^+$ or $OH^-$ on the edges of the crystals. The negative electric charge determines the type of interaction between clay layers and their cation absorption capacity.

There are many types of clay minerals as a result of the variations in composition produced by isomorphic substitutions that take place in the tetrahedral layer, where $Si^{4+}$ is replaced by $Al^{3+}$, and in the octahedral layer, where $Al^{3+}$ is replaced by $Mg^{2+}$ and $Fe^{2+}$.

To compensate the excess negative charge and maintain the neutrality of the structure, monovalent ($Na^+$ and $K^+$) and divalent ($Ca^{2+}$ and $Mg^{2+}$) cations are incorporated, in a new layer called the **interlaminar layer**.

Layer charge controls intrinsic soil properties, such as cation exchange reactions, specific surface area and degree of hydratation. The value of the layer charge allows various groups of minerals to be distinguished. These can be basically differentiated by the type and characteristics of interlayer cation or cations and their incorporation in either anhydrous or hydrated form. Layer charge in the mica group is approximately 1; in the illite group it falls to 0.8, and in the smectite group it drops to values lower than 0.6.

Clay minerals tend to replace Si or Al with other elements within the crystalline network. This property, known as **isomorphic substitution**, is produced when an ion belonging to the clay layer is replaced by another ion of the same size but with a lower valence (normally $Al^{3+}$ is replaced by $Si^{4+}$, and $Mg^{2+}$ by $Fe^{3+}$), producing a charge defect on the surface of that layer, and a slight deformation in the network, because the size of the ions is not identical. This substitution leads to an increase in negative charge on the surface of the clay. To compensate this charge defect, as well as to maintain electric neutrality and satisfy the broken bonds on the edges of the crystals, clays attract exchangeable cations and anions to their surfaces and, in some cases, to the unit cell. The sum of all the exchangeable cations that a mineral can absorb is known as the **cation exchange capacity** or ionic exchange capacity. The maximum amount of exchangeable cations for each type of clay is constant and it is expressed in milli-equivalents per 100 grams of dry clay at 110°C ( = cmol(+)kg$^{-1}$).

Table 2.11 shows average values of these properties for the main clay minerals.

## Geotechnical properties and mineralogical composition

From the geotechnical point of view, clays are considered problematic materials as their behaviour depends both on their mineralogical and chemical composition and on environmental conditions. These factors often change; e.g. modification of the chemical composition of water may produce reactions within the mineral structure and changes in the geotechnical properties of the soil. The mineralogical composition of clays is the factor with most influence on geotechnical properties such as plasticity, strength, compressibility, and change in volume.

The amount of water adsorbed by clay minerals depends on the cation exchange capacity and the specific surface area. Water molecules are joined at the particle surface by bipolar bonding surrounding them with a film of water. The weakness of the bipolar bonds allows displacement of the particles when pressure is applied. Figure 2.63 shows the position of different types of clays on the Casagrande plasticity chart. The lowest plasticities correspond to the kaolinites and the highest to the smectites, with sodium montmorillonites being the highest in this particular group.

**Activity** is defined as the ratio between the plasticity index and the clay fraction ($PI/\%$ particles <2 µm). It is a dimensionless indicator of the mineralogical composition of clays. Clays are classified according to their activity as:

— Active: with an activity ratio higher than 1.25.
— Normal: with an activity ratio between 1.25 and 0.75.
— Inactive: with activity lower than 0.75.

Both the clay fraction content and the type of predominant clay mineral influence the **strength of soils**. Shear strength decreases as clay content increases, as shown in Figure 2.64.

The capacity for water adsorption at the edges of the layers and interlayer areas produces **changes in volume** in

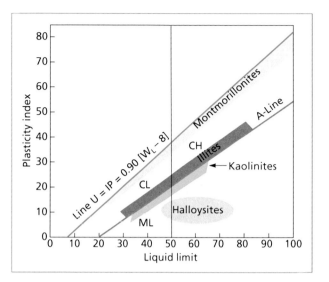

Figure 2.63   Mineralogical composition and plasticity (Day, 1999).

clays. Figure 2.65 shows the most characteristic tendencies for some of the more typical clay minerals.

The swelling of clay minerals by water adsorption is due to various factors: surface adsorption of water molecules, exchange cation hydration, (intercrystalline expansion), osmotic expansion, interlayer charge in sodium montmorillonites, and pressure at the edges of magnesium smectites (saponites).

Mineralogical composition also influences how compressible clays are. Figure 2.66 shows the most significant tendencies, with smectites being the most compressible and kaolinites the least.

In their natural surroundings, soils are mixed, in both mineralogical composition and particle size distribution and this determines the complexity of their behaviour. However, the influence of mineralogy on the geotechnical properties of soil is so marked that even the presence of small percentages of certain minerals, such as smectites (above 10%), can have a significant influence on soil behaviour.

## Microfabric of clayey soils

Another factor influencing the geotechnical behaviour of soils is their fabric. The soil **fabric or microfabric** is defined as the spatial arrangement of particles, groups of particles, pores, discontinuities and other elements present in the soil. The fabric is directly related to the degree of orientation of the soil components, their porosity and density. It conditions other properties which are of great importance in geological engineering, such as strength, compressibility and permeability. **Microfabric analysis** is usually performed with

| Table 2.11 | PHYSICAL-CHEMICAL PROPERTIES OF CLAY MINERALS | | | |
|---|---|---|---|---|
| | Cation exchange capacity | Specific surface (m²/g) | | |
| Mineral | (meq/100 g) | External | Internal | Total |
| Smectite | 80–150 | 50 | 750 | 800 |
| Illite | 10–40 | 25 | 2 | 27 |
| Kaolinite | 1–10 | 15 | 0 | 15 |

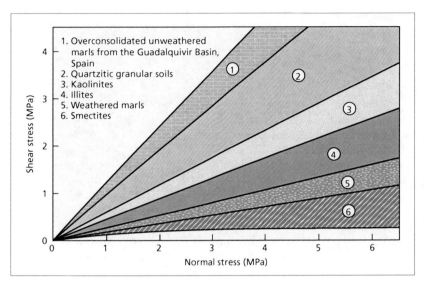

*Figure 2.64* Influence of mineralogy on strength (Tsige *et al.*, 1995).

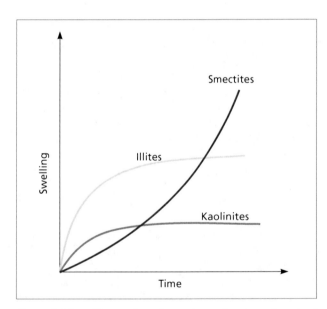

*Figure 2.65* Change in volume depending on mineralogical composition.

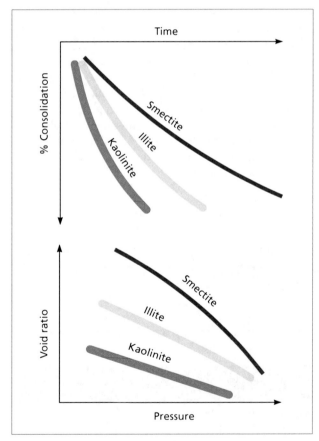

*Figure 2.66* Influence of mineralogical composition on compressibility.

a Scanning Electron Microscope (SEM) and a Transmission Electron Microscope (TEM).

While the spatial distribution of the particles constituting the coarse fraction of soils basically depends on the shape, size and degree of packing, in fine fractions the microfabric depends on physico-chemical properties, especially the degree of interaction between clay crystals. This interaction is due to the attraction and repulsion forces developed on the surface of the clay minerals. Attraction between particles is due to the Van der Waals forces while repulsion forces are due to the negative charges present on clay surfaces and in the double diffuse layer.

The electric charge on clay surfaces varies according to the different parameters of the system (pH, degree of isomorphic substitution, concentration of dissolved salts, temperature, etc.). This means that in certain conditions the edges of the particles may acquire as many negative charges as positive ones, with a variable degree of interaction.

Depending on the condition of a soil suspension, especially the concentration of electrolytes, clay minerals may adopt the following basic associations: **face-to-face, face-to-edge,** where the face has a negative charge and the edge a positive charge, and **edge-to-edge,** in cases of edges with positive and negative charges (Figure 2.67).

In suspensions with high concentrations of electrons, where the pH is relatively high (>8.2), edge-to-face interactions predominate because the positive charge is maintained at the edges and the negative charge on the clay surface. In conditions like these, with high concentrations of ions, the net electrical forces between adjacent particles are predominantly attractive, leading to the phenomenon of **flocculation,** which gives rise to an open structure (flocculated structure) with large voids, typical of marine soils (Figure 2.68).

On the other hand, when electrolyte concentration is low, clay minerals tend to have a negative charge, both at the surface and at the edges. In this case, the double diffuse layer will increase and therefore electrical repulsion forces will predominate between the adjacent particles in the phenomenon known as **dispersion** (Figure 2.69). This produces a dense elongated structure (dispersive structure) in which the clay particles are not in contact due to the predominance of repulsive forces. This dispersive structure is characteristic of freshwater lacustrine and river deposits.

Between these two types of structure (flocculated and dispersive) there are multiple ways in which clay particles can be spatially organized, since various factors intervene in the interaction. These factors include mineralogical and chemical

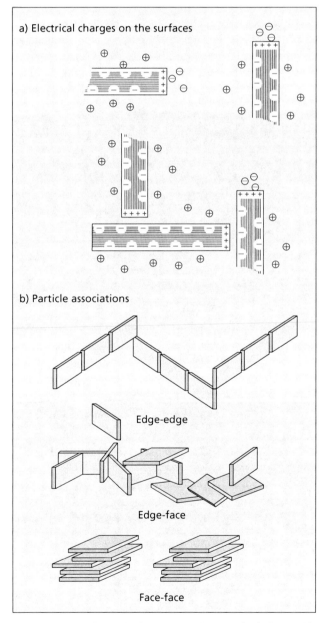

Figure 2.67  Electrical charges in clays and their particle associations.

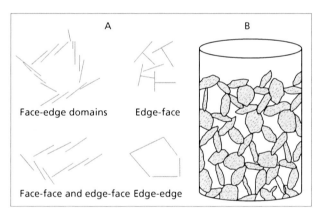

Figure 2.68  Basic structural associations in flocculated clays. A) Flocculation forms. B) Flocculated structure in an aqueous medium.

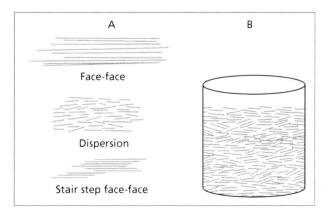

Figure 2.69  Basic structural associations in dispersed clays. A) Dispersed forms. B) Dispersed structure in an aqueous medium.

composition, particle size, concentration of solids and dissolved salts, water turbulence, biological activity, temperature, and sedimentation rate. As a result, there is a wider and more complex variety of particle associations than just the two described above.

Depending on the type of basic particle association and the different environmental factors that intervene in sedimentation, natural clay soils have various types of microfabric. The following are some of the most characteristic:

— **Turbostratic or regular aggregation:** continuous compact clay matrix; very dense structure with no preferred orientation; characteristic of overconsolidated marine sediments (*Figure 2.70*).
— **Laminar or oriented:** homogeneous matrix formed of clay particles oriented in a preferred direction; compact anisotropic structures (*Figure 2.71*).
— **Honeycomb:** open structures formed by clay particle floccules bonded by adhesion forces. These have a large quantity of intercommunicating pores and are characteristic of saline environments and sensitive soils (*Figure 2.72*).
— **Skeletal:** metastable organization of clay aggregates and fragments joined by large connectors and abundant pores. Typical of weathered and collapsible soils (*Figure 2.73*).
— **Oolitic or nodular:** made up of nodules or spherical aggregates which may be densely packed; characteristic of continental environments rich in Fe oxides (*Figure 2.74*).

As well as solid particles, other elements are also present in the microfabric: pores, discontinuities, microfissures, shear surfaces, particle bonds and cementing agents.

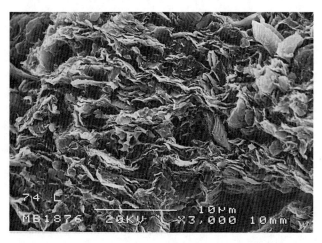

*Figure 2.70* Turbostratic microfabric in marls, Guadalquivir Basin, Spain.

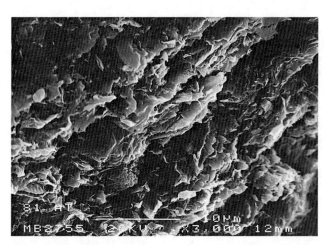

*Figure 2.71* Laminar microfabric in marls, Guadalquivir Basin, Spain.

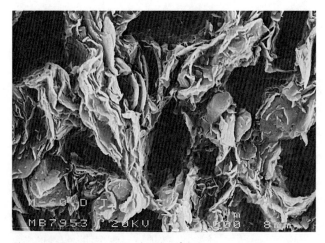

*Figure 2.72* Honeycomb microfabric.

*Figure 2.73* Skeletal microfabric.

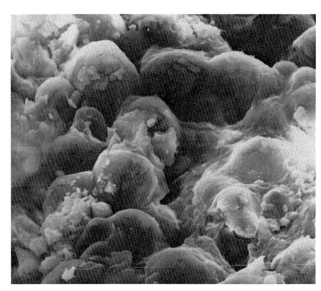

*Figure 2.74* Oolitic microfabric in volcanic clays, Tenerife, Spain.

## Geotechnical properties and microfabric

The geotechnical properties of soil are related to its structure or microfabric. This is the result of different geological and environmental processes acting on it throughout its geological history. Properties such as porosity and anisotropy have their origins in the orientation and reordering of particles (*Figure 2.75*). Other relationships between microfabric and geotechnical properties are shown in *Table 2.12*.

Collapsibility and sensitivity are closely related to the flocculation state, with very open structures typical of saline sediments and residual soils. **Sensitivity** (*St*) is defined as the ratio of undrained shear strength ($S_u$) in undisturbed state to the strength with the same water content in a remoulded state, and indicates the loss of shear strength in a soil that has been remoulded:

$$St = S_{u\ (undisturbed)} / S_{u\ (remoulded)}$$

Based on this ratio, clay soils are classified as:

— Non-sensitive: $St \approx 1$
— Slightly sensitive: $St = 1–2$
— Moderately sensitive: $St = 2–4$
— Highly sensitive: $St = 4–8$
— Extra sensitive: $St = 8–16$
— Quick clays: $St > 16$

Sensitive soils have an open and meta-stable microfabric. The most characteristic are the **quick clays**, where original intergranular cementation and particle interaction is lost by leaching when it comes into contact with fresh water. This phenomenon may also be seen in some residual soils.

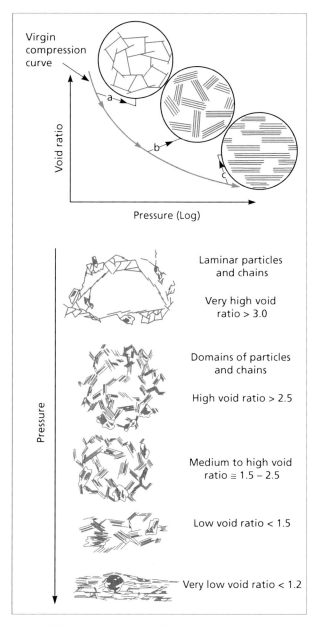

*Figure 2.75* Rearranging of particles and associated void ratio as a function of consolidation pressure (Bennet and Hulbert, 1986).

**Microfissures and microdiscontinuities** also form part of soil microfabric. These are frequent in overconsolidated soils and in laminar and turbostratic-type fabrics. Intergranular **cementing agents** (carbonates, sulphates, etc.) in soils may have an influence on stress properties producing a considerable increase in their cohesion. The microfabric may also undergo changes, in both natural and man-made conditions. For example: changes in the chemical composition of water, external loads, remoulding, or compaction can modify the original particle arrangement and as a result their geotechnical behaviour.

*Table 2.12* **MICROFABRIC OF CLAYS AND ENGINEERING GEOLOGICAL PROPERTIES**

| Fabric type | Porosity | Strength | Collapsibility | Susceptibility | Engineering geological implications |
|---|---|---|---|---|---|
| Turbostratic or regular | Low | Very high | No | No | — Characteristic of overconsolidated clays.<br>— Discontinuity surfaces at depth. |
| Laminar or oriented | Very low | Dependent on orientation | No | No | — Failure planes according to preferred orientations.<br>— Discontinuities at depth. |
| Honeycomb | High | High when unweathered. Very low when remoulded | Possible | Very high | — Instabilities and earthflows.<br>— Quick clays. |
| Skeletal | Very high | Low | Possible | High | — Residual and weathered soils.<br>— Rapid weathering on slopes.<br>— Unstable conditions. |
| Oolitic or nodular | Low | High | Low | High | — Unusual behaviour according to soil index properties. |

# Summary

The main properties of clay soils can be identified and interpreted on the basis of their mineralogical composition and microfabric. Complex or unfavourable geotechnical behaviour can be explained by a variety of factors that participate in their formation. In addition to composition and fabric, geological and human factors are very significant. However, most of the properties associated with soils considered unfavourable in the geotechnical context originate from their mineralogy and microfabric.

Another factor to be taken into account with clay soils is that they are unstable over time. Possible modifications in the environment, both natural and man-made, bring about important changes in the structure and fabric of clays, and this, together with other properties, alters their strength, deformability, and expansivity.

## 2.8 Engineering geological characteristics of sediments

Sedimentary deposits are formed by geomorphological and climatic processes, of which weathering and transportation are the most significant. These deposits are the product of different types of sedimentation and their geotechnical characteristics are related to the conditions under which these sediments were formed. The particle size and shape of sedimentary materials will reflect their transportation. Understanding the geomorphological and climatic factors that have affected transportation and deposition of a sediment allows predictions to be made of its arrangement and geometry, its physical properties and other aspects of interest in engineering geology. For these purposes, the following types of deposits can be recognised, according to their main engineering geological characteristics:

— Colluvial deposits.
— Alluvial deposits.
— Lacustrine deposits.
— Coastal deposits.
— Glacial deposits.
— Desert and arid climate deposits.
— Evaporitic deposits.
— Tropical soils.
— Volcanic soils.

Much has been written on this subject and the following notes are for guiding the further reading given in the references.

# Colluvial deposits

These deposits are transported by gravity, by freeze-thaw action and by water. They are the product of *in situ* weathering of rocks, that are then transported as material scoured from the slopes or deposited as a result of solifluction. These deposits are often associated with masses whose stability is easily disturbed, especially by excavation, and their composition depends on the parent rock. They consist of generally coarse, angular heterometric fragments, lumped together in a clayey silt or sandy matrix. They are not generally very thick, but can vary considerably.

Colluvial deposits can be considered as potentially unstable materials in most cases. Their strength is low, especially in contact with the underlying rock, or when high pore pressures develop as a result of heavy rain. Identifying these materials is fundamental in any engineering geological study, and is therefore a priority in site investigations. Their presence may imply a geotechnical problem. *Figures 2.76* and *2.77* show a typical ground profile of colluvial deposits.

# Alluvial deposits

These materials are transported and deposited by river water. Their particle size varies from clay to coarse gravel and boulders, and many of its particles have rounded edges. They are distributed in layers, with a distinct classification and

Figure 2.77   Colluvial deposits, Tenerife, Spain.

varying density. They are very common in temperate climates in riverbeds, valleys, plains, alluvial fans, terraces and paleochannels.

Their soil distribution is highly anisotropic, with very variable geotechnical properties closely related to particle size; coarse particles (sand and gravel) are typical of channel deposits whereas silts, clays and organic deposits (e.g. peat) are typical of the flood plains on either side of alluvial channels. Their continuity is irregular and they may have a high content of organic matter in certain environments. Permeability depends on particle size, and the water table is generally high. Because of their heterogeneity and anisotropy, a large number of site investigations are needed to describe these materials adequately.

Alluvial deposits are a good source of aggregates (Chapter 12). *Figure 2.78* shows a ground profile of these soils. An example can be seen in *Figure 2.79*.

# Lacustrine deposits

These materials are deposited in lakes and are generally fine-grained sediments, predominantly silts and clays. Organic matter may predominate, especially in marshy areas with peat bogs. They are frequently structured in fine layers. Salt precipitates are present in saline waters.

The main geotechnical problem is related to the high proportion of organic matter, which generally makes these soils very soft. Quick clays may also be found. *Figure 2.80* includes a typical ground profile.

# Coastal deposits

These materials are formed in the inter-tidal zone by the mixed action of land and marine environments and influenced by currents, waves and tides. Storm beaches can contain boulders and cobbles but away from these fine sands and

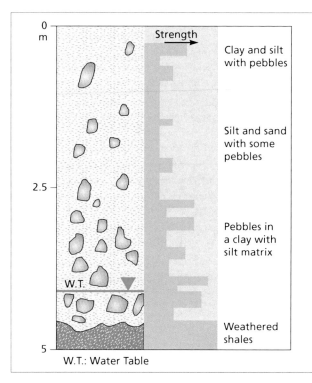

Figure 2.76   Typical ground profiles of colluvial deposits.

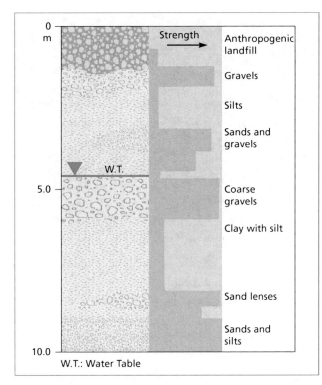

*Figure 2.78* Typical ground profile of alluvial deposits.

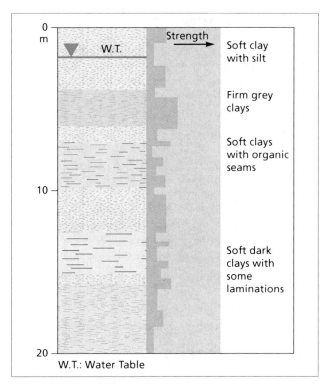

*Figure 2.80* Typical ground profile of lacustrine deposits.

*Figure 2.79* Alluvial fan deposits, Quebrada de Purmamarca, Argentina.

silts predominate and may contain abundant organic matter and carbonates. Very fine sediments, mud and organic matter are characteristic of deltaic and estuarine areas. The consistency of these deposits is highly variable, although they are generally soft to very soft and although they may contain crusts of dried material the main characteristic of coastal soils is that they are highly compressible. Another type of deposit present in coastal areas is dunes which are mobile and very unstable. *Figure 2.81* shows a typical ground profile.

## Glacial deposits

Glacial deposits (tills) are transported and deposited by ice or meltwater. They are formed under the ice to produce extremely dense and hard deposits, from material carried on top of the ice and pushed in front of the ice of an advancing glacier, and by meltwater flowing from a glacier. Their composition can be very heterogeneous and highly erratic in its distribution. Fluvio-glacial deposits contain fractions ranging from coarse gravels to clays. They are slightly stratified, with grain sizes decreasing the further they are from the glacier front. Those of lacustrine-glacial origin, however, have finer fractions, with a predominance of clays and layered structures typical of **varved clays.**

Heterogeneity and anisotropy are the common characteristics of these deposits, where fine clays may be found together with coarse gravels and large boulders (*Figure 2.82*). Their geotechnical properties are therefore extremely variable. Glacial soils can vary considerably in their strength and stiffness and some are sensitive. Solifluction and palaeo-slope instability are common in areas subjected to peri-glacial conditions. Site investigations are complex and different techniques have to be used e.g. bore hole drilling and geophysics, using a considerable number of testing points. Thickness is very variable and often considerable. *Figure 2.83* shows a typical profile for these soils.

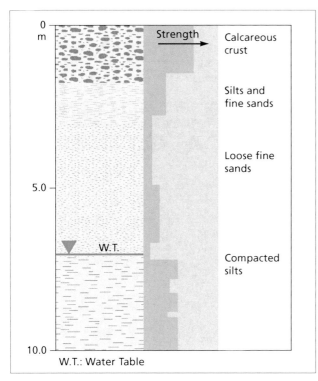

Figure 2.81   Typical ground profile of coastal deposits.

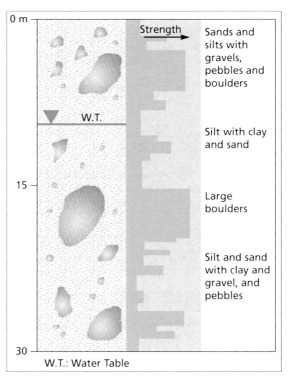

Figure 2.83   Typical ground profile of glacial deposits.

Figure 2.82   Moraine debris.

# Deserts and arid climate deposits

The properties of soils in arid areas are conditioned by a series of factors, such as depth of desiccation, accumulation of salts and transport by wind, all of which have important implications for engineering and the environment. Amongst the most significant properties are:

— A very low moisture content giving unsaturated soils with relatively high suction (i.e. pore pressures less than atmospheric pressure).
— A low organic matter content, making arid soils too poor for agricultural purposes.
— The development of salt-rich crusts; the loss of moisture through surface evaporation leads to soluble solids such as salt being precipitated as mineral cement.
— Many arid soils have been deposited by the wind and are therefore poorly graded with a very loose structure.

From a geological engineering point of view, the main problems encountered include:

— Swelling of clays.
— Collapse of loose soils.
— High sensitivity to erosion.

— Chemical attack of concrete and metal by salts, chlorides and sulphates.
— Volumetric changes in anhydrites.
— Slope failures on steep slopes.

The main features of these soils are shown in *Figure 2.84*.

## Evaporitic deposits

These deposits are formed by the chemical precipitation of salts, chlorides or sulphates. They are typical of arid or desert conditions, and also of shallow lacustrine and coastal environments (*Figure 2.85*). Common characteristics include:

— Chemical reactions with concrete which may lead to its deterioration and destruction.
— Easily soluble, especially the chlorides.
— Changes in volume as anhydrites form gypsum.
— Formation of surface crusts.
— Risk of collapse as a result of dissolution and karstification.

## Tropical soils

In regions with tropical climates, high water content and high temperatures cause intense chemical weathering that produces well-developed residual soils. Their geotechnical

*Figure 2.85* Salt deposits ("*salares*") in the desert area of La Puna, NW Argentina.

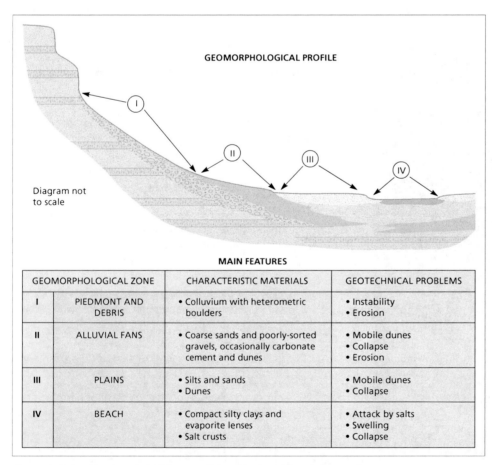

*Figure 2.84* Characteristic features of deposits in arid and desert climates.

# SOIL MECHANICS AND ENGINEERING GEOLOGY OF SEDIMENTS

Figure 2.86  Tropical soils (volcanic latosols) affected by a landslide, Hainan Island, China.

behaviour is controlled by mineralogical composition, microfabric and the geochemical environmental conditions. When high iron and aluminium contents are present, **laterites** can be formed (*Figure 2.86*). If drainage is poor, **"black soils"** may form, which are rich in smectites. If there is good drainage **red clays** will form, rich in halloysites. *Figure 2.87* shows a typical profile of these soils.

Crusts are common in tropical soils. These have better geotechnical properties on the surface than at depth. They tend to form aggregations of silt- and sand-sized clay particles which do not reflect their clayey nature when analysed for particle size and plasticity analysis and they are highly sensitive to desiccation. The most typical soils include:

— **Red soils:** These form on slopes and in mountainous areas. In good drainage conditions they are rich in halloysites. Their geotechnical properties change with desiccation and particle aggregation.
— **Black soils:** These are found in low-lying areas and on plains, and predominantly contain smectite minerals. Swelling and inadequate drainage may cause problems.
— **Crusted soils:** These behave well in geotechnical terms. Depending on the predominant type of mineral, they form laterites (Al), ferricretes (Fe), silcretes (Si) or calcretes (Ca).

## Volcanic soils

Volcanic soils may be residual, resulting from the *in situ* weathering of underlying materials to form silty sands and clays, or they may be formed by accumulations of pyroclastic material from volcanic eruptions. When transported by water, they may show alluvial or lacustrine characteristics.

Minerals from volcanic rocks are highly unstable and are quickly weathered to alteration products and clays, the most common being halloysites, allophanes (with an amor-

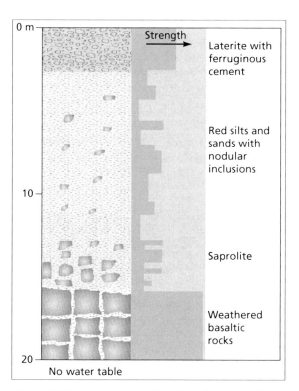

Figure 2.87  Typical ground profile of tropical climate soils.

phous structure) and smectites. The predominance of any one of these minerals depends on drainage conditions and the geochemical environment. Volcanic clays tend to form oolitic fabrics and clay aggregations (*Figure 2.74*), giving particle size distributions and plasticity values corresponding to soils of greater particle size. Smectitic soils are expansive, with high plasticity. Residual soils may be highly sensitive, with very unstable behaviour under rapid increases in pore pressure conditions or cyclic loading due to earthquakes, triggering landslides and earthflows.

Another group of volcanic soils is formed from pyroclastic deposits. These are made up of particles of various sizes, ranging from ash (<2 mm) to lapilli (2–64 mm), or even larger fragments. They accumulate in stratified layers according to the direction of the wind, or the outflow of pyroclasts during an eruption, forming very open structures with very low density and high porosity. When the ashes consolidate or compact together they form tuffs, which can be weathered and be collapsible under relatively low loads. If the pyroclasts are still molten at the moment of sedimentation, they agglomerate into compact tuff.

Layers of lapilli and ash are common. They are associated with cinder cones and have variable dips and stratification (*Figure 2.88*). Their density is very low and they are potentially collapsible. If the particles become strongly bound together through the welding and compaction of vitreous materials during deposition and cooling, their strength increases, giving high internal friction angles and high apparent cohesion. When an incandescent lava flow covers one of these pyroclastic deposits or residual soils, rubefaction of the surface takes place, giving a compact red soil layer.

*Figure 2.89* shows weathering profiles of residual volcanic soils.

The composition of lacustrine deposits that may form in volcanic regions often contains smectites, organic matter, and biogenic materials. These conditions are typical of the clays in the Valley of Mexico, where composition is allophanic, with a high content in salts, organic matter, and fossil remains, giving their high plasticity and compressibility.

## 2.9 Problematic soils

Problem soils generally belong to one of the following groups:

— Soils forming part of a natural environment that has been altered by human action, such as open cast mining and quarrying, tunnelling, or excavation of foundations. Many of the common problems in geological engineering derive from the soil behaviour in these situations including those associated with Made Ground.

— Soils posing special problems because of their own particular condition and the action of nature, without any human intervention. They include the earth-flow of clayey soils caused by intense rain on slopes, leading to mudslides, or liquefaction in sandy silt soils from earthquakes. However, they may cause serious problems if, e.g. construction work is carried out in a valley at the foot of a slope with risk of avalanches (as has happened many times in Peru, Colombia and Central America), or if an urban development is built on deposits which are susceptible to liquefaction (such

*Figure 2.88* Pyroclastic deposits, Tenerife, Spain.

| RESIDUAL SOILS ON BASALTS ||||||| RESIDUAL SOILS ON PYROCLASTS |||||
|---|---|---|---|---|---|---|---|---|---|---|---|
| LC | DESCRIPTION | T (m) | R (Ω·m) | SPT RQD | DW | LC | DESCRIPTION | T (m) | R (Ω·m) | SPT RQD | DW |
|  | Reddish or brown silty clays | 0.5-2 | 10-80 | SPT 40-50 | VI |  | Reddish or ochre silty clays and weathered lapillis | 0-0.5 | 10-150 | 20-40 | VI-V |
|  | Brown sandy silts | 0-1 |  |  |  |  | Highly weathered yellowish or reddish lapillis | 1 | 100-200 | 40-50 | IV |
|  | Basaltic gravels in a clayey matrix | 0-2 | 40-150 | SPT Re | V |  |  |  |  |  |  |
|  | Basaltic scoria | 0-1 | 60-500 | RQD < 20 | IV |  | Scoria and pyroclasts | — | 100-700 | Re | II-I |
|  | Highly scoriated basalts | — | 200-500 | RQD 20-40 | II |  |  |  |  |  |  |
|  | Massive basalt | — | 1,000 | RQD > 70 | I |  |  |  |  |  |  |

LC: lithological column; T: thickness; R: apparent resistivity; SPT: no of blows SPT; RQD: rock quality designation; DW: degree of weathering; Re: Rebound.

*Figure 2.89* Typical lithological columns of residual volcanic soils, Tenerife.

as Niigata in Japan, or Anchorage, Alaska). Such soils can be considered problem soils and are dealt with in more detail further on in this section.

In general, the most common problems posed by problematic soils in geological engineering are related to the following:

— **Bearing capacity**, where the ground has to be capable of supporting increments in stress induced by engineering work, without reaching the limit of pre-established safety levels, i.e. the curve of intrinsic resistance or load-settlement ratio for a certain factor of safety (Figure 2.90).
— **Deformability**, where the foundations have to be able to withstand displacements without any significant structural consequences. These displacements are the result of deformations induced in the ground by loads transmitted to it by the foundations. Usually a maximum settlement ($S_{max}$) and an appropriate maximum angular distortion ($\Delta/L$) that must not be exceeded are established (depending on the type of structure) (Figure 2.91).
— **Long term stability**, where strength and deformability conditions must remain unchanged over time, or at least not fall below the appropriate values established. For example, an excavation close to a pre-existing structure may produce new movements that affect nearby foundations (Figure 2.92).

Special problems may arise from natural causes or be brought about by human activity interfering with and changing the balance of nature, either once or several times. Climatic action, which may be periodic but is often intense, can affect engineering projects in a very different way from the surrounding area. In this respect, it is important to highlight problems typically found in the following soils:

— Swelling and shrinking clays.
— Dispersive soils.
— Saline and aggressive soils.
— Collapsible soils.
— Permafrost.
— Soft sensitive soils.
— Soils susceptible to liquefaction

The behaviour of some of these soils, such as swelling and shrinking clays and permafrost, is influenced by climatic variations, as well as by their properties. Others, such as those susceptible to liquefaction, are influenced by geological processes like earthquakes.

## Swelling and shrinking clays

This group includes clayey soils with a mineralogical structure and microfabric which allows water absorption, producing significant changes in volume. Water molecules penetrate the crystal network between the silica sheets which are joined with weak bonds that are reduced or eliminated in the process, so that the crystal network ends up occupying a greater apparent volume, without any chemical reactions having taken place. If conditions then change (for example, through prolonged desiccation or drainage), the water molecules may leave the crystal network, causing reduction or shrinkage.

The capacity for volume change in these materials is therefore conditioned by the clay content and its mineralogy, structure and microfabric.

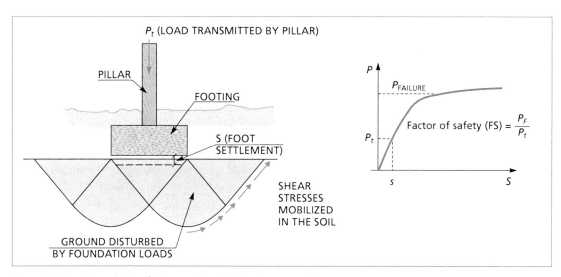

Figure 2.90  Foundation failures related to bearing capacity.

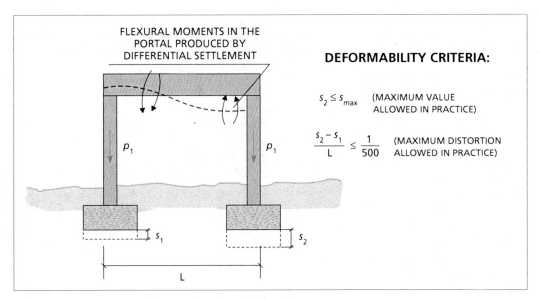

*Figure 2.91*  Foundation failures related to deformability.

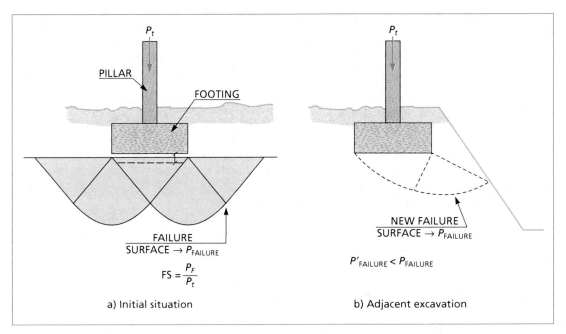

*Figure 2.92*  Stability problems due to change in stress conditions over time.

**Swelling** is the increase in volume through absorption of water, and **shrinkage** the reduction of volume through elimination of water.

In addition to geological factors, volume change (swelling or shrinkage) is conditioned by the following factors:

— **Climatic variations**, which determine either the presence of water necessary for expansion or conditions for evaporation that induce shrinkage. Changes in volume are reflected in buildings constructed on swelling clays. If the foundations are subjected to movements, cracking may occur. This may reoccur regularly if the movements are induced by seasonal or cyclical changes in climatic conditions (*Figure 2.93*).

— **Vegetation**, which may have a local influence by changing the moisture content of the foundation ground, with resulting changes in volume. Both vegetation and the action of roots may trigger this phenomenon.

*Figure 2.93* Building affected by severe cracks due to expansive clays, Jaen, Spain.

— **Hydrological changes** in general, produced both by climatological action and variations in the water table, e.g. due to the exploitation of aquifers, dam construction or deep excavations.

Clay soil can therefore be potentially expansive in the following conditions:

— A clay must be present that has the appropriate mineralogy and microfabric. Carbonates may cement the soil structure and prevent or reduce swelling, but the destruction of diagenetic bonds (e.g. when soil is remoulded or compacted) causes the cementing action to disappear, making the minerals susceptible to the action of water.
— The soil moisture content must vary for whatever reason, leading to swelling in rainy periods followed by shrinkage in periods of drought.

Because of this, potentially expansive soils are found in regions with arid and semi-arid climates, such as those around the Mediterranean (Spain, Italy, Turkey, Israel, Morocco, Tunisia, etc.), South Africa, the southern part of North America (Arizona, Texas, Northern Mexico), and the north of South America (Colombia, Ecuador, Peru).

Given that swelling is related to clay content, it is common practice to use clay characterization parameters to evaluate and grade the possible expansivity of a soil. Four grades of expansivity (I to IV) are generally considered, as shown in *Table 2.13*. Other laboratory tests of a higher quality are also carried out, specifically to determine swelling potential:

a) **Lambe's test:** A calibrated piston measures the pressure exerted by the soil (remoulded) after it has been moistened inside a mould (*Figure 2.94*).
b) **Swelling pressure test:** The maximum pressure attained by an undisturbed sample in an oedomer when swelling is prevented during and after wetting.

*Figure 2.94* Lambe's apparatus for testing soil expansivity.

c) **Free swell test:** This is the maximum variation in the thickness of an undisturbed sample in an oedomer when it is moistened and free to expand vertically.

The index values of these three tests, used to establish the level of potential swelling are included in *Table 2.13* and *Figure 2.95*.

Swelling clays are usually present in volcanic areas. They may form alluvial or lacustrine deposits, and sometimes appear as layers between basaltic lava flows.

Other swelling phenomena also occur when anhydrite (dehydrated calcium sulphate) is hydrated, passing to the dihydrate form (gypsum) from water absorption. This is particularly important when encountered in tunnels.

There are other soils which also cause problems of swelling, derived from freezing ground water.

## Dispersive soils

The composition and micro-fabric of **dispersive soils** are determined by the repelling forces between the fine particles (clays) which are greater than the attraction forces. In these

| Table 2.13 | SWELLING POTENTIAL AND AVERAGE VALUES OF GEOTECHNICAL PROPERTIES | | | | | |
|---|---|---|---|---|---|---|
| Grade | Swelling potential | Fine fraction (%) | Liquid limit (%) | Lambe index (kPa) | Swelling pressure (kPa) | Free swelling (%) |
| I | Low | <30 | <35 | <80 | <25 | <1 |
| II | Low to medium | 30–60 | 35–60 | 80–150 | 25–125 | 1–4 |
| III | Medium to high | 60–95 | 50–65 | 150–230 | 125–300 | 4–10 |
| IV | Very high | >95 | >65 | >230 | >300 | >10 |

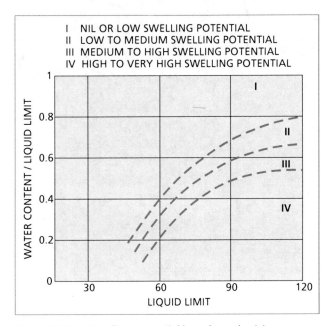

Figure 2.95   Swelling potential based on plasticity.

conditions and where water is present the soils flocculate, i.e. the particle aggregates break up into smaller particles that are more easily flushed away by water flowing at a certain velocity, producing soil erosion.

The particle aggregates, or floccules, are made up of clays, usually with a high proportion of dissolved salts (above 12% in the water occluded in the soil pores).

Dispersion may cause soil erosion in roads and dam embankments. Two methods can be used to recognize dispersion phenomena: one is a physical method consisting of a double particle size analysis by sedimentation test, with or without a particle dispersant. The other method is chemical, and consists of calculating the Na, Ca, Mg and K ions content and making a relative comparison of them (Figure 2.96). For the double particle size analysis test the **dispersion index** $I_{dis}$ is defined as the ratio between the percentage of particles of less than 0.005 mm in the test with demineralized water, and the same parameter obtained in a normal test with dispersant. If this index is greater than 50% the soil is considered to be very stable with regard to dispersion; between 50 and 30–35% there is intermediate or marginal stability, and below 30–35% the material is dispersible.

## Saline and aggressive soils

These soil types usually contain more than 15% of their ionic exchange capacity saturated with sodium ions, and also appreciable quantities of other soluble salts. Their pH in saturated solution is around 8.5 or less.

Saline soils associated with high levels of evaporation, and therefore concentrations of salts, such as those found in the Middle East, may also show swelling characteristics, but slight changes in the saline composition may change the swelling potential to collapse potential, depending on the original density of the clays.

An extreme case of saline soils is found in the Dead Sea depression, where the Karameh Dam (Jordan) was built on calcareous silts with fine layers of aragonite. These soils are well cemented and the aragonite is stable due to the composition of the interstitial water, which is strongly saline. However, the dam contains fresh water, seepage of which may bring about a change in the salinity of the interstitial water of the foundations in the long term, with consequences which are not yet apparent.

Many saline soils can be very aggressive to concrete, particularly if the water table is near the foundation. Generally, below 0.02% of sulphates (measured by the $SO_3$ content) there is no aggressive potential. Table 2.14 gives reference values for soils and water for levels of attack on concrete.

## Collapsible soils

These soils are characterized by their very open and loose soil structure. However, they remain stable in dry or very dry

# SOIL MECHANICS AND ENGINEERING GEOLOGY OF SEDIMENTS

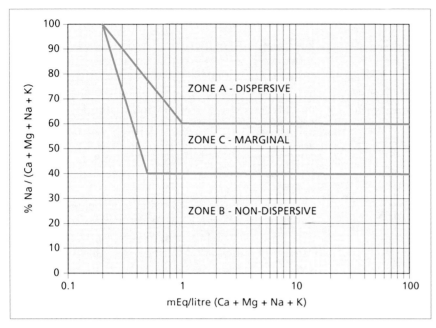

Figure 2.96  Dispersion potential as a function of chemical components (Sherard et al., 1976).

| Table 2.14 | SULPHATE ATTACK ON CONCRETE | |
|---|---|---|
| Attack potential | Water mg $SO_4^=$/l | Soil mg $SO_4^=$/kg dry soil |
| Low | 200–600 | 2,000–3,000 |
| Moderate | 600–3,000 | 3,000–12,000 |
| High | >3,000 | >12,000 |

| Table 2.15 | COLLAPSE POTENTIAL | |
|---|---|---|
| Grade of collapse | Dry unit weight (kN/m³) | Collapse potential (%)* |
| Low | >14.0 | <0.25 |
| Low to medium | 12.0–14.0 | 0.25–1.0 |
| Medium to high | 10.0–12.0 | 1.0–5.0 |
| High to very high | <10.0 | >5.0 |
| * Settlement induced by collapse in saturated conditions referred to the initial height of the specimen. | | |

environments. When they are first deposited, by water or wind, they have no cohesion, however they can be slightly cemented by sulphate crystals, or by the filling of their voids with finer particles, which in a dry state can give them considerable strength.

The behaviour of these soils varies according to their water content. If this increases, the original structure may be destroyed, leading to a significant decrease in apparent volume (collapse) and consequent settlement, as well as the possible transport of particles by water flowing at a certain velocity. In the Central Valley of California subsidence of more than 4 m has been recorded due to the gradual infiltration of water following the introduction of irrigation on soils of this type. At times, if the surface areas have formed crusts (e.g. from carbonate deposits) the dissolution of sulphate ions will take place below the surface, forming cavities or sinkholes which end up caving in when surface crusts are broken.

The first indication of the potential a soil has for collapse comes from its dry unit weight. This potential is then confirmed by **collapse tests** (Table 2.15). These tests are carried out in an oedometer, subjecting the samples to a certain load and measuring settlement after flooding the sample.

Other possible types of collapsible soils include:

a) Volcanic materials like **soft tuffs and pyroclastic low density agglomerates.** These have very open structures and weak contacts between the particles. Due to their very low density, they may collapse under moderate loads and from flooding.

b) Non-compacted **man-made fills** produce a weak structure with water remaining in the contacts between particles, forming menisci due to the suction created by the difference in pressure between the atmosphere and pore water. These menisci introduce intergranular forces which compress the particles and cause considerable strength at normal moisture levels. Saturation in water eliminates the menisci, reduces intergranular forces and brings about collapse or a decrease in volume, a phenomenon that has led to great problems with different fills. In this type of soil, the probing or penetration test (see Chapter 5) can clearly distinguish between natural soil (more than 25–30 blows for 20 cm penetration) and insufficiently compacted fill material (5–15 blows for 20 cm penetration).

One of the best known collapsible soils is **loess**, a material which has been deposited by wind. Loess is found between parallels 30° and 55° in both hemispheres (Siberia, Ukraine, Romania, Australia, Argentina, Uruguay, the North American Midwest, etc.). There are cases of whole valleys which have been flooded, with the aim of provoking collapse by flooding and making the ground more stable and dense so that in terms of foundations, its behaviour becomes acceptable. In central Spain gypsum bearing silts have been mechanically worked and compacted to reach optimum dry densities in the order of 17.5 kN/m$^3$, which means they are transformed into a much denser soil than the original.

## The action of ice and permafrost

Frost penetration underneath the soil surface is accompanied by diverse physical phenomena. One the most important is the increase in water volume of the soil as it freezes, which may destroy the structure of the soil or the rock. The most significant effect is usually the accumulation of ice lenses which produce expansion of the ground in winter and softening in summer.

As the proportion of soil with particle sizes below 0.02 mm increases, so does its susceptibility to ice action. If this fraction is greater than 3% and the coefficient of uniformity ($C_u = D_{60}/D_{10}$) is in the order of 15, the soil is sensitive to the effects of frost. When this fraction exceeds 10%, the coefficient of uniformity must fluctuate at around 5 for the soil to be sensitive to this phenomenon.

Permafrost is found in wide areas of Canada, Alaska and Siberia, with permanently frozen soils reaching depths that depend on the thermal conductivity of the ground and on the climatic conditions. Below the surface, which is generally very hard, the soil may have a weak structure because frozen water increases in volume, destroying the bonding and cementation between particles. While the ice exists, the soil is hard, but if for some reason the temperature in the ground rises (i.e., when a centrally-heated building is constructed on it), the pore ice liquefies and converts the ground, which is weak, to mud with very low strength. Due to the low bearing capacity of soils in these conditions, building foundations in these areas are supported on piles.

## Soft sensitive soils

River mouths, flood plains and some coastal areas with softer rocks are covered by fine silt and clay deposits, which are very soft, saturated, and usually contain organic matter (4–5%). In these materials the water content is very high (60–140%) and the structure very weak (dry unit weight from 7.0–14.0 kN/m$^3$), depending on the type of deposit, the amount of organic matter, and the particle size distribution, amongst other factors. As a result, they are very deformable and very soft, with a compression index from 0.4 to more than 1.0.

This high degree of deformability also implies that resistance to undrained shear stresses is very low, from about 15 to 50 kPa, although on the ground surface (because of salt deposits or cyclical effects of the water table) there may be some crust affecting the top meters where the shear strength may be much higher.

When water content is above the liquid limit (fluid state) the structure may be governed by the chemistry of the interstitial water. The identification of soft soils with piezocones is a straightforward process and an evaluation of their deformability can be adequately studied using experimental embankments. *Figure 2.97* shows the relative settlements measured under embankments on different soft soils none of which have been subjected to ground treatment.

In addition, soft soils and muds may be liable to thixotropy, causing them to lose their initial strength through remoulding, for example, due to landslides or the sinking of piles nearby. Large movements involving sensitive marine clays (quick clays) have occurred in Norway, sometimes triggered by small excavations no deeper than 2–3 m.

## Soils susceptible to liquefaction

Loss of soil strength due to cyclical loading or earthquake shaking is usually referred to as **liquefaction**. Loose saturated, predominantly silty sandy soils can undergo rapid increases in pore pressures (because of the absence of drainage and collapse of the fabric) and the pore pressures can reach total normal stress. When this happens, effective stress is practically zero, and the grains lose frictional contact with each other, shear strength disappears, and the material behaves like a liquid. This can permit vertical and horizontal movements of the ground to occur, which can result in landslides and extensive settlements rapidly developing.

The presence of loose, silty sandy soils with low permeability has been associated with some major earthquake disasters, because the rapid, cyclical repetition of tangential forces annuls effective stresses during earthquake shaking. In the 1964 earthquake at Niigata, Japan, dozens of buildings collapsed, although they had been designed to be earthquake

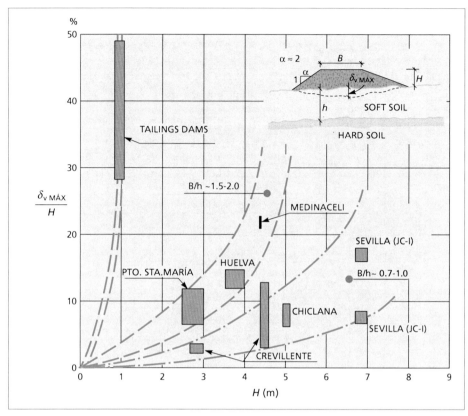

*Figure 2.97* Settlement measured in embankments at different sites in Spain.

resistant, because they were founded on deposits liable to liquefaction. This led to settlements of many metres and to the overturning and rotation of buildings. The same year, and for the same reasons, there were huge landslides in the vicinity of Anchorage, Alaska. Affected buildings were carried some 200 m from their original position.

Liquefaction is studied in more detail in Section 14.6 of Chapter 14.

# Recommended reading

Atkinson, J. (2007). The mechanics of soils and foundations. 2nd ed. Taylor and Francis. The Netherlands.
Barnes, G. (2000). Soil mechanics: principles and practice. 2nd ed. Palgrave MacMillan. New York.
Craig, R.F. (1987). Soil mechanics. 4th ed. Chapman & Hall.
Lambe, T.W. and Whitman, R.V. (1979). Soil mechanics. John Wiley & Sons. New York.
Mitchel, J.K. (1993). Fundamentals of soil behaviour. 2nd ed. John Wiley & Sons. New York.
Terzaghi, K. and Peck, B. (1948). Soil mechanics in engineering practice. John Wiley & Sons, New York.

# References

Bennet, R.H. and Hulbert, M.H. (1986). Clay microstructure. Int. Human Resources Dep. Co. Boston, Houston, London.
Burland, J.B. (1988). Behavior and design of foundations. MSc Course on Soil Mechanics. Imperial College, London.
Capper, P.L., Cassie, W.F. and Geddes, J.D. (1974). Problems in engineering soils. E.F.N. Spon. London.
Day, R.W. (1999). Geotechnical and foundation engineering. McGraw-Hill.
González de Vallejo, L.I., Jiménez Salas, J, A. y Leguey Jiménez, S. (1981). Engineering geology of the tropical volcanic soils of La Laguna, Tenerife. Engineering Geology, 17, pp. 1–17.
Hansbo, S. (1957). A new approach to the determination of the shear strength of clay by the fall-cone test. Swedish Geot. Inst. Pub. No. 14, Stockholm.
Lambe, T.W. and Whitman, R.V. (1979). Soil Mechanics. John Wiley & Sons. New York.
Lancellotta, R. (1995). Geotechnical Engineering. Taylor & Francis. The Netherlands.

Lupini, J.F., Skinner, A.E. and Vaughan, P.R. (1981). The drained residual strength of cohesive soils. Geotechnique 31, no 2, pp. 181–213.

Mayne, P.W. and Kulhawy, F.H. (1982). $K_o$-OCR relationships in soil. Journal of the Geotechnical Engineering Division. ASCE, vol. 108, GT6, pp. 851–872.

Powers, J.P. (1992). Construction dewatering. New methods and applications. 2nd ed. John Wiley & Sons, N.Y.

Schmertmann, J.M. (1955). The undisturbed consolidation of clay. Trans. ASCE, vol. 120, pp. 1201–1227.

Sherard, J.L., Dunnigan, L.P., Decker. R.S. and Steele, E.F. (1976). Pinhole test for identifying dispersive soils. Journal of the Geotechnical Eng. Division, ASCE 102(GT1), pp. 69–85.

Skempton, A.W. (1954). The pore pressure coefficients A and B. Geotechnique 4(4), 143–147.

Skempton A.W. (1970). The consolidation of clays by gravitational compaction. Quart. J. Geol. Soc. 125, 373–412. London.

Tsige, M., González de Vallejo, L.I., Doval, M. and Oteo, C. (1995). Microfabric of Guadalquivir blue marls and its engineering significance. Procc. 7th Int. Congress of Eng. Geol. IAEG. Lisbon. Balkema. Vol. II., pp. 655–704.

# 3

# ROCK MECHANICS

1. Introduction
2. Physical and mechanical properties of rocks
3. Stress and strain in rocks
4. Strength and deformability of intact rock
5. Discontinuities
6. Strength and deformability of rock masses
7. *In situ* stress
8. Rock mass classifications

## 3.1 Introduction

### Definition, objectives and scope

**Rock mechanics** includes the study, in theory and practice, of the properties and mechanical behaviour of rock materials in response to the forces acting on them within their physical environment.

Rock mechanics developed as a consequence of the use of the geological medium for surface and underground engineering projects and mining activities. Fields of applications of rock mechanics include those where rock is the structure (e.g. excavation of tunnels, adits and slopes), those where rock is the support for man-made structures (e.g. building or dam foundations) and those where rock itself is used as building material (e.g. rockfills, rock embankments).

Rock mechanics is closely related to other disciplines such as structural geology, for the study of tectonic processes and structures affecting rocks, and soil mechanics, for the study of weak and weathered rocks.

Rocks usually have **discontinuities** or weakness surfaces which divide a **rock mass** into blocks of **intact rock** (*Figure 3.1*); both components are studied in rock mechanics, the discontinuity planes being the main difference with soil mechanics and giving a rock mass its discontinuous and anisotropic character.

Characterizing and describing rocks and rock masses and their mechanical and deformational behaviour is a complex process, because their characteristics and properties vary considerably and many different factors govern them.

The main aim of rock mechanics is to understand the behaviour of rocks and to be able to predict how they will behave in response to internal and external forces acting on them. Any excavation or construction work carried out on rock will modify its original conditions, and as a result the rock may be deformed and/or fail. At a microscopic level, mineral particles are displaced and this may cause failure planes in response to new state of stress. At rock mass scales, deformations and failure usually occur along discontinuity planes.

Understanding the **stresses** and **strains** a rock material may be subjected to under certain conditions allows its mechanical behaviour to be assessed for the design and planning of engineering or structural work. The relationship between these two parameters describes the behaviour of the different types of rock and rock masses, which in turn

*Figure 3.1* Rock mass. Blocks of Bunter sandstone separated by discontinuities.

depends on their material properties and the prevailing natural conditions affecting them.

The strength and deformability characteristics of **intact rock** are controlled by its **physical properties**: mineral composition, density, structure and fabric, porosity, permeability, durability and hardness; these are determined by the genesis of the rock and by the geological and tectonic conditions and processes that have affected it over time (*Figure 3.2*). In the case of **rock masses**, the mechanical behaviour is also influenced by **geological characteristics**: lithology and stratigraphy, geological structures, tectonic joints and diagenetic discontinuities and *in situ* state of stress. At both scales the mechanical response also depends on other factors, such as hydrogeological and environmental conditions, climatic and meteorological phenomena acting on the geological medium and causing weathering processes which modify the original properties of rocks and rock masses.

The mechanical state and behaviour of rock masses are the result of a combination of all these factors to a different extent depending on each situation. For example, at and near ground level, discontinuities and weathering processes play a significant role in the mechanical characteristics and behaviour of the rock mass, while at greater depths the existing state of stress, and the corresponding *in situ* stress

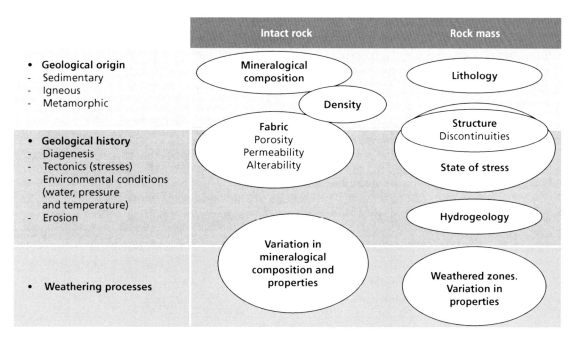

Figure 3.2    Geological control of the properties of intact rock and of the rock mass.

magnitudes, are the most important factors conditioning the mechanical response and failure.

The study of **geological structure** and **discontinuities** is a fundamental aspect of rock mechanics, provided that pre-existing planes of weakness control the deformation and failure processes in the rock mass at depths where most engineering work takes place.

How far the blocks of **intact rock** affect the overall behaviour of the rock mass will depend on the relative properties of the rock material and the discontinuities, on the number, type and characteristics of the discontinuities, and on the scale of the project in question. For example, failure processes in rock masses consisting of blocks of hard intact rock with high strength properties are clearly determined by discontinuities; however, in jointed rock masses with weak intact rock, differences between the strength and stiffness of discontinuities and intact rock may not be so relevant. To evaluate these aspects in the design of an engineering project, the dimensions of the project must be considered in relation to both the rock mass structure and the separation between the discontinuities (Figure 3.3).

Compared with geological processes, engineering works modify the **state of stress** the rock mass is subjected to in a very short space of time. This may lead to mutual interactions between the structures and the relief or redistribution of natural stresses. It is therefore essential to know the prior state of stress to be able to assess how it will be influenced by the proposed work.

Rock mass strength is reduced by the presence of **water,** which generates pressure inside it and alters its

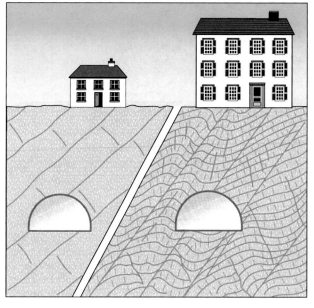

Figure 3.3    Dimensions of engineering works in relation to the rock mass structure and spacing of discontinuities.

properties, hindering both surface and underground excavations. A study of the permeability and flow characteristics of the rock mass is essential to evaluate the influence of water; any assessment of rock mass properties should take possible ground water into account.

As pointed out in Chapter 1, in terms of the effect of engineering works on ground behaviour, it is important to consider the differences between **geological time** and time on a human scale. Certain natural processes, which would take hundreds or thousands of years to occur in a natural state, are accelerated by engineering works: the weathering of excavated rock surfaces, the relief of natural stress and opening up of discontinuities, or changes in water flow. All these lead to reduced rock mass strength within very short periods of time (a few years or even months). To evaluate these effects the evolution of rock material properties over time has to be investigated, as well as the geological, environmental and mechanical conditions to which it is subjected.

The most significant time-dependent process is **weathering**, which causes the disintegration and decomposition of rock material, particularly clay rocks. Other processes, such as swelling, may take place in particular types of rock as a result of stress relief or chemical reactions, e.g. the conversion of anhydrite to gypsum after hydration.

Some weak or heavily jointed rock materials may show **rheological behaviour** and creep may occur; once a certain deformation level is reached as a result of an applied load, loss of strength is only a question of time.

The factors outlined above are the basis of the study of rock mechanics applied to geological engineering and are dealt with in more detail in this chapter.

To predict the response of a rock mass to an action involving a change in its original state and conditions, its overall properties and behaviour have to be investigated using appropriate geological engineering and geotechnical **methods of study and research**. Expertise in geology and field surveys is essential for evaluating the mechanical conditions of rocks.

The tools used in rock mechanics to determine the geomechanical properties required to predict rock and rock mass behaviour include the data from *in situ* and laboratory tests, analyses, the application of empirical strength criteria and physical and mathematical models. Given the complexity of the geological medium, **experience** is always crucial for interpreting and evaluating correctly the results from such studies.

**Laboratory tests** are used to quantify the physical and mechanical properties that define the intact rock behaviour:

— The nature of the rock.
— Strength and resistance to failure.
— Short and long-term deformation.
— The effects of water on behaviour.
— The effects of weathering on behaviour.
— Time-dependent behaviour.

*In situ* **tests** measure the properties and condition of rock masses in their natural state at a representative scale, and allow *in situ* simulation of the possible effects on the rock mass of construction and engineering work at that scale.

# Rock and soil

**Rock** is a hard natural aggregate of mineral particles connected by strong cohesive forces, and is usually considered to form a continuous system. Geological classification is based on the proportion of different minerals forming the rock, its granular structure, its texture and its origin.

According to the accepted useage for geological engineering purposes, **soils** are natural aggregates of mineral grains joined to the surfaces of adjacent particles by contact forces that can be broken by gentle mechanical means or stirring and agitation in water.

Unlike soils, the composition, characteristics and properties of intact rocks are often highly variable. As a result, these natural materials show heterogeneous and anisotropic

*Figure 3.4*    Alternating rock materials with varying composition and structure on the slope of a volcanic rock mass, showing different degrees of weathering, strength and mechanical behaviour.

characteristics, making it very difficult to study their behaviour in the laboratory, due largely to the problem of obtaining representative samples at a working scale. In addition, rocks are affected by geological and environmental processes that lead to fracturing, alteration and weathering.

Some of the main distinguishing characteristics of the physical and mechanical properties of intact rocks are:

— Generation of failure mechanisms and planes during deformation.
— Higher deformation modulus than soils.
— Lower permeability than soils.

In terms of *in situ* conditions or characteristics, other difference with soils is that rock masses are affected by tectonic joints and other planes of weakness, and are subjected to natural tectonic related stress, whereas soils, being formed at shallower depths, have relatively low *in situ* state of stress due to lithostatic forces.

Uniaxial compressive strength is a widely-used criterion in geological engineering for setting the **limits between soil and intact rock**, i.e. the maximum stress an axially-loaded cylindrical sample in the laboratory can bear before it fails. So-called hard soils and soft rocks are found in the transition zone. Limits suggested by different classifications and authors have gradually fallen to 1 or 1.25 MPa, because the strengths of certain very soft rocks have values of this order, and these values are currently considered to be appropriate (*Tables 3.7* and *3.10*).

Rocks can be grouped, in a simple classification, into three groups based on composition, the geometrical relationship of particles (texture) and genetic characteristics:

— Sedimentary rocks: detrital and non-detrital.
— Igneous rocks: plutonic and volcanic.
— Metamorphic rocks.

*Figure 3.5* Highly weathered clay rich rock with characteristics common to both rocks and soils.

# Rock masses

As described at the beginning of this chapter, rock masses contain discontinuities, or planes of weakness, which separate blocks of intact rock. The study of rock mass behaviour should include the study of the properties of both the intact rock and the discontinuities.

A "blocky" structure means that the properties and behaviour of rock masses have a **discontinuous character**. The presence of systematic discontinuities with a particular orientation, such as bedding planes or lamination, also implies **anisotropic behaviour**, where the mechanical properties change depending on direction: e.g. there may be drastic variations in the strength of a stratified rock mass between directions parallel to the bedding planes and those perpendicular to it. Another characteristic of rock masses is their **heterogeneity** or the variability of their physical and mechanical properties in different zones of the rock mass (*Box 3.2*).

It should be noted that on a microscopic scale, and even as a laboratory sample, intact rock is also discontinuous, anisotropic and heterogeneous due to the presence of lamination, micro-cracking, the preferred orientation of minerals, and such like. However, in geotechnical terms intact rock is considered, in many rock mechanics applications, to be continuous and isotropic in relation to the rock mass as a whole.

In most cases, discontinuity surfaces form planes of weakness that govern the geomechanical behaviour of rock masses by conditioning the strength of the formation as a whole and the zones and mechanisms of deformation and failure. The control exerted by discontinuities is definitive in hard rock masses, such as granites or quartzites, where the strength of intact rock blocks is much stronger than the strength of planes separating them. In soft rock masses such as mudstones, shales or marls, the difference between the strength of both components may not be very significant; in such cases, the behaviour of the overall rock mass may even be determined by the intact rock properties.

Mechanical behaviour may also be controlled by the presence of particular discontinuities affecting the rock mass, such as faults, dykes or lithological separation surfaces, rather than by systematic sets (*Figure 3.6*).

When excavations or foundations are made in the ground, the initial conditions and the forces acting on the rock masses are modified, both internally, due to the weight or intrinsic properties of the materials themselves, and externally (body and surface forces, respectively); in addition, pore pressures change because the flow and water tables are modified. These produce modifications to the state of stress that, together with the strength and deformational characteristics and properties of the rock materials,

## Box 3.1

## Rock to soil transition

Soils originate from the weathering and disintegration of sedimentary, igneous or metamorphic rocks, as a result of the action of external geological processes and climatic phenomena. When the soil formed by rock decomposition remains in its original place it is called a **residual soil**, and when it is moved it is called a **transported soil**. These processes start as soon as rocks on the earth's surface undergo alteration and mechanical fragmentation caused by physical or chemical weathering. Transported soils are the result of several processes or stages:

— Disintegration and remobilization of the original constituents through alteration and weathering of the parent rock.
— Displacement of the material by different transport agents, like wind and water.
— Accumulation of the material in low energy areas, by sedimentation, controlled by the mechanical and physical-chemical forces operation, and by the biological characteristics of the environment.
— Transformation through diagenesis into a new compacted material, with reduced porosity, the addition of new substances and mineralogical changes.

The sedimentary cycle is completed when the sediments are transformed into sedimentary rocks (lithification).

The boundary between soil and rock is sometimes difficult to define and to determine. Photo a) shows a clear division between transported soil and rock while photo b), in contrast, shows a gradation between residual soil formed by weathering and alteration *in situ* and the parent rock, so there is no clear dividing line between the two materials.

a)

b)

a) Clear division between soil and transported rock (photo courtesy of R. Mateos). b) Continuous transition between weathered rock and residual soil. (Field of view 8–10 m wide)

control the mechanical response and deformation and failure models.

The geological factors determining the behaviour and mechanical properties of rock masses are:

— Type and properties of the intact rock.
— Geological structure and discontinuities.
— State of stress the material is subjected to.
— Degree of weathering.
— Hydrogeological conditions.

The type of rock and its degree of weathering determine the strength properties of the intact rock. The geological structure of a rock mass defines its zones and planes of weakness, its zones of stress concentration and its zones prone to weathering and water flow. Stresses acting on the rocks determine the deformation models and the mechanical behaviour of the rock mass as a whole; the state of stress is a result of geological history, although knowledge of this is not enough for a quantitative evaluation of stress to be made.

## Box 3.2

### Intact rock, discontinuities and rock mass

**Intact rock** is rock material with no joints or discontinuities. Although intact rock is considered to be a continuum, its behaviour is often heterogeneous and anisotropic related to its fabric and mineral micro-structure. In mechanical terms, it is characterized by its specific weight, strength and deformability.

A **discontinuity** is any plane of mechanical or sedimentary origin that separates or isolates blocks of intact rock within the rock mass. The tensile strength of discontinuity planes is generally very low or even zero and their mechanical behaviour is characterized by their shear strength or the strength of any existing fill material.

The **rock mass**, as a whole, comprises the blocks of intact rock and the different types of discontinuity that bound them, taken together as a whole. Mechanically, rock masses are discontinuous, anisotropic and heterogeneous. For practical purposes, their tensile strength can be considered to be zero.

- **Anisotropy:** the presence of planes of weakness with preferred orientations (e.g. bedding and joint sets) implies different properties and mechanical behaviour, depending on the direction in question. The orientation of the forces exerted on an anisotropic rock mass will result in an anisotropic state of stress.
- **Discontinuity:** the presence of discontinuities (e.g. bedding planes, joints, faults, dykes) breaks the continuity of the mechanical properties of the rock blocks, so that the geomechanical and hydraulic behaviour of the rock mass as a whole is discontinuous and conditioned by the nature, frequency and orientation of the discontinuity sets.
- **Heterogeneity:** the areas within a rock mass with different lithologies, degrees of alteration and weathering or water content may have very different properties and mechanical behaviour.

Discontinuities and blocks of intact rock together make up the rock mass structure and control its overall behaviour, with one or other of these dominating, depending on their relative properties and scale considered.

Apart from the intrinsic properties of the rock mass related to the characteristics of the intact rock and the discontinuities, which largely define its strength, other factors can also affect the mechanical behaviour of the rock mass. These include:

— Non-discontinuous tectonic and sedimentary structures in the rock mass (e.g. folds).
— The natural stress the rock mass is subjected to (*in situ* state of stress).
— Hydrogeological and geo-environmental factors.

Isotropic and homogeneous intact fresh rock at macroscopic scale. Volcanic tuff.

Jointed rock mass showing various sets of discontinuities and zones with different degrees of weathering.

*Figure 3.6* Fault affecting a limestone rock mass.

An important aspect of the study of rock masses is the influence of weathering processes on certain types of low-strength rocks, such as marls, mudstone and shale. Their properties vary considerably over time if they are exposed to the atmosphere or the action of water, or to changes in state of stress, all factors which are normally related. Account should be taken that with engineering construction work on or in this type of material, strength may be reduced with time, until the limit of stability is reached.

## 3.2 Physical and mechanical properties of rocks

### Rock characteristics

In solid mechanics, materials are normally assumed to be continuous, isotropic, elastic and linear, but this is not true of rock. Both intact rock and jointed rock masses have a wide variety of physical and mechanical properties and characteristics that imply a differential behaviour.

On the smallest scale, i.e. that of intact rock, the study of rock properties and how they vary deals with the chemical composition of the heterogeneous crystal aggregates and amorphous particles forming the rocks; e.g. sandstone may be cemented by quartz or calcite, or granite may contain variable quantities of quartz. The rock fabric or petrofabric, the result of its genesis and geological history, shows its preferred anisotropy in the orientation of grains and crystals, foliation or schistosity planes; pores, microcracks and recrystallizations give the rock its discontinuous and non-linear character, and an irregular distribution of minerals and rock components makes

*Figure 3.7* Lithological and structural characteristics and environmental conditions determine the great variability in physical and mechanical properties of rock masses. The photograph on the left shows a weak rock mass susceptible to weathering, with lithologies of varying competence and structure in horizontal layers with few tectonic joints. The rock mass on the right is formed by competent hard rock, with thin folded strata affected by intense fracturing. (Field of view of 40 m wide).

# Box 3.3

## Physical and mechanical properties of rocks

The physical properties of rocks are the result of their mineralogical composition and fabric and of their geological, deformational and environmental history, including alteration and weathering processes. The great variability of these properties is reflected in their different mechanical response to forces as defined by the strength of the material and its deformation modulus: fresh granite subjected to large loads behaves in a brittle, elastic way, whilst marl or mudstone may show ductile behaviour under medium or low loads.

It is therefore the physical properties of rocks that determine their mechanical behaviour, as shown in the figures in this box. These properties can be quantified using specific techniques and laboratory tests (*Table 3.1*).

Photo A (view through petrological microscope)

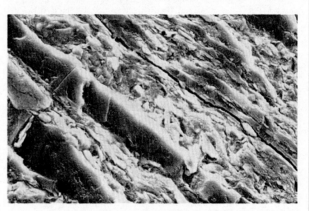

Photo B (electronic microscope image)

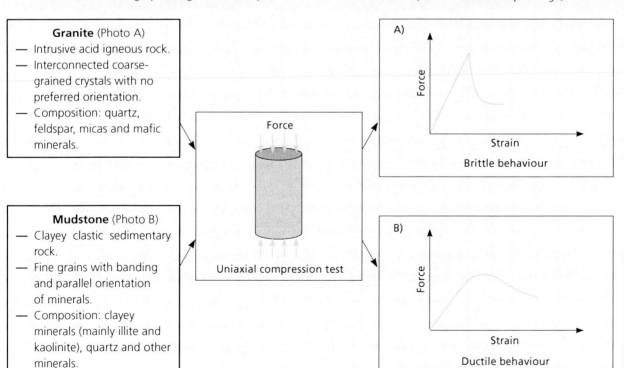

**Granite** (Photo A)
— Intrusive acid igneous rock.
— Interconnected coarse-grained crystals with no preferred orientation.
— Composition: quartz, feldspar, micas and mafic minerals.

**Mudstone** (Photo B)
— Clayey clastic sedimentary rock.
— Fine grains with banding and parallel orientation of minerals.
— Composition: clayey minerals (mainly illite and kaolinite), quartz and other minerals.

Uniaxial compression test

A) Brittle behaviour

B) Ductile behaviour

it a heterogeneous medium. Physical and chemical alteration and weathering processes also modify the rock composition, with the generation of new minerals with different properties.

On a larger scale, i.e. that of the rock mass, the intact rock is generally considered to be a continuous isotropic material, but the aspects outlined above must be taken into account when studying certain types of rock materials, e.g. those with lamination or schistosity; in these cases, to evaluate the influence of these "defects", the scale or scope of the project must also be considered. For certain geological engineering applications, e.g. selecting a storage site for radioactive waste, laboratory-based tests and detailed investigation of the properties and characteristics of the intact rock as described above are required; but these are less important when studying jointed hard rock masses with discontinuous behaviour, where the blocks of intact rock can be considered homogeneous and isotropic.

The characteristics and mechanical behaviour of hard rock masses depends on factors such as the degree of fracturing and weathering, the presence of water, the type and orientation of the discontinuities and the size of the blocks. How important discontinuities such as bedding planes, joints or faults are, is also relative to the scale of the project; if the discontinuities do not affect the behaviour of the rock mass significantly, or if they only affect it to a small extent in relation to the overall scale of the project or structure in question, then the rock medium can be considered to be continuous; however, if the weak planes or zones are large enough to affect the behaviour of the rock mass on the site, a separate study of them must be carried out.

# Physical properties of intact rock

To identify and describe the basic properties of rocks, a series of quantitative parameters are used for an initial classification for geotechnical purposes. These are known as index properties and with the **mineralogical composition** and **fabric**, they indicate the properties and mechanical behaviour of the intact rock in the first instance. They are listed in Table 3.1, with the methods for evaluating them.

The geological description of a rock includes its name, mineralogy, texture, type of cementation and degree of alteration. The petrographic description is made from macroscopic observation of the samples and from microscopic analysis, to determine the composition, texture, fabric, degree of alteration, existing microcracks, and porosity; it includes thin section analysis, optical and electronic microscopy and X-ray diffraction techniques.

The preferred mineral orientation, hardness or crystalline structure may determine the reaction or mechanical

| Table 3.1 | PROPERTIES OF INTACT ROCK AND HOW THEY ARE DETERMINED | |
|---|---|---|
| | **Properties** | **Determination methods** |
| Identification and classification properties | Mineralogical composition. Fabric and texture. Grain size. Colour. | Visual description. Optical and electronic microscopy. X-ray diffraction. |
| | Porosity ($n$) | Laboratory techniques. |
| | Unit weight ($\gamma$) | |
| | Water content. | |
| | Permeability (permeability coefficient, $k$) | Permeability test. |
| | Durability. Alterability (alterability index) | Alterability tests. |
| Mechanical properties | Uniaxial compressive strength ($\sigma_c$) | Uniaxial compression test. Point load test. Schmidt hammer. |
| | Tensile strength ($\sigma_t$) | Direct tension test. Indirect tension test. |
| | Sonic wave velocity ($V_p$, $V_s$) | Laboratory measurement of elastic waves velocity. |
| | Strength (parameters $c$ and $\phi$) | Triaxial compression test. |
| | Deformability (static or dynamic elastic deformation modules: $E$, $\nu$) | Uniaxial compression test. Sonic velocity test. |

response of the rock to changes in external forces. Many relevant rock properties for engineering depend on the structure of the mineral particles and how they are linked.

The physical or **index properties** of rocks are determined in the laboratory. Porosity, unit weight, permeability, alterability, strength and the elastic wave propagation velocity have the most significant influence on an understanding of the mechanical behaviour to be expected. Some of these properties are directly related to the strength and deformational characteristics of the rocks and are used to classify them.

**Porosity** is the ratio between the rock pore volume, $V_v$, and the total volume $V$ (solid particles + pore spaces or voids): $n(\%) = V_v/V$. This is the property that most affects strength and mechanical characteristics, as it is inversely proportional to strength and density, and directly proportional to deformability. In crystalline, igneous or metamorphic rocks, pores may be microcracks or cracks in the intact rock. In general, porosity decreases with depth and the age of the rocks.

The value of $n$ varies between 0% and 90%, with normal values ranging from 15%–30%. Carbonate bioclastic sedimentary rocks and volcanic rocks may have very high porosity values, the same as altered or weathered rocks. Table 3.2 shows porosity data for some rocks.

**Effective porosity** is the ratio between the interconnected pore void volume and the total volume $V$ of the sample rock; it can be obtained from the dry and saturated weights of the sample:

$$n_e = (W_{sat} - W_{dry})/(\gamma_w V)$$

where $\gamma_w$ = unit weight of water.

Pores in rocks are often not interconnected, so that real porosity is greater than effective porosity. The void ratio is defined as the ratio of the volume of void space, $V_v$, to the volume of solid particles, $V_{sol}$: $e = V_v/V_{sol}$.

The **unit weight** of the rock depends on its components and is defined as the weight per unit volume. The units used are those of force per unit volume. Care should be taken because in geotechnical literature "density" $\rho$ ($\rho$ = mass/volume) is sometimes referred to as specific or unit weight; when working with weight ($\gamma = \rho g$) it should be made clear that units of force (i.e. mass × acceleration), not mass, are being used; i.e. 1 $kg_{mass}/m^3$ × 1 $m/s^2$ = 1 $N/m^3$, making 1 $kg_{mass}/m^3$ × 9.81 $m/s^2$ = 9.81 $N/m^3$ the unit weight of 1 $kg_{mass}/m^3$ on earth. Water has a density of 1,000 $kg_{mass}/m^3$ so giving 1 $m^3$ a weight of 9.81 kN.

Unlike soils, the specific weight values of rocks vary widely. Table 3.2 gives average values for some rocks.

**Permeability** is the water-transmitting capacity of a rock. Most rocks have low or very low permeability. Water infiltrates and flows through intact rock through pores and cracks, and the permeability is determined by how these are interconnected and other factors, such as the degree of weathering, anisotropy and the state of stress the material is subjected to.

The permeability of a rock is measured by the coefficient of permeability or hydraulic conductivity, $k$, expressed in m/s, cm/s or m/day.

Darcy's law states that the rate of flow $Q$ per unit area is proportional to the gradient of the potential head, $i$, measured in the direction of flow:

$$Q = kiA$$

For most types of rock, the flow in intact rock can be considered to follow Darcy's law:

$$q_x = k(dh/dx)A$$

| Table 3.2 | TYPICAL VALUES FOR UNIT WEIGHT AND POROSITY OF ROCKS | |
|---|---|---|
| Rock | Unit weight (kN/m³) | Porosity (%) |
| Andesite | 22–23.5 | 10–15 |
| Amphibolite | 29–30 | – |
| Basalt | 27–29 | 0.1–2 |
| Chalk | 17–23 | 30 |
| Coal | 10–20 | 10 |
| Diabase | 29 | 0.1 |
| Diorite | 27–28.5 | – |
| Dolomite | 25–26 | 0.5–10 |
| Gabbro | 30–31 | 0.1–0.2 |
| Gneiss | 27–30 | 0.5–1-5 |
| Granite | 26–27 | 0.5–1.5 (0.9) |
| Greywacke | 28 | 3 |
| Gypsum | 23 | 5 |
| Limestone | 23–26 | 5–20 (11) |
| Marble | 26–28 | 0.3–2 (0.6) |
| Mudstone | 22–26 | 2–15 |
| Quartzite | 26–27 | 0.1–0.5 |
| Rhyolite | 24–26 | 4–6 |
| Salt | 21–22 | 5 |
| Sandstone | 23–26 | 5–25 (16) |
| Schist | 25–28 | 3 |
| Shale | 25–27 | 0.1–1 |
| Tuff | 19–23 | 14–40 |

where $q_x$ is the volume of flow in direction $x$ (volume/time), $h$ is the hydraulic head, $A$ is the section perpendicular to direction $x$, and $k$ is hydraulic conductivity.

The coefficient of permeability $k$ also equals:

$$k = K(\gamma_w/\mu)$$

where $K$ is the intrinsic permeability (dependent only on the characteristics of the physical medium), $\gamma_w$ is the unit weight of the water and $\mu$ is the kinematic viscosity of the water (see Chapter 4, Section 4.2).

Table 3.3 includes permeability coefficient values for certain rocks. As it is difficult to estimate and evaluate this parameter, values are indicated in orders of magnitude.

**Durability** is the resistance of rock to weakening and disintegration processes. This property is also described as alterability, defined in this case as the tendency of the intact rock components or structures to failure.

The properties of intact rock are changed by a number of processes including hydration, dissolution and oxidation. In some rocks, such as volcanic rock, mudstones and shales, which contain significant quantities of clay minerals, exposure to air or the presence of water degrades the strength properties, which may mean that the strength of these rocks may be overestimated for engineering applications such as surface excavations, tunnels and embankments, unless their medium-term behaviour, in contact with the atmosphere, is taken into account. Rock durability increases with density and decreases with water content.

Durability is measured with the **Slake Durability Test** (SDT), which subjects previously fragmented rock samples to standard 10 minute drying and wetting cycles in the laboratory. Figure 3.8 shows the test apparatus. After the fragments of rock have been oven-dried, they are weighed and placed in a drum with a 2 mm external mesh. The drum is placed in a bath filled with water to a level below the drum axis and

Figure 3.8   Slake durability test apparatus.

then rotated for ten minutes. The samples remaining in the drum are taken out, oven-dried and weighed and the process is repeated. The **slake-durability index**, $I_D$, represents the proportion of dry-weight rock remaining in the drum after one or two disintegration cycles ($I_{D1}$, $I_{D2}$), and may vary from 0% to 100%:

$$I_D(\%) = \frac{\text{Dry weight after 1 or 2 cycles}}{\text{Initial weight of sample}}$$

Standard classification is based on the value of $I_{D2}$ (Table 3.4). In cases of very weak clay rocks which give $I_{D2}$ values lower than 10% after the second cycle, the index corresponding to the first cycle, $I_{D1}$, is recommended (Table 3.5).

There are other laboratory tests for assessing durability which also involve weakening and disintegrating the rock by simulating weathering processes using wetting/drying, heating/cooling, freezing/thawing cycles. The results of strength tests also give indirect qualitative information on the durability of the rocks.

**Uniaxial compressive strength** or uniaxial strength is the maximum stress the rock can carry under uniaxial

| Table 3.3 | TYPICAL PERMEABILITY VALUES FOR INTACT ROCK |
|---|---|
| **Rock** | **$k$ (m/s)** |
| Granite | $10^{-9}$–$10^{-12}$ |
| Limestone and dolomite | $10^{-6}$–$10^{-12}$ |
| Metamorphic rocks | $10^{-9}$–$10^{-12}$ |
| Mudstone | $10^{-9}$–$10^{-13}$ |
| Salt | $<10^{-11}$–$10^{-13}$ |
| Sandstone | $10^{-5}$–$10^{-10}$ |
| Schist | $10^{-7}$–$10^{-8}$ |
| Shale | $10^{-11}$–$10^{-13}$ |
| Volcanic rocks | $10^{-7}$–$10^{-12}$ |

| Table 3.4 | DURABILITY CLASSIFICATION BASED ON THE $I_{D2}$ INDEX |
|---|---|
| **Durability** | **% weight retained after 2 cycles** |
| Very high | >98 |
| High | 95–98 |
| Medium-high | 85–95 |
| Medium | 60–85 |
| Low | 30–60 |
| Very low | <30 |

Table 3.5  DURABILITY CLASSIFICATION BASED ON THE $I_{D1}$ INDEX

| Durability | % weight retained after 1 cycle (1) | (2) |
|---|---|---|
| Extremely high | – | >95 |
| Very high | >99 | 90–95 |
| High | 98–99 | 75–90 |
| Medium-high | 95–98 | – |
| Medium | 85–95 | 50–75 |
| Low | 60–85 | 25–50 |
| Very low | <60 | <25 |

(1) Gamble, 1971 (in: Goodman, 1989)
(2) Frankling and Chandra, 1972 (in: Johnson and De Graff, 1988)

Table 3.6  STRENGTH VALUES FOR FRESH INTACT ROCK

| Fresh rock | Uniaxial compressive strength (MPa) Average values | Range of values | Tensile strength (MPa) |
|---|---|---|---|
| Andesite | 210–320 | 100–500 | 7 |
| Amphibolite | 280 | 210–530 | 23 |
| Anhydrite | 90 | 80–130 | 6–12 |
| Basalt | 80–200 | 60–350 | 5–25 |
| Diabase | 240–350 | 130–365 | 55 |
| Diorite | 180–245 | 120–335 | 8–30 |
| Dolerite | 200–300 | 100–350 | 15–35 |
| Dolomite | 60–200 | 50–350 | 5–25 |
| Gabbro | 210–280 | 180–300 | 14–30 |
| Gneiss | 60–200 | 50–250 | 5–20 |
| Granite | 70–200 | 50–300 | 7–25 |
| Greywacke | 100–150 | 80–220 | 5.5–15 |
| Gypsum | 25 | 10–40 | 1–2.5 |
| Limestone | 60–140 | 50–200 | 4–30 |
| Marble | 120–200 | 60–250 | 6–20 |
| Marl | 30–70 | 20–90 | – |
| Mudstone | 20–40 | 10–90 | 1.5–10 / 0.5–1* |
| Quartzite | 200–320 | 100–500 | 10–30 |
| Salt | 12 | 5–30 | – |
| Sandstone | 55–140 | 30–235 | 5–20 |
| Schist | 30–60 | 20–160 | 2–5.5 |
| Shale | 40–150 | 30–200 | 7–20 |
| Siltstone | – | 35–250 | 2.7 |
| Tuff | – | 10–46 | 1–4 |

(*) Along lamination planes.
Data selected from Rahn (1986), Walthan (1999), Obert and Duvall (1967), Farmer (1968).

compression, measured on an unconfined cylindrical specimen in the laboratory and is given by:

$$\sigma_c = \frac{F_c}{A} = \frac{\text{Compressive force applied}}{\text{Area of application}}.$$

The strength value provides information on the engineering properties of the rock. Table 3.6 shows uniaxial compressive strength values for different types of rock.

The approximate compressive strength can also be estimated from correlation with indexes obtained with simple field tests, such as the **Point Load Test (PLT)** or the **Schmidt hammer test** (described in Chapter 5, Section 5.5). With the values obtained using either of these two methods, rock can be also classified according to strength (Table 3.7).

**Field indexes** provide an initial approximation of the rock strength value. Table 3.7 shows the criteria for their identification and the strength value corresponding to each one.

**Tensile strength** is the maximum stress a material can sustain in uniaxial tension before fracture occurs. It is obtained by applying tensile forces to a cylindrical rock specimen in the laboratory (the test is described in Section 3.4):

$$\sigma_t = \frac{F_t}{A} = \frac{\text{Tensile force applied}}{\text{Area–section of sample}}$$

The $\sigma_t$ value for intact rock usually ranges from 5–10% of the uniaxial compressive strength value (10% for most brittle rocks and closer to 5% for soft rocks, such as mudstones and claystones); it can be 14–16% in some sedimentary rocks (Duncan, 1999).

The **velocity of propagation of elastic waves** as they pass through the rock depends on the density and the elastic properties of the material. Measuring this provides information on certain characteristics, such as porosity. The laboratory test to determine this velocity is described in Section 3.4.

The longitudinal or compression wave velocity $V_p$ is used as a classification index; this value has a linear correlation with uniaxial compressive strength $\sigma_c$ and gives an indication of intact rock quality. This velocity in rocks ranges from 1,000 to 6,000 m/s. For weathered rocks, values below 900 m/s are obtained; i.e. fresh granite may give values

| Table 3.7 | | SOIL AND ROCK STRENGTH CLASSIFICATION FROM FIELD INDICES | |
|---|---|---|---|
| Class | Description | Field identification | Approx. range to uniaxial compressive strength (MPa) |
| $S_1$ | Very soft clay | Easily squeezed between fingers | <0.025 |
| $S_2$ | Soft clay | Easily penetrated several centimetres by thumb. | 0.025–0.05 |
| $S_3$ | Firm clay | Can be penetrated several centimetres by thumb with moderate effort. | 0.05–0.1 |
| $S_4$ | Stiff clay | Readily indented by thumb but penetrated only with great difficulty. | 0.1–0.25 |
| $S_5$ | Very stiff clay | Readily indented by thumbnail. | 0.25–0.5 |
| $S_6$ | Hard clay | Indented with difficulty by thumbnail. | >0.5 |
| $R_0$ | Extremely weak rock | Indented by thumbnail. | 0.25–1 |
| $R_1$ | Very weak rock | Crumbles under firm blows with point of geological hammer and can be peeled by a pocket knife. | 1–5 |
| $R_2$ | Weak rock | Can be peeled by a pocket knife with difficulty, shallow indentations made by firm blow with point of geological hammer. | 5–25 |
| $R_3$ | Moderately strong rock | Cannot be scrapped or peeled with a pocket knife, specimen can be fractured with single firm blow of geological hammer. | 25–50 |
| $R_4$ | Strong rock | Specimen requires more than one blow of geological hammer to fracture it. | 50–100 |
| $R_5$ | Very strong rock | Specimen requires many blows of geological hammer to fracture it. | 100–250 |
| $R_6$ | Extremely hard rock | Specimen can only be chipped with geological hammer. | >250 |

(ISRM, 1981).

of up to 6,000 m/s, but in weathered granite the velocity decreases proportionally to half or a third of this, depending on the degree of weathering, and in very decomposed rock the values will be less than 700–800 m/s. Table 3.8 shows some velocity of propagation values for elastic compressional waves. The wide ranges of velocity for many of the rocks are mainly due to variations in density (variation in porosity or mineral composition).

# Rock classification for geotechnical purposes

**Geological or lithological classifications** (Table 3.9) are fundamental in geological engineering; they provide information on the mineralogical composition, texture and fabric of rocks and also on the structural isotropy or anisotropy in rocks of a particular origin, e.g. massive rocks compared with laminated or foliated rocks. Thus, the terms igneous or metamorphic correspond to rocks with a specific structure, texture, composition, grain size, etc. These factors condition the physical and strength properties of the rock and are used for sub-dividing the main rock groups.

How lithology is related to specific geological processes is another important factor to consider when studying the behaviour of rock materials; e.g. how easily carbonate or evaporitic rocks dissolve, how clay rocks alter and swell and how creep processes develop in saline rocks.

However, lithological classifications in themselves are not enough in geological engineering since there may be considerable variation in the physical and mechanical properties of similar lithologies, e.g. in strength; moreover, these classifications do not give any quantitative information on properties.

Rock classification for engineering purposes is a complex task as the properties used as input parameters for design calculations must be quantified. The qualitative terms used for rock, such as hard/strong or soft/weak, should be defined by the value of their corresponding uniaxial compressive strength, i.e. 50 to 100 MPa for hard rock and 5 to 25 MPa for weak rock. Geotechnical classification is difficult because of the wide variety of rock properties and the

# ROCK MECHANICS

| Table 3.8 | PROPAGATION VELOCITY FOR LONGITUDINAL WAVES IN ROCKS |
|---|---|
| Fresh rock | Propagation velocity of $V_p$ waves (m/s) |
| Basalt | 4,500–6,500 |
| Conglomerate | 2,500–5,000 |
| Diabase | 5,500–7,000 |
| Dolerite | 4,500–6,500 |
| Dolomite | 5,000–6,000 |
| Granite | 4,500–6,000 |
| Gabbro | 4,500–6,500 |
| Gneiss | 3,100–6,000 |
| Gypsum | 3,000–4,000 |
| Limestone | 2,500–6,000 |
| Marble | 3,500–6,000 |
| Marl | 1,800–3,200 |
| Mudstone | 1,400–3,000 |
| Quartzite | 5,000–6,500 |
| Salt | 4,500–6,000 |
| Sandstone | 1,400–4,200 |
| Schist | 4,200–4,900 |
| Shale | 3,500–5,000 |

| Table 3.9 | GENERAL ROCK CLASSIFICATION BY ORIGIN | |
|---|---|---|
| Rock classification by origin | | |
| Sedimentary rocks | Detrital: | quartzite, sandstone, mudstone, silt, conglomerate |
| | Chemical: | evaporites, dolomitic limestone |
| | Organic: | limestone, coal, corals |
| Igneous rocks | Plutonic: | granite, gabbro, diorite |
| | Volcanic: | basalt, andesite, rhyolite |
| Metamorphic rocks | Massive: | quartzite, marble |
| | Foliated or schistose: | shale, phyllite, schist, gneiss |

limitations on the methods and procedures for determining them. Rocks in a wet condition, e.g. saturated, are usually weaker than when dry.

**Uniaxial compressive strength** is the most frequently measured property of rocks and is the basis of classification in rock mechanics. *Table 3.10* shows different classifications based on this parameter.

A value also used for the mechanical classification of intact rock is the ratio between the elasticity modulus $E$ (described in Section 3.3) and the uniaxial compressive strength, $\sigma_c$, which usually varies according to lithology. *Figure 3.9* shows this engineering classification on the basis of **modulus ratio** for different types of rock; for most rocks the ratio lies between 200 and 500.

A qualitative classification of rocks can be made from the **degree of weathering** of the intact rock, which gives some idea of the mechanical or geotechnical characteristics. Weathering increases the porosity, permeability and deformability of rock material and reduces its strength. Chapter 6, Table 6.4, shows a classification with four categories

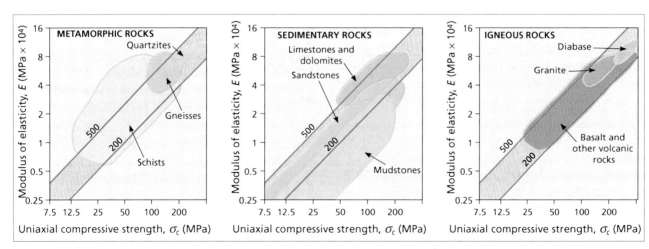

Figure 3.9  Intact rock classification based on the modulus ratio $E/\sigma_c$. Three categories are established based on this ratio: high (>500), medium (200–500) and low (<200). (Modified from Deere and Miller, 1966; in Attewell and Farmer, 1976).

| Uniaxial compressive strength (MPa) | ISRM (1981) | BS 5939 (1999) | Bieniawski (1973) | Examples |
|---|---|---|---|---|
| <1 | | | Soils | |
| <1.25 | Very weak | Very weak | Very low | Salt, mudstone, silt, marl, tuff, coal |
| 1.25–5 | | Weak | | |
| 5–12.5 | Weak | Moderately weak | | |
| 12.5–25 | | Moderately strong | | |
| 25–50 | Moderately strong | | Low | Schist, shale |
| 50–100 | Strong | Strong | Medium | Schistose metamorphic rocks, marble, granite, gneiss, sandstone, porous limestone |
| 100–200 | Very strong | Very strong | High | Hard igneous and metamorphic rocks, highly cemented sandstone, limestone, dolomite |
| >200 | | Extremely strong | Very high | Quartzite, gabbro, basalt |
| >250 | Extremely strong | | | |

Table 3.10 ROCK CLASSIFICATION BY UNIAXIAL COMPRESSIVE STRENGTH (UCS)

for rock, ranging from fresh to decomposed. Weathering processes in intact rock are dealt with later in this section.

## Rock mass classification

The classification of a rock mass is based on one or more factors that determine its likely mechanical behaviour:

— Properties of the intact rock.
— Type and frequency of discontinuities; these define the degree of fracturing, the shape and size of the rock blocks, the hydrogeological properties, and such like.
— Degree of weathering.
— *In situ* state of stress.
— Presence of water.

The great variability of these factors and the discontinuous and isotropic nature of the rock masses make it difficult to establish general geotechnical or geomechanical classifications that are valid for different types of rock mass.

**Geomechanical classifications** are the most useful in rock mechanics, with Bieniawski's Rock Mass Rating (RMR) and Barton's Tunnelling Quality Index (Q) the most widely used (see Section 3.8). These give quality ratings for the rock mass, depending on the properties of the intact rock and of the discontinuities, and provide estimated values for the overall strength properties.

Other classifications use different parameters that are more or less representative of the properties of the rock mass as a whole. The one shown in *Table 3.11* is based on the **degree of fracturing**, as measured by the Rock Quality Designation (RQD) index, and classifies the rock mass at different quality levels (the RQD index is defined in Chapter 5, Box 5.1).

The number of discontinuity sets and the size and shape of the intact rock blocks are also used as rating parameters to give information on configuration and the degree of fracturing (see Chapter 6, Tables 6.10, 6.11 and 6.12).

The **velocity index** $(V_{situ}/V_{lab})^2$ or $(V_F/V_L)^2$ is another parameter used for rock mass classification. This index, used as a quality criterion, gives the ratio of the velocity of longitudinal waves measured *in situ* in the rock mass in a given direction to the velocity measured in laboratory specimens (*Table 3.12*). For a high quality rock mass, with few closed discontinuities, this ratio should be around 1.0, decreasing in value as the degree of fracturing increases and the quality of the rock mass diminishes. This index can be correlated in a given direction with the RQD in that direction (*Figure 3.10*):

$$RQD = 100 \left(\frac{V_F}{V_L}\right)^2$$

The **degree of weathering** is also used for the quality rating of rock masses as it influences their strength and mechanical properties. Chapter 6, Table 6.14 shows six different degrees of weathering for rock mass rating, ranging from "fresh" (grade I), with no sign of weathering, to "residual soil" (grade VI), where the rock mass has been turned into soil and its original fabric destroyed.

| Table 3.11 | ROCK MASS CLASSIFICATION BY RQD |
|---|---|
| RQD % | Quality |
| <25 | Very poor |
| 25–50 | Poor |
| 50–75 | Fair |
| 75–90 | Good |
| 90–100 | Excellent |

| Table 3.12 | VELOCITY INDEX AND ROCK MASS QUALITY |
|---|---|
| Velocity index $(V_F/V_L)^2$ | Rock mass quality |
| <0.2 | Very poor |
| 0.2–0.4 | Poor |
| 0.4–0.6 | Fair |
| 0.6–0.8 | Good |
| >0.8 | Very good |

$V_F$ = compressive wave velocity in the field
$V_L$ = compressive wave velocity of intact rock sample
Coon and Merritt, 1970; in Bieniawski, 1984.

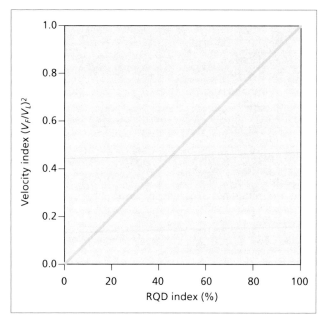

Figure 3.10   Theoretical correlation between RQD and the velocity index.

# Weathering of rock

## Weathering processes

Weathering is the disintegration and/or decomposition of geological materials. It includes any physical or chemical modifications of the materials' characteristics and properties. The final products of rock weathering processes are soils; these may remain as **residual soils** in their original location on the source rock or they may be transported as sediment. **Transported soils** (alluvial, aeolian or glacial) can be either lithified to form new rocks or remain as soils. The contact between soil and rock may be clearly defined or gradual, which is characteristic of residual soils (see Box 3.1). The extent to which rock either intact or as a rock mass is weathered has a significant influence on its physical and mechanical properties.

Weathered rock materials can be broadly defined as in transition between rock and soil; they present a wide range of geotechnical properties and characteristics which blend with those of soils and rocks, depending on the degree of weathering. The terms "soft" or "weak" rock are sometimes used to infer weathered materials, although not all soft rocks (e.g. mudstones, siltstones, marls) are a product of only weathering.

When materials with a significant degree of weathering are classified according to their mechanical behaviour, the problem is whether they should be considered as rocks or as soils; in the first case their properties are undervalued while in the second case they are overvalued. Some authors prefer to distinguish soils from rocks by their level of compaction or cementation, by their structure and by their durability.

Weathering processes are controlled by climatic conditions and variations in temperature, humidity and rainfall. These factors determine the type and intensity of the physical and chemical transformations that affect rocks at and near ground level.

Physical actions produce mechanical fracturing of rock. The most important actions controlled by climatic factors, especially temperature and humidity are:

— **Ice formation:** the volume of water held in pores and cracks increases as the temperature drops and ice forms, which can lead to the rock fracturing.
— **Solar radiation or "insolation":** in arid climates with marked thermal differences over short periods of time, can induce fracturing as a result of the stresses from successive dilation and contraction of the rocks.
— **Salt formation:** salt crystallization in pores or cracks in the rocks produces failure and disintegration as the crystals expand.
— **Hydration:** when clays and sulphates are hydrated their volume increases, producing considerable deformation which may cause fracturing in the rock.
— **Capillarity and thermal expansion:** fissured and sheet structure minerals, such as micas and gypsum, allow water to penetrate, and with temperature changes this may cause structural failure, as the water expands.

When water is present, temperature controlled chemical processes take place. These occur more quickly and intensely in wet regions than in dry ones and lead to the formation of new minerals or compounds from existing ones. The most important processes are:

— **Dissolution:** the decomposition of minerals by the action of water, leading ultimately to disintegration of the material. Dissolution is a chemical reaction; the dissolution of calcium carbonate causes cracks, joints and existing voids to open up in carbonate and other soluble rocks.
— **Hydration:** the formation of new minerals or chemical compounds from the absorption of water; e.g. gypsum from anhydrite.
— **Hydrolysis:** the decomposition of a mineral or chemical compound by the action of water. The level of hydrolysis depends on to what extent the element ions attract the water molecules.
— **Oxidation and reduction:** the formation of new minerals from a combination of a host mineral with oxygen, either by losing one or more electrons and fixing oxygen (oxidation), or by losing oxygen and fixing electrons (reduction).

Which of these actions occur depends on the climatic characteristics of the area. In cold climates and mountainous elevations with average rainfall, physical weathering will predominate, basically controlled by ice, but in warm tropical climates with high rainfall there will be more chemical activity. *Figure 3.11* shows the different types of predominant weathering and their intensity depending on temperature and rainfall.

Weathering processes affect both intact rock and the rock mass as a whole.

## Weathering of intact rock

Physical weathering in intact rock causes exfoliation along planes in a preferred direction and the opening up of microdiscontinuities by the action of ice or salt growth, and changes in volume due to variations in temperature or water content. Chemical weathering produces the dissolution of soluble minerals and the formation of new minerals through oxidation, reduction and hydration processes. The results of chemical weathering range from discolouration of the intact rock to decomposition of silicates and other minerals, although some minerals, such as quartz, are much more resistant to such action than others. Dissolution processes play a very important role in the chemical alteration of rocks, such us carbonate, gypsiferous and evaporite materials, especially rock salt.

The action and effects of weathering differ according to the type of rock, and are directly related to its mineral composition and structure.

Although chemical weathering is usually more intense and leads to decomposition and mineralogical changes in rock, physical weathering causes the rock to disintegrate, weakening its structure as the minerals and contacts between particles are broken down; this increases the surface area exposed to the atmosphere and allows water to infiltrate.

Some minerals are more prone to chemical weathering than others; common minerals are listed below with the least stable at the top and increasing in stability with descending order:

— Olivine
— Feldspar (Ca-rich)
— Pyroxene
— Amphibolite
— Feldspar (Na-rich)
— Biotite
— Feldspar (K-rich)
— Muscovite
— Quartz

How easily rocks are weathered depends on how great the difference is between the original pressure and

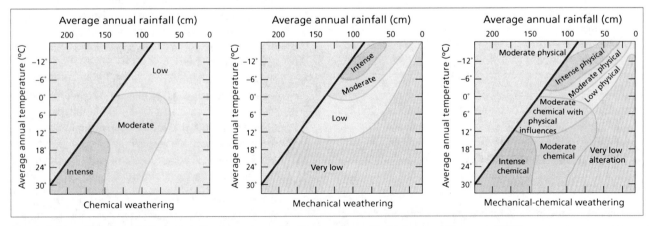

*Figure 3.11* Relationship between climate and weathering processes (Embleton and Thornes, 1979).

temperature conditions when they were formed and current environmental conditions. Quartz is the mineral that forms at temperatures closest to environmental temperatures ($\approx 300°C$).

In general, most silicates (particularly feldspars and micas) weather to clay minerals. Under certain environmental conditions such as tropical or humid climates, they decompose into iron and aluminium oxides and hydroxides. The type of clay minerals which result from weathering depends on the original minerals, the chemistry of the water content and its pH.

**Clay rocks** are the most susceptible to physical weathering processes which affect their physical and mechanical properties. However, these rocks often remain stable in mineralogical terms and do not experience chemical weathering because they are formed at temperatures and pressures similar to those at the surface, although some clay rocks may contain weathered accessory minerals. **Igneous** and **metamorphic rocks** can be chemically unstable at the surface as they have formed under very different temperature and pressure conditions and have undergone intense chemical and mineralogical changes. Their resistance to weathering depends on their mineralogy and they are generally more resistant to physical weathering than sedimentary rocks.

Reduced strength is the most important effect of the rock decomposition caused by chemical weathering. A small increase in the water content or porosity of the rock may mean a considerable reduction in strength and deformation modulus. This means that granite may range from more than 250 MPa when fresh, to half this value with weathering and to less than 100 MPa if weathering is intense. The sonic velocity also decreases from more than 5000 m/s in fresh granite to less than half if it is moderately weathered and to below 800 m/s if it has been altered to residual soil.

Durability and alterability tests are carried out in the laboratory (e.g. the slake durability test) to determine how rocks are altered by weathering. Strength tests (e.g. the PLT and uniaxial compression test), also provide qualitative information on the resistance of the rock to weathering. The extent to which intact rock components are weathered can be determined from mineralogical analysis. Visual descriptions and standard indexes are used for a qualitative classification of intact rock based on its **degree of weathering** (Chapter 6, Table 6.4).

## Weathering of rock masses

The action of weathering processes on a rock mass affects its blocks of intact rock and also its existing discontuinities. Mechanical or physical weathering may cause discontinuities already present in the rock mass to open further or new ones to form when the grain contacts break or mineral forming the intact rock fracture. Discontinuities are preferred pathways for water, and contribute to increase chemical and physical weathering.

Residual soils are the final product of *in situ* weathering in rock masses. Weathered rock masses whose vertical profile retains their rock structure, even though the intact rock composition has been altered and is weaker than the fresh rock, are sometimes called **saprolites** or regoliths; as well as the original discontinuities, such profiles may also contain other planes of weakness due to preferential weathering. Depending on the degree of weathering, they may either preserve the original blocks of intact rock or be reduced all to soils.

Weathering processes affect different lithologies to different extents, increasing in intensity with longer exposure to atmospheric agents. *Figure 3.12* shows schematically the variation with depth in the degree of weathering in rock masses of different geological origins.

The depth of weathering depends on the type of rock, the climate and the amount of exposure to weathering. Clay rocks, porous sandstone and weak limestone often weather to a greater depth than granites or metamorphic rocks. In humid tropical climates soils produced by the weathering of rock masses may be as much as 20 or 30 m thick, or more.

**Unloading** processes as a result of erosion are an important agent of rock mass weathering. When the lithostatic pressure is reduced, the rock masses expand along fracture planes generated parallel to the ground surface, producing a layered structure. This phenomenon is more important in materials with pre-existing planes of weakness, such as mudstones or shales.

**Clay** or mudstone rocks are frequently found at the surface, outcropping when erosion reduces the overburden; this process triggers a relaxation of the materials and the widening of discontinuities that normally affect them, allowing water to penetrate the weakness planes and even the intact rock. Several metres below the surface, the joints remain closed and weathering processes do not take place. *Figure 3.13* shows a clay rock mass that has been considerably altered by unloading and physical weathering.

**Carbonate rock masses** are weathered both at their surface and internally along fractures and bedding planes, leading to the formation of cracks and cavities from dissolution. These processes cause irregular karstic relief (*Figure 3.14*) in which hard rocks are found side by side with soft clay infill, cavities and sinkholes (*Figure 3.15*). Karstic processes and forms are also found in evaporitic rock masses.

When **igneous rocks** such as granites and diorites are exposed, decompression cracks appear parallel to the surface, facilitating chemical weathering (*Figure 3.16*) and ultimately leading to the alteration of the feldspars and micas in the intact rock into clay minerals, while the quartz remains as a sand. Discontinuities are preferred zones of alteration, and

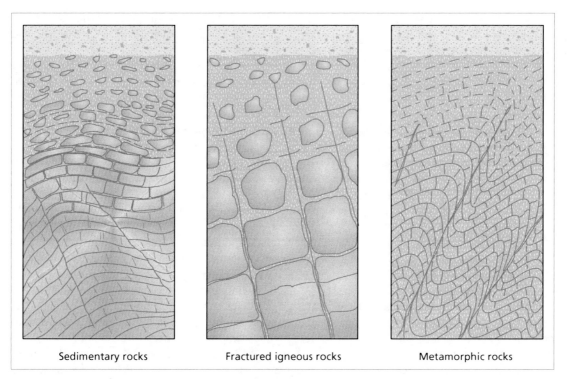

*Figure 3.12* Typical weathering profiles resulting in residual soils in sedimentary, igneous and metamorphic rocks.

*Figure 3.13* Weathered clayey rock mass (area shown 10 m wide).

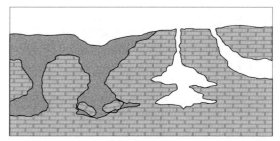

*Figure 3.15* Weathering of carbonate rock masses.

*Figure 3.14* Karstic morphology.

*Figure 3.16* Unloading, stress relief fractures parallel to the surface in a granitic rock mass.

spheroidal or "onion skin" weathering occurs on the blocks of intact rock between them, leaving the core less altered.

Weathering of **basaltic rock masses** is most likely to occur along fracture planes although the intact rock blocks may show spheroidal weathering and their minerals may alter to clays and iron oxide (Figure 3.17).

In **metamorphic rocks**, such as gneisses and amphibolites, the feldspars and pyroxenes tend to alter more rapidly than the amphiboles, while quartz survives. Gneisses typically show banding and the bands with minerals most liable to weathering are preferred zones for chemical alteration, which produces areas of weakness in a rock mass.

Schists, slates and phyllites show a marked tendency to split along the schistosity planes and cleavage, creating weak areas prone to weathering. Despite the presence of hard minerals, alteration takes place more easily because of the penetration of water and ice.

The **degree of weathering** in a rock mass can be assessed from the degree of fracturing using the RQD index (described in Chapter 5, Section 5.4), since the number of discontinuities indicates how prone the rock mass is to weathering. The sonic velocity value of waves through the rock mass also indicates its degree of weathering as explained above. Qualitative classification of rock masses according to their degree of weathering is based on visual descriptions and standard indexes (see Chapter 6, Section 6.5).

# Groundwater

## Permeability and water flow

Rocks are described as permeable or impermeable depending on whether they can transmit water at rates relevant to the engineering work in question. This depends on how porous they are and how the pores are inter-connected.

Permeability can be defined as the capacity for water to flow through the interconnected pores and spaces of the rock medium and is represented by the **coefficient of permeability**, expressed as a velocity. Normal values for rocks range from 1 m/day to 1 m/year.

The permeability of intact rock is intergranular; water flows through interconnected pores and micro-cracks in the rock, and this is called **primary permeability**. In rock masses the water flows along the discontinuities; this is called **secondary permeability**. The permeability of intact rock is normally negligible compared to that of a jointed rock mass; sandstones and other porous rocks where water can flow through intact rock are the exceptions. Rock masses affected by karstification have the highest permeability coefficients, due to wide open discontinuities and cavities produced by the dissolution of carbonate or other soluble materials. Lava flows can be very permeable if they are fractured or with cavities, and pyroclastic deposits, usually very porous, can also have very high permeability if the voids are connected.

In permeable rock masses a groundwater level or **water table** is established and below it the interconnected pores and/or discontinuities are filled with water. The shape of this surface normally adapts to the topography and it may reach the ground surface at certain points because of changes in lithology or topography, or geological features such as faults. There may be fluctuations in the water table caused by heavy or persistent rain, pumping or periods of drought. Usually only a small percentage of rainwater infiltrates the ground and only a part of this, perhaps even none, reaches the water table, depending on the quantity of infiltrated water, the rock mass permeability coefficient and the depth of the water table.

Subsurface water exerts pressure on the pores and discontinuities of rocks and rock masses.

The flow in a jointed rock mass depends on the number and aperture of discontinuities, which in turn depends on the vertical stress exerted with depth, and on how the discontinuities are interconnected. Vertical stress increases with depth and below a certain level discontinuities, in theory, do not exist or are closed, and the permeability of the rock mass is the same as that of the intact rock, or primary permeability. Section 3.6 includes some aspects of permeability in rock masses. Figure 3.18 shows data on primary and secondary permeability for different types of rock.

The permeability of intact rock is measured in the laboratory and the permeability of the rock mass is determined from in situ tests in boreholes (see Chapters 4 and 5).

## Effects of water on the properties of rock masses

Water is a geological "material" that coexists with rocks and influences their mechanical behaviour and their response to applied forces. Water flows through preferred areas,

Figure 3.17   Volcanic rock mass affected by intense weathering to soil along joint planes.

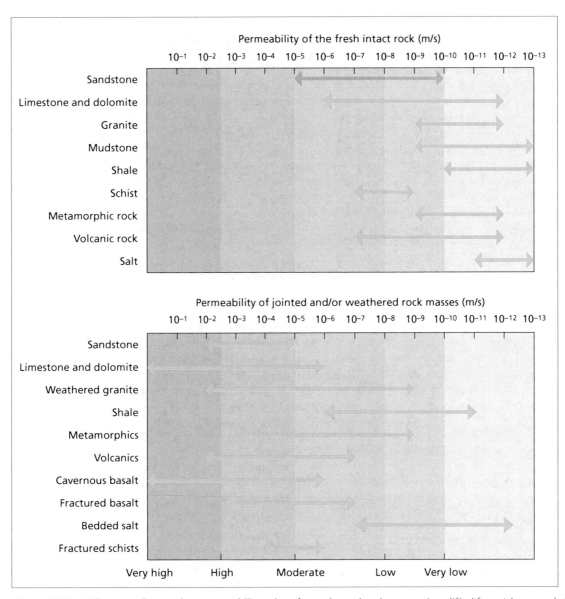

Figure 3.18　Primary and secondary permeability values for rocks and rock masses (modified from Isherwood, 1979; in Hudson and Harrison, 2000).

i.e. superficial weathered areas, and pathways, i.e. major discontinuities and faults. The most important effects of water are that it:

— Transmits a pore pressure that reduces the strength of the intact rock in porous rocks.
— Fills discontinuities in rock masses and similarly affects their strength.
— Causes physical and chemical alteration in intact rock that can weaken it and the rock masses it forms.
— Is an erosive agent.
— Promotes chemical reactions that in turn may change its composition.

The presence of groundwater produces hydrostatic pressure, exerted on the rocks with the same magnitude in all directions.

Water can affect the mechanical behaviour of the two components of the rock mass: intact rock and discontinuities. The significance of groundwater in rocks is normally less at an inter-granular level than in soils, as the strength of intact rock is generally greater than that of soil; but in porous rocks the principle of effective stress holds true and the presence of water reduces the normal stresses acting on the mineral particles. The intact rock strength will therefore be lower if its pores are filled with water. In discontinuities, water exerts hydrostatic pressure and this reduces normal stresses between their walls, reducing their shear strength. In both cases, water present in the pores or discontinuities reduces the strength of the rock mass as a whole. The effects of water on the strength of intact rock and of discontinuities are also covered in Sections 3.4 and 3.5.

Groundwater also reduces the strength of rock masses by internal erosion in soft and clayey materials, transporting fine material and causing voids in the structure of the rock mass. In soluble materials like carbonates and gypsum, dissolution processes lead to the opening up of discontinuities and the creation of cavities.

In soft and clay or sandy-clay rocks, the presence of pore water reduces the cohesive and frictional strength properties and increases deformability. The frictional properties of the discontinuities may also be reduced if these are altered or filled with clay.

## 3.3  Stress and strain in rocks

## Force and stress

Solid mechanics assumes that the behaviour of materials is ideal: homogeneous, continuous, isotropic, linear and elastic. Unlike artificial materials such as steel or concrete (which are made using known ingredients and well defined, repeatable processes and contain known defects), rocks have structural "defects" due to the natural variations in their mineralogical composition and to geological history, which collectively influence their mineral orientation, their porosity and microfissures, or their degree of weathering. At larger scale, rock masses contain different types of discontinuities and weathered or tectonized areas. These characteristics reflect their heterogeneous, discontinuous and anisotropic physical and mechanical properties and determine the mechanical response of both intact rock and rock masses to the forces acting on them.

Applying new forces, or modifying the magnitude or distribution of existing ones, produces changes in the mechanical state of rock and leads to a series of internal effects, including displacement, strain and change in the state of stress. Laboratory tests apply forces to produce failure of the rock material and define its strength and deformational properties.

The **mechanical state** of a system is characterized by:

— The position of each part of the system, defined by its coordinates.
— The forces acting on and between the parts of the system.
— The velocity with which each part changes position.

The difference between two mechanical states will therefore be defined by displacement, strain and changes in the state of stress.

**Displacement** $u$ is the change in position of particle $s$ defined by a vector $u = p' - p$. The displacement field in a system will be homogeneous if the displacement vectors of each particle are equal in magnitude and direction (Figure 3.19).

**Strain** $\varepsilon$ indicates the variation in the distance or space between two particles in two different mechanical states, and is expressed as the ratio of the change in distance to the initial distance between the particles: $\varepsilon = (l_i - l_f) / l_i = \Delta l / l_i$. This parameter is adimensional and compares situations in two different mechanical states.

The **state of stress** of a system is the result of the forces acting on it. If the forces vary, the state of stress associated with the planes under consideration also changes.

**Forces** are the main factors defining the state and mechanical behaviour of any system. Two types of forces act on any rock body (Figure 3.20): gravitational or body force, $F = mg$ (although $g$ depends on the position of the body in the earth's gravitational field, a constant value, $g = 9.80$ m/s$^2$ is assumed), and surface forces, which are exerted on the body by surrounding materials; these act on the contact surfaces between the adjacent parts of the rock system and are transmitted to any point within the body, e.g. tectonic forces exerted on rocks. Both these forces are closely related, as the body forces are conditioned by the distribution and spatial variation of the surface forces.

Surface forces are classified as compressive (positive) and tensile (negative). They are represented respectively by vectors pointing inwards or outwards from the point of application. A force is a vector quantity described by its magnitude, orientation and direction of application.

If a force is exerted on a plane, it may take any direction in relation to the plane; if it is perpendicular it is called normal force, and if it is parallel it is called tangential or shear force. The first type may be compressive or tensile, but not the second. For shear forces, a sign convention has to be defined: normally this is positive if the force vector and the related vector on the other surface of the plane are anti-clockwise, and negative in the opposite case (Figure 3.21).

The effect a force has depends on the total area over which it is applied; using forces in calculations is therefore not appropriate for determining how they influence material

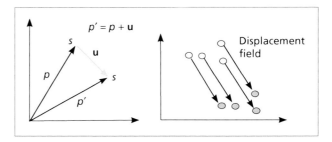

Figure 3.19  Displacement vector and displacement field.

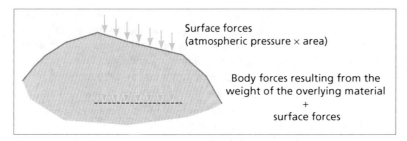

Figure 3.20   Forces acting on a rock body.

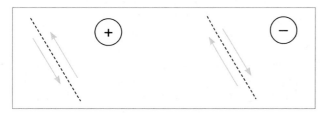

Figure 3.21   Conventional signs for shear forces.

behaviour. If the total force refers to area A of the plane it is acting on, it is expressed as stress, a parameter independent of the area of application: $\sigma = F/A$.

Force is measured in SI or equivalent systems of units: newtons (N) or dynes, kiloponds (kp) or ton-forces (t); the SI unit of stress is $N/m^2$ or Pa.

**Stress** is defined as the intensity of internal forces set up in a body under the influence of applied surface forces; it is a quantity that cannot be directly measured, since the physical parameter measured is force. If the force acts uniformly over a surface, the stress indicates the intensity of the forces that act on the plane.

Stress does not vary depending on the area under consideration if the forces on the surface are uniformly distributed. If this is not the case, stress will vary over different areas of the surface. Taking an infinitesimal area $\Delta A$ inside a body of rock in equilibrium, the magnitude of the resultant stress on the area will be:

$$\sigma = \lim_{\Delta A \to 0} \frac{\Delta F}{\Delta A} = \frac{dF}{dA}$$

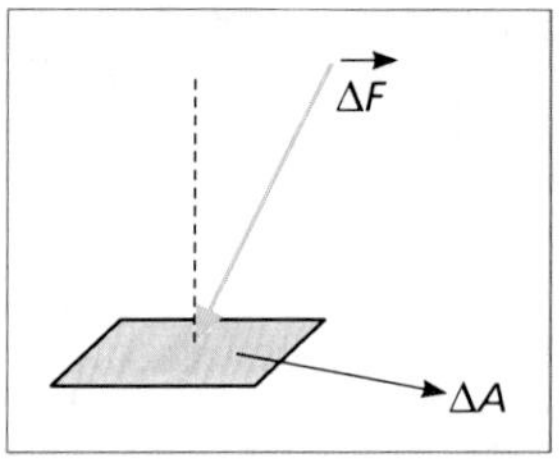

Figure 3.22   Force over area.

As force is a vector quantity, the above can be expressed as a vector equation:

$$\vec{\sigma} = \lim_{\Delta A \to 0} \frac{\Delta \vec{F}}{\Delta A} = \frac{d\vec{F}}{dA}$$

Stress is also a vector quantity as it is the product of a vector, $\vec{\Delta F}$, by a scalar, $1/\Delta A$. The notation $\vec{\sigma}$ represents the magnitude and direction of the vector. The notation $\sigma$ or $|\vec{\sigma}|$ represents magnitude only and is the scalar of $\vec{\sigma}$. The stress vectors can be summed vectorially if they are related to the same plane.

**Stress on a plane** can be fully represented by the stress vector, with a magnitude equal to the ratio of force to the area of the plane and the direction parallel to the direction of the force acting on the plane (Figure 3.23). Just like forces, compressive stresses are positive and tensile stresses are negative.

Stress, like any vector, may be broken down into normal and shear components, $\sigma_n$ and $\tau$, related to any plane; these components depend on the orientation of the plane selected. In the same way, stress may be broken down into two components, $\sigma_x$ and $\sigma_y$, parallel to the x and y axes of an orthogonal coordinate system.

## Stress on a plane

The state of stress at any given point is defined by the force per unit area with reference to two perpendicular planes x, y, through the point.

If a continuous homogeneous material is subjected to a uniform force field over an infinitesimally small square area at rest (Figure 3.24), the resulting stresses on the sides of the square, or in other words the force per unit area exerted by the surrounding material on the sides of the square must be in equilibrium. A normal and a shearing component will act on each side.

If the square is referenced with respect to a system of x and y axes, the stress components on the x plane

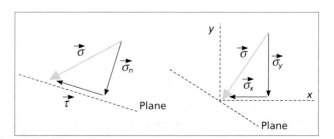

Figure 3.23   Stress on a plane.

(perpendicular to the x axis) are $\sigma_x$ and $\tau_{xy}$, and on the y plane (perpendicular to the y axis) they are $\sigma_y$ and $\tau_{yx}$.

For equilibrium, the resultant of the forces acting in the directions x and y must be equal to zero. In addition, rotational equilibrium requires the moments must also be equal to zero:

$$\tau_{xy} = \tau_{yx} \qquad (1)$$

This means that the state of stress in two dimensions is determined by three components: $\sigma_x$, $\sigma_y$ and $\tau_{xy}$. Although the state of stress does not depend on the direction of the system of coordinates selected, its components do.

Once the state of stress at any given point is known from its $\sigma_x$, $\sigma_y$ and $\tau_{xy}$ components, stresses can be calculated on any plane in any direction that passes through the point. If the state of stress of the plane is determined referring to a randomly chosen axes system, the values of the normal and shearing components will depend on the axes selected.

The orientation of any plane P within the square can be specified with the cosines of the angles between the perpendicular to the plane and the x and y axes. These are the **direction cosines** of unit length perpendicular to P: $l = \cos \alpha$ and $m = \cos \beta$ (Figure 3.25). In other words, the direction cosines of any line that passes through the origin of the coordinate system used are the coordinates of a point situated on the line at a unit distance from the origin. Thus for the perpendicular to a plane parallel to the x axis, the direction cosines will be $l = 0$ and $m = 1$.

## Calculation of x and y stress components on a plane

On a plane AB with perpendicular OP inclined at angle $\theta$ in relation to the axis Ox, the $p_x$ and $p_y$ components of vector p, parallel to OP, can be determined considering the equilibrium of the triangular area OAB, for which the sum of the forces acting in any direction must be zero. The force equilibrium equations are (Figure 3.26):

$$p_x AB = \sigma_x OB + \tau_{yx} OA$$

$$p_x AB = \sigma_x AB \cos\theta + \tau_{yx} AB \sin\theta$$

$$p_x = \sigma_x \cos\theta + \tau_{yx} \sin\theta \qquad (2)$$

Similarly, in direction y:

$$p_y = \sigma_y \sin\theta + \tau_{xy} \cos\theta \qquad (3)$$

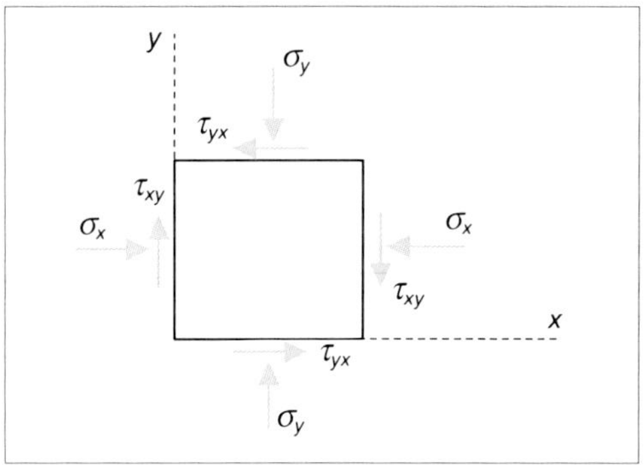

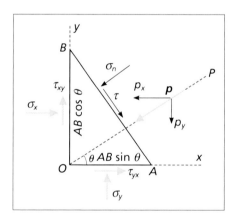

Figure 3.24  State of stress components in two dimensions.

Figure 3.26  Stresses on a plane.

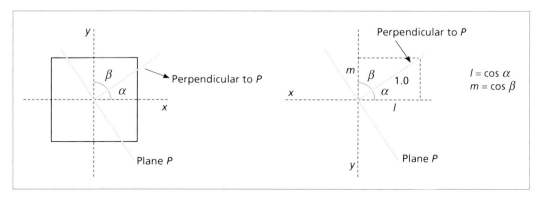

Figure 3.25  Direction cosines.

## Calculation of the normal and shear stresses acting on a plane

Equilibrium equations derived for *Figure 3.26*, as a function of normal and shear stresses, $\sigma_n$ and $\tau$, acting on the plane AB, and considering (1), (2) and (3), will give:

$$\sigma_n = \sigma_x \cos^2\theta + 2\tau_{xy} \sin\theta \cos\theta + \sigma_y \sin^2\theta$$

$$\sigma_n = \frac{1}{2}(\sigma_x + \sigma_y) + \frac{1}{2}(\sigma_x - \sigma_y)\cos 2\theta + \tau_{xy} \sin 2\theta \quad (4)$$

$$\tau = (\sigma_y - \sigma_x)\sin\theta\cos\theta + \tau_{xy}(\cos^2\theta - \sin^2\theta)$$

$$\tau = \frac{1}{2}(\sigma_y - \sigma_x)\sin 2\theta + \tau_{xy} \cos 2\theta \quad (5)$$

The expressions above give the values of the normal and shear stresses on any plane that passes through O (application point of stresses $\sigma_x$, $\sigma_y$, $\tau_{xy}$, $\tau_{yx}$) and whose normal forms an angle $\theta$ with respect to the axis Ox.

## The stress ellipse

If the principal stresses $\sigma_1$ and $\sigma_3$ are parallel to the x and y axes respectively, then equations (2) and (3) will give:

$$p_x AB = \sigma_1 AB \cos\theta$$

and

$$p_y AB = \sigma_3 AB \sin\theta$$

Depending on the direction cosines *l* and *m* of line OP in *Figure 3.26*, the above expressions (for $\alpha = \theta$ and $\beta = 90 - \theta$) are:

$$p_x = \sigma_1 l \quad \text{and} \quad p_y = \sigma_3 m$$

These expressions give the x and y components of the stress on any plane, where *l* and *m* are the direction cosines, and the direction of $\sigma_1$ and $\sigma_3$ coincides with those of the x and y axis respectively. From *Figure 3.25* it can be seen that if a plane parallel to the x-axis, i.e. normal to $\sigma_3$, is rotated through 90° to be parallel to the y-axis and normal to $\sigma_1$, the length of *m* and *l* will change. The vector they define varies from its shortest length when co-incident with $\sigma_3$ to its longest length when co-incident with $\sigma_1$. Their length defines an ellipse the equation for which is:

$$l^2 + m^2 = 1$$

Thus with reference to Fig. 3.26:

$$p_x^2/\sigma_1^2 + p_y^2/\sigma_3^2 = 1$$

where $|\vec{\sigma}_1|$ and $|\vec{\sigma}_3|$ are the largest and smallest radii respectively, and the major and minor axes are parallel to the x and y axes (*Figure 3.27*). Any point on the ellipse has x and y coordinates of the same magnitude as $p_x$ and $p_y$, and any radius vector from the origin to the point is a stress vector *p*.

The ellipse represents the two-dimensional state of stress at a point for the plane containing the principal stresses. On each of the infinitely many planes perpendicular to the ellipse that pass through the centre point, a pair of parallel and opposing stresses, $\sigma_{ab}$ and $\sigma_{ba}$, act obliquely to the plane, which can be decomposed into normal and shearing components. It is only on the two planes perpendicular to the principal stresses, $\sigma_1$ and $\sigma_3$ (the principal planes of stress) that shear stresses do not act. The locus of all these pairs of stresses forms the ellipse and those of greatest and smallest magnitude are the principal stresses $\sigma_1$ and $\sigma_3$.

## Calculating the $\sigma_n$ and $\tau$ components from $\sigma_1$ and $\sigma_3$

When the magnitude and direction of the principal stresses $\sigma_1$ and $\sigma_3$ are known, the normal and shear stresses can be calculated for any plane with a given orientation. For two dimensions, the force equilibrium for the plane in *Figure 3.28* will be:

$$\sigma_n A = \sigma_1 \cos\theta\, A\cos\theta + \sigma_3 \sin\theta\, A\sin\theta =$$

$$= \sigma_1 A \cos^2\theta + \sigma_3 A \sin^2\theta$$

$$\tau A = \sigma_1 \sin\theta\, A\cos\theta - \sigma_3 \cos\theta\, A\sin\theta$$

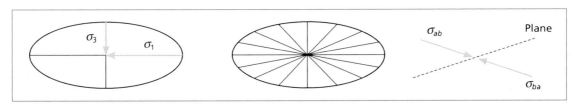

*Figure 3.27* Stress ellipse.

# ROCK MECHANICS

## Box 3.4

### Principal stresses

At any point subjected to stress, three mutually orthogonal planes are found where shear stresses are zero. Such planes are called **principal planes of stress** and the normal stresses acting on them are the **principal stresses.** Of the three stresses, the greatest is $\sigma_1$, the intermediate is $\sigma_2$ and the smallest is $\sigma_3$: $\sigma_1 > \sigma_2 > \sigma_3$. Assuming that stresses at a point are due only to gravitational forces, the horizontal plane and all the vertical planes that pass through that point will be principal planes of stress. If $\sigma_1 = \sigma_2 = \sigma_3$, the state of stress is called isotropic or hydrostatic, the same as for fluids.

All self-supporting walls of surface and underground excavations, are principal planes of stress on which shear stresses do not act.

In contrast to what occurs with shear stress, there is no orientation in space for a surface on which the normal stress is zero; in other words, the sum of the principal stresses on a body under unchanging conditions always has the same value regardless of the orientation of the principal stresses: $\sigma_1 + \sigma_2 + \sigma_3 = $ constant.

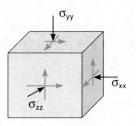
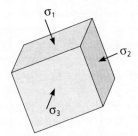

Stress components referred to a system of x, y and z axes and principal stress components.

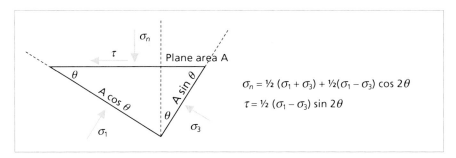

*Figure 3.28* Normal and shear stresses on a plane.

Applying trigonometric identities:

$$\sigma_n = \frac{1}{2}(\sigma_1 + \sigma_3) + \frac{1}{2}(\sigma_1 - \sigma_3)\cos 2\theta \qquad (6)$$

$$\tau = \frac{1}{2}(\sigma_1 - \sigma_3)\sin 2\theta \qquad (7)$$

Equations (6) and (7) provide a complete description of the state of stress on a plane, where the angle $\theta$ and principal stresses are known. The maximum shear stress is $(\sigma_1 - \sigma_3)/2$, found on sections at 45° to the principal planes. Maximum normal stress and maximum shear stress are exerted on sections at 45° to each other.

### Mohr's circle

Equations (6) and (7) define a circle in $\tau - \sigma_n$ space. This graphic representation of the state of stress at a given point is called Mohr's Circle (*Figure 3.29*). The principal stresses, $\sigma_1$ and $\sigma_3$, define the position and diameter of the circle on the $\sigma_n$ axis. Any point on the circumference of the circle represents the state of $\tau$ and $\sigma_n$ stress on a plane whose normal forms the angle $\theta$ with the direction of the major principal stress $\sigma_1$. The radius of the circle represents the maximum value of the shear stress $\tau$. From the diagram, given the stresses $\sigma_1$ and $\sigma_3$, the values of $\sigma_n$ and $\tau$ can be calculated graphically for any plane; in the same way, the magnitude and direction of the principal stresses can be obtained from $\sigma_n$ and $\tau$ (*Figure 3.30*).

Mohr's circle allows different state of stress to be represented, as shown in *Figure 3.31*.

### The effect of pore pressure

Water exerts hydrostatic pressure, $u$, of the same magnitude in all directions. If water is present in the rock, this stress counteracts the perpendicular stress component but has no effect on the shear component. This means that the effective stress acting perpendicularly to a plane will be the total stress minus the stress $u$:

$$\sigma'_n = \sigma_{n\,total} - \sigma_{water} = \sigma_n - u$$

In the Mohr diagram, the effect of this is that the stress circles move to the left a distance equal to the value of the pore stress or pressure $u$ (*Box 3.5*).

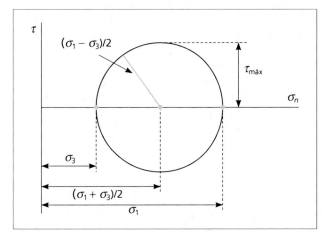

Figure 3.29    Mohr's circle of stress.

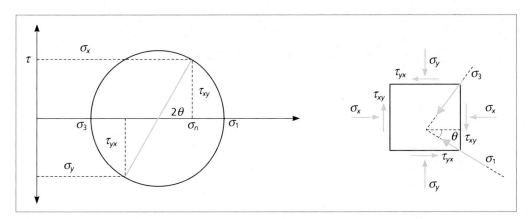

Figure 3.30    Diagram with the Mohr's circle showing the stresses acting on the vertical plane of the figure on the right (orthogonal to the plane of the page), whose normal makes the angle $\theta$ with the direction of $\sigma_1$.

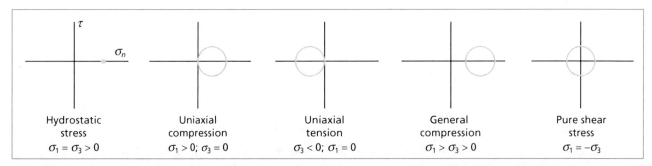

Figure 3.31:   Mohr's circles for different states of stress.

# ROCK MECHANICS

## Box 3.5

### Graphical and analytical methods for calculating normal and shear stresses acting on a plane

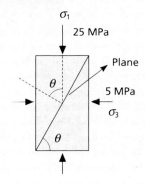

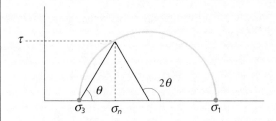

### Method a)

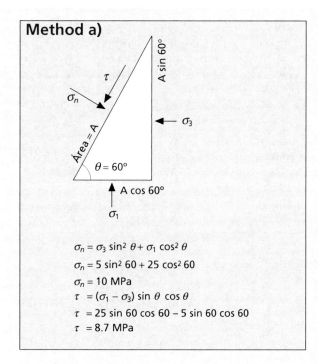

$\sigma_n = \sigma_3 \sin^2 \theta + \sigma_1 \cos^2 \theta$
$\sigma_n = 5 \sin^2 60 + 25 \cos^2 60$
$\sigma_n = 10$ MPa
$\tau = (\sigma_1 - \sigma_3) \sin \theta \cos \theta$
$\tau = 25 \sin 60 \cos 60 - 5 \sin 60 \cos 60$
$\tau = 8.7$ MPa

If $\sigma_{water} = 2$ MPa

$\sigma'_1 = 23$ MPa
$\sigma'_3 = 3$ MPa
$\sigma'_n = 8$ MPa

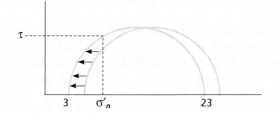

## Method b)

Graphical construction of Mohr's circle:

$\sigma_n = 10$ MPa  and  $\tau = 8.7$ MPa

## Method c)

From the expressions:

$\sigma_n = \tfrac{1}{2}(\sigma_1 + \sigma_3) + \tfrac{1}{2}(\sigma_1 - \sigma_3) \cos 2\theta = 10$ MPa
$\tau = \tfrac{1}{2}(\sigma_1 - \sigma_3) \sin 2\theta = 8.7$ MPa

If there is fluid pore pressure inside the rock, the fluid will carry part of the stress applied to cause failure. The "effective" stress on the solid particles in the rock will be the difference between the total stress applied and that supported by the fluid:

$\sigma'_1 = \sigma_{1\,total} - \sigma_{water}$
$\sigma'_3 = \sigma_{3\,total} - \sigma_{water}$
$\sigma'_n = \sigma_{n\,total} - \sigma_{water}$

## Stress in three dimensions

If instead of a plane, where the stress is defined by a vector, a point inside a rock body is considered, then infinite planes with different orientations will pass through it. If the stress vectors for each of these planes is determined, this will define the state of stress for that point, represented by a second-order tensor.

In other words, to quantify the state of stress at a given point its state of stress must be defined, by defining the forces per unit area acting on three orthogonal planes passing through the point. The state of stress is not altered by the choice of reference system of axes but its components are.

If an infinitesimal area $\Delta A$ is taken around a point O inside a rock mass in equilibrium, and $\Delta F$ is the resultant force acting on the plane (Figure 3.32), the magnitude of the resultant stress on point O, or the stress vector, $\sigma_R$, is given by:

$$\sigma_R = \lim_{\Delta A \to 0} \frac{\Delta F}{\Delta A}$$

Its normal and tangential components on the plane containing the point are given by:

$$\sigma_n = \lim_{\Delta A \to 0} \frac{\Delta N}{\Delta A} \quad \tau = \lim_{\Delta A \to 0} \frac{\Delta T}{\Delta A}$$

If the normal to the $\Delta A$ surface is oriented parallel to one of the axes, e.g. to the x axis, the stress components acting on that surface can be referred to the x, y and z axes. Unlike normal stress, tangential stress is not so readily defined as generally it does not coincide with the direction of any of the axes and so it must be referred to two components. The plane stress here is therefore given by three components:

$$\sigma_{xx}, \tau_{xy} \text{ and } \tau_{xz}$$

with the first subscript indicating the direction of the normal to the plane (or the plane on which the component acts), and the second the direction in which the stress component acts.

In the same way, for the other two directions y, z, the stress components acting on the planes normal to them are:

$$\sigma_{yy}, \tau_{yx} \text{ and } \tau_{yz}$$
$$\sigma_{zz}, \tau_{zx} \text{ and } \tau_{zy}$$

The stress matrix with the nine components is defined by:

$$[\sigma] = \begin{bmatrix} \sigma_{xx} & \tau_{xy} & \tau_{xz} \\ \tau_{yx} & \sigma_{yy} & \tau_{yz} \\ \tau_{zx} & \tau_{zy} & \sigma_{zz} \end{bmatrix}$$

The state of stress at a point is defined by nine independent stress components, 3 normal and 6 tangential. If the equilibrium of the cube in Figure 3.32 is considered, the following conditions must be met:

$$\tau_{xy} = \tau_{yx}, \quad \tau_{xz} = \tau_{zx}, \quad \tau_{yz} = \tau_{zy}$$

so that only six stress components are needed to determine the state of stress at any given point:

$$[\sigma] = \begin{bmatrix} \sigma_{xx} & \tau_{xy} & \tau_{xz} \\ \tau_{yx} & \sigma_{yy} & \tau_{yz} \\ \tau_{zx} & \tau_{zy} & \sigma_{zz} \end{bmatrix}$$

The stress tensor corresponding to the principal stresses is:

$$[\sigma] = \begin{bmatrix} \sigma_1 & 0 & 0 \\ 0 & \sigma_2 & 0 \\ 0 & 0 & \sigma_3 \end{bmatrix}$$

If there is pore pressure, $u$, only the perpendicular components of the tensor will be modified since hydrostatic pressure does not act on tangential components; if tangential

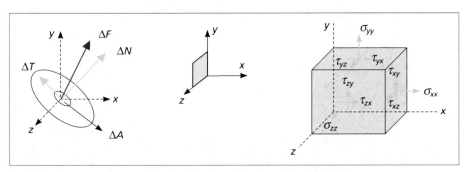

Figure 3.32   Stresses in three dimensions.

components or principal stresses are present, the stress tensors will be:

$$[\sigma] = \begin{bmatrix} \sigma_{xx} - u & \tau_{xy} & \tau_{xz} \\ \tau_{yx} & \sigma_{yy} - u & \tau_{yz} \\ \tau_{zx} & \tau_{zy} & \sigma_{zz} - u \end{bmatrix}$$

$$[\sigma] = \begin{bmatrix} \sigma_1 - u & 0 & 0 \\ 0 & \sigma_2 - u & 0 \\ 0 & 0 & \sigma_3 - u \end{bmatrix}$$

### Stress ellipsoid

The three-dimensional state of stress at a given point is represented by an ellipsoid. In the same way as the equations for the two-dimensional stress ellipse were deduced above, if the principal stresses $\sigma_1$, $\sigma_2$ and $\sigma_3$ are considered parallel to the x, y, z axes it can be written as:

$$l = p_x/\sigma_1, \quad m = p_y/\sigma_2, \quad n = p_z/\sigma_3$$

and as:

$$l^2 + m^2 + n^2 = 1$$

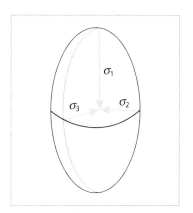

Figure 3.33    Stress ellipsoid.

it follows that:

$$p_x^2/\sigma_1^2 + p_y^2/\sigma_2^2 + p_z^2/\sigma_3^2 = 1$$

The three planes cut in the ellipsoid in Figure 3.33 are the principal stress planes. The planes containing the $\sigma_1$ and $\sigma_2$ stresses and the $\sigma_1$ and $\sigma_3$ stresses respectively are the two ellipses representing the state of stress on any plane perpendicular to the ellipse shown.

The different state of stress can be defined by the shape of the ellipsoid or by the relative values of the stresses acting on a point at its centre:

**Shape of the ellipsoid**

Uniaxial: $\sigma_1 \neq 0$; $\sigma_2 = \sigma_3 = 0$
Biaxial: $\sigma_1 \neq 0$; $\sigma_3 \neq 0$; $\sigma_2 = 0$
Triaxial: $\sigma_1 \neq 0$; $\sigma_2 \neq 0$; $\sigma_3 \neq 0$

**Relative stress value**

Axial: $\sigma_1 > \sigma_2 = \sigma_3$
Polyaxial: $\sigma_1 \neq \sigma_2 \neq \sigma_3$
Hydrostatic: $\sigma_1 = \sigma_2 = \sigma_3$

Hydrostatic stress is represented by a sphere. Figure 3.34 shows these states of stress in laboratory specimens.

## Strength and failure

### Basic concepts

Stresses generated by the application of forces on rocks may produce deformation and failure, depending on the strength of the rocks and on other conditions external to the rock material itself.

Deformation indicates a change in the shape or configuration of a body as a result of displacements of the rock particles when supporting the load. As very small displacements are difficult to measure, deformation is expressed by comparing the deformed and original states, and thus has no units. Longitudinal strain or elongation, $\varepsilon$, was defined above as the difference in length between two particles in two different mechanical states, expressed as:

$$\varepsilon = (l_i - l_f)/l_i = \Delta l / l_i$$

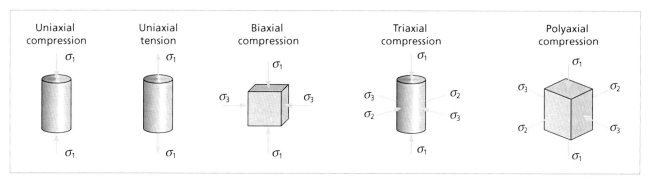

Figure 3.34    Different states of stress applied to laboratory specimens.

Volumetric strain or dilation is the ratio between the change in volume of a body and its original volume:

$$\Delta = (V_i - V_f)/V_i = \Delta V/V_i$$

While stress refers to the condition of the rock at a given moment and depends on the forces applied, strain compares conditions at two times and refers only to the configuration of the bodies.

Strength is defined as the stress that the rock can bear under a given set of deformation conditions. **Peak strength**, $\sigma_p$, is the maximum stress the rock can sustain (*Figure 3.35*) and is reached for a specific strain known as peak strain. **Residual strength**, $\sigma_r$, is the lowered value of the strength of certain rocks when there is considerable post-peak strain, after peak strength is exceeded. It is difficult to analyse and predict the deformations that the rock will reach before either peak strength or residual strength is reached; this is an important problem in geological engineering with major practical consequences.

In natural conditions, strength depends on the intrinsic properties of the rock, its cohesion and angle of friction, and on external factors, such as the magnitude of the stresses acting on it, the loading and unloading cycles it has sustained and the presence of water. For this reason, **strength is not a single intrinsic value** and so it is important to know the range of values and variations for rock materials under certain conditions.

In intact rock, compressive strength is the most characteristic and most frequently measured property because it is easy to obtain samples and carry out laboratory tests. In contrast, in rock masses strength cannot be measured directly, because of the great difficulties in carrying out *in situ* large scale testing, and either empirical criteria or numerical modelling must be used.

When a certain load or force is applied, if the stress generated exceeds the strength of the rock material, unacceptable deformation is reached and failure occurs.

**Failure** occurs when the rock can not sustain the forces applied, and the stress reach a maximum value corresponding to the peak strength of the material. Although it is generally assumed that failure occurs when peak strength is reached, this is an over-simplification which is not always true. Similarly, rock failure does not necessarily coincide with the initiation of fracture planes. **Fracture** is the formation of planes of separation within the rock as the cohesion between particles is broken and new surfaces are formed. In the process, cohesive forces are lost and only frictional forces remain.

Failure may occur in different ways depending on the rock strength and the relationships between applied stresses and the resulting strain: it is generally described as being either **brittle failure** (instant and violent) or **ductile failure** (progressive). These concepts are dealt with later in this chapter.

The failure phenomenon is accompanied by the generation of fracture surfaces through the rock. The direction of these will depend on:

— The direction in which the forces are applied.
— The anisotropies present in the rock material either at microscopic level (preferred orientation of the minerals, presence of oriented microcracks) or at macroscopic level (schistosity or lamination surfaces).

In a jointed rock mass, failure may take place through the intact rock or along pre-existing discontinuities; a mixture of both may also occur.

## Failure mechanisms

Failure in rocks is a varied and complex process in which different types of phenomena and numerous factors are involved acting together. Analysis of failure is more complex in rocks than in soils; when rock is subjected to compressive stresses, microscopic tension cracks and shear planes are produced which advance inside it. The different failure modes in rocks are briefly described below; *Figure 3.36* illustrates some examples at a rock mass scale.

**Shear stress failure.** This takes place when a specific rock surface is subjected to shear stresses that are high enough to cause one face of the surface to slip relative to the other. Examples are failure along discontinuities in rock slopes (*Figure 3.36a*) and collapse of a tunnel roof along vertical discontinuities bounding stiff side walls. It is the most common type of failure and the most important.

**Flexure failure.** This occurs when a section of rock is subjected to a series of variable stresses, failing in the area

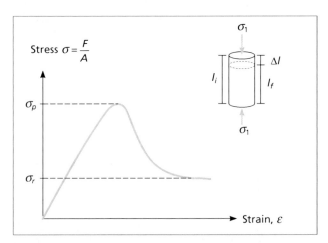

*Figure 3.35* Complete stress-strain curve showing peak strength and residual strength values.

where tensional stresses accumulate. This can occur in the capping beams of underground galleries or in the roofs of karstic cavities (Figure 3.36b).

**Direct tension failure.** This type of failure takes place when a certain section of the rock is subjected to pure or almost pure tension due to the configuration and/or structure of the rock mass. In fact, such failure by itself occurs very seldomly. An example would be the tensional state of stress generated in some parts of the larger failure surface of a slope (Figure 3.36c).

**Compression failure.** Uniaxial compression does not often occur naturally or in engineering projects; an example of this would be the failure of support pillars in a mining excavation (Figure 3.37).

**Failure from collapse.** Failure brought about by mechanical collapse occurs when rock fabric fails and the rock is transformed into a soil-like material. This particular type of failure can occur in very porous rocks, such as low-density volcanic rocks, cemented chalky sandstones and similar weak materials.

## Stress-strain relationships in rock

The **stress-strain behaviour** of a body is defined by the relationship between the stress applied and the resulting deformation, usually expressed as strain, and describes how it is deformed and how the behaviour of the rock material varies as the load is applied; in other words, how the strength of the material varies for specific amounts of deformation, considering:

— Behaviour before failure is reached.
— How the failure occurs.
— Behaviour after failure.

All these conditions are studied with laboratory tests where compressive forces are applied and the stress-strain curves are recorded. The rocks show non-linear relationships between the forces applied and the resulting deformation at

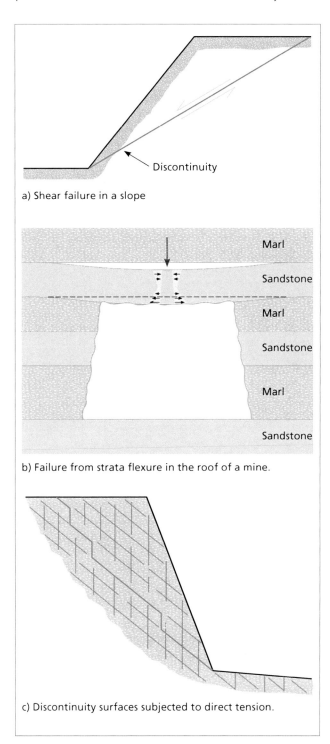

Figure 3.36   Failure mechanisms.

Figure 3.37   Compression failure in the pillar of a early Christian church excavated in volcanic tuff, Cappadocia, Turkey.

specified stress levels, and different types of the σ-ε curve are obtained for different types of rock.

If peak strength is exceeded (that is, if strain exceeds peak strain), the following may occur:

— Rock strength is drastically reduced and may even fall to values approaching zero; this represents **brittle behaviour**, which occurs e.g. in glass (*Figure 3.38*, curve 1). This behaviour is typical of hard, high-strength rocks. Brittle behaviour implies an almost instantaneous loss of strength in the rock with little or no plastic deformation.
— The strength of the rock decreases to a certain value, after significant deformation is reached. This is **brittle-ductile** or partly brittle **behaviour** (*Figure 3.38*, curve 2), which occurs when the components of fabric can be rearranged e.g. as when shearing rock discontinuities or over-consolidated clays.
— Deformation continues to increase with no loss of strength (strength remains constant after large-scale deformation). This is known as **ductile behaviour** (*Figure 3.38*, curve 3) and is seen in certain materials where fabric is continually remobilised during failure, e.g. as in rock salt.

In ductile behaviour, peak and residual strength are the same. When deformation occurs with no further change in strength it is called ductile deformation. Brittle behaviour is characterized by significant difference between peak and residual strength and as there is a sudden reduction in strength, there is practically no difference between peak deformation and the deformation corresponding to the residual strength.

If an axial compressive force is gradually applied to an unconfined rock specimen in a laboratory test it will produce an axial deformation that can be measured by gauges installed on the specimen and converted to axial strain. The stress-strain curve can be drawn from the corresponding log of stresses and deformations during the test (*Figure 3.39*). Before reaching peak strength, the rising part of the curve usually shows linear or elastic behaviour for most rocks. In the elastic domain, strain is proportional to stress and satisfies the ratio:

$$E = \sigma/\varepsilon_{ax}$$

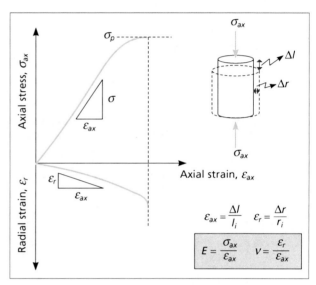

Figure 3.39  Stress-strain curves obtained from the uniaxial compression test.

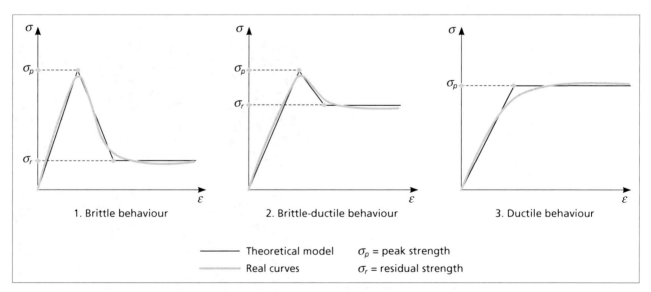

Figure 3.38  Stress-strain behaviour models.

where $E$ is the constant of proportionality known as **Young's modulus** or modulus of elasticity, $\sigma$ is the stress and $\varepsilon_{ax}$ is the axial strain (i.e. in the same direction as the force applied).

**Poisson's ratio** is another constant that defines the elastic behaviour of rock material:

$$\nu = \varepsilon_r / \varepsilon_{ax}$$

where $\varepsilon_r$ is the radial strain of the rock specimen being tested.

Both these elastic constants are obtained from the uniaxial compressive strength test described in Section 3.4.

If the force causing elastic deformation on a specimen is removed, it will return to its original configuration (Figure 3.40). Beyond a certain level of deformation the rock cannot sustain elastic behaviour and a point is reached at which ductile or plastic deformation starts to occur and the linear relationship between stress and strain is no longer valid. This is the **yield point**, which is shown in the inflexion of the stress-strain curve, and the corresponding strength is written as $\sigma_y$ (it should not be confused with the normal stress component along the $y$ axis, $\sigma_{yy}$). Beyond this point the rock may still withstand considerable deformation before its ultimate strength is reached. In brittle rocks, the $\sigma_y$ and $\sigma_p$ values are either very close or coincide, but this does not occur in rocks with ductile behaviour (Figure 3.40). The difference between these values is very important for studying the behaviour of certain types of rock as it indicates the capacity of the rock to continue withstanding loads once it exceeds its elastic limit and before inadmissible deformations occur.

Once the elastic limit is exceeded, deformation of the rock does not recover even if the entire applied load is removed.

It is important to know the value of $\sigma_y$ and the deformations associated with this stress level when the design and planning of engineering work and structures involves weak or soft rocks, which may undergo non recoverable plastic deformations for stresses below peak strength. Once beyond yield even a small incremental increase in load may cause progressive failure of the material. Even if the load remains constant, weathering processes and the passage of time may cause permanent loss of strength.

Elastic or plastic behaviour depends on the intrinsic strength properties of the rock and the conditions under which stresses are applied, especially the value of the confining pressure and temperature, and the presence of pore water.

The effect of confining pressure $\sigma_3$ on rock may cause its behaviour to change from brittle to ductile. The value of $\sigma_3$ for which this variation is produced is called brittle-ductile transition pressure; once this is reached, the rock behaves plastically and deforms without any further increase in stress. This transition pressure is very high for the ranges of stress at the usual depths where engineering work takes place, but for some clays and evaporitic rocks, this pressure is quite low (<20 MPa at ambient temperature; Goodman, 1989).

What has been explained so far refers to models of behaviour which can be reproduced in the laboratory to study the "immediate" or short term deformational response of the rock to conditions of applied stress; the influence of time on the long term behaviour of the rock under certain stress or strain conditions is not considered. However, some kinds of rock may exhibit rheological behaviour and undergo time-dependent processes such as **creep** (increasing strain under constant stress) and **relaxation** (decreasing stress under constant strain).

This is very important because material can evolve and even fail under constant, maintained, long term load or strain conditions. The most illustrative example of creep is in

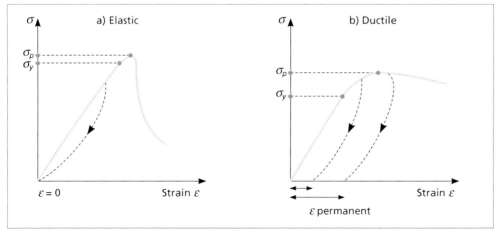

Figure 3.40   Model curves showing elastic behaviour, with recoverable deformations once the load is removed, and ductile behaviour, with permanent deformation once the yield point is exceeded.

evaporites e.g. rock salt. In creep processes, the material shows viscous behaviour, i.e. slow, continuous, time-dependent deformation which is also influenced by the moisture content. Many rocks exhibit visco-elastic behaviour when stress is applied, through a combination of instantaneous (elastic) and longer term rheological deformations.

*Figure 3.41* shows a theoretical model of a creep curve in which the different deformation phases are shown as a function of time.

When the load is applied initially, immediate elastic deformation is produced, followed by primary creep (I), in which deformation slows down with time if conditions remain constant (transient creep). In some rocks, the primary creep curve may evolve into so-called secondary creep (II), where there is a gradual increase in deformation until its range becomes constant (steady state creep). If the acting stresses are close to the peak stress, secondary creep may develop into tertiary creep (III), where the range of deformation increases with time, until failure eventually takes place (accelerating creep).

Creep can be generated by either flow or cracking mechanisms operating at a microscopic level. Apart from evaporitic salts, other materials may also show this rheological behaviour and creep may occur under high pressures or temperatures that remain steady over prolonged periods of time, as occurs in deep mines and tunnels. Overconsolidated mudstones or metamorphic shales may suffer deformation from creep along planes of weakness and/or deterioration when exposed to changing environmental conditions. Evaporitic rocks and compacted mudstones undergo creep when subjected to relatively low levels of stress. Also, poorly cemented hard rocks with low porosity may present primary creep due to micro-cracking.

## Strength criteria

The mechanical behaviour of rock mainly depends on its strength and on the forces applied. These forces create a particular state of stress, defined by the acting principal stresses: $\sigma_1$, $\sigma_2$ and $\sigma_3$. Deformation and/or failure of the rocks depend mainly on the magnitude and direction of these stresses. Rocks fracture under conditions of differential stress, and a specific relationship between the principal stresses produces a specific state of deformation. If these stress-strain relationships are known, how the material will behave under a given state of stress can be predicted.

In theory, this prediction could only be made if the rocks and the rock masses were isotropic and homogeneous, but even rocks that appear to be isotropic and homogeneous, such as granites, have variable physical and mechanical properties, and a behaviour that is difficult to predict.

The **behaviour law** of a material is defined as the relationship between the stress components that indicate the state of strain that the material undergoes. It is a broader concept than that of strength or failure criterion since it refers to the relationship between the stresses throughout the whole process of deformation of rock.

In theory, a behaviour law should lead to the following predictions:

— Peak strength value of the material.
— Residual strength.
— Yield point strength.
— The point at which generation of the fracturing begins.
— Deformations of the material.
— Energy of the deformation and failure process.

As it is practically impossible to obtain the specific laws that govern the behaviour, strength and failure of specific rock materials (both intact rock and rock masses), a series of **strength criteria** are used that are obtained empirically from laboratory tests and other practical experience. These criteria are mathematical expressions representing simple models which allow the strength of the material to be estimated on the basis of the stresses applied and the strength properties operating, in order to predict when failure will occur:

$$\text{strength} = f(\sigma_1, \sigma_2, \sigma_3, K_i)$$

where $\sigma_1$, $\sigma_2$ and $\sigma_3$ are the principal stresses in three directions in space and $K_i$ is a representative set of parameters for the rock.

A peak strength criterion is therefore an expression providing the combination between the stress components at which peak strength of the material is reached, and a plasticity criterion or elasticity limit criterion is the ratio between the stress components that is satisfied at the onset of permanent deformations.

Failure criteria are based on stresses because it is easier and quicker to measure these than other parameters,

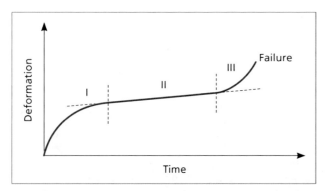

*Figure 3.41* Time dependent deformation curve of creep.

## Box 3.6

## Models for stress-strain behaviour in rocks

Rock behaviour can be classified as:

— **Brittle**, with elastic deformation; typical of strong hard rocks.
— **Brittle-ductile**, with elastic and irrecoverable plastic deformations; typical of soft weak rocks.
— **Ductile**, with predominantly plastic (permanent) deformations; e.g. rock salt.

These behavioural models are reflected in the deformation and failure mechanism shown by rock specimens when loaded. While brittle rocks fail along clearly-defined failure planes and generally instantaneously, ductile materials fail gradually, with a greater deformation range, generating numerous failure planes. Any rock can exhibit brittle, brittle-ductile or ductile behaviour depending on the period over which it is loaded and the magnitude of the stresses involved. The examples of rocks likely to behave in this manner refer to their behaviour in normal civil and mining engineering environments.

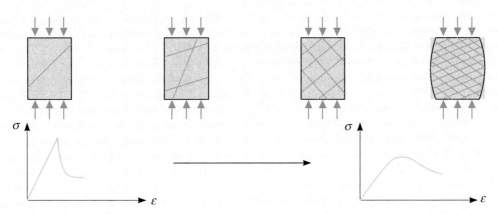

Brittle behaviour. Elastic deformation → Ductile behaviour. Plastic deformation

Stress-strain curves showing brittle and ductile behaviour obtained from uniaxial compression tests with servo-controlled testing machine in Carboniferous shales.

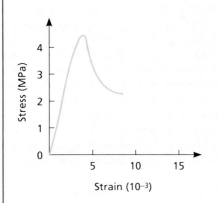
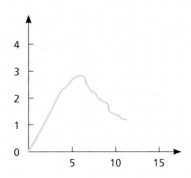
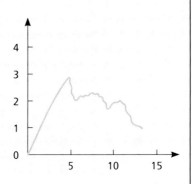

such as strain ($\varepsilon$) or strain energy released during the loading process; however, failure criteria can be based on these parameters if the relevant quantities can be measured:

$$\text{strength} = f(\varepsilon_1, \varepsilon_2, \varepsilon_3, K_i)$$

The use of servo-controlled testing machines allows strength tests to be carried out with a control variable other than stress (as described in Section 3.4).

As well as considering the principal stresses, strength criteria can also be established in terms of the normal and shear stresses acting on a plane (*Figure 3.42*):

$$\sigma_1 = f(\sigma_2, \sigma_3, K_i) \quad \text{or} \quad \tau = f(\sigma_n, K_i)$$

Because intermediate stress $\sigma_2$ has little influence on the peak strength of the materials with respect to minimum stress $\sigma_3$, the criteria in general are expressed as follows:

$$\sigma_1 = f(\sigma_3)$$

For porous rocks, strength or failure criteria should be established in terms of effective stress, given that this is what controls the stress-strain behaviour.

For the **peak strength** of the material, the failure criterion most widely used in rock mechanics is the linear failure criterion proposed by Coulomb at the end of the eighteenth century (*Figure 3.43*):

$$\tau = c + \sigma_n \tan \phi$$

where $c$ is the cohesion and $\phi$ the internal friction angle, which are parameters of rock material defined in Section 3.4. This criterion expresses the shear strength along a plane in a triaxial state of stress, giving the linear relationship between normal and shear stresses acting at the moment of peak or ultimate failure.

Unlike soils, the mechanical behaviour of rocks is non-linear; therefore, despite having the advantage of simplicity, linear failure criteria are not entirely appropriate for rocks in that they may give erroneous data when assessing the state of strain in the rock, especially for low stress. The Mohr-Coulomb criterion does not fit the real behaviour of the rock material either in intact rock, rock masses or discontinuities. Tests have shown that rock mass strength increases less with the increase in normal confining pressure than what is obtained by applying a linear law.

Therefore, non-linear failure criteria are more suitable in rock mechanics. Incorrect use of the linear model may give rise to serious errors, by overvaluing or undervaluing the mechanical capacities of the rock material.

*Figure 3.44* shows the envelopes corresponding to a linear and a non-linear failure criterion. At point 1, the corresponding state of stress shows stable conditions (no-failure) for both criteria. If it is assumed that pore pressure exists, the state of stress will be displaced to position 2, which will continue to be stable if the criterion adopted is linear; for the non-linear envelope, however, the state of stress at point 2 is not admissible, indicating that the rock strength has been exceeded.

The most widely used criteria in rock mechanics for the limit of elasticity, or **plasticity criteria**, include those of Drucker-Prager, Von Mises and Tresca, with different expressions which relate the stresses acting at the moment

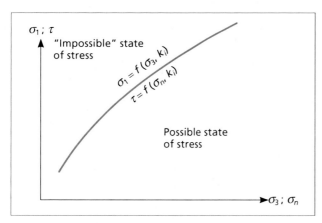

Figure 3.42   Diagram of a general failure criterion in two dimensions.

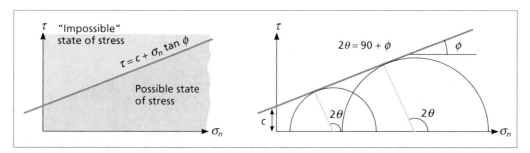

Figure 3.43   Mohr-Coulomb linear failure criterion.

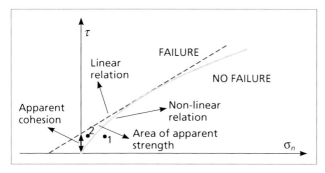

Figure 3.44 Diagram showing linear and non-linear failure criteria. When extrapolated to low stresses, the linear criterion gives a zone of apparent strength and an apparent cohesion value.

permanent or plastic deformations of the material occur. Plasticity criteria are usually expressed in terms of invariants of stress or deviatoric stresses, as the plastification process in isotropic materials is independent of the axes considered.

Strength or failure criteria can be expressed for intact rock, for discontinuity planes or for rock masses. Those most widely used in rock mechanics are described in the sections that follow.

## 3.4 Strength and deformability of intact rock

### Strength and strength parameters

The mechanical behaviour of rocks is defined by their strength and deformability. **Strength**, as already defined, is the resistance of a rock to deformation under a specific stress regime. The strength measured in unconfined rock specimens is called **uniaxial compression strength** (UCS) and its value is used in geotechnical rock classifications. The intact rock strength depends on the rock type and characteristics such us the mineralogical variation, grain size distribution, micro-cracks, within each rock type. Table 3.6 shows typical values for this parameter in different types of rock obtained from the uniaxial compression test (described below). Laboratory tests on brittle rocks generally give higher results than the real strength values.

Fracturing of the intact rock is a complex process. Rocks fail when fractures planes are generated, as the peak strength is exceeded, and slide past each other. So compression tests indirectly measure the **shear strength** of rocks. Fracture planes might be expected to develop in a direction parallel to the application of load; however, the lowest strength is obtained for the direction in which the greatest shear stress is exerted, forming a specific angle in relation to the direction of the applied load. According to the Mohr circle, failure in isotropic rocks will occur when $2\theta = 90° + \phi$ or $\theta = 45° + \phi/2$ (Figure 3.43), although this expected result is not always obtained in laboratory tests. In the case of isotropic intact rock, in theory the compressive strength will always be the same if the same state of stress is applied under the same environmental conditions of water pressure and temperature.

Strength depends on the cohesive and frictional forces of the rocks. **Cohesion**, c, is the bonding force between the mineral particles making up the rock. The **angle of internal friction**, $\phi$, between two surfaces of the same rock is the minimum angle of inclination of a surface which causes a superimposed block of similar material to slide; for most rocks, this angle varies between 25° and 45°. The frictional force depends on the angle of friction and on the normal stress, $\sigma_n$, acting on the plane in question.

Rock strength is not a single value; apart from its c and $\phi$ values, it depends on other conditions, such as the magnitude of the confining stresses, the presence of pore water or the rate of load application. Even in apparently isotropic and homogeneous rocks, the c and $\phi$ values may vary, depending on the level of cementation or variations in mineral composition.

Table 3.13 shows characteristic cohesion and friction values for intact rock. Both parameters are determined from the triaxial compression laboratory test, described later in this section.

### Effects of anisotropy and pore pressure on strength

When rock is anisotropic, its compressive strength for the same state of stress varies according to the angle $\beta$ ($\beta = 90-\theta$) between the planes of anisotropy and the direction of the load applied; as a result, values may differ considerably (Figure 3.45).

The direction most favourable to failure corresponds to the angle $\theta = 45° + \phi/2$, so the rock strength will be minimum if the planes of weakness have this orientation. Similarly, strength will be maximum for orientations of $\theta = 90$ and $\theta = 0°$, where shear stresses are zero. The strength value will vary for the $\theta$ angle values lying between these extremes. Figure 3.46a) shows the theoretical strength curve of an anisotropic rock, with the curved section corresponding to failure along the planes of weakness and the straight section corresponding to failure through the rock material. Figure 3.46b) illustrates real curves obtained in the laboratory for different $\theta$ angle values.

| Table 3.13 | TYPICAL VALUES OF c AND φ FOR FRESH ROCK | |
|---|---|---|
| Rock | Cohesion c (MPa) | Basic friction angle $\phi_b$ (degrees) |
| Andesite | 28 | 45 |
| Basalt | 20–60 | 48–55 |
| Diabase | 90–120 | 40–50 |
| Diorite | 15 | 50–55 |
| Dolomite | 22–60 | 25–35 |
| Gabbro | 30 | 35 |
| Gneiss | 15–40 | 30–40 |
| Granite | 15–50 | 45–58 |
| Greywacke | 6–10 | 45–50 |
| Gypsum | – | 30 |
| Limestone | 5–40 | 35–50 |
| Marble | 15–35 | 35–45 |
| Marly limestone | 1–6 | 30 |
| Mudstone | 3–35 | 40–60 |
|  |  | 15–25* |
| Quartzite | 25–70 | 40–55 |
| Sandstone | 8–35 | 30–50 |
| Schist | 25 | 25–30 |
|  | 2–15* | 20–30* |
| Shale | 10–50 | 40–55 |
|  | <10* | 15–30* |
| Tuff | 0.7 | – |

(*) In bedding or foliation planes.
Data taken from Walthan (1999), Rahn (1986), Goodman (1989), Farmer (1968), Jiménez Salas and Justo Alpañés (1975).

This variability of the compressive strength in intact rock means that a representative value for $\sigma_1$ cannot be assigned with complete certainty. The minimum value is often taken to provide a safety margin; however, where there is no risk of failure along the planes of anisotropy, an appropriate strength value should be used within the real working context.

Intact rock strength depending on the direction of anisotropy can be assessed in two ways:

— By laboratory tests on specimens with differently oriented planes of weakness.
— By applying the Mohr-Coulomb empirical failure criteria.

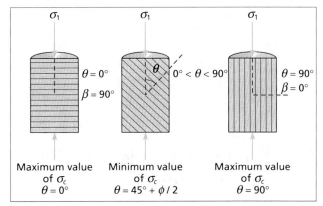

Figure 3.45   Rock strength varies according to the θ angle considered. Minimum strength value in a rock specimen with lamination planes or foliation is found when the failure occurs along these planes of weakness. Maximum strength corresponds to θ angles of 0° and 90°.

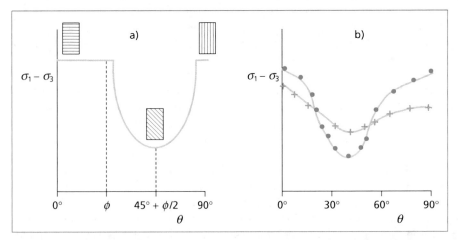

Figure 3.46   Variation in compressive strength with the angle of application of the load. a) Theoretically, when θ values are near 90° or between 0° and φ, failure cannot take place along a pre-existing plane of weakness, it will occur across these planes. b) Curves obtained from laboratory tests on samples with different θ angle values.

Anisotropic rocks are difficult to test because their strength varies, and many tests are required to obtain representative parameters for the whole strength range.

In porous intact rock, strength is reduced by **pore pressure,** which acts against the normal stress that resists failure, complying with the principle of effective stress:

$$\sigma'_n = \sigma_n - u$$

This mainly affects permeable porous rocks where water can infiltrate and may become saturated. Many rocks may be considered to be practically dry, although in conditions where water is present, saturation is only a question of time.

## Failure criteria

The strength of **isotropic intact rock** can be assessed using the Mohr-Coulomb and Hoek-Brown criteria. The main difference between the two is that the first is a linear criterion and the second is non-linear and more suited to real rock behaviour. In recent years other failure criteria have been developed, although they are not so well-known and are not used so often. Sheorey (1997) details the main failure criteria found in the literature on rock mechanics. Griffith's criterion, dating from 1921 (Jaeger and Cook, 1979), is based on the study of glass and is a classic in solid mechanics; although it is not wholly appropriate for application to rock, it has been very useful for studying how existing microcracks influence material tensile failure.

## Mohr-Coulomb criterion

At its simplest this criterion expresses shear strength along a plane in a triaxial state of stress; the ratio between normal and shear stresses acting at the moment of failure is given by the equation:

$$\tau = c + \sigma_n \tan \phi$$

where:

$\tau$ and $\sigma_n$ are the normal and shear stresses on the failure plane
$c$ and $\phi$ are the cohesion and internal friction angle of the intact rock

The criterion can also be expressed as a function of the principal stresses $\sigma_1$ and $\sigma_3$ (Figure 3.47):

$$\sigma_1 = \frac{2c + \sigma_3 \left[\sin 2\theta + \tan\phi \left(1 - \cos 2\theta\right)\right]}{\sin 2\theta - \tan\phi \left(1 + \cos 2\theta\right)}$$

from which the strength on any plane defined by $\theta$ can be obtained. For the critical failure plane $\theta = 45° + \phi/2$, the equation above will now be:

$$\sigma_1 = \frac{2c \cos\phi + \sigma_3 (1 + \sin\phi)}{(1 - \sin\phi)}$$

If $\sigma_3 = 0$, $\sigma_1$ will be the uniaxial compressive strength of the rock:

$$\sigma_1 = \sigma_c = \frac{2c \cos\phi}{1 - \sin\phi}$$

By extension (Figure 3.47b) the criterion also gives the tensile strength value:

$$\sigma_t = \frac{2c \cos\phi}{1 + \sin\phi}$$

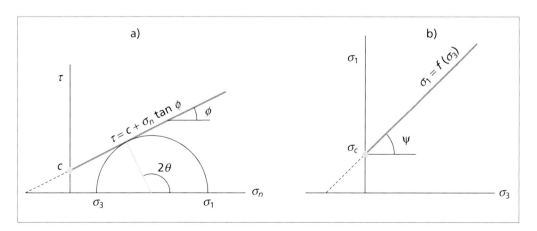

Figure 3.47   Mohr-Coulomb envelopes in terms of shear and normal stresses (a) and principal stresses (b). Failure does not take place for states of stress below the envelopes.

The Mohr-Coulomb criterion implies that shear fracture will occur when peak strength of the material is reached. The main advantage of this criterion is that it is so simple. However, it does have following drawbacks:

— Rock strength envelopes are not linear: experiments have shown that rock strength increases less with the increase in normal confining stress than the result obtained if a linear law is used; this may lead to errors when using Mohr-Coulomb criterion, especially for low confining stresses (*Figure 3.44*).
— According to this criterion, the direction of the fracture plane does not always coincide with the experimental results.
— This criterion overvalues tensile strength.

However, if this linear failure criterion is used to evaluate intact rock strength, the following recommendations should be followed:

— It should be assumed that the value for cohesion is around 10% of the uniaxial compressive strength of the intact rock.
— The value taken for the internal friction angle should be appropriate to the operative level of stress; this is taken either from specific tests or from tables (*Table 3.13*).

## Hoek-Brown's criterion

A non-linear criterion is more appropriate for assessing intact rock strength, with failure represented graphically as a curve. Hoek and Brown (1980) proposed a non-linear empirical failure criterion, valid for assessing isotropic intact rock strength in triaxial conditions:

$$\sigma_1 = \sigma_3 + \sqrt{m_i \sigma_{ci} \sigma_3 + \sigma_{ci}^2}$$

where $\sigma_1$ and $\sigma_3$ are the major and minor principal stresses at failure, $\sigma_{ci}$ is the uniaxial compressive strength (UCS) of the intact rock, and $m_i$ is a constant that depends on the properties of the intact rock.

The value of $\sigma_{ci}$ should be determined from laboratory tests, or, failing that, indirectly from point load tests (PLT). It can also be estimated from *Table 3.7*. The constant $m_i$ should be determined by triaxial testing of core samples (i.e. of intact rock) wherever possible, or estimated from a qualitative description of the rock material as described by Hoek and Brown (1997). This parameter depends on the internal frictional characteristics of the intact rock and it has a significant influence on the strength characteristics of rock.

When it is not possible to carry out triaxial tests, an estimate of $m_i$ can be obtained from *Table 3.14*.

The failure envelope can be drawn using the above equation (*Figure 3.48*). *Figure 3.49* shows the relationship between normalized stresses $\sigma_1$ and $\sigma_3$ for intact rock.

Expressed dimensionlessly, in terms of normalized stresses in relation to $\sigma_{ci}$, the criterion is:

$$\frac{\sigma_1}{\sigma_{ci}} = \frac{\sigma_3}{\sigma_{ci}} + \sqrt{m_i \frac{\sigma_3}{\sigma_{ci}} + 1}$$

The uniaxial compressive strength of the rock is given by the previous expression, substituting $\sigma_3 = 0$, and the tensile strength is obtained by resolving $\sigma_1 = 0$ and $\sigma_3 = \sigma_t$:

$$\sigma_t = \frac{1}{2} \sigma_{ci} \left( m_i - \sqrt{m_i^2 + 4} \right)$$

The expression of the failure criterion as a function of shear and normal stresses is:

$$\tau = A \sigma_{ci} \left( \frac{\sigma_n - \sigma_t}{\sigma_{ci}} \right)^B$$

where $\sigma_t$ is the tensile strength and $A$, $B$ are constants depending on the value of $m_i$.

## Deformability

Deformability refers to the property the rock has for changing its shape in response to forces acting on it. Depending on the intensity of the force exerted, how it is applied and the mechanical characteristics of the rock, deformation can be permanent or elastic; if it is elastic the rock will return to its original shape if the load is removed. The stress-strain relationships of rocks and plastic and elastic behaviour models are described in Section 3.3 of this chapter. The deformability of a rock is expressed by its elastic constants, $E$ and $v$.

$$E = \sigma / \varepsilon_{ax} \text{ (units of stress)}$$
$$v = \varepsilon_r / \varepsilon_{ax} \text{ (dimensionless)}$$

**Young's modulus**, $E$, defines the linear elastic relationship between the stress applied and the deformation produced in the direction of its application. The Poisson ratio, $v$, defines the relation between radial and axial strain. Both constants are obtained from the **uniaxial compression test** and define the characteristics of "static" elastic deformation of the rock. Hard rock with brittle behaviour gives a greater

## Table 3.14  VALUES OF THE CONSTANT $m_i$ FOR INTACT ROCK

| Rock type | Class | Group | Texture | | | |
|---|---|---|---|---|---|---|
| | | | Coarse | Medium | Fine | Very fine |
| SEDIMENTARY | Clastic | | Conglomerates (21 ± 3) Breccias (19 ± 5) | Sandstones 17 ± 4 | Siltstones 7 ± 2 Greywackes (18 ± 3) | Claystones 4 ± 2 Shales (6 ± 2) Marls (7 ± 2) |
| | Non-Clastic | Carbonates | Crystalline Limestone (12 ± 3) | Sparitic Limestones (10 ± 2) | Micritic Limestones (9 ± 2) | Dolomites (9 ± 3) |
| | | Evaporites | | Gypsum 8 ± 2 | Anhydrite 12 ± 2 | |
| | | Organic | | | | Chalk 7 ± 2 |
| METAMORPHIC | Non Foliated | | Marble 9 ± 3 | Hornfels (19 ± 4) Metasandstone (19 ± 3) | Quartzites 20 ± 3 | |
| | Slightly foliated | | Migmatite (29 ± 3) | Amphibolites 26 ± 6 | Gneiss 28 ± 5 | |
| | Foliated* | | | Schists 12 ± 3 | Phyllites (7 ± 3) | Slates 7 ± 4 |
| IGNEOUS | Plutonic | Light | Granite 32 ± 3 Granodiorite (29 ± 3) | Diorite 25 ± 5 | | |
| | | Dark | Gabbro 27 ± 3 | Dolerite (16 ± 5) Norite 20 ± 5 | | |
| | Hypabyssal | | | Porphyries (20 ± 5) | Diabase (15 ± 5) | Peridotite (25 ± 5) |
| | Volcanic | Lava | | Rhyolite (25 ± 5) Andesite 25 ± 5 | Dacite (25 ± 3) Basalt (25 ± 5) | |
| | | Pyroclastic | Agglomerate (19 ± 3) | Breccia (19 ± 5) | Tuff (13 ± 5) | |

Hoek and Marinos, 2000.
Values in parenthesis are estimates.
The range of values quoted for each material depends upon the granularity and interlocking of the crystal structure—the higher values being associated with tightly interlocked and more frictional characteristics.
(*) Values for intact rock specimens tested normal to bedding or foliation planes. The value of $m_i$ will be significantly different if failure occurs along a weakness plane.

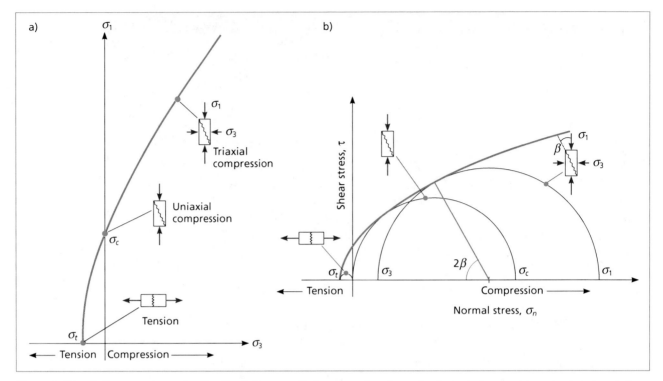

Figure 3.48    Failure envelopes for the Hoek-Brown criterion, as a function of principal stresses (a) and normal and shear stresses (b) for different stress conditions within intact rock.

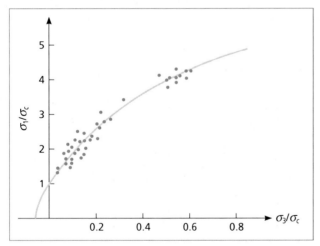

Figure 3.49    Representation of the peak strength envelope of intact rock in terms of normalised stresses.

Young's modulus and a smaller Poisson ratio than a soft rock with ductile behaviour.

Rocks rarely show ideal linear elastic behaviour, so that the $E$ and $v$ values vary. If an axial load is applied to a specimen of ideal elastic, isotropic, homogeneous material, its volume will not vary, although deformation occurs. For a specimen 10 cm long and 5 cm in diameter, and assuming there is axial deformation of 4% of the specimen length, the Poisson ratio is:

$$v = \varepsilon_r / \varepsilon_{ax} = \left[ (r_i - r_f)/r_i \right] / \left[ (10 - 9{,}6)/10 \right]$$

where $r_i$ is the initial radius and $r_f$ the final radius of the specimen (before and after deformation). As the volume remains constant, the value of $r_f$ can be calculated to obtain the value of $v$, which will be 0.5. This is the value for ideal elastic materials. Rocks always have lower values, generally ranging from 0.15 to 0.30.

The values of $E$ and $v$ can also be obtained from the velocity of elastic waves $V_p$ and $V_s$, measured in laboratory using **sonic velocity**, corresponding in this case to "dynamic" values for the elastic "constants". The dynamic Young's modulus is greater than the static one: $E_d > E$.

Table 3.15 includes values for both the static and dynamic Young's modulus and for the Poisson ratio for different rocks. It shows the range for these parameters as generally experienced, sometimes very wide due to the great variety of physical properties involved (e.g. porosity, mineral structure, cementation) and to the anisotropic character of certain rocks (e.g. where there is lamination or schistocity).

| Table 3.15 | GUIDE TO ELASTIC CONSTANTS FOR ROCKS | | |
|---|---|---|---|
| Intact rock | Static elasticity modulus, $E$ (GPa) | Dynamic elasticity modulus, $E_d$ (GPa) | Poisson ratio, $v$ |
| Andesite | 30–40 | | 0.23–0.32 |
| Amphibolite | 13–92 | 46–105 | |
| Anhydrite | 1.5–76 | | |
| Basalt | 32–100 | 41–87 | 0.19–0.38 (0.25) |
| Diabase | 69–96 | 60–98 | 0.28 |
| Diorite | 2–17 | 25–44 | |
| Dolomite | 4–51 | 22–86 | 0.29–0.34 |
| Gabbro | 10–65 | | 0.12–0.20 |
| Gneiss | 17–81 (53–55) | 25–105 | 0.08–0.4 (0.20–0.30) |
| Granite | 17–77 | 10–84 | 0.1–0.4 (0.18–0.24) |
| Greywacke | 47–63 | 23–107 | |
| Gypsum | 15–36 | | |
| Limestones | 15–90 (29–60) | 8–99 | 0.12–0.33 (0.25–0.30) |
| Marble | 28–72 | | 0.1–0.4 (0.23) |
| Marl | 4–34 | 10–49 | |
| Mica-schist | 1–20 | | |
| Mudstone | 3–22 | 10–70 | 0.25–0.29 |
| Quartzite | 22–100 (42–85) | | 0.08–0.24 (0.11–0.15) |
| Salt | 5–20 | | 0.22 |
| Sandstones | 3–61 | 5–56 | 0.1–0.4 (0.24–0.31) |
| Schist | 6–39 (20) | | 0.01–0.31 (0.12) |
| Shale | 5–30 | | |
| Siltstone | 53–75 | 7–65 | 0.25 |
| Tuff | 3–76 | | 0.24–0.29 |

Maximum and minimum values. Average values in brackets.
Data taken from Rahn (1986), Johnson and De Graff (1988), Goodman (1989), Walthan (1999) and Duncan (1999).

# Strength and deformability laboratory tests

Most building materials, such as metal and concrete, are so uniform and homogeneous that, once in place, the mechanical properties of these materials are virtually the same as those obtained in laboratory tests. This is not the case with rocks, so test results must be interpreted taking into account their limitations and how representative they are. Even apparently isotropic and homogeneous rocks show preferred anisotropic directions and variations which affect the results of laboratory tests.

Experimental methods for determining rock strength and deformability are independent of the failure criteria adopted in each case; their aim is to establish the relationship between stresses and strains during loading and failure processes, the stresses to which the rock is subjected at the moment of failure and its strength parameters. These laboratory tests are the uniaxial compression, triaxial compression and tensile strength tests.

If a statistically representative number of tests are carried out, characteristic values for the strength parameters of a rock can be obtained from the force applied at the moment of failure. Appropriate tests will give the stress-strain curves that typify behaviour (a behavioural model or law), which must be studied to define the deformational properties of rock.

*Table 3.16* lists the laboratory tests used to obtain the strength and deformability parameters of intact rock. *Figure 3.50* shows a diagram of strength tests.

Laboratory tests are carried out using cylindrical rock specimens. As borehole cores are normally used, the specimens are usually small. The tests must be carried out systematically and the results should be statistically representative of the rock examined. It is very important to define clearly what is to be measured and evaluated.

The resulting values will depend on the type of rock and on its condition (e.g. mineralogy, grain size and cementation, micro-cracks, porosity and degree of weathering) and on the test conditions (shape and volume of the specimen, how it is prepared, moisture content, temperature, load velocity, direction of load application and the stiffness of the testing machine).

## Uniaxial compression test

This test determines the unconfined uniaxial strength, or uniaxial compressive strength (UCS), $\sigma_c$, of the rock and its elastic constants: Young's modulus, $E$, and the Poisson ratio, $v$. The results are used to classify rock by **strength** and to determine its **deformability.** The relationship of the stresses applied in the tests is $\sigma_1 \neq 0; \sigma_2 = \sigma_3 = 0$.

### Procedure

An axial force is gradually applied to a cylinder of rock until failure occurs (*Figures 3.51*, *3.52* and *3.53*). In conventional testing machines, the control variable is force, applied with controlled magnitude and velocity. During the test, the axial deformations produced in the specimen are measured, and the axial stress-strain curves $\sigma - \varepsilon_{ax}$ of the specimen are calculated and logged. The radial or circumferential deformations in the specimen can also be measured to obtain the $\sigma - \varepsilon_r$ curve.

The ISRM (1979) gives a series of recommendations when preparing specimens:

— Specimens must be cylindrical with a ratio of $L/D = 2.5 - 3$, and with $D > 54$ mm. Diameter $D$ should be at least 10 times greater than the largest grain size of the rock.
— The ends of the specimen must be flat and parallel, and perpendicular to the axis of the cylinder.

To characterize intact rock, at least five tests must be carried out.

| Table 3.16 | LABORATORY TESTS FOR STRENGTH AND DEFORMABILITY | |
|---|---|---|
| | **Tests** | **Parameters obtained** |
| Strength | Uniaxial compression | Uniaxial compressive strength, $\sigma_c$ |
| | Triaxial compression | Cohesion ($c$), peak internal friction angle ($\phi_p$) and residual friction angle ($\phi_r$) |
| | Direct tension | Tensile strength, $\sigma_t$ |
| | Indirect tension | Tensile strength, $\sigma_t$ |
| Deformability | Uniaxial compression | Static deformation modulus, $E$ and $v$ |
| | Sound velocity | Dynamic deformation modulus, $E_d$ and $v_d$ |

ROCK MECHANICS

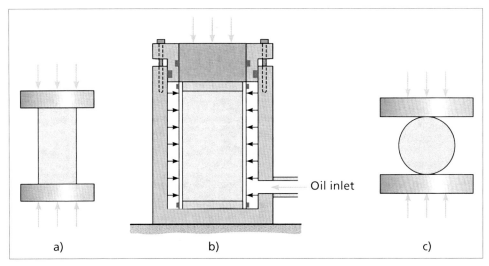

*Figure 3.50* Diagrams of strength tests: a) uniaxial, b) triaxial, c) indirect tensile strength (Brazil test).

*Figure 3.51* Equipment for uniaxial compression tests.

### Interpretation

*Figure 3.54* shows an example of stress-strain curves obtained in this test. The curves show one rising portion until peak strength, $\sigma_c$, is reached and a falling portion showing loss of strength. The uniaxial compressive strength is the value of the maximum force supporting the specimen divided by the area to which force is applied. This parameter depends, to a certain extent, on the shape and size of the specimen, its moisture content, and the regime and velocity of the load applied.

Although it is assumed that failure in the rock due to compression occurs when peak strength is reached, it has been demonstrated experimentally that the failure process and the generation of microcracks begin at stress levels before that of the peak, at between 50% and 95% of uniaxial compressive strength, $\sigma_c$ (Brady and Brown, 1993).

On the rising portion of the $\sigma - \varepsilon_{ax}$ curve there is a part where the relationship between the load applied and the deformation produced is linear and it can be assumed that Hooke's law is valid: $E = \sigma/\varepsilon =$ constant. Young's modulus $E$ is a constant in linear elastic materials where the deformations are recoverable.

A high percentage of rock materials are relatively elastic, or behave relatively elastically; i.e. when they are subjected to a load, displacement occurs but when the load is removed, the displacement returns to zero; i.e. deformation disappears. However, only some rocks display truly linear elasticity or approach this behaviour; for the other materials the deformation modulus $E$ varies during the test and is not a constant for the material. Even the behaviour of a particular type of rock varies depending on environmental conditions.

The static elastic constants of the rock can be obtained from the stress and strain values of the specimen in its elastic field: $E = \sigma/\varepsilon_{ax}$ and $v = \varepsilon_r/\varepsilon_{ax}$ (Box 3.7).

After peak strength is reached, the rock may continue to maintain the load, and lose its strength gradually. The post-peak portion of the stress-strain curve for the

Figure 3.53 Fractured specimens from uniaxial compression tests.

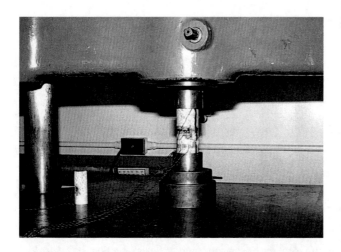

Figure 3.52 Specimens prepared for uniaxial compression tests, with electrical strain gauges and micrometer gauges to measure axial and radial displacements.

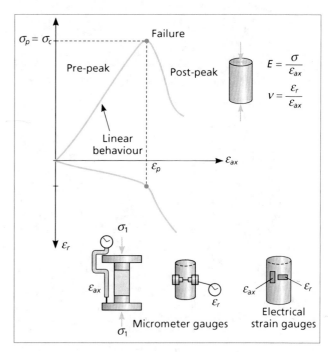

Figure 3.54 Stress-strain curves $\sigma - \varepsilon_{ax}$ and $\varepsilon_{ax} - \varepsilon_r$ obtained from a compression test.

## Box 3.7

### Calculating the elastic constants for the rock: Young's modulus, $E$, and Poisson's ratio, $v$

Young's modulus can be determined in the following ways:

— Average modulus, $E_m$, or slope of the straight portion of the curve.
— Tangent modulus, $E_t$, or slope of the curve at a particular point (generally at 50% of peak strength).
— Secant modulus, $E_s$, or slope of the straight line that joins the origin of the curve to peak strength.

The first two give more representative values and, moreover, the results usually coincide. In the example shown in graph d) the values measured are:

$$E_m = 34 \times 10^3 \text{ MPa}; \quad E_t = 34 \times 10^3 \text{ MPa}$$
$$E_s = 25.5 \times 10^3 \text{ MPa}$$

The value of the Poisson ratio measured for the straight portion of the curve, $\varepsilon_r - \sigma_{ax}$ is $v = 0.40$.

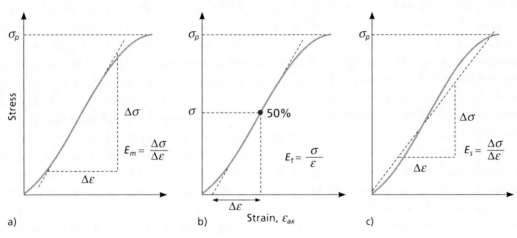

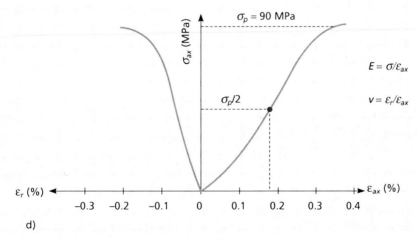

$\sigma_{ax}$ = Axial force/Initial area of the base of the specimen
$\varepsilon_{ax}$ = Axial strain
$\varepsilon_r$ = Radial or transversal strain

specimen can only be recorded if stiff or servo-controlled testing equipment is used. Recording this portion to find out how the rock behaves after failure is important for some engineering applications, for example designing excavations in soft rock.

### Factors affecting the measurement of uniaxial compression strength in rocks

The results of the laboratory tests are affected by the type and condition of the rock samples and by the test conditions, especially:

— The shape and volume of the specimen.
— The preparation and grinding of the loaded ends of the sample.
— The direction the load is applied in (for anisotropic rocks).
— The rate of loading.
— The moisture content and degree of saturation of the sample.

The distribution of stresses in a specimen varies with its **geometry**. *Figure 3.55* shows the effects of the length/diameter ratio, L/D, on test results. Variation is mainly due to friction between the ends of the specimen and the platens. Uniaxial compressive strength decreases as the **volume** of the specimen is increased.

The concave effect that usually appears at the start of the test, before the elastic portion of the stress-strain curve, can be considerably reduced if the **ends of the specimen** are exactly parallel.

The direction of the load applied in anisotropic rocks and its effects have already been dealt with in this Section. A rock when wet is invariably weaker than when dry, for a variety of reasons. Finally, to minimize the influence of the rate of loading the ISRM (1979) recommends using loads ranging from 0.5 to 1 MPa/s; this corresponds approximately to a 5–10 minute period before peak strength is reached (for hard materials in general). Rapid application may cause sudden failure and lead to overestimation of the strength of the material.

### Recording the complete stress-strain curve

In a compression test, both the specimen and the testing machine are deformed with each load increment, and during the test both store strain energy proportional to their stiffness.

Whether or not the complete stress-strain curve of a rock material can be recorded depends on the relative

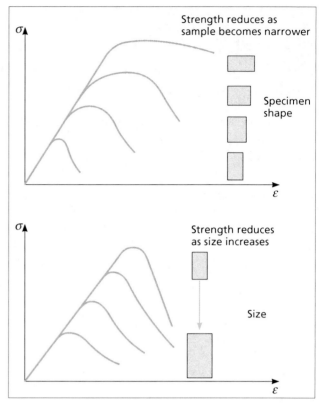

Figure 3.55  Variation in uniaxial compressive strength with respect to sample shape and size.

stiffness of the specimen and of the testing machine. The stiffness, $K$, of an elastic member is defined as the force necessary to cause a unitary displacement, $s$, in the direction of load application $P$:

$$K = P/s$$

which, as a function of stress and strain, can be expressed as:

$$K = EA/l$$

where $E$ is Young's modulus, $A$ is the area of application of the load $P$, and $l$ is the length of the body (either machine or specimen) in the direction of load application.

The amount of strain energy, $W$, stored in an elastic body on the application of a load is defined (*Figure 3.56*):

$$W = 1/2\ Ps \quad \text{or} \quad W = P^2/2K$$

The lower the stiffness value of the testing machine, $K_m$, the greater will be the elastic energy stored in the machine during the application of load.

If $K_m < K_{specimen}$, when peak strength of the specimen is reached, the deformation energy stored in the machine, $\Delta w_m$, is suddenly released and the specimen cannot absorb it. The testing machine is "soft" with respect to the specimen, sudden failure takes place (Figure 3.56a) and the post-peak portion of the curve cannot be recorded correctly. The test will give the stress-strain relationship up to peak strength but will not give any information on the rock's characteristics once this has been exceeded.

In contrast, if $K_m > K_s$, the machine is "stiff" with respect to the specimen, which is able to gradually absorb the energy released by the machine, $\Delta w_m < \Delta w_s$, and the post-peak portion of the curve can be recorded correctly (Figure 3.56b). In this case, the machine-specimen system is stable.

The record of the post-peak curves allows a study of the whole fracture mechanism of a rock. Once peak strength is reached, propagation of the fracture is "stable" when energy has to be supplied to the specimen for failure to continue (Class I, Figure 3.57), and "unstable" when energy must be withdrawn to prevent sudden failure (Class II). Classification of the post-peak region of the curve is based on these two types of behaviour.

In very brittle homogeneous rocks, the post-peak strain curve cannot be recorded, even with stiff machines, and **servo-controlled testing machines** are used. These are controlled by feed-back from the sample so that sample response to change in load is used to control the next increment of change of load. This is achieved by measuring selected variables throughout the test which are compared instantaneously and electronically with programmed values, so that the system reacts and either applies or withdraws pressure until the load is adjusted to a pre-set condition (Figure 3.58).

This system allows deformation to be used as a control variable in the test, and a complete record of the post-peak curve can be obtained for almost any type of rock. Brady and Brown (1993) and Hudson and Harrison (2000) describe the servo-control system and how it is applied to compression testing in rocks.

## Triaxial compression test

This test represents the condition of rocks *in situ* subjected to confining stresses, and does so by applying uniform hydraulic pressure around the specimen, making it possible to obtain the strength envelope of the rock; from this, its strength parameters are obtained: cohesion $c$ and friction $\phi$. The triaxial compression test is the most widely used of the multi-axial compression tests. The relationship between the stresses applied to the specimen is:
$\sigma_1 > \sigma_2 = \sigma_3 \neq 0$.

### Procedure

The test is carried out on specimens similar to those used in uniaxial compression tests. A specimen is placed in a metal cylinder or cell and then pressurised by fluid inside the cell. The specimen is surrounded by a flexible impermeable jacket to isolate it from this fluid.

At the beginning of the test, axial load and confining pressure are applied simultaneously so that both stresses are equal. Once the required level of confining pressure is reached, axial load is applied until failure of the specimen occurs. The confining pressure must be kept constant throughout the test.

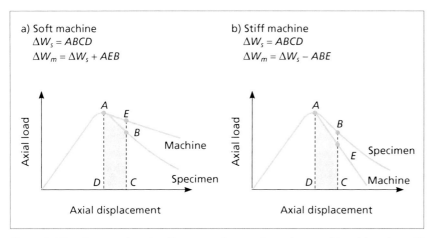

Figure 3.56  Post-peak unloading curve for a soft testing machine (a) and a stiff testing machine (b) with respect to the specimen stiffness (modified from Brady and Brown, 1993).

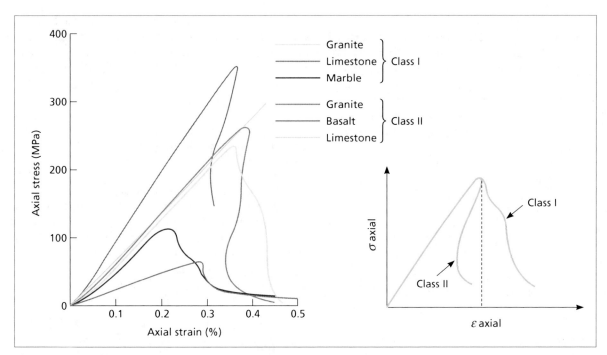

*Figure 3.57*  Class I and Class II stress-strain behaviour for uniaxial compression. Curves for different types of rock (Wawersick and Fairhurst, 1970).

*Figure 3.58*  Servo-controlled machine for compression tests.

The data recorded during the experiment are: load or axial stress $\sigma_1$, axial deformation, the angle of the fracture plane and, if required, the angle formed by the planes of anisotropy relative to the direction of the axial load.

*Figure 3.59* shows a diagram of a triaxial cell with the components needed to carry out the test and the strain gauges attached to the specimen to record the deformation as local strains.

When axial loading starts the specimen shortens and flattens (due to the confining pressure) until it starts to "dilate" as a result of the internal cracking of the material (*Figures 3.60* and *3.61*). Dilation begins in the elastic field and continues in the post-peak region of the test; it decreases as confining pressure increases, and may become non-existent in tests with high $\sigma_3$ values.

### Interpretation

The results of triaxial compression tests depend basically on the characteristics of the rock and on the confining pressure applied.

Peak strength will be different in each case and will increase as $\sigma_3$ increases. *Figure 3.62* shows curves obtained in triaxial tests for different values of confining pressure.

Interpretation of test results is based on the application of the Mohr-Coulomb failure criterion. The Mohr circles and failure envelope can be plotted from the $\sigma - \varepsilon$ curves obtained

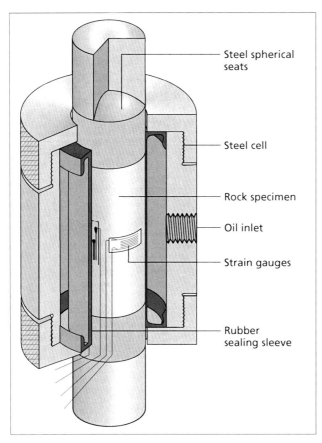

*Figure 3.59*  Diagram of a triaxial cell (Hoek and Brown, 1980).

for different values of $\sigma_3$, to give c and $\phi$ values of the material tested (*Box 3.8*).

## Factors affecting triaxial compression test results

Results of the triaxial test on identical samples of the same rock are controlled by **confining pressure**. Any increase in this (see *Figure 3.62*) will lead to:

— Increase in peak strength (although this increase is not generally linear).
— Transition from brittle to ductile behaviour in the specimen and its deformation mechanisms.
— Flattening and widening of the peak portion of the curve.
— Reduction of the post-peak portion of the curve up to the point where residual strength is reached; with high confining pressures, this may disappear.

The brittle-ductile transition pressure of the rock is defined as the confining pressure at which a change takes place from brittle to ductile deformation mechanisms; this is

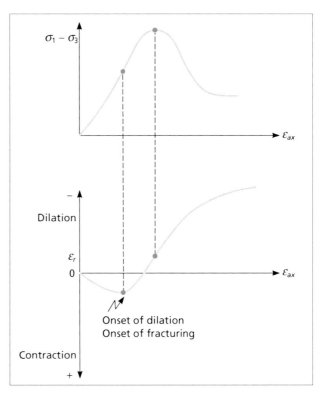

*Figure 3.60*  Volumetric strain in the triaxial compression test.

shown by the near horizontal stress-strain curves after peak strength, typical of ductile behaviour. In most hard rocks, this can be considered to occur when $\sigma_1 > 3.5\sigma_3$.

*Figure 3.61b)* shows the influence of confining pressure on the dilation of the specimens as a result of internal cracking: the "amount" of dilation is reduced as pressure is increased. In *Figure 3.61a)*, for the curve $\sigma_3 = 2$ MPa, residual values of the material are reached after a noticeable peak strength; for the curve $\sigma_3 = 5$ MPa this tendency is less noticeable and residual values close to those of the peak are reached; finally, in the case of the curve $\sigma_3 = 10$ MPa, there is no differentiated peak strength, and hardening occurs because the brittle-ductile transition pressure has been exceeded.

In permeable rocks, **pore pressure**, $u$, counteracts the influence of confining pressure so that the mechanical response of the rock is controlled by the effective pressure: $\sigma'_3 = \sigma_3 - u$. For the same value $\sigma_3$, an increment in $u$ leads to a reduction in the peak strength of the rock and to more brittle behaviour (*Figure 3.63*); i.e. the effect is the opposite of that produced by an increase in confining pressure.

An increase in **temperature** in triaxial tests generally causes a decrease in peak strength and in the brittle-ductile transition pressure.

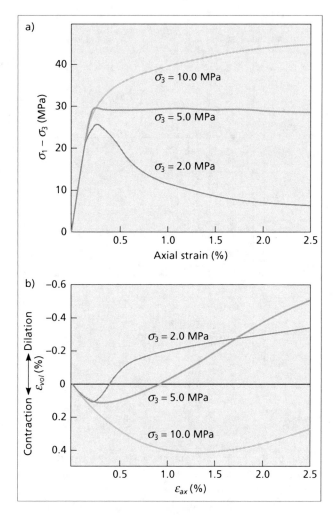

Figure 3.61   Results of triaxial compression test on oolitic limestone, with volumetric strain measurements (Elliot, 1982; in Brady and Brown, 1993).

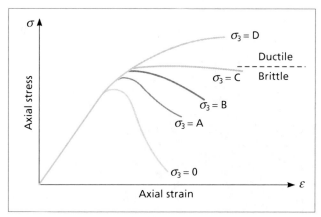

Figure 3.62   Stress-strain curves from triaxial tests on rock under different confining pressures increasing from 0 to D. Above a certain $\sigma_3$ value, rock behaviour changes from brittle to ductile.

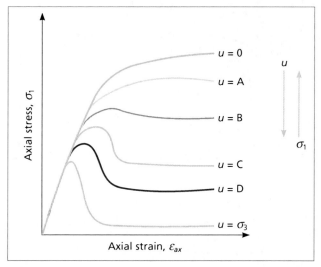

Figure 3.63   Influence of pore pressure, $u$, on rock behaviour with a constant confining pressure $\sigma_3$. Pore pressure increases from 0 to $u = \sigma_3$.

The presence of pre-existing **microcracks** in rocks influences both test results and the stress-strain curve model.

Confining pressure has no influence on the orientation of the failure plane.

## Tensile strength tests

### Direct tensile test

This test measures directly the uniaxial tensile strength of a rock cylinder. Each end of the specimen is held firmly in place and a uniaxial tensile force is applied in the direction of the greatest length of the specimen until failure occurs. Traction is applied to two metal caps bonded to each end of the cylinder with resin or cement. The specimen can also be trimmed wider at each end to match the traction system.

The L/D ratio of the specimen should be 2.5 to 3, and the diameter not less than 54 mm. The cylinder bases should be flat and parallel, and perpendicular to the length of the specimen. For preparing and trimming the specimen, the same specifications should be followed as those used in compression tests. Tensile force is applied continuously and uniformly, within a range between 0.5 and 1.0 MPa/s, so that failure occurs after a few minutes. The tensile strength $\sigma_t$ is calculated by dividing the force applied at the moment of failure by the circular area of the specimen. At least five tests are recommended to obtain a representative value for tensile strength (ISRM, 1981).

## Box 3.8

### Example of the calculation of strength parameters c and $\phi$ from triaxial tests

The peak strength values, $\sigma_p$, are taken from the $\sigma_1 - \varepsilon_{ax}$ curves obtained from each test and used to draw the corresponding Mohr circles on a diagram in ($\tau - \sigma_n$) space. The tangent to the circles is drawn, representing the failure envelope of the material being tested. The typical cohesion and friction values of the material are read directly from this line.

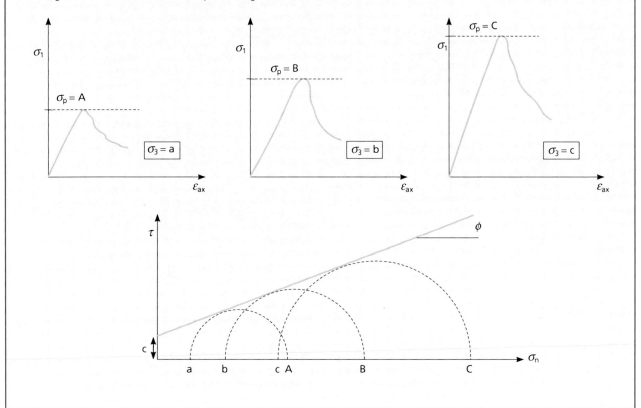

These tests are difficult to carry out due to the problems involved in cutting the specimen and ensuring that the metal caps are perfectly stuck to it.

### Indirect tensile or Brazil test

This measures the uniaxial tensile strength of a rock specimen indirectly. It is assumed that failure is produced by traction when the rock is subjected to a biaxial state of stress, with a tensile principal stress and a compressive stress of magnitude no greater than 3 times that of the tensile stress.

A vertical compressive load is applied to a disc or cylinder of rock, placed horizontally between two platens through which the force is transmitted, until failure occurs. The platens used to transmit the loads may be either flat or spherical and concave and must be exactly parallel (*Figure 3.64*).

The load applied is within a range where the failure of the rock occurs in 15 to 30s; the ISRM (1981) recommends a range of 200 N/s. The guidelines given in previous sections should be followed for the preparation and trimming of the specimens.

Compressive load produces a complex stress distribution within the specimen. Tensile strength is obtained using the formula:

$$\sigma_t = 2P/\pi DL$$

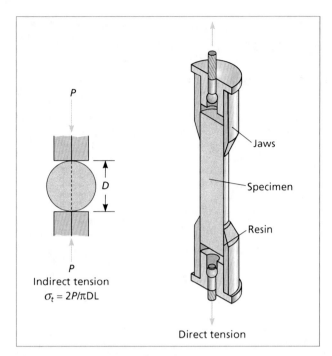

Figure 3.64   Diagram of tensile tests.

where:

P = load causing failure.
D = diameter of specimen.
L = length of specimen.

## Sonic velocity

This test measures the velocity of longitudinal and shear elastic waves, $V_p$ and $V_s$, through a specimen of dry or saturated rock. Their velocity is related to the strength and deformability of the material, and from it are calculated the dynamic elastic deformation moduli: $E_d$ and $v_d$.

The test consists of transmitting longitudinal waves using ultrasonic compression and measuring the time taken for the waves to pass through the specimen. In the same way, shear waves are transmitted via sonic pulses and their arrival times recorded. The corresponding velocities, $V_p$ and $V_s$, are calculated from these times. The transmitter or generator of the compressive force and of its pulses is fixed to one end of the specimen; at the other end is the receiver, which measures how long the waves take to pass through the length of the rock sample. The receiver may also be placed on one side of the specimen to vary the distance the waves have to travel. Specimens can be cylindrical or rectangular blocks, and their minimum dimension should be at least 10 times the wave length (ISRM 1981).

Shear wave velocity $V_s$ is approximately two thirds the velocity of the longitudinal waves $V_p$. The **dynamic elastic moduli** of the rock, $E_d$ and $v_d$, are obtained from the formulae:

$$E_d = \rho V_p^2 \frac{(1-2v_d)(1+v_d)}{(1-v_d)} \quad E_d = 2\rho V_s^2(1+v_d)$$

$$v_d = \frac{(V_p/V_s)^2 - 2}{2\left[(V_p/V_s)^2 - 1\right]}$$

where $\rho$ is the density of the rock (kg/m³) and $V_p$ and $V_s$ are the velocities of the longitudinal and shear waves (m/s):

$$V_p = \left[\frac{E_d}{\rho}\frac{1-v_d}{(1+v_d)(1-2v_d)}\right]^{1/2}$$

$$V_s = \left[\frac{E_d}{\rho}\frac{1}{2(1+v_d)}\right]^{1/2}$$

$$\frac{V_p}{V_s} = \left[2\frac{(1-v_d)}{(1-2v_d)}\right]^{1/2}$$

The value of the dynamic deformation modulus $E_d$ is greater than that obtained from uniaxial compression tests, since rapid application of low magnitude stresses results in the rock having a purely elastic behaviour.

As well as having a linear correlation with the deformability of the rock, the $V_p$ value is also indicative of rock quality, as described in Section 3.2, because it is related to many properties including porosity and uniaxial compressive strength (Figure 3.66). The Poisson ratio does not have a well-defined relationship with $V_p$.

## Limitations of laboratory tests

Laboratory testing is necessary for determining the properties of intact rock, and is a crucial part of rock mechanics. The

Figure 3.65   Sonic velocity test apparatus.

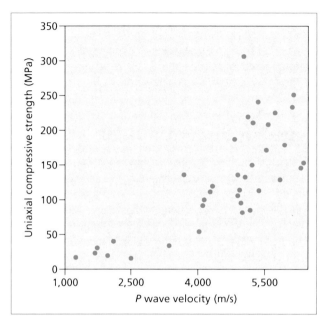

*Figure 3.66* Relationship between wave velocity and uniaxial compressive strength (modified from Johnson and De Graff, 1988).

type and number of tests to be carried out depends mainly on the aim of the research and the nature of the project. The size, number and location of the test samples depend on the kind of geological engineering problem posed and also on financial considerations.

Laboratory tests by themselves do not give the properties of rock masses, although they provide values that can be extrapolated to and correlated with properties of a rock mass. Their advantage is that they are cheaper and easier to perform than *in situ* tests, and a large number of tests can be carried out under different conditions. However, laboratory tests and their results have certain limitations when it comes to extrapolating data to a rock mass scale:

— **Representability.** The samples tested correspond to isolated points on the rock mass and are not representative of the whole area under study, or of the variability of environmental factors conditioning the behaviour of the materials; it is therefore essential to carry out a statistically representative programme of sampling and testing.

  In addition, the environmental conditions of rocks in the field (e.g. confining pressure, temperature, chemical composition of pore water) are difficult to reproduce in the laboratory.

— **Scale.** Small portions of material are tested to characterize and predict the behaviour of larger volumes. The difference between these scales means conversion factors and corrections have to be used to extrapolate results to the scale of a rock mass.

— **Velocity.** Deformation and failure processes are generally reproduced in the laboratory in a matter of minutes and at most days, while such processes in nature may be the result of actions taking place over much longer periods of time.

If the influence of other factors related to laboratory tests is also taken into account, e.g. the type and characteristics of the testing machine and preparation of the specimens, it is easy to understand the limitations and difficulties associated with characterizing rock mass properties from such tests.

The same limitations are present in *in situ* testing, although to a lesser extent: results are only applicable to the area affected by the test. Their great advantage, however, is that they are carried out on the rock mass itself.

## 3.5 Discontinuities

### Influence on rock mass behaviour

Planes of discontinuity define the strength, deformational and hydraulic properties, and the general behaviour of rock masses. The discontinuities make the rock mass discontinuous and anisotropic, meaning it is weaker and more easily deformable, which makes it very difficult to assess its mechanical behaviour in the context of engineering work. Discontinuities allow water flow and provide preferred planes for weathering and fracture (*Figure 3.67*). It is essential to describe and characterize discontinuities in a study of the mechanical and hydrogeological behaviour of the rock mass. The stability of excavations and foundations in rock, for example, depends on the direction and strength of the discontinuities. *Figures 3.3* and *3.68* show different examples of how they can affect engineering projects.

*Figure 3.67* Joint sets in shales. Foundation of the Llyn-Brenig Dam, U.K. (height shown 4 m).

In engineering work for excavations or foundations, the relative **orientation** of discontinuities may determine whether the ground is stable or not, as shown in Figure 3.68. In surface excavations the stability of a slope depends on its orientation in relation to discontinuities; in arch dams, the presence of discontinuities parallel to the direction of the resultant force transmitted by the dam and the water may cause problems of stability; in tunnels, discontinuities with pronounced dips running parallel to the tunnel axis are equally unfavourable. Orientation is even more important if there are other factors, such as a large number of joints, of close spacing and low angles of friction.

When various sets of discontinuities are present, in different directions, they will determine the level of fracturing in the rock mass and the shape and size of blocks of intact rock.

**Shear strength** is the most important aspect to consider when determining the strength of jointed hard rock masses. A description of the physical and geometrical characteristics of the planes is required to evaluate the shear strength, as laboratory or field tests alone do not always give satisfactory results.

Discontinuities are grouped into families or sets characterized by their average representative values for their orientation and strength characteristics. Discontinuities in the same family are parallel or sub-parallel to each other (Figure 3.69). Single macro-discontinuities are sometimes present, running through the whole rock mass in addition to the other different sets; these should be studied on an individual basis.

## Types of discontinuities

The term discontinuity refers to any plane of separation or weakness in a rock mass; its origin may be sedimentary (bedding or lamination planes), diagenetic or tectonic (joints and faults). In Table 3.17 the different types of discontinuities have been grouped as "systematic", when they appear in sets, and "singular", when there is a single plane running through the rock mass; the latter type is usually more continuous and persistent than systematic discontinuities and may be up to several kilometres long in the case of faults. While sets are classified statistically by their average orientation and by their general characteristics, singular discontinuities require individual description and treatment. They may even influence and control the mechanical behaviour of a mass to a greater extent than the systematic discontinuities.

**Joints** are the most usual discontinuity planes in rock masses. These are fracture or failure surfaces in the rock along which there has been little or no displacement. They affect all types of rock and can be classified by their origin:

— Joints of tectonic origin, associated with folds and faults. Joints associated with folds have a characteristic arrangement (Figure 3.70). Joints associated with faults are parallel to the fault surface and they become less frequent the further they are from the fault.
— Joints in igneous rock, caused by contraction of the igneous body during or after its emplacement. These have a characteristic arrangement in three mutually orthogonal sets. An example of joints caused by

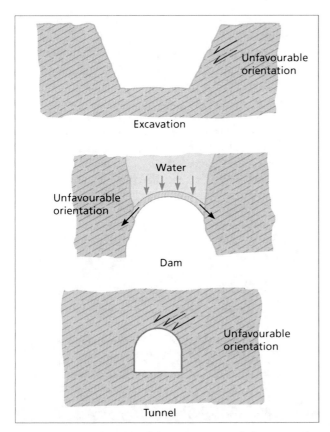

Figure 3.68  Influence of the orientation of discontinuities in relation to engineering work.

Figure 3.69  Inclined bedding in flysch.

## ROCK MECHANICS

### Table 3.17  TYPES OF DISCONTINUITIES

| Discontinuities | Systematic | Singular |
|---|---|---|
| Planar | — Bedding planes<br>— Lamination planes<br>— Joints<br>— Foliation planes | — Faults<br>— Dykes<br>— Discordances |
| Linear | — Intersection of planar discontinuities<br>— Lineations | — Axes of folds |

Figure 3.71   Columnar jointing in basalt, Canary Islands, Spain.

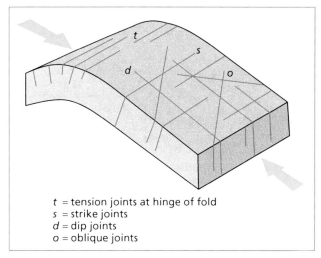

*t* = tension joints at hinge of fold
*s* = strike joints
*d* = dip joints
*o* = oblique joints

Figure 3.70   Joint sets associated with folds (Blyth and de Freitas, 1984).

Figure 3.72   Horizontal bedding planes with a high persistence in a limestone rock mass.

contraction during cooling is the columnar jointing that forms in basaltic lavas from tensile cracking (*Figure 3.71*).

— Relaxation joints, the result of a reduction in lithostatic load on the rock mass. These are arranged subparallel to the topographic surface and become less frequent with depth.

**Bedding planes** are the surfaces between the beds in sedimentary rocks (*Figure 3.72*). These systematic discontinuities extend over a wide area, with spacing generally ranging from a few centimetres to several metres.

**Lamination planes** are systematic discontinuities found in sedimentary rocks and are the surfaces separating the layers or smallest megascopic levels in the sedimentary sequence (*Figure 3.73*). These surfaces are more significant in fine-grained rocks and are characterized by very close spacing, of a few millimetres or centimetres.

**Foliation planes** have tectonic origin and occur in rocks that have undergone considerable deformation. They are arranged perpendicularly to the maximum compressive stress operating at the time of their formation. The smaller the grain of the rock, the more likely these systematic discontinuities are to develop, with high frequency and millimetric spacing (*Figure 3.74*).

**Lithological contact surfaces** are singular separation planes between different lithologies in a rock mass. In unfolded sedimentary rocks, they can be less important to the behaviour of the rock mass as a whole than other features, and are simply considered as bedding. In folded sedimentary rocks they can be surfaces of syn-tectonic shear and very important to the behaviour of a rock mass. Contact surfaces in igneous rocks are very important, especially in dykes and dyke rocks (*Figure 3.75*).

**Faults** are singular discontinuities corresponding to failure or fracture planes that show relative displacement

*Figure 3.73* Lamination in gypsum.

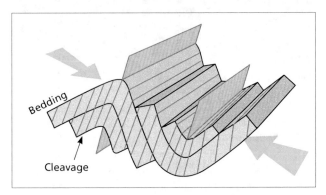

*Figure 3.74* Clavage related to folds. The arrows mark the maximum shortening direction (Price, 1981; in Blyth and De Freitas, 1984).

between the blocks (*Figure 3.76*). The size of the faults may vary from a few metres to hundreds of kilometres. They may be associated with areas of weakness known as fault zones or breccia where a clearly defined fracture plane cannot always be distinguished.

# Characteristics of discontinuities

The description of a set of discontinuities must include the following characteristics and geometric parameters: orientation,

*Figure 3.75* Dyke intruding pyroclasts and lava flows (cliff height: 100 m).

spacing, persistence, roughness, aperture, filling, seepage and wall strength. Some of these, such as roughness, aperture, filling and wall strength, will determine mechanical behaviour and the strength of the discontinuity planes.

These parameters are described and measured in the field. Chapter 6 describes the procedure for collecting data in the field, along with examples, classifications and tables for assessing the different factors involved, complementing what is described below.

The spatial **orientation** of a discontinuity is defined by its dip and dip direction. The average orientation of each set is determined from representative statistical values. Graphic representation of the discontinuities and their orientation will give a general overview of the rock mass geometry. Block diagrams give a three-dimensional representation of the plane distribution, making it easier to visualize the orientation of fracturing and how it will affect an engineering project or structure (*Figure 3.77*).

Spatial orientation cannot normally be determined in boreholes; the special techniques required are only used very occasionally.

**Spacing** is the average perpendicular distance between discontinuity planes in the same set. It defines the size of blocks of intact rock and affects the overall behaviour of a rock mass. With small spacing, the strength of the rock

a)

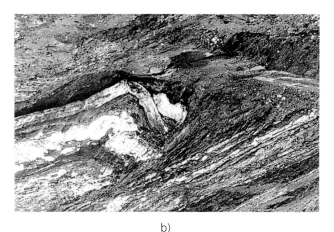

b)

Figure 3.76   Types of fault: a) Normal fault in Muschelkalk materials; b) Reverse fault in Carboniferous mudstone.

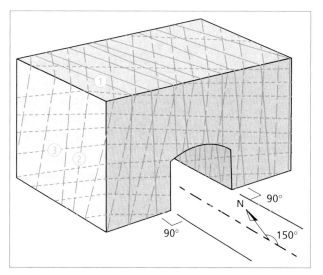

Figure 3.77   Block diagram showing sets of discontinuities (ISRM, 1981).

mass will be considerably reduced, and in extreme cases will lead to behaviour similar to granular, non-cohesive materials.

The spacing between discontinuities plays a very important role in the permeability of the rock mass. If the apertures of individual discontinuities are comparable, the hydraulic conductivity associated with a particular set is generally inversely proportional to its spacing.

**Continuity** is the area of the discontinuity. This determines to a large extent whether or not the intact rock will be involved in failure processes in the rock mass, and how far it affects the overall strength parameters of a rock mass.

Continuity can be represented in diagrams, as shown in Chapter 6, Figure 6.9.

The **roughness** of a discontinuity plane largely determines its shear strength (Figure 3.78); the rougher it is, the stronger it will be. Any irregularities make movement along the discontinuities more difficult during the shear tangential displacement processes.

The waviness and roughness of the planes may control the possible directions of the displacement and define the shear strength for different directions. The strength may vary considerably, depending on whether the direction of the movement coincides with that of the roughness or is transversal to it.

**Aperture** is the perpendicular distance separating the discontinuity walls when there is no fill. This may be very different in different areas of the rock mass: although the aperture may be very large at the surface, it will be become smaller with depth and may even close up completely. The influence of the aperture on shear strength is significant even in closed discontinuities because it modifies the effective stresses acting on the walls.

Discontinuities may sometimes appear with a **fill** of soft clayey material or a rock material different from that of the walls. The physical and mechanical properties of the fill,

Whether there is fill or not, discontinuities provide a preferred path for water **seepage** through the rock mass (secondary permeability). Water reduces the shear strength by reducing effective stresses acting on the discontinuity planes.

Finally, the **compressive strength** of the discontinuity wall, which depends on the type of intact rock and the degree of weathering of the walls, affects the shear strength and deformability of the discontinuity, whether or not fill is present, but especially when there is no fill.

Due to surface weathering, uniaxial compressive strength is usually lower than that of the intact rock.

## Shear strength of discontinuity planes

The study of the mechanical behaviour of discontinuities is based on the relationship between the shear stress applied and the shear displacement generated as a result. This ratio, $\tau/\mu$, is the **stiffness** of the discontinuity, in stress/length units (MPa/mm). Typical discontinuity behaviour curves are very similar in general form to those for intact rock, and always fail along a pre-existing plane (*Figure 3.79*).

The strength of discontinuity planes is obtained from Mohr-Coulomb's failure criterion and is determined in the laboratory using the direct shear test. Triaxial tests also give shear strength values if carried out on specimens cut so that failure takes place along the pre-existing discontinuity plane, i.e. ideally with angles of 25° to 40° between the plane and the direction of the vertical compressive stress. Shear strength can also be estimated from the *in situ* direct shear test (described in Chapter 5, Section 5.5).

The shear strength of discontinuities basically depends on the friction of the planes and, to a lesser extent, on cohesion. **Roughness** or irregularity of the discontinuity walls is one of the most influential factors on frictional strength, especially in discontinuities subjected to low normal stresses.

Peak **shear strength**, $\tau'_p$, in **smooth discontinuities** is given by Mohr-Coulomb's expression (*Figure 3.79*):

$$\tau'_p = c' + \sigma'_n \tan \phi'_p$$

a)

b)

*Figure 3.78*    a) Smooth planar discontinuity with high persistence. b) Rough undulating discontinuities produced by bedding planes.

such as its shear strength, deformability and permeability, may vary considerably and control how the discontinuity behaves; if the fill is made of soft or weathered materials, its strength may vary significantly in the short-term if the moisture content of the fill varies or if displacement takes place along the joints.

The main characteristics of the fill are its nature, thickness, shear strength and permeability.

where $\sigma'_n$ is the normal effective stress on the discontinuity plane, $c'$ is cohesion and $\phi'_p$ is the peak friction angle in terms of effective stress. Various factors controlling shear strength in discontinuities have already been mentioned (normal stress, roughness, wall strength, type, thickness and properties of the filling), but the above expression only considers normal stress and the strength properties of the plane of weakness;

so that although it is easy to apply and often used, it is still a simplification.

Patton (1966) proposed a bilinear failure model based on the **influence of the roughness** or irregularities usually present in discontinuities. The irregularity of a discontinuity plane can be defined by the roughness angle $i$ which is added to the basic angle of friction $\phi_b$ to obtain the total $\phi_p$ value of the surface:

$$\phi_p = \phi_b + i$$

$i$ is the angle of the irregularity in relation to the discontinuity plane and very influential on the geomechanical behaviour of the discontinuities; in fact, the main objective of describing and measuring roughness ($i$) is to estimate the shear strength of the planes. The value of $\phi_p$ is usually between 30°–70°; the $\phi_b$ angle is generally between 20°–40° and the $i$ angle may vary from 0°–40°.

According to *Figure 3.80*, if the discontinuity has no cohesion:

$$\tan \phi = \tau^*/\sigma_n^*$$
$$\tau^* = \tau \cos i - \sigma_n \sin i$$
$$\sigma_n^* = \sigma_n \cos i + \tau \sin i$$

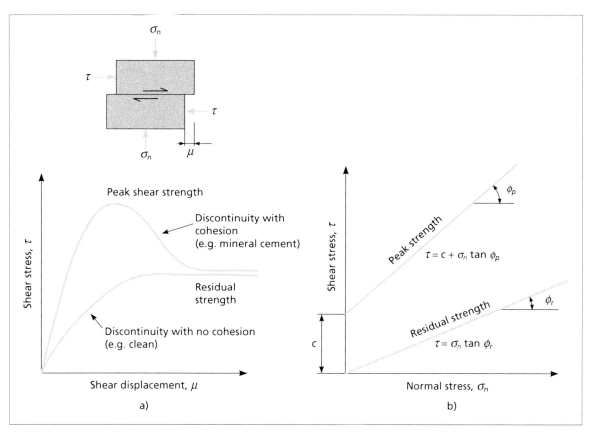

Figure 3.79   a) Typical shear stress $\tau$ - shear displacement $\mu$ curves for planar discontinuities. b) Theoretical shear strength of a planar discontinuity.

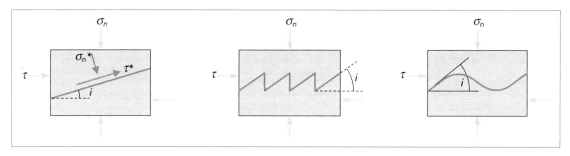

Figure 3.80   Influence of the roughness angle on discontinuity shear strength.

from which:

$$\tau/\sigma_n = \tan(\phi + i)$$

When shear stress is exerted on a discontinuity subjected to low normal stresses, displacement takes place along the plane, and **dilation** of the discontinuity walls occurs which open up and move apart as the $i$ angle has to be exceeded for displacement to take place; at this point, effective friction ($\phi_b + i$) will come into operation (Figure 3.81) and the value of $\tau_p$ (taking $c = 0$) is expressed by:

$$\tau'_p = \sigma'_n \tan(\phi'_b + i)$$

As shear displacement progresses, the sharpest edges may break and "smooth out" the roughness; the two surfaces come into contact, with the $\phi_b$ value prevailing. If the stress $\sigma_n$ on the plane is increased, a value is reached which prevents dilation and the irregularities have to fail for displacement to take place, at which point the inclination of the curve $\tau$-$\sigma_n$ is approximately the same as the value of the residual friction angle $\phi_r$. For high normal stresses:

$$\tau_p = \sigma_n \tan\phi_r$$

The inflection point of Patton's bilinear criterion corresponds to a specific value of $\sigma_n$.

Several authors have developed empirical criteria based on the Patton criterion for failure along **rough discontinuity planes** depending on the normal and shear stresses acting on them, including Barton and Choubey or Ladanyi and Archambault (Figure 3.82), with the former more widely used.

Figure 3.83 shows the results of shear tests on rough discontinuities for different $\sigma_n$ values. In the top curve in graph a), where $\sigma_n = 0$, there will be dilation, and shear strength will be practically nil because there is no effective friction (graph b). If the $\sigma_n$ value is increased, the corresponding curves show decreased dilation or opening, and increased shear strength.

The above is valid when the direction of shear displacement is perpendicular to the irregularities of the discontinuity walls. If it is parallel to them, the roughness will have no effect on the strength of the plane (Figure 3.84).

## Barton and Choubey criterion

This empirical criterion (Barton and Choubey, 1977), deduced from analysis of discontinuity behaviour in laboratory tests,

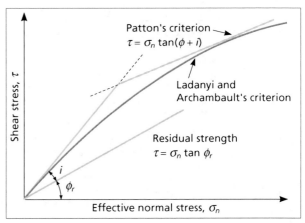

Figure 3.82  Representation of Patton's linear criterion and Ladanyi and Archambault's non-linear criterion for estimating the strength of rough discontinuity planes as a function of normal stress.

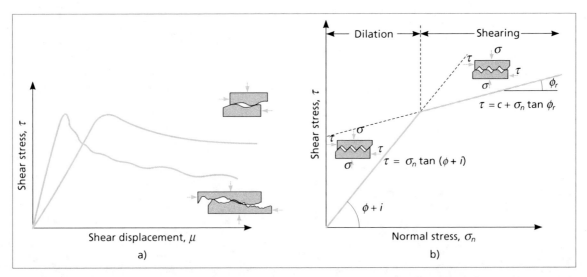

Figure 3.81  a) Typical shear stress $\tau$ - shear displacement $\mu$ curves for rough discontinuities. b) Bi-linear failure criterion for rough discontinuities.

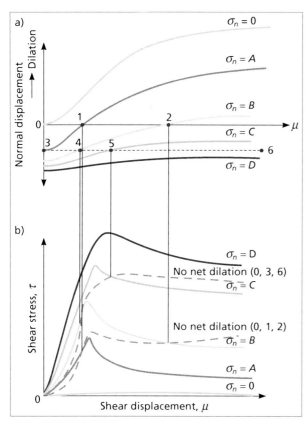

*Figure 3.83* Curves corresponding to shear strength tests on rough joints for different $\sigma_n$ values increasing from 0 to $D$; a) normal displacement-shear displacement; b) shear stress-shear displacement (Goodman, 1989).

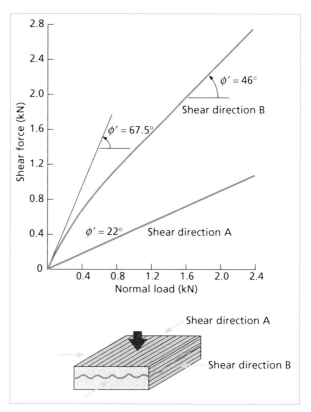

*Figure 3.84* Influence of roughness on discontinuity strength depending on shear direction (Brown et al., 1977; in Brady and Brown, 1993).

allows peak shear strength in rough discontinuities to be estimated, considering the roughness of the joint and the compressive strength of the joint surface in relation to the applied effective normal stress. It is expressed as:

$$\tau' = \sigma'_n \tan\left[ JRC \log_{10}\left(\frac{JCS}{\sigma'_n}\right) + \phi'_r \right]$$

where:

— $\tau'$ and $\sigma'_n$ are the shear and normal effective stresses on the discontinuity plane.
— $\phi'_r$ is the residual friction angle.
— JRC is the joint roughness coefficient of the discontinuity.
— JCS is the joint wall compression strength of the discontinuity.

According to the above expression, discontinuity strength depends on three components: a frictional component, $\phi'_r$, a geometrical component given by the JRC parameter, and an "asperity" component controlled by the ratio JCS/$\sigma'_n$. The asperity and geometrical components represent roughness, $i$, and the value for roughness generated dilation is nil for high normal stresses, when JCS/$\sigma'_n$ = 1; typical values are normally between 3 and 100. The overall friction angle is given by ($\phi_r + i$) and is generally not higher than 50°. As $\sigma_n$ increases so frictional strength increases but dilation decreases.

Because of roughness, very high friction angles are obtained with the Barton and Choubey equation for very low normal stresses acting on a rough discontinuity. It should therefore not be used when JCS/$\sigma'_n$ > 50; in this case, a constant friction angle, independent of the load should be taken, with a $\phi_p$ value equal to:

$$\phi_p = \phi_r + 1.7\, JRC$$

### How to estimate the residual friction angle, $\phi_r$

Generally, the joint wall is weathered and so the angle of residual friction will be lower than the angle of the fresh rock, $\phi_b$. The formula used to estimate this is:

$$\phi_r = (\phi_b - 20°) + 20\frac{r}{R}$$

where $R$ is the Schmidt hammer rebound number, described in Chapter 5, Section 5.5, measured on a fresh dry surface; $r$ is the rebound number on the surface of the joint wall in its natural state, wet or dry; and $\phi_b$ is the basic friction angle of the rock, which can be obtained from the literature (see Table 3.13).

If the discontinuity walls are fresh, $\phi_r = \phi_b$. Typical $\phi_b$ values in planar unweathered discontinuities are in the order of 25° to 37° for sedimentary rocks, 29° to 38° for igneous rocks and 21° to 30° for metamorphic rocks.

### Joint Wall Compression Strength, JCS

If the joint walls are not weathered, the uniaxial compressive strength value of the intact rock, $\sigma_c$, is taken. If the wall is weathered, as usually happens, the JCS value can be obtained from the results of the Schmidt hammer on the joint wall, using the expression:

$$\log_{10} JCS = 0.00088\, \gamma_{rock}\, r + 1.01$$

where JCS is in MN/m² and $\gamma_{rock}$ in kN/m³.

Where no direct measurements are available, a ratio of JCS/$\sigma_c$ = 0.25 may be used, or field index testing may be used.

### Joint Roughness Coefficient, JRC

The JRC coefficient depends on the roughness of the discontinuity walls and varies from 1–20. It can be obtained from:

- The standard roughness profiles (Figure 3.85). The roughness of the joint walls should be classified in advance, taking both macro and micro scales into account (according to the roughness profiles in Chapter 6, Figure 6.11).
- The tilt test, described in Chapter 5, Section 5.5. For this test, fragments of rock or borehole cores are used. Angle $\alpha$ is determined (the angle at which one of the rock fragments begins to move in relation to the others) and the following expression is applied:

$$JRC = \frac{\alpha - \phi_r}{\log\left(\dfrac{JCS}{\sigma_n}\right)_{test}}$$

### Scale effects

The JRC and JCS parameters depend on the scale used, as shown in Figure 3.86. JRC depends on the magnitude and amplitude of irregularities, which can be divided in waviness (large scale undulation, metres) and roughness or asperities (small scale, cm, mm). As the scale is increased, the value of $i$ is lowered (due to the influence of waviness, and if the discontinuity is allowed to dilate (for low normal stresses), the value of $\phi_p$ falls; if dilation does not take place, the effect of scale is lost. The JRC values obtained empirically correspond to 10 cm long measuring lengths. To analyse the behaviour of longer joints, the values for other scales have to be corrected. The JCS compressive strength value, and therefore the JCS/$\sigma'_n$ component, decreases with an increase in scale. To counteract these effects, Bandis et al. (1981) established the following ratios to obtain parameters for joints of real length $L_n$ ($L_0$ = 10 cm):

$$JCS_n = JCS_0 (L_n/L_0)^{-0.03\, JCS_0}$$

$$JRC_n = JRC_0 (L_n/L_0)^{-0.02\, JRC_0}$$

The strength of the joint on a real scale can then be estimated from the expression (Barton, 1990):

$$\tau = \sigma'_n \tan\left[ JRC_n \log_{10}\left(\frac{JCS_n}{\sigma'_n}\right) + \phi_r + i \right]$$

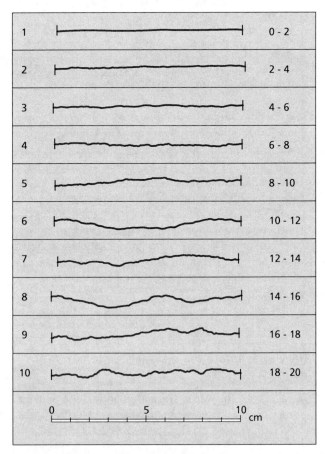

Figure 3.85   Standard profiles for estimating joint roughness coefficient (JRC) (Barton and Choubey, 1977).

where $i$ is the large scale angle of waviness of the discontinuities.

# ROCK MECHANICS

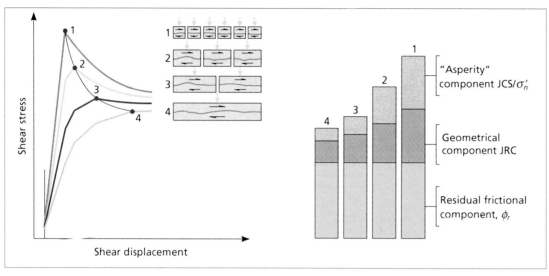

*Figure 3.86* Effect of block size on the shear strength components of joints (Bandis et al., 1981).

Other aspects of the **scale effect** when estimating shear strength in discontinuities are described in Section 3.6.

## Discontinuities with infilling

In discontinuities filled with clay or other materials (e.g. from weathering, shear failure of the walls or deposited by water), the shear strength of the planes is conditioned by the type and thickness of the filling material. If it is thick, shear failure will generally take place through the fill, and the strength of the discontinuity plane will be the strength of the fill. If the fill is hard and consolidated, failure may occur along the contact between the rock and the fill.

The properties of the fill, e.g. shear strength, deformability and permeability, may vary considerably and control how the discontinuity behaves. In contrast to clean discontinuities, fills usually have cohesion. The type of fill is critical and, in general terms, may be:

— Clayey materials.
— Breccia or angular rocky fragments with a higher or lower proportion of clay matrix.
— Crystallized materials (e.g. calcite, gypsum).

Depending on how thick the filling is, the roughness of the plane (a definitive parameter in the shear strength of clean discontinuities) may have no effect on shear strength. *Table 3.18* gives values for the cohesive and frictional strength parameters for discontinuities with fill.

## Direct shear strength laboratory test

This test provides residual and peak shear strength of discontinuities as a function of the normal stress applied to the plane. Shear stresses are applied to a rock sample that includes the discontinuity under study until relative displacement between the two parts takes place. The normal load applied, $\sigma_n$, remains constant throughout the test. Both peak and residual values of the strength parameters of the discontinuity, $c$ and $\phi$, can be obtained from the data obtained on shear stress and shear displacements.

According to the ISRM (1981) at least five tests on five specimens of the same discontinuity plane should be carried out to determine the shear strength.

### Procedure

The testing equipment includes a shear box that can be dismantled and separated into two halves. One part of a prepared sample or specimen containing the discontinuity plane is placed in each half, so that the discontinuity plane coincides with where the two halves of the box join. The samples are fixed in each half of the box with mortar or resin. The surface to be tested should be placed parallel to the direction of the shear force applied, and should preferably be square with a minimum area of 2500 mm. The top and bottom halves of the shear box should be far enough apart to allow for vertical contraction of the discontinuity when the sample is loaded normally.

The discontinuity plane should be affected by weathering as little as possible and maintain the natural conditions of moisture content and roughness it has in the rock mass. The height of each of the two parts of the sample separated by the discontinuity should be $\geq 0.2\ L$, where L is the length of the side of the sample.

When the sample has been placed in the shear box, normal stress is applied perpendicular to the discontinuity

| Table 3.18 | SHEAR STRENGTH OF FILLED DISCONTINUITIES | | | | |
|---|---|---|---|---|---|
| | | Peak strength | | Residual strength | |
| Rock | Description | Cohesion (kPa) | Friction angle (°) | Cohesion (kPa) | Friction angle (°) |
| Basalt | Clayey basaltic breccia with rocky fragments. | 240 | 42 | | |
| Clay shale | Clay filling. | 60 | 32 | | |
| | Clay in bedding planes. | | | 0 | 19–25 |
| Diorite | Clay filling. | 0 | 26.5 | | |
| Dolomite | Clayey filling ≈ 15 cm thick. | 41 | 14.5 | 22 | 17 |
| Granite | Clay filled faults. | 0–100 | 24–25 | | |
| | Sandy loam fault filling. | 50 | 40 | | |
| | Shear zone, broken granites, disintegrated rock and clay gouge. | 240 | 42 | | |
| Greywacke | 1–2 mm clayey filling in bedding planes. | | | 0 | 21 |
| Limestone | 6 mm clay filling. | | | 0 | 13 |
| | 1–2 cm clay filling. | 100 | 13–14 | | |
| | < 1 mm clay filling. | 50–200 | 17–21 | | |
| | 2 cm marl filling. | 0 | 25 | 0 | 15–24 |
| Schist and quartzite | Clayey filling 10–15 cm thick. | 30–80 | 32 | | |
| | Thin clayey filling in bedding planes. | 610–740 | 41 | | |
| | Thick clayey filling in bedding planes. | 380 | 31 | | |
| Slate | Laminated and altered. | 50 | 33 | | |
| Data from various authors and tests performed under different conditions (Barton, 1974; Hoek and Bray, 1981). | | | | | |

surface until the required value is reached. A shear force is then applied, by hydraulic or mechanical means, on the sides of the shear box until displacement by shearing along the plane occurs.

The test becomes more complicated if the discontinuity is full of soft saturated material; in this case, the filling has to be consolidated and the water pressure dissipated before the shear test is run (ISRM, 1981).

The **Hoek cell** is a portable shear apparatus for carrying out field or laboratory tests using the same procedure described above (Figures 3.87 and 3.88). It allows rapid tests to be carried out using borehole cores containing a discontinuity.

## Interpretation

The peak values of normal and shear stresses are obtained by dividing the forces applied by the section of the sample that remains in contact:

$$\tau_p = P_{shear}/A; \quad \sigma_n = P_{normal}/A$$

During the test, the shear stress values and the normal and shear displacement values are measured (in rough discontinuities, displacements perpendicular to the plane will occur as any irregularities have to be overcome for shear displacement to take place). The corresponding $\tau$-shear displacement and $\tau$-normal displacement curves can then be plotted. From these curves, the values of $\tau_{peak}$ and $\tau_{residual}$ are obtained, which are represented on a diagram, $\tau - \sigma_n$,

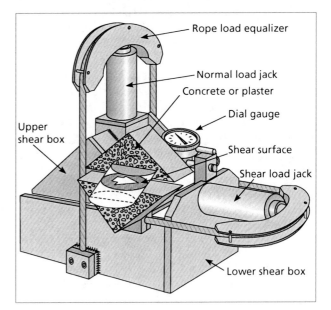

Figure 3.87   Hoek shear box for measuring discontinuity shear strength (Hoek and Bray, 1981).

and the values of $\phi$ and c corresponding to shear and residual strength can be read directly (Box 3.9). Care should be taken with rough surfaces to compare results from samples sheared in the same direction across the surface.

### Influence of scale

Test results are influenced by the scale of the test, i.e. the size of the sample being tested; this is called the **scale effect**. Shear strength in discontinuities depends mainly on the roughness and waviness of the planes, and therefore on the area tested. In the laboratory, only a small portion of the joint is tested, whereas with *in situ* tests roughness can be considered on a much larger scale (Figure 3.89). The effect of scale on shear strength is more marked the greater the roughness, and diminishes as the scale of testing is increased. The above is applicable if normal stresses are low and the discontinuity is allowed to dilate or open during the test; if this is not the case, the influence of scale is less. Peak shear strength falls as the test area increases. For joints filled with clay material, the scale effect may be non-existent.

To summarize the above: when tests are carried out on a larger scale, the roughness angle, *i*, is lower, and the value of $\phi_p$ and of the shear strength is therefore lower; the scale effect on discontinuities is also described in Section 3.6.

## Permeability and water pressure

The **permeability** of a discontinuity depends on its aperture and the type of fill. Aperture is conditioned by the state of stress of the rock mass; permeability therefore depends on *in situ* stresses. The value of *k* in a smooth clean discontinuity is given by:

$$k = a^2 g/12\eta \quad \text{or} \quad k = a^2 \gamma_w/12\mu$$

where:

$k$ = permeability coefficient (cm/s)
$g$ = gravity acceleration.
$a$ = discontinuity aperture.
$\gamma_w$ = unit weight of water.
$\eta$ = kinematic viscosity coefficient of water (0.0101 cm$^2$s$^{-1}$ at 20°).
$\mu$ = dynamic viscosity coefficient of water (0.01005 g · s$^{-1}$cm$^{-1}$ at 20°).

If the discontinuity is rough, the "hydraulic" aperture ($a_h$) will be less than the "real" or "mechanical" one (*a*), and both are related (according to Lee *et al.*, 1996; in Singhal and Gupta, 1999):

$$a_h = a^2/\text{JRC}^{2.5}$$

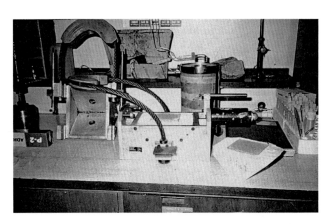

Figure 3.88  Hoek shear box.

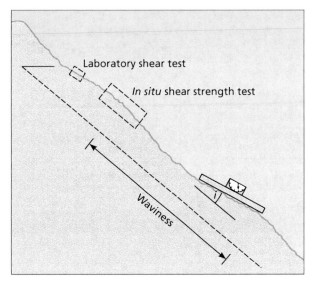

Figure 3.89  Different scales of discontinuity roughness (ISRM 1981). Waviness can be characterized by the angle *i*. Roughness on a millimetric scale can only be determined in laboratory tests.

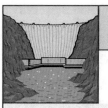

## Box 3.9

### Calculating the strength parameters c and ϕ for discontinuities

The shear stress values $\tau_{peak}$ and $\tau_{residual}$ are measured from the $\tau$-shear displacement curves obtained for each test. These values are plotted on a graph, $\tau - \sigma_n$, for the $\sigma_n$ values corresponding to the different tests, and lines fitted to the points, to give the cohesion and friction values typical of residual and peak shear strength for the discontinuity being tested. Because the procedure is based on the Mohr-Coulomb linear failure criterion, the points on the graph must be fitted to a straight line. In the case of rough discontinuities, the relation between $\tau$ and $\sigma_n$ will be bilinear, as shown in *Figure 3.81b*.

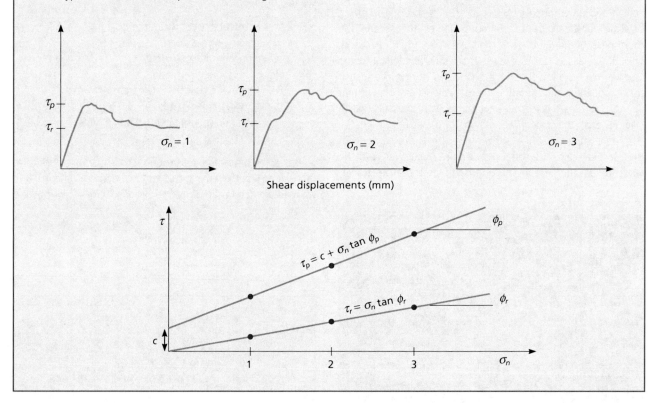

where JRC is the joint roughness coefficient (according to the description earlier in this section).

For a set of discontinuities, permeability also depends on the spacing between the planes. The permeability coefficient or hydraulic conductivity of a system of smooth clean discontinuities, with spacing $b$, can be estimated from the following empirical expressions:

$$k_f = a^3 g / 12 \eta b \quad \text{or} \quad k_f = a^3 \gamma_w / 12 \mu b$$

The relation between the permeability coefficient, the joint aperture and spacing is shown in *Figure 3.90*.

The presence of water in discontinuities reduces their shear strength; the **pressure** exerted by the water is directly opposed to the normal stress component on the joint, reducing the effective stress (*Figure 3.91*).

Using the Mohr-Coulomb criterion, the water pressure value, $u$, needed to produce shear displacement in a discontinuity is:

$$u = \sigma_n + \frac{c - \tau}{\tan \phi}$$

or, in relation to the principal stresses:

$$u = \sigma_3 + (\sigma_1 - \sigma_3)\left(\cos^2 \theta - \frac{\sin \theta \cos \theta}{\tan \phi}\right) + \frac{c}{\tan \phi}$$

where $\theta$ is the angle formed by the perpendicular to the discontinuity plane with the major principal stress and $\phi$ the friction angle of the discontinuity. The value of $u$ will be the minimum value calculated with the above equation, either when $c = 0$ and $\phi = \phi_b + i$, or when $c \neq 0$ and $\phi = \phi_r$.

## 3.6 Strength and deformability of rock masses

### Rock mass strength

The strength of rock masses depends as much on the strength of the intact rock as it does on the discontinuities, both of which are extremely variable, and on the geo-environmental conditions the rock mass is subjected to, such as natural stresses and hydrogeological conditions. The presence of tectonized or weathered areas or those with a different lithological composition implies weak and anisotropic zones with different mechanical behaviour and strength characteristics. These factors make evaluating rock mass strength very complex.

Strength can be assessed in terms of the maximum stress the rock mass can bear under certain conditions, and in terms of its strength parametres, $c$ and $\phi$, which are often required for the analysis and calculation of engineering projects.

Depending on the degree of fracturing of the rock mass, its behaviour and strength properties are defined by:

— The strength of the intact rock (isotropic or anisotropic)
— The shear strength of one discontinuity set.
— The shear strength of two or three discontinuity sets (as long as they are representative of the rock mass)
— The overall strength of a rock block system with isotropic behaviour.

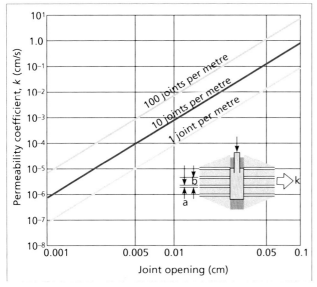

Figure 3.90 Influence of joint opening and spacing on the permeability coefficient of a set of smooth planar parallel discontinuities (Hoek and Bray, 1981).

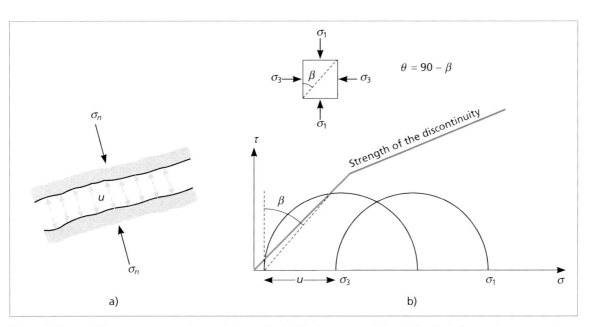

Figure 3.91 a) Water pressure acting on joint walls. b) Displacement of the Mohr circle for total stresses when pore pressure ($u$) is subtracted to give the magnitude of effective stresses.

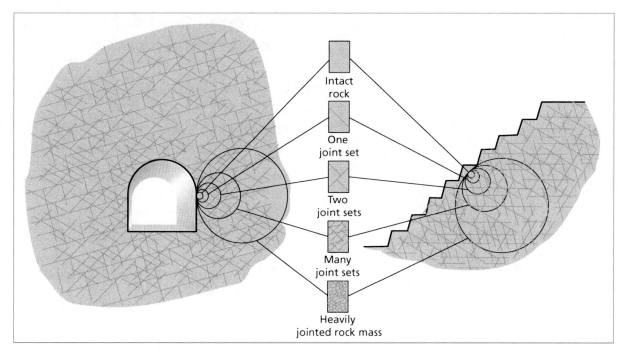

*Figure 3.92* Diagram showing the transition from intact rock to heavily jointed rock mass with increasing sample size (Hoek and Brown, 1997).

*Figure 3.92* shows the transition between the different situations described. In shallow and deep excavations, the excavation work, stability and mechanical behaviour problems are directly related to the rock material's strength and the presence of discontinuities. The strength of the intact rock or a discontinuity plane can be calculated with laboratory tests or *in situ*. With respect to the rock mass, their dimensions and natural conditions cannot be reproduced in the laboratory, and there are no appropriate methods available for estimating its strength *in situ*. This is why the strength of rock masses must be estimated using indirect methods.

Once the elements that control the rock mass strength have been established (e.g. one or more discontinuity sets, the intact rock, the rock mass as a whole, specific weak areas), it can be evaluated using the following procedures:

— Empirical methods based on experience and laboratory tests.
— Indirect methods based on quality indexes (geomechanical classifications).
— Mathematical models and back analysis.
— Physical models.

**Failure or strength criteria** are the basis of empirical methods. They are used to assess rock mass strength based on applied stresses and material properties, with the following results:

— The response of the intact rock to different stress conditions.
— Forecasting the influence of discontinuities on the behaviour of the rock mass.
— Forecasting the global behaviour of the rock mass.

**Quality indexes** defined from geomechanical classifications can be used to estimate strength, by establishing correlations between different rock mass classes and the rock mass strength parameters, c and $\phi$, (see *Table 3.20*). The classifications are described in Section 3.8 and Chapter 10.

**Mathematical models** can be used to estimate strength through numerical modelling of the behaviour of the rock mass, its physical and mechanical properties, its behaviour law and factors influencing it (e.g. stresses or water pressure). These models are specially useful for **back analysis**, using numerical modelling of deformations and failure process in a real rock mass (where the failure characteristics and mechanisms are known) to obtain the strength parameters corresponding to the rock mass failure or to a specific deformation level.

**Physical models** use scale models built with natural or artificial materials (e.g. plaster components, blocks of rigid material, a mixture of sand and clay with a binder material), subjecting them to different loads to observe their behaviour.

The above methods allow the approximate strength of the rock masses to be obtained, depending on the quantity and quality of the available information and data and how representative these are. Empirical criteria and mathematical models based on back analysis give the most representative results; the determination of the relevant values for c and $\phi$, the characteristic rock mass strength parameters, is the most questionable point. From the procedures mentioned above, only mathematical and physical models take the deformational behaviour of the rock masses into account.

## Failure criteria for isotropic rock masses

### Hoek-Brown criterion

This failure criterion is valid for isotropic rock masses and takes into account the determining factors for failure of rock on a large scale, such as non-linearity with stress level, influence of rock type and state of the rock mass, relationship between the compressive and tension strength or reduction of the angle of internal friction with increased confining stress.

The criterion was initially developed to be applied to **unaltered, jointed rock masses with hard intact rock**, assuming that the intac rock blocks are in contact with each other and that the strength of the mass is controlled by the strength of the jointing or discontinuities. The strength of the mass is defined by the expression (Hoek and Brown, 1980):

$$\sigma_1 = \sigma_3 + \sigma_{ci}\sqrt{m\frac{\sigma_3}{\sigma_{ci}} + s}$$

where:

$\sigma_1$ and $\sigma_3$ are the major and minor principal stresses at failure.

$\sigma_{ci}$ is the uniaxial compressive strength of the intact rock material.

$m$ and $s$ are rock mass constants chosen to reflect the properties of the rock mass, and the type, frequency and characteristics of the discontinuities. (Thus an unbroken solid cylinder of rock has $s = 1.0$ making $\sigma_1$ for unconfined failure, when $\sigma_3 = 0$, equal to its unconfined compressive strength, $\sigma_{ci}$).

The value of $\sigma_{ci}$ should be obtained in uniaxial compressive strength laboratory tests or else it can be estimated from the PLT test (described in Chapter 5, Section 5.5) or from field indexes (Table 3.7). The $m$ and $s$ values can be obtained from the RMR index (described in Section 3.8), taking into account if the rock mass is undisturbed or disturbed in terms of its properties (Hoek and Brown, 1988):

— For undisturbed rock masses not affected by blasting:

$$m = m_i \exp\frac{RMR - 100}{28} \qquad s = \exp\frac{RMR - 100}{9}$$

— For disturbed rock masses or affected by blasting:

$$m = m_i \exp\frac{RMR - 100}{14} \qquad s = \exp\frac{RMR - 100}{6}$$

where $m_i$ is the value corresponding to the intact rock obtained from triaxial compression tests for appropriate $\sigma_3$ value ranges (the values for different types of rocks are included in Table 3.14). If the rock mass is completely fresh (RMR = 100) $m = m_i$ and if it is unbroken by discontinuities $s = 1$.

Table 3.19 gives the values for constants $m$ and $s$ depending on rock type and rock mass quality. Values are included for undisturbed and disturbed rock masses.

The normal recommendation is to use values corresponding to disturbed rock mass conditions. However, there is some confusion when selecting the $m$ and $s$ values, since the "disturbed" classification refers both to rock mass disturbed by excavation or blasting and to a weathered rock mass. For this reason, using the different available methods is recommended, so these parameters can be adjusted by judgement as far as possible.

Neither the Hoek-Brown criterion nor the expressions for calculating $m$ and $s$ give representative values for weathered or poor quality rock masses. As a result, a new expression was developed, also valid for **poor quality jointed rock masses** with soft and weathered materials, introducing the concept of the **generalised Hoek-Brown criterion** for jointed rock masses (Hoek, 1994):

$$\sigma_1 = \sigma_3 + \sigma_{ci}\left(m\frac{\sigma_3}{\sigma_{ci}} + s\right)^\alpha$$

where $m$ is a reduced value of the intact rock constant $m_i$, and $s$ and $\alpha$ are constants depending on the properties of the rock mass.

The uniaxial compressive strength is obtained by setting $\sigma_3 = 0$:

$$\sigma_c = \sigma_{ci} \cdot s^\alpha$$

From the equation of the generalised criterion, the shape of the principal stress curve $\sigma_1$ against $\sigma_3$ could be adjusted by means of the variable coefficient $\alpha$.

The equivalent Mohr envelope corresponding to this criterion is expressed (Figure 3.93):

$$\tau = A\sigma_{ci}\left(\frac{\sigma_n - \sigma_{tm}}{\sigma_{ci}}\right)^B$$

### Table 3.19 APPROXIMATE RELATIONSHIPS BETWEEN ROCK MASS QUALITY AND MATERIAL CONSTANTS m AND s

**Empirical failure criterion**

$$\sigma_1 = \sigma_3 + \sqrt{m\sigma_{ci}\sigma_3 + s\sigma_{ci}^2}$$

$\sigma_1$ and $\sigma_3$: major and minor principal effective stress
$\sigma_{ci}$: uniaxial compressive strength of intact rock
$m$ and $s$: empirical constants of rock mass

Disturbed rock mass $m$ and $s$ values
Undisturbed rock mass $m$ and $s$ values

| | Material constants: $m$ and $s$ | Carbonate rocks: dolomite, limestone, marble | Lithified argillaceous rocks: mudstone, siltstone, shale, slate | Arenaceous rocks: sandstone and quartzite | Fine-grained igneous crystalline rocks: andesite, dolerite, diabase, rhyolite | Coarse-grained igneous and metamorphic crystalline rocks: amphibolite, gabbro, gneiss, granite, norite, quartz-diorite |
|---|---|---|---|---|---|---|
| **Intact rock samples** Laboratory size specimens free from discontinuities. RMR = 100 Q = 500 | m s m s | 7.0 1.0 7.0 1.0 | 10.0 1.0 10.0 1.0 | 15.0 1.0 15.0 1.0 | 17.0 1.0 17.0 1.0 | 25.0 1.0 25.0 1.0 |
| **Very good quality rock mass** Undisturbed rock with unweathered joints at 1 to 3 m. RMR = 85 Q = 100 | m s m s | 2.40 0.082 4.10 0.189 | 3.43 0.082 5.85 0.189 | 5.14 0.082 8.78 0.189 | 5.82 0.082 9.95 0.189 | 8.56 0.082 14.63 0.189 |
| **Good quality rock mass** Fresh to slightly weathered rock, slightly disturbed with joints at 1 to 3 m. RMR = 65 Q = 10 | m s m s | 0.575 0.00293 2.006 0.0205 | 0.821 0.00293 2.865 0.0205 | 1.231 0.00293 4.298 0.0205 | 1.395 0.00293 4.871 0.0205 | 2.052 0.00293 7.163 0.0205 |
| **Fair quality rock mass** Several sets of moderately weathered joints spaced at 0.3 to 1 m. RMR = 44 Q = 1 | m s m s | 0.128 0.00009 0.947 0.00198 | 0.183 0.00009 1.353 0.00198 | 0.275 0.00009 2.030 0.00198 | 0.311 0.00009 2.301 0.00198 | 0.458 0.00009 3.383 0.00198 |
| **Poor quality rock mass** Numerous weathered joints at 3 to 50 cm, some gouge. Clean compacted waste rock. RMR = 23 Q = 0.1 | m s m s | 0.029 0.000003 0.447 0.00019 | 0.041 0.000003 0.639 0.00019 | 0.061 0.000003 0.959 0.00019 | 0.069 0.000003 1.087 0.00019 | 0.102 0.000003 1.598 0.00019 |
| **Very poor quality rock mass** Numerous heavily weathered joints spaced <5 cm, with gouge. Waste rock with fines. RMR = 3 Q = 0.01 | m s m s | 0.007 0.0000001 0.219 0.00002 | 0.010 0.0000001 0.313 0.00002 | 0.015 0.0000001 0.469 0.00002 | 0.017 0.0000001 0.532 0.00002 | 0.025 0.0000001 0.782 0.00002 |

Hoek and Brown, 1988.

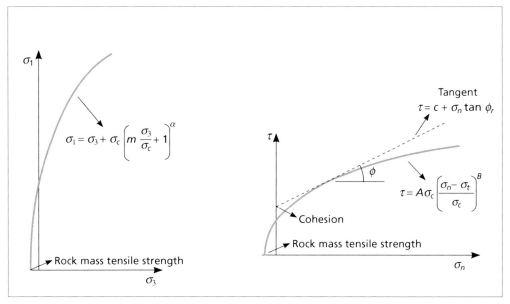

Figure 3.93   Hoek and Brown generalized failure criterion in terms of major and minor principal stresses and in terms of shear and normal stresses.

where A and B are material constants, $\sigma_n$ is the normal stress at the point of interest and $\sigma_{tm}$ is the tensile strength of the rock mass, given by:

$$\sigma_{tm} = \frac{s\,\sigma_{ci}}{m}$$

The Hoek-Brown criterion is not appropriate for anisotropic rock masses with preferred discontinuity or failure directions, or for rock masses with widely spaced discontinuities in relation to the scale of the engineering structure, i.e. when the study area is not representative of the geological structure of rock mass as a whole (see Figure 3.92).

## Obtaining the coefficients $m$, $s$ and $\alpha$

The Rock Mass Rating of Beniawski does not consider many of the characteristics of the rock masses which can only be obtained from geological observations in the field, particularly for weak rock masses, so it is not appropriate to include it with the Hoek-Brown criterion when applied to poor or very poor quality rock masses.

To widen the application range of the generalized criterion to poor quality rock masses, and to use "more geological" parameters to evaluate their strength, Hoek (1994) and Hoek et al. (1995) proposed the **Geological Strength Index** (GSI) which evaluates rock mass quality depending to some extent on the degree and characteristics of fracturing, geological structure, block size and discontinuity alteration.

The GSI replaces the RMR initially adopted for evaluating the generalized Hoek-Brown parameters $m$, $s$ and $\alpha$.

Initially drawn up for 4 classes of rock mass, the GSI has evolved into new applications for different geological conditions, in parallel to the evolution of the Hoek-Brown criterion. The latest updates extend the range of conditions for weak and sheared, poor and very poor quality rock masses (Hoek et al., 1998; Hoek and Marinos, 2000).

The Geological Strength Index (Figure 3.94) classifies the rock mass based on geological information, combining structural characteristics and surface conditions. In spite of its advantages compared with the use of the RMR index for poor and very poor quality rock masses, the GSI is still a merely qualitative classification of rock masses and a simplification of the geological conditions which these present in the natural setting.

From this index, the values of $m$, $s$ and $\alpha$ can be obtained (Hoek et al., 2002):

$$m = m_i \exp\left(\frac{GSI-100}{28-14D}\right)$$

$$s = \exp\left(\frac{GSI-100}{9-3D}\right)$$

$$\alpha = \frac{1}{2} + \frac{1}{6}\left(e^{-GSI/15} - e^{-20/3}\right)$$

These equations give smooth continuous transitions for the entire range of GSI values between very poor quality rock masses (GSI<25) and hard rocks.

The parameter $D$ is a factor that depends on the degree of disturbance to which the rock mass has been

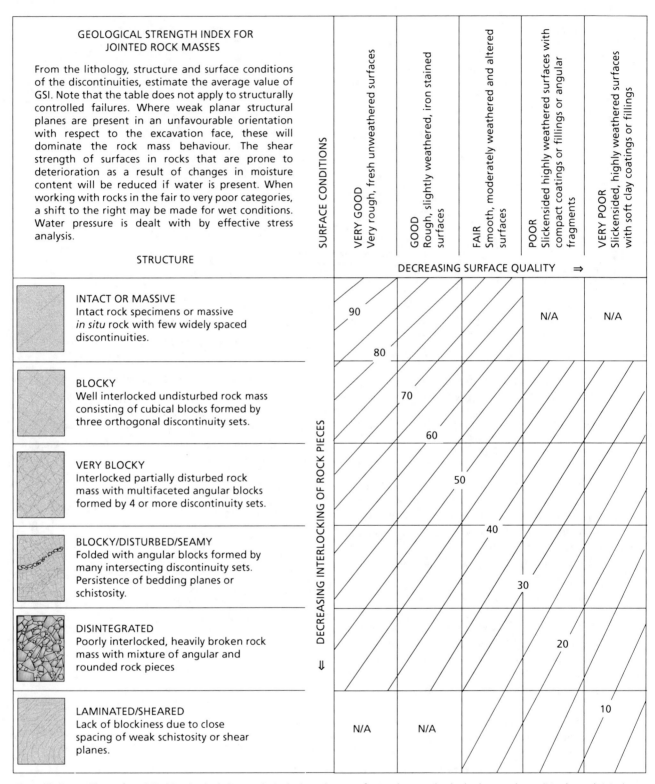

Figure 3.94  Chart for GSI (Geological Strength Index) estimates from the geological observations (Hoek and Marinos, 2000).

subjected by blast damage or stress relaxation. It varies from 0 for undisturbed *in situ* rock masses to 1 for very disturbed rock masses.

The GSI can be obtained from the RMR (Section 3.8) through the following correlation, where a value of 15 should be assigned to the groundwater conditions of the rock mass and a value of 0 to the adjustment parameter for joint orientation:

$$GSI = RMR_{(89)} - 5$$

When applying the Hoek-Brown criterion and interpreting the strength values obtained, the following should be taken into account:

— The criterion is only valid for rock masses with isotropic behaviour.
— The GSI is based on qualitative estimates and on simple models which seldom coincide with actual conditions.
— Results must be contrasted with other methods and wherever possible with field data and back analysis.

## Obtaining the rock mass strength parameters c and $\phi$

Determining the equivalent Mohr-Coulomb parameters for a rock mass, i.e. friction angle and cohesive strength, from the Hoek-Brown criterion expressions is not straightforward: the main difficulty is that as it is a non-linear criterion, the values of the two parameters are not constant, but a function of the normal stress value $\sigma_n$.

The values of $c$ and $\phi$ can be obtained by fitting a linear Mohr-Coulomb envelope, by least squares methods, to the curve of the generalised Hoek-Brown criterion for a range of minor principal stress values defined by $\sigma_t < \sigma_3 < \sigma_{3max}$, so that the equations for the angle of friction and cohesive strength can be obtained (Hoek et al., 2002). The value of $\sigma_{3max}$, the upper limit of confining stress over which the relationship between the Hoek-Brown and the Mohr-Coulomb criteria is considered, has to be defined for each individual case. Guidelines for selecting these values when applying the equations for slopes and shallow or deep tunnels are given in Hoek et al. (2002).

This procedure has been incorporated into the Windows program RocLab (www.rocscience.com) which includes tables and charts for estimating the uniaxial compressive strength of the intact rock, the material constant $m_i$ and the GSI index. An example of the application of this program is shown in Chapter 10, Box 10.2.

The Mohr-Coulomb shear strength $\tau$, for a given normal stress $\sigma_n$, is found by substitution of the values of $c$ and $\phi$ in the equation:

$$\tau = c + \sigma_n \tan\phi$$

## Mohr-Coulomb criterion

The advantage of the Mohr-Coulomb criterion is its simplicity. However, it does have the disadvantages mentioned above for intact rock (Section 3.4), and above all in relation to the non-linear stress-strain behaviour of the rock masses, so it is not an appropriate criterion to use to estimate the strength of a rock mass. However, in some cases the criterion can be used for hard rock masses where failure occurs along clearly identifiable discontinuity surfaces, taking into account that representative values for the rock mass should be taken for the cohesion and for the angle of friction. Beniawski (1979) proposed some indicative values for both these parameters, depending on the quality of the rock mass given by the RMR index (defined in Section 3.8) (Table 3.20).

## Considerations on the shear strength calculation of rock masses using Mohr-Coulomb, Hoek-Brown and Q parameters

The conventional addition of cohesion and tangent of friction angle in continuum models, assumed by the linear Mohr-Coulomb failure criterion and the non-linear Hoek & Brown failure criterion, cannot be appropriate when considering that the cohesive component of strength (representing the complex small-strain fracturing of the intact rock) may be a precursor to the mobilization, at larger strains, of the non-linear frictional strength (Hajiabdolmajid et al., 2002; Barton and Pandey, 2010).

In both of the standard criteria, it is assumed that the cohesive and the frictional strength components are mobilized instantaneously and simultaneously, i.e., they are assumed to be additive, considering the shear strength as $c + \tan\phi$. However, these approaches with typical strength parameters have not been successful in predicting the type and extent of failure in deep underground openings in hard rock masses, or borehole breakouts in deep boreholes. A constitutive model introducing cohesion weakening and frictional strengthening (CWFS), making the mobilization of the strength components a function of plastic strain, and assuming that the cohesion

| Table 3.20 | APPROXIMATE VALUES FOR c AND $\phi$ OF THE ROCK MASS DEPENDING ON QUALITY | | | | |
|---|---|---|---|---|---|
| Rock class | I | II | III | IV | V |
| RMR | >80 | 61–80 | 41–60 | 21–40 | <20 |
| Cohesion (MPa) | >0.4 | 0.3–0.4 | 0.2–0.3 | 0.1–0.2 | <0.1 |
| Friction angle | >45° | 35°–45° | 25°–35° | 15°–25° | <15° |

must be overcome before the frictional strength can be mobilized, was presented by Hajiabdolmajid et al. (2002). The comparison of this model with that of H&B (where c and $\phi$ rock mass shear strength parameters are estimated by the H&B formulations using the Roclab application), resulted in more realistic results.

In fact the following simple approach can be used to estimate the always very uncertain shear strength of a rock mass as "c then tan $\phi$" for continuum analysis, by using the Q parameters (Barton, 2002):

— Cohesion component (CC):

$$c = \left(\frac{RQD}{J_n} \times \frac{1}{SRF} \times \frac{\sigma_c}{100}\right)$$

— Frictional component (FC):

$$\phi = \tan^{-1}\left(\frac{J_r}{J_a} \times J_w\right)$$

This simple approach has been applied in some underground mining excavations and compared to the H&B approach (Barton and Pandey, 2010). The results have shown that the failure mode is highly unlikely to be governed by the c + tan $\phi$ formulation, the "c then tan $\phi$" approach being more realistic. Modelling with cohesion softening and friction hardening criteria has been shown by increasing numbers of researchers to better match break-out resulting from stress fracturing (e.g. Edelbro, 2008).

The complexity and highly anisotropic behaviour of most rock masses must be carefully considered when shear strength parameters have to be calculated. The use of the H&B—Roclab procedure based on a simplified continuum and the assumption that the mobilization of the cohesion and frictional strength components is simultaneous, may lead to unrealistic rock mass shear strength parameters, and therefore unrepresentative results of breakout and stress-induced failure.

It is difficult to quantify the geomechanical behaviour of a rock mass, even when simplified to an isotropic continuum medium. As mentioned above, the determination of the values of the rock mass strength parameters, c and $\phi$, needs always to be questioned. The calculation of input parameters for numerical modelling of rock masses using single RMR, Q or GSI values, given the actual complexity and variations of most of rock masses, suppose a gross simplification that may lead to the misinterpretation of the rock mass behaviour and strength.

Also, even ignoring the failure mechanism complexity, it must be considered that, in spite of the extent and use of numerical models considering a jointed medium and discontinuity properties (such as UDEC), given the difficulty of obtaining realistic input data inevitably means that the results must be interpreted based on experience.

# Failure criteria for anisotropic rock masses

## Stratified rock masses (with a single discontinuity set)

The strength of a stratified rock mass, assuming that there is no other set of discontinuities affecting its behaviour, may be considered as similar to that of a transversally anisotropic intact rock. The strength of the rock mass depends on the direction of the applied stresses in relation to the direction of the bedding planes and will vary in value from a minimum corresponding to the strength of these planes to a maximum corresponding to the intact rock strength. In cases where the intact rock is weak and anisotropic, as e.g. in clay or marl rocks, there may be little difference between strengths.

Where failure occurs along the stratification planes, the strength of the rock mass can be found using the Mohr-Coulomb criterion:

$$\sigma_1 = \frac{2c + \sigma_3\left[\sin 2\theta + \tan\phi(1-\cos 2\theta)\right]}{\sin 2\theta - \tan\phi(1+\cos 2\theta)}$$

where $\theta$ is the angle the perpendicular to the failure plane forms with the major principal stress $\sigma_1$. For values of $\theta$ around 90° or lower than the angle of friction of the plane, $\phi$, failure is not possible along the stratification planes and the rock mass will failure through the intact rock. If the intact rock is isotropic, strength can be evaluated applying the Hoek-Brown criterion (s = 1):

$$\sigma_1 = \sigma_3 + \sigma_{ci}\sqrt{(m\sigma_3/\sigma_{ci}) + s}$$

## Jointed rock masses (many discontinuity sets)

The strength of fractured competent rock masses affected by 2 or 3 sets of discontinuities at right angles to each other, will depend on the strength of the discontinuities and the angle of incidence of the load applied in relation to the weakness planes. For each possible direction, one of the sets will control the strength of the rock mass, and its strength as a whole is obtained by integrating the strength curves of each set (Figure 3.95). In the curve corresponding to 3 sets of discontinuities, it can be seen that the rock mass cannot fail through the intact rock, as a joint set orientation will prevail in any direction. The strength of the rock mass becomes more uniform as the number of sets increases.

Supposing that four sets of discontinuities are present in the rock mass, each with the same strength parameters and arranged spatially at 45° from each other, the strength of the rock mass will be represented by the superimposition of the strength curves $\sigma_1 - \beta$ ($\beta = 90 - \theta$) of each set, as shown

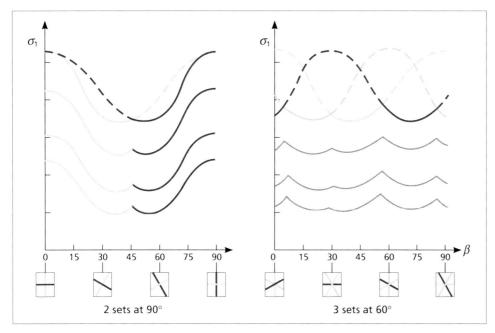

*Figure 3.95* Strength curves for rock masses with 2 and 3 sets of discontinuities (Hoek and Brown, 1980).

in *Figure 3.96*. Failure will always occur along one of the four sets, depending on the stress orientation. The behaviour of the rock mass for each orientation is determined by the lowest strength of the individual curves of each set.

As can be seen in the strength curve in *Figure 3.96*, isotropic behaviour can be assumed for a rock mass with enough discontinuities. Hoek and Brown (1980) consider a rock mass to be isotropic, in the context of designing underground excavation, if 4 or more sets of discontinuities are present. Where this is the case, the failure criteria for isotropic rock masses should be used.

## Summary

*Table 3.21* gives the failure criteria applicable to the different cases described above for evaluating rock mass strength.

# Rock mass deformability

The deformability of a rock mass is defined by the relationship between the applied stress and the strain produced, and is defined through its modulus of deformation, which relates stress to resulting deformations. As is the case with the other properties of the rock mass, deformability exhibits an anisotropic and discontinuous character in nature, and so it is not easy to measure, and one of the most difficult problems to solve in rock mechanics. It depends on the deformability of both the intact rock and the discontinuities, and it is always lower than that of the intact rock. Intact rock deformability is given by its Young modulus, $E_i = \sigma/\varepsilon$,

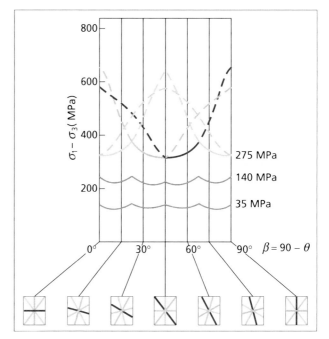

*Figure 3.96* Hypothetical strength curves for rock masses affected by 4 sets of discontinuities (Brady and Brown, 1993).

and for discontinuities it is given by discontinuity stiffness, $k = \sigma/\mu$. (MPa/mm). Deformations in each case are given by $\sigma/E_i$ and $\sigma/k$. For a rock mass with a single family of discontinuities with spacing $S$, the deformation normal to the planes will be the sum of the deformations for intact

Table 3.21 ROCK MASS FAILURE CRITERIA AND DATA NEEDED TO APPLY THEM

| Rock mass characteristics | Failure along discontinuity planes | Failure through intact rock |
|---|---|---|
| Massive rock mass with no discontinuities | Not possible | Hoek and Brown ($m_i$ for intact rock and $s = 1$) |
| Stratified rock mass | Mohr-Coulomb ($c$ and $\phi$ of discontinuities) | Hoek and Brown ($m_i$ for intact rock and $s = 1$) |
| Rock mass with two sets of discontinuities | Mohr-Coulomb ($c$ and $\phi$ of discontinuities) | Hoek and Brown ($m_i$ for intact rock and $s = 1$) |
| Rock mass with three orthogonal sets of discontinuities | Hoek and Brown ($m$, $s$ and $\alpha$ of the rock mass) | Generally not possible. If it is possible, follow criteria for 2 sets |
| Rock mass with four or more sets of discontinuities | Hoek and Brown ($m$, $s$ and $\alpha$ of the rock mass) | Not possible |

rock and discontinuities, $\sigma/E_i$ and $\sigma/k_n$, where $k_n$ is the normal stiffness. Assuming the rock mass behaves as an elastic continuum, then the deformation moduli are related through the following formula (Goodman, 1989):

$$1/E = (1/E_i) + (1/k_n S)$$

where $E$ is the deformation modulus for the rock mass.

When these components are known, the influence the spacing has on the rock mass deformability modulus can be assessed.

Estimating deformability directly is a problem as it is difficult to carry out real-scale compression tests on rock masses and to test representative volumes in the laboratory. Methods for evaluating rock mass deformability can be classified as direct or indirect. The first group includes *in situ* tests, and the second group includes geophysical methods and a series of empirical correlations.

The most appropriate methods are *in situ* tests, although the different methods have not been sufficiently compared and contrasted to determine how representative they are. In addition, the anisotropic character of deformability and the influence of test method on the results obtained means that values often vary for the same rock mass.

## *In situ* deformability tests

Laboratory tests on small specimens can not predict the properties of the rock masses; *in situ* tests are necessary, but they are expensive and time consuming.

Using *in situ* tests to evaluate deformability, just as for other properties of the rock mass, presents problems of knowing how representative the results are and how they can be extrapolated, especially in relation to the scale of the test. Extrapolation of results from rock mass scale tests, and also of laboratory tests, can only be considered valid if the volume of the sample tested is equal or larger than the "representative elemental volume" (REV). This scaling effect is discussed later in this chapter.

The most commonly used tests to calculate deformability are the dilatometer, plate bearing and flatjack tests, which are described in Chapter 5, Section 5.5. The first of these is performed in excavations and adits, the second on adit walls, and the third in boreholes. The dilatometer is the only method available for gauging rock mass deformability at depth. Dilatometer tests, also called "pressuremeter" tests, generally provide deformation moduli considerably smaller than plate bearing tests. Some other tests, less representative and less commonly used, are shown in Table 3.22, which summarises the characteristics of different deformability tests.

An example of a comparison of results given by different tests and techniques is given in Table 3.23, with data from a well-documented case study, and from which a relationship between the *in situ* and the laboratory deformation moduli of the rock mass is obtained: $E_{situ}/E_{lab} = 0.35$.

Due to the variability of rock mass deformability, it is advisable to consider a possible range of variation defined by a maximum and minimum value related to the value of $E$ calculated *in-situ*: $E_{min} = 0.4E$ and $E_{max} = 1.6E$.

## Geophysical methods

These methods are used to assess the "dynamic" deformability of a rock mass, and are based on the velocity of propagation of elastic longitudinal, $V_p$, and shear waves, $V_s$, through the rock mass. The static deformability constants can be calculated from the dynamic constants by correlations, which in general are imprecise. For massive rocks of low porosity, the static and dynamic values are quite close, while for heavily jointed rock masses, large differences are found.

The velocity of propagation of longitudinal waves depends on the type of rock material, the degree of weathering and fracturing of the rock mass, the existing state of stress and the hydrogeological conditions.

The dynamic rock mass deformation moduli, $E_d$ and $v_d$, are obtained from formulae included in Section 3.4 (sonic wave velocity test).

## Table 3.22  IN-SITU TESTS FOR CALCULATING ROCK MASS DEFORMABILITY

| Type | Size | Place | Advantages | Disadvantages |
|---|---|---|---|---|
| Flat jack | Several m$^3$ | On adit walls, in tunnels, etc. | Large samples. Well-known stress distribution. Multiphase. | Expensive. Special work necessary. Few tests. |
| Plate bearing test | Up to 1 m$^3$, according to plate size | In adits, tunnels, shafts, etc. | Large samples. Undisturbed conditions. | Poorly known stress distribution. Hypotheses necessary when interpreting results. Expensive. Special work. Few tests. |
| Dilatometer | Several dm$^3$ | In boreholes. | Can be performed at a depth. Good method for obtaining distribution of the rock mass deformability with closed joints. Low cost. | Small samples relative to the rock mass. Unknown stress distribution. Requires additional hypotheses. |
| Radial jack | Several m$^3$, according to jack size | In special adits. | Large samples. Especially suited to pressurized adits. | Very expensive. Usually only one test. |
| Large scale triaxial test | Several m$^3$ | Close to the surface, in adits, tunnels, etc. | Large samples. Well-known stress distribution. | Very expensive. Special work. Very few tests. Not without sample disturbance. |
| Loaded pillar | Several m$^3$ | In special adits. | Large samples. Well-known stress distribution. | Very expensive. Special work. Very few tests. |

## Table 3.23  DEFORMATION MODULI FOR A PHYLLITE ROCK MASS

| Method | Number of tests | Range (GPa) | Mean (GPa) | Standard deviation (GPa) |
|---|---|---|---|---|
| Small flat jack | 9 | 25.2–47.9 | 31.8 | 6.9 |
| Goodman jack | 6 | 6.0–20.0 | 12 | 6.2 |
| Tunnel relaxation | 4 | 9.7–39.6 | 20 | 13.4 |
| *Petite sismique* | 25 | 12.3–21.5 | 15.4 | 4.6 |
| RMR quality index | 7 | 15.1–22.4 | 20.1 | 2.6 |
| Mean *in-situ* value | – | – | 19.8 | – |
| Laboratory tests | 7 | 46.0–69.0 | 56 | 11.9 |

Bieniawski, 1984.

The value of the dynamic modulus of deformation $E_d$ is greater than the modulus obtained through *in situ* static tests.

No representative results have been obtained despite attempts to determine the relationship between the static deformation modulus of a rock mass and the modulus of the intact rock measured in the laboratory, using what is known as the velocity index $(V_{p\text{-situ}}/V_{p\text{-lab}})^2$ or $(V_F/V_L)^2$ (the square of the quotient of the longitudinal wave velocity as measured in the field and as measured in the laboratory in intact rock samples).

Barton (1995, 2006) presents an empirical expression to estimate the static deformation modulus of the rock mass from the P wave velocity. This is described in the following section.

## Empirical correlations

Empirical criteria relate the rock mass properties to index properties. Several authors have proposed empirical correlations for evaluating a rock mass deformation modulus from other parameters, although at present these relationships are not sufficiently sensitive. Values obtained in this way must be considered as an approximate average value of the deformation modulus, and the anisotropic character of rock mass deformability must be taken into account when applying them.

Based on the **RQD** quality index (described in Chapter 5, Section 5.4) or on **seismic wave velocity** indices, correlations have been established with the rock mass deformation modulus and with the **modulus reduction ratio** (or modulus ratio: quotient of the static modulus of the rock mass obtained from *in situ* tests and the intact rock modulus measured in laboratory: $E_{situ}/E_{lab}$ or $E/E_i$). Figures 3.97 and 3.98 and Table 3.24 show these relationships, although in general these correlations are not precise and are not sufficiently sensitive and contrasted, since they do not consider some of the factors affecting the rock mass deformation modulus. Figure 3.97 shows the non-linear relations between RQD and the ratio $E/E_i$, from data covering the range $0 \leq RQD \geq 100\%$. It must be remembered that:

— when RQD = 100% the rock may still have fractures within it (they just happen to be at distances greater than 100 mm in the direction of measurement), so $E$ may be smaller than $E_i$
— RQD is only one of the factors affecting the deformation modulus of the rock mass
— conditions of discontinuities are not considered.

According to Bieniawski (1984) deformation moduli calculated *in situ* are between 0.2 and 0.6 of the value measured in the laboratory for the intact rock, depending on the rock quality. This author also mentions results obtained by Heuze, in 1980, with a value of 2.5 as a ratio between the lab-obtained deformation modulus and the *in situ* values.

The most widely used empirical relationships for estimating the deformation modulus of an isotropic rock mass are those based on classification indices such as the **RMR** (Rock Mass Rating), **Q** (Tunnel Quality Index) and **GSI** (Geological Strength Index). Below are some correlations proposed by various authors (Table 3.25):

— Bieniawski (1978) established a relationship between the rock mass *in situ* deformation modulus, $E$ (GPa) and the RMR index (described in Section 3.8) using *in situ* test results; this correlation is not valid for rock masses, with RMR<50–55 (Figure 3.99):

$$E = 2RMR - 100$$

— Serafim and Pereira (1983) deduced an expression depending on the value of the RMR index, valid for RMR<50, and particularly for values of $E$ between 1 and 10 GPa. For poor to very poor quality rock masses, excessively high values are obtained:

$$E = 10^{\left(\frac{RMR-10}{40}\right)}$$

The formulae above do not take laboratory data into account and are independent of the deformational properties of the rock mass.

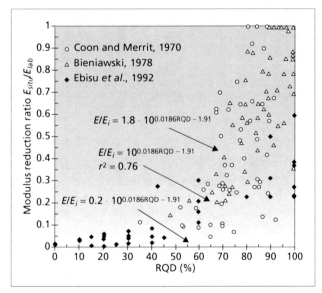

Figure 3.97  Relations between modulus reduction ratio and RQD (Zhang and Einstein, 2004).

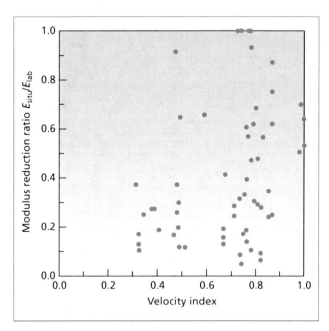

Figure 3.98  Relation between modulus reduction ratio and velocity index (modified from Coon and Merritt, 1970).

| Table 3.24 MODULUS REDUCTION RATIO $E/E_i$ AND RELATIONSHIPS WITH OTHER PARAMETERS | | |
|---|---|---|
| $E/E_i$ | $E/E_i$ and RQD | $E/E_i$ and sound velocity |
| $E = E_i/2.5$ (Heuze, 1980)<br><br>$E = 0.2$ to $0.6\ E_i$, according to rock quality (Bieniawski, 1984)<br><br>$E = jE_i$<br>($j$ = mean spacing between discontinuities)<br>(Kulhawy and Goodman, 1980) | Acceptable correlation for good-quality rock masses (Coon and Merritt, 1970)<br><br>$E/E_i$ and $E$ depend on:<br>— RQD, orientation and spacing of discontinuities.<br>— Discontinuity properties and stiffness.<br><br>$E/E_i = 10^{0.0186RQD-1.91}$<br>(Zhang and Einstein, 2004). | $E/E_i$ and $(V_F/V_L)$: non representative results; poor correlation for good-quality rock masses (Coon and Merritt, 1970)<br><br>There is a correlation between the $E/E_i$ ratio and S-wave length.<br><br>Correlation between $E$ and S-shear wave frequency $f$:<br>$E = 0.054f - 9.2$ (Schneider, 1967; Bieniawski, 1984).<br><br>$E_d > E$ in jointed rocks.<br><br>$E_d/E \leq 13$. |

$E$ = Rock mass deformation modulus.
$E_i$ = Intact rock deformation modulus (laboratory).
$E_d$ = Rock mass dynamic deformation modulus.
$(V_F/V_L)$ = Velocity index (ratio between the longitudinal wave velocity measured in the field and on an intact rock sample in laboratory).
$V_F$ varies with rock type, weathering degree, jointing degree, in-situ state of stress and hydrological conditions.

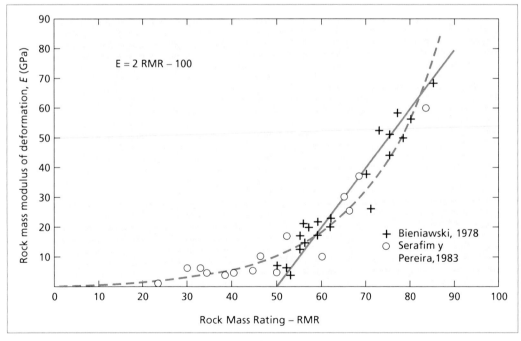

Figure 3.99   Relationship between the *in situ* deformation modulus and the RMR index (Bieniawski, 1984).

— Hoek and Brown (1997) propose a change in Serafim and Pereira's formula based on observations and back-analysis of the behaviour of underground excavations in poor quality masses, where the non-validity of the formula has been verified because higher values of $E$ were obtained than for the intact rock in the laboratory, $E_{lab}$. This formula takes into account the rock's intact compressive strength and replaces the

Table 3.25 **EMPIRICAL EXPRESSIONS FOR ESTIMATING THE DEFORMATION MODULUS IN ISOTROPIC ROCK MASSES**

| Criterion | Application | |
|---|---|---|
| $E = 2\,RMR - 100$ (GPa) (Bieniawski, 1978) | — Good quality rock masses, RMR > 50–55<br>— Not valid for low quality rock masses | — Does not take laboratory data into account<br>— $E$ does not depend on the values of $\sigma_{ci}$ or $E_i$ |
| $E = 10^{(RMR-10)/40}$ (GPa) (Serafim & Pereira, 1983) | — Fair-low quality rock masses, 10 < RMR < 50<br>— Especially indicated for values 1 < $E$ < 10 GPa<br>— Too high values are obtained for low-very low quality rock masses | |
| $E = \left(1 - \dfrac{D}{2}\right)\sqrt{\dfrac{\sigma_{ci}}{100}}\,10^{(GSI-10)/40}$<br>($\sigma_{ci}$ in MPa; $E$ in GPa)<br>(Hoek et al., 2002) | — Indicated for weak or soft rock masses with low or very low quality (GSI < 25) and intact rock with $\sigma_{ci}$ < 100 MPa<br>— D (disturbance factor) = 0 (undisturbed rock mass) to 1 (fully disturbed) | — $E$ is a function of the value of $\sigma_{ci}$ |
| $E = 100{,}000\left(\dfrac{1 - D/2}{1 + e^{(75+25D-GSI)/11}}\right)$<br>($E$ in MPa)<br>(Hoek & Diederichs, 2006) | — Valid for GSI = 20 to 80<br>— D = 0 (undisturbed rock mass)<br>— D = 1 (fully disturbed) | — Properties of the intact rock are not considered<br>— Recommended when reliable data for the intact rock are not available |
| $E = 10\,Q_c^{1/3}$<br>($E$ in GPa)<br>(Barton, 1995, 2006) | — Indicated for jointed or fractured rock masses | — Intact properties of the rock are considered by means of $\sigma_{ci}$ |
| $E = 10^{(V_p - 0.5)/3}$<br>($V_p$ in km/s; $E$ in GPa)<br>(Barton, 2006) | | — Rock mass properties are considered ($V_p$ picks up density and strength) |

$E$ = Rock mass empirical static deformation modulus.
$E_i$ = Intact rock deformation modulus measured in laboratory.
RMR = Rock Mass Rating.
GSI = Geological strength index.
Q = Quality index.
$Q_c = Q \cdot \sigma_{ci}/100$ ($\sigma_{ci}$ in MPa).
$\sigma_{ci}$ = Intact rock uniaxial compressive strength.
$V_p$ = P wave velocity.

- These correlations have not yet been sufficiently verified.
- They give approximate, merely indicative values.
- In general, they overestimate the value of the rock mass deformation modulus.
- They do not take the possible anisotropic nature of the *in situ* deformation modulus into account.
- For the rock mass, a range of values is recommended between $0.4E$ and $1.6E$.

RMR parameter with the GSI index, which is obtained from the rock mass characteristics:

$$E = \sqrt{\dfrac{\sigma_{ci}}{100}}\,10^{\left(\dfrac{GSI - 10}{40}\right)}$$

where $\sigma_{ci}$ is in MPa and $E$ in GPa.

*Figure 3.100* shows the relationship between its parameters. It is well suited for low quality weak or soft rock masses, where the intact rock has a compressive strength below 100 MPa. In soft rock masses, intact rock properties have a fundamental influence on the deformation process of the rock mass, and this is considered by the above formula.

# ROCK MECHANICS

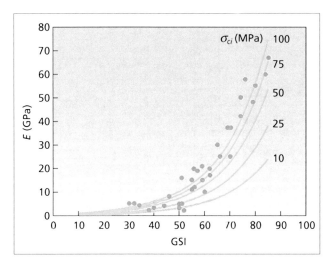

*Figure 3.100* Relationship between the GSI index, the rock mass deformation modulus and the intact rock uniaxial compressive strength (Hoek and Brown, 1997). Points show *in-situ* moduli obtained by Serafim and Pereira (1983) and Bieniawski (1978).

Hoek *et al.* (2002) further introduced the factor D in the expression above:

$$E = \left(1 - \frac{D}{2}\right)\sqrt{\frac{\sigma_{ci}}{100}} \, 10^{\left(\frac{GSI - 10}{40}\right)}$$

where D varies from 0 for undisturbed *in situ* rock masses to 1 for very disturbed rock masses.

— Barton (1995, 2006) proposes an expression to calculate the modulus of deformation of hard fractured isotropic rock masses from the value of the Q index, which considers the intact rock properties:

$$E = 10 Q_c^{1/3}$$

where $E$ is in GPa and $Q_c = Q\sigma_c/100$ (Q = Quality Index; $\sigma_c$ in MPa).

From this, by applying the expression which relates the P wave velocity $V_p$ and the Q index, $V_p = \log Q_c + 3.5$ (km/s), a new empirical expression is deduced which can be used as a general estimation method of the static modulus of deformation:

$$E = 10^{(V_p - 0.5)/3}$$

where $V_p$ is in km/s and $E$ is in GPa.

*Figure 3.101* shows the $V_p$–$Q_c$–$E$ correlations, with depth and porosity corrections.

— Hoek and Diederichs (2006) propose the following expression depending on the GSI index and of what is known as the "disturbance factor" $D$ ($D = 0$: undisturbed or undamaged rock mass; $D = 0.5$: partially disturbed; $D = 1$: fully disturbed rock mass), from a reliable data series from *in situ* tests performed in China and Taiwan (*Figure 3.102*):

$$E = 100{,}000 \left( \frac{1 - D/2}{1 + e^{(75 + 25D - GSI)/11}} \right)$$

This equation provides an upper bound for $D = 0$, and is recommended for confined rock masses where disturbance is insignificant, while for surface or shallow conditions the effects of rock mass disturbance can be taken into account by means of the disturbance factor.

## Permeability and water pressure

When assessing the hydrogeological behaviour of a rock mass, the most important parameters to take into account are permeability and water pressure. Except when the intact rock is permeable, water flows most easily along discontinuities, and pressures exerted can vary quite quickly as water flows easily through joints.

**Permeability** for a rock mass of low-permeability intact rock is governed by the degree of fracturing, the interconnection of discontinuities and their permeability. Evaluating permeability is complicated because it is very variable, even in areas close together in the same rock mass, and can be dominated by one or two open and continuous discontinuities. Multiple joint sets, single discontinuities such as faults, weathered or tectonized areas, are all anisotropic areas in the permeability of the rock mass.

The permeability of a rock mass can be assessed *in situ* with permeability tests (described in Chapter 5, Section 5.5), with the Lugeon test being the most characteristic. Table 5.10 in Chapter 5 shows rock mass classification according to the results obtained in this test.

There are some empirical formulae that can be used to obtain the approximate permeability of jointed rock masses in simple cases. The permeability of a set of discontinuities depends on its aperture, fill and spacing (*Figure 3.90*). As mentioned in Section 3.5, the permeability coefficient of a family of smooth, clean discontinuities, assuming laminar flow, can be estimated using the following empirical formula:

$$k = ga^3 / 12\eta b$$

where $g$ is the acceleration of gravity, $a$ is the aperture of the discontinuities, $\eta$ is the kinematic viscosity coefficient (0.0101 cm²/s for pure water at 20°C) and $b$ is the spacing between the discontinuities.

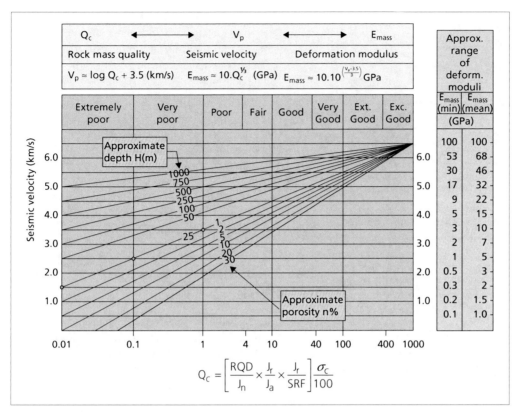

*Figure 3.101* Correlations between normalised $Q_c$ index, static modulus of deformation $E$ (from in situ tests and back-analysis) and seismic velocity $V_p$, with depth and porosity corrections (Barton, 1995).

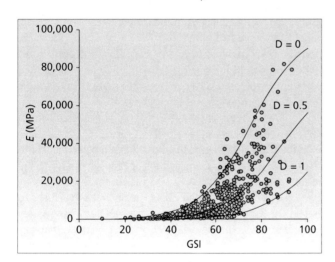

*Figure 3.102* Relationship between the GSI index and the rock mass deformation modulus for Chinese and Taiwanese data (Hoek and Diederichs, 2006).

This value of $k$ is the maximum attainable by a rock mass with low-permeability intact rock affected by one set of discontinuities (the intact rock permeability is disregarded because it is very low compared to that of the discontinuities).

If the discontinuities are filled, the permeability coefficient will be lower, and the permeability of the intact rock can influence the rock mass permeability; in this case, the rock mass permeability is given by:

$$k = (a/b)k_r + k_i$$

where $k_r$ is the permeability coefficient of the fill material, and $k_i$ of the intact rock, and $a$ and $b$ are as defined above.

When there is permeable intact rock, the value of $k_i$ can be calculated through laboratory tests or using tables from the literature (*Table 3.3*).

For a rock mass with three orthogonal sets of discontinuities, with equal spacing and opening, its permeability will be:

$$k = (2a^3 \gamma_w / 12 \mu b) + k_i$$

where $\gamma_w$ is the unit weight of the water and $\mu$ the dynamic viscosity coefficient of the water (0.01005 g · s$^{-1}$cm$^{-1}$ at 20°C).

The flow and drainage patterns of a jointed rock mass depend on the orientation of the different sets and their permeability. A rock mass with two or more sets of discontinuities has anisotropic permeability; in general and in practice, the highest value of $k$ for each set is taken.

**Water pressure** inside a rock mass is independent from its permeability, but dependent on the flow pattern. Anisotropy of the permeability distribution is of great importance for evaluating water pressures.

When there is water inside the rock mass, whether in the intact rock or in the discontinuities, rock mass strength must be assessed in terms of effective stresses, subtracting water pressure from the total normal stress acting on it. However, the following aspects must be considered:

— In jointed rock masses with low permeability and high strength intact rock, water pressure has its greatest influence on discontinuities and weathered or weak areas.
— In heavily jointed rock masses, even with low permeability intact rock, water plays an important role in reducing strength, because the intensity of jointing causes the behaviour of the rock mass to approximate to that of an isotropic granular medium.
— The role of water pressure must also be considered in heavily weathered rock masses, and in those with soft intact rock. In these cases, the drained strength can be so low that further reduction from water pressure may be of minor significance.

Water pressure is measured by direct methods like piezometers (described in Chapter 5, Section 5.6), or by indirect methods, from the rock mass flow net or by measuring the water table. Chapter 2, Section 2.3, describes the process for constructing the flow net and estimating water pressures. Chapter 9, Box 9.1, shows an example of a flow net for a slope. If there is no data available for drawing the flow net, pressure can be estimated by considering the weight of the column of liquid above the point of interest (in the case of an unconfined aquifer):

$$u = \gamma_w h$$

The importance of evaluating water pressure lies mainly in its application to calculating the stresses that act on rock masses. An increase in water pressure can in itself cause failure along a discontinuity plane.

## Scale effect

Due to their variability, there is always a certain degree of uncertainty attached to determining the properties of intact rock, discontinuities and rock masses. Physical properties of rocks generally depend on the samples tested (heterogeneity), the direction considered when evaluating them (anisotropy), and the volume considered in the test (scale effect).

**Scale effect** is the most important result of the heterogeneous and discontinuous nature of rock masses. Extrapolation of test results to rock mass scale can only be assumed to be valid if the volume of tested material is representative of the whole. *Figure 3.103* represents the scale effect in rock masses, illustrating how samples vary as their size increases, involving successively the intact rock, the intact rock and a single discontinuity, several sets of discontinuities, and the rock mass as a whole. Samples taken from the same rock mass, subjected to the same loads under the same conditions, will show variation in their properties according to their size. At laboratory test level, the scale effect becomes apparent when different-sized samples of the same rock mass give variable, scattered results (*Figure 3.104*). According to the ISRM, scale effect exists when sets of samples with different sizes collected from the same rock mass, show different statistical distributions with size for the same property.

On the basis of experimental data, it has been observed that above a certain sample volume test results are independent of size. This volume, the smallest that is considered representative of the rock mass behaviour for a certain property, is called **representative elemental volume (REV)** (*Figure 3.104*). The REV can vary significantly according to the property being considered, and from one rock mass to another. For some properties like deformability this volume can be very large, beyond what can reasonably be tested in a laboratory. In general, it is not possible to define this volume, and when it is possible, it is almost always too big to test.

The REV is usually greater than the volume of rock normally used for testing. In most cases only a few large-scale tests can be carried out to characterize the rock mass, and sometimes none. In general, results from small-scale tests, laboratory or *in situ*, have to be extrapolated to the rock mass. The scale effect must also be considered when applying

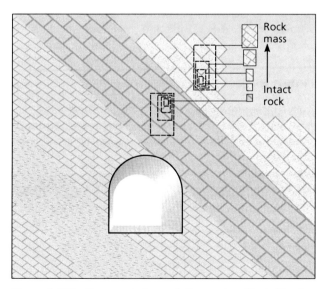

*Figure 3.103* Representation of the scale effect (Cunha, 1990).

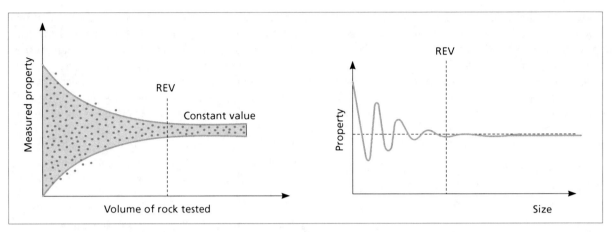

*Figure 3.104* Properties of rocks vary according to the volume considered. In general, the scatter of measured values diminishes as size increases.

empirical failure criteria, especially for those deduced mainly from laboratory tests results.

Experimental data show that for many rock mass properties, such as deformability and strength, an increase in sample size reduces test result scatter even though in some cases the mean value does not vary. This means that the smaller the sample size, the greater the number of tests needed to obtain the same reliability of the estimated mean value.

Because of its complexity, the scale effect has not been sufficiently studied, but it is expected that the current safety factors used in different geological engineering applications will gradually be replaced with a qualitative and quantitative understanding of how sample size influences test results. Establishing this relationship will enable geotechnical design parameters to be selected on a more scientific basis (Cunha, 1990).

Scale affects the measurement of properties such as strength, deformability, permeability and *in situ* state of stress of rock masses, and also the measurement of the deformability and strength of intact rock and discontinuities.

### Intact rock

Scale effect in intact rock is due to mineralogical and structural heterogeneity and variability. Compressive **strength** decreases as sample size increases, as can be seen in the examples shown in *Figures 3.105* and *3.106*. Hoek and Brown (1980) and Barton (1990) have proposed the following ratio between the compressive strength of a sample with diameter $d$ and that of a sample with diameter 50 mm, valid for values of $d$ between 10 and 200 mm:

$$\sigma_c = \sigma_{c50} \left(\frac{50}{d}\right)^A$$

where $A$ equals 0.18 or 2 depending on the author.

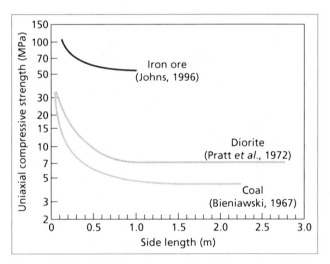

*Figure 3.105* Variation of strength with sample size for several types of rock (Bieniawski, 1984).

In relation to **deformability**, the mean deformation modulus appears to be largely independent of size, but deviations in results decrease as the sample volume increases.

### Discontinuities

The scale effect related to the mechanical behaviour of discontinuities has been partly described in Section 3.5. It depends mainly on the roughness of the planes and the surface area considered in the test. In laboratory tests, only small scale roughness is represented, while for *in situ* tests, waviness or larger-scale roughness can be considered. This variation in area is mainly responsible for scale effect when estimating shear strength of rough planes, as it influences, among other properties, the concentration of effective stresses at the points of contact of the joint walls during shear. Although the literature gives contradicting opinions, Bandis et al. (1983) and Cunha (1990) have established the following conclusions about the influence of scale on the evaluation of the mechanical behaviour of rough discontinuities. As the size of the tested area increases:

— The shear strength, $\tau$, decreases.
— The shape of the stress-shear displacement curve varies from brittle to ductile.
— The peak shear displacement increases.
— The shear stiffness of the joint, $k_s$, decreases.
— The scatter in the values of $\tau$ and $k_s$ decreases.

Figure 3.107 illustrates these effects, as it clearly shows how strength decreases as the area being tested increases, and the variation in the shape of the curves (see also Figure 3.86).

The scale effect on shear strength of joints is larger the rougher a joint is, and decreases as the scale of the test increases. If normal stresses on the discontinuity are high, and the shear process takes place without dilation, the influence of scale can be annihilated. Figure 3.108 shows the influence of effective normal stress on the shear strength of discontinuities in laboratory tests and in situ tests of different-sized samples. For laboratory tests with low normal stresses, sample size has a definite influence, resulting in a high scatter of results, whereas its influence is considerably lower for in situ tests of larger areas. At high normal stresses, scatter in both laboratory and in situ data is not so significant, because the effects of roughness on the joint shear strength are suppressed; the values for $\tau$ and therefore for $k_s$, increase as normal stress on discontinuities increases.

## Rock mass

As a larger volume of the rock mass is considered, the discontinuities play a more significant role in its strength and deformability, as can be seen in Figure 3.103. For a given rock mass the representative elemental volume (REV) will depend on joint spacing, with dimensions several times larger than the spacing.

Compressive **strength** decreases as the sample size considered increases, in the same way as it does in intact rock and with individual discontinuities, on which the strength of the rock mass depends. However, different combinations of failure modes can modify the behaviour of the rock mass, so that for the time being no general criterion can be established.

Due to the near impossibility of evaluating a rock mass strength from in situ tests, failure criteria are frequently

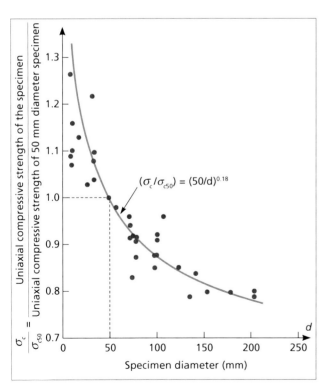

Figure 3.106 Variation of uniaxial compressive strength with sample size for several types of rock. Strength values normalised to a 50 mm diameter sample to eliminate the influence of variation in test characteristics (Hoek and Brown, 1980).

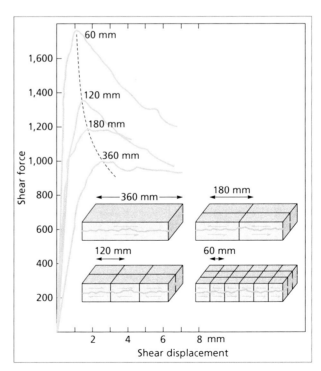

Figure 3.107 Influence of block size on the shear stress-displacement behaviour of discontinuities (Bandis, 1980; in Cunha, 1990).

applied. These criteria must be interpreted according to the scale of application. Thus, the Hoek-Brown failure criterion takes this effect partly into account by including the $m$ and $s$ rock mass constants, which depend on the degree of fracturing and other characteristics of the rock mass.

In relation to **deformability**, theoretical and experimental studies coincide in showing how, for an equal degree of fracturing or joint frequency, mean deformation modulus and sample size are largely independent. However, scattering of results decreases as the volume tested increases, as occurs in the intact rock.

Deformability depends on the degree of fracturing and the deformational properties of discontinuities and intact rock. When fracturing is more intense, the deformability of the rock mass increases, because of the influence of the growing number of discontinuities. *Figure 3.109* shows the results of biaxial tests carried out on samples with different degrees of fracturing and block sizes. The stress-strain curves obtained show a decrease in Young's modulus, $E$, for the smallest block sizes and, contrary to what might be expected, a continued increase in strength as failure mechanisms within the sample change and its broken pieces begin to rotate. In relation to the influence of sample size, *Figure 3.110* shows elasticity modulus values obtained in laboratory and *in situ* tests (dilatometer and large-scale flat jack). As the volume tested increases, and so a greater number of discontinuities is considered, the results scatter and the mean value of $E$ decrease.

However, it is also important to consider the influence of the type of test carried out. The mean deformation modulus measured with a dilatometer for volumes of 10, 20 and 30 $m^3$ does not vary, but the scatter of results decreases as size increases. Load bearing test results show a variation in deformation modulus according to test size on different rock types, with no single law for scale effect; in some cases the deformation modulus becomes smaller and in others it increases with the increase in the load area. *Figure 3.111* shows different procedures for estimating deformability according to the spacing of the rock mass joints.

Another question discussed in the literature is the contradictory theoretical effect of the size and the depth of the test sample both being increased: on one hand, the deformation modulus decreases, as more discontinuities are included, but on the other hand it should increase, as the tested area affects deeper parts where rock is less weathered. In unweathered rock masses, at great depths, with widely spaced closed joints, the intact rock and the rock mass should show the same deformability, independent of scale.

Concerning the extrapolation of laboratory-obtained deformability data to rock mass scale, a series of correlations

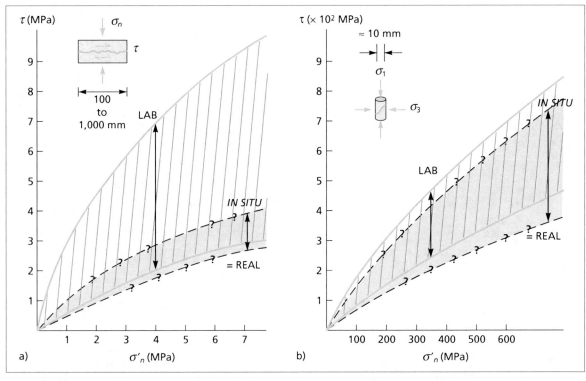

*Figure 3.108* Scale effects are expected under high and low stress. a) Laboratory and *in-situ* direct shear tests with 100 to 1,000 mm long samples, with low normal stress (engineering scale). b) Triaxial tests with 10 mm diameter samples, with high normal stress (crustal scale). Scale effect curves are estimates. (Barton, 1990).

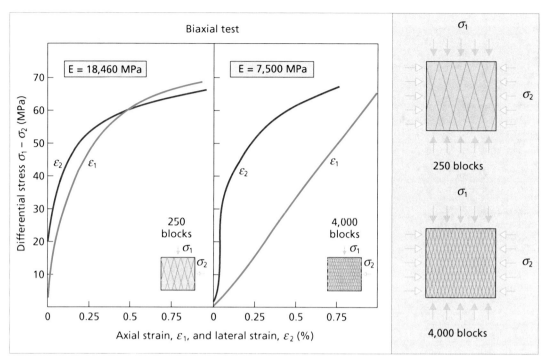

*Figure 3.109* Effect of block size on shear strength of interlocked assemblies of blocks (modified from Barton, 1990).

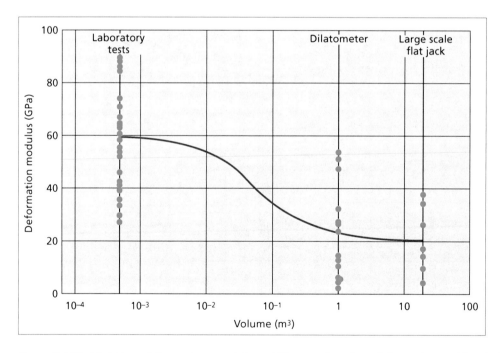

*Figure 3.110* Variation of the rock mass deformability as a function of the scale considered: the curve tends to stabilise for volumes approaching REV (Cunha and Muralha, 1990).

exist (already described), but their limitations and the specific characteristics of the rock mass in question must be taken into account. These correlations are generally applicable to hard rock masses.

In general terms, it can be said that to reduce the scale effect and obtain representative results a large number of small-scale tests or a smaller number of larger-volume tests (close to REV) will have to be performed; this last condition

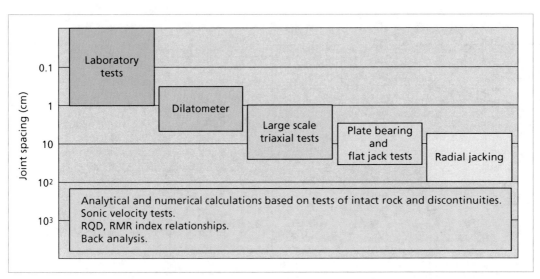

*Figure 3.111* Procedures for estimating the deformability of a rock mass depending on joint spacing (modified from Natau, 1990).

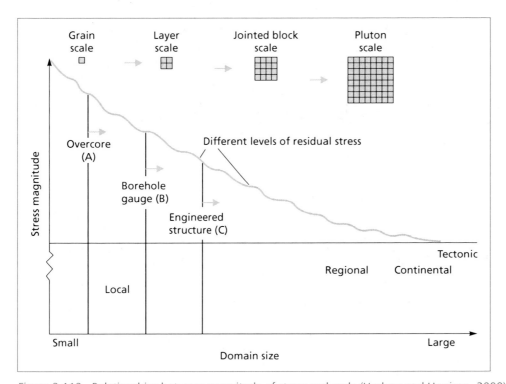

*Figure 3.112* Relationships between magnitude of stress and scale (Hudson and Harrison, 2000).

is difficult to achieve because to evaluate deformability, and most other rock mass properties, volumes of several cubic metres would have to be tested.

The scale effect in the measurement of **state of stress** (Section 3.7) is difficult to evaluate because few experiments and results are representative at regional scale (stress measurements are frequently affected by local effects), and because there are significant differences between the methods used. On the other hand, with the exception for anisotropic state of stress in rock masses associated with different structural, tectonic or topographic effects, stress is probably the factor least affected by the scale effect, as depth is the fundamental variable determining stresses. Some authors propose a general law whereby the magnitude of stresses decreases as the volume of the area considered increases (*Figure 3.112*), however according to Martin et al. (1990), and based on a test programme in granite undertaken for Atomic Energy of Canada Limited to study the scale effect in the calculation of *in situ* stresses, there is no significant evidence of volume being an influence even when varying over five orders of magnitude.

In practice, the different testing methods must be considered depending on the volume involved, ranging from overcoring methods in boreholes, to measurements in shafts or adits several metres in diameter. Methods that involve larger volumes, such as hydraulic fracturing or convergence measurements in underground excavations, are the most representative. *Figure 3.113* shows an example of the effect of scale on the results obtained from hydraulic fracturing as a function of core diameter.

**Permeability** of rock masses is dependent upon the number of discontinuities, their aperture, spacing and interconnection. Determining it through *in situ* testing in boreholes is only valid if the boring goes through a representative number of discontinuities, and so the test volume may have no influence on the result. Measuring permeability in small volumes can give a completely erroneous idea of the real value, and depending on the tested area, results can be very variable (*Figure 3.114*); if the results of small-scale permeability tests are extrapolated to predict water flow in the rock mass, the values calculated may sometimes exceed real values by one or more orders of magnitude.

The REV is therefore a fundamental concept when determining permeability and flow in rock masses. Sometimes, the limitations of borehole tests can be overcome with excavations or adits in the rock mass, where flow can

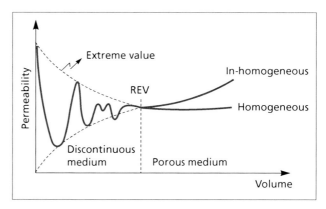

Figure 3.114 Scale effect on the measurement of permeability in rock masses (Bear, 1972; in Hudson and Harrison, 2000).

be measured directly. However, the cost of these works limits them to specific engineering projects.

## 3.7 *In situ* stress

## Origin and types of *in situ* stress

Geological materials are subjected to natural stresses as a result of their geological history. Stress inside a rock mass is generated by external applied forces and by its own weight. The resulting state of stress is usually complex. The heterogeneous, discontinuous and anisotropic nature of rock masses also means that the state of stress can vary significantly between adjacent areas. The state of stress at any point is defined with the stress tensor, as is explained in Section 3.3.

Tectonic forces are the main cause of stresses stored in rocks which can be released in many different ways, from earthquakes and fault movements to rock bursts, fracturing and deformations in underground excavations. An understanding of these forces is essential for many different fields such as oil, gas or geothermal exploration and production, mining and underground work, or seismic hazard studies and earthquake prediction. Engineering projects affect *in situ* states of stress, adding new forces or modifying the distribution of existing forces, inducing new states of stress. The most important applications of the study of natural stresses in geological engineering are tunnels and underground excavations in general, where the stability depends on the magnitude and direction of stresses, and where it is essential to know the *in situ* stress.

The Earth's crust is subjected to different types of stress, whose effects are reflected in the movement of tectonic plates, the deformation of rocks or in the sudden release

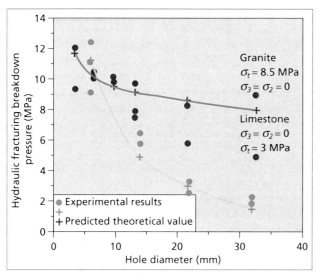

Figure 3.113 Variation of the hydraulic fracturing breakdown pressure as a function of borehole diameter for two rock masses (modified from Haimson, 1990).

of energy as in earthquakes. The state of stress of the crust is due to various causes, the main ones being:

— Tectonic stress.
— Gravitational stress.
— Non-renewable stress.

**Tectonic stress** is responsible for the movement of lithospheric plates and is the main source of stresses present in rocks. Knowledge of the geology may make it possible to determine crustal areas subjected to stress; their orientation and magnitude can be estimated by applying methods of structural geology analysis.

**Gravitational or lithostatic stress** is a result of the weight of geological materials. If at any given point there are no other acting stresses, the state of stress is defined by gravitational forces produced by the overlying and confining materials, and principal stresses are thus vertical and horizontal: $\sigma_1 = \sigma_V$, $\sigma_2 = \sigma_3 = \sigma_H$. Vertical stress at any point due to overlying material loading is $\sigma_V = \rho g z$, where $\rho$ is the material density, $g$ the force of gravity (9.8 m/s²) and $z$ the material depth or thickness. In general terms the magnitude of this lithostatic stress is in the order of 0.027 MPa/m (1 MPa ≈ 40 m) (Figure 3.115).

This vertical compressive stress creates horizontal lateral stresses as the rock tries to expand along directions transversal to the vertical loads. In elastic bodies transversal expansion can be expressed using Poisson's ratio:

$$v = \varepsilon_r / \varepsilon_l$$

where $\varepsilon_r$ is the radial strain and $\varepsilon_l$ is the longitudinal strain. If the rock is not free to expand transversally ($\varepsilon_r = 0$), a radial or transversal stress appears, with the value:

$$\sigma_{radial} = \sigma_H = (v / 1 - v) \sigma_V$$

Taking an average value of the Poisson ratio, $v = 0.25$ (values usually lie between 0.15 and 0.35), the ratio $\sigma_H / \sigma_V = K$ will be approximately equal to one third:

$$\sigma_H = 0.33 \sigma_V \quad \text{or} \quad K \approx 1/3$$

If the behaviour of the rock is not elastic and creep or plastic deformation occurs, the material cannot sustain shear stresses over geological time, and the horizontal stress component will in time become equal to the vertical component, producing a hydrostatic stress field, where $\sigma_H = \sigma_V$ and $K = 1$. This situation occurs at great depths.

Lateral density variations in materials can modify the distribution of gravitational loads in the crust. In shallow areas, stresses can also be modified through topographical effects: differences in topographical elevation give rise to non-uniform distribution of forces. Erosion processes free the

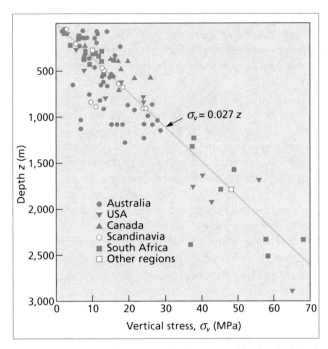

*Figure 3.115* Variation in vertical stress with depth (Hoek and Brown, 1980).

underlying materials from loads, generating decompression stresses which may even fracture the rock, as is the case with horizontal fracturing in igneous rocks, or the sub-vertical fractures parallel to deep valley slopes caused by erosion.

**Non-renewable stress** can be generated through thermal effects on the rock materials, through bending or buckling of the crust or as a result of variations in the radius of curvature of a tectonic plate; in theory this can generate very great stresses in the crust, although as they are non-renewable, they are released through creep or brittle failure in relatively short geological periods.

The mechanism that produces **thermal stress** is the expansion or contraction that a homogeneous rock undergoes when slowly heated or cooled. The relationship between the strain $\varepsilon$ and the temperature variation $\Delta t$ is:

$$\varepsilon = \alpha \Delta t$$

where $\alpha$ is the thermal expansion coefficient. If the rock can not freely expand or contract, stresses will be generated. One effect this mechanism causes in rock masses is the appearance of fractures and areas with anisotropic state of stress in dykes, lavas or intruded materials with different thermal properties from those into which they are intruded (Figure 3.116). Stress can also appear at a crystal or mineral particle scale if these have different elastic constants or thermal coefficients.

Large-scale bending or buckling of the crust resulting from differential loads in subduction areas also creates

Figure 3.116  Columnar jointing in basalt, Giant's Causeway, N. Ireland.

non-renewable stresses. What are called **membrane** stresses originate from variation in a plate's radius of curvature as it moves across the surface of the Earth, which is not truly spherical.

To sum up, the most important crustal stresses are of **tectonic origin**, due to the forces that act on lithospheric plates, which are responsible for the processes of subduction and collision along transcurrent faults. The action of these forces produces stresses in the order of 25 MPa in the lithosphere, and although these are relatively small, due to the lithosphere's viscoelastic properties they undergo an **amplifying effect** from acting constantly over very long periods of time (millions of years). This effect means the rock's strength is exceeded, producing tectonic deformations. Non-renewable stresses have little influence because they act over short geological periods and dissipate quickly.

**Residual stresses** remain stored or accumulated in rock even when external forces, such as tectonic forces, have stopped acting. Residual stresses include those generated at an inter and intra-crystalline scale due to different thermal or elastic properties of the crystals.

**Induced stresses** are those generated through modification or redistribution of the *in situ* state of stress, e.g. when a slope or a tunnel is excavated. This aspect is important for the design and construction of underground excavations.

## Geological and morphological factors which influence the state of stress

The geological and morphological conditions of an area can modify regional gravitational stress fields and the direction and magnitude of stresses, creating stress anisotropies, or anisotropic states of stress, mainly due to the following factors:

— Presence of faults, folds, dykes and other structural anisotropies.
— Various processes of loading or unloading: e.g. erosion, sedimentation, and glacial processes.
— Deep valleys and areas with intense relief.
— Volcanic processes.

Certain geological structures indicate the direction of stresses, and can therefore show possible anisotropic stress fields, as well as giving the orientation of the stress tensor from geological field observations (Figure 3.117). For example dykes are usually oriented at right angles to the minor principal stress $\sigma_3$, and the alignment of volcanoes can also indicate the direction of $\sigma_3$.

The value $\sigma_V = 0.027$ MPa/m for the vertical gravitational stress is valid for flat or featureless areas, where the principal stress directions are horizontal and vertical, both on the surface and underground. However, this generalization does not hold in certain circumstances. In mountain areas, with valleys and slopes, the direction and magnitude of stresses near the surface is determined by the morphology: one of the principal stresses is normal to the slope and is zero, while the other two principal stresses are contained in the plane of the slope (Figure 3.118).

In deep valleys in mountain areas this topographical effect is very strong, and stress anisotropies occur because of the magnitude of the stresses concentrated in the slopes (Figure 3.119). Considerable stresses due to these causes have been measured most frequently in:

— Valley slopes more than 500 m deep and with gradients greater than 25°.
— Valley slopes in soft rock more than 300 m deep.

In areas where $\sigma_V = \sigma_1$ and $\sigma_H = \sigma_3$, the magnitude of stresses and their inter-relationship can also be affected by

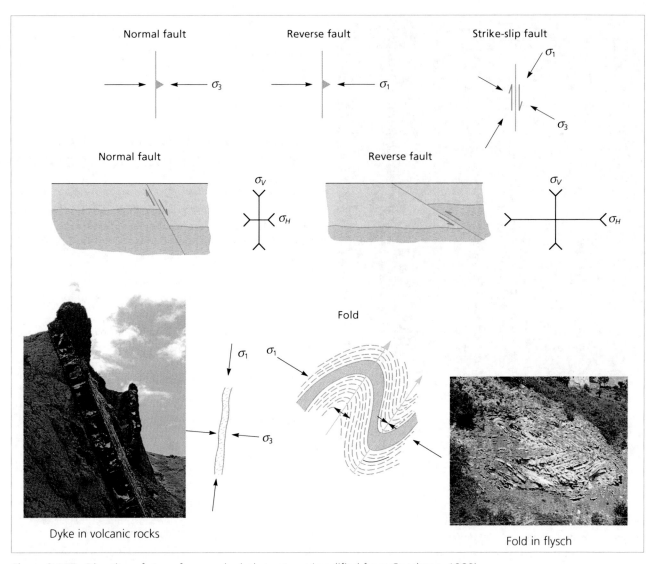

*Figure 3.117* Direction of stress from geological structures (modified from Goodman, 1989).

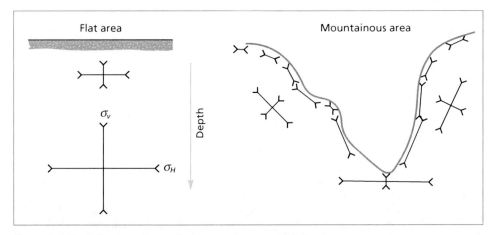

*Figure 3.118* Influence of morphology on the stress distribution.

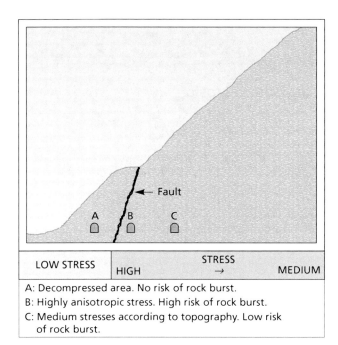

A: Decompressed area. No risk of rock burst.
B: Highly anisotropic stress. High risk of rock burst.
C: Medium stresses according to topography. Low risk of rock burst.

*Figure 3.119* State of stress in an underground excavation inside a valley slope affected by a fault (Selmer-Olsen, 1977).

geological factors. As already explained, in elastic materials $K = 1/3$, and at great depths hydrostatic conditions occur where $K = 1$; but this relationship is not valid in the range of depths of less than 500 m where most engineering work is carried out, where the value of $K$ can be greater than 1 (*Figure 3.120*).

Although the main cause of elastic and hydrostatic conditions not occurring are tectonic stresses, the effects of **erosion** can also give rise to values of $K$ greater than 1. If initially the rock has a hydrostatic or elastic state of stress, where $\sigma_V = \gamma z$, erosion will have the effect of reducing $z$ quite rapidly, with the isostatic adjustment to $\sigma_H$ lagging behind, and therefore increasing the value of $K$, reaching the condition where $\sigma_H > \sigma_V$ (*Box 3.10*).

## Methods for measuring *in situ* stress

Determining the state of stress of rock masses is one of the most complex problems in rock mechanics. Principal stresses at any point are defined by their direction and magnitude,

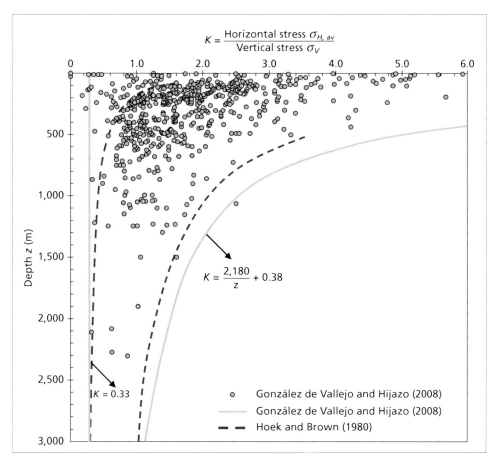

*Figure 3.120* Variation in the relationship $\sigma_H/\sigma_V$ with depth (González de Vallejo and Hijazo, 2008).

## Box 3.10

### Variation in the relationship $\sigma_H/\sigma_V$ due to erosion

The equation $K = \nu/(1-\nu)$, valid for materials with elastic behaviour, does not hold for near surface areas of the crust which have undergone a process of unloading as the result of erosion. When part of the overburden is eliminated from an area the value of the vertical stress $\sigma_V$ is lowered and the state of stress and relationships between the acting stresses are modified.

Taking $\sigma_{V0}$ and $\sigma_{H0}$ as the initial horizontal and vertical stresses on a point at depth $z_0$:

$$\sigma_{V0} = \gamma z_0$$
$$\sigma_{H0} = \sigma_{V0} K_0 = \gamma z_0 K_0$$

after a thickness of overburden $\Delta z$ is removed by erosion, the new values of stresses $\sigma_V$ and $\sigma_H$ will be:

$$\sigma_V = \gamma z_0 - \gamma \Delta z$$

$$\sigma_H = \gamma z_0 K_0 - \gamma \Delta z \, \nu/(1-\nu)$$

and the new value of $K$ for $z = z_0 - \Delta z$ will be:

$$K(z) = \frac{\sigma_H}{\sigma_V} = \frac{\gamma z_0 K_0 - (\gamma \Delta z \, \nu/(1-\nu))}{\gamma z_0 - \gamma \Delta z}$$

$$K(z) = K_0 + \left[\left(K_0 - \frac{\nu}{1-\nu}\right)\Delta z\right]\frac{1}{z}$$

The erosion tends to increase the value of $K$, as the horizontal stress becomes greater than the vertical one for depths of less than a specific value.

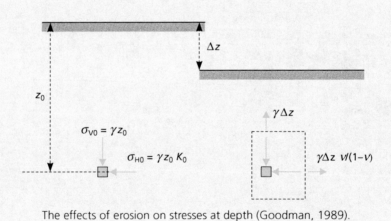

The effects of erosion on stresses at depth (Goodman, 1989).

---

which can only be measured *in situ*. The most used measurement methods are:

— Overcoring.
— Flat jack.
— Hydraulic fracturing.

There are other methods that only measure the direction of the stress:

— Tectonic structural analysis.
— Earthquake focal mechanisms analysis.
— Observing the orientation of failure planes or breakouts on borehole walls.
— Stress relief measurement on outcrops.

### Measuring the direction of stresses by geological methods

The analysis of tectonic structures, mainly stylolitic bedding and joints filled with growth-crystallizations, is used to interpret **paleostress** fields and estimate stress directions and trajectories.

Another method is based on the analysis of seismic waves generated by an earthquake. By identifying the **focal mechanism** the direction of the stresses acting during the earthquake can be calculated (Figure 3.121).

Fractures that form symmetrically by spalling of the borehole wall due to the stress concentration around the borehole, can also indicate stress directions, as they develop approximately parallel to the maximum in situ horizontal stress direction (Figure 3.122). Borehole breakouts (hole ovalisation) in the borehole wall form in a preferential direction perpendicular to the maximum horizontal stress.

In rock outcrops the stress directions can be determined by analyzing the stress relief produced when a large diameter core is overcored. This technique is shown in Box 3.11.

## Estimating stress magnitude from empirical relationships

When enough stress measurement data is available, relationships between horizontal and vertical stresses, and their variation with depth, can be established. Figures 3.115, 3.120 and 3.123 show several examples of these relationships. Empirical data show a good correlation between maximum vertical stress and depth, with a gradient between 0.025 MPa/m for acid rocks and 0.035 MPa/m for alkaline rocks. However, there is great regional variation in horizontal stresses, and in most cases they are seen to be greater than vertical stresses, with an average value of 10 MPa at shallow depths.

As indicated previously, $\sigma_H/\sigma_V$ tends towards unity at great depths (below 1,000 m), whereas for depths of less than 500 m the $K$ values present great dispersion.

Chapter 10, Section 10.4, contains other empirical methods for estimating the magnitude of stresses, including example calculations.

## Instrumental methods for measuring orientation and magnitude of stress

The most used methods for measuring in situ stress can be classified as follows:

- **Overcoring methods**
- **Pressurization methods**

The overcoring methods are based on determining the in situ rock stress relieved when overcoring a borehole, by measuring the strains produced at the bottom of the borehole or on its walls. An ideal, elastic, homogeneous and isotropic material is assumed for the application of overcoring methods described below. Stresses are calculated applying the theory of elasticity to convert the measured strains into stresses, when the modulus of elasticity and Poisson's ratio are known; these two parameters must be calculated in the laboratory. The methods described are:

— Doorstopper method.
— USBM method.
— CSIR triaxial cell method.

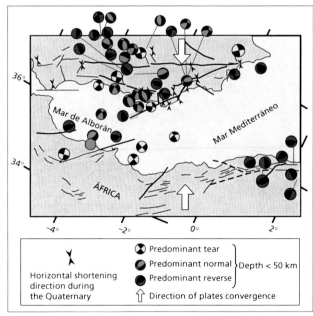

Figure 3.121 Focal mechanisms in the Mar de Alborán area, between the Iberian Peninsula and the African plate. The different types of mechanisms for each earthquake are shown. Also shown is the direction of convergence between the European plate to the north and the African plate to the south, as well as the Quaternary shortening directions obtained from seismological and geological data (courtesy of J. Martínez-Díaz).

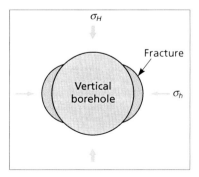

Figure 3.122 Fractures induced in a borehole when horizontal stress exceeds the rock strength, producing fracture surfaces perpendicular to the direction of the minimum horizontal stress.

### Box 3.11

### Determining stress direction through stress relief methods in outcrops

1. Measurement of main directions of tectonic structures in the area (Photo A).
2. Strain gauges are fixed to an outcrop with no fractures, oriented according to structural directions (Photo B).
3. Drilling a slot around the area instrumented with strain gauges, approximately 20 cm in diameter and 30 cm in length (Photo C).
4. Measuring strains (Photo D).
5. Extracting the core and measuring the elastic constants in the laboratory.

Photo A: Structural measurements and site preparation

Photo B: Placing the strain gauges

Photo C: Drilling

Photo D: Measuring strains

The pressurization methods are based on applying fluid pressure to produce rock fracture or generate mechanical dilation. The methods described are:

— Flat jack.
— Hydraulic fracturing.

The stress tensor has nine components (*Figure 3.124*), six of which are independent: 3 normal stresses and 3 shear stresses. Therefore, 6 measurements are needed to define the state of stress at any point. Of all the methods mentioned above, only the triaxial cell can supply the 6 components in a single test; USBM-type cells can supply 3 components (defining the two-dimensional state of stress) and the flat jack test gives the stress acting in a certain direction; the hydraulic fracturing test supplies principal stresses in the three dimensions.

A detailed description of stress measurement methods can be found in Hudson and Cornet (2003) and in the ISRM Suggested Methods (http://www.isrm.net).

## The doorstopper method

This method gives the direction and magnitude of stresses in a plane perpendicular to the borehole axis. To determine the complete state of stress, three boreholes are necessary, bored in different directions inside an adit or tunnel.

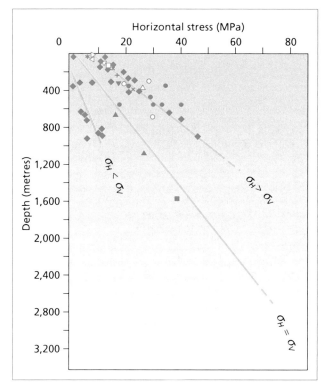

Figure 3.123  Relationships between maximum horizontal stress and depth (Herget, 1988).

Figure 3.125 shows a diagram of the measurement procedure. First, a borehole around 60 mm is drilled; the head of the measuring instrument (doorstopper) is stuck to the bottom of the borehole; this head contains strain gauges to measure unit strains in three directions, and an initial reading is taken at this stage. Next, rods and measurement equipment are taken out, leaving the doorstopper stuck to the borehole bottom; drilling is then continued using the same diameter, and a rock core sample is extracted with the doorstopper head stuck to it; stresses in the sample are relieved when it is extracted, and final strain measurements are then taken. These strains then have to be reversed to obtain the stresses that caused them.

## USBM and CSIR triaxial cell methods

Both methods are based on measuring strains on borehole walls that are a result of the stress release that occurs when overcoring. USBM type cells can estimate stresses in two dimensions and CSIR type cells in three dimensions. The borehole is overcored using a larger diameter crown, so that stresses are released in the ring that forms around the original borehole. Transducers inside the USBM gauge measure strains along three diameters in the borehole. If the reference axes are parallel and perpendicular to the borehole, the measurements can supply the three components of the two-dimensional state of stress in a plane perpendicular to the borehole axis: $\sigma_{xx}$, $\sigma_{yy}$, $\tau_{xy}$. To measure three-dimensional stresses, measurements have to be taken in three different boreholes drilled inside an adit in three different directions; otherwise, the CSIR method must be used, with 9 or 12 transducers.

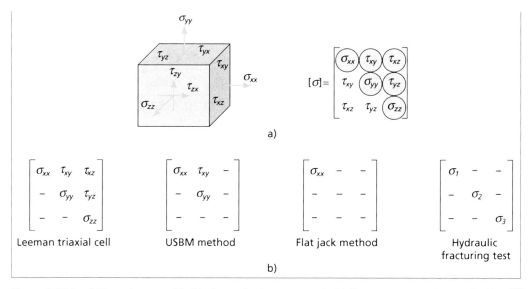

Figure 3.124  a) Stress tensor with 6 independent components. b) Stress components supplied by different measurement methods.

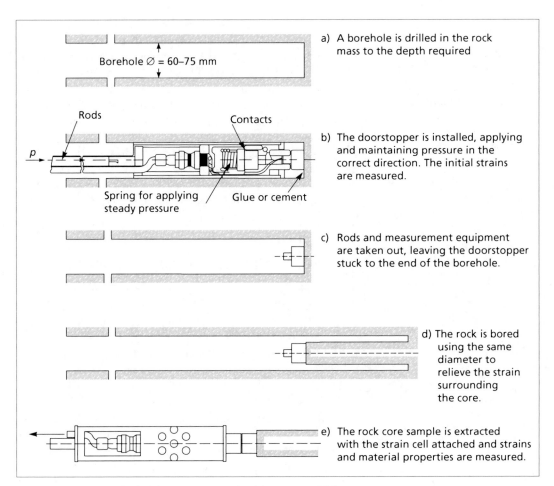

Figure 3.125 Steps in the doorstopper method (Leeman, 1970).

Using the CSIR method, cylindrical probes are placed in boreholes to measure strains on their walls in different directions, as the result of stresses released by overcoring. Figure 3.126 shows the measurement procedure. After a borehole has been drilled to the depth at which measurements are to be taken, a smaller hole is drilled from the bottom of the original borehole, and a measurement torpedo is installed in it. This torpedo is usually equipped with three strain gauge rosettes at three different places, pointing in three different directions. The smaller hole is then overcored using the original borehole diameter, relieving stresses from the resulting hollow cylinder of rock, on which the induced strains are measured. Figure 3.126 shows a triaxial cell. The complete state of stress can be estimated using this method.

## Flat jack test

This test is carried out on a wall inside a rock mass. The procedure is described in Chapter 5, Section 5.5. This method provides the stress perpendicular to the flat jack installed in a slot cut into the wall (Figure 3.127 and Figure 5.85).

To determine the state of *in situ* stress of the rock mass, several tests must be carried out on differently oriented slots, and the probable stress distribution in the area must also be known, to be able to calibrate results. This technique is simple and can also be used to estimate the deformation modulus of the rock, whereas when other methods are used, this has to be calculated in laboratory tests.

## Hydraulic fracturing test

The objective of the hydraulic fracturing test is to measure the state of *in situ* stress through a borehole. The test obtains the magnitude and direction of the maximum and minimum principal stresses on a plane perpendicular to the borehole. At present, it is the only reliable technique available for deep drilling: it can be used at depths exceeding 1,500 m.

### Method

The test is performed by injecting a fluid under pressure, generally water, into a section of a borehole (usually around 1 m in length) that has been isolated using special packers,

# ROCK MECHANICS

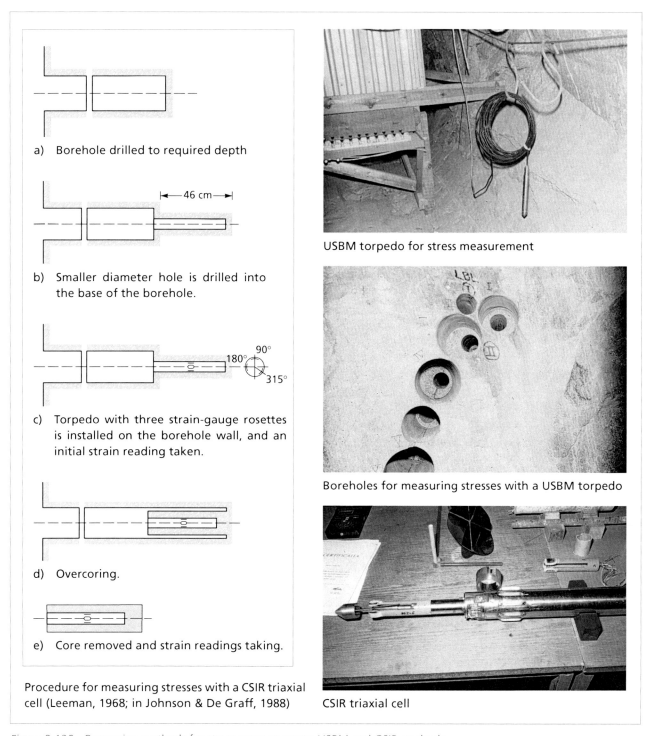

Figure 3.126 Overcoring methods for stress measurements: USBM and CSIR methods.

until a fracture is initiated in the rock. The pressure needed to generate, propagate, sustain and reopen fractures is measured. These pressures are related to the magnitude of the principal *in situ* stresses existing in the drilling area.

It is assumed that the axis of the borehole coincides with one of the principal stress directions (e.g. vertical), and the other two are calculated using the orientation of the fracture plane induced by the hydraulic fracturing. In general,

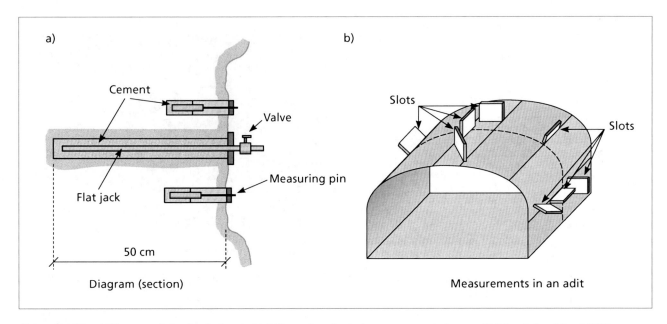

*Figure 3.127* a) Diagram of the flat jack test. b) Differently-oriented measurements in an adit (Kim & Franklin, 1987).

this assumption is valid for vertical boreholes, and the vertical stress in the area is calculated from the overburden weight: $\sigma_V = \gamma z$. However, it is questionable how precise this method is if the direction of drilling deviates more than 15 degrees from one of the principal stress directions.

This method is more exact when applied to non-porous materials which show elastic, homogeneous, continuous, isotropic behaviour, since elasticity theory is assumed. The section used to perform the test must be checked to make sure it has not been previously fractured. This can be done by inspecting the drilling cores or by inserting a four-armed calliper or a TV camera into the borehole. When drilling, rock samples must be taken to monitor discontinuities and their orientations and to perform laboratory tests.

### Testing

The rock fractures if the stresses induced by the injected fluid reach the tensile strength $\sigma_t$ of the rock surrounding the borehole. The injection pressure that generates the initial fracturing, $P_f$, is called fracture initiation pressure or breakdown pressure.

After fracturing has started, $P_f$ is maintained for a short time, enough for the failure to propagate, and then the injection circuit is shut in instantaneously. A small fall in pressure occurs, and the instantaneous shut-in pressure, $P_s$, is then measured. Finally, the circuit is opened again until atmospheric pressure is reached. *Figure 3.128* shows recorded pressure-time and flow rate-time curves.

The fracture is reopened with further pressure cycles, recording the fracture reopening pressure, $P_r$, and new

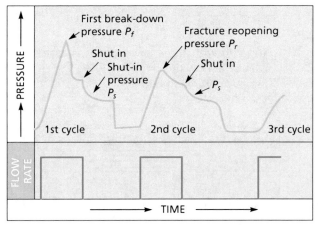

*Figure 3.128* Pressure and flow rate versus time curves from a hydraulic fracturing test (Kim & Franklin, 1987).

additional values of the shut-in pressure $P_s$. The test is considered to be complete when similar values of $P_s$ are obtained in several cycles.

*Figure 3.128* shows the evolution of the test in three cycles. In the first, the pressure $P_f$ corresponds to the highest peak, with a subsequent pressure drop produced as a result of the opening of the fracture, with the injection circuit being shut in instantaneously at that point. From that time there is pressure stabilization corresponding to the shut-in pressure $P_s$.

The initial development of the fracture is very quick, and little work is required to propagate it, due to the high stress concentration at its propagating tip. The pressure needed to keep the fracture open is $P_s$, which is equivalent to the normal stress acting on the fracture, $\sigma_n$, since the rock's tensile strength has already been exceeded and now $\sigma_t = 0$.

In a vertical borehole in an area where principal stresses are vertical and horizontal, a vertical fracture will be generated if $\sigma_1 = \sigma_V$, and the values of $P_f$ and $P_s$ will be:

$$P_f = \sigma_n + \sigma_t$$
$$P_s = \sigma_n \quad (\sigma_t = 0)$$

If $\sigma_1 = \sigma_H$, a horizontal fracture will form, and the values of $P_f$ and $P_s$ will be:

$$P_f = \sigma_V + \sigma_t \quad (\sigma_n = \sigma_V \text{ in this case})$$
$$P_s = \sigma_V = \gamma z \quad (\sigma_t = 0)$$

The hydraulic fracture generated in this test is always vertical and perpendicular to the smaller horizontal stress, $\sigma_h$, regardless of the magnitude of $\sigma_V$ (Figure 3.122). However if $\sigma_V$ is the minimum principal compressive stress, the fracture that forms on the borehole wall will change direction towards a horizontal fracture as soon as the failure propagates beyond the local stress field created by the drilling pressurization system (Herget, 1988).

### Calculating in situ stress

Tensile failure of the rock may occur when the fluid pressure reaches the minimum shear stress around the borehole plus the tensile strength of the rock:

$$P_f = 3\sigma_3 + \sigma_t - \sigma_2$$

where $\sigma_2$ and $\sigma_3$ are the intermediate and minimum principal stresses. If $\sigma_V = \sigma_1$ in the test area, the principal stresses in the horizontal plane will be $\sigma_H = \sigma_2$ and $\sigma_h = \sigma_3$, and the fracture plane generated will be parallel or sub-parallel to the borehole axis and perpendicular to the direction of the minimum horizontal stress $\sigma_h$. In this case, the following equations will give the horizontal stresses (Kim & Franklin, 1987):

$$\sigma_h = P_s$$
$$\sigma_H = \sigma_t + 3P_s - P_f - P_0 \text{ for the initial pressurization cycle}$$
$$\sigma_H = 3P_s - P_r - P_0 \text{ for subsequent pressurization cycles}$$

$P_r$ is the fracture propagation pressure, or reopening pressure, and $P_0$ is the initial pore pressure as measured with a piezometer, $P_0 = \gamma_w z$.

The rock's tensile strength $\sigma_t$ is measured in the laboratory or *in situ* comparing the fracture initiation pressure with subsequent fracture reopening pressures.

The vertical stress is assumed to be equal to the overburden pressure, $\sigma_V = \gamma z$. The maximum horizontal stress $\sigma_H$ has the same orientation as the fracture plane, and $\sigma_h$ is oriented perpendicular to that plane. The directions of the stresses $\sigma_H$ and $\sigma_h$ can be obtained by observing and measuring the plane orientation through visual inspection of the borehole using a TV camera, through reflected acoustic signal pictures or with impression packers equipped with magnetic compasses (Figure 3.130).

Figure 3.129 Equipment used for hydraulic fracturing tests.

Figure 3.130 Impression of induced fractures in the borehole.

## Box 3.12

### Example of a hydraulic fracturing test in a deep borehole

Several hydraulic fracturing tests have been performed in an 800 m deep borehole. The purpose of these tests was to determine in-situ stresses as part of the design of a mine shaft. The equipment used for the tests is shown in *Figure 3.131*.

The figure below shows the results of a test on a section at a depth of 427 m. The first two pressure cycles did not produce fracture propagation, as is shown by the rapid fall in pressure. During the third attempt, a fracture was created at a breakdown pressure $P_f$ of approximately 7.6 MPa, as shown by the peak in the pressure-time graph and by the brief increase in the corresponding flow rate. Pressure stabilized to the fracture propagation pressure $P_r$. The shut-in pressure $P_s$ was then estimated at 5.5 MPa from the pressure-time graph.

The pore pressure measured with the piezometers in the tested section was 3.5 MPa, corresponding to $P_0$. The tensile strength of the rock, a Carboniferous siltstone, was determined through laboratory tests to be 2.4 MPa, and unit weight 26 kN/m³.

### Calculating stresses

The pressure $P_H$ corresponding to the weight of the column of water of height $H$ inside the borehole, between the test section and the pressure gauge installed on the circuit, must be added to the value of $P_f$ and $P_s$; in this case, $H = 428$ m giving $P_H = 4.28$ MPa.

$$P_f = 7.6 \text{ MPa} + 4.28 \text{ MPa} = 11.88 \text{ MPa}$$
$$P_s = 5.5 \text{ MPa} + 4.28 \text{ MPa} = 9.78 \text{ MPa}$$

Assuming $\sigma_1 = \sigma_V$

$$\sigma_V = 26 \cdot 427 = 11.102 \text{ kPa} = 11.1 \text{ MPa}$$
$$\sigma_h = \sigma_3 = P_s = 9.78 \text{ MPa}$$
$$\sigma_H = \sigma_2 = \sigma_t + 3P_s - P_f - P_0$$
$$\sigma_H = 2.4 + (3 \cdot 9.78) - 11.88 - 3.5 = 16.36 \text{ MPa}$$

which gives:

$$K = \sigma_H / \sigma_V = 16.36/11.1 = 1.47$$
$$\sigma_H / \sigma_h = 16.36/9.78 = 1.67$$

The results show a strong stress anisotropy, with $K \approx 1.5$, and a maximum horizontal stress 1.67 times the minimum horizontal stress. These results agree with the geological situation at the test site, a synclinal structure almost reaching overturning. The stress directions were not obtained using instruments, but were deduced from an analysis of geological structure, judging the maximum principal stress direction to be N–S.

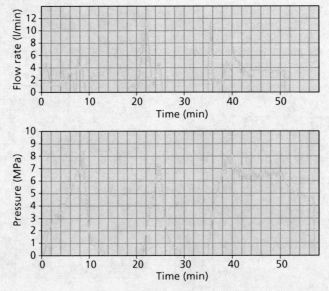

Flow rates and pressures in the hydraulic fracturing test performed at a depth of 427 m.

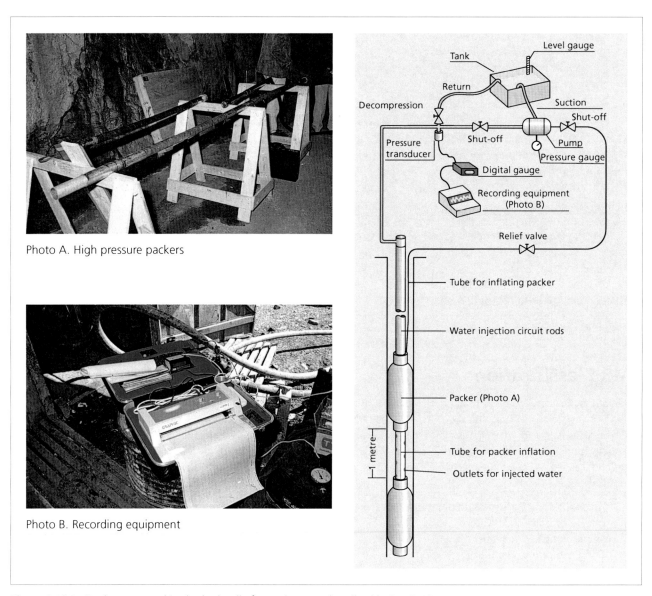

Figure 3.131   Equipment used in the hydraulic fracturing test described in Box 3.12.

In order to obtain representative stress data, several tests must be carried out at different depths in the borehole, so that variation curves $\sigma_H$ and $\sigma_h$ with depth can be plotted.

## 3.8   Rock mass classifications

Section 3.2 dealt with the classification of rock masses for geotechnical purposes, aimed at obtaining geomechanical parameters to use when designing engineering projects. Rock masses are discontinuous media with complex geomechanical behaviour, which when idealised can be studied and classified according to their suitability for different uses. **Geomechanical classifications** were drawn up for this purpose, and through direct observation of rock mass characteristics and the use of simple tests, provide **quality indexes** related to the geomechanical parameters of the rock mass and their suitability for applications including tunnel and slope support and excavability.

The rock mass characteristics considered in the various classifications include:

— Strength of rock material
— Rock Quality Designation (RQD)

- Spacing of discontinuities
- Orientation of discontinuities
- Condition of discontinuities
- Geological structure and faults
- Seepage and water
- State of stress

The most frequently used geomechanical classifications nowadays are the RMR and the Q classifications. Both are applied for characterising rock masses and their properties and for tunnelling applications.

RMR rock mass classification is included in this section as the most widely used system for classifying rock masses. The Q system is described in Chapter 10, Section 10.5, as a specialised classification for tunnelling; SRC rock mass classification is also included in the same section as a specific classification for tunnelling under high tectonic stress in weak rocks. Chapter 10 also describes the tunnel support systems estimated from RMR and Q classifications.

The SMR classification applied to slopes is included in Chapter 9.

## RMR Classification

Developed by Bieniawski in 1973, and updated in 1979 and 1989, this is a rock mass classification system that relates quality indexes with rock mass geotechnical parameters and with excavation and support parameters in tunnels. This classification takes the following geomechanical parameters into account:

- Uniaxial compressive strength of intact rock.
- Degree of fracturing (RQD).
- Spacing of discontinuities.
- Condition of discontinuities.
- Groundwater conditions.
- Orientation of discontinuities.

The influence of these parameters in the geomechanical behaviour of a rock mass is expressed using the **RMR (Rock Mass Rating) quality index**, which varies from 0 to 100.

To apply the RMR classification, the rock mass is divided into areas or sections with similar geological characteristics according to field observations, where data and measurements are taken relating to the properties and characteristics of the intact rock and its discontinuities. For systematic data collection, log books such as the one shown in Chapter 6, Figure 6.2, are used. The RMR index for each area is calculated using the procedure described in Table 3.26.

Once the scores obtained from applying the five classification parameters are calculated, an adjustment for discontinuity orientation is made, and a numerical value obtained which finally allows the rock mass to be classified.

This classification differentiates five classes, with their geotechnical meaning explained in Table 3.27; each rock mass class is assigned a quality and geotechnical characteristics.

Thus, a rock mass classified as Very Good (Class I), will be hard, slightly jointed, with no important seepage and slightly weathered, and so will present very few problems of strength and stability. It can be inferred that its bearing capacity will be high, it will allow excavation of steep slopes and no support or reinforcement measures will be needed in tunnels.

Figure 3.132 shows examples of rock masses belonging to different classes according to RMR index.

Table 3.26 (D) also describes the stand-up time and unsupported tunnel span characteristics of the rock mass, and the influence of the discontinuity orientation in relation to the tunnel. Both these aspects considered in Bieniawski's classification are discussed in Chapter 10, Section 10.5.

## Geomechanical classifications in practice

Geomechanical classifications are a procedure for characterizing rock masses from field and borehole data, and are applied mainly to tunnels, because of the difficulty of studying rock masses at depth; this aspect is discussed in Chapter 10. But they are also applied to describe rock masses in general, and to provide a geotechnical rock mass classification. Calculating the RMR or Q indices allows an estimate of rock mass **strength and deformability parameters** (see Section 3.6), and of their expected behaviour when excavated (Chapter 10, Sections 10.5 and 10.6).

To apply them, field observations and measurements must be made, which are the basis and the systematic practice of the classification, as indicated in Figure 3.133. The rock mass classes obtained refer to conditions before excavation, and their description must specify if any adjustments have been applied, for discontinuity orientation or any other specific adjustments for tunnels, as described in Chapter 10. Other aspects to be highlighted are significant geological structures or areas, such as faults, folds or other tectonic structures, discordances, heavily weathered areas or areas with water infiltration.

The main advantage of geomechanical classifications is that they provide an easy and simple estimate of the mechanical parameters of the rock mass. However, their over-simplification must be taken into account when they are applied to weak, soft and weathered rock masses, for which in general they overestimate strength and mechanical properties, and do not consider certain important aspects such as rock mass deformability. When applying these classifications, their limitations must be considered, and the results always interpreted carefully, based on a knowledge of the properties and geomechanical behaviour of the different types of rock mass.

### Table 3.26  ROCK MASS RATING SYSTEM

#### A. Classification parameters

| 1 | Strength of intact rock material (MPa) | Point-load strength index | >10 | 10–4 | 4–2 | 2–1 | Uniaxial compressive strength (MPa) | | |
|---|---|---|---|---|---|---|---|---|---|
| | | Uniaxial compressive strength | >250 | 250–100 | 100–50 | 50–25 | 25–5 | 5–1 | <1 |
| | | Rating | 15 | 12 | 7 | 4 | 2 | 1 | 0 |
| 2 | RQD | | 90%–100% | 75%–90% | 50%–75% | 25%–50% | <25% | | |
| | Rating | | 20 | 17 | 13 | 8 | 3 | | |
| 3 | Spacing of discontinuities | | >2 m | 0.6–2 m | 0.2–0.6 m | 60–20 mm | <60 mm | | |
| | Rating | | 20 | 15 | 10 | 8 | 5 | | |
| 4 | Conditions of discontinuities (see E) | | Very rough surfaces. Not continuous. No separation. Unweathered wall rock. | Slightly rough surfaces. Separation <1 mm. Slightly weathered walls. | Slightly rough surfaces. Separation <1 mm. Highly weathered walls. | Slickensided surfaces. or Gouge <5 mm thick or Separation 1–5 mm. Continuous. | Soft gouge >5 mm thick. Separation >5 mm. Continuous. | | |
| | Rating | | 30 | 25 | 20 | 10 | 0 | | |
| 5 | Ground water | Inflow per 10 m tunnel length | None | <10 litres/min | 10–25 litres/min | 25–125 litres/min | >125 litres/min | | |
| | | (Joint water press)/(Mayor principal stress) | 0 | 0.0–0.1 | 0.1–0.2 | 0.2–0.5 | >0.5 | | |
| | | General conditions | Completely dry | Damp | Wet | Dripping | Flowing | | |
| | Rating | | 15 | 10 | 7 | 4 | 0 | | |

#### B. Rating adjustment for discontinuity orientations (see F)

| Strike and dip orientations | | Very favourable | Favourable | Fair | Unfavourable | Very unfavourable |
|---|---|---|---|---|---|---|
| Rating | Tunnels and mines | 0 | –2 | –5 | –10 | –12 |
| | Foundations | 0 | –2 | –7 | –15 | –25 |
| | Slopes | 0 | –5 | –25 | –50 | –60 |

#### C. Rock mass classes

| Class | I | II | III | IV | V |
|---|---|---|---|---|---|
| Description | Very good | Good | Fair | Poor | Very poor |
| Rating | 100–81 | 80–61 | 60–41 | 40–21 | <20 |

(continued)

Table 3.26  ROCK MASS RATING SYSTEM (CONT.)

**D. Meaning of rock classes**

| Class number | I | II | III | IV | V |
|---|---|---|---|---|---|
| Average stand-up time | 20 yrs for 15 m span | 1 yr for 10 m span | 1 week for 5 m span | 10 hrs for 2.5 m span | 30 min for 1 m span |
| Cohesion of rock mass | >400 kPa | 300–400 kPa | 200–300 kPa | 100–200 kPa | <100 kPa |
| Friction angle of rock mass | >45° | 35°–45° | 25°–35° | 15°–25° | <15° |

**E. Guidelines for classification of discontinuity conditions**

| | | | | | |
|---|---|---|---|---|---|
| Length (persistence) | <1 m | 1–3 m | 3–10 m | 10–20 m | >20 m |
| Rating | 6 | 4 | 2 | 1 | 0 |
| Separation (aperture) | None | <0.1 mm | 0.1–1.0 mm | 1–5 mm | >5 mm |
| Rating | 6 | 5 | 4 | 1 | 0 |
| Roughness | Very rough | Rough | Slightly rough | Smooth | Slickensided |
| Rating | 6 | 5 | 3 | 1 | 0 |
| Infilling (gouge) | None | Hard filling | Hard filling >5 mm | Soft filling <5 mm | Soft filling >5 mm |
| Rating | 6 | 4 | 2 | 2 | 0 |
| Weathering | Unweathered | Slightly weathered | Moder. weathered | Highly weathered | Decomposed |
| Rating | | 6 | 5 | 3 | 1 | 0 |

**F. Effect of discontinuity strike and dip orientation in tunnelling**

| Strike perpendicular to tunnel axis | | | | Strike parallel to tunnel axis | | Dip 0°–20° Irrespective of strike |
|---|---|---|---|---|---|---|
| Drive with dip | | Drive against dip | | | | |
| Dip 45–90° | Dip 20–45° | Dip 45–90° | Dip 20–45° | Dip 45–90° | Dip 20–45° | |
| Very favourable | Favourable | Fair | Unfavourable | Very unfavourable | Fair | Fair |

(Bieniawski, 1989).

Table 3.27  ROCK MASS QUALITY ACCORDING TO THE RMR INDEX

| Class | Quality | RMR rating | Cohesion (MPa) | Friction angle |
|---|---|---|---|---|
| I | Very good | 100–81 | >0.4 | >45° |
| II | Good | 80–61 | 0.3–0.4 | 35°–45° |
| III | Fair | 60–41 | 0.2–0.3 | 25°–35° |
| IV | Poor | 40–21 | 0.1–0.2 | 15°–25° |
| V | Very poor | <20 | <0.1 | <15° |

## Class I (RMR = 81-100) and Class II (RMR= 61-80) rock masses

Cretaceous dolomite. Very good quality. Two main sets of discontinuities.

Granite. Good quality. Several sets of weathered discontinuities.

## Class III rock masses (RMR= 41-60)

Ordovician slate. Fair quality. High fracturing degree. Weathering degree: III.

Ordovician quartzite. Fair quality. High fracturing degree. Very hard intact rock.

## Class IV (RMR= 21-40) and Class V (RMR ≤ 20) rock masses

Ordovician quartzite. Poor quality. Weathered and heavily jointed rock mass.

Palaeozoic slate. Very poor quality. Heavily jointed. Degree of weathering: V.

*Figure 3.132* Examples of classes of rock mass.

**1. Geological analysis of the rock mass**

| | |
|---|---|
| Identifying lithological units | → Lithological description |
| Structural analysis | → Structural data |
| Litho-structural zoning | → Identifying zones |
| Hydrogeological conditions | → Hydrogeological data |
| Geomorphological conditions | → Geomorphological data |

⇒
— Production of detailed geological sections and maps.
— Litho-structural zoning.

**2. Obtaining geomechanical data**

Selecting geomechanical sites: identifying outcrops representative of the different litho-structural zones.

Collecting geomechanical data in each geomechanical field site according to the field log.

⇒
— Fill in field log reports at sites.
— Geotechnical core borehole logging.

**3. Calculating the RMR, Q, SMR or SRC** [(1)]

Calculating RMR index at each geomechanical field site.

Calculating Q, SRC or SRM index, depending on their application.

Geomechanical sectors defined according to rock mass classification.

⇒
— Correlation with geotechncial properties of the rock mass.
— Application to tunnels, slopes and foundations.

[(1)] These indexes are described in Chapters 3 (RMR), 9 (SMR) and 10 (Q and SRC).

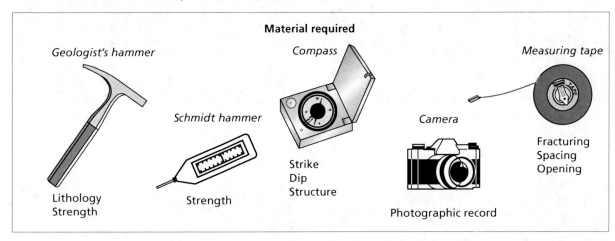

*Figure 3.133* Procedure for applying geomechanical classifications.

# Recommended reading

Bieniawski, Z.T. (1989). Engineering rock mass classifications. John Wiley & Sons.

Brady, B.H.G. and Brown, E.T. (1993). Rock mechanics for underground mining. 2nd ed. Kluwer Academic Publishers.

Goodman, R.E. (1989). Introduction to rock mechanics. John Wiley & Sons.

Hudson, J.A. and Cornet, F.H. (2003). Rock stress estimation. Int. Soc. for Rock Mechanics. Suggested methods and associated supporting papers. Special issue of the Int. J. of Rock Mech. and Min. Sci., vol. 4, issues 7–8.

Hudson, J.A. and Harrison, J.P. (2000). Engineering rock mechanics. An introduction to the principles. Pergamon Press.

ISRM, 1981. Rock characterization. Testing and monitoring. Int. Soc. for Rock Mechanics. Suggested methods. Brown, E.T. (ed.). Commission on testing and monitoring, ISRM. Pergamon Press.

# References

Attewell, P.B. and Farmer, I.W. (1976). Principles of engineering geology. Chapman and Hall, London.

Bandis, S.C., Lumsden, A.C. and Barton, N. (1981). Experimental studies of scale effects on the shear behaviour of rock joints. Int. J. of Rock Mech. and Min. Sci., Abstracts, vol. 18, pp. 1–21. Elsevier.

Bandis, S.C., Lumsden, A.C. and Barton, N. (1983). Fundamentals of rock joint formation. Int. J. of Rock Mech. and Min. Sci., vol. 20, no. 6. Elsevier.

Barton, N. (1990). Scale effects or sampling bias? In: Scale effects in rock mechanics. Cunha, A.P. (ed.). Balkema.

Barton, N. (1995). The influence of joints properties in modeling jointed rock masses. Keynote lecture. Proc. 8th ISRM Congress. Fuji, T. (Ed.). pp. 1023–1032. Balkema.

Barton, N. (2002). Some new Q-value correlations to assist in site characterization and tunnel design. Int. J. Rock Mech. & Min. Sci., 39/2, pp. 185–216.

Barton, N. (2006). Rock quality, seismic velocity, attenuation and anisotropy. Taylor and Francis, London.

Barton, N. and Choubey, V. (1977). The shear strength of rock joints in theory and practice. Rock Mechanics, vol. 10, 1/2, pp. 1–54.

Barton, N. and Pandey, S.K. (2010). Numerical modelling of two stoping methods used at HZL's Zawar Mines in India using FLAC 3D and degradation and mobilization of c and $\phi$ based on Q-parameters. Int. J. of Rock Mech. and Min. Sci. (in press).

Bieniawski, Z.T. (1973). Engineering classification of jointed rock masses. Transactions, S. African Inst. of Civil Engineers, vol. 15, no. 12, pp. 335–344.

Bieniawski, Z.T. (1978). Determining rock mass deformability: experience from case histories. Int. J. of Rock Mech. and Min. Sci., vol. 15, pp. 237–248.

Bieniawski, Z.T. (1979). The geomechanics classification in rock engineering applications. Proceedings of the 4th International Conference on Rock Mechanics. Montreux. Balkema, vol. 2, pp. 41–48.

Bieniawski, Z.T. (1984). Rock mechanic design in mining and tunnelling. Balkema.

Bieniawski, Z.T. (1989). Engineering rock mass classifications. John Wiley and Sons, Inc.

Blyth, E. and De Freitas, M. (1984). Geology for engineers. Edward Arnold, London.

Brady, B.H.G. and Brown, E.T. (1993). Rock mechanics for underground mining. 2nd ed. Kluwer Academic Publishers.

Coon, R.F. and Merritt, A.H. (1970). Predicting in situ modulus of deformation using rock quality indexes. Am. Soc. Test. Mater. (ASTM), Spec. Tech. Publ. 477, pp. 154–173.

Cunha, A.P. (1990). Scale effects in rock mechanics. In: Scale effects in rock masses. Cunha, A.P. (ed.). Balkema

Cunha, A.P. and Muralha, J. (1990). About LNEC experience on scale effects in the deformability of rock masses. In: Scale effects in rock masses. Cunha, A.P. (ed.). Balkema

Duncan, C.W. (1999). Foundations on rock. 2nd ed. E & F.N. Spon.

Edelbro, C. (2009). Numerical modelling of observed fallouts in hard rock masses using an instantaneous cohesion-softening friction-hardening model. Tunnelling and Underground Space Tech., 24, 4, pp. 398–409.

Embleton, C. and Thornes, J.B. (1979). Process in geomorphology. Arnold, London.

Farmer, I.W. (1968). Engineering properties of rocks. Spon Ltd., London.

González de Vallejo, L. and Hijazo, T. (2008). A new method of estimating the ratio between in situ rock stresses and tectonics based on empirical and probabilistic analysis. Engineering Geology, 101, 185–194.

Goodman, R.E. (1989). Introduction to rock mechanics. John Wiley & Sons.

Haimson, B.C. (1990). Scale effects in rock stress measurements. In: Scale effects in rock masses. Cunha, A.P. (ed.). Balkema.

Hajiabdolmajid, V., Martin, C.D. and Kaiser, P.K. (2000). Modelling brittle failure. Proc. 4th North American Rock Mechanics Symp. Girard, Liebman and Breeds Eds., pp. 991–998. Balkema.

Herget, G. (1988). Stresses in rock. Balkema.

Hoek, E. (1994). Strength of rocks and rock masses. Int. Soc. for Rock Mechanics New Journal 2, (2), pp. 4–16.

Hoek, E. and Bray, J.W. (1981). Rock slope engineering. 3rd ed. The Institution of Mining and Metallurgy, London.

Hoek, E. and Brown, E.T. (1980). Underground excavation in rock. The Institution of Mining and Metallurgy, London.

Hoek, E. and Brown, E.T. (1988). The Hoek-Brown failure criterion. A 1988 update. Rock engineering for underground excavations. Proc. 15th Canadian Rock Mechanics Symp. Curran (ed.). University of Toronto.

Hoek, E. and Brown, E.T. (1997). Practical estimates of rock mass strength. Int. J. of Rock Mech. and Min. Sci. Elsevier. vol. 34, no. 8, pp. 1165–1186.

Hoek, E., Carranza-Torres, C. and Corkum, B. (2002). Hoek-Brown criterion – 2002 edition. Proc. NARMS-TAC Conference, Toronto, 267–273.

Hoek, E. and Diederichs, M.S. (2006). Empirical estimation of rock mass modulus. Int. J. of Rock Mech. and Min. Sci., 43, pp. 203–215.

Hoek, E., Kaiser, P.K. and Bawden, W.F. (1995). Support of underground excavations in hard rocks. Balkema.

Hoek, E. and Marinos, P. (2000). Predicting tunnel squeezing. Tunnels and Tunnelling Int. Part 1, v. 32:11, pp. 45–51.

Hoek, E., Marinos, P. and Benissi, M. (1998). Applicability of the Geological Strength Index (GSI) classification for very weak and sheared rock masses. The case of the Athens Schist Formation. Bull. Eng. Geol. Env. 57(2), 151–160.

Hudson, J.A. and Cornet, F.H. (2003). Rock stress estimation. Int. Soc. for Rock Mechanics. Suggested methods and associated supporting papers. Special issue of the Int. J. of Rock Mech. and Min. Sci., vol. 4, issues 7–8.

Hudson, J.A. and Harrison, J.P. (2000). Engineering rock mechanics. An introduction to the principles. Pergamon Press.

ISRM (1979). Suggested methods for determining the uniaxial compressive strength and deformability of rock materials. ISRM Commission on standardization of laboratory and field tests. Int. J. of Rock Mech. and Min. Sci., Geomechanical Abstracts, vol. 16.

ISRM (1981). Rock characterization. Testing and monitoring. Int. Soc. for Rock Mechanics. Suggested methods. Brown, E.T. (ed.). Commission on testing and monitoring, ISRM. Pergamon Press.

Jaeger, J.C. and Cook, N.G.W. (1979). Fundamentals of rock Mechanics. 3rd ed., Chapman and Hall, London.

Johnson, R.B. and De Graff, J.V. (1988). Principles of engineering geology. John Wiley & Sons.

Kim, K. and Franklin, J.A. (1987). Suggested methods for rock stress determination. Int. J. of Rock Mech. and Min. Sci. Geomechanical abstracts. 24–1, pp. 53–73.

Leeman, E.R. (1970). Experience throughout the world with the CSIR "doorstopper" rock stress measurement equipment. 2nd Congress ISRM, Belgrade. vol. 2 pp. 4–6.

Martin, C.D., Read, R.S. and Chandler, N.A. (1990). Does scale influence in situ stress measurements? Some findings at the underground research laboratory. In: Scale effects in rock masses. Cunha, A.P. (ed.). Balkema.

Natau, O. (1990). Scale effects in the determination of the deformability and strength of rock masses. In: Scale effects in rock masses. Cunha, A.P. (ed.). Balkema.

Patton, F.D. (1966). Multiple modes of shear failure in rock. Proceedings 1st Congress on Rock Mechanics, ISRM, Lisbon. vol. 1, pp. 509–513.

Rahn, P.H. (1986). Engineering geology. An environmental approach. Elsevier.

Selmer-Olsen, R. and Broch, E. (1977). General design procedure for underground openings in Norway. Rockstorage 77. 1st Int. Symp. on Storage in Excavated Rock Caverns, Sweden. vol. 2 (11–22).

Serafim, J.L. and Pereira, J.P. (1983). Considerations of the geomechanical classification of Bieniawski. Proceedings Int. Symp. on Engineering Geology and Underground Construction, Lisbon. Balkema.

Serrano, A. and Olalla, C. (1998). Ultimate bearing capacity of an anisotropic discontinuous rock mass. Parts 1 and 2. Int. J. of Rock Mech. and Min. Sci., vol. 35, 3, pp. 301–348.

Sheorey, P.R. (1997). Empirical rock failure criteria. Balkema.

Singhal, B.B.S. and Gupta, R.P. (1999). Applied hydrogeology of fractured rock masses. Kluwer Academic Publishers.

Walthan, A.C. (1999). Foundations of engineering geology. E & F.N. Spon.

Wawersick, W.R. and Fairhurst, C. (1970). A study of brittle rock failure in laboratory compression experiments. Int. J. of Rock Mech. and Min. Sci., vol. 7, no. 5. pp. 561–575.

Zhang, L. and Einstein, H.H. (2004). Using RQD to estimate the deformation modulus of rock masses. Int. Journal of Rock Mechanics and Mining Sciences, 41, pp. 337–341.

# 4

# HYDROGEOLOGY

1. Hydrogeological behaviour of soils and rocks
2. Hydrogeological parameters
3. Flow. Darcy's law and fundamental flow equations in porous media
4. Evaluation methods for hydrogeological parameters
5. Solution methods
6. Chemical properties of water

## 4.1 Hydrogeological behaviour of soils and rocks

Aquifers are permeable geological formations which can store and transmit water. There is a wide range of natural formations with very different capacities for storing and transmitting water; in hydrogeological terms, these formations can be divided into four main groups (*Figure 4.1*):

— **Aquifers**: can store and transmit water (gravels, sands, limestone materials, etc.); these are formations with a **high drainage capacity** where wells and boreholes can be drilled to satisfy human needs for water supplies, agriculture, livestock or industry.
— **Aquitards**: can store very large quantities of water but transmit it with difficulty; these are often classified as semi-permeable formations (silts, silty or clayey sands) with **medium to low drainage capacity**; they cannot be used to produce flow rates for water supply requirements but they play a very important natural role as water transmitting elements capable of vertical recharge over large surface areas.
— **Aquicludes**: can store large quantities of water but cannot transmit it, and **drain with great difficulty**; the water occupies the formation pores and cannot be released (clays, plastic clays or clay silts); in classical hydrogeology they are classified as impermeable, but in engineering geology this concept is less precise, as even very limited drainage can cause problems in some projects.
— **Aquifuges**: formations that cannot store or transmit water; among them are hard rock materials, such as granite and gneiss, and even very compacted and non-karstified limestone; these materials are impermeable unless there are fractures and discontinuities which allow flow.

## Types of aquifers and their behaviour

First of all, to establish some basic points of reference, it should be noted that when a well is pumped the water level falls in it and in the part of the aquifer surrounding it; this drawdown is more significant in the well itself and less so the further away from the well the measurement is taken. The distance between the pumping well and the area of the aquifer where its influence is nil is known as the radius of influence (*Figure 4.2*). A large pumping cone is formed around the well, with its surface being the transient, or dynamic, piezometric surface, and the water level at a point on this surface is known as the pumping, or dynamic level. The static level is what exists in the aquifer before pumping starts.

With these points established, the types of aquifers which exist naturally are described below, as well as the way they react in different hydraulic and structural situations and how they behave when pumping takes place. There are basically three types of aquifer materials in terms of their texture, which are shown in *Figure 4.3*:

— The so called **porous aquifers**, with permeability resulting from intergranular porosity; these include

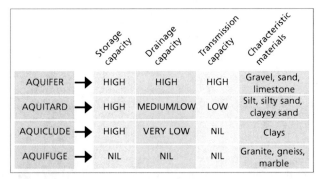

Figure 4.1   Hydrogeological behaviour of geological formations.

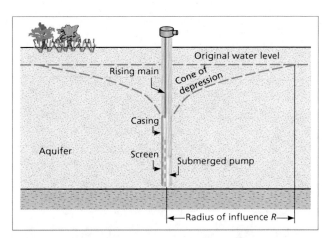

Figure 4.2   Cone of depression when pumping a catchment well.

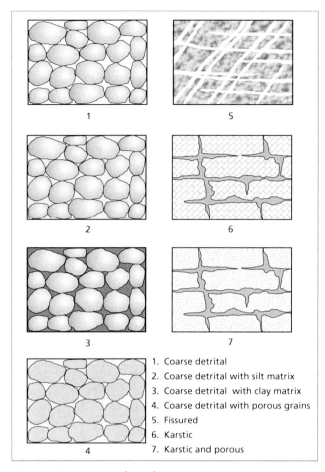

Figure 4.3   Types of aquifer by texture.

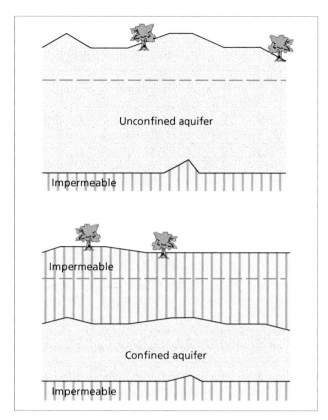

Figure 4.4   Diagrams of unconfined and confined aquifers.

gravels, sands and in general all detritic materials of at least sand grain size. The texture of the medium is made up of grains which allow water to be stored and to circulate through the intergranular spaces; these pores may be filled with very fine granular material, reducing the capacity of the medium for water storage and transportation, or they may be filled with clay materials, reducing these characteristics to almost nothing. Sometimes, the grains themselves are made up of porous material, which provides even better properties for water storage. Because of their genesis, granular media are usually very homogeneous on reduced scales.

— The second type are those where permeability is due to discontinuities and fissures, caused either mechanically or by dissolution, which form **karstic and fissured aquifers** and include limestone, dolomite, granite, basaltic formations, etc., with the first two of these the most important. Karstification is a dissolution process caused by the action of water in carbonated, previously fissured formations; karstic aquifers are not very homogeneous on a small scale, but can be considered more homogeneous if the working scale is large enough.

— The third type are aquifers where permeability is due to a combination of the two causes outlined above, and sometimes called "double aquifers", making them **karstic, fissured and porous**. Typical of these are the calcarenites.

In general, porous aquifers are the most homogeneous, within the characteristic heterogeneity of all aquifers. Both these and the karstic-porous ones are able to store a large quantity of water per unit of aquifer volume and so they tend to be slow aquifers which react to pumping with inertia. Their radius of influence tends to be shorter than in karstic and fissured ones; karstic and fissured aquifers, however, are less homogeneous and tend to produce the most varied test results; they have less water storage capacity and so can be considered as quick aquifers, where the radii of influence are longer than in other aquifers.

Aquifers, depending on their hydraulic and structural circumstances, may work in three different ways (Figures 4.4 and 4.5):

— **Unconfined aquifers**, where the water level is found below the upper boundary of the permeable formation. They release water by dewatering, i.e. the water

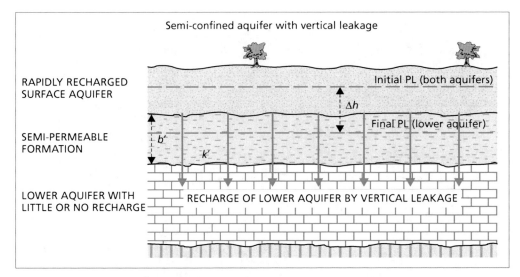

*Figure 4.5*   Diagram of semi-confined aquifer (PL = Piezometric Level).

they discharge is what they have stored; this volume of water is high compared to the aquifers described below and for this reason they have considerable inertia. The pumping cones are normally elongated i.e. with a significant level of drawdown in the well and a short radius of influence.

— **Confined aquifers**, isolated in the sub-soil, bounded by impermeable materials sometimes on all sides. The water level can be above the upper boundary of the aquiferous formation; in fact, they are usually under pressure due to the weight of the overlying materials. The water they release comes from the decompression, when water is taken from the aquifer. In the strictest sense, completely confined aquifers do not exist naturally, since there are no absolutely impermeable materials capable of isolating them, although in practical terms there are many which could be considered as such, including those permeable formations which remain confined over a large area and appear as outcrops on the surface at either end (this case will be described in more detail below). These are quick aquifers, with very little inertia due to their low water storage capacity which immediately react or transmit the influences of pumping. The pump cones tend to be shallower than in unconfined aquifers, but with a longer radius of influence.

— **Semi-confined aquifers**, where the surrounding materials are not all impermeable; this means that the upper or semi-confining layer is made up of semi-permeable formations, which allow the water from other higher aquifers to pass through to the semi-confined aquifer. As a result the speed of reaction of these aquifers to pumping is more moderate than in

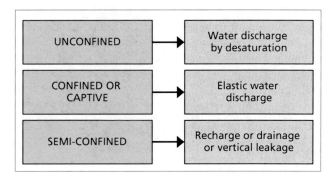

*Figure 4.6*   Type of aquifer by structure and mechanism.

the confined ones and the average values of the radii of influence fall between unconfined and confined. A semi-confined aquifer is, in fact, an integrated physical system including an upper and rapidly recharged aquifer, a semi-permeable layer or aquitard and a lower semi-confined aquifer; the difference in levels between the upper and lower aquifers results in a vertical transference of water which recharges the lower aquifer.

*Figure 4.6* lists the types of aquifer which exist naturally, by structure and mechanism.

These considerations give a qualitative idea of the shape of the cones of influence and the reaction speed of different aquifers to pumping. For example, the slower aquifers, i.e. those which transmit the influence of pumping more slowly and which also have deep cones of influence but with short radii, will be those with a granular porous texture when they work as unconfined systems. In contrast, the aquifers which will feel the effects of specific pumping more quickly, with shallower pumping cones and long radii, will be the karstic and fissured ones, functioning as confined aquifers.

# Piezometric level

The movement of water within geological formations is conditioned by the capacity of its elemental particles to do work, i.e. by their energy. This energy depends both on the force field to which the fluid is subjected and, on the other hand, on the agent sensitive to that field, and on the type of energy itself or the way the formation has stored its capacity to do work.

The three types of energy which may affect the water moving within a permeable formation matrix, assuming there are no temperature variations, are:

— Potential energy: $E_h = mgz$
— Kinetic energy: $E_k = 1/2 \cdot mv^2$
— Pressure energy: $E_p = pV$

where $g$ represents the intensity of the gravitational field and $m$ the mass or agent sensitive to the gravitational field, $z$ is the height with respect to a reference elevation, $v$ is the velocity of the fluid (water), $p$ is the pressure the water is subjected to and $V$ is volume.

By the principle of energy conservation, this will remain constant along a flow line and therefore:

$$E = E_h + E_k + E_p = \text{constant}$$

or

$$mgz + \frac{1}{2}mv^2 + pV = \text{constant}$$

extracting common factor $m \cdot g$:

$$mg\left(z + \frac{v^2}{2g} + \frac{p}{\rho g}\right) = \text{constant}$$

where $\rho$ is the water density.

Considering the same fluid and gravitational field (the earth's), gives:

$$z + \frac{v^2}{2g} + \frac{p}{\rho g} = \text{constant}$$

which is the expression of Bernouilli's theorem where all the terms have length dimensions:

$$[z] = L$$

$$\left[\frac{v^2}{2g}\right] = \frac{L^2 T^{-2}}{LT^{-2}} = L$$

$$\left[\frac{p}{\rho g}\right] = \frac{MLT^{-2}L^{-2}}{ML^{-3}LT^{-2}} = L$$

As a result, the terms of the equation are called heads:

$z \rightarrow$ position head

$\dfrac{v^2}{2g} \rightarrow$ velocity head

$\dfrac{p}{\rho g} \rightarrow$ pressure head

The term $v^2/2g$ or velocity head, in the case of water movement within a permeable medium, may be ignored compared with the pressure head and position head, given the low water velocity value in these media; the capacity of the water to do work at point A of an aquifer will therefore be given by its piezometric level or the sum of the position and pressure heads:

$$h_A = z_A + \frac{p_A}{\gamma_w}$$

where $\gamma_w$ is the unit weight of the water.

The piezometric surface is also known as the piezometric head or level or hydraulic head. The piezometric surface remains constant with depth in a standing body of water or in a saturated permeable medium when this is homogeneous and isotropic and contains no flow.

As shown in *Figure 4.7*, point A is at atmospheric pressure and therefore its piezometric surface, $h_A$, is equal to its position head $z_A$. Point B has a piezometric level $h_B$, which is to be compared with $h_A$.

If $p_B$ is the pressure at point B and $z_B$ its position head:

$$h_A = z_A$$

$$h_B = z_B + \frac{p_B}{\gamma_w} = z_B + \frac{(z_A - z_B)\gamma_w}{\gamma_w} = z_A = h_A$$

i.e. the piezometric level is the same at points A and B.

---

The PL at any given point in an aquifer is the height the water level reaches above a horizontal reference, or datum, when this level is in contact with atmospheric pressure.

It is the sum of two measurements:
 – Position or elevation head
 – Pressure head

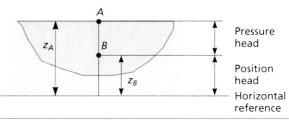

*Figure 4.7*  Piezometric level.

As a result, the **piezometric level** at any point in an aquifer can be defined as the water level compared with the level the water reaches when this point is freed to atmospheric pressure. This height is the sum of two components: the position head of the point compared to a reference point, or datum, and the pressure head equivalent to the level the water would reach at atmospheric pressure.

The term **water table** is used very often, mainly in geotechnics. In technical publications and engineering projects there is often some misunderstanding about the way this term is used.

What is often referred to is the depth at which water can be found in an excavation, well or bore hole. This concept, however, is not much use, since it does not give any idea of the working capacity of a water particle at any given point. The measurement between the ground surface and the point at which the water is encountered in boreholes and wells on a site may not be comparable with the level the water reaches in an open excavation to that depth. The term water table is also used by some as the **piezometric level** associated with unconfined aquifers, but often only the free **water surface** is referred to, i.e. the upper surface of the saturated area of the unconfined aquifer. This concept is also insufficient, above all for geological engineering purposes, because the calculation of the ground pore pressures cannot be based exclusively on it, especially if there are confined aquifers at depth. The problem is usually solved by thinking in terms of **hydraulic head**, comparable to piezometric level. Often practitioners use the terms water table and piezometric level as synonymous and interchangeable.

The recommendation is to use the term piezometric surface with its associated concept and whenever the term water table is used, the reader will have to deduce the exact meaning the writer intends.

## Water movement in aquifers

In the soil and subsoil there are various zones where the movement of the water has singular characteristics and its own laws. Four zones can be differentiated: soil, non-saturated zone, capillary zone and saturated zone. *Figures 4.8* and *4.9* show these zones and give an indication of the relative piezometric levels for each.

### Water movement in the soil

A wet zone is found in the aquifer formation nearest the surface and in contact with the atmosphere, depending on the season, characterized in general by its high porosity and abundance of organic material. This is normally known as "soil" and in it the water movement is characterized by the phenomena of storativity, evaporation and transpiration.

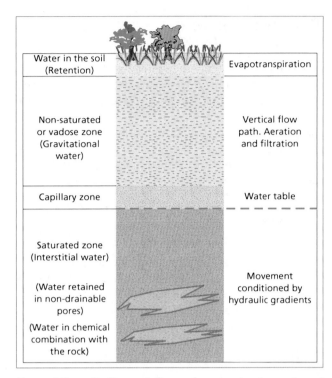

Figure 4.8   Profile and flow of water in the soil and subsoil.

The soil can store a quantity of water or moisture available for use by the plants which depends on its field capacity (the maximum moisture content it can hold), permanent wilting point (minimum moisture content needed for the plants to be able to survive), average root depth and apparent density. When there is precipitation, this reservoir is recharged. If the reservoir is "full" i.e. if the soil is at its field capacity, the excess either becomes surface run-off or drains by gravity to the saturated zone in an infiltration process. If there is no rainfall, the plants use up the water in the reservoir by evotranspiration during their life cycle until it is exhausted and they wilt, or until there is more rainfall, which recharges the reservoir. This is an absorption zone and therefore, due to the negative pressures, the piezometric surface is lower than the position head.

### Movement of water in the non-saturated zone

The water flow in the non-saturated zone is by gravity, and therefore it falls vertically. In fact, although the vertical component is very important, depending on its position within this zone and local circumstances it may be altered and have other components, but in general terms, it should be assumed that the water moves vertically downwards. There is also absorption in this zone, and due to the negative pressures the piezometric surface is lower than the position head.

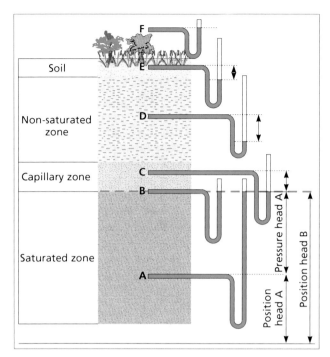

*Figure 4.9* Piezometric levels in soil and subsoil. F = water at atmospheric pressure. E = water pressure < atmospheric. D = water pressure < atmospheric. C = water pressure < atmospheric. B = water = atmospheric = water table. A = water pressure > atmospheric.

## Movement of water in the capillary fringe

This is a transition zone between the non-saturated and saturated zones due to the absorption by the part of the aquifer formation situated immediately above the saturated zone. In the upper part of the capillary fringe the air pockets may stop the descending movement while in the lower part the water flow is very similar to the water movement in the saturated zone. Due to absorption, the piezometric surface is below the position head (*Figure 4.9*).

## Movement of water in the saturated zone. Equipotential lines

This is the lower zone where the water completely saturates the aquifer pores. In this area the piezometric surface is never lower than the position head. In the highest part, near the base of the capillary fringe, the piezometric level and the position head are coincident and the pressure head is zero. Further down in the saturated zone, the piezometric level is maintained, but the head losses are compensated by the increment in the pressure head. These statements are valid assuming that there are no vertical flows within the formation.

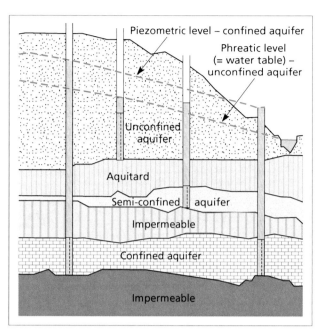

*Figure 4.10* Types of aquifers, water table and piezometric levels.

The **piezometric surface** is the geometric location of points of equal piezometric level within the formation. It should be differentiated from the **phreatic surface**, which is the geometric location of the points of the aquifer where the pressure head is nil and which coincides with the water surface in unconfined aquifers (*Figure 4.10*). In the phreatic surface there may and in fact do exist different piezometric levels, depending on the pressure head. In unconfined aquifers, under conditions where there are no vertical flows, piezometric and phreatic surfaces are coincident terms.

The water in the saturated zone moves from points of higher piezometric level to points of lower piezometric level, i.e. from zones of greater to zones of lesser energy. As a result the water in the saturated zone may move in horizontal and vertical directions, rising or falling, independently from the elevation level of the formations, in response only to the energy at each point in the space.

The piezometric surfaces in confined aquifers are higher than the roof of the aquifer formation, except sometimes near pumping wells which produce a considerable drop in the water level. In unconfined aquifers the piezometric surface coincides with the phreatic surface or the surface of the geometric location of aquifer points at atmospheric pressure.

The piezometric level is normally obtained by direct measurements at points of the aquifer and the piezometric surface is defined by drawing lines of equal piezometric level. These **equipotential lines** are perpendicular to the impermeable borders of the aquifer and parallel with the lines of

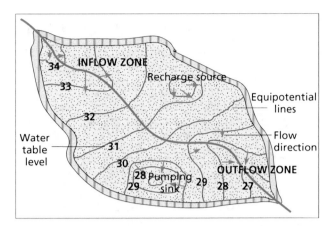

*Figure 4.11   Equipotential lines and flow lines.*

recharge or discharge. The **flow lines** at each point are perpendicular to the equipotential lines in ground that is isotropic. On maps of piezometric surfaces the direction of flow is normally represented by arrows, oriented from greater to lesser piezometry (*Figure 4.11*).

The study of piezometric surfaces allows basic data on groundwater movement to be obtained. The situation of the piezometric surface of an aquifer varies with time as a function of its storage capacity and it is important to remember that it represents the existing situation at any given moment. Whenever the piezometric surface of an aquifer is mentioned it is essential to indicate the date this refers to.

When the piezometric suface intersects the ground surface in an unconfined aquifer this produces a spring or the discharge of water into a river. When the flow is towards a water course whether this is a river, canal, fracture, etc, this is considered as a **drainage** along the length of that line; however if the flow moves away from that line, it means the opposite. In the first case the river is said to be effluent, since it drains the aquifer and in the second case to be influent, as it recharges the aquifer. Closed contours indicate localized areas of recharge or discharge, although these areas are not always represented by closed contours.

## 4.2 Hydrogeological parameters

The water storage and transmission capacity of an aquifer must be quantified to be able to evaluate the geological formations from the hydrogeological point of view. Therefore, four basic parameters should be identified to describe how geological formations can contain and transmit water. These paired and interdependent parameters are porosity and storage coefficient, and permeability and transmissivity, and they must be identified to define the characteristics of the aquifer.

## Porosity

Porosity is the ratio between the pore volume and the total rock volume. It is a dimensionless parameter and depends only on the rock or soil composition, i.e. on its characteristic texture, independently of its geometric shape or thickness of the formation or its natural hydraulic character.

Depending on the type of formation, the pores may be due to intergranular spaces, in detrital formations, or to cracks and fissures in the case of fissured or karstic rocks; the concept of porosity may be linked to either of these.

The porosity of a formation may be only texture related or may also depend on the characteristics of the fluid moving within it. The first case is **total porosity**, $n$, referring to the total pore volume, independently of whether or not the water can circulate:

$$n = \frac{\text{pore volume}}{\text{total volume}}$$

The second concept is **effective porosity**, $n_e$, also known as kinematic porosity, which is the ratio of the part of the pore volume where the water can circulate to the total volume of a representative sample of the medium, and which is related not only to the texture of the formation, but also to the characteristics of the fluid:

$$n_e = \frac{\text{volume of connected pores}}{\text{total volume}}$$

The total porosity of a detrital formation depends on the grain shape, grain size distribution and type of grain packing. It may be useful to point out here that in their natural state the grains in detrital formations may be more or less compacted, due to factors such as formational genesis, sedimentary environment, lithostatic pressure or overburden. The type of grain packing may range from cubic, with most inter-particle space, to rhombohedral, the most compacted.

If the medium is considered to be made up of similar sized spheres, as shown in *Figure 4.12*, cubic packing would leave intergranular spaces giving a total porosity of 47.64%, and rhombohedral packing would give 25.95%. The size of the spherical grains does not affect the porosity since in percentile terms the volume of voids compared with the total volume will always be the same.

If the medium is heterometric, with different sized grains, the finest grains will occupy the voids between the coarser ones, reducing the total porosity value; the same thing happens if the shape of the grains is variable and angular.

In their natural state, detrital formations tend to be closely packed, with a varied granular distribution depending

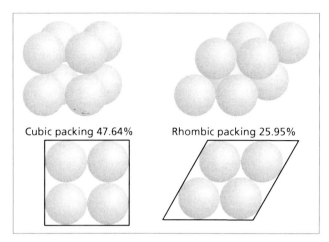

*Figure 4.12* Maximum and minimum packing of a homogeneous granular medium.

on each case and grains which are not perfectly spherical with, in most cases, differently shaped grains with a large number of angular edges.

Real porosity, which is important in geological engineering studies, is that of the formation in its natural state. In general terms, the porosity measured in decompressed core samples tends to give much higher values than those measured *in situ* for the same formation.

## Storage coefficient

This coefficient represents the aquifer's capacity to release water. It is defined as the volume of water which a prism with unit base and height of the aquifer is able to release when the piezometric surface falls 1 m. It is therefore a physical parameter referring to the volume released per unit volume of the aquifer and, just like porosity, it is dimensionless.

The way in which a permeable geological formation releases water varies according to its natural situation and the state of its piezometric surfaces. This means that formations or aquifers can be differentiated depending on whether they function as confined or unconfined.

Confined aquifers (*Figure 4.13*) are located structurally between impermeable layers, with the piezometric surface above the level of the upper boundary of permeable material. At the upper boundary of the aquifer the pressure is higher than atmospheric pressure, and this means that if a well connects the surface with the aquifer, the water level will rise up through the borehole to a point of equilibrium, where the piezometric surface will be the sum of the position and pressure heads. A fall in the piezometric level of the aquifer, so that it is lower than this upper boundary, will result in a variation in pressure state in the physical medium, which will result in an elastic release of water.

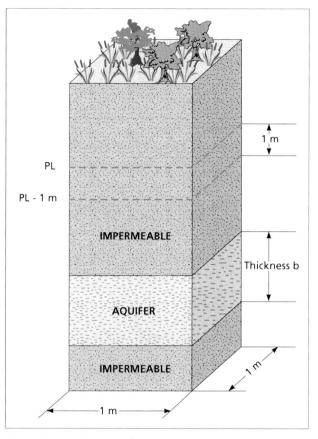

*Figure 4.13* Diagram for calculating storage coefficient from the elastic response of the aquifer.

When the level falls, two changes occur:

— As the pressure the water is subjected to is lowered, the water expands.
— As the internal pressure of the water on the aquifer is lowered, the aquifer compresses.

In fact, in confined aquifers, the water is released by elastic mechanisms due to the combined effect of the water expansion and the vertical decompression of the granular structure.

The storage coefficient of a confined aquifer or storage coefficient by elastic release, $S$, can be deduced in general terms by considering the volumes of water released in these two phenomena. As shown in *Figure 4.13* (a unit base prism) if the piezometric level falls 1 m the quantity of water discharged by elastic release will be equivalent to the storage coefficient.

• Water decompression:

If the pressure decreases by $\Delta p$, the original volume of water in the aquifer $V_W$ will increase by the value $\Delta V_W$. This value $\Delta V_W$ will be the water released in the decompression i.e. from the expansion of the water contained in the aquifer. The relationship between the pressure which produces the decompression and the relative variation in volume caused is given by the volumetric elasticity modulus of the water:

$$B = \frac{\Delta p}{\frac{\Delta V_W}{V_W}} \qquad \Delta V_W = \frac{1}{B}\Delta p \cdot V_W$$

In the above equation, the following can be substituted:

$1/B = \beta$ (water compressibility)
$\Delta p = \rho g = \gamma_w$ (decrease in pressure as piezometric level falls 1 m)
$V_W = n_e V_A$

Given that the volume of existing water in the aquifer, $V_W$ in the prism considered, is equal to the volume of the aquiferous material of the prism $V_A$ multiplied by the effective porosity $n_e$ of the aquifer, it can be established that:

$$\Delta V_W = \beta \gamma_w n_e V_A$$

As shown in *Figure 4.13*, $V_A$ is the product of the unit base by the thickness of the aquifer, so that:

$$\Delta V_W = \beta \gamma_w n_e b$$

- Aquifer decompression:

If the pressure decreases by $\Delta p$, the original volume of the aquifer, $V_A$, will be reduced by the value $\Delta V_A$. This value $\Delta V_A$ is equivalent to the water which will be released from the decompression of the aquifer. The relationship between the pressure producing the decompression and the relative variation in the volume caused will be given by the elasticity modulus of the permeable formation:

$$E = \frac{\Delta p}{\frac{\Delta V_A}{V_A}} \qquad \Delta V_A = \frac{1}{E}\Delta p \cdot V_A$$

In this equation the following can be substituted:

$1/E = \alpha$ (vertical compressibility of the formation)
$\Delta p = \rho g = \gamma_w$ (increase in pressure when piezometric surface falls 1 m)
$V_A = b$ where b is thickness of aquifer:

$$\Delta V_A = \alpha \gamma b$$

The total release of water or storage coefficient by elastic release, $S$, is the sum of the two volumes calculated:

$$S = \Delta V_W + \Delta V_A = \beta \gamma_w b n_e + \alpha \gamma_w b$$
$$S = \gamma_w b (\alpha + n_e \beta)$$

$S$ is conceptually dimensionless as has been shown:

$$[\gamma] = \left[\frac{\text{Force}}{\text{Volume}}\right] = \frac{M \cdot L \cdot T^{-2}}{L^3} = \frac{M \cdot T^{-2}}{L^2}$$

$[b] = L$

$$[\alpha] = \left[\frac{\text{Surface}}{\text{Force}}\right] = \frac{L^2}{M \cdot L \cdot T^{-2}} = \frac{L}{M \cdot T^{-2}}$$

$$[\beta] = \left[\frac{\text{Surface}}{\text{Force}}\right] = \frac{L^2}{M \cdot L \cdot T^{-2}} = \frac{L}{M \cdot T^{-2}}$$

$$[S] = \frac{M \cdot T^{-2}}{L^2} L \left(\frac{L}{M \cdot T^{-2}} + \frac{L}{M \cdot T^{-2}}\right) =$$

$$= \frac{M \cdot T^{-2}}{L} \cdot \frac{L}{M \cdot T^{-2}} \rightarrow \text{dimensionless}$$

When the water level falls in unconfined aquifers the same phenomena of elastic release occur, but in addition to the water released by this concept there is also the existing water in the pores, i.e. the water desaturated from draining a 1 m high prism per base unit of the aquifer. This value coincides with the effective porosity. As a result, taking $S'$ to be the storage coefficient of an aquifer functioning as unconfined:

$$S' = S + n_e$$

which logically is dimensionless.

The storage coefficient by elastic release $S$ is very small compared to the storage coefficient by desaturation or effective porosity, so that it is usually assumed that with no significant errors:

$$S' = n_e$$

This means that the storage coefficient in unconfined aquifers coincides with the effective porosity $n_e$ or storage coefficient by drainage.

# Permeability

Permeability is the parameter which allows a formation's water transfer capacity to be evaluated depending on its texture, without relating it to structure or geometric form. Two further concepts can be defined:

— Permeability linked to the textural characteristics of the physical medium as well as to the fluid it transmits, is called **effective permeability** or **hydraulic**

**conductivity**, and represented by the coefficient of permeability, $k$.

— What is known as **intrinsic permeability**, $K$, only depends on the internal characteristics of the permeable medium.

The effective permeability, $k$, is defined as the discharge which passes through a unit section of the aquifer, normal to the flow, under a piezometric gradient equal to 1.0. It depends on the characteristics of both the physical medium (aquifer) and the fluid (water) flowing through it.

The effective permeability dimensions are:

$$[k] = \frac{L^3 T^{-1}}{L^2} = LT^{-1}$$

The most usual units are cm/s and m/day (the use of m/day is normal practice in hydrogeology).

The intrinsic permeability and effective permeability are only related by the parameters able to define the fluid characteristics.

$$k = K\frac{\gamma_w}{\mu} \quad \text{or} \quad k = K\frac{g}{\eta}$$

where:

$k$ = effective permeability $[LT^{-1}]$
$K$ = intrinsic permeability
$\gamma_w$ = water unit weight $[ML^{-2}T^{-2}]$
$\mu$ = dynamic fluid viscosity $[ML^{-1}T^{-1}]$
$\eta$ = kinematic viscosity $[L^2T^{-1}]$
$g$ = gravity acceleration $[LT^{-2}]$

and as a result:

$$[K] = \left[k\frac{\mu}{\gamma}\right] = \frac{LT^{-1} \cdot ML^{-1}T^{-1}}{ML^{-2}T^{-2}} = L^2$$

A relationship able to define the intrinsic permeability in general terms depending on the medium has not been found, in spite of the efforts of researchers. In any case, it is not only the size of the component particles of the medium which influences this, but also their shape and surface, and generalizations of these cannot be made. Different authors have established the following relationships:

$$K = c \cdot d_e^2 \qquad \text{Hazen}$$

$$K = c \cdot m^n \cdot d_e^2 \qquad \text{Slichter}$$

$$K = \lambda \left(\frac{m - 0.13}{(1-m)^{1/3}}\right)^2 d_e^2 \qquad \text{Terzaghi}$$

The variable $d_e$ is known as the effective diameter which in general terms is the $d_{10}$ of the grain size curve of the sample: i.e. the sieve size which allows 10% of the weight of the granular material sample to pass through. The coefficients $c$, $m$, $n$ and $\lambda$ depend on the geometric characteristics and friction of the grains, so that in any case the following is acceptable:

$$K = \text{constant} \cdot d_e^2 \quad \text{and therefore:} \quad K = \text{constant} \cdot d_{10}^2$$

for the same fluid at the same temperature.

Permeability, in contrast with porosity, does depend on the grain size; as has been seen, the value $d_{10}$ influences the permeability value.

## Transmissivity

Transmissivity, $T$, is the parameter defined to evaluate an aquifer's water transmission capacity, taking into account the aquifer texture and the fluid characteristics and also structural and geometric characteristics. It is defined as the product of the hydraulic conductivity $k$ and the thickness of the aquifer $b$:

$$T = kb$$

Aquifers which are very permeable but not very thick may, even in spite of their excellent textural characteristics, not be suitable for efficient water transmission, and therefore they are considered as having low transmissivity.

## 4.3 Flow. Darcy's law and fundamental flow equations in porous media

### Darcy's law

This was formulated in 1856, after considerable research and many experiments. It states that the discharge $Q$ which passes through a permeable medium is proportional to the cross sectional area normal to flow of the permeable medium, $A$, and the piezometric gradient between the entry and exit of the flow in the permeable medium, $i$.

The constant of proportionality is the permeability of the medium, which considers the fluid characteristics, i.e. the effective permeability, or Darcy's conductivity or hydraulic conductivity. As a result and in general terms:

$$Q = -kAi$$

where:

$Q$ = discharge $[L^3T^{-1}]$
$k$ = Darcy's permeability $[LT^{-1}]$
$A$ = cross sectional area to flow $[L^2]$
$i$ = piezometric gradient.

The gradient of a scalar is a vector, defined by its magnitude and direction. The magnitude is the directional derivative of the scalar function; the direction is normal to the contour level of the scalar function.

When the flow moves from a higher piezometric level zone to a lower one, which in this case is the scalar function, the flow and the gradient have different directions, which justifies the negative sign in the second term in Darcy's equation.

Taking $h$ to be the piezometric level and $x$ the space in the flow direction, the equation can be expressed in a differential form, taking into consideration the variation in piezometric levels along the whole flow through the porous medium:

$$Q = -kA\frac{dh}{dx}$$

In the case shown in *Figure 4.14*, the experiment maintains levels $h_A$ and $h_B$ constant and as $k$ and $A$ are constant for the whole permeable medium, the simplest form of Darcy is established:

$$Q = kA\frac{h_A - h_B}{L}$$

Darcy's law is only valid for laminar flow regime. To evaluate whether the regime is really laminar the Reynolds number must be considered, which represents the ratio of inertial to viscous forces in a fluid:

$$R_e = \frac{\rho v d}{\mu}$$

where:

$v$ = fluid velocity
$d$ = mean particle size; $d_{50}$ is considered in granular media and $2_e$ in fissured media, with $e$ being the mean width of the fissures.
$\rho$ = fluid density.
$\mu$ = dynamic viscosity.

The Reynolds number is dimensionless.

$$[R_e] = \frac{ML^{-3}L}{MLT^{-2}TL^{-2}} \rightarrow \text{dimensionless}$$

In any case, homogeneous units must be used to calculate it; e.g. $v$ in cm/s, $d$ in cm, $\rho$ in g/cm³ and $\mu$ in dyne · seg/cm².

A Reynolds number value in the range 1–10 denotes a laminar regime where Darcy's law can be applied, although it is particularly reliable for values lower than 4 (*Figure 4.15*). The regime will be completely turbulent for values ranging from 60 to 180 and for intermediate values can be assumed to be quasi-turbulent. In fact, within this last range of parameters, Darcy's law does not hold for the flow.

In the equation:

$$Q = -kA\frac{dh}{dx}$$

the permeability $k$ depends on both the characteristics of the medium and of the fluid, i.e. it refers to the effective permeability or hydraulic conductivity, which is also known as Darcy's permeability.

## Darcy's velocity and real velocity

The velocity $v$ with which the fluid passes through the permeable medium can be easily found:

$$Q = vA; \quad v = \frac{Q}{A}$$

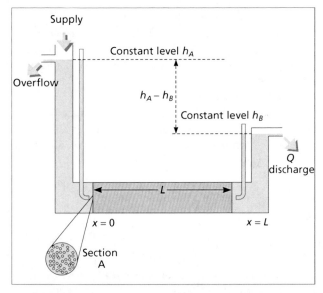

Figure 4.14   Darcy's law.

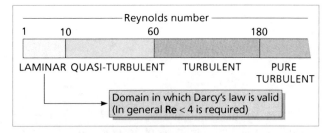

Figure 4.15   Flow regime.

where $Q$ is the discharge and $A$ the section.

Given that:

$$Q = -k \cdot A \cdot \frac{dh}{dx} \Rightarrow v = -k \cdot \frac{dh}{dx}$$

This velocity is called Darcy velocity or flow velocity and refers to the fluid rate if this is through an open unobstructed section with value $A$.

When the fluid crosses section $A$ it finds this section mainly occupied by the aquifer particles so that in real terms the available section is related to $A$ by the effective porosity $n_e$, equal to $An_e$ (Figure 4.16). In this case the discharge rate is the same, but the real velocity $v_R$ of the fluid between the particles becomes:

$$Q = v_R A n_e \Rightarrow v_R = \frac{Q}{An_e} \Rightarrow v_R = \frac{v}{n_e}$$

i.e. that the real velocity of the fluid is the quotient of the Darcy velocity and the effective porosity.

Although it is simple, Darcy's law is widely applicable. It is the basis, along with the law of continuity, for calculating different flow equations in permeable media in permanent and transitory regimes, and on its own it can also cover an important range of hydrodynamic applications in permeable media.

## Generalization of Darcy's law

Darcy's law expressed in one dimension becomes:

$$v = -k \left[ \frac{dh}{dl} \right]$$

Generalized in three dimensions, the vector $v$ will have three components:

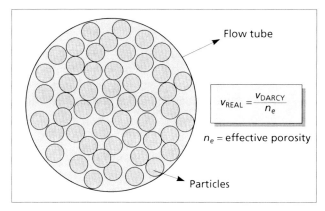

Figure 4.16   Darcy's velocity and real velocity.

$$v_x = -k \frac{\partial h}{\partial x}; \quad v_y = -k \frac{\partial h}{\partial y}; \quad v_z = -k \frac{\partial h}{\partial z}$$

and can be synthesised with the generalized Darcy expression for **homogeneous and isotropic media**:

$$\bar{v} = -k \, \overline{\text{grad}} \, h$$

where:

$$\bar{v} = (v_x, v_y, v_z)$$

$$\overline{\text{grad}} \, h = \left( \frac{\partial h}{\partial x'}, \frac{\partial h}{\partial y'}, \frac{\partial h}{\partial z'} \right)$$

($k$ = scalar)

Where there is an **anisotropic medium** the permeability is a second order tensor, as a permeability value will exist for each point of the space and for each of the directions of the space at that point. This means that the permeability tensor will have three vectorial components and nine scalar components:

$$[\bar{\bar{k}}] = \begin{bmatrix} k_{xx} & k_{xy} & k_{xz} \\ k_{yx} & k_{yy} & k_{yz} \\ k_{zx} & k_{zy} & k_{zz} \end{bmatrix}$$

$$\bar{v} = -\bar{\bar{k}} \, \overline{\text{grad}} \, h$$

and the components of the velocity vector will be expressed as:

$$v_x = -k_{xx} \frac{\partial h}{\partial x} - k_{xy} \frac{\partial h}{\partial y} - k_{xz} \frac{\partial h}{\partial z}$$

$$v_y = -k_{yx} \frac{\partial h}{\partial x} - k_{yy} \frac{\partial h}{\partial y} - k_{yz} \frac{\partial h}{\partial z}$$

$$v_z = -k_{zx} \frac{\partial h}{\partial x} - k_{zy} \frac{\partial h}{\partial y} - k_{zz} \frac{\partial h}{\partial z}$$

Making the axes coincide with the main directions of the anisotropic medium, the scalar components of the permeability tensor will be:

$$[\bar{\bar{k}}] = \begin{bmatrix} k_{xx} & 0 & 0 \\ 0 & k_{yy} & 0 \\ 0 & 0 & k_{zz} \end{bmatrix}$$

where:

$$v_x = -k_{xx} \frac{\partial h}{\partial x}; \quad v_y = -k_{yy} \frac{\partial h}{\partial y}; \quad v_z = -k_{zz} \frac{\partial h}{\partial z}$$

are the components of the velocity vector which will be used in the deduction of the flow equations as shown below.

## Continuity equation for steady flow

The concept of steady flow implies that the water which flows into a closed space is equal to the water which is going to flow out of it without any variation in the water stored in the space and that, therefore, there are no variations in the piezometric level; i.e. the level is time-independent, or stationary, and the water flowing in is the same as the water flowing out.

Figure 4.17 shows a REV (representative elementary volume) referring to an ideal porous material cube, representative of the properties of the medium. The inflow on each of its sides should produce the same quantity of outflow from the other sides. The sum of the inflows should be equal to the sum of the outflows.

The elementary cube has dimensions $\Delta x$, $\Delta y$, $\Delta z$ and volume $\Delta V = \Delta x \cdot \Delta y \cdot \Delta z$. On each side there is an inflow $Q$ which will produce a unit discharge (discharge per unit area) as quotient of discharge $Q$ and the surface area of the side. It is assumed that if a unit flow $v$ flows in through one side, then $v + \Delta v$ will flow out through the opposite side.

According to Taylor's solution, the incremented function is equal to the non-incremented function plus the function derivative for the increment and a series of terms which are higher level infinitesimals, and which can be ignored depending on the circumstances; i.e.:

$$f(x+\Delta x) = f(x) + \frac{\partial f}{\partial x}\Delta x + \frac{\partial^2 f}{\partial x^2}\frac{\Delta x^2}{2!} + \frac{\partial^3 f}{\partial x^3}\frac{\Delta x^3}{3!} + \cdots$$

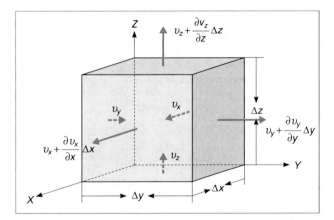

Figure 4.17  Water balance in an elementary porous cube (REV).

which is approximated by:

$$f(x+\Delta x) = f(x) + \frac{\partial f}{\partial x}\Delta x$$

The balance of the water in the elementary volume can be calculated by applying the Taylor series solution, ignoring infinitesimals higher than the first order and adding the unit discharge components in each direction. This gives the following balance between the sides of the REV:

$$\text{side } yz: \left(v_x + \frac{\partial v_x}{\partial x}\Delta x - v_x\right)\Delta y \Delta z \Rightarrow \frac{\partial v_x}{\partial x}\Delta x \Delta y \Delta z$$

$$\text{side } xz: \left(v_y + \frac{\partial v_y}{\partial y}\Delta y - v_y\right)\Delta x \Delta z \Rightarrow \frac{\partial v_y}{\partial y}\Delta y \Delta x \Delta z$$

$$\text{side } xy: \left(v_z + \frac{\partial v_z}{\partial z}\Delta z - v_z\right)\Delta x \Delta y \Rightarrow \frac{\partial v_z}{\partial z}\Delta z \Delta x \Delta y$$

To respect the conditions of mass in equal to mass out these three quantities must sum to zero. Therefore:

$$\frac{\partial v_x}{\partial x}\Delta x \Delta y \Delta z + \frac{\partial v_y}{\partial y}\Delta y \Delta x \Delta z + \frac{\partial v_z}{\partial z}\Delta z \Delta x \Delta y = 0$$

$$\frac{\partial v_x}{\partial x} + \frac{\partial v_y}{\partial y} + \frac{\partial v_z}{\partial z} = 0$$

which is the equation of continuity in steady, or stationary, flow.

## Laplace equation

The Laplace equation is the first of the differential equations in partial derivatives which govern the flow through porous materials (permeable in general, when the medium can be assumed to be porous). It refers to the flow in a permanent (i.e. unchanging) regime conditions and in the absence of springs and drains.

This equation, like all flow equations in porous materials, is obtained by applying the continuity equation and Darcy's law together. The flow is studied accepting that the ingoing mass should be equal to the outgoing, and that within the porous medium both are conditioned by Darcy's approach. Thus:

$$\frac{\partial v_x}{\partial x} + \frac{\partial v_y}{\partial y} + \frac{\partial v_z}{\partial z} = 0 \quad \text{(continuity)}$$

$$v_x = -k_{xx}\frac{\partial h}{\partial x}; \quad v_y = -k_{yy}\frac{\partial h}{\partial y}; \quad v_z = -k_{zz}\frac{\partial h}{\partial z}$$

substituting:

$$\frac{\partial}{\partial x}\left(-k_{xx}\frac{\partial h}{\partial x}\right)+\frac{\partial}{\partial y}\left(-k_{yy}\frac{\partial h}{\partial y}\right)+\frac{\partial}{\partial z}\left(-k_{zz}\frac{\partial h}{\partial z}\right)=0$$

If the medium is considered to be homogeneous (i.e. its permeability is the same at any point of the space) and isotropic (i.e. the permeability is the same in any direction of the space) then:

$$k_{xx}=k_{yy}=k_{zz}=k$$

and:

$$\frac{\partial^2 h}{\partial x^2}+\frac{\partial^2 h}{\partial y^2}+\frac{\partial^2 h}{\partial z^2}=0$$

which is Laplace's equation (permanent or unchanging regime, homogeneous and isotropic medium and absence of springs and drains).

There are two characteristic types of boundary conditions:

— Dirichlet conditions: piezometric level known at a boundary.
— Neumann conditions: flow known at a boundary.

The resolution of the equation is only possible by analytical methods in simple physical systems, as is the case in flows towards wells and some problems of flow between trenches or open pit excavations. For more complex real cases approximate numerical methods should be used or, in some cases, graphical resolution methods such as flow nets.

## Poisson's equation

Laplace does not include the existence of springs and drains in his approach. However, even in permanent regime problems, the pumping and recharge (springs and drains) associated with real problems have to be included.

Poisson's equation for flow in porous media in permanent regime, integrating pumping and recharge, can be developed with the help of the diagram in *Figure 4.18*.

Recharge $R(x, y)$ is considered by unit area and unit time. Applying stationary flow conditions:

$$\frac{\partial v_x}{\partial x}\Delta x(b\Delta y)=\frac{\partial v_y}{\partial y}\Delta y(b\Delta x)=R(x,y)\Delta x\Delta y$$

Applying Darcy with homogeneity and isotropy hypothesis:

$$k_{xx}=k_{yy}=k_{zz}=k$$

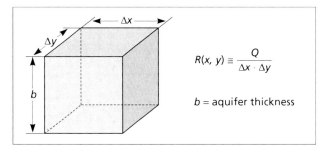

Figure 4.18   Diagram to deduce Poisson's equation.

and with:

$$T=kb$$

Poisson's equation is obtained:

$$\frac{\partial^2 h}{\partial x^2}+\frac{\partial^2 h}{\partial y^2}=-\frac{R(x,y)}{T}$$

in a permanent unchanging regime, for the presence of springs and drains in homogeneous and isotropic permeability.

In this equation:

— if $R(x, y) = 0$, it coincides with Laplace's equation
— if $R(x, y)$ is positive = recharge
— if $R(x, y)$ is negative = discharging, e.g. pumping.

All the above for the resolution of Laplace's equation is also valid for Poisson's equation. However, the latter is more realistic and complete and adapts better to most real problems which may occur. It establishes simplistically that the sum of the ingoing and outgoing water in a closed domain must be equal to the pumping or recharging affecting this domain, in permanent regime conditions (with no variation in the piezometric levels in the domain) and in a homogenous and isotropic medium in terms of permeability.

## Flow equation in transitory regime

In a permanent regime equilibrium solutions are obtained for a specific phenomenon. This is not how or when this equilibrium is reached, but only calculating equilibrium is reached.

If a system is in equilibrium, to change this situation requires certain actions on the system such as pumping or recharging. The system evolves until it reaches a new state of equilibrium. Until this new state is reached, a changing set of situations occur over time, which is what is meant by transitory regime. The levels vary and as a result water is stored in the system or discharged from it. In this case, the equation of continuity in a transitory, or non-steady regime is applied to the domain as follows:

Outgoing volume = Incoming volume − Discharged volume (in time interval $\Delta t$)

In the interval $\Delta t$ the level varies $\Delta h$ as shown in *Figure 4.19*. The volume of rock emptied is expressed by $\Delta h \cdot \Delta x \cdot \Delta y$, and the volume of water contained in this volume of rock is expressed by $S \cdot \Delta h \cdot \Delta x \cdot \Delta y$.

The volume discharged in the time interval $\Delta t$ will be $S \cdot \Delta h \cdot \Delta x \cdot \Delta y/\Delta t$. Putting this term into the continuity equation where springs and drains are present, i.e. including the term of the volume discharged in the time unit in the recharging term gives:

$$\frac{\partial v_x}{\partial x}\Delta x (b\Delta y) + \frac{\partial v_y}{\partial y}\Delta y (b\Delta x)$$
$$= R(x,y,t)\Delta x \Delta y - S\frac{\partial h}{\partial t}(\Delta x \Delta y)$$

applying generalized Darcy as in the equations above:

$$v_x = -k_{xx}\frac{\partial h}{\partial x}$$
$$v_y = -k_{yy}\frac{\partial h}{\partial y}$$
$$v_z = -k_{zz}\frac{\partial h}{\partial z}$$

and again assuming homogeneous and isotropic conditions ($k_{xx} = k_{yy} = k_{zz} = k$), squared discretization ($\Delta x = \Delta y$) and $T = kb$, gives:

$$\frac{\partial^2 h}{\partial x^2} + \frac{\partial^2 h}{\partial y^2} = \frac{S}{T}\frac{\partial h}{\partial t} - \frac{R(x,y,t)}{T}$$

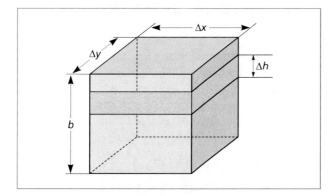

*Figure 4.19*    Diagram to deduce the general transient flow equation.

the general flow equation in a transitory non-steady regime, as with the presence of springs and drains and homogeneous and isotropic permeability.

The terms of this equation have the following meanings:

$\dfrac{\partial^2 h}{\partial x^2} + \dfrac{\partial^2 h}{\partial y^2}$: sum of inflows through the lateral sides

$\dfrac{S}{T}\dfrac{\partial h}{\partial t}$: volumes discharged in the time unit

$\dfrac{R(x,y,t)}{T}$: pumping or recharging in the time unit.

In fact, this equation shows that the balance of the inflows and outflows discharges entering and exiting in a domain due to piezometric gradients, plus the external recharging of the system, such as pumping and recharging, must be equal to discharges filling or voiding the domain.

As can be seen:

if $\dfrac{S}{T}\dfrac{\partial h}{\partial t} = 0$, Poisson's equation for a permanent regime with springs and drains is obtained;

if $\dfrac{S}{T}\dfrac{\partial h}{\partial t} = 0$ and $\dfrac{R(x,y,t)}{T} = 0$, Laplace's equation is obtained.

The general flow equation and its resolution for each specific case with the appropriate boundary conditions is widely used in hydrogeology and geological engineering problems.

## 4.4 Evaluation methods for hydrogeological parameters

Evaluating hydrogeological parameters such as permeability, transmissivity, porosity or storage coefficient is one of the basic tasks of a hydrogeological investigation. These parameters are the basis for calculating drainage, leakages, pumping discharges, transit time for substances transported by groundwater, etc. Maximum reliability is required when determining these parameters which in fact define the characteristics of the aquifers and their capacity for storing and transmitting groundwater.

Basically there are three methods: pumping tests, injection tests and tracers. Laboratory methods can also be used on samples extracted from the aquifer, although the other methods are generally considered to be more reliable and also have the advantage of a greater significant spatial validity.

### Pumping tests

Pumping tests are undoubtedly the most complete and reliable method for calculating the hydrogeological parameters of an aquifer. They consist in pumping a well, at first with constant discharge, and analyzing the drawdown in the piezometric levels, both in the pumping well and in wells or piezometers nearby.

There are two basic types of methods: pumping tests in a permanent unchanging regime (steady flow) and pumping tests in transitory or variable regime (non-steady flow). In the former the piezometric levels do not vary with time and the drawdown in the well area is interpreted as a result of the constant pumping. In variable regime what is analyzed is the evolution of the levels in the pumping well and observation bore holes throughout the test.

## Pumping tests in permanent regime. Deduction of the equations

The general equation for bi-dimensional flow under confined aquifer and steady flow conditions is given by:

$$\frac{\partial h}{\partial x^2} + \frac{\partial^2 h}{\partial y^2} = 0$$

where $h$ is the piezometric level.

This means that the aquifer may be considered homogeneous, isotropic and infinite, with no existing vertical recharge.

In the case of radial flow towards a point where the pumping well is sited, the equation can be transformed into polar coordinates, taking the well axis as the coordinates axis, which will be a point when projected on the plane $xy$. The transformation is carried out with:

$$x = r\cos\alpha$$
$$y = r\sin\alpha$$

Cancelling the terms in $\alpha$ gives the equation:

$$\frac{\partial^2 h}{\partial r^2} + \frac{1}{r}\frac{\partial h}{\partial r} = 0$$

which as it depends only on $r$ may be written:

$$\frac{1}{r}\frac{d}{dr}\left(r\frac{dh}{dr}\right) = 0$$

and therefore:

$$\left(r\frac{dh}{dr}\right) = \text{constant}$$

To calculate the constant, the outflow from the well is taken as equal to that crossing a cylindrical surface of constant height $b$ and radius $r$ from the well.

In this case, according to Darcy:

$$Q = 2\pi r b k \frac{dh}{dr}$$

then:

$$\text{constant} = r\frac{dh}{dr} = \frac{Q}{2\pi b k} = \frac{Q}{2\pi T}$$

and therefore:

$$dh = \frac{Q}{2\pi T}\frac{dr}{r}$$

Integrating using the radius $r$ and the radius of influence $R$, i.e. the distance at which the drawdown is nil, the level will be seen to vary between the value $h$ at $r$ and the value $h_0$ representative of levels of repose on a regional scale at $R$:

$$\int_h^{h_0} dh = \frac{Q}{2\pi T}\int_r^R \frac{dr}{r}$$

giving **Thiem's solution**:

$$h_0 - h = \frac{Q}{2\pi T}\ln\frac{R}{r}$$

where:

$h_0$ = initial piezometric level
$h$ = piezometric level at distance $r$
$Q$ = pumping discharge
$T$ = transmissivity
$R$ = radius of influence
$r$ = distance from pump well.

If the aquifer behaves as if unconfined, there would be a fundamental variation with respect to the previous case; the constant thickness $b$ represented by the height of the cylinder will be variable and equal to level $h$ since, as the aquifer is unconfined, the saturated level corresponds to the piezometric level referred to the bottom of the permeable formation. Therefore:

$$Q = 2\pi r h k \frac{dh}{dr}$$

then:

$$h\,dh = \frac{Q}{2\pi k}\frac{dr}{r}$$

An analogy of the case above, but with $h_0$ conceptually being the initial saturated thickness, gives:

$$\int_h^{h_0} h\,dh = \frac{Q}{2\pi k}\int_r^R \frac{dr}{r}$$

and finally:

$$h_0^2 - h^2 = \frac{Q}{\pi k}\ln\frac{R}{r}$$

which is **Dupuit's solution**.

All these formulas start off from the assumption that transmissivity is constant at any point of the space; i.e. they assume a homogeneous and isotropic medium.

**Thiem's equation** can also be reached by simpler methods, without using the general equation. This mathematical development is included below as it is easy to use and understand.

As shown in *Figure 4.20*, in a confined aquifer where a well is pumping at a constant discharge $Q$ and there are two observation boreholes (no 1 and no 2), with water level stabilized in the whole pumping cone, it can be established that the discharge from the well is equal to that passing through an ideal cylindrical cross section with radius $r$ from the well and thickness of aquifer $b$.

The depressions in the pumping well and in the piezometers no 1 and no 2 will be $d_p$, $d_1$ and $d_2$ and the heads $h_p$, $h_1$ and $h_2$ respectively. The piezometers are placed at distances $r_1$ and $r_2$ from the pumping well; at the distance $r$, the depression $d$ and the head $h$.

According to Darcy's law, the discharge across the porous cylindrical surface will be equal to the product of the permeability of the aquifer by flow area and by the hydraulic gradient existing between the two sides of the cylindrical surface considered, i.e.:

$$Q = kAi$$

where $A = 2\pi rb$ is the flow area, $i = dh/dr$ is the gradient:

$$Q = k2\pi rb\frac{dh}{dr} \qquad kb = T$$

$$Q = 2\pi Tr\frac{dh}{dr} \qquad dh = \frac{Q}{2\pi T}\cdot\frac{dr}{r}$$

To integrate this last expression, the integration limits must be defined. When $h$ varies between a generic level $h$ and the initial level of the aquifer $h_0$, $r$ will vary between a generic radius $r$, in which the head value is $h$, and a distance $R$ (radius of influence) where the drawdown is nil and therefore $h$ is equal to $h_0$:

$$\int_h^{h_0} dh = \frac{Q}{2\pi T}\int_r^R\frac{dr}{r} \quad \text{and} \quad h_0 - h = \frac{Q}{2\pi T}\ln\frac{R}{r}$$

or

$$d = \frac{Q}{2\pi T}\ln\frac{R}{r}$$

The deduction of **Dupuit's formula** by this procedure is analogous as shown in *Figure 4.21*:

$$Q = kAi$$

where $A = 2\pi rh$ is the flow area, $i = dh/dr$ is the gradient:

$$Q = k2\pi rh\frac{dh}{dr} \quad\Rightarrow\quad h\,dh = \frac{Q}{2\pi k}\cdot\frac{dr}{r}$$

and integrating with the already known limits of Dupuit's formula:

$$h_0^2 - h^2 = \frac{Q}{\pi k}\ln\frac{R}{r}$$

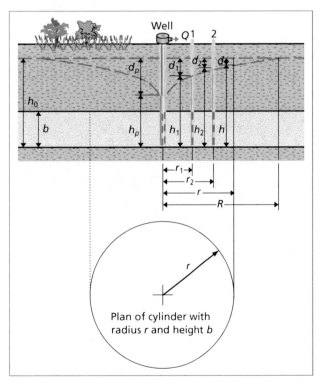

Figure 4.20   Diagram to deduce Thiem's formula.

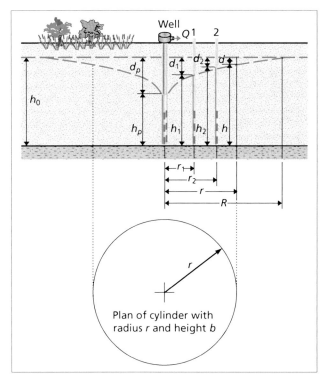

*Figure 4.21*   Diagram to deduce Dupuit's formula.

## Thiem's method. Confined aquifer in a permanent regime

If the following conditions are physically met in the pumping test:

— permanent regime i.e. steady flow
— no exterior recharging exists
— homogeneous and isotropic aquifer in terms of its permeability $k$
— the aquifer is infinite
— the diameter of the pumping well is zero
— the well fully penetrates the permeable formation
— the pumped water produces an immediate drawdown and does not flow back into the aquifer
— the flow of water towards the well is radial with no vertical components
— the pumped discharge $Q$ is constant

then, as has already been seen, Thiem's equation for permanent regime and confined aquifers is obtained (*Figure 4.20*):

$$d = \frac{Q}{2\pi T} \ln \frac{R}{r}$$

which particularized for the distances $r_1$ and $r_2$ and subtracting is transformed into:

$$d_1 - d_2 = \frac{Q}{2\pi T} \ln \frac{r_2}{r_1}$$

and in decimal logarithms and dividing by $2\pi$ gives:

$$d_1 - d_2 = 0.366 \frac{Q}{T} \log \frac{r_2}{r_1}$$

By analogy, if it is considered that there is no head loss in the well and the value of the radius of influence (the distance to the point where the depression caused by pumping is practically zero) is taken as $R$:

$$dp = \frac{Q}{2\pi T} \ln \frac{R}{r_p} \quad \text{or} \quad dp = 0.366 \frac{Q}{T} \log \frac{R}{r_p}$$

which is Thiem's equation stated above, where $dp$ is the drawdown of the well and $r_p$ is the radius of the well. This indicates simply that the differences between the depressions $(d_1 - d_2)$ in two observation boreholes, located at distances $r_1$ and $r_2$ from the well pumping at constant discharge $Q$, is the product of $Q/2\pi T$ by the Napierian or natural logarithm of the inverse quotient of the distances, $r_2/r_1$.

This type of test obviously does not allow the calculation of the value of the storage coefficient, $S$, given that as the head does not vary with time and no drainage of the aquifer occurs. However, it is very useful for calculating the transmissivity, $T$, in areas which have been pumping continuously over a long period and where the regime can be considered as permanent.

The method for calculating $T$ is as follows: if a generic observation point is considered at a distance $r_i$ from a well pumping at a constant discharge $Q$, and where a drawdown $d_i$ has been produced with respect to the static level before pumping began, then it can be established that:

$$d_i = 0.366 \frac{Q}{T} \log \frac{R}{r_i}$$

$$d_i = 0.366 \frac{Q}{T} \log R - 0.366 \frac{Q}{T} \log r_i$$

If $\log r_i$ is taken as variable, using a semi-logarithmic scale, this formula will give a line:

$$y = mx + n$$

where:

$$y = d_i$$
$$m = -0.366 \frac{Q}{T}$$
$$x = \log r_i$$
$$n = \text{constant} = 0.366 \frac{Q}{T} \log R$$

i.e. the plot will be as shown in *Figure 4.22*. This line:

— has a slope value $m = -0.366\, Q/T$
— intersects the *x* axis at the value $R$ = radius of influence
— intersects the *y* axis at the value of the pumping well drawdown when there are no head losses
— when there are head losses, these are represented by the deviation from the ideal line.

The practical method for calculating the slope is to measure the difference between the coordinates for each logarithmic cycle on the *x*-axis: the slope will be this difference:

$$m = \frac{-\Delta y}{\Delta x} = \frac{-\Delta d}{\Delta \log r_i}$$

As a result, if a logarithmic cycle is taken:

$$\Delta \log r_i = \log 10^n - \log 10^{n-1} = \frac{\log 10^n}{\log 10^{n-1}} = \log 10 = 1$$

and:

$$m = -\Delta d$$

then:

$$T = 0.366 \frac{Q}{\Delta d}$$

## Dupuit's method. Unconfined aquifer in a permanent regime

Where the aquifer is unconfined, the flow is no longer radial (*Figure 4.23*). In this case, a correction known as Dupuit's correction for unconfined aquifers is applied to the value of

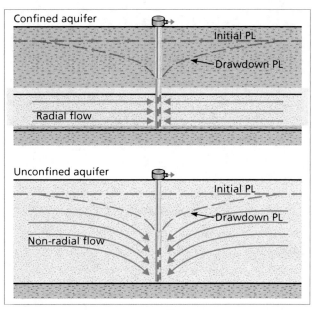

*Figure 4.23*  Diagram of pumping in confined and unconfined aquifers.

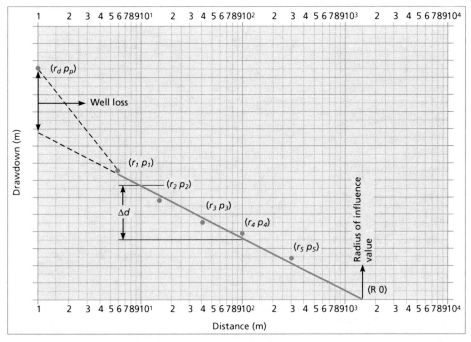

*Figure 4.22*  Thiem's solution.

the drawdown measured *in situ*. This correction is calculated from Dupuit's formula:

$$h_0^2 - h^2 = \frac{Q}{\pi k}\ln\frac{R}{r}$$

and

$$h_0^2 - h^2 = (h_0 - h)(h_0 + h) = d(2h_0 - d)$$
$$= 2h_0\left(d - (d^2/2h_0)\right)$$

then

$$d - \frac{d^2}{2h_0} = \frac{Q}{2\pi T}\ln\frac{R}{r}$$

given that $kh_0 = T$.

This formula is the same as Thiem's except that it must be drawn with a correction to the scale, i.e. with a correction in the observed drawdown:

— observed drawdown = $d$
— corrected drawdown = $d - (d^2/2h_0)$, where $h_0$ is the initial saturated thickness.

Once the drawdown values in the well and piezometers have been corrected, the procedure is the same as described for Thiem's method for a confined aquifer, but using the corrected drawdown. However, it is not worth making this correction when the drawdown is less than 10% or 15% of the saturated thickness $h_0$.

### De Glee's method. Semi-confined aquifer in a permanent regime

The last case is when there is a uniform vertical recharge through semi-permeable ground i.e. a semi-confined aquifer. The conditions are shown in *Figure 4.24* (see also *Figure 4.5*). The solution of the equations gives De Glee's formula:

$$d = \frac{Q}{2\pi T}K_0\left(\frac{r}{B}\right)$$

where:

$r$ = distance to the observation point
$B$ = leakage factor (defined below)

$K_0(r/B)$ is a function which has no analytical solution, so that it is tabulated; it is shown in *Figure 4.25*, where the values of $K_0$ can be obtained depending on $r$ and $B$.

In this formula the following conditions are considered:

— there is an existing higher, rapidly recharged aquifer
— the starting level is the same in the higher and lower aquifers

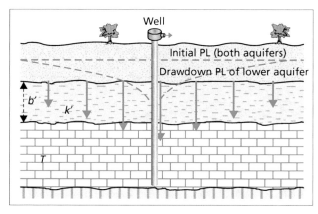

Figure 4.24 Diagram of pumping in semi-confined aquifer.

— the higher aquifer does not discharge water through the well
— as the level of the lower one falls, a gradient is created towards it, which makes the higher aquifer recharge it through the semi-permeable formation.

The leakage factor is given by:

$$B = \sqrt{\frac{T \cdot b'}{k'}}$$

where:

$b'/k'$ = hydraulic resistivity
$k'$ = semi-permeable vertical permeability
$b'$ = semi-permeable thickness
$T$ = transmissivity of the lower aquifer.

When the ratio $r/B$ is less than 0.1, the function $K_0(r/B)$ takes values which can be replaced by $\ln 1.12B/r$. In this case:

$$d = \frac{Q}{2\pi T}\ln\frac{1.12B}{r}$$

and there is no need to use tables or charts.

Finally, evidently how close the results of the test are to reality will depend on how far the physical reality of the test meets the mathematical conditions imposed to solve the general equation.

### Tests in a transitory regime. Deduction of the equations

The general equation for the bi-dimensional flow in confined aquifer and transitory (non-steady) regime conditions is given by:

$$\frac{\partial^2 h}{\partial x^2} + \frac{\partial^2 h}{\partial y^2} = \frac{S}{T} \cdot \frac{\partial h}{\partial t}$$

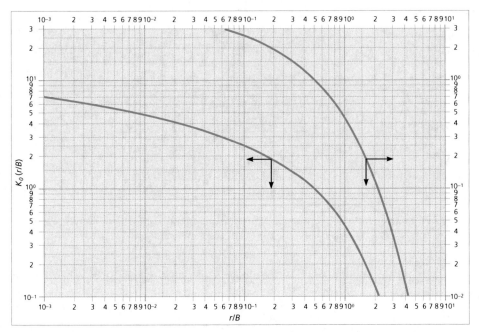

*Figure 4.25* Chart of well function for a semi-confined aquifer at steady state (arrows indicate the axes reading for each curve).

implying that the aquifer is homogeneous and isotropic and that there is no vertical recharge.

Where there is radial flow towards a point where the pumping well is sited, the equation may be transformed into polar coordinates, taking the well axis as y-axis, which will be a point projected on the plane $xy$. The transformation is carried out using:

$$x = r\cos\alpha$$
$$y = r\sin\alpha$$

In the substitution and operation the terms in $\alpha$ are cancelled out giving:

$$\frac{\partial^2 h}{\partial r^2} + \frac{1}{r}\cdot\frac{\partial h}{\partial r} = \frac{S}{T}\cdot\frac{\partial h}{\partial t}$$

which is the general flow equation in porous media, expressed in polar coordinates.

The solution, worked out by Theis in 1935, gave rise to the hydraulics of wells in transitory regime. To do this, first the change of variable is carried out:

$$u = \frac{r^2 S}{4Tt}$$

and the following conditions applied:

— initial conditions:

$$h(r,0) = h_0 \quad \text{for } r > 0$$

i.e. the piezometric head in the aquifer is uniform before pumping starts and equal to $h_0$

— boundary conditions:

$$h = h_0 \quad \text{for } r \to \infty \quad \text{being } t > 0$$

($t$ = time since start of pumping)

$$\lim_{r \to 0}\left(r\frac{\partial h}{\partial r}\right) = \frac{Q}{2\pi T} \quad \text{Darcy's law}$$

The solution for this equation and these conditions is:

$$h_0 - h = d = \frac{Q}{4\pi T} W(u)$$

where $W(u)$ is what Theis called the well function for confined aquifer and transitory regime:

$$W(u) = \int_u^\infty \frac{e^{-u}}{u} du$$

The Theis solution refers to the general case where the permeability $k$ remains constant at any point in the space.

A simplification of this is the **Jacob method**, which involves developing in series $W(u)$.

$$W(u) = -0.577216 - \ln u + u - \frac{u^2}{2 \cdot 2!} + \frac{u^3}{3 \cdot 3!} - \frac{u^4}{4 \cdot 4!} + \cdots$$

When $u < 0.03$, all the terms except the first two can be ignored. However it is usual practice to ignore them from $u < 0.1$. Therefore:

$$d = \frac{Q}{4\pi T}(-0.577216 - \ln u)$$
$$= \frac{Q}{4\pi T}(\ln 0.562 - \ln u) = \frac{Q}{4\pi T}\ln\frac{0.562}{u}$$
$$d = \frac{Q}{4\pi T}\ln\frac{2.25Tt}{r^2 S}$$

or

$$d = 0.183\frac{Q}{T}\log\frac{2.25Tt}{r^2 S} \quad \text{(Jacob)}$$

### Theis method. Confined aquifer in a variable regime

In this case the solution of the fundamental equation is more complicated as the term $(S\partial h)/(T\partial t)$ appears, and as the level of $(h)$ varies with time, it is not cancelled out.

For the confined aquifer and taking the same limiting conditions as for the permanent regime, Theis formula is obtained, shown above:

$$d = \frac{Q}{4\pi T}\int_u^\infty \frac{e^{-u}}{u}du$$

where:

$d =$ drawdown in a well at distance $r$
$u = (r^2 S)/(4Tt)$
$S =$ storage coefficient
$t =$ time from the start of pumping, taking repose conditions.

This integral has no analytical solution, so that the well function $W(u)$ is tabulated.

In *Table 4.1* and *Figure 4.26* the $W(u)$ values can be obtained depending on $u$ and $1/u$ respectively. Therefore:

$$d = \frac{Q}{4\pi T}W(u) \quad (1) \qquad T = \frac{QW(u)}{4\pi d} \quad (3)$$

$$u = \frac{r^2 S}{4Tt} \quad (2) \qquad S = \frac{4Tt}{r^2 \cdot 1/u} \quad (4)$$

The test field data are represented in charts of the type:

$$d - \log t; \quad d - \log r^2/t; \quad d - \log r^2$$

The first of these is the simplest and most widely used and is commented on below. Logarithms are taken in equations (1) and (2):

$$\log d = \log\frac{Q}{4\pi T} + \log W(u)$$

$$\log t = \log\frac{r^2 S}{4T} + \log\frac{1}{u}$$

If a standard curve is available (on tracing paper) which represents $W(u)$ as a function of $1/u$, it can be seen that the curve of field $d - \log t$ and the standard curve $W(u) - 1/u$ only differ in constant quantities for $x$ and $y$ axes:

$$\frac{Q}{4\pi T} \text{ and } \frac{r^2 S}{4T} \text{ (in double logarithmic representation)}$$

which is in fact the same curve, represented on different axes. *Figure 4.27* may help to clarify this point.

The method consists in laying the tracing paper on top, making the two curves coincide. A simple point is taken on the standard curve (e.g. $W(u) = 1$, $1/u = 10$) and the values for $d$ and $t$ corresponding to the field curve can be seen.

When these values $W(u)$, $1/u$, $d$ and $t$ have been obtained, formulae (3) and (4) can be used to obtain values for $T$ and $S$. The value of $S$ can only be found when there are values for $h$ from a piezometer. The value of $T$ can be found using the levels in pumping wells and in piezometers.

### Jacob's method. Confined aquifer in a variable regime

Jacob's formula as has been deduced is given by the expression:

$$d = 0.183\frac{Q}{T}\log\frac{2.25Tt}{r^2 S}$$

Taking:

$$\frac{r^2}{2.25T} = t_0$$

obtains:

$$d = 0.183\frac{Q}{T}\log\frac{t}{t_0}$$

$$d = 0.183\frac{Q}{T}\log t - 0.183\frac{Q}{T}\log t_0$$

## Table 4.1 VALUES FOR WELL FUNCTION $W(u)$

| k | $k \times 10^{-14}$ | $k \times 10^{-12}$ | $k \times 10^{-10}$ | $k \times 10^{-8}$ | $k \times 10^{-6}$ | $k \times 10^{-4}$ | $k \times 10^{-2}$ | $k \times 10^{0}$ |
|---|---|---|---|---|---|---|---|---|
| 1.0 | 31.6590 | 27.0538 | 22.4486 | 17.8435 | 13.2383 | 8.6332 | 4.0379 | 0.2194 |
| 1.5 | 31.2535 | 26.6483 | 22.0432 | 17.4380 | 12.8328 | 8.2278 | 3.6374 | 0.1000 |
| 2.0 | 30.9658 | 26.3607 | 21.7555 | 17.1503 | 12.5451 | 7.9402 | 3.3547 | 0.04890 |
| 2.5 | 30.7427 | 26.1375 | 21.5323 | 16.9272 | 12.3220 | 7.7172 | 3.1365 | 0.02491 |
| 3.0 | 30.5604 | 25.9552 | 21.3500 | 16.7449 | 12.1397 | 7.5348 | 2.9591 | 0.01305 |
| 3.5 | 30.4062 | 25.8010 | 21.1959 | 16.5907 | 11.9855 | 7.3807 | 2.8099 | 0.006970 |
| 4.0 | 30.2727 | 25.6675 | 21.0623 | 16.4572 | 11.8520 | 7.2472 | 2.6813 | 0.003779 |
| 4.5 | 30.1549 | 25.5497 | 20.9446 | 16.3394 | 11.7342 | 7.1295 | 2.5684 | 0.002073 |
| 5.0 | 30.0495 | 25.4444 | 20.8392 | 16.2340 | 11.6289 | 7.0242 | 2.4679 | 0.001148 |
| 5.5 | 29.9542 | 25.3491 | 20.7439 | 16.1387 | 11.5336 | 6.9289 | 2.3775 | 0.0006409 |
| 6.0 | 29.8672 | 25.2620 | 20.6569 | 16.0517 | 11.4465 | 6.8420 | 2.2953 | 0.0003601 |
| 6.5 | 29.7872 | 25.1820 | 20.5768 | 15.9717 | 11.3665 | 6.7620 | 2.2201 | 0.0002034 |
| 7.0 | 29.7131 | 25.1079 | 20.5027 | 15.8976 | 11.2924 | 6.6879 | 2.1508 | 0.0001155 |
| 7.5 | 29.6441 | 25.0389 | 20.4337 | 15.8286 | 11.2234 | 6.6190 | 2.0867 | 0.00006583 |
| 8.0 | 29.5795 | 24.9744 | 20.3692 | 15.7640 | 11.1589 | 6.5545 | 2.0269 | 0.00003767 |
| 8.5 | 29.5189 | 24.9137 | 20.3086 | 15.7034 | 11.0982 | 6.4939 | 1.9711 | 0.00002162 |
| 9.0 | 29.4618 | 24.8566 | 20.2514 | 15.6462 | 11.0411 | 6.4368 | 1.9187 | 0.00001245 |
| 9.5 | 29.4077 | 24.8025 | 20.1973 | 15.5922 | 10.9870 | 6.3828 | 1.8695 | 0.000007185 |

To choose an appropriate value for $W(u)$ given $u$, where in this table $u = k \times$ power 10, read the table as follows: for $u = 5 \times 10^{-10}$ select the value 5 from the left column ($k = 5$) and move across the table to the column for $k \times 10^{-10}$, to give $W(u)$ for $u = 5 \times 10^{-10} = 20.84$. Likewise, for $u = 5 \times 10^{0}$, the value for $W(u) = 0.001$.

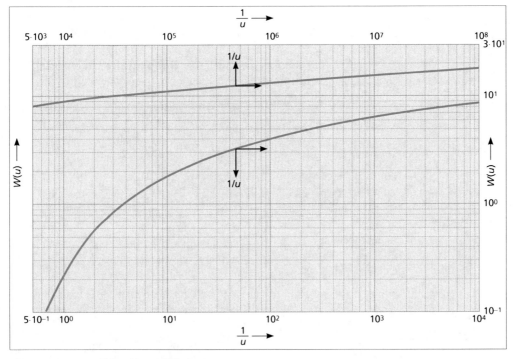

Figure 4.26   Well function $W(u)$ chart.

which represented on semi-logarithmic paper is a line with form $y = mx + n$ (Figure 4.28), where:

$$y = d$$

$$m = 0.183 \frac{Q}{T}$$

$$x = \log t$$

This line intersects the x axis in:

$$d = 0 = 0.183 \frac{Q}{T} \log t - 0.183 \frac{Q}{T} \log t_0$$

where: $t = t_0$.

Looking back to what was explained for cases in permanent regime or steady flow, $T$ is obtained by:

$$T = 0.183 \frac{Q}{m} \quad \text{or} \quad T = 0.183 \frac{Q}{\Delta d}$$

given that $m = \Delta d$ = drawdown over a logarithmic cycle.

To calculate the storage coefficient the point is determined where the line intersects the x axis; this value is made equal to $t_0$ ($t_0$ = point of intersection with 0x) which as seen above is:

$$t_0 = \frac{r^2 S}{2.25 T} \quad \text{and} \quad S = \frac{2.25 T t_0}{r^2}$$

As regards when the Jacob method can be used it should be noted that the values of $u$ are lower the smaller the distance $r$ is from the pumping well, and the less time $t$ has passed since the start of the test. For this reason Jacob is always applicable in the pumping well, while at the observation points it is only applicable after a specific time lapse, which is calculated as follows:

$$u < 0.1 \quad \frac{r^2 S}{4 T t_v} < 0.1 \quad t_v > \frac{10 r^2 S}{4 T} \quad t_v > 2.5 \frac{r^2 S}{T}$$

where $t_v$ is the Jacob validity time.

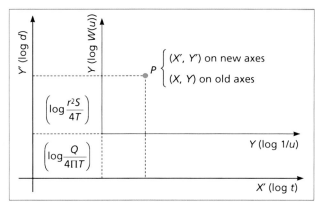

Figure 4.27  System of coordinate axes change to apply the Theis curve-matching method.

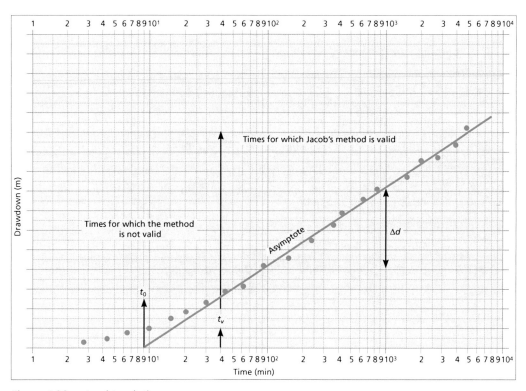

Figure 4.28  Jacob's solution.

## Jacob-Cooper method. Correction of saturated thickness. Unconfined aquifers in a transitory regime

When dealing with unconfined aquifers the Theis and Jacob methods are valid, whenever the depression is of minor importance with respect to the saturated thickness of the aquifer; if this is not the case, the correction should be made by substituting the value of the observed drawdown, $d$, by $(d - d^2/2h_0)$.

In fact the problem is more complicated, but the correction shown gives an acceptable approximation.

## Hantush method. Semi-confined aquifer in a variable regime

In the case of semi-confined aquifers, where the vertical recharge system has already been commented on in the permanent regime tests, the solution of the general flow equation is the Hantush formula:

$$d = \frac{Q}{4\pi T} W\left(u, \frac{r}{B}\right)$$

where the function $W(u, r/B)$ is also tabulated. Figure 4.29 shows the function:

$$\left[W\left(u, \frac{r}{B}\right), \frac{1}{u}\right]$$

analogous to the Theis function but with derivations for each value of $r/B$.

Overlaying the field curve $d - \log t$ to coincide with the $r/B$ derived from the above which fits best, gives, in exactly the same way as Theis, the values for $W(u, r/B)$, $1/u$, $d$, $t$, and $r/B$, making it possible to obtain the transmissivity, $T$, the storage coefficient $S$ and the vertical permeability of the semi-confining formation, $k'$.

# Injection tests

In general terms, the geological ground conditions at the site of any engineering project must be studied. When water is present on the site, the water itself and also its movement (flow) usually cause problems. If water movements have not been predicted and controlled, seepage, piping, erosion, dissolution, volume changes, uplift or artesian pressures and other water related problems may occur. The aim is to eliminate or control excess water which complicates or even prevents usual engineering working conditions, or which changes the ground properties making them unsuitable for carrying out the planned construction works.

When considering the methods normally used to solve these problems, it is extremely important to understand the laws which govern groundwater flow. One of the most important parameters to be studied is Darcy's **coefficient of permeability**, $k$. This coefficient is often calculated from injection tests. Apart from what has already been explained about pumping tests, note that although these tests are the most reliable and offer most parameters, the traditional use of injection tests in geological engineering means that

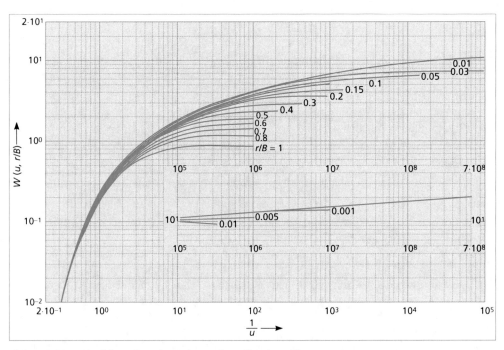

Figure 4.29   Well function charts in semi-confined aquifers.

they can be considered as key techniques for investigating permeability of the ground.

Injection tests, using the Lugeon or Lefranc methods, are described in Chapter 5, Section 5 and should be consulted to understand the analysis made here. Injection tests are of interest mainly because they are *in situ* tests. However, it should be noted that most formulae for calculating permeability by injection methods start from Thiem's equation for steady flow in a confined aquifer, and when tests are carried out the set of limitations and boundary conditions imposed to solve the flow equation in steady flow when deducing Thiem's formula should be maintained: i.e. confined aquifer, saturated medium, steady flow, homogeneous and isotropic medium, etc.

It is difficult, or almost impossible, to reach the steady state when water is injected into a borehole for only a few minutes. In the strictest sense these tests do not obey the theory upon which they are based. They follow a standardized methodology and for this reason tend to be a good method for comparing permeability at different depths in a borehole test or between contiguous borehole tests drilled in the same rock or soil formation. However, in most cases the permeability values obtained are not very reliable (see Chapter 11, Section 6).

The **Lugeon test** is the most common injection test although often badly carried out. This is an injection test using water under pressure in boreholes in rock masses where permeability is generally the result of jointing or fissuring from tectonic stress or dissolution phenomena. This simple test gives an overall idea of the permeability of the rock mass and provides immediate qualitative measurements, and by applying specific formulae, a value of the coefficient $k$ is obtained.

Even if the traditional nature of this test is recognized, and the undoubted advantages it has over laboratory methods in core samples, its low reliability has to be admitted. Pumping tests are without any doubt the most reliable method for obtaining permeability parameters in saturated formations, although they are the most expensive and for this reason are not widely used in geological engineering. But the guarantee their data offer compensates amply for any extra costs incurred. On the other hand, in injection tests, very low volumes of water are injected into the formation being tested, which means that the results are spatially very limited. Pumping tests last a significant time (1 to 3 days is usual) and during this time a large quantity of water is extracted from zones of the formation separated from the pumping well, testing a large volume of the physical medium so that the results are highly representative in spatial terms.

## Tracer tests

This type of tests consists basically in injecting a tracer at a specific point of the aquifer and observing when it reaches another given point of the same aquifer, measuring the transit time between the two points as exactly as possible. In this way the flow velocity in the direction between the injection and arrival points can be obtained, so that the permeability or even the effective porosity can be estimated if other data of the aquifer are known, such as the hydraulic gradients, depending on the test type and method used.

The use of tracers is a particularly specialized technique, widely used in hydrogeology and geological engineering contexts. Tracers are commonly used, e.g. for measuring flow rates in turbulent rivers or where discharges are low; they are also often used when investigating leakage and filtrations in dams or tunnels.

The problems have to be approached within real dimensions from a hydrogeological point of view. This means that distances between injection and collection points have to be appropriate for the permeability, effective porosity and hydraulic gradients of the aquifer; if not, the transit times may be so long, or the tracer may be so diluted that the test is invalid and is considered to be non-effective, when in fact it is not the method that is at fault, but rather an inadequate test approach for the aquifer conditions.

Different types of tracers are commonly used: radioactive isotopes with a low half-life period, colorants, soluble salts which are non-interactive with the formation or even, in specific circumstances, solids in suspension. Each type has its advantages and disadvantages, which are summarized below.

**Radioactive tracers** can be detected at very low concentrations and tests can be carried out at concentrations lower than the normally permitted maximum for drinking water. These allow precise field measurements to be made, even inside bore holes, without the need to take samples. Very small quantities of the tracer are needed, which makes field operations easier as they are able to trace large volumes of water. They disappear by disintegration, after several half-life periods have passed.

The typical drawbacks of this kind of product are that the detection method is expensive and complex, that specialized equipment and human resources are required and that radioactive isotopes are difficult to obtain and must be requested from the corresponding authority.

The most commonly used **colorants** include fluorescein or some of its more soluble derivatives and are easily detected with a photocolorimeter at very low concentrations. These are very suitable for karstic aquifers but are not much use in detrital materials. One drawback of these is that they are easily retained in the ground.

**Soluble salts,** mainly NaCl, $NH_4Cl$ and $CaCl_2$, are very popular. The easiest to use is NaCl, but the danger is that it may alter the clay fraction of the formation, changing the permeability, so one of the other two must be used. The chloride ion behaves as a practically ideal tracer, but should be used in aquifers where it already exists in low proportions in the natural water; it can only be used if the natural content is lower than 100 ppm. It is detected by chemical analysis and where the concentration of salts in the aquifer is low, salinometers can be used, so that it is not necessary to extract samples from the test bores.

**Solid tracers in suspension** are generally substances which remain in suspension in the water such as oat bran, starch grains, yeast, coloured pollen or even sawdust, but they can only be used when water flows through large cracks or karstic conduits. In porous media they are very easily retained by filtration. The injection can be done in two ways: instantaneous or continuous. In the former, a specific volume of the tracer is injected as an instantaneous discharge, while in the second the tracer is injected continuously throughout the whole time span of the test.

Calculating the tracer transit times from the point of injection to the collection point is not immediate; various observations have to be made. Due to the effect of hydrodynamic dispersion in the aquifer, the tracer will reach the collection point gradually, not instantaneously. With instantaneous injection, the concentration of tracer will increase at the collection point until it reaches its maximum and then will be reduced until it returns to the original aquifer values, so that a concentration/time curve similar to a Gaussian bell curve is obtained and the transit time can be estimated as corresponding to the maximum of the curve, if this is symmetrical. With continuous injection, the concentration of the tracer increases progressively from the aquifer base levels until the level stabilizes, estimating the transit time as that corresponding to the sum of the base concentration and 50% of the increase due to the tracer.

There are many ways of carrying out tests with tracers to estimate the hydrogeological parameters of the formation. Some of the most common are described below.

### Well-to-well tracer injection method

Two wells are used, sited at distance $d$. The tracer is injected into one of them and the other is pumped at constant discharge $Q$. Considering the radial flow as a result of the pumping, when the tracer reaches the pumping well and is collected it can be supposed that any water particle within the circumference of the radius $d$ has also arrived. At that point all the water contained within an aquifer cylinder with base $\pi d^2$ and height of the saturated aquifer $h_0$ will have been pumped. As the water pumped from the moment when the tracer was injected is $Qt_T$, with $t_T$ being the transit time, it can be established that:

$$\pi d^2 h_0 n_e = Qt_T$$

from which the value of the effective porosity can be obtained:

$$n_e = \frac{Qt_T}{\pi d^2 h_0}$$

Some limitations must be taken into account when this test is being run:

— the aquifer is supposed to be homogeneous and isotropic
— the aquifer must be confined, or unconfined with drawdown of less than 10–15% of the initial saturated thickness
— the pumping must be started before the injection until a steady flow is reached for practical purposes
— the volume of water added with the tracer must be relatively small in comparison with the volume extracted in the transit time.

Finally, it must be considered that the greater the distance between the injection well and the collection well, the greater the reliability will be of the effective porosity value obtained, but the more difficult it will be to carry out the test satisfactorily, because of the effect of the dilution of the tracer and because of the large volume of water which has to be pumped.

### Injection method in a single well

This involves injecting a quantity of tracer into a well and waiting a specified time $t_E$, for the natural flow of the aquifer to displace it a specific distance. After the waiting time has elapsed, pumping at constant discharge $Q$ is started in the same well where the tracer was injected and the time it takes to reappear $t_R$ is calculated.

After the tracer is added, the flow will have taken it away from the well to the distance $D = vt_E$, with $v$ being the real velocity of the flow in the aquifer. When pumping at discharge $Q$ during time $t_R$ until the tracer is recovered, the water will have been extracted from an aquifer cylinder with radius $D$ and the aquifer height $h_0$. As a result it can be established that:

$$v = Qt_R = \pi D^2 h_0 n_e$$

where:

$$D = vt_E = \frac{ki_N t_E}{n_e}$$

with $k$ and $i_N$, the permeability and the natural gradient of the aquifer, which should be known.

Substituting, the following expression is obtained:

$$Qt_R = \pi(ki_N t_E)^2 \frac{h_0}{n_e}$$

from which $k$ can be obtained if $n_e$ is known, or $n_e$ can be obtained if $k$ is known.

## 4.5 Solution methods

There are many problems in geological engineering which need a very rigorous and practical calculation of the flow in natural materials under specific boundary conditions. The laws governing the groundwater flow in permeable media have been outlined in the preceding sections, where emphasis was placed on the difficulty involved in resolving the characteristic differential equations for each phenomenon. To be able to solve these satisfactorily depends on the complexity of the real problem and especially on the boundary conditions. The problems may appear in permanent (steady) and transitory (non-steady) regimes and there are three classic methods used to solve them: **analytical methods, flow nets and numerical methods**.

Analytical methods have many limitations and can only solve problems in a permanent regime that have a simple configuration of the physical medium, and very few cases in transitory regime. Flow nets can be used to solve permanent regime problems with very simple physical medium configurations and with perfectly delimited constant level boundary conditions. Numerical methods, with approximate solutions, are the most complete, and allow an approach to problems of permanent and transitory regimes in two and three dimensions with variable hydrogeological parameters in space and even in time, with different boundary conditions; in fact they are the only method available for analyzing complex systems.

Although numerical methods require some specialization in mathematics and computer science, the applications of flow and transport models currently available do not need much specialization in areas outside hydrogeology and geological engineering, and these techniques are being used more and more in general.

For a clear understanding of the difference between the three methods an example is included below using each one. As can be seen in Figure 4.30, there are two open pits or trenches, A and B where constant levels: $h_A = 15$ m and $h_B = 8$ m. The trenches are in an unconfined aquifer with permeability $k = 2$ m/day, they penetrate the permeable formation completely and are separated by distance $L = 26$ m. The phenomenon is studied calculating the flow rates between the trenches (per linear meter of trench) and the shape of the piezometric surface.

## Analytical methods

The aim is to find an expression able to calculate the flow rate per meter of trench and the piezometric surface equation. In the example shown in Figure 4.30, the flow per meter of open pit does not pass through a constant section $b \cdot 1$, but instead through a variable section $h \cdot 1$. The thickness of the aquifer, as it is unconfined, is not constant, but depends on the saturated thickness at each point situated at distance $x$ from the trench A, given by piezometric level $h$.

The discharge $Q$ which flows through a generic point $x$ for a 1 m of length of the trench is:

$$Q = -kA\frac{dh}{dx}$$

where:

$A$ = flow area = $h \cdot 1$
$k$ = coefficient of permeability

$$Q = -kh\frac{dh}{dx}$$

then:

$$\frac{Q}{k}dx = -hdh$$

When $x$ varies between 0 and $L$, $h$ varies between $h_A$ and $h_B$:

$$\frac{Q}{k}\int_0^L dx = -\int_{h_A}^{h_B} h\,dh$$

$$\frac{Q}{k}[x]_0^L = -\left[\frac{h^2}{2}\right]_{h_A}^{h_B}$$

$$\frac{QL}{k} = \frac{1}{2}\left(h_A^2 - h_B^2\right)$$

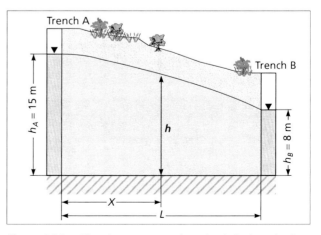

Figure 4.30  Flow between trenches. Analytical method.

and the discharge per meter of trench will be given by:

$$Q = \frac{k}{2L}\left(h_A^2 - h_B^2\right)$$

which in the case in question, for $k = 2$ m/day, $h_A = 15$ m, $h_B = 8$ m and $L = 26$ m, gives:

$$Q = 6.19 \text{ m}^3/\text{day}$$

The piezometric surface equation can be obtained by finding the relationship which would give the piezometric surface value $h$ as a function of $x$. To do this, the integration between trench $A$ and trench $B$ has to be set to be able to eliminate $Q$ from both equations.

So, in the case of the unconfined aquifer:

$$\frac{Q}{k}dx = -h\,dh$$

Integrating between 0 and $x$:

$$\frac{Q}{k}\int_0^x dx = -\int_{h_A}^h h\,dh; \quad \frac{Qx}{k} = \frac{h_A^2 - h^2}{2}$$

Integrating between $x$ and $L$:

$$\frac{Q}{k}\int_x^L dx = -\int_h^{h_B} h\,dh; \quad \frac{Q}{k}(L-x) = \frac{h^2 - h_B^2}{2}$$

Eliminating $Q$ by quotient between both equations:

$$\frac{L-x}{x} = \frac{h^2 - h_B^2}{h_A^2 - h^2}; \quad \left(\frac{L}{x} - 1\right)\left(h_A^2 - h^2\right) = h^2 - h_B^2$$

and operating and simplifying:

$$h^2 = \frac{h_A^2 - h_B^2}{L}x + h_A^2$$

The shape of the piezometric surface is independent of the permeability and thickness of the aquifer, and its equation is a parabola which depends only on the boundary conditions.

As can be seen, the deduction of the piezometric surface equation by analytical methods is a simple, easily solved problem and therefore easily applied.

## Flow nets

Solution methods based on constructing flow nets are very common in geological engineering. In the past there were no adequate computers or numerical methods available and the solution of flow problems by analytical methods had the well known difficulties of boundary conditions and heterogeneity of the physical medium, which often made them impossible to apply.

Flow nets and the general application of this method were explained in Chapter 2, Section 3; however, there are some additional points to be made here. The construction of a flow net to solve Laplace's equation graphically is relatively simple for problems with stable limits and also for movement in confined aquifers, but it is more difficult for those which refer to movement in unconfined aquifer conditions if the position of the water table is not known.

A flow net is a two-dimensional graphic with two families of a special type of curve: the **flow lines** which show the direction of the ground water flow and the **equipotential lines** which join up the points of constant head. The use of flow nets is limited to investigating two-dimensional sections which represent the movement in the porous medium and to analyzing three-dimensional problems with axial or radial symmetry.

In a homogeneous, isotropic medium where Darcy's law is valid, the flow lines are orthogonal to the equipotential lines. A flow net can be imagined as a network of two families of mutually orthogonal lines arranged in such a way that as a rule the flow lines end in equipotential lines and vice versa, partly delineating the movement domain. The exception to this rule appears when there is a seepage surface. To be able to design a flow net the flow must satisfy criteria of independence with time and homogeneity, in the plane of the net, and Darcy's law must also be valid.

Flow nets can be constructed to study movement in a vertical section and on a horizontal plane. Most applications are to investigate seepage through embankments, dams, diaphragm walls and similar structures, with vertical sections. Flow nets can also be constructed to study horizontal movements using maps of piezometric surfaces; in this case, due to the variations in transmissivity and the lack of parallelism between the plane of flow movement and that of the net, it is unusual to be able to construct a network of square grids; the deviations in the squared grid can be interpreted as variations in recharge, discharge or transmissivity.

There is not only one way to construct a flow net, because as the potential ($\phi$) and flow ($\psi$) functions are continuous, there are many choices to be made between the flow lines and equipotentials; in practice a few lines are chosen as representative of each family. However, the ratio between the number of flow tubes and the number of potential drops is a constant for each problem and is deduced from the formula for discharge.

The flow net is defined when the number of flow lines or equipotential lines is specified. It is recommended to limit the number of flow tubes to four or five, but in fact, a fractional number may be used for either flow tubes or potential drops.

To calculate the discharge below a hydraulic structure in a system which can be represented in a vertical section, the flow net is drawn and refined until it is "squared". The equipotentials will always be perpendicular to the impermeable boundaries and markedly parallel to the constant level boundaries; the opposite will be true for the flow lines. When the net is completed there will be a number $N_f$ of flow channels or tubes, each limited by two flow lines, and a number $N_d$ of drops in potential between the constant potential boundaries. Darcy's law is applied then to calculate $q$, the discharge through each of the flow tubes:

$$\Delta q = k \Delta b \left( \frac{\Delta h}{\Delta l} \right)$$

where:

$\Delta b$ is the width of the tube
$\Delta l$ is the length between two potential drops
$\Delta h$ is the potential drop in the square element.

As $\Delta b = \Delta l$, by construction $\Delta q = k \Delta h$. The total discharge, $Q$, which passes through the flow tubes $N_f$ will be:

$$Q = N_f \Delta q = N_f k \Delta h$$

On the other hand:

$$\Delta h = \frac{H}{N_d}$$

where $H$ is the total drop in piezometric levels or head loss between the two constant level boundaries which limit the problem, therefore:

$$Q = k \frac{N_F}{N_d} H$$

which allows the calculation of the transference discharge between boundaries.

From the values obtained for $h$, the flow net allows the pore pressure and velocity in the porous medium to be determined. In fact, in each curvilinear square discrete average values for $h$ are obtained instead of a continuous variation, which would be the result of an analytical solution. The smaller the subdivision of the elements of the net the more exact the discrete values obtained for $h$ will be. As the potential lessens along the length of each flow line, this allows $h$ to be calculated at any point. At the start of each flow line, the piezometric level is determined from the reference plane chosen.

The reduction between this point and the one where the values of the pressure and velocity are to be found is calculated as a function of the numbers of drops in potential between the two points and of the loss per drop, i.e. $H/N_d$. Once $h$ has been calculated, the pressure values, $p$, are deduced immediately from the value of $z$ and the function of $h = z + p/\gamma_w$.

The problem set is solved from the flow net in the same way as by analytical methods. *Figure 4.31* shows the net for the problem of the open pits or trenches. The two constant level boundaries are easy to identify, as they are the levels at 15 m and 8 m defined in the problem for trenches $A$ and $B$ respectively.

The base layer of the aquifer is an impermeable boundary and the equipotentials run perpendicular to it. The most difficult point is to draw the upper surface since this is an unconfined aquifer; it has to be adjusted between levels $A$ and $B$ into a smooth parabola shape, typical of an unconfined aquifer, as seen above. As shown in *Figure 4.31*, four flow tubes are drawn of the same thickness which go from $A$ to $B$ keeping the same proportions. Then the equipotentials are drawn perpendicular to the boundaries and to the flow lines so that the grid is square. There are four flow tubes and nine level drops with defined values which are a ninth part of the total drop $H$. As a result, the discharge will be:

$$Q = k \frac{N_f}{N_d} H = 2 \frac{4}{9}(15-8) = 6.22 \, m^3/day$$

which turns out to be very similar to the value obtained by the analytical method.

## Numerical methods

These methods consist basically of discretizing the physical medium into a defined set of intervals, setting the place and characteristics of the system boundaries, assigning the corresponding piezometric levels and hydrogeological parameters

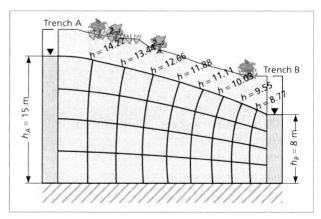

*Figure 4.31*    Flow between trenches. Flow net method.

to each interval and setting out the flow equations for each interval. This gives a set of equations with the same number of equations and unknowns as there are intervals, which is solved by computer. The solution to this system of equations is the piezometric level in each discrete interval.

*Figure 4.32* shows two open pits or trenches penetrating right through an unconfined aquifer. The physical space existing between the two trenches A and B, with length L, is divided into intervals of equal width $\Delta x$, so that $(n-1)\Delta x = L$.

Each band of width $x$ will have a control node on the right and the left, so that there will be $n$ nodes. Each generic node $i$ will have representative piezometric level $h_i$ and average permeability $k_i$. As has already been shown, the discharge which passes through a generic point $x$ for a 1 m length of trench $A = h \cdot 1$ is:

$$Q = kA\frac{dh}{dx}$$

$$Q = kh\frac{dh}{dx}; \quad \frac{Q}{k}dx = hdh$$

Integrating between $i-1$ and $i$:

$$\frac{Q}{k_{i-1}}\int_{x_{i-1}}^{x_i}dx = \int_{h_{i-1}}^{h_i}hdh; \quad \frac{Q}{k_{i-1}}[x_i - x_{i-1}]_{x_{i-1}}^{x_i} = \left[\frac{h^2}{2}\right]_{h_{i-1}}^{h_i}$$

$$\frac{Q}{k_{i-1}}x = \frac{h_i^2}{2} - \frac{h_{i-1}^2}{2}$$

$$\frac{2Qx}{k_{i-1}} h_i^2 - h_{i-1}^2$$

Analogously integrating between $i$ and $i+1$:

$$\frac{2Qx}{k_i} = h_{i+1}^2 - h_i^2$$

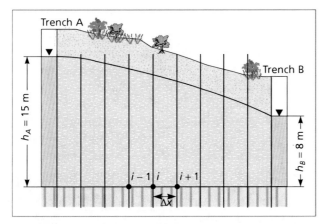

*Figure 4.32* Flow between trenches. Numerical method.

Eliminating $Q$ and $x$ by quotient between both equations:

$$\frac{\frac{2Qx}{k_{i-1}}}{\frac{2Qx}{k_i}} = \frac{h_i^2 - h_{i-1}^2}{h_{i+1}^2 - h_i^2}; \quad \frac{k_i}{k_{i-1}} = \frac{h_i^2 - h_{i-1}^2}{h_{i+1}^2 - h_i^2}$$

where:

$$B_i = \frac{k_i}{k_{i-1}} \Rightarrow B_i h_{i+1}^2 - B_i h_i^2 - h_i^2 + h_{i-1}^2 = 0$$

$$h_{i-1}^2 - (B_i + 1)h_i^2 + B_i h_{i+1}^2 = 0$$

This equation represents a system of $n - 2$ linear equations, by varying $i$ between $i = 2$ and $i = n -1$. There are $n - 2$ unknowns, given that $h_1 = h_A$ and $h_n = h_B$ which are the known levels in trenches A and B. The set of equations is as follows:

$$i = 2: \quad h_1^2 - (B_2 + 1)h_2^2 + B_2 h_3^2 = 0$$

$$i = 3: \quad h_2^2 - (B_3 + 1)h_3^2 + B_3 h_4^2 = 0$$

$$i = 4: \quad h_3^2 - (B_4 + 1)h_4^2 + B_4 h_5^2 = 0$$

......

......

$$i = n-2: \quad h_{n-3}^2 - (B_{n-2} + 1)h_{n-2}^2 + B_{n-2}h_{n-1}^2 = 0$$

$$i = n-1: \quad h_{n-2}^2 - (B_{n-1} + 1)h_{n-1}^2 + B_{n-1}h_n^2 = 0$$

The system solution allows the levels at each node to be found, which is a numerical approximation to the piezometric surface equation.

The system outlined is not linear and so its solution is more complicated. Normally an iterative algorithm is used based on the real behaviour of unconfined aquifers in hydrodynamic problems. In fact, in unconfined aquifers the flow area varies with the saturated thickness at any given point and this varies with the piezometric level. If the saturated thickness varies, the transmissivity varies for the same permeability value. The method consists in giving iterations to the configured model for confined aquifer, which is linear and simple to calculate, with the transmissivities modified depending on the levels obtained at the end of each iteration. This allows the real thicknesses entered in each iteration to configure the system as an unconfined aquifer. When the results between two successive iterations satisfy an error criterion, the simulation is considered to be complete, giving the last levels calculated as the result. In fact, the only modification needed is to equalize the cell thicknesses at the mean piezometric levels of the two adjacent nodes after each iteration. On the other hand, it makes no sense to talk of aquifer thicknesses functioning as unconfined, since they depend on the piezometric level. This is why to deal with variable thicknesses they have to be deduced from the difference between the piezometric levels and the (variable) base level of the aquifer.

The example problem given may be solved with a numerical model, carried out for that purpose, and the following results are obtained:

| Initial simulation values | Values of simulated levels | |
|---|---|---|
| | Node | Simulated piezometric level (m) |
| | 1 | 15 |
| Total number of nodes = 10 | 2 | 14.42941 |
| General permeability = 2 m/day | 3 | 13.83626 |
| General thickness = 15 m | 4 | 13.21767 |
| Level $h_A$ = 15 m | 5 | 12.57014 |
| Level $h_B$ = 8 m | 6 | 11.88925 |
| | 7 | 11.16937 |
| | 8 | 10.4031 |
| | 9 | 9.580376 |
| | 10 | 8 |

Here it can be seen that the simulated levels reproduce the equation of the piezometric surface obtained by analytical methods with acceptable accuracy; to obtain the discharge rates as well, Darcy can be applied between nodes to find the mean. This will give:

$$Q = 6.19 \text{ m}^3/\text{day}$$

which is the same value as was obtained by analytical methods.

It should be pointed out that in this case only 10 nodes have been taken, and as a result there is some discrepancy in the solutions, mainly in the nodes near trench $B$. A more accurate result will be obtained by simulating a higher number of nodes and with more demanding error criteria.

This method may seem slower and more difficult, but there are easy to use computerized models available to configure the physical medium and simulate all types of options. In this context the "Modflow" model, used to simulate the flow of groundwater through aquifers, is outstanding, with a basic downloadable version available (McDonald and Harbaugh, 1988; http://water.usgs.gov/nrp/gwsoftware/modflow2000/modflow2000.html).

## 4.6 Chemical properties of water

The chemical quality of the groundwater and its contamination processes have a significant effect on many problems related to geological engineering. These can be different physical or chemical processes which occur between the groundwater and its supporting medium, whether this medium is natural (geological formations) or constructed (compacted infill, foundations, diaphragm walls, etc.). Sometimes engineering geologists or hydrogeologists have to study and solve problems of the groundwater contamination itself, for environmental surveys or in relation to facilities for discharging potentially contaminating waste.

This section attempts to offer guidelines for understanding processes linked to quality and contamination of groundwater, its modifying phenomena, the influence of human activities and the mechanisms for introducing and spreading contaminants.

## Chemical quality of groundwater

The natural chemical quality of groundwater is conditioned by its source and by the natural chemical and physical processes which develop along its flow path. Generally, the source of groundwater is rainwater infiltration and this contains gases and ions which are found dissolved in the atmosphere. As soon as the water begins to infiltrate, it starts to mineralize. There is a close relationship between the lithology of an aquifer and the groundwater mineralization. In the case of aquifers composed of sedimentary rocks the mineralization of the waters depends on the factors shown in *Table 4.2*.

The main constituents of groundwater are:

— major: $HCO_3^-$, $SO_4^{2-}$, $Cl^-$, $NO_3^-$, $Na^+$, $K^+$, $Ca^{2+}$, $Mg^{2+}$, $CO_2$, $O_2$ and $SiO_2$
— minor: $NO_2^-$, $PO_4^{3-}$, $BO_3^-$, $F^-$, $S^{2-}$, $Br^-$, $I^-$, $NH_4^+$, $Fe^{2+}$, $Mn^{2+}$, $Li^+$, $Sr^{2+}$, $Zn^{2+}$
— trace elements: $Al^{3+}$, $Ti^{4+}$, $Co^{2+}$, $Cu^{2+}$, $Pb^{2+}$, $Ni^{2+}$, $Cr^{3+}$.

Sometimes there may be organic matter as the result of organic substances in the infiltration water.

The natural quality of water varies considerably; in some aquifers very pure water is found, with very low ionic content around 3 mEq/l while in others there is important dissolution of salts and the ionic content reaches 20 or 30 mEq/l.

In general terms groundwater is better protected against possible contamination than surface water, but once the contaminating agent has been incorporated into the groundwater flow it is very difficult and expensive to detect its presence, and therefore to put measures into place to control it before it affects productive wells and boreholes.

| Table 4.2 | FACTORS AFFECTING WATER CHEMISTRY IN SEDIMENTARY ROCK AQUIFERS | | | |
|---|---|---|---|---|
| **Lithological factors** | **Water factors** | **Rock/water system factors** | **External factors** | |
| — Rock/water contact surface.<br>— Grain size.<br>— Soluble solid phases.<br>— Mobile pore and locked pore water. | — pH.<br>— Eh.<br>— Saturation.<br>— Vapor pressure from dissolved gases. | — Permeability.<br>— Length of flow path and contact time.<br>— Flow regime.<br>— Ionic change, diffusion and adsorption. | — Temperature.<br>— Pressure.<br>— Biological activity. | |

For planning and control activities is essential to understand the contaminant routes based on the geological, hydrogeological and chemical conditioning factors.

# Physical-chemical processes. Water-aquifer interaction

Groundwater moves from a recharge zone, where rainwater infiltrates, to a discharge zone, where it flows out. The contaminant transport processes are different in the non-saturated and the saturated zone. In both cases they depend on the water flow. In a non-saturated zone the movement of water, and therefore also of the contaminant, is vertical, towards the phreatic surface, while in the saturated zone, the movement of the water is preferentially horizontal or at least conditioned by the piezometric gradients of the aquifer.

Once the contaminant has reached the groundwater flow it moves in its same direction and at the same speed, producing dispersion and dilution mechanisms.

When the contaminant reaches the ground the infiltration mechanism begins and a series of physical and chemical changes take place. These changes are closely linked to the lithology of the terrain, the thickness of the non-saturated zone and how long the contaminant remains in the aquifer. The changes include surface phenomena, precipitation-dissolution and oxidation-reduction. The slow flow rate of groundwater means that contamination of aquifers is also slow, but depending on the conditions of the medium and of the contaminants, processes take place which affect the contamination rate or may even reduce it. The most usual processes are described below:

## Physical processes

— **Mechanical filtration**: this occurs with particles in suspension such as sands, silts, clays, algae, microorganisms, etc.; the smaller the average size of the pores and the more uniform their distribution, the more effective the filtration.

## Chemical processes

— **Surface phenomena**: the most important are ionic changes and molecular adsorption which depends on the colloidal nature, inorganic clays and organic substances such as humus. **Molecular adsorption** is the main mechanism for retaining from the water synthetic organic compounds, heavy metals, anions such as orthophosphate, cations such as ammonium and potassium, etc. Other surface phenomena may be established related to the volatility of some substances, the acidity level, heavy metals, organic materials, etc.

— **Ionic interchange**: this is the precipitation of iron and manganese oxides along with the continuous formation of bacterial growth around the surface of gravel and fissures. Generally this takes places in the transit zone between a reduction and an oxidation medium. This process allows the sorption (retention) capacity of the ground to be maintained, but means its permeability is reduced.

— **Neutralization**: base acid reactions, important in very acid or very basic residues.

— **Precipitation-dissolution**: directly related to the degree of saturation and the equilibrium constants. Precipitation mainly affects the $Fe^{2+}$ and $Mn^{2+}$ originated in aerobic processes and which are then found dissolved in the water. It is due to the oxidation at pH between 6 and 8 of the ions mentioned above, forming the compounds $Fe(OH)^3$ and $Mn(OH)^3$ which are insoluble. The precipitation of these compounds drags down many metallic ions by co-precipitation such as Cu, Pb, Zn, As, W, V and $F^-$. Calcium, magnesium, bicarbonate and sulphates may undergo this process. Some trace constituents such as arsenic, boron, cadmium, cyanide, iron, lead, mercury, etc. have a considerable capacity for this type of process.

— **Oxidation-reduction**: this is produced mainly during the infiltration process in the non-saturated medium and the bacteria and other microorganisms existing in the ground play a very important role. This mechanism is important in reducing the contamination by inorganic and nitrogenous products, which as they oxidize produce $CO_2$ and $NO_3^-$ respectively. However, where there is little or no oxygen, anaerobic processes may occur which form $CO_2$, $CH_4$, $SH_2$, S, $Fe^{2+}$, $Mn^{2+}$,

$NH_4^+$, $N_2$ and $NO_2^-$ by the reduction of $NO_3^-$, $SO_4^{2-}$ and iron and manganese compounds present in the soil. These cause a kind of contamination which is generally detected from the presence of a bad smell and colour.
- **Adsorption-desorption**: the adsorbed elements may revert to a solution when they come in contact with water with a lower concentration of these elements.
- **Ground retention capacity**: this is the property the ground has to reduce the circulation speed of the contaminants either temporarily or permanently. This happens mainly when they are large, complex molecules. The ground has a specific retention capacity under certain conditions and so if the compounds retained are not decomposed by other actions, the process is detained when the maximum retention capacity is reached. Retention implies a greater permanence period and this allows other processes to occur.
- **Dilution processes**: these are closely linked to the way contamination is produced and are less important when the contamination is localized than when it is widespread.

# Contamination of groundwater

The contamination of groundwater is the deterioration of its natural quality due to human activities or actions which make it unsuitable for the use for which it was intended. These activities give their name to the type of contamination: urban, agricultural, livestock, industrial, over-pumping, etc. The concept of localized or widespread contamination depends on the area affected, generally urban, industrial or livestock farming in the first case and agricultural in the second. As mentioned above, groundwater is better protected from contamination than surface water but once the contaminating agent has entered the groundwater flow it is difficult to stop it before it affects wells, boreholes or springs.

Groundwater is contaminated when waste products from various activities reach the aquifer, causing the presence or increase of specific substances characteristic of these activities. These substances, which show the existence of contamination either by being present, or by exceeding a defined concentration or by a combination of both of these, are called **contamination indices**.

The contaminants of groundwater are very varied and from different sources. In general terms they are all substances from various different activities which seep into groundwater flow causing deterioration of the water and make it unsuitable for its intended or potential use.

**Chemical contaminants** include a wide range of ions and chemical substances, such as the normal ions present in the water, but in higher quantities than would be present in its natural state, and compounds which do not occur naturally in the water. The ions ammonium and nitrate, which are the result of the oxidization of organic material, should also be mentioned; although they are sometimes not detected in contaminated water in significant quantities due to their instability, when they are detected this indicates that there is a source of contamination nearby.

Trace element concentrations of heavy metals, which are sometimes detected in water, increase their content as a result of the incidence of industrial activities. Some are very dangerous such as lead, which is cumulative, and hexavalent chromium, which is toxic and very persistent. Chemical contaminants include also dangerous toxic compounds such as cyanides, detergents, fats, pesticides, etc.

**Biological contaminants** basically include bacteria and viruses but generally their effect is lessened by the purifying capacity of the soil and subsoil.

**Radioactive contaminants** are not often found because of the strict controls on radioactive substances. Of the six radio nuclids with greatest toxicity, mobility and radioactive period (H-3, Sr-90, I-129, Cs-137, Ra-226 and Pu-239), five are produced in the processes for nuclear energy, and if they are incorrectly stored, can reach aquifers.

# Anthropogenic activities

Human activities and actions which may damage water resources can be urban, agricultural, livestock or industrial. The damage caused will depend on the characteristics of the waste, on the place, type, intensity and duration of the discharge and on the specific conditions of the receptor medium.

## Urban activities

Urban activities are all those which are the result of human settlement. The number of urban activities varies depending on the type of population concerned.

In general terms two types of waste are produced: urban waste water and urban solids or refuse.

**Liquid urban waste** or waste water is the return discharge of the water used in urban activities and generally includes domestic waste water, water from street cleaning and industrial waste water. The composition of this type of waste varies from one place to another and even over an annual or daily period, but in general it is characterized by the presence of dissolved salts, organic matter, causing colour and smell, solids in suspension and micro-organisms. The possible presence of heavy metals or foreign substances originating in industries connected to the sewerage system network also has to be taken into account.

What often happens is that waste is discharged into a surface water course; the repercussions on the groundwater depend on the dilution level and the river-aquifer relationship. In coastal areas the discharge is often directly into the sea or through an underwater discharge pipe. Another way of eliminating this waste is to discharge it onto the ground surface, taking advantage of the purifying capacity of the soil, either as irrigation or to recharge the aquifer.

**Solid urban waste** is usually made up of garbage or household refuse, waste from road cleaning, etc. This type of waste can be incinerated, recycled to make compost or accumulated in a waste disposal dump. In the last case a liquid residue or **leachate** will form, as a result of the compacting of the waste itself and precipitation. This leachate varies in content, since it depends on the composition of the rubbish, which varies considerably from country to country, or even from one district to another within the same city. The leachate will have high chemical and biochemical oxygen demand values. There will also be a high content of dissolved solids and concentrations of chlorides, sulphates, bicarbonates, sodium, potassium, calcium and magnesium. Leaching is often found in a reducing medium, so that it often also contains ammonium, iron, manganese and zinc. It may also contain more dangerous products if solid industrial waste is dumped from industries in built-up areas.

Groundwater which is contaminated from an urban solid waste disposal will show an increase in mineralization and temperature, presence of colour, unpleasant smell and a high content of $NH_4^+$, DQO and $DBO_5$, with presence of $Fe^{++}$ and absence of $NO_3^-$.

### Industrial activities

These are all activities involved in the production of raw materials or manufacture of consumer goods for direct human use or activities.

Industry may produce solid, or more usually, liquid waste. Smoke emissions into the atmosphere may constitute a secondary risk for water supplies.

The substances these contain are both inorganic and organic. The inorganic ones include dangerously toxic heavy metals, high concentrations of dissolved solids and acids and alkali which provide acidity and causticity.

There is a wide variety of organic substances present in industrial waste. These include by-products of petrol and coal, synthetic organic substances like phenols, detergents, gypsums, oils, etc. and natural organic substances such as sugars, dairy products or distillery by-products.

The main cause of this type of contamination is waste disposal, especially of liquids. These are disposed of, treated to a greater or lesser extent or even untreated, through public water courses, dumped on the soil or injected directly into the aquifer. Other causes of contamination are leaks in storage facilities or pipelines, and accidental spills in the transport of dangerous substances.

Solid waste is dumped along with other urban solid waste after being stockpiled on the ground near factories. Sometimes, depending on the type of waste, they are deposited in controlled waste disposal for toxic and dangerous residues.

### Actions on water

Two types of human actions on water may unintentionally cause a reduction in quality: from exploitation and from recharge.

Recharging consists in injecting a specific volume of water into the aquifer through injection wells. This may be waste water; the treatment of this water and the place where it is injected may lower the quality of the aquifer. However, sometimes recharging is used as a corrective measure, injecting better quality water and altering the flow.

## Mechanisms of ground water contamination

An aquifer is contaminated by the introduction and spread of substances or ions which, when they occur in higher than ideal quantities, cause deterioration of water quality. There are many different ways of introducing these agents into aquifers, but generally they are the following:

### From the ground surface

This occurs in the following situations:

— Intentional dumping on the ground.
— Leaks in pipelines and surface installations.
— Accidental industrial spills.
— Accidents in the transport of toxic substances.
— Through careless handling during loading and unloading.

The contamination mechanism is the seepage of the contaminants from the ground surface and their movement depends on the water flow.

The lithology of the soil and its thickness play a very important role in the spread of contamination as it can act as a physical filter where the solids in suspension and micro-organisms are retained, which is very important in the physical and chemical processes which take place in the non-saturated zone. The filter effect does not exist when the lithology of the non-saturated zone is karstic.

### From the non-saturated zone

This occurs in the following situations:

— Leaks in industrial sewage systems.
— Leaks in pipelines and underground tanks.
— Injections of waste into the ground.

The spreading mechanisms from the non-saturated zone are similar to when the contaminant has been dumped on the ground surface, but the distances to the saturated zone may be considerably reduced.

### From the saturated zone

Injection wells are a quick way of contaminating the aquifer, since the contaminant is fed directly into it, with no type of previous treatment. If the contaminants are fed in under pressure, this causes their dispersion within the radius of the well, both upstream and downstream.

### From other water sources connected to the aquifer

This happens basically when there is a hydraulic connection between a contaminated river and an aquifer and this drains to the river.

## Recommended reading and references

Fetter, C.W. (2001). Applied hydrogeology. 4th ed. Prentice Hall International. London.

Freeze, R.A. and Cherry, J.A. (1979). Groundwater. Prentice Hall. U.S.A.

McDonald, M.G. and Harbaugh, A.W. (1988). A modular three-dimensional finite difference ground-water flow model: U.S. Geological Survey Techniques of Water Resources Investigations, Book 6, Chapter A1.

Rushton, K.R. (2003). Groundwater hydrology: conceptual and computational models. Wiley.

Schwartz, F.W. and Zang, H. (2003). Fundamentals of ground water. John Wiley & Sons.

Todd, D.K. and Mays, L.W. (2005). Groundwater hydrology. Wiley.

# PART II
# METHODS

# 5

# SITE INVESTIGATION

1. Planning and design
2. Preliminary investigations
3. Engineering geophysics
4. Boreholes, trial pits, trenches and sampling
5. *In situ* tests
6. Geotechnical instrumentation

## 5.1 Planning and design

### Aims and importance

Site investigation is essential for any engineering project where the ground will be excavated and used as the base for foundations or as a source of materials. Before construction starts, the site is investigated to define the geological and geotechnical models, and specify the parameters and properties of the ground in engineering geological terms (*Figure 5.1*).

The results of site investigations are crucial for accurate budget estimates for any construction project. A considerable proportion of the overall cost increase for public works, often over 50%, is the result of inadequate site investigations. This is estimated to lead to delays in at least a third of all projects (Tyrell *et al.*, 1983), mainly from inadequate planning and misinterpretation of data. There is also a tendency to dismiss the cost-effectiveness of investing in site investigations, which means they are generally under-financed. This all results in uncertainty, delays and extra costs prior to construction, and it is generally agreed that **sooner or later, the cost of site investigations have to be paid for.**

There are no fixed rules for estimating the appropriate cost of geotechnical investigations as every project has its own set of characteristics. These depend on the type of ground and the scale of the project, and also how the geological complexity influences the construction conditions. As a rough guide, for large projects investigation should be budgeted as 15–20% of the total project cost, and around 10% or less for smaller projects. If the project is considerably affected by geological complexity then this percentage may be much higher.

The **general aim of site investigation** is to identify and quantify ground conditions that may affect the viability, design or construction of an engineering structure. Once the scale and nature of the project is known, the **objectives of site investigation** should be to:

— Establish the viability of the site for the proposed works in terms of its geological, geotechnical and geoenvironmental conditions.
— Establish which sites or corridors offer the most favourable conditions.
— Identify any geological hazards and instability problems present in the area.
— Determine the geotechnical properties necessary for the structural design and construction and environmental acceptability.

Site investigations should be carried out at each stage of the project as follows:

— **Preliminary and desk-based studies:** conceptual definition of the project and viability studies based on information obtained.
— **Preliminary design:** site selection; study of possible solutions and approximate cost estimates.
— **Design:** detailed site description, design, cost estimates, completion dates and technical specifications.
— **Construction:** design; on-site control and geotechnical monitoring.
— **Operation:** monitoring and control of ground-structure behaviour.

The separate scheduling of these stages means that site investigation can be programmed in a work sequence where data collection is cumulative, both in intensity and detail, and each stage complements the preceding one (*Tables 5.1* and *5.2*).

### Planning site investigations

In view of how important site investigations are in technical and financial terms, it is essential that they are adequately planned and matched to the project specifications. Each project requires its own particular site investigation planning and so there are no general rules. For planning purposes the following factors should be considered:

— Information available from previous studies.
— Aims of the project; structural loads acting on the ground.
— Regional and local geological conditions.
— Site access and physiographical characteristics.
— Suitable site investigation methods.
— Cost estimates and completion deadlines.

Site investigation surveys must be planned and supervised by engineering geology or geotechnical experts and

# SITE INVESTIGATION

Structural field data

Measuring deformations in saline rocks

**1. To define the geological conditions of the site**
- Type and description of geological materials.
- Orientation and description of discontinuities.
- Rock and/or soil mass characterisation.

**3. To quantify the engineering geological data and geomechanical properties of the materials on the site**
- Geomechanical characterization and rock and/or soil mass classification.
- Strength and deformation properties of soils, intact rock and rock mass.

**2. To identify geological problems that may affect design and construction**
- Faults and fracture zones, particular geological structures.
- Seepage zones.
- Soft and expansive materials.
- Aggressive materials.
- Abrasive rocks and problematic soils.

**4. To provide engineering geological criteria for the design and construction**

Inclined rotary core drilling

Slope stabilization with an anchored wall

*Figure 5.1*     Aims of site investigations.

should be based on data obtained from available information, aerial photo interpretation and field surveys. They include the following:

— Geological description of the area.
— Main lithological groups.
— Geomorphological and hydrogeological conditions.
— Identification of possible engineering geological problems and geohazards.
— Main engineering geological properties and data to be obtained.
— Environmental conditions for fieldwork.

*Table 5.1* **AIMS OF SITE INVESTIGATION AND RELATED ENGINEERING GEOLOGICAL STUDIES**

| Project stage | Objectives |
|---|---|
| Preliminary studies | — Identification of geological hazards.<br>— Preliminary assessment of geological-geotechnical conditions.<br>— Identification of geological factors that may affect project viability. |
| Preliminary design | — Site selection.<br>— Geomechanical classification of materials.<br>— Preliminary geotechnical solutions. |
| Design | — Detailed geomechanical characterization.<br>— Geomechanical parameters for design of excavations, foundations, etc. |
| Construction | — Monitoring and ground control.<br>— On-site design and remedial measures. |
| Operation | — Ground-structure response control.<br>— Monitoring. |

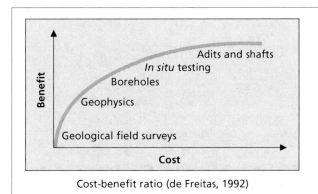

| Activity | Cost (C) | Benefit (B) | B/C |
|---|---|---|---|
| Literature review | Low | Very high | 2.7 |
| Field surveys | Low to medium | Very high | 2.7 to 1.6 |
| Laboratory tests | Low to medium | High to low | 2.3 to 0.6 |
| Preliminary site investigations | Medium to high | High to low | 1.4 to 0.4 |
| Site investigations for project design | High | High | 1.0 |

Cost-benefit ratio (de Freitas, 1992) — Cost-benefit indices (Fookes, 1997)

*Figure 5.2*  Cost-benefit ratio of site investigations.

Depending on the above data, different techniques or methods will be selected, taking the following criteria into account:

— Decision on and scope of methods for obtaining geomechanical parameters required and their limitations.
— Schedule of work and completion dates.
— Cost-benefit ratio of the investigation techniques used.

*Figure 5.2* shows the ratio between the result, or benefit, of the method and its cost. Surface geological techniques and field surveys show the highest ratios. Geological field work is both highly cost-effective and also essential for determining which investigation methods to use and their characteristics. *Table 5.3* shows the influence of geological factors on site investigation planning.

Other aspects to be taken into account are the location, depth and number of surveys. These must be carried out at representative geological and geotechnical points; the number and depth of the surveys will also depend on the project aims, the scope of the methods used, and the size of the area affected by the engineering activity. Methods that complement each other are used to cross-correlate and so help define the resulting engineering geological properties. These aspects are dealt in Part III of this book.

Investigation programmes should strike a balance between technical objectives and time/cost related estimates. It is therefore essential that they take logistical aspects into account, such as site access and climatological conditions.

Site investigation design can be summed up as an art requiring sound geological or geotechnical engineering criteria and appropriate experience to be able to combine the following successfully:

— Adequate planning
— Appropriate investigation methods
— Reliable and representative data
— Relevant results
— Clearly written reports

## Table 5.2  SITE INVESTIGATION PROCEDURES

| Project stages | Main activities | Site investigations |
|---|---|---|
| Preliminary studies | Literature review and desk-based study | — Topography, relief and land use.<br>— Hydrology and hydrogeology.<br>— Regional geological maps.<br>— Geological history.<br>— Seismicity and other geological hazards. |
| | Aerial and remote sensing interpretation | — Aerial photographs and remote sensing.<br>— Geomorphology.<br>— Lithological and structural characterisation.<br>— Geo-hazards.<br>— Geological mapping. |
| | Walk-over survey and preliminary geological reconnaissance | — Identification of soils and rocks.<br>— Faults and structures.<br>— Hydrogeological data and drainage.<br>— Geomorphology, slope stability, subsidence, collapses, flooding, land use, etc.<br>— Geo-environmental problems.<br>— Accesses and sites for borehole drilling, trial excavations and geophysical surveys. |
| Preliminary design | Engineering geological mapping (scales 1:5,000–1:10,000) | — Lithostratigraphy and structure.<br>— Geomorphology and Hydrogeology.<br>— Classification and properties of materials. |
| | Hydrological and hydrogeological data | — Identification of karstic areas and areas susceptible to flooding and runoff.<br>— Regional and local hydrological data. |
| | Basic site investigations [1] | — Boreholes and trial excavations.<br>— Geophysical prospecting.<br>— Laboratory tests. |
| Design | Detailed site investigations [2] | — Boreholes<br>— *In situ* and laboratory testing. |
| | Detailed geotechnical mapping (scales 1:500–1:2,000) | — Detailed geological-geotechnical mapping.<br>— Geomechanical properties, classification and zoning. |
| Construction | Geotechnical validation | — Detailed geotechnical mapping.<br>— Stability of excavations and tunnels.<br>— Ground control and improvements.<br>— Foundations and ground reinforcement work. |
| | Ground structure monitoring and control | — Monitoring equipment installation and instrumental readings.<br>— *In situ* testing.<br>— Quality control. |
| Operation | Monitoring | — Ground-structure behaviour monitoring. |

[1] Basic investigations refer mainly to exploratory investigations and identification tests.
[2] Detailed investigation involves intrusive investigations for each structure and for the whole area affected by the project, followed by *in situ* and laboratory testing.

### Table 5.3 INFLUENCE OF GEOLOGICAL AND GEOMORPHOLOGICAL CONDITIONS ON SITE INVESTIGATION PLANNING

| Geological and geomorphological factors | Typical features | Factors to be considered |
|---|---|---|
| Sedimentary and metamorphic rocks of sedimentary origin | — Relatively uniform formations over large areas<br>— Well-defined and stratified structures<br>— Rocks of marine origin more uniform and continuous than those of continental origin. | — Greater reliability in extrapolation of geological data.<br>— Number of site investigation points usually lower than those for metamorphic igneous rocks.<br>— Use of borehole drilling is highly effective. |
| Extrusive igneous rocks | — Stratiform structures<br>— High heterogeneity and lithological anisotropy. | — Great number of boreholes is usually required.<br>— Limitations in the use of geophysical methods. |
| Intrusive igneous rocks | — Non-homogeneous lithological and geometric conditions. | — Difficult to identify geometric boundaries of intruded bodies.<br>— Highly effective use of geophysics. |
| Tectonic structures | — Great continuity.<br>— Soft and breccia materials filling.<br>— Anisotropic conditions at both sides of the structure | — Structural and geological mapping are fundamental.<br>— Use of boreholes and geophysics essential. |
| Low relief | — Predominance of alluvial deposits, heterogeneous soils and weathered rocks.<br>— Poor drainage. | — Relatively low drilling costs.<br>— Highly effective use of geophysics and penetrometers. |
| Moderate to high relief | — Lithological and structural relief control.<br>— Rocks, colluvial and alluvial deposits. | — High drilling costs.<br>— Accesses can be very difficult. |
| High relief and mountain regions | — Hard rocks.<br>— Lithological and structural relief control.<br>— Colluvial deposits and slope instabilities. | — Adverse conditions for site investigations.<br>— Difficulty of access.<br>— High drilling costs.<br>— Weather limitations.<br>— Extensive use of photointerpretation and geological mapping. |

## 5.2 Preliminary investigations

The aim of preliminary investigations is to acquire a general engineering geological understanding of the project area or site. Much of this first phase is desk-based, and includes examining available information, results of aerial photo interpretation and walk-over surveys. From these preliminary studies, site investigation can be planned and geological factors assessed, including any geohazards and environmental restrictions which may affect the viability of the project.

## Desk-based study

Before any field work is carried out, all available information relating to the project and the construction site should be examined. This includes a literature review, publications and reports, both on the project itself and its site, together with

| Table 5.4 | SOURCES OF INFORMATION FOR DESK STUDIES |
|---|---|
| **Subject** | **Information** |
| Topography | — Topographical maps.<br>— Aerial photographs. |
| Photointerpretation and remote sensing | — Aerial photographs.<br>— Satellite images. |
| Geology | — Geological maps.<br>— Geological surveys and reports.<br>— Aerial photographs.<br>— Soil survey and edaphological maps. |
| Geotechnics | — Geotechnical publications.<br>— Geotechnical reports.<br>— Geotechnical maps. |
| Hydrogeology and hydrology | — Hydrogeological maps.<br>— Topographical maps.<br>— Aerial photographs.<br>— Well and borehole data.<br>— Hydrogeological reports.<br>— Flood risk maps. |
| Meteorological data | — Rainfall and temperature records. |
| Seismological data | — Earthquakes and seismic codes. |
| Mining and quarrying | — Industrial rocks information.<br>— Mining and quarrying records.<br>— Inventories and maps. |
| Land use | — Urban and land use planning. |
| Environmental and natural resources | — Environmental maps and environmental impact studies. |
| Pre-existing constructions and services | — Industrial and building records. |

geological maps and aerial photos. The analysis of documents and reports on various other projects carried out in the area, such as roads, tunnels, hydraulic works and building foundations is also important. Possible information sources are summarized in *Table 5.4*.

# Aerial photo and remote sensing interpretation

Aerial photo interpretation is one of the most widely used working methods in geology. As well as its other applications, it is used as a basic tool in geological mapping, lithological surveys and geomorphological studies. Aerial photo interpretation is also extremely useful in geological engineering, especially during the preliminary phases of a project.

Since the 1970s, remote sensing has been based on the development of satellite based multispectral sensors. This technique has evolved considerably and now provides many applications for geological investigation and engineering.

The following sections summarize the most commonly used applications of both methods.

## Aerial photo interpretation

Aerial photo interpretation is an essential technique for carrying out preliminary studies and geological-geotechnical field surveys, with the following advantages:

— Large areas can be surveyed, so the characteristics of both the project site and the surrounding area can be easily analysed.
— As it is extensive, regional factors which may be relevant to the project can be taken into account.
— It is quicker and cheaper to use than other survey methods.

Types of aerial photographs include:

— Colour photos. These are available for most projects and give the most realistic view of the ground. They are very useful for identifying materials and structural features.
— Black and white photos. These have higher definition than colour photos and so different types of vegetation, wet areas and water are clearly defined. They can be used to identify geological structures and to analyse drainage networks and geomorphological data.
— Infra-red or false-colour photos. These produce distortions in tonality that make it easier to identify vegetation (which appears red), water, wetlands and drainage networks. They are very useful for identifying landslides, anomalies in the drainage network and karstic zones.

The scale of the photo is an important factor to consider. In preliminary studies the most useful scales for geological engineering are those ranging from 1/30,000 to 1/5,000. Scales used as back-up for field work range from 1/10,000 to 1/5,000.

Before analysing an aerial photograph or arranging a flight over a particular area, the prevailing conditions should be considered, such as the time of year, time of day (which both affect light quality), length of shadows and cloud cover.

In geological engineering the most characteristic applications of aerial photo interpretation include:

— Selecting and comparing possible sites and corridors.
— Planning site investigations.
— Identifying construction materials.
— Identifying hydrological features and drainage networks.
— Identifying landslides, karst and mining subsidence.
— Identifying different types of lithology, their contacts, structural and geomorphological characteristics; the presence of alluvial materials and other recent deposits.
— Locating man-made landfills and access routes, and land use in the area.

*Figures 5.3* and *5.4* show two examples of aerial photographs used to identify landslides and karstic structures.

## Remote sensing

Remote sensing uses a series of techniques to collect information without being in physical contact with the object of study. The interaction of electromagnetic energy with the surface of the earth is recorded by sensors installed on airborne or spaceborne mobile platforms, such as aircraft or satellites.

*Figure 5.3*  Aerial photograph of Benamejí landslide, S. Spain; original scale 1:5,000.

*Figure 5.4*  Railway line affected by landslides and large karstic open fractures; original scale 1:4,000.

The **electromagnetic (EM) spectrum** is divided into several spectral regions. Electromagnetic energy can be absorbed by atmospheric gases in specific regions of the spectrum, producing areas considered as "opaque" which cannot be used in remote sensing for studying the ground surface. EM regions that are less influenced by atmospheric absorption are called atmospheric transmission windows. These are the visible, near thermal infrared and microwave regions used in remote sensing.

All objects emit and/or reflect electromagnetic radiation as a result of their interaction with their own or external sources of energy. The absorption, emission, diffraction and reflection of EM energy by an object depend on the wavelength of the incoming EM radiation and on its material composition. The spectral reflectance signature or spectral response curve of a material depends on the intensity of the EM radiation throughout a range of wavelengths. The characteristics of this signature depend primarily on the object's atomic and molecular structure, surface characteristics and temperature. How much radiation is reflected, absorbed or transmitted by an object depends on its surface characteristics and varies in different EM regions.

The earth's surface is mostly soil, rock, water and vegetation. High spectral resolution remote sensors can identify each material as they absorb or reflect energy in unique ways throughout the different EM wavelengths. Knowing the spectral properties of different ground targets is crucial for selecting the most suitable EM regions to study them. The optimum discrimination level between two materials is shown when there are significant differences in their reflectivity and/or emissivity in a specific EM region. In general, the spectral features of a material are shown as absorption bands that can be used to identify the material.

The spectral signature of **soils** depends on the spectral response of their mineralogical and organic components, their grain size, surface texture, moisture content and atmospheric humidity. The reflectivity of dry soils characteristically increases with wavelength, especially in the visible and near infrared regions. The higher the moisture content of the soil, the lower the reflectivity value. Soil reflectivity is also conditioned by chemical composition. The characteristic spectral signature of certain **minerals** in the reflected infrared region means that iron minerals, clays, carbonates, phosphates and silicates can be distinguished, i.e., that the lithologies can be differentiated by their spectral features.

**Water** masses totally absorb incident radiation in the short and near infrared wavelengths. Variations in the spectral curve of water are detected in the shortest wavelengths of the visible region (blue and green) and are basically related to the depth of the water mass, the content of materials in suspension (sediments and chlorophyll) and the roughness of the water surface.

The analysis of the spectral characteristics of **vegetation** for geological applications is interesting because particular variations in the typical vegetation reflectivity curve may indicate anomalies in the plant phenology, which could indicate changes in the chemical composition of soil/rock at depth. The correlation of plant chemistry with geology is known as geobotany and biogeochemistry.

## Remote sensing systems

Remote Sensing uses EM energy from natural and man-made sources. **Passive systems** are based on registering the natural EM radiation reflected or transmitted by objects. The energy source is the sun or the earth and the sensor registers the signal reaching it. **Active systems** produce their own energy that is transmitted in a specific direction, and record the part reflected back by the targets. The most common active system is **radar**.

Radiation is registered by sensors that capture the reflected or transmitted EM radiation from the ground surface at a particular wavelength interval. The most common sensors are digital cameras, scanning radiometers and radar or lidar systems.

Most remote sensing systems operate using **multispectral scanners (MSS)** that record EM radiation in the visible and infrared regions. The information is recorded in discrete electromagnetic bands by scanning the ground in consecutive lines along or perpendicular to the flight direction (along-track scanning or across-track scanning). Radiation from each ground element (pixel) is recorded simultaneously in a specific range of wavelengths by a series of internal detectors that transform it into an electric signal that is then converted into digital data. This digital value can then be stored or transmitted to earth. These systems are used on aircraft and natural resource satellite platforms such as Landsat, SPOT, IRS, MODIS.

**Hyperspectral sensors** are multispectral scanners that record ground surface radiation in hundreds of very narrow contiguous spectral bands throughout the visible, near-infrared, and mid-infrared ranges of the EM spectrum. The very high spectral resolution allows the spectral signature of the observed material to be generated and targets to be discriminated based on their spectral response. These sensors offer a complete spectral capacity for discriminating diagnostic spectral features of ground surface materials, with important implications for carrying out direct lithological identification of rocks and soils (*Figure 5.5*).

Portable **field spectrometers** register radiation from objects within a specific spectral range. In contrast to other systems, their field of vision is fixed, so the device has to be repositioned to measure a different object. They can be used for calibrating multispectral data recorded from aircraft or satellites.

**Radar systems** transmit a microwave signal towards the target and register the backscatter from the signal. The strength of the backscatter is measured to discriminate between different targets and the time delay between the transmitted and reflected signals determines the distance to the target (or **range**). Because microwaves can penetrate clouds, haze and dust, these systems can generate ground

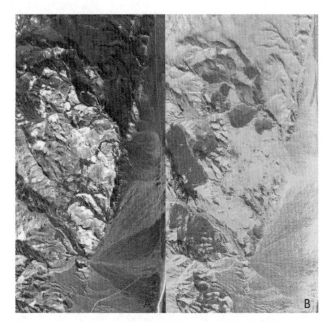

*Figure 5.5* Images recorded by the Modis/Aster sensor over Death Valley National Park (USA). A) Combination of visible bands. B) Combination of thermal infra-red bands. Note how the capacity for discriminating different lithologies depends on the spectral region used (NASA data processed by AIG; in Photogrammetric Engineering & Remote Sensing, 2000).

surface images in any meteorological conditions, except heavy rain clouds. They are particularly useful for studying areas with adverse atmospheric conditions, such as the tropics and polar regions. Synthetic aperture radar (SAR) is the system most used in airborne and spaceborne platforms.

Land observation satellites are usually near-polar and sun-synchronous, meaning the angle of solar illumination remains constant for a given latitude. The first satellite designed specifically to monitor the Earth's surface, Landsat-1, was launched by NASA in 1972. **LANDSAT** satellites have provided a continuous record of the ground surface up to the present. They register images in seven spectral bands with 30 m of spatial resolution in the visible and infra-red EM regions, except the thermal infrared band which is 120 m. They supply an enormous collection of data for research using long term monitoring and detection of change.

The **SPOT** satellite, dating from 1986, has two operational modes: a high spatial resolution single-channel panchromatic mode or a coarser spatial resolution three-channel multispectral mode. Its oblique viewing capability provides imagery for stereoscopic coverage which can be analysed as a three dimensional model.

The launch of NASA's Terra satellite in December 1999 included advanced **ASTER** sensors in the visible, near- and thermal infra-red regions.

**MODIS** provides almost daily morning (EOS-Terra) and afternoon (EOS-Aqua) views.

Hyperspectral sensors, until recently only airborne, are now also functional in satellites and can generate much more information than the multispectral scanners on LANDSAT or SPOT. **EO-1**, launched in 2000, had a standard 9 band sensor (ALI) and a 220 band hyperspectral sensor (Hyperion). ESA launched in 2001 the **Proba** satellite which included CHRIS (Compact High Resolution Imaging Spectrometer) with 200 narrow bands.

The **SEASAT**, launched in 1978, was the first satellite to carry a spaceborne SAR sensor. Other satellites with radar-type sensors are the **ERS-1** and **ERS-2** (the first European natural resources satellite), **JERS**, **RADARSAT** and **ENVISAT**.

## Applications in engineering geology

The first contribution by remote sensing to geological applications was in the field of structural geology. Satellite images have introduced a new dimension into regional structural studies by providing synoptic views of large areas in identical sunlight conditions. The mosaics obtained from this type of imagery give exceptional overviews of the structural layout of continental areas and allow detection of extensive linear features, which frequently correspond to the surface expression of a set of fractures. The traces of these structures may show volcanic alignments, intrusions, dykes, fractures, topographical crests or depressions and watersheds. They are very useful for studying seismic hazards and fault activity. To improve image interpretation the selection of image spectral bands and the image registration date must be taken into account. Images recorded in winter are generally more suitable because the low angle of solar illumination enhances the topographical features.

Near infra-red bands provide better information for distinguishing lithologies of rocks and soils. The spectral characteristics of certain minerals allow the identification of iron oxides and hydroxides, and clays formed from weathering of the underlying rock, which may indicate the presence of potentially altered areas. The detection of jarosite and other alteration minerals is critical for mapping heavy metal contamination in mining areas (*Figure 5.6*). The greatly improved information from hyperspectral data acquired both from airborne and spaceborne sensors has generated renewed interest in using such data for mineral identification and other applications. Hyperspectral imaging means that surface mineralogy can be mapped directly, and even that the relative importance of different minerals present within the observed area can be determined. An important application has been developed to map smectites responsible for expansive soils. Spectroscopic identification shows high correlations with mineralogical x-ray diffraction analyses and geotechnical engineering tests.

# SITE INVESTIGATION

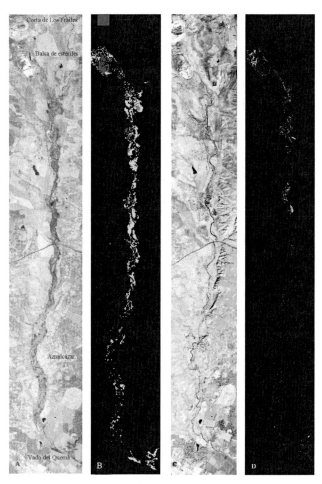

Figure 5.6   Mapping of mud spill at the Aznalcóllar mine on the Guadiamar river, S. Spain, before and after cleaning operations. A) Image taken July 1998. B) Digital classification (in yellow: remaining pyritic mud; in red: salt crusts). C) Image taken June 1999. D) Digital classification (in yellow: remaining pyritic mud). Daedalus ATM images. (Courtesy of C. Antón-Pacheco).

One of the principal characteristics of satellite images is the return interval of observations for areas covered. This multi-temporal capacity makes them extremely useful in studies of dynamic processes such as sediment deposition in delta areas. Bands registered in the visible region provide information on the turbidity of coastal waters or on river-transported materials in sea water (Figure 5.7). They are also useful for controlling and monitoring certain climatic phenomena, such as droughts and flooding. The photograph in Figure 5.8 shows the area affected by flooding of the Fitzroy River in Australia. Thermal infra-red images are critical in regions with recent volcanic activity as they are able to detect temperature differences and thermal anomalies in the observed areas.

Spatial resolution has improved considerably with the latest generation of satellites. Images with 10, 5 and 0.5 m

Figure 5.7   Landsat image recorded in February, 2000, showing the plumes of the Guadalquivir and Tinto rivers, S. Spain.

resolution are systematically recorded, enabling detailed mapping to be carried out in scales up to 1:10,000. MODIS images supply information that is particularly relevant to global climate change research, the movement of glaciers and polar ice caps, changes in coastlines, changes in the vegetation cover and volcanic eruptions. The use of geographic information systems (GIS) greatly facilitates the overall handling of multi-temporal and multi-sensor data.

Radar sensors can be transported on satellite, airborne or terrestrial platforms. These sensors acquire synthetic aperture radar (SAR) images either of large areas, up to 100 km × 100 km in the case of spaceborne radars, or small areas for ground-based radar. The latter is used to monitor specific sites where instabilities such as landslides or subsidence are expected, covering a monitoring surface of few square kilometres (see Chapter 13, Figure 13.29). Advanced differential interferometry techniques (A-DInSAR), and Persistent Scatterer Interferometry (PSI), are techniques used for SAR image analysis providing displacement and subsidence maps over the study area and estimating the temporal evolution of the displacement for every detected ground target with millimetric precision (see Chapter 13, Figure 13.42).

A combination of terrestrial airborne LIDAR and GPS is an innovative monitoring tool that can produce extremely accurate ground elevation models and can even measure ground elevation through a cover of trees.

## The walk-over survey

The walk-over survey is one of the most important tasks carried out at the preliminary stage. It should be implemented

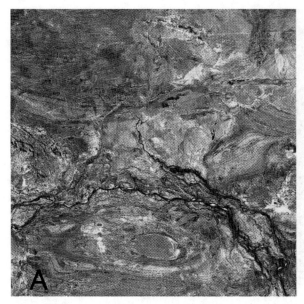 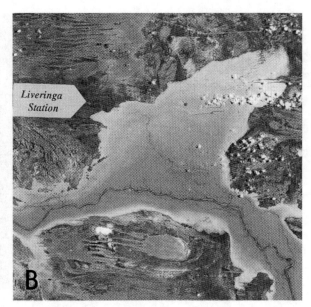

*Figure 5.8* A) Landsat image of the Fitzroy river basin in 1989. B) Image taken in 1993 after flooding. The area shown covers 50 x 50 km. (Australian Centre for Remote Sensing; in Landsat Data Users Notes, 1993).

after all relevant available information has been collected, including aerial photo interpretation and a geological synthesis and basic topography. This information allows the planning and design of the site investigations required at the next stage.

The following field studies should be carried out:

## Geological data

— Types of materials, lithology and composition, lithological contacts, bedding and sedimentary structures.
— Geological structure and tectonic contacts, degree of fracturing, systematic and isolated discontinuities (joints, foliation, stratification), fault zones, areas with tectonic and neotectonic activity and definition of the basic sequence of geological events.
— Surface deposits and materials affected by weathering.
— Geomorphological conditions, morphological evolution and ground surface processes.

## Geotechnical description of soils

The geotechnical field description of soils should be based on the following sequence:

— The soil composition is determined using the Unified Soil Classification System, differentiating soils according to particle size distribution (Chapter 2, Section 2.2).
— The colour corresponding to that observed on site may indicate important properties; e.g. a yellowish-red colour indicates intense weathering and the presence of iron oxides; a dark brownish-green colour and black indicate the presence of organic material.
— Soil structure is described as: homogeneous when it includes soil with uniform characteristics; stratified when there are different soil layers; banded if residual layers of soil are present; laminated when the soil layers are less than 3 mm thick.
— The density of granular soils and the consistency of cohesive soils can easily be determined in the field with simple tests (*Tables 5.5* and *5.6*).
— Note that soil colour and structure are sensitive to the environment under which they are observed and so it is relevant to record the age of the exposure or sample described (hours, days, weeks) and the moisture conditions under which it was exposed (open to the atmosphere, sealed bag etc).

A soil could be described as e.g.: a fine sandy clay, light grey in colour, with low compressibility, firm and homogeneous when seen fresh at natural moisture content.

## Geomechanical description of rock masses

This includes three areas: characterization of the rock matrix or intact rock, description of discontinuities and characterization of the properties of the rock mass. The data-collection processes and sequence are described in Chapter 6.

## Hydrogeological and hydrological data

— Water tables, inspection of wells and springs.
— Location of aquifers, permeable and impermeable materials, areas of swampland, etc.
— Location of recharge and drainage areas.

| Table 5.5 | DENSITY AND FIELD DESCRIPTION OF GRANULAR SOILS | |
|---|---|---|
| Density | Relative density (%) | Field assessment |
| Loose | <35 | Easily penetrated by hand-held 12.5 mm diameter steel bar. |
| Medium dense | 35–65 | Easily penetrated by a steel bar with a hammer of 2–3 kg weight. |
| Dense | 66–85 | Easily penetrated to a depth of 30 cm by a steel bar and hammer. |
| Very dense | >85 | Only penetrated a few centimetres by a steel bar and hammer. |

| Table 5.6 | CONSISTENCY AND FIELD DESCRIPTION OF COHESIVE SOILS IN THE FIELD AND GUIDE TO STRENGTH | |
|---|---|---|
| Consistency | Uniaxial compressive strength (MPa) | Field identity test |
| Very soft | 0–0.025 | Soil squeezes through the fingers when the hand is closed. |
| Soft | 0.025–0.050 | Easily moulded with the fingers. |
| Firm | 0.05–0.1 | Moulded with strong pressure from the fingers. |
| Stiff | 0.1–0.15 | Dented with strong pressure from the fingers. |
| Very stiff | 0.15–0.2 | Slightly dented with strong pressure from the fingers. |
| Hard | >0.2 | Slightly dented with a pencil point. |

### Ground instabilities

— Signs of landslides or rock falls.
— Areas of intense erosion.
— Areas affected by subsidence, sinking processes and cavities.

### Carrying out site investigations

— Establishing land ownership, permission for access and presence of underground utilities (cables, pipes, etc.).
— Location of paths and access routes for carrying out site investigations, especially boreholes.
— Availability of water and electricity, and authorization for use.
— Selection of possible sites for boreholes, trial excavations, geophysical surveys and field tests.
— Assessment of risks to personnel, others and the environment.

### Observation of structural damage

Inspection of buildings, bridges, tunnels, embankments, walls and other structures in the vicinity which show some sort of structural damage. Attention must be paid to the appearance of cracks and other signs of deformation, such as leaning walls and much can be learned from mapping these brittle fractures, their relative ages and the direction in which movement has occurred across them.

## Preliminary site investigation report

The following reports should be presented:

— Engineering geological report based on the preliminary studies.
— Potential problems and engineering geological-geotechnical factors that may affect the project objectives.
— Proposal for detailed site investigations.

## 5.3 Engineering geophysics

Geophysical prospecting covers a series of techniques for investigating the earth's interior based on variations detected in its physical parameters and their correlation with geological characteristics. Such techniques are non-destructive and are used in investigations covering large areas to complement on-site tests and direct investigation techniques, such as borehole drilling and trial pits. Their application in geological engineering requires specialists because familiarity with the geotechnical characteristics of the materials involved is required. These techniques are used to determine the following: the thickness of fills or overburden; the excavability of materials; the position of the water table; the location of cavities or other anomalies in the ground; calculation of the volume of borrow pits; structure of the substratum; the geomechanical properties of materials; location of faults or landslide surfaces; the thickness of weathered rock; rates of fracturing and location of underground structures (sewers, piles, galleries, etc.).

The different geophysical methods of surveying the substratum are grouped according to the physical parameters under investigation. These may be: electric (resistivity), seismic (velocity of seismic wave propagation), electromagnetic (electrical conductivity and magnetic permeability), gravity (density), magnetic (magnetic susceptibility) and radioactive (natural or induced radiation levels). A distinction is made in the field between surface and borehole techniques and they are usually described separately, although they are based on the same theory.

## Surface geophysics

### Electrical methods

These methods are frequently used in geological engineering to study the response of the ground when continuous electric currents (DC) are passed through it. The physical parameter tested is resistivity ($\rho$), and a final interpretation is made based on the geological characteristics of the test area. **Resistivity** is an intrinsic property of rocks and soils that depends on lithology, microstructure and, above all, water content; it is therefore not an isotropic property of the ground but a function of the direction in which it is measured. Table 5.7 shows some resistivity values in soils and rocks.

Archie's formula expresses the relation between rock resistivity, $\rho$, water contained in the pores, $\rho_w$, and porosity, $\varphi$:

$$\rho = a\varphi^{-m}S^{-n}\rho_w$$

where $S$ is saturation and the terms $a$, $m$, and $n$ are experimental coefficients. The formula above is often used with average values:

$$\rho = (\varphi S)^{-2}\rho_w$$

Resistivity in the substratum is measured in the following steps (Figure 5.9):

— A continuous current of intensity I is passed through the ground by two electrodes, A and B, connected to an energy source.
— The difference in potential ΔV generated by passing the current is measured between two electrodes, M and N.
— The resistivity of the depth of the ground affected by the current is measured.

The resulting resistivity does not correspond to a specific lithological unit but defines the materials affected by the current as a whole. This is known as **apparent resistivity** ($\rho_a$):

Table 5.7 RESISTIVITY VALUES OF COMMON GEOLOGICAL MATERIALS

| Materials | Resistivity $\rho$ ($\Omega$ m) |
|---|---|
| Marls | 50–5,000 |
| Limestones | 300–10,000 |
| Shales | 100–1,000 |
| Granites | 300–10,000 |
| Clays | 1–20 |
| Sands | 50–500 |
| Conglomerates | 1,000–10,000 |
| Sandstones | 50–5,000 |
| Alluvium | 50–800 |

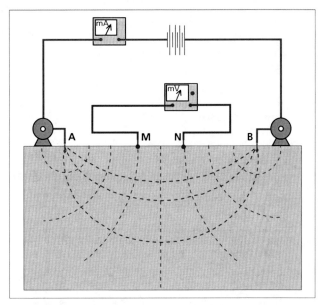

Figure 5.9   Measuring ground resistivities using electrical methods.

$$\rho_a = K\frac{\Delta V}{I}$$

where $K$ is the constant in the geometric array of the device at each measurement, depending on the distances between the electrodes AM, MB, AN and NB.

$$K = \frac{2\pi}{\left(\dfrac{1}{AM} - \dfrac{1}{MB}\right) - \left(\dfrac{1}{AN} - \dfrac{1}{NB}\right)}$$

Modification of the electrode array provides numerous possibilities for investigation. The most frequently used are normalized arrays, the most common being the Schlumberger,

Wenner and dipole-dipole methods (Figure 5.10). The first is symmetrical, with the potential MN electrodes inside aligned to the AB current, with a separation between M and N that is less than 1/5 of that of A-B. In the second, the array is the same, except that the distances A-M, M-N, and N-B are equal. In the third, the potential dipole (MN) is situated laterally to the current dipole (AB). The equipment used is similar for different arrays and consists of a box containing batteries, current and potential electrodes, connection cables and resistivity-meters (Figure 5.11).

The most common techniques in investigation are vertical electric sounding (VES), for testing the distribution of resistivities at depth, and 2-D resistivity pseudo-sections, for testing lateral resistivity by means of electrical tomography.

## Vertical Electric Sounding (VES)

This technique is carried out by progressively separating the electrodes A and B, with respect to the central point, passing a current through each position and measuring the difference in potential generated between the potential electrodes M and N. In each case, the calculated resistivity is that of the whole of the material affected by the passage of the current and is known as apparent resistivity $\rho_a$. A greater distance between electrodes A and B implies a greater thickness of material affected by the current. Therefore, representation of $\rho_a = f(AB/2)$ will give the variation in apparent resistivity with depth (Figure 5.12a). The VES curve is interpreted by applying inversion procedures to give a geological model made up of a series of layers, with real resistivity $\rho$ and thickness $e$. This will give, within a predetermined margin of error, an apparent resistivity curve similar to the one obtained from the measuring device (Figure 5.12b).

## 2-D Electrical imaging surveys: pseudo-sections and tomography

With this technique, the entire set-up (A, B, M and N) is moved laterally. A current is applied in each position and the difference in potential generated is measured. This is carried out using any of the electrodic arrays already referred to: Schlumberger, Wenner, dipole-dipole, or one of the many variants. The resistivity obtained in each case is apparent resistivity $\rho_a$, and variations observed are lateral because of the type of displacement (Figure 5.13).

### Pseudosections

When sounding at different depths and levels are made simultaneously with any electrode device (Figure 5.14) the distribution of $\rho_a$ can be represented on a cross section called a pseudo-section (Figure 5.15).

### Electrical tomography

This is a model of the distribution of real resistivities in the subsoil obtained by an inversion process from the apparent resistivities detected in the pseudo-sections. The complexity of the inversion process depends on the methods used: e.g. finite differences, finite elements, the application of distorted grids to correct topography, or others.

## Seismic methods

Seismic methods are used to study the propagation of artificially-produced seismic waves through the ground and establish their relationship with the geological structure and lithology. Propagation velocity depends on elastic constants and the density of the medium. Different geological masses have different seismic wave transmission velocities and it is at the contacts, or separation surfaces, between them that waves are refracted, reflected or diffracted (Snell's law).

Seismic refraction is a basic geological engineering technique for studying energy that returns to the surface

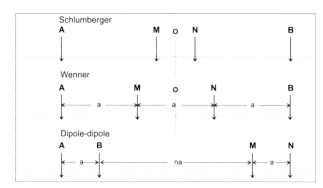

Figure 5.10   Schlumberger, Wenner and dipole-dipole electrode arrays.

Figure 5.11   Vertical electrical sounding (VES) using an Ambrogeo DataRes resistivity-meter.

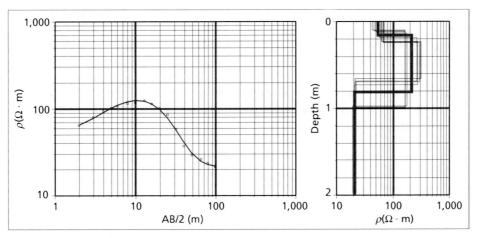

*Figure 5.12*  Example of vertical electrical sounding (VES).

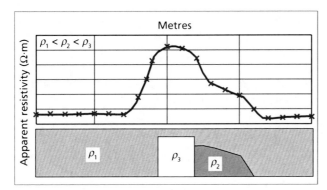

*Figure 5.13*  Example of a resistivity pseudo-section.

after it has undergone total refraction at surface boundaries in the subsoil (*Figure 5.17*).

The normal subsoil model gives velocities that increase with depth (soil–weathered rock–fresh rock); there are exceptions where velocity inversions occur (low velocity buried beds).

## Seismic refraction

The most commonly used seismic method is seismic refraction. Longitudinal sections equipped with an array of sensors (geophones) are repeated at regular, known intervals. Energy released by a "shot" (usually a blow with a 6–8 kg hammer) reaches the sensors, generating a seismic disturbance recorded on a seismograph. The sections are usually 25–100 m long, and the geophones are placed less than 5 m apart, to guarantee detailed results. There are usually at least three shot points, at the beginning, middle and end of each profile. For profiles more than 60 m long, five shot points would be used.

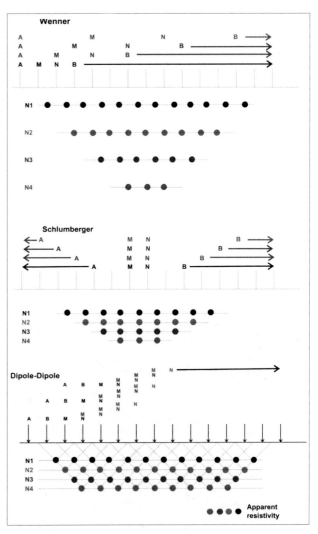

*Figure 5.14*  2D resistivity measurements (pseudo-sections) with different electrode arrays.

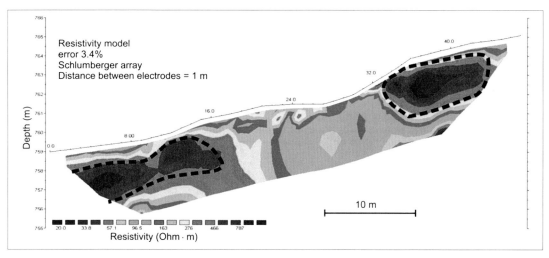

*Figure 5.15* Resistivity model resulting from the inversion of apparent resistivities obtained with a Schlumberger device; electrodes spaced at a distance of 1 m and 7 investigation levels.

*Figure 5.16* Field surveying of a 2-D resistivity section using 24 electrodes with a DMT Resecs system.

The time the elastic waves take to reach the geophones is measured to give the propagation velocity value and thickness of the different materials they pass through. *Figure 5.18* shows one type of seismograph, and *Figure 5.19* shows an example of a seismogram.

The travel time between the precise moment of the shot and the arrival of the first disturbance is measured for each geophone. The first waves to arrive are direct waves, but at a certain point (the critical distance), the refracted waves travelling through the lower levels of the substratum arrive first. The longer distance travelled by these waves is compensated by their greater velocity (*Figure 5.20*).

The travel time graph is the linear function relating the arrival time of the first wave with the distance it has travelled. Each refractor has a travel time graph and the slope and ordinate at origin of each arrival, are used to calculate the

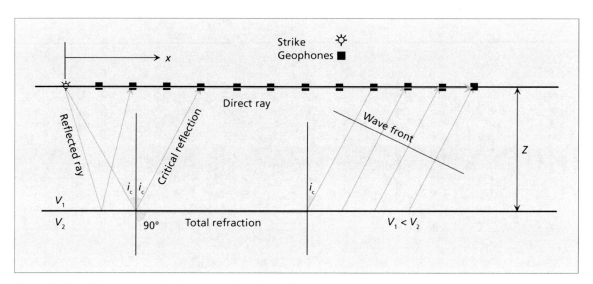

*Figure 5.17* 2-D seismic refraction surveying using a hammer as source.

velocity of the medium and the depth at which the refraction surface is found (*Figure 5.21*). The line passing through the origin to the arrival of the refracted rays corresponds to the arrival of direct waves.

Refractors are not usually flat and so the arrival times of the signal from the refractor need not be perfectly aligned. There are several methods for obtaining the depth and velocity below each geophone, based on deviations from the theoretical curve that are observed for arrival times at a geophone when the outward time is measured and return time is read (*Figure 5.22*).

The transmission velocity of seismic waves is a good indicator of the geotechnical characteristics of materials. Tables of velocities for different rock materials frequently appear in the literature, although there is a significant divergence of velocity values. This is due to the variability in lithology or internal structure, to the percentage of pores or cavities, and to fluid saturation (*Figure 5.23*). As the materials disintegrate and the degree of alteration increases, the velocity is reduced.

The **degree of weathering** of rocks is clearly a factor conditioning seismic wave propagation velocity; an unaltered

Figure 5.18  Seismic investigation in an urban setting with a 24 channel DAQ Link seismograph.

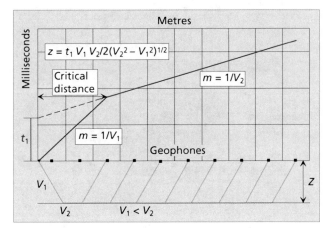

Figure 5.20  Examples of arrival times of P waves at different geophones.

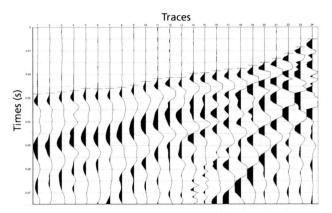

Figure 5.19  Seismic traces obtained with a 24-channel seismograph corresponding to a seismic refraction section (shot located at 2 m after geophone 24).

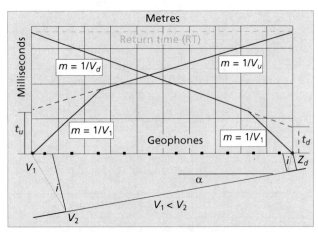

Figure 5.21  Interpretation of travel time graphs obtained in a seismic refraction profile.

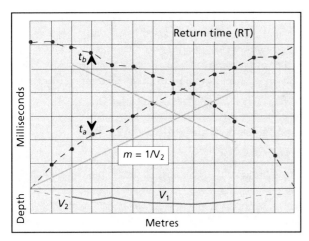

*Figure 5.22* Irregularities in the alignment of arrival times at different geophones.

| P wave velocity in km/s |
| 0 | 1 | 2 | 3 | 4 | 5 | 6 |
|---|---|---|---|---|---|---|
| Air |
| Water |
| Ice |
| Soil |
| Sands |
| Clays |
| Schists |
| Sandstones |
| Limestones |
| Dolomites |
| Salt |
| Gypsum |
| Anhydrite |
| Granite |
| Gneiss |
| Basalt |

*Figure 5.23* Transmission velocity of longitudinal seismic P waves in different materials.

rock such as a fresh granite may have a velocity of 5,000 m/s, but if it is highly altered this may drop to as low as 1,000 m/s or less.

Seismic refraction is used in geological engineering to determine the depth of made ground, superficial drift, substratum structure, material rippability and borrow pit volumes.

## Seismic reflection

The seismic reflection method has not been used much in geological engineering, although it is increasingly being used for defining deep geological structures such as in tunnelling projects, deep landslides, underground storage.

Seismic reflection involves generating seismic waves by applying an appropriate energy source (with a hammer, gun, weight drop or explosives) to an array of geophone sensors aligned on a cross section. Then the arrival times are calculated after they have been reflected from the interfaces of contrasting lithological layers, faults or discontinuity surfaces. The ray paths of the primary waves can be reconstructed from the arrival times of the longitudinal waves at the geophones and the different horizon velocities. This enables the structural arrangement of different seismic horizons in the section to be defined. How clearly these reflectors are observed depends on a reflection coefficient based on the amplitude of the incident wave and its reflection, the difference in density between the material above and below the reflector, and the ratio of the propagation velocities of P waves between both materials. Acoustic impedance $Z$ = density × P-wave velocity, and the greater the impedance contrast between materials at the discontinuity or lithological contact, the more clearly the reflector will be observed.

The generation/transmission of seismic waves is associated with other types of waves produced by phenomena such as surface conditions, random environment noise, multiple reflections and diffractions. These mask the results as they are recorded at the same time as primary waves. This effect can be partially reduced by correcting the signals during interpretation.

The advantage of seismic reflection over other geophysical techniques is that it allows multiple horizons to be represented graphically with a single shot without losing any significant accuracy with depth.

## Seismic investigations using surface waves

Surface waves are generated by the heterogeneity of the earth's surface. They are propagated in two ways: Rayleigh waves and Love waves. Rayleigh waves have a vertically polarized movement with an elliptical retrograde particle vibration. Love waves have a shear movement and are horizontally polarized perpendicular to propagation. Both types of wave are dispersive, i.e., they contain groups of waves of different wavelengths, each one travelling at a different velocity and with a different penetration depth (*Figure 5.24*). The long waves travel at a greater velocity than shorter waves and can penetrate further into the subsoil. Each wavelength is characterized by a particular phase velocity. The waves most commonly used are Rayleigh waves because the vertical component is the most frequently detected with conventional geophones.

The aim of these techniques is to obtain velocity values for Rayleigh waves ($V_{rayleigh}$) which have maximum velocities

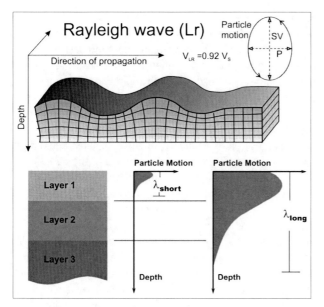

*Figure 5.24* Diagram of particle vibration and the propagation mode and dispersive nature of Rayleigh waves.

close to those of internal S waves ($V_{rayleigh} = 0.92\ V_s$). There two most common survey techniques are: refraction microtremor (ReMi, or passive seismic) and spectral analysis of surface waves (SASW).

### Refraction microtremor (ReMi) technique

This technique consists of carrying out 10 or more registers over a period of time (30 seconds, with a sampling interval equal to or less than 2 msecs.) along a line or or set-up with 24 or more geophones in an area where seismic noise is present. Once the recorded data are stored, transformation to the frequency domain is carried out using the p/f transform and the slowness-frequency spectrum is calculated (*Figure 5.25 A*). A tendency to increased velocity, with wavelengths characteristic of Rayleigh waves, can be distinguished in this spectrum. Stacking of 10 or more spectra allows a group of points to be defined (Vphase/frequency), that delineate a dispersion curve (*Figure 5.25B*). Finally, a vertical profile of $V_{rayleigh}$ below the central point of each station is calculated, using a 1-D modelling technique (*Figure 5.25C*).

The ground can be characterized by the $V_{rayleigh}$ distribution at depth (see the NEHRP-UBC classification, BSSC, 1998); if P wave velocity is available, dynamic modules can also be calculated.

With this method, the penetration depth depends on the type of geophones used and the separation between them, but it is generally much greater than depths reached with conventional seismic refraction (depths of more than 40–50 m are often reached with twenty four 10 Hz geophones at 5 m intervals, and depths of over 100 m can be reached with 4.5 Hz geophones). One basic advantage is a signal does not need to be generated; the signal is the background noise, which makes this technique especially useful for work in built-up areas.

### Spectral analysis of surface waves (SASW)

This technique also uses the dispersion of Rayleigh waves to obtain a profile of their velocity at depth, but, unlike ReMi, a discrete energy source, such as striking the ground is necessary. The set-up in the field e.g. with 2 or 4 geophones separated by distance d, responds to the ground being struck in symmetrical positions. The sensors are then separated at different intervals and the velocities of each phase are calculated. From estimations of the time/phase differences and the power spectrum for each wavelength, a dispersion curve similar to that of ReMi can be calculated. Finally, a $V_{rayleigh}$ profile at depth is obtained below the centre of the device. The penetration depth is half the separation between the sensors. The disadvantage of this method compared with ReMi is that the data collection and processing is more complicated.

## Electromagnetic methods

Electromagnetic methods study the response of the ground when electromagnetic (EM) fields are propagated through it. Due to both the many different ways EM fields can be generated or detected and the diversity of their characteristics, electromagnetic methods give rise to a larger number of applications than any other geophysical method. They can be broadly classified into two groups:

— Techniques in which conduction currents predominate; these can be classified in most cases by the position of the energy source:

- From a nearby induction source; these are commonly known as electromagnetic methods and can be subdivided into frequency-domain EM methods (FDEM) or time-domain EM methods (TDEM).
- From a distant induction source, very low frequency (VLF).

— Techniques in which displacement currents predominate: geo-radar or ground penetration radar (GPR).

### Frequency domain electromagnetic prospecting

Electromagnetic impulses are transmitted from a transmitting coil to a receiver on the ground. The penetration

# SITE INVESTIGATION

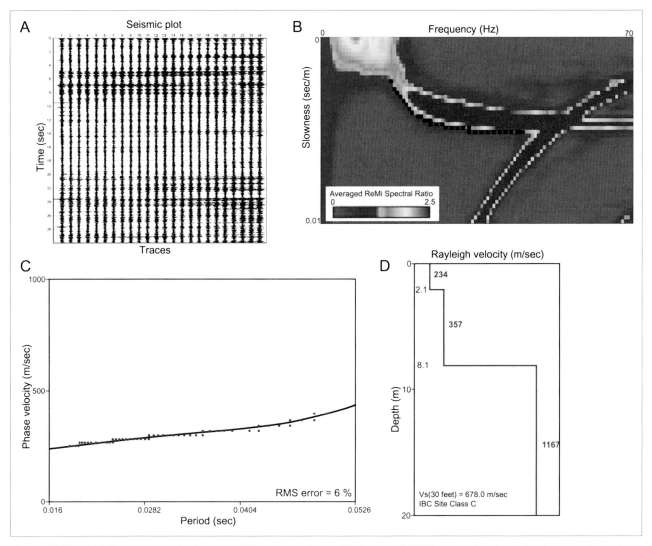

*Figure 5.25* A) Microtremor seismic traces (30 seconds long). B) Averaged REMI spectral ratio and selected points along the spectrum base. C) Dispersion curve. D) V Rayleigh's model and IBC (International Building Code) ground classification.

depth depends on the transmission frequency, which usually ranges from 100 Hz–10 kHz, and the distance between the transmitter and the receiver. The method is operated by placing a transmitting point and a receiving point at a specified distance from each other (generally 5–50 m) and moving them along the profiles at regular intervals. The result for each measurement point is obtained at the point halfway between the transmitter and receiver, at a depth depending on the frequency used and whether the coils are horizontal or vertical. Multi-frequency equipment is normally used; this allows several successive measurements to be taken at the same point and the frequency of each one modified so the ground can be investigated at different depths (*Figure 5.26*).

## Time domain electromagnetic prospecting

With this technique, variations over time in the secondary magnetic field generated are recorded while the transmitter is shut down. This eliminates noise and allows the transmitting coil to be used as a receiver, or the receiver coil to be placed inside the transmitter.

## Very low frequency

This technique differs from the others as the transmitting source is some distance away. The primary field is generated by low frequency radio antennas which may be situated hundreds or thousands of kilometres away. As well as being very accurate, the great advantage of these systems

*Figure 5.26* Right: Geonics EM34 electromagnetic equipment showing the transmitting (orange) and receiving (white) coils. Left: investigation on a landfill site using this equipment.

is that the setting up of heavy primary field generating equipment is avoided as only lighter coils are used to pick up the resulting field. The waves transmitted are in the VLF band (3–30 kHz) and each transmission has a specific frequency.

## Ground Penetration Radar (GPR)

Ground Penetration Radar (GPR) uses reflection to give continuous high resolution profiles, similar to those obtained by seismic reflection. Its main advantages are the speed with which data can be collected, and its versatility, as the antennas can be exchanged for others with different frequencies. Its main disadvantage is an excessive dependence on the surface characteristics of the ground where it is applied.

GPR radiates short impulses of electromagnetic energy by means of a transmitting antenna, with frequencies at present between 50 Mhz and 1.5 Ghz. When the radiated wave detects heterogeneity in the electromagnetic properties of the materials (contacts between materials, fractures, cavities, areas of different geomechanical quality, metallic elements, etc.), part of the energy is refracted back to the surface and part is transmitted to greater depths. The reflected signal is amplified, transformed to the audio frequency spectrum and recorded. This provides a continuous profile showing the total travel time of a signal as it passes through the subsoil, is reflected by a heterogeneity and returns to the surface. This double trajectory (TWT-Two Way Time) is measured in nanoseconds (1 ns = $10^{-9}$ s).

Selection of antenna frequency for a particular study is conditioned by a compromise between resolution and penetration. High frequencies have better resolution at low depths while low frequencies penetrate further but have lower resolution.

Georadar equipment consists of four main elements: transmitter, receiver, control unit and recording unit (*Figure 5.27*). Normal operative procedure involves recording profiles by moving the antennas along a path while maintaining a constant distance between them.

The interpretation of georadar recordings, or radargrams, is normally based on characterization of the texture, range, continuity and termination of the reflections. *Figure 5.28* shows an example of a georadar cross-section.

When an investigation using georadar is planned, the following factors have to be borne in mind: contrast in the electrical properties of the materials, penetration and resolution (which depend on the electrical properties of the ground and the frequency of the antenna used), background noise (the

*Figure 5.27* Investigation with Geo-Radar using 900 Mhz antennas.

# SITE INVESTIGATION

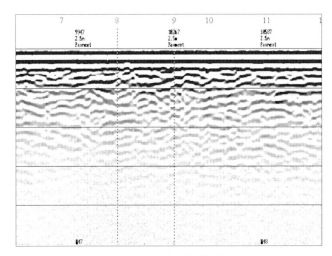

*Figure 5.28*  Cross-section obtained using geo-radar.

*Figure 5.29*  LaCoste & Romberg Model G gravity-meter.

equipment is very sensitive to the influence of metallic structures, radio waves, electricity power lines, etc.) and the water table.

Applications of georadar include the detection of cavities and contacts between materials and location of metallic structures, electric cables and pipelines.

## Gravity methods

Gravity methods are based on the study of differences between average values of the earth's gravitational field at a certain point and the theoretical value that point ought to have (i.e. a gravity anomaly). Anomalies are caused by heterogeneity in the density of the subsoil, and are positive or negative depending on whether a body of greater or lesser density than that of the surroundings is present at that point. The unit of measurement is the milligal (1 mgal = $10^{-3}$ cm/s$^2$) or the gravimetric unit (gu = $10^{-4}$ cm/s$^2$) and measurements

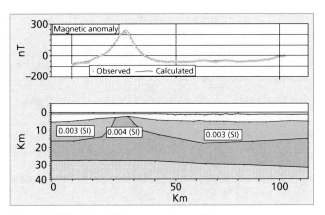

*Figure 5.30*  Modelling a gravity profile.

are taken with a gravimeter (*Figure 5.29*). The accuracy of normal models is 0.01 mgal and of micromodels 0.001 mgal. Gravimeters do not give direct measurements of gravity, and the average values have to be corrected, adding a "family name" to the anomaly to show that a particular correction has been made (Free Air and Bouguer Anomalies). *Figure 5.30* shows the interpretation of a gravimetric profile.

Gravimetric methods are useful for localizing any phenomenon in which density variation is a fundamental characteristic. In geological engineering gravimetric methods have many applications, including detecting cavities and calculating their volume, locating galleries, detecting areas with significant losses of fines leading to reduced density, identifying areas where ground treatment has increased the density.

This technique is generally applied using longitudinal sections, with measurement points arranged linearly or on an evenly-spaced grid. The distance between the measuring points will depend on the scale and depth of the anomaly being investigated.

In geological engineering, microgravity has important applications in the investigation of small-scale gravimetric anomalies. The measuring points are placed 1 m or less apart and equipment sensitivity is 1 µgal ($10^{-6}$ gal).

## Magnetic methods

These are used for studying local variations in the earth's magnetic field and give absolute measurements of the vertical component of the magnetic field. Anomalies are due to differences in the magnetic susceptibility of soils and rocks, and to the presence of permanently magnetized minerals. The results obtained are usually interpreted qualitatively as they cannot be interpreted quantitatively directly from field data. In geological engineering their usefulness is very limited, the main applications being the location of features such as metal pipes under the ground, lithological interfaces, faults, dykes and mineralized masses.

The main advantage of magnetometry is that it is cheap and very quick, taking as little as 30 s for readings at each station. The area under investigation is normally covered with a grid of regularly spaced measuring points at intervals that vary depending on the reason for the investigation. As a general rule, the separation between measuring points should be a maximum of half the smallest horizontal extent of the body or anomaly being investigated.

Currently, the most commonly used equipment is the proton magnetometer (*Figure 5.31*).

Field work can be seriously affected by the presence of power lines, railways, moving vehicles or very heterogeneous ground.

# Borehole geophysics

The application of geophysical techniques inside boreholes is a useful tool for measuring certain physical properties of the geological formations intersected. The information obtained complements core logging and geophysical results at the surface.

## Geophysical logging

Borehole logging records physical properties of the ground, such as density, porosity, level of saturation, etc., from information supplied by electrical, nuclear or acoustic methods. As well as physical parameters, information is obtained on mechanical properties and *in situ* characteristics of materials. Use of this technique is highly recommended in all deep boreholes. Logs, or diagraphs, are obtained by lowering a sonde to the bottom of the borehole and taking measurements, either continuously or at intervals, as it is raised. The equipment has four parts: the measuring instrument or sonde, the connecting cable and apparatus for lowering and raising it, the battery and the control and logging unit (*Figure 5.32*).

These techniques only allow the areas around the boreholes to be investigated, which means that results cannot be extrapolated to other areas; the advantage, however, is that depths to hundreds of metres can be investigated. The equipment used in geotechnics allows boreholes with diameters as small as 50–150 mm to be logged. Logging with several sondes allows correlation between them. Depending on the physical parameter measured, logs can be classified as follows:

— Electric: measuring electrical resistivity, spontaneous potential and electrical conductivity.
— Nuclear or radioactive: natural gamma, spectral gamma, neutron-neutron or neutron-gamma and gamma-gamma.

*Figure 5.31* Magnetic investigation using a Scintrex proton gradiometer.

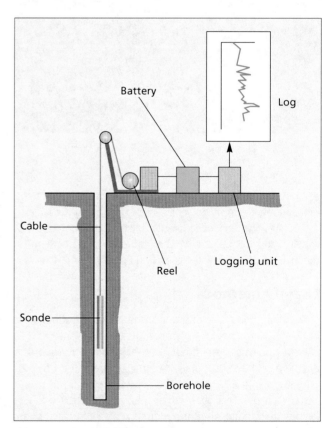

*Figure 5.32* Equipment for geophysical testing of boreholes (Clayton *et al.*, 1995).

- Sonic and acoustic.
- Fluid: temperature, conductivity and flow velocity.
- Geometric: calipers, dip meters and television recording.

**Electric sondes** provide information on: the **electrical resistivity** of the ground surrounding the borehole, which depends mainly on the salinity of the pore water and the pore size and interconnection; the **spontaneous potential** (SP), which responds to differences in electrical potential between contrasting formations arising from differences in the salinity of fluids or minerals; the **electrical conductivity** of the ground.

Measuring electrical resistivity requires an uncased borehole filled with liquid. The records obtained provide qualitative information on the lithological sequence, which facilitates correlations between boreholes. The sonde measures the resistivity of the ground when a current is passed artificially between one electrode, A, in the borehole and another, B, at the surface. The SP register is a passive measurement of the difference between the electrical potential of an electrode, M, placed inside the borehole, and another, N, at the surface. The presence of these potentials can be attributed to natural causes. The response of electrically conductive formations to an induced magnetic field is obtained from the combined results of the induction or conductivity sonde and the continuous log of conductivity of the ground around the borehole.

**Radioactive sondes** may be passive or active. Passive sondes measure the natural radioactive emissions from the ground surrounding the borehole; active sondes register the ground's response to being bombarded with gamma rays or a stream of neutrons. The **natural gamma sonde** measures gamma radiation emissions. Gamma rays are electromagnetic waves with frequencies above $10^{10}$ Mhz which are emitted spontaneously by radioactive elements present in rocks rich in clay materials. Radiation is mainly due to the radioactive isotope $K^{40}$ and to uranium and thorium isotopes. K forms part of the crystalline structure of micas, illites, smectites and other clay minerals, so natural gamma ray registers are used as qualitative indicators of the clay content in sedimentary formations. This sonde can be used in both cased and uncased boreholes.

The **gamma-gamma sonde** is mainly used for estimating ground density. An artificial source of radioactive isotopes that emits gamma rays (radium-226, cesium-137 and cobalt-60) is used to bombard the ground, and the gamma rays that return moments later with a certain loss of energy are logged. This value is inversely proportional to the density of the formation intersected. Before proceeding to calculate density, the level of natural gamma radiation in the ground must be deducted from radiation received.

The **neutron-neutron sonde** emits a stream of neutrons, and those returning moments later with loss of energy (thermal neutrons) are measured.

The **neutron-gamma sonde** measures the emission of gamma rays produced as the thermal neutrons are absorbed by atoms. They are highly sensitive to the presence of hydrogen atoms. With the neutron-neutron sonde, the amount of water present in the ground can be determined. If the ground is saturated, the neutron log gives a direct measurement of its porosity. When interpreting the neutron log, the diameter of the borehole, type of liquid in its interior, type of casing tube, lithology of the materials and degree of saturation in the ground should be taken into account.

**Sonic or acoustic** logging measures the propagation velocity and the attenuation characteristics of the elastic waves in the formation intersected by the borehole. These can be correlated with the mechanical properties and degree of fracturing of the material.

**The temperature sonde** records the temperature of the borehole fluid depending with depth. This provides information on the points or areas where water enters or leaves a borehole. The **conductivity sonde** measures the electrical conductivity of the borehole fluid. By logging the flow velocity, the presence of different zones of hydraulic head intersected by the borehole can be determined.

Among the most commonly used geometrical measures are: **caliper logging,** which gives a continuous graphical log of the borehole diameter as well as data on roughness or irregularities of the walls from lithological changes, gaps, fractures or dissolved areas, etc.; individual fractures can even be identified if the log is sufficiently detailed. It is also used for correlating the results of other types of logging. The **diplog** gives the dip and direction of dip in discontinuities intersected by the borehole from "micro" electrical devices arranged in such a way that discontinuities in walls are logged diametrically. From this, the scale and direction of any deviation in the borehole can be determined. Lithological interfaces, discontinuities, fractures, cavities, etc. can be observed from recordings of borehole walls, either without water or with clean water, provided by **television logging.**

## Seismic logging inside boreholes

This is carried out by introducing a triaxial sonde (i.e. one that can operate in 3 principal directions orthogonal to each other) into a previously cased borehole. The sonde logs the arrival times of the P and S waves, used to calculate transmission velocities and the dynamic deformation moduli of the ground. These constants depend on the velocity of the longitudinal ($V_p$) and transversal ($V_s$) elastic waves, and the density of the material, $\rho$ (see Section 5.5 and Chapter 3, Section 3.6).

Calculating the P wave velocity from seismic refraction from the surface is a common practice that makes use of travel time graphs of seismic profiles. As it is difficult to locate the arrival of S waves on the seismograms, down-hole and cross-hole techniques are used inside the boreholes to improve reception and identification. Expertise is needed to handle the sensors, and the generation of seismic waves for these techniques to work properly inside the borehole and pick up the transversal or shear waves. *Figure 5.33* shows as cartoons the principles of equipment used to generate shear waves. The following are the most commonly used investigation techniques:

### Cross-hole

This is carried out between two or three boreholes close together. A triaxial sonde is lowered into one or two of them to different depths to act as a receiver and the third is activated as a transmitter, also at a variable depth. The result gives a cross section of the different velocities of the ground between the boreholes.

### Down-hole and up-hole

This takes place in a single borehole. A triaxial sonde is placed at different heights, spaced at regular intervals. It receives seismic waves from a source either at the top of the borehole (down-hole) or from the bottom (up-hole). The impulses at ground level can be generated by a lateral blow on an element fixed to the ground and immobilised with a weight. This gives a cross section of ground velocities.

The geophone used has three components, two arranged horizontally and orthogonally in relation to each other, and a vertical one. This arrangement allows the arriving S waves to be identified by comparing seismograms received in the same component but resulting from impulses in opposite directions. Once the arrival time of P and S waves is identified, velocities $V_p$ and $V_s$ can be calculated from representation of the time-distance curves (travel time graphs), and these velocities can be used to determine Young's modulus and Poisson's coefficient. In geological engineering these techniques are normally used when designing underground work and foundations.

## Seismic tomography

Tomography is a geophysical investigation method used inside boreholes. It gives an image of the spatial distribution of seismic wave propagation velocity in the section of ground affected.

Seismic tomography involves the generation of seismic impulses inside boreholes and at ground level by mechanical means. Signals are received by geophones installed at numerous points in the interior of the borehole and/or on the surface. The ground's response to many seismic impulses from multiple points is studied by measuring wave arrival times. The section of ground affected by the test is divided into pixels. Propagation time between a point of emission and of reception will be equal to the sum of the travel time in each pixel. This depends in turn on the velocity and distance travelled in each of them. If as many equations (traces) are available as unknowns (velocities and spaces) a map can be drawn up of the velocity distribution in the section. The size, and therefore the number, of pixels will depend on the number of traces carried out (*Figure 5.35*).

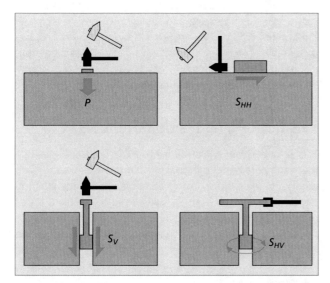

Figure 5.33  Examples of percussive methods for generating P and S waves, $S_{HH}$ and $S_V$ waves and S waves polarised on the horizontal and vertical plane (Clayton *et al.*, 1995).

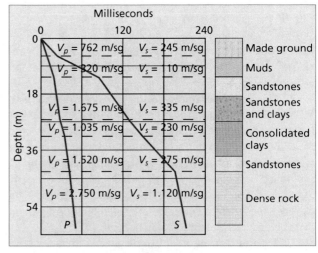

Figure 5.34  Propagation velocity of P and S waves in a cross-hole test.

# SITE INVESTIGATION

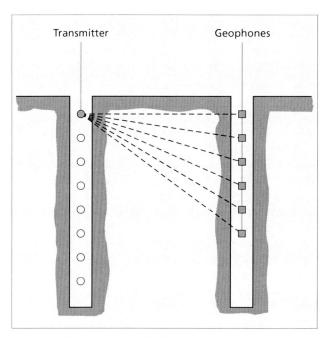

*Figure 5.35* Example of investigation using seismic tomography.

The following conclusions can be drawn from experiences based on these techniques:

— Anomalies with a low transmission velocity, such as cavities, are more difficult to locate than high velocity zones, such as blocks of sound rock.
— Propagation velocity values of seismic waves can be used to compare the properties of different materials, but should not be used as absolute values in calculations for engineering purposes.
— The lower the contrasts in velocity, the more reliable the interpretation of the ground will be.
— Planar structures, like faults, can be studied in detail using this technique.

## 5.4 Boreholes, trial pits, trenches and sampling

### Borehole drilling

Geotechnical boreholes are normally small in diameter and are made with lightweight, versatile, easily transported equipment. They can be drilled to a depth of around 100 m, after which heavier equipment is used. They can drill through any type of material and samples can be extracted for testing and the ground tested inside them. Boring procedures depend on the type of material and the type of sampling and testing to be done. The most common procedures are rotary, auger and percussion drilling.

### Rotary drilling

Rotary drilling can penetrate any type of soil or rock at any angle of inclination and to considerable depths (*Figures 5.36* and *5.37*). They are not usually deeper than 100 m for geotechnical purposes although they may be as deep as 500 m. Core extraction is a continuous process and can give a very high percentage of core recovery in relation to the length drilled, depending on the drilling system used. Some types of materials, such as gravels and boulders, or fine sands below the water table, are difficult to recover with rotary drilling, due to erosion by the drilling fluid.

Rotary drilling uses the following elements housed in the barrels: the drill head, core barrel, core catcher and drilling bit.

The drill head is the part that joins the core barrel, where the sample is collected, and the drill rods that transmit the rotary movement and force exerted by the drilling machine. The core catcher houses a catcher spring that opens like an iris when core passes through it and grips core to prevent it from

*Figure 5.36* Lightweight roller—mounted rotary drill.

*Figure 5.37* Inclined rotary coring rig.

slipping as the core barrel is withdrawn from the hole. The actual perforation is done by the drill bit. The cutting edge may be tungsten carbide (*Figure 5.38*) or diamond (*Figure 5.39*). Tungsten carbide bits are used for softer rocks and soils. Diamond bits must be used for hard or very hard rocks.

Core barrels (*Figure 5.40*) may be single or double tubes (*Figure 5.41*). In the **single tube** core barrel, drilling fluid passes over the whole surface of the sample. This effect, and the rotation of the tube, may lead to the disintegration of partially cemented soils or softer rocks. For this reason, the single tube is used when high recovery is not required. If high recovery is required, a **double tube** core barrel is used, where water runs down the annulus between the tubes, and contacts with the sample only at the base of the tube where it joins the core bit. The inner tube is mounted on bearings so that it is almost stationary while the outer tube rotates. The damaging effect of drilling fluids may be further reduced by using a **triple tube** core barrel; the sample here is sheathed in a third tube housed inside the double tube. This tube can extend a short way beyond the bit of the rotating outer tube and pierces the ground with a cutting shoe that retracts or extends depending on how compact the ground is. These barrels are generally used where the drilling depth does not exceed 100 m. For greater depths a **wireline** system is more appropriate for brining sample barrels to the surface, as this considerably reduces operating time and offers better performance (*Figure 5.42*).

*Table 5.8* shows the ratio between different types of drilling diameters and core sizes, with the most common diameters being NX (75.5 mm) or greater.

*Figure 5.38* Tungsten carbide bit.

*Figure 5.39* Impregnated diamond bit.

*Figure 5.40* Core barrels and coring bits.

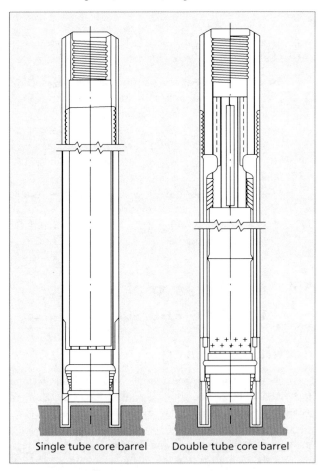

*Figure 5.41* Core barrel sections.

Rotary core drilling can be carried out with a flushing medium such as water or bentonite slurry or compressed air, even though there may be water or mud present on the drill rig. The flow is generally direct, with downward flow through the rods; it can also be reverse, in which case a special system of rods is required. To obtain the best results and performance, the operating techniques must be adapted to the type of material being drilled and the most suitable types of drill, barrel and core bit for it selected. Rotation velocities, pressures exerted on the bit and the operating torque must also be appropriate for the material being drilled, making the success of such drilling very dependent on the skill and experience of the driller and the condition of the equipment used.

In deep boreholes, any deviation must be controlled from the planned direction, caused by the tendency of the borehole to follow the dip of different layers or strata.

## Auger drilling

Auger drilling is suitable for relatively soft and cohesive soils and unsuitable for hard or consolidated soils. Its advantages include low cost, portability and rapid installation of the equipment.

Boreholes of this type often do not allow precision better than ±0.50 m in defining the depth of different layers encountered. Samples obtained from auger drilling will be disturbed, although, as described below, it is possible to obtain undisturbed samples with certain types of sampling devices.

*Figure 5.42*  Wire line system.

### Table 5.8  DIAMETERS OF CORE BITS, CASING AND CORES

| System | Core bits | | | Casing | | | |
|---|---|---|---|---|---|---|---|
| | Size (mm) | Perforation diameter (mm) | Core diameter (mm) | Size (mm) | External diameter | Internal diameter | Weight (kg/m) |
| Craelius metric standard | 36 | 36 | 22 | 35 | 35 | 29 | 1.4 |
| | 46 | 46 | 32 | 44 | 44 | 37 | 3.5 |
| | 56 | 56 | 42 | 54 | 54 | 47 | 4.4 |
| | 66 | 66 | 52 | 64 | 64 | 57 | 5.2 |
| | 76 | 76 | 62 | 74 | 74 | 67 | 6.3 |
| | 86 | 86 | 72 | 84 | 84 | 77 | 7.2 |
| | 101 | 101 | 84 | 98 | 98 | 89 | 10.5 |
| | 116 | 116 | 86 | 113 | 113 | 104 | 12.4 |
| | 131 | 131 | 101 | 128 | 128 | 119 | 13.8 |
| | 146 | 146 | 116 | 143 | 143 | 134 | 15.4 |
| American Standard: Diamond Core Drill manufact Assoc. of USA | EX | 37.7 | 21.4 | – | – | – | – |
| | AX | 48.0 | 30.1 | EX | 46.0 | 38.1 | 4.1 |
| | BX | 60.0 | 42.0 | AX | 57.2 | 48.4 | 4.5 |
| | NX | 75.5 | 54.7 | BX | 73.0 | 60.3 | 9.0 |
| | HX | 99.2 | 76.2 | NX | 88.9 | 76.2 | 11.8 |
| | 23/4" × 3 7/8" | 98.4 | 68.3 | 4" | 129.0 | 102.0 | 16 |
| | 4' × 5$^{1/2}$" | 139.6 | 100.8 | 6" | 187.0 | 154.0 | 30 |
| | 6" × 7$^{3/4}$" | 196.9 | 151.6 | 8" | 239.0 | 203.0 | 39 |
| | Wire line | | | | | | |
| | AQ | 48.0 | 27.0 | EX | 46.0 | 38.1 | 4.1 |
| | BQ | 60.0 | 36.5 | AX | 57.2 | 48.4 | 4.5 |
| | NQ | 75.7 | 47.6 | BX | 73.0 | 60.3 | 9.0 |
| | HQ | 96.0 | 63.5 | NX | 88.9 | 76.2 | 11.8 |

Auger drilling can be done by hand for shallow depths (2–4 m) and small diameters (2–5 cm), or by power equipment for depths of up to 40 m with diameters from 7 to 20 cm. It is usually done for preliminary survey purposes (*Figure 5.43*).

There are two types of augers: hollow and flight. Flight augers have a helical thread. With the hollow type, undisturbed samples can be obtained without having to extract the flight to the surface. The shaft of the flight is slightly larger than for a flight only drill, so that it can contain a central casing that ends in a small drill bit at the helical head of the flight and turns simultaneously with it. These are removed from the inside of the shaft and a sampler inserted (*Figure 5.44*).

## Percussion drilling

Percussion or **shell and auger** drilling allows soils with a firm or very firm consistency to be bored and is therefore used in both granular and cohesive soils. Boreholes of this type can reach depths of up to 30 or 40 m, although depths of 15 to 20 m are the more usual. Boring is carried out by driving a series of steel tubes into the ground with a 120 kg hammer that drops from a height of 1 m (*Figure 5.45*). The number of blows needed to penetrate each 20 cm section must be counted systematically in order to determine the compactness

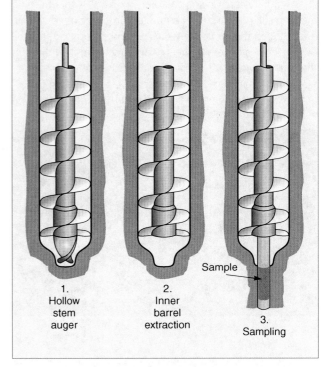

Figure 5.44  Sampling using augers.

Figure 5.43  Auger drilling rig.

Figure 5.45  Shell and auger rig and tools, showing a clay cutter and open drive samplers.

of the soil. The barrel used, which may have outside diameters of 91, 128, 178 or 230 mm, acts as a shoring mechanism during the extraction of samples, which is carried out using clay cutters and shells (Figure 5.46).

## Special boreholes

In addition to the methods already described, drilling is sometimes carried out with drill bits, impact hammers or a **rotary percussive drill**, or by using **tricone** rollers (Figure 5.47).

Figure 5.46  Open tube sampler U – 100 type used in percussion drilling for sampling cohesive soils with diameter 100 mm and length approximately 450 mm.

Figure 5.47  Tricone rollers.

These techniques are known as **destructive drilling** because, instead of a core sample, they produce rock chips and pulverized material which is expelled up the borehole by the drilling fluid. Methods used will depend on the type of ground to be drilled and the purpose of the investigation. Destructive drilling is used to penetrate boulders and loose blocks. Rotary percussive drilling can be used to detect cavities in limestones, volcanic rocks or abandoned mines.

## Number and depth of boreholes

Deciding on the number and depth of boreholes is one of the basic questions in planning site investigations, and is dependent on several factors. These aspects are dealt with in Part III (foundations, slopes, dams, etc.) As a guideline, a borehole should reach the level of the deepest substratum that may be affected by any structural action on the ground (loads, seepages, deformations, etc.). The number made will depend on both the aims and scope of the site investigation and the extent to which each borehole is representative of the area involved.

## Borehole data presentation

The results of borehole investigations are presented on field data sheets together with geotechnical data obtained from logging (Figures 5.59 and 5.60). These are described below. It is usual for a country to adopt a certain standard method of description, e.g. Eurocode 7, or the ASTMS of the United States, and these should be followed when requested to do so.

## Trial excavations

Trial pits, trenches, and shafts are examples of trial excavations made mechanically or manually which allow observation of the ground to a certain depth and permit *in situ* testing and sampling to be carried out (Figures 5.48 and 5.49).

Their main advantage is that they provide direct access to the ground, so that lithological variations, structures and discontinuities can be observed directly and samples of considerable size can be taken for testing and analysis.

**Trial pits** are one of the methods most frequently used in geotechnical surveys. They are cheap and quick to make and are a common feature of any type of site investigation, although they have some limitations:

— They are not normally deeper than 4 m.
— The presence of water restricts their use.
— The ground usually has to be excavated mechanically.
— All the safety regulations on the excavation site have to be observed to prevent the sides collapsing, and it is necessary to check beforehand that there are no underground installations such as pipelines and cables that could be disturbed.

*Figure 5.48* Trial pit excavation showing soil structure.

*Figure 5.49* Trench excavation. Note the unsafe conditions of the excavation: depth is greater then the height of a person, slopes are unstable and no support has been provided.

Results obtained from this type of investigation are recorded on field data sheets and should include data on depths, continuity of different strata, discontinuities, descriptions of the lithology, seepages, location of samples and photographs (*Figure 5.50*).

# Geotechnical sampling

The main object of sampling from boreholes or trial pits is to obtain materials for testing in the laboratory that are representative of the ground properties. The samples can be classified into two groups:

**Disturbed samples:** these have undergone modifications in their structure and water content but preserve their mineralogical composition. Disturbed samples are usually taken from trial excavations and boreholes. These samples are suitable for classification, mineralogical and compactness tests.

**Undisturbed samples:** in theory these have not undergone any alteration in their structure or water content and care is taken to achieve this as far as possible, but all samples from depth will expand on being released from their surrounding in-situ stress. When obtained from boreholes they are extracted in appropriate samplers. Samples taken from trial excavations can be cut out as a block or taken from tube samplers that are either pressed or driven into the ground. Such samples are needed for tests of strength, deformability and permeability and for soil analysis. Rock cores can be covered with a thin layer of molten wax for protection against changes in water content or transportation and handling damage.

**Water samples:** these are obtained for chemical analysis from different water levels detected during drilling. The most common laboratory analysis include pH, salt content and the detection of contaminating elements. Samples should not be taken immediately after drilling. Any residue produced by the boring process should be allowed to disappear; this may include particles in suspension and the remains of injected water or slurry used during drilling. The water sample is collected in clean bottles (plastic or glass depending on the analyses required) which are washed out with the same water before being filled. Each sample is labelled with the date and data identifying the borehole and depth and refrigerated if necessary.

## Borehole samplers

The type of sampler will depend on the drilling system used for extraction. The following are the most commonly used:

**Rotary core samplers:** In rotary drilling, the barrels themselves and their bits are used as samplers (*Figure 5.40*). Cores extraction from single barrels will be disturbed after being subjected to rotary movement. With double barrels, the outer barrel turns but the inner remains static, allowing undisturbed samples to be obtained. The ends of undisturbed samples must be coated and sealed with paraffin wax the moment they are extracted. For a more careful extraction split inner tuber core barrels can be used as well as triple- tube barrels.

**Driven tube samplers:** In percussion and auger boring the barrel is replaced by a tube sampler which is inserted under pressure or driven into the ground. **Open tube samplers** may

# SITE INVESTIGATION

| | |
|---|---|
| **TRIAL PIT DATA SHEET** | **PROJECT:** ................................................ <br> **TRIAL PIT NUMBER:** C-1 <br> **LOCATION:** ................................................ <br> **DATE:** ................................................ **COORDINATES** <br> **SUPERVISOR:** ................................................ X: <br> **MACHINERY:** ................................................ Y: <br> **SUPPORT:** ................................................ |

**DEPTH:** 2.40 m (large rock block prevents further excavation)

**DESCRIPTION:**
- From 0.00 m to 0.20 m: layer of concrete (pavement).
- From 0.20 to 0.40 m: reddish brown clayeysands. Incipient root system (organic topsoil).
- From 0.40 to 1.00 m: dark brown silty clays with centimetric sub-angular limestone pebbles (non-compacted fill).
- From 1.00 to 2.40 m: non-compacted fill made up of calcarenite blocks, 0.20 to 1.20 m in size, in a brownish red silty clay matrix.

**TOPSOIL:** 0.20 m of reddish brown clayey sands.

**WATER TABLE:** Not present.

**COLUMN**

0.0 m — Pavement / Topsoil / Non-compacted fill of centimetric pebbles
1.0 m
2.0 m — Non-compacted fill of calcarenite blocks

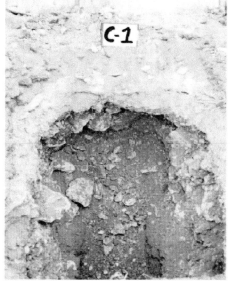

Methods of description will vary according to accepted national standards.

*Figure 5.50* Trial pit record.

be thick or thin-walled and are permanently open at the lower end, whereas closed drive samplers can be open or temporarily closed (*Figure 5.51*). Samplers used in standard penetration tests (SPT) are of the thick-walled type. **Shelby tubes** are a type of thin-walled sampler (*Figure 5.52*). Thick-walled open samplers are driven in and thin-walled ones inserted under pressure. **Piston samplers** are gently and smoothly pushed into the ground, and used for obtaining better quality undisturbed samples from soft and very soft soils (*Figure 5.53*).

## Trial pit and trench samples

Disturbed or undisturbed samples can be taken during the excavation of a trial pit, trench or shaft. Disturbed samples are extracted manually or with shovels and put into watertight plastic bags. The quantity of samples taken will depend on the soil particle size and the type of tests to be carried out. For index tests, 2 or 3 kg of material are usually sufficient. However, if CBR tests are to be carried out (see Chapter 12, Box 12.1), the minimum quantity required is 20 kg. In sands and gravels these quantities are doubled or tripled depending on grain size, and may reach over 100 kg in cases of large boulders or rock fragments (such as those found in colluvial or alluvial deposits). Undisturbed samples are obtained from trial excavations in two ways:

**Block samples.** In this procedure a block is cut out of the ground and removed manually. It is immediately sealed with warm wax and wrapped in cheesecloth (*Figures 5.54* and *5.55*).

**Tube samplers.** An open tube sampler is driven into the sides or base of an excavation, either manually in the case of soft soils, or mechanically with the excavator shovel in firmer soils. The ends of the tube are sealed with paraffin wax (*Figure 5.56*).

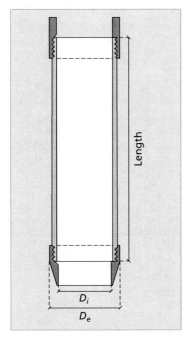

Figure 5.51   Section of an open tube sampler. $D_i$ = internal diameter; $D_e$ = external diameter.

Figure 5.52   Shelby-type thin wall open tube sampler.

Figure 5.53   Hydraulically operated piston sampler used for soft soil sampling.

Figure 5.54   Excavation of block sample.

Figure 5.55   Block samples sealed and wrapped for protection.

# Sample sizes

The size of undisturbed samples is conditioned by the requirements of the laboratory tests. The most common diameters range from 55–100 mm. For uniaxial compressive tests a diameter of approximately 55 mm may be sufficient, while for oedometer tests a minimum diameter of 80 mm is advisable. If three samples are needed from the same plane for the triaxial test a diameter of at least 100 mm will be required. Because some disturbance at each end of a sample is inevitable, its minimum length should be sufficient to provide a middle section that is as long and as intact as possible. When transporting undisturbed samples it is essential to avoid heat, knocks and vibrations. Once in the laboratory, they should be stored in a chamber of 100% relative humidity.

# Borehole logging

Borehole logging consists of a description of cores and samples obtained from boreholes, together with other data relating to the drilling operations. This task must be carried out by a specialist in engineering geology who is able to control the drilling process and make a detailed description of the samples obtained. These descriptions will follow either national or prescribed standards.

It is essential to include the following data in the **description of the drilling records**:

— Basic information: project name, reference number, location and date; names of contractor, supervisor and driller; borehole number, coordinates, inclination and orientation.
— Drilling method: rig, type of drilling, diameter, characteristics of drilling tools, flushing medium, type of flow (normal or reverse) and other technical characteristics.
— Progress made during the drilling process: operations, distance advanced in metres, rate of advance, recovery, loss and infiltration of fluids, instability of the walls, breakdowns, water tables, number of blows needed for driven samplers, types of tests carried out, etc.

**Engineering geological logging** consists of the recording and description of drill cores obtained from mechanical boreholes. Cores should be arranged in order in wooden or paraffin-waxed boxes and labelled, with markers to indicate the depth of any change in lithology or the presence of other significant structural features, such as faults, fractures and cavities. The empty spaces that remain where samples have been extracted should be marked, and characteristics of the samples indicated (undisturbed sample, paraffined sample, SPT, etc.) (Figures 5.57 and 5.58).

An engineering geological description of the cores can be done at the same time as drilling or immediately afterwards. It should not be delayed because certain types of materials undergo alterations that modify their properties

Figure 5.56   Open tube samples in trial pits sealed with paraffin wax.

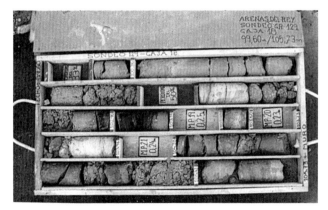

Figure 5.57   Soil cores obtained by rotary drilling labelled and stored in a core box.

Figure 5.58   Rock cores obtained by rotary drilling labelled and stored in a core box.

## Box 5.1

## RQD calculation

RQD is an index of the ratio between the sum of the lengths of core fragments longer than 10 cm and the total length of the core run:

$$RQD = \frac{\sum \text{lengths of cores fragments} \geq 10\,\text{cm}}{\text{total length of core run}} \times 100$$

Only fresh or hard pieces of core are considered for estimating RQD. Those showing significant weathering (from Grade IV and above) are eliminated; in these cases, RQD is considered to be 0%. It is recommended that the operational run length should not exceed 1.5 m. The minimum core diameter on which the index should be calculated is 48 mm. The length of the core piece is measured along its central axis, using fragments with at least one complete diameter.

The procedure for measuring RQD is shown below, as well as the description of rock quality based on this index. It must be remembered that RQD is a function of the direction in which it is measured, which can result in different values for RQD being obtained from the same rock mass.

| RQD % | Quality |
|---|---|
| <25 | Very poor |
| 25–50 | Poor |
| 50–75 | Fair |
| 75–90 | Good |
| 90–100 | Very good |

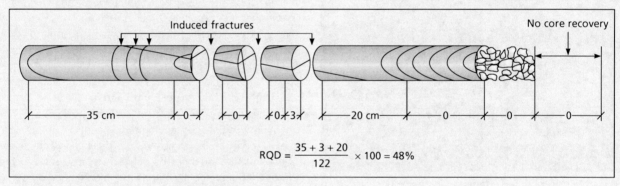

$$RQD = \frac{35 + 3 + 20}{122} \times 100 = 48\%$$

Modified from Norbury et al., 1986 (in Clayton et al., 1995).

---

(such as loss of water in soils). The following procedure should be followed:

— Systematic description: Observed nature and composition, lithology, grain size, colour, texture, degree of weathering, consistency and soil penetration resistance (with a pocket penetrometer), etc according to the standards being used.
— In rock materials: Description of discontinuities (type, spacing, roughness, fills) and the percentage of sample recovery.
— RQD index (described in Box 5.1) and $N_{30}$ index representing the number of fractures for every 30 cm of core.

— Data from tests performed inside the borehole.
— Photographs of the cores stored in boxes, taken so that the spacers dividing the core runs and indicating depth, and the colours, textures and fractures in the cores can be clearly identified, together with the box number, the bore hole number and its depth.

In addition the following data should be recorded:

— Depth and type of samples obtained.
— Depth of water table.

Figures 5.59 and 5.60 show examples of borehole logs.

# SITE INVESTIGATION

| COMPANY | | | | | BOREHOLE LOGGING IN SOILS | | | | | | |
|---|---|---|---|---|---|---|---|---|---|---|---|
| | | | | | PROJECT: | | | | | | |
| BOREHOLE N°: | | | | | SITUATION:<br>DRILL TYPE:<br>FLUSH:<br>INCLINATION:<br>AZIMUTH: | | | COORDINATES: | X:<br>Y:<br>Z: | | |
| DATE: | | | | | DEPTH: 12 m | | | SHEET: | | | |

| DEPTH (m) | SECTION LENGTH (m) | CASING DEPTH (m) | WATER LEVEL | LEGEND | DESCRIPTION | SAMPLES | SPT NUMBER | ATTERBERG LIMITS | | WATER CONTENT (%) | U.S.C.S. CLASSIF. |
|---|---|---|---|---|---|---|---|---|---|---|---|
| | | | | | | | | LL (%) | PI (%) | | |
| | 0.25 | | | | Organic topsoil. | | | | | | |
| 0.90 | 0.65 | | | | Well-graded sandy gravels with blackish angular pebbles. | | | | | | |
| 1.80 | 0.90 | | | | Sandy gravels similar to above with sandy-silty-clays that become more abundant with depth | | | | | | |
| 2.00 | 0.20 | | | | | | | | | | |
| 2.60 | 0.60 | | | | Silts with some sands with pebbles occasionally whitish. Medium dense. | US | 5-7-8-9 | 33.0 | 15.9 | 16.8 | CL |
| 3.20 | 0.60 | | | | | SPT | 5-8-9-12 | | | | |
| | | | | | Clayey sandy silts with angular pebbles 2-3 cm in size. Dark brown and loose, with nodules of cohesive material. | | | | | | |
| | 3.25 | | | | | | | | | | |
| | | | | | Sandy silts with scattered sub-rounded pebbles that can reach 5 cm in size. Loose and brownish-white. | | | | | | |
| 6.45 | | | | | | | | | | | |
| 7.00 | 1.15 | | 7.30 | | Dark brown clayey silt with high content of organic material with gravel pebbles ranging from 0.5 to 1 cm in size. | US | 3-4-5-8 | 24.8 | 8.5 | 16.2 | ML |
| 7.60 | | | | | | SPT | 4-7-7-6 | | | | |
| | 3.40 | | | | Loose sandy silts, light brown colour | | | | | | |
| | | | | | Firm light brown silts with some sands | | | | | | |
| 11.00 | | | | | | | | | | | |
| 11.50 | 0.50 | | | | Hard clayey marls with a few scattered bluish-coloured pebbles. | MI | 15-20-30-40 | 30 | 20 | 8.5 | CL |
| 12.00 | 0.50 | | | | | SPT | 25-50-R | | | | |

OBSERVATIONS:

US: Undisturbed sample  
DS: Disturbed sample  
SPT: Standard Penetration Test  
NSO: No sample obtained  
TP: Telltale paraffin-treated  
WT: Water table

Methods of description will vary according to accepted national standards.

*Figure 5.59* Borehole log in soils.

| COMPANY | BOREHOLE LOGGING IN ROCKS | |
|---|---|---|
| | PROJECT: | |
| BOREHOLE Nº: | SITUATION:<br>DRILL TYPE:<br>FLUSH:<br>INCLINATION:<br>AZIMUTH: | COORDINATES: |
| DATE: | DEPTH: 12 m | SHEET: |

| DEPTH (m) | SECTION LENGTH (m) | CASING DEPTH (m) | WATER LEVEL | LEGEND | CORE DESCRIPTION | SAMPLES | CORE RECOV. (%) (20, 40, 60, 80, 100) | RQD (%) (20, 40, 60, 80, 100) | FRACT. N/30 cm (1, 2, 3, 4, 5) | DISCONTINUITIES |
|---|---|---|---|---|---|---|---|---|---|---|
| 0.6 | 0.60 | | | | Organic topsoil composed of blackish clayey organic silts. | | 100 | | | Soil |
| 1<br>2<br>3 | 2.4 | | | | Heterometric conglomerates with a reddish matrix. Sub-rounded pebbles of up to 5 cm maximum size with average size 1–2 cm. | 3.00 | 100 | 60 | | Wavy-smooth discontinuities, 80° dip |
| 4<br>5 | 2.0 | | 5.0 | | Red to brown sandstone with little matrix and gravels. | NSO<br>3.60 | 82.5 | 23<br>43 | | Two families: one wavy-rough with $\beta = 80°$ and another flat-rough with $\beta = 20°$ |
| 6<br>7 | 2.0 | | | | Red sandstone with little matrix and thin seams of conglomerates with pebbles up to 1 cm. | 5.35<br>TP<br>5.80 | 100 | 76 | | Flat-smooth |
| 8<br>9<br>10<br>10.5 | 3.5 | | | | Conglomerates with reddish matrix. Sub-rounded pebbles of up to 7 cm in size, with average size 3–4 cm. | 7.80<br>TP<br>8.30<br>10.2 | 100<br>100 | 90<br>100 | | Wavy-smooth discontinuities, 80° dip<br>No discontinuities |
| 11<br>12 | 1.5 | | | | Very fine-grained red sandstones. | TP<br>10.8 | 100 | 100 | | No discontinuities |

OBSERVATIONS:

UD: Undisturbed sample
DS: Disturbed sample
SPT: Standard Penetration Test

NSO: No sample obtained
TP: Telltale paraffin- treated
WT: Water table

Methods of description will vary according to accepted national standards.

*Figure 5.60* Borehole log in rocks.

## 5.5 *In situ* tests

*In situ* tests are very important for establishing the geomechanical properties of soil and rock materials. They permit the measurement of parameters that determine the geomechanical behaviour of soils and rocks, such as strength, deformability and permeability.

The main advantage of *in situ* tests is that they are more representative than laboratory tests of the ground conditions of the proposed work site. This is because they involve a much greater volume of material in its natural state. However, the scale of *in situ* tests means they do not represent the whole volume of rock or soil mass and this should be taken into account when results are interpreted and extrapolated.

### Standard penetration test (SPT)

SPT is a dynamic penetration test carried out inside boreholes. It gives a **penetration resistance** value $N$ that can be correlated with geotechnical parameters such as relative density, friction angle, and settlements in granular soils. The penetration resistance ($N$) is the number of blows required to drive the split spoon sampler for the last 30 cm of penetration. A disturbed sample is also obtained from this test for carrying out identification tests in the laboratory.

The SPT can be done in practically all types of soil, even in very disturbed rock, although it is recommended for use in granular soils; the difficulty in obtaining undisturbed samples in these types of soils makes SPT even more relevant.

SPT is usually carried out as the borehole progresses at a frequency of about one every 2 to 5 m, or more, depending on the characteristics of the ground.

The procedure followed consists of carefully cleaning the bottom of the borehole of debris to the depth required for the test, removing the drilling rod and installing in its place a standard-sized sampler. This has three components: the drill shoe, the split spoon sampler and the coupling head attached to the rod (Figure 5.61). It should be driven 60 cm into the ground and the number of blows to drive in each 15 cm section should be counted. Blows are produced by a 63.5 kg hammer falling from a height of 76 cm onto an anvil plate (Figure 5.62). The readings from the first and last 15 cm section should not be considered, in the first case because of soil disturbance and wall collapse, and in the second because of possible over-compaction. The penetration resistance $N$ is the number of blows required to drive in the split spoon for the two middle 15 cm sections (30 cm total) of penetration. If the advance or penetration of the sampler is very slow, or rebounds due to the strength of the ground, and more than 100 blows are needed for the sampler to advance 15 cm, the test is discontinued.

SPT results may be affected by the following factors:

— Quality of the borehole: cleanliness and stability of the borehole wall.
— Rod length and borehole diameter may condition the weight driven in and generate friction with the borehole walls.
— Hammer device: this can be manual or automatic, with significantly different results for each method; automatic devices awre preferable as they guarantee the same impact force every time.

When SPT is carried out below the water table the following correction is applied to low permeability soils (silts and fine sands):

$$N = 15 + [(N' - 15)/21]$$

valid for $N' > 15$, where $N$ is the corrected value and $N'$ the value measured.

Figure 5.61    SPT split spoon sampler, drilling rods and anvil.

Figure 5.62    SPT equipment.

The extensive use of SPT has led to the establishing of a series of correlations with different geotechnical parameters:

— Depth correction using the ratio between relative density and $N$ value (*Figure 5.63*).
— Friction angle in granular soils ($\phi$) applicable below a depth of 2 m (*Figure 5.64*).
— Compactness in granular soils (Table 2.6 in Chapter 2).

## Probing penetrometers

Probing tests are simple and inexpensive tests that estimate soil penetration resistance according to depth. Different layers of soil can be correlated with available geological information from boreholes or trial pits in nearby areas. These tests are frequently used in geotechnical studies for building and road foundations or railway infrastructures. Useful information obtained from probing is the assessment of the depth of the rock head or resistant layer, the thickness of Made Ground and anthropogenic fill, and the presence of buried obstructions (old foundations, tanks etc).

A metal cone attached to a series of rods is driven into the ground. The hammering equipment consists of a hammer, an anvil plate and guide tracks. The anvil transmits the energy received to the cone via the rods, which are progressively coupled together and extended as the test advances. The hammer has a velocity equal to zero at the moment its fall is initiated. The anvil is firmly attached to the rods; its diameter is equal to or greater than 100 mm and less than or equal to half the diameter of the hammer. There are several variations of this test, depending on the force of the blow. The different types of equipment used depend on the character of the ground.

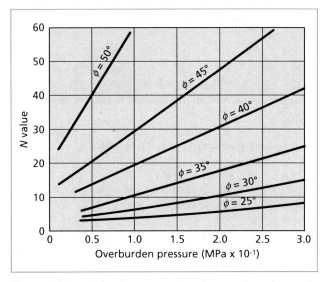

*Figure 5.64*   Ratio between $N$ value and $\phi$ in sands (de Mello, 1971).

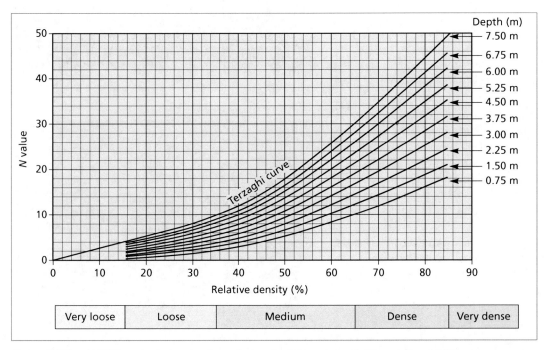

*Figure 5.63*   $N$ (SPT) and relative density ratio as a function of depth (Thornburn, 1963).

# Swedish weight penetrometer

This test is also known as the "Borros Test". It can be carried out at shallow depths, usually less than 15 m, but in some cases may exceed 25 m (*Figures 5.65* and *5.66*). The hammer, which weighs 63.5 kg, falls from a height of 50 cm. The drop point may be square or conical. The number of blows $N_B$ are recorded every 20 cm. Rebound is considered to be encountered when more than 100 blows are needed to drive in 20 cm of tubing.

As an approximation, $N$ can be estimated as equal to $N_B$ when the value of $N_B$ is between 8 and 12. For higher values, $N_B$ is somewhat greater than $N$. Two correlations (not strictly equivalent) applicable to sandy soils have been proposed, Dahlberg (1974):

$$\log(N_B) = 0.035\, N + 0.668 \pm 0.044$$

$$N = 25 \log (N_B) - 15.16 \pm 1.16$$

## Cone penetration test (CPT)

The cone penetration test measures the soil's reaction to the continuous hydraulic insertion of a cone into the ground, *Figure 5.67*. In this type of static penetration test the cone resistance ($q_c$) and the lateral friction ($f_s$) are recorded. If a pore pressure sensor or **piezocone** is installed, pore pressures ($u$) are also recorded and the test is known as **CPTU**.

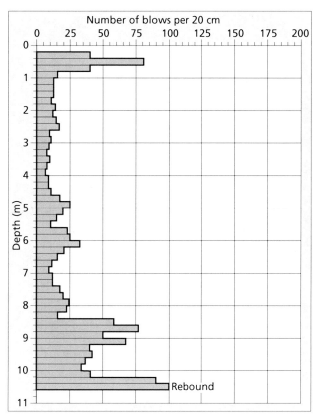

Figure 5.66   Borros test log.

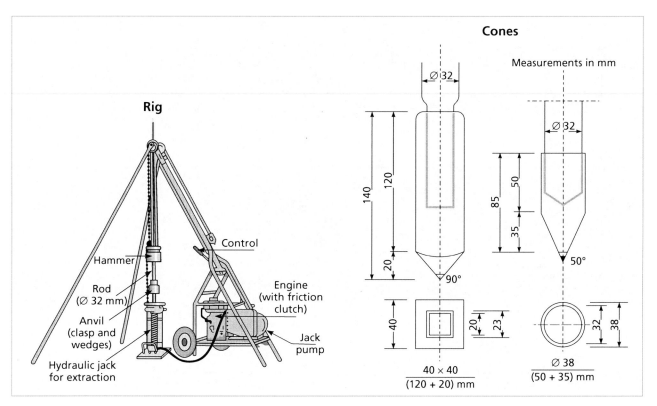

Figure 5.65   Swedish weight penetrometer rig and equipment.

Figure 5.67   Static cone penetration test equipment. (CPT)

Parameters $q_c$, $f_s$ and $u$, measured during the test, are represented graphically in relation to depth. *Figure 5.68* shows how the presence of thin silty or sandy layers sandwiched between less permeable beds can be detected from the resulting pore pressure peaks, as can less permeable layers between sandy strata. From this, the ground profile can be interpreted.

CPT tests are performed in granular soils and cohesive soils with soft consistencies as the equipment is obstructed and damaged by the presence of boulders, gravels, cemented soils and rock. CPT results are used for foundation design and provide a continuous depth – resistance profile of the ground. From data obtained, correlations with other geotechnical parameters can be established, for example:

— With the internal friction angle for granular soils; although there is no simple or general correlation, *Figure 5.69* gives some indicative values.
— With Young's modulus ($E$), for granular soils, Schmertmann (1978) proposed the following expression:

$$E = 2.5 q_c$$

where $q_c$ is the force required to push the cone divided by the plan area of the cone.

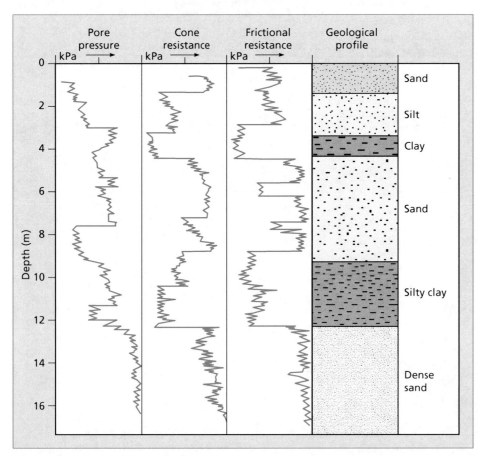

Figure 5.68   Example of CPTU test log.

# Field vane test

This test is usually carried out at the bottom of a borehole, although it can also be performed from the ground surface. The vane test best is used to determine the undrained shear strength of soft cohesive soils.

The vane shear test consists of four steel blades welded onto a central rod (Figure 5.70). This is driven down at the bottom of the borehole to a depth 5 times the height H of the blades, normally 50–100 mm. Then the blades are rotated at a constant velocity of 0.1°/s (6°/min), and the maximum torque T at which the soil fails is measured. As it is a quick test, it is carried out in undrained conditions so that the shear strength is without drainage, which is equivalent to the undrained shear strength (sometimes called "cohesion") of the material (for $\phi = 0°$). The remoulded strength, or residual strength value once the soil has failed, should also be measured. If used in the base of a hole separation between testing points as the borehole progresses should be at least 0.5 to 0.7 m.

The undrained shear strength, $S_u$, depends on the maximum torque and soil sensitivity, $S_t$:

$$S_u = 2T/[\pi D^2(H + D/3)]$$
$$S_t = S_{u(peak)}/S_{u(residual)}$$

where $T$ is the maximum torque needed for soil failure, $H$ the length of the vane blades, $D$ the diameter of the vane, $S_{u(peak)}$ the undrained peak shear strength and $S_{u(residual)}$ the residual strength.

# Schmidt hammer test

This test is used to obtain an approximate estimation of the uniaxial compressive strength of rocks. Its main application is for intact pieces of rock, but it can also be used on discontinuities.

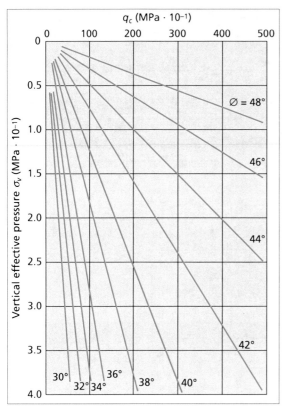

Figure 5.69  Cone resistance ($q_c$) and angle of internal friction ($\phi$) for non-cemented sands (Robertson and Campanella, 1983).

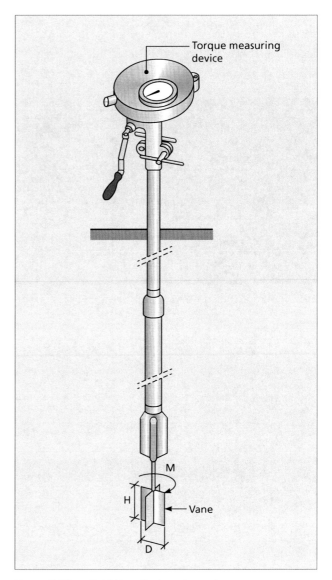

Figure 5.70  Field Vane Test equipment.

The L-type sclerometer consists of a cylindrical metal device containing a spring that drives a rod (and its hammer) out of the cylinder. When the hammer strikes the rock surface its rebound is measured (*Figure 5.71*). The surface to be tested should be free of cracks and fissures and must be cleaned before testing by removing the patina of weathered rock. Pressure is then applied to the hammer until the spring is released. The instrument must be placed perpendicular to the test plane. The spring rebound value depends on the hardness or strength of the rock. This value is indicated on a scale on the side of the apparatus. At each measuring point there should be 10 hammer blows, provided they do not damage the surface; the 5 lowest values are then discarded and the average value is taken of the rest.

The resulting rebound values are correlated with a chart with the uniaxial compressive strength, which depends on the unit weight of the rock and the inclination of the hammer and test plane. *Box 5.2* shows an example of its application. Uniaxial compressive strength laboratory tests results should be available to calibrate measurements with the Schmidt hammer test and to establish correlations.

## Point load test

The test measures the Point Load Strength Index of rock samples, $I_s$, which is used as an index for strength classification and for correlation with the uniaxial compressive strength of the rock. The test consists of breaking a rock specimen by applying a point load, and it can be performed in a laboratory testing machine or with portable equipment (*Figure 5.72*).

The specimens used can be cylindrical cores, cut blocks or irregular lumps. Results are more reliable if tests are conducted on core specimens, diametrically or axially loaded, with $L/D > 1.0$ and $L/D = 0.3–1.0$ respectively (*Figure 5.73*).

The test procedure consists of inserting a rock specimen between the two conical steel platens of the test machine. The distance $D$ between the platen points is recorded. Then the load is steadily increased until failure

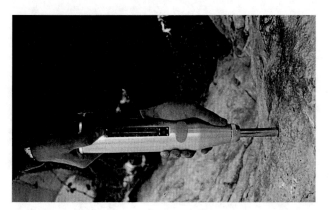

Figure 5.71   Schmidt hammer or sclerometer test.

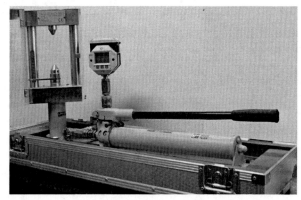

Figure 5.72   Point load test apparatus.

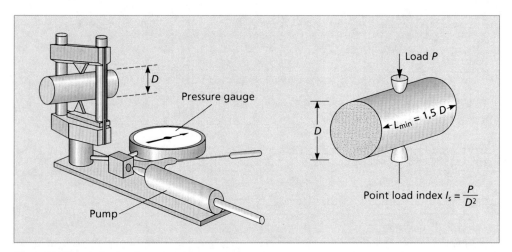

Figure 5.73   Point load test equipment and test.

## Box 5.2

### Uniaxial strength assessed using the Schmidt hammer

Calculate the uniaxial compressive strength expected of a rock with a density of 27 kN/m³ using the Schmidt hammer on specimens of the intact rock.

Hammer rebound values: 49, 46, 45, 45, 44, 50, 48, 46, 43, 44 (the measurements are obtained with the hammer applied perpendicularly to a vertical fresh rock surface).

Method: mean rebound value is calculated from Schmidt tests. For every 10 values, the five lowest are discarded and the mean value is obtained from the remaining five (for 12 values, 6 are discarded). In this case, the values that remain are: 46, 46, 48, 49, 50; the average rebound value is therefore 48.

This mean rebound value is plotted on the x axis of the graph shown below until it reaches the density value of the rock being studied, depending on the direction of the hammer. From this point, a horizontal line is plotted until it intersects the y axis, giving the expected uniaxial compressive strength value of the rock, which in this case is 125 MPa.

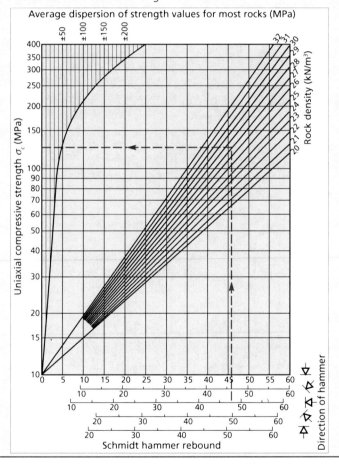

occurs within 10–60 s, and the failure load $P$ is recorded. At least 10 tests must be conducted per material sampled. The tests should be rejected as invalid if the fracture surface passes through only one loading point (ISRM, 1985).

When a rock sample is anisotropic (shaly, bedded, schistose) it should be tested both parallel and perpendicular to the planes of anisotropy, so the greatest and least strength values are obtained.

The uncorrected point load strength is calculated from the failure load:

$$I_s = P / D_e^2$$

where $D_e$ is the equivalent diameter: $D_e^2 = D^2$ for diametral tests and $D_e^2 = 4A/\pi$ for axial, block or lump tests ($A$ = cross sectional area of a plane between the platen contact points).

$I_s$ varies as a function of $D$ (for diametral tests) and of $D_e$ (for the other types of test), so that a size correction must be applied to obtain a unique Point Load Strength value for the rock sample: the size-corrected Point Load Strength Index, $I_{s(50)}$, which is defined as the value of $I_s$ that would have been measured by a diametral test with $D = 50$ mm (ISRM, 1985). To avoid this size effect, diametral tests are best conducted on samples having a diameter at or close to $D = 50$ mm; in other cases the size correction must be applied as explained in Box 5.3.

From the results obtained in the tests, the mean value of $I_{s(50)}$ is calculated, and it can be used directly as an index for rock strength classification, and also to predict the uniaxial compressive strength. Box 5.3 includes an example of the calculation of strength using the Point Load Test.

## Shear strength test on discontinuities

Measuring the shear strength of a discontinuity plane can be performed in adits or galleries or at ground level by cutting out rock blocks with dimensions that may vary between 40 × 40 cm and 100 × 100 cm, although the usual size is 50 × 50 cm (Figures 5.74 and 5.75). The base of the block sample is the discontinuity plane to be tested. The testing procedure is carried out in two stages. First, a normal load is applied to the discontinuity on the block sample and the vertical displacement produced is measured; the normal load then remains constant throughout the test. Then, a tangential load is applied until failure occurs along the discontinuity plane. The load value is recorded, together with tangential and normal displacements. The normal load is applied to the block sample with a jack and is distributed uniformly. Tangential stress is applied through a jack inclined at an angle, which may vary according to the characteristics of the rock and the geometry of the discontinuity. The test is usually carried out in adits, where the sides and roof act as reactions for the jacks. If the test is carried out at the surface, the jacks are supported by an anchored kentledge.

Three or four block samples are tested in each test. Different normal loads and increasing tangential stress are applied to each block sample until failure is reached. It is also usual practice to use the same block sample and carry out several shear failures on it by subjecting it to increasing normal stress. The results are shown on a tangential stress-normal stress graph, $\tau - \sigma_n$, where each block sample tested

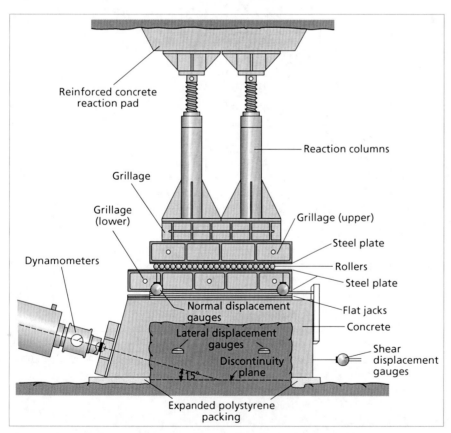

Figure 5.74    Typical arrangement of equipment for *in situ* direct shear test (ISRM, 1981).

## Box 5.3

## Uniaxial strength calculated using the Point Load Test (PLT)

The correlation between the $I_s$ index, obtained from the PLT test and the uniaxial compressive strength of a rock, is for 50 mm diameter cores. For specimens with a different diameter, a size correction must be applied to obtain the corrected Point Load Strength Index, $I_{s(50)}$.

The size correction must be applied using the formula:

$$I_{s(50)} = F \times I_s$$

where $F$ is a size correction factor, which can be obtained from the chart in this box or from the expression (with $D_e$ in mm):

$$F = (D_e/50)^{0.45}$$

which can be applied irrespective of the degree of anisotropy of the specimen and the loading direction (ISRM, 1985).

Once the mean value of $I_{s(50)}$ for the sample tested is calculated, after excluding the two highest and lowest values from 10 or more valid tests, it can be used to predict the uniaxial compressive strength expected of the material in that direction. The relationship between $\sigma_c$ and the point load strength index can be expressed as:

$$\sigma_c = f \times I_{s(50)}$$

where $f$ is the conversion factor, ranging in most cases between 25 and 20 depending on the type of rock (higher for hard, strong rocks and lower for soft rocks), with even lower values for some shales and mudstones; however, $f$ can vary between 15 and 50, especially for anisotropic rocks (ISRM, 1985).

The PLT's accuracy in predicting the uniaxial compressive strength depends on the ratio between the $\sigma_c$ and the tensile strength. For most brittle rocks, the ratio is approximately 10. For soft mudstones and claystones, however, the ratio may be closer to 5. This implies that PLT results might have to be interpreted differently for the weakest rocks. The accuracy of the estimate also depends on the number of tests conducted.

Point load strength often replaces uniaxial compressive strength, $\sigma_c$, when there is sufficient confidence in the relationship between PLT and $\sigma_c$ obtained from direct tests of UCS, since when properly conducted PLT is as reliable and much quicker to measure.

### Example:

A series of diametral tests are conducted on cores with a diameter $D = 6.5$ cm, and gave the following values for $I_s = P/D_e^2$ (MN/m$^2$): 2.1, 2.4, 1.7, 1.9, 2.2, 1.6, 2.3, 2.1, 1.8, 1.9. From the chart below, the size correction factor for $D = 65$ mm is $F = 1.125$, and thus the corrected values for $I_{s(50)} = P/D_e^2$ (MN/m$^2$) are: 2.362, 2.7, 1.912, 2.137, 2.475, 1.8, 2.587, 2.362, 2.025, 2.137. The $I_{s(50)}$ mean value, once the two highest and lowest values have been excluded, is 2.25, and this value can be entered in the expression for estimating the uniaxial compressive strength, $\sigma_c = f \times I_{s(50)}$. If a value of 23–24 is assumed for the conversion factor $f$, because the rock tested is a hard, fresh material, then $\sigma_c = 52$–54 MN/m$^2$ or MPa.

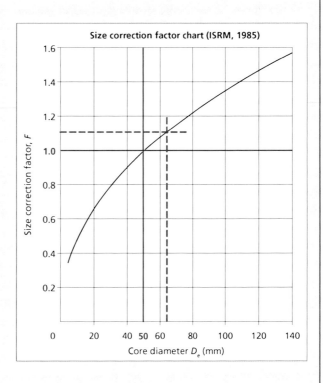

Size correction factor chart (ISRM, 1985)

*Figure 5.75*   Rock specimen preparation and equipment for *in situ* shear strength test in rock.

is represented by a point. The 3 or 4 points obtained are joined to give the curve defining the cohesion and friction angle of the discontinuity tested. In shear tests, both peak and residual strength parameters can be determined; in the second case, after failure is reached, the operation is continued until large shear displacements occur; the displacement orientation of the upper half of the sample block can be reversed during the procedure, if necessary.

Shear strength in discontinuities can also be calculated in the field from cores or samples containing a discontinuity plane, using a **Hoek cell** (Chapter 3, Figures 3.87 and 3.88).

## Tilt test

This test is used to estimate the angle of friction of discontinuities or basic angle of friction for smooth discontinuities to allow calculation of the residual angle of friction and joint roughness coefficient (JRC).

A sample rock block containing a non-cohesive discontinuity plane is required to estimate the angle of friction in discontinuities. The rock sample is placed on an adjustable testing plane, separated along the surface where the roughness is to be measured. Then the plane is slowly tilted until the sample starts to slide. As soon as this occurs, the angle of the support plane is measured in relation to the horizontal, $\alpha$, (Figure 5.76). The procedure should be repeated with several rock samples. The value of $\alpha$ is a function of the ratio between the shear stress and normal stress acting on the discontinuity:

$$\alpha = \arctan(\tau/\sigma_n) = \phi$$

The JRC value of the discontinuity is obtained from $\alpha$. This is needed to apply the Barton and Choubey criterion, to estimate the shear strength of rough discontinuities:

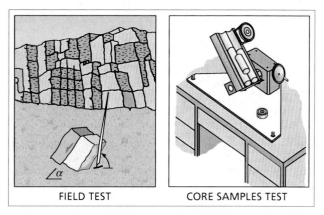

*Figure 5.76*   Tilt test (Barton, 1981).

$$JRC = (\alpha - \phi_r)/(\log(JCS/\sigma_n))$$

The test can also be carried out in the field or the laboratory using three cylindrical core samples. Two parallel cores are placed together on a horizontally supported surface with a third core lying on top of them. The support base is gradually tilted until the upper core slides along the two lower samples. This gives the angle $\alpha'$ and $\phi$ can be obtained using cylindrical cores of the same diameter, because the friction angle is $\alpha'$. The following expression is used to assess the basic friction angle:

$$\phi_b = \arctan(\tan\alpha' \cos 30°)$$

The basic angle of friction of the material, $\phi_b$, corresponds to the strength of smooth, flat, non-weathered discontinuities. From this parameter and data obtained with the Schmidt hammer, the residual angle of friction, $\phi_r$, can be calculated using Barton and Choubey's expression, as explained in Chapter 3, Section 3.5.

# SITE INVESTIGATION

Figure 5.77    Pressuremeter probe used for soil deformability tests.

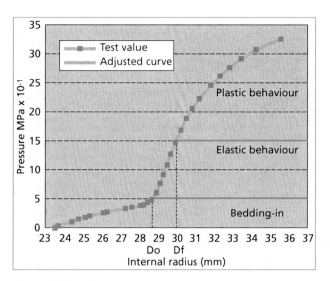

Figure 5.78    Pressuremeter test curve.

elastic behaviour of the soil; and a plastic or irrecoverable deformation phase leading up to soil failure. From this curve it is possible to calculate the yield pressure $P_F$, or pressure at which the material ceases to behave elastically, and the limit pressure $P_L$, or pressure at which the ground shears and no longer accepts pressure increases. Finally, the pressiometric deformation modulus, $E_P$, is obtained from the following expression:

$$E_P = (1 + \nu) M \cdot r$$

where $\nu$ is Poisson's ratio obtained in laboratory tests, $M$ is the stiffness of the ground, calculated from the slope of the elastic section of the pressiometric curve, and $r$ is the borehole radius.

## Pressuremeter test

The pressuremeter test is carried out to assess deformability inside a borehole. Equal increments of radial pressure are applied to an expandable cylindrical membrane (the pressuremeter) in the borehole. Dilation induced in the surrounding ground is measured after each pressure step. Once the maximum allowable pressure is reached, deflation of the pressuremeter is carried out step by step and deformations during deflation are measured. Pressure is applied through a rubber sheath, using water or gas. Most equipment works with pressures of less than 10 MPa but some allow pressures up to 20 MPa to be applied (Figure 5.77).

Depending on the type and characteristics of the soil being tested, the following phases can be identified in the pressure-deformation curve obtained (Figure 5.78): an initial phase, when the probe enters into contact with the sides of the borehole; a linear elastic phase, corresponding to the

## Plate loading test on soils

The plate loading test can be carried out in a trial trench or borehole, or at ground surface if this has adequately prepared. The test procedure is to apply a vertical load to a smooth, rigid plate to determine the deformations produced. The plate size can vary between 30 cm and 100 cm and be either circular or rectangular. The load at each step is usually maintained until the settlement increment is lower than 0.01 mm, with a 5 minute interval between readings. The load in the last step tested should be 3 times greater than the projected working load of the structure planned. Several loading and unloading cycles can be carried out in one test. The load is exerted with hydraulic jacks. These are either anchored or load against some suitable reaction e.g. a heavy truck (Figures 5.79 and 5.80).

This test is mainly applied to granular soils and to study shallow foundations. The parameters measured during the test are: time, applied load and settlement, shown on load-settlement and time-settlement diagrams (Figure 5.80).

*Figure 5.79* Plate loading test equipment in soils.

By applying Boussinesq's theory Young's modulus, $E$, can be obtained from the following expression:

$$E = 1.5(P_s/S)r$$

where $r$ is the radius of the plate, $P_s$ is the average pressure under the plate and $S$ is the plate settlement.

In roads, railways and raft foundations, the **ballast coefficient** $K_s$ is used, corresponding to a coefficient of proportionality defined by the following ratio:

$$K_s = P/S$$

where $S$ represents vertical displacements (settlements) under pressure $P$.

## Dilatometer test

The dilatometer test is carried out to assess rock mass deformability inside a borehole. It is an adaptation of the pressuremeter test and works on the same principles. Test results give load-displacement curves used to determine the dilatometric deformation modulus. Unlike soils, however, rock masses are discontinuous and anisotropic, which to a large extent conditions their deformability. This is why the dilatometric test usually measures deformation in six directions across three diameters.

Increasing pressure is applied through an elastic sheath housed inside a borehole (*Figure 5.81*). Once a linear section is obtained in the load-displacement curve, pressure is released. This cycle is repeated one to three times per test. Higher pressures are reached in each successive cycle. These will depend on the strength and deformational characteristics

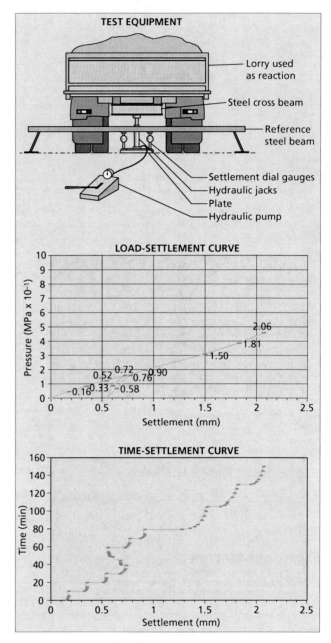

*Figure 5.80* Plate load test arrangement and results.

of the rock. A series of pressure-displacement curves are obtained from test results (*Figure 5.82*) and the following stages of deformation can be distinguished: adaptation of sheath to borehole, elastic deformation, one or several cycles of loading and unloading pressure, plastic deformation and failure. The dilatometric deformation modulus $E_D$ of the rock, for both loading and unloading pressures, is given by the following expression:

$$E_D = (1 + v) M \cdot r$$

where $v$ is the Poisson's ratio, $M$ the stiffness of the rock, corresponding to the slope of the elastic section of the dilatometric curve of the test, and $r$ the borehole radius.

This test is very useful in highly fractured rock masses, soft or deformable rocks, in general where good samples are difficult to obtain, or where the elastic properties of the rock must be obtained *in situ*.

## Plate loading test on rock

This test is usually carried out in galleries or tunnels. Young's modulus E can be obtained from the parameters measured (loading, settlement, displacements and time). The test is used for special rock foundations such as concrete dams.

The test procedure consists of loading a plate placed on the rock to be tested and measuring the displacements produced when the loads are applied. The load can be horizontal (if applied to the side of the tunnel) or vertical (if applied to roof and floor). Loads are applied with a jack and a hydraulic pump to reach higher pressures, using the opposite sides of the tunnel for reaction (*Figures 5.83* and *5.84*). The load area usually ranges from 30 × 30 cm to 100 × 100 cm, although a smaller plate often has to be used because of the high strengths found in rock masses. In each test, several loading and unloading cycles are carried out, and Young's modulus is obtained in both the loading and unloading cycles, according to the following expressions:

$$E = [qL(1 - v^2)]/z \quad \text{for square plates}$$
$$E = [\pi qd(1 - v^2)]/(4z) \quad \text{for circular plates}$$

where $v$ is Poisson's ratio, $q$ the load applied, $z$ the plate settlement and $d$ and $L$ the diameter and length of the plate.

## Flat jack test

The flat Jack test is carried out on the sides of excavations, e.g., galleries and tunnels, to obtain the deformation modulus and occasionally *in situ* stress in hard continuous rock masses. The results can be considered representative for up to several metres inside the rock mass from the test surface.

Before starting the test, reference point markers are inserted into the rock surface and the distance between them is measured (*Figures 5.85* and *5.86*). A groove is cut between the points with a saw or by overlapping drill holes. The groove tends to close up and the points are moved due to stress release. The point displacements are measured immediately after the groove is cut, and after a specified period of time, normally one to three days later. A flat jack is cemented into the groove and pressure is applied until the

*Figure 5.81* Dilatometer test equipment.

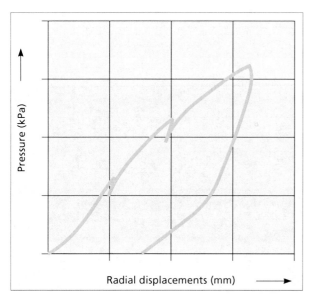

*Figure 5.82* Dilatometric test curve.

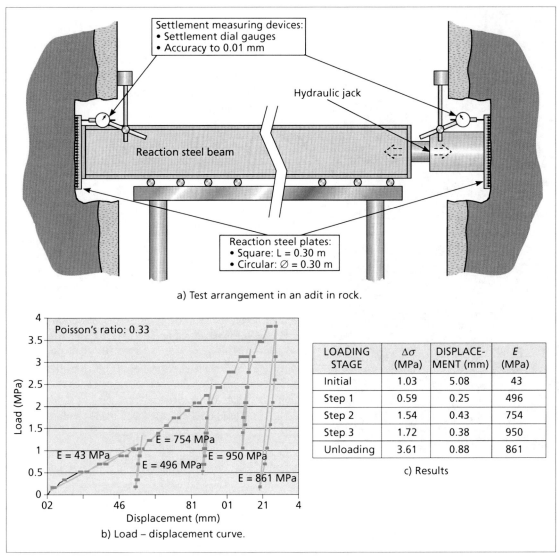

*Figure 5.83* Plate loading test in rock.

*Figure 5.84* Plate load test in rock: equipment and installation.

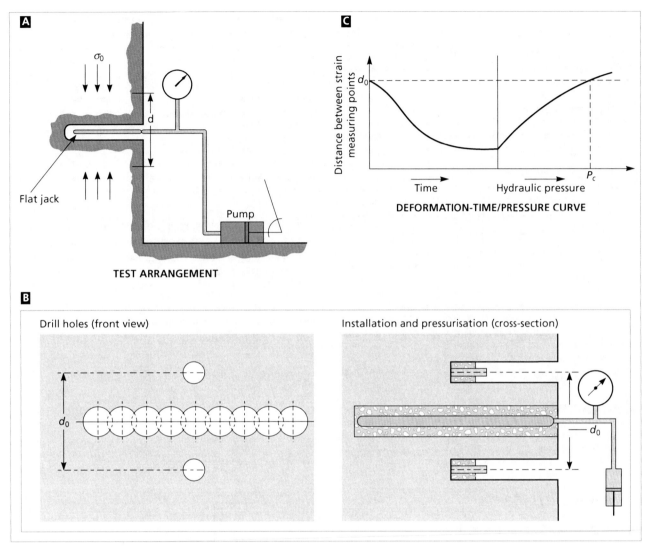

*Figure 5.85* A) Flat jack test. B) Groove excavation procedure and flat jack installation C) Deformation/distance curve results (A and C: Kim and Franklin, 1987; B: Brady and Brown, 1985).

*Figure 5.86* Flat jack test in a rock drift. A) Equipment. B) Taking readings.

distances between the points return to what they were originally. This applied pressure is considered equal to the normal initial *in situ* stress $\sigma_o$ for the groove. During the test the pressures and deformations are recorded to obtain the elastic deformation modulus of the rock mass.

### Seismic methods

Deformation moduli can also be obtained by seismic methods, as dynamic deformation moduli, for both soils and rock masses. These methods, described in Section 5.3, estimate the dynamic deformability from the velocity of longitudinal elastic or compression waves $V_p$, and transversal or shear waves $V_s$. Longitudinal wave velocity depends on the type of material, its degree of weathering and fracturing, the state of stress and the hydrogeological conditions.

The expressions relating these parameters with dynamic moduli are:

$$E_d = V_p^2 \rho[(1 + v_d)(1 - 2v_d)/(1 - v_d)]$$
$$v_d = 1/2[(V_p/V_s)^2 - 2]/[(V_p/V_s)^2 - 1]$$

where $E_d$ is Young's modulus, $v_d$ Poisson's ratio and $\rho$ the density of the material.

## Measuring *in situ* stress

Methods of measuring *in situ* stresses are described in Chapter 3, Section 3.7. A list of these procedures is included in *Table 5.9*.

## Permeability tests

### Permeability tests on soils

*In situ* tests to estimate the hydrogeological parameters are described in Chapter 4, Section 4.4.

The **Lefranc Test** is one of the most common permeability tests in geotechnical investigations. This test is used to estimate the permeability coefficient of permeable or semi-permeable granular soils lying below the water table or, occasionally, in highly fractured rocks. The test is performed inside boreholes and can be done either during or after the drilling operation.

The procedure consists of filling the borehole with water and measuring the flow rate needed to keep the water level constant **(constant head test)**, or by measuring the velocity of drop or change in the water level **(falling and rising head tests)**. Intake flow rate measurement must be carried out every 5 minutes so that the water level at the top of the borehole is kept constant for 45 minutes. If there is very high intake, this must be measured every minute during the first 20 minutes and every 5 minutes thereafter up to a total of 45 minutes.

Before time and flow are measured, the borehole is filled with water and checked to ensure that all the air has been expelled, and that the water level and the velocity of drop have been stabilized. For subsequent calculations the height of the water table has to be known.

*Figure 5.87* shows the factors that should be taken into account in order to obtain the permeability coefficient $k$, defined by the expression:

$$k = Q/(C\Delta h)$$

where $Q$ is the rate of flow (m³/s), $\Delta h$ the height of the water level above the initial piezometric level (m) and $C$ is a shape factor defined by the expression:

$$C = 4\pi/[(2/L)\log(L/r) - (1/2H)]$$

| Table 5.9 | METHODS FOR *IN SITU* STRESS MEASUREMENTS |
|---|---|
| **Method** | **Test location** |
| Geological | At outcrops |
| Geophysical | Focal mechanism from seismic data |
| Doorstopper technique | In boreholes and drifts |
| Triaxial cells | In boreholes and drifts |
| Hydrofracture | In boreholes |
| Flat jack | In tunnels and drifts |

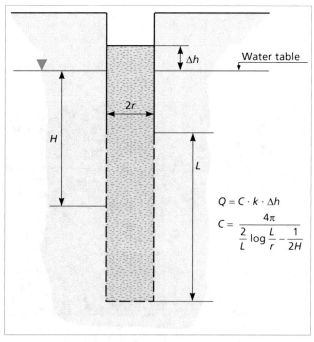

*Figure 5.87* Lefranc type constant head permeability test.

where $L$ is the length of the section tested (m), $r$ the radius of the borehole (m) and $H$ the distance from the midpoint of the tested section to the water table (m).

The section of borehole tested is always the length between the end of the casing and the bottom of the borehole.

## Permeability tests on rock

Permeability is one of the rock mass properties showing the highest variation within the same rock formation. This means that when the permeability of a rock mass is quantified it is more appropriate to refer to an order of magnitude (power of 10) than to precise values. In sound rock masses, permeability may be very low, $10^{-8}$–$10^{-10}$ cm/s, although if the rock mass is formed by a porous matrix, such as sandstone, values may reach $10^{-3}$ cm/s. Permeability in a jointed rock mass may reach $10^{-2}$ and $10^{-3}$ cm/s. The **Lugeon Test** is the most commonly used to estimate permeability in a fractured rock mass.

This test is carried out inside the borehole, to estimate the rock mass permeability. The method consists of pumping water at constant pressure (1 MPa) into a borehole and measuring the water intake for a 10 minute period. 5 m sections of borehole are generally tested, with the section being tested isolated from the rest of the borehole by means of two "packers", which is why the test is also referred to as the *packer test* (*Figures 5.88* and *5.89*). If the test is carried out at the bottom of the borehole (the last 5 m), only one packer is needed, as in the test originally defined by Lugeon in 1933.

Pressure is applied in successive stages at 0, 0.1, 0.2, 0.5 and 1 MPa, respectively, and is kept constant for

Figure 5.89  Inflatable packers and drill rods used for the Lugeon test.

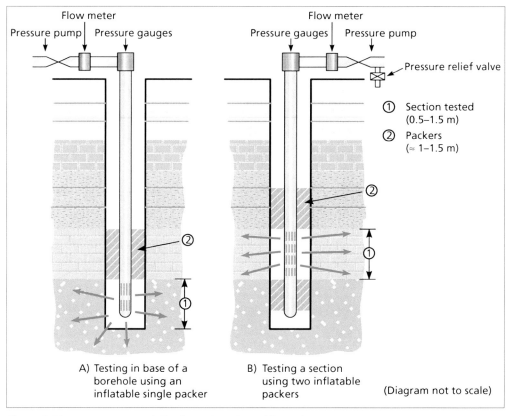

Figure 5.88  Lugeon test. A) Single packer B) Double packer.

10 minutes at each stage. The 1 MPa stage should always be reached, except in weak rocks where hydraulic fracturing may occur before this pressure is reached. Permeability values obtained at 0.5 and 1 MPa cannot be extrapolated linearly for greater pressures.

The unit of measurement for this test is the **lugeon** which corresponds to the absorption of 1 litre of water per metre of borehole per minute, under 10 bars of pressure during 10 minutes. A lugeon unit (LU) is equivalent to a permeability coefficient of $10^{-5}$ cm/s (LU = 1 l/m × min = $10^{-5}$ cm/s).

The results of this test can be represented graphically (*Figures 5.90* and *5.91*). *Table 5.10* gives a classification of rock masses, according to their permeability. These concepts are also dealt with in Chapter 11, Section 11.7.

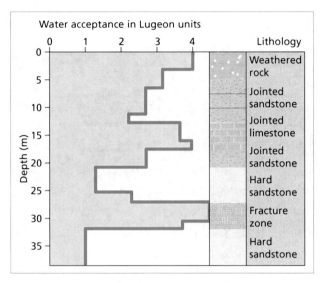

*Figure 5.90*   Example of Lugeon test results.

| Table 5.10 | ROCK BEHAVIOUR ACCORDING TO PERMEABILITY | |
|---|---|---|
| Rock mass watertightness | Lugeon units | Pressure (MPa) |
| Very high | 0–1 | 1 |
| High | 1–3 | 1 |
| Low | >3 | 1 |
| | 1.5–6 | 0.5 |
| Very low | >3 | 1 |
| | >6 | 0.1 |

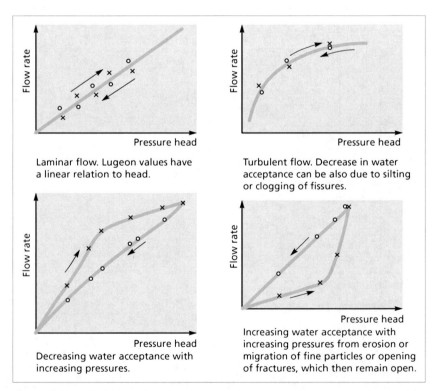

*Figure 5.91*   Typical results of Lugeon test.

# 5.6 Geotechnical instrumentation

The purpose of geotechnical instrumentation is to determine the behaviour and characteristics of the ground to be able to predict how it will react to loading, movements, thrusts and other actions, both natural and induced by human activity. This section describes common instrumentation in geological engineering.

An instrumentation programme involves selecting the scale of dimensions to be measured and choosing the instruments accordingly. Elements to consider include:

— Surface movements.
— Underground movements.
— Displacements from cracks opening and between different points.
— Pore pressures.
— Lateral earth pressure on retaining structures.

The frequency of readings and data collection depend on the dimensions to be measured and the speed of the process studied. Readings may be manual or automatic. Manual readings are more appropriate when the number of sensors or logging points is small, where data is collected at weekly or longer intervals, or where readings are taken at easily accessible points.

The choice of the data collection system is conditioned by the number of sensors and their characteristics, their location, the situation and accessibility of the site, the frequency of readings, the amount of data involved and how quickly this has to be processed and interpreted.

## Displacement measurements

### Displacements between points in close proximity

To check displacement between points in close proximity situated either on surface or underground, the following methods are used:

#### Mechanical reading systems

Sensors used include: tape extensometer, measuring tapes, calipers and fleximeters. A tape extensometer (Figure 5.92) is recommended for distances of over 2 m. For shorter distances other systems are used depending on the precision required: a metal measuring tape is used for a low level of precision (millimetres), calipers for medium precision (tenths of millimetres), and the fleximeter for high precision (hundredths of millimetres) (Figure 5.93). Relative displacements on the surface of excavations and structures are measured.

#### Electrical reading systems

These are essential for the automation of data logging processes, or when the points to be checked are inaccessible. Figure 5.93 shows a diagram of the equipment. Types of measuring sensor include the potentiometer, LVDT and vibrating wire.

For ranges of several centimetres the use of potentiometers is preferred, for those of a few millimetres, either potentiometers or LVDT, and for ranges of tenths of a millimetre, vibrating wire sensors.

### Displacement between widely distanced surface points

#### Geodesic methods

These methods enable horizontal and vertical movements to be measured with an average degree of precision to centimetres. The systems used include:

— Triangulation: measurements of angles from two or more fixed points
— Trilateration: measurements of distances from three or more fixed points
— Polygonation: measurements of angles and distances from at least three fixed points.

#### Levelling

This method measures vertical movements, with a precision of up to 1 mm in stretches of 1 km. Measurement of these movements is carried out in relation to a series of fixed

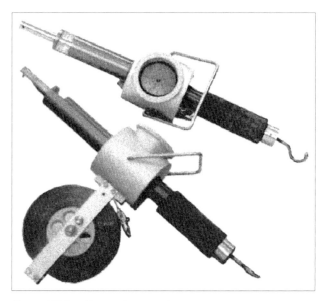

Figure 5.92  Tape extensometer.

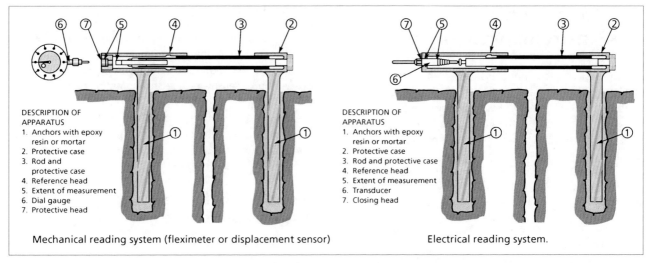

*Figure 5.93* Rod extensometer used on the ground surface.

reference bases. The procedures for reading and processing the data are quick and straightforward.

## Collimation

This involves measuring horizontal movements perpendicular to the collimation plane. Horizontal movements of the control points are measured with respect to a fixed vertical collimation plane. Precision is high, to millimetres, and the reading and processing of the data is quick and simple.

In all three systems it is important to make sure that the topographical or reference bases are fixed and situated away from unstable areas.

## Deep displacements

### Inclinometers

Inclinometers are one of the main methods for investgating landslides and for general control of transversal displacements in boreholes. Inclinations at various points inside a borehole are measured by means of a sonde that transmits an electric signal proportional to the angle of slope (*Figures 5.94* and *5.95*). Transversal displacements in a borehole can be recorded and quantified from differences between measurements taken at different points and the times at which these are taken. *Figure 5.96* shows an example of readings taken by an inclinometer in which two failure surfaces are identified at depths of 7.5 and 17 m.

Inclinometers may have an electrical resistance, vibrating wire sondes or servoaccelerometers. This last type can measure rotation precisely to $2 \times 10^{-4}$ rad. It is important to make sure that the inclinometer is installed to below any areas of possible movement so it can straddle them.

*Figure 5.94* Inclinometer equipment.

### Extensometers

This instrument can measure movements between two points, one at the top of the borehole and the other inside it, where the extensometer is anchored. Displacements at the

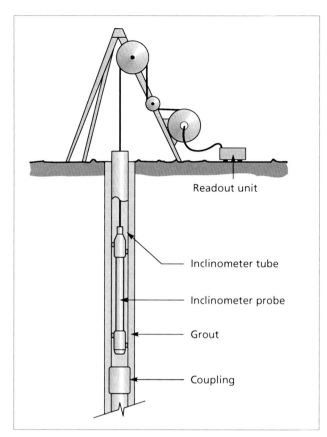

*Figure 5.95* Installation of inclinometer sonde (Soil Instruments Ltd.).

anchor points are transmitted to the top of the borehole by wires and rods. Measurement is by either mechanical or electrical procedures.

The rod extensometer is used for lengths under 40 m (*Figures 5.97* and *5.98*) and the wire extensometer for lengths over 60 m.

## Displacements in shafts and boreholes

Other methods used to detect displacements or estimate the depth of sliding surfaces include:

— Observing the deformation in large diameter bores or shafts with segmented casing where the depth of the deformation or failure can be measured approximately.
— Inserting a "telltale", a piece of metal tubing 25–40 cm long, into the bottom of a cased borehole to measure the depth at which the tube is intercepted because the borehole is deformed or blocked by ground displacements (*Figure 5.99*).

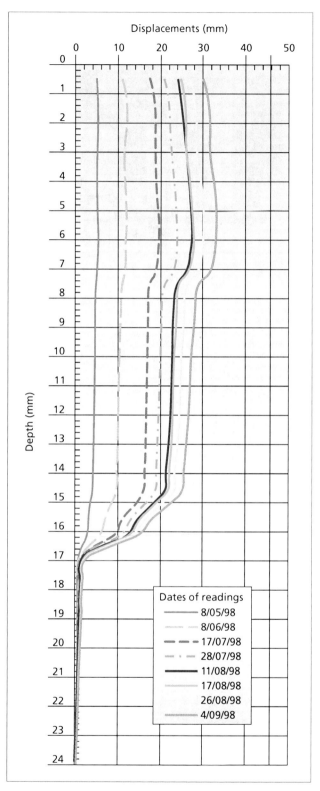

*Figure 5.96* Inclinometer readings.

Figure 5.97   Rod extensometer.

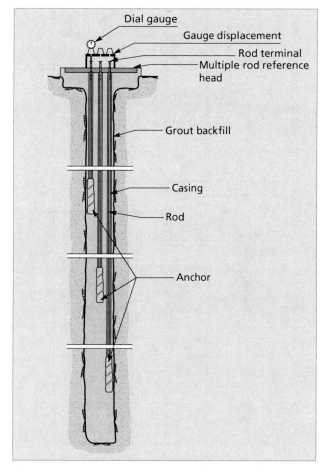

Figure 5.98   Rod extensometer with three anchor points in a borehole.

# Pore pressure and water level measurements

## Standpipes

A standpipe is a perforated pipe open at both ends, that can be inserted to the whole depth of a borehole or to shallower depths if the remainder of the borehole is to be sealed on completion (Figure 5.100A). The water level is measured, generally several hours after the borehole has been completed and then over a period of days or even longer. This measurement of water levels carried out both during and after drilling, provides essential information on the type of aquifer, and related hydrogeological and geotechnical conditions.

Standpipes create a zone of vertical hydraulic continuity in layered ground where such continuity does not exist naturally. The water level recorded, especially in deep standpipes, might be some unknown sum of the piezometric pressures encountered along the length of the hole. These conditions must be taken into account to avoid misinterpretations of the water table and piezometric heads (Chapter 4, Section 4.1).

## Open piezometers

A section of borehole is isolated with bentonite plugs and a perforated pipe open at the top is installed in the isolated section to measure the water height, or piezometric head, corresponding to that section (Figure 5.100B). In the Casagrande type, a porous tip is embedded in sand or gravel at the pressure measurement level.

## Closed piezometers

A sensing system or transducer is installed at a point previously isolated inside the borehole. From here, pore pressures are transmitted to a data reading device outside the borehole (Figure 5.100C).

There are three types of transducers: pneumatic, electrical resistance or vibrating wire (Figure 5.101). Pneumatic transducers are installed between the sensor and the data reading device and are appropriate for distances of less than 200 m when automatic measurements are not required. The precision of electrical resistance transducers is lost with temperature variations. Transducers with a vibrating wire allow the signal to be transmitted as far as 1000 m without any loss of precision.

Because their response time is short, these piezometers are used on low permeability ground. They allow pore pressure readings to be taken in various sections or at different

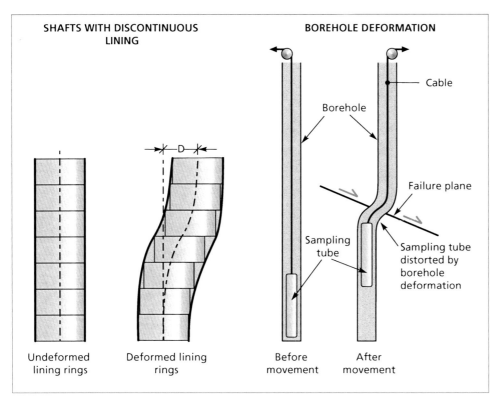

Figure 5.99   Deformations in shafts and boreholes.

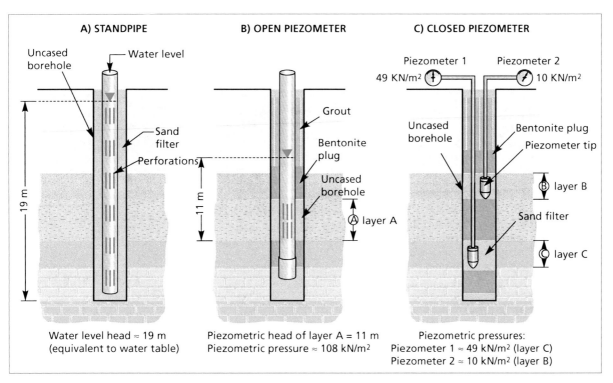

Figure 5.100   Standpipes and piezometers.

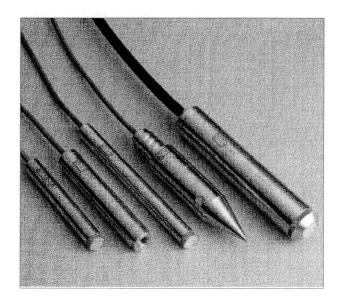

Figure 5.101  Vibrating wire piezometers.

Figure 5.102  Total pressure cell.

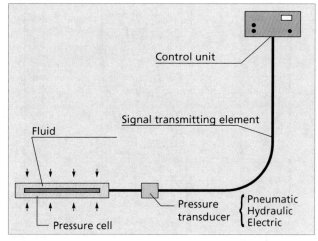

Figure 5.103  Total pressure cell equipment.

Figure 5.104  Vibrating wire pressure cells.

levels in the same borehole, and have the advantage of being less affected by possible ground movements. However they are more expensive than open stand pipe piezometers.

## Stress measurements

Measurements of stress caused by loads and lateral earth pressures, in the ground and in structures, are taken using total pressure cells. Load cells are used to measure stresses or loads transmitted to anchorage points.

### Total pressure cells

These cells consist of two welded steel plates with a fluid such as oil inside them (Figures 5.102 and 5.103). Ground pressure exerted on the cells is transmitted through the fluid to a pressure transducer, which may be pneumatic, hydraulic or electric.

Total pressure cells are used on experimental embankments for pre-loading control, on permanent structures such as behind retaining walls and tunnel linings.

### Load cells

These cells are installed at anchorage points to measure stress transmitted to the ground as well as the stress of the anchor itself (Figure 5.104). Measurement points are situated between the anchorage head and the ground. The cells are of various types:

— Mechanical: deformations are measured directly with a strain gauge.
— Hydraulic: deformations are measured with cells with an oil chamber which transmits the load to a transducer.
— Electrical: deformations in cylindrical metal cells are transmitted to electric sensors.

## Recommended reading

Clayton, C.R.I., Matthews, M.C. and Simons, N.E. (1995). Site investigation. Blackwell Science.
Colwell, R.N. (Ed) (1999). Manual of remote sensing. American Society of Photogrammetry. Sheridan Press.
Day, R.W. (1999). Geotechnical and foundation engineering. McGraw-Hill.
Joyce, M.D. (1982). Site investigations practice. E. and F.N. Spon.
McDowell, P.W., Barker, R.D., Butcher, A.P. et al. (Eds) (2002). Geophysics in engineering investigations. Geological Society, Engineering Geology Special Publications, 19.
Reynolds, J.M. (1997). An introduction to applied and environmental geophysics. John Wiley and Sons Ltd.

## References

Barton, N. (1981). Shear strength investigations for surface mining. 3rd Int. Conf. on Stability Surface Mining. Vancouver.
Brady, B.H.G. and Brown, E.T. (1985). Rock mechanics for underground mining. George Allen and Unwin, London.
ISRM (1981). Rock characterization. Testing and monitoring. Int. Soc. for Rock Mechanics. Suggested methods. Brown, E.T. (Ed.). Commission on testing and monitoring, ISRM. Pergamon Press.
BSSC – Building Seismic Safety Council (1998). 1997 Edition. NEHRP Recommended provisions for seismic regulation for new buildings, FEMA 302/303, developed for the Federal Emergency Management Agency, Washington, D.C.
Clayton, C.R.I., Matthews, M.C. and Simons, N.E. (1995). Site investigation. Blackwell Science.
Dahlberg, R. (1974). A comparison between the results from Swedish penetrometers and standard penetration test. Results in sand. ESOPT, 2:2.
de Freitas, M.H. (1992). Site investigation. MSc Lecture Notes. Universidad Complutense de Madrid. (Unpublished)
de Mello, V.F.B. (1971). The standard penetration test. Proc. 4th Pan-American Congress on Soil Mechanics and Foundation Engineering. Vol. I, pp. 1–86. Puerto Rico.
Fookes, P.G. (1997). Geology for engineers: the logical model; prediction and performance. The First Glossop Lecture. Geological Society of London. Ql. Jl. Engineering Geology, Vol. 30, no. 4, pp. 293–424.
ISRM – Int. Soc. of Rock Mechanics. Commission on Testing Methods (1985). Suggested method for determining point load strength. Int. J. Rock Mech. Min. Sci. and Geomech. Abstr. 22, 51–60.
Kim, K. and Franklin, J.A. (1987). Suggested methods for rock stress determination. Int. Journal Rock Mech. Min. Sci. and Geomech. Absts., 24–1, pp. 53–74.
Landsat Data Users Notes (1993). EOSAT. Vol. 8, no. 2. http://landsat.org.
Photogrammetric engineering and remote sensing (2000). Amer. Soc. of Photogrammetric and Remote Sensing, Vol. 66, no. 4.
Robertson, P.K. and Campanella, R.G. (1983). Interpretation of cone penetration test. Part I. Sand. Canadian Geotechnical Journal, 20, 4, pp. 718–733.
Schmertmann, J.K. (1978). Guidelines for cone penetration test performance and design. U.S. Dept of Transportation, Federal Highway Administration, Offices of Research and Development. Report No. FHWA-TS-78-209.
Thornburn, S. (1963). Tentative correction chart for the standard penetration test in non-cohesive soils. Civ. Eng. and Public Works 58; 683: 752–753.
Tyrell, A.P., Lake, L.M. and Parsons, A.W. (1983). An investigation of the extra costs arising on highway contracts. TRRL Supplementary Report SR 814, Transport and Road Research Laboratory. U.K.

# 6

# ROCK MASS DESCRIPTION AND CHARACTERISATION

1. Methodology
2. Description and zoning
3. Intact rock characterisation
4. Description of discontinuities
5. Rock mass parameters
6. Rock mass classification and characterisation

## 6.1 Methodology

The description and the characterisation of outcrops of rock during the early stages of a site investigation are essential for any geological engineering study of the properties and geotechnical characteristics of rock materials and masses.

A study of outcrops provides information needed to evaluate the geotechnical behaviour of rock masses, and allows the more advanced stages of investigation to be planned and their results to be interpreted. Rock mass characterisation can be a complex task, owing to the great diversity of conditions and properties found in nature; this is particularly so in areas where rock and soil materials are present together and in areas with heavily jointed, tectonised or weathered materials. Descriptions should include all aspects and parameters that can be observed, deduced or measured from the outcrops.

Descriptions of rock masses for geotechnical purposes require additional observations and measurements from those made for geological purposes alone. It is therefore essential to establish a system of standard working procedures in order to assist communication between all professionals concerned with descriptions of rock mass outcrops, and so reduce as much as possible the effect of subjectivity. This is achieved through systematic methods of observation and use of standardised terminology, taking the following into account:

— All factors should be examined systematically and in logical sequence.
— No basic information on the outcrop should be omitted.
— The descriptions should transmit a clear mental picture of the materials and mass, enabling the most relevant information to be deduced from them.
— The amount of data collected should be statistically representative.

Field characterisation of a rock mass is a progressive exercise that begins with the identification of general ground conditions and their objective description, and the identification and classification of the materials making up the rock mass. At a later stage, more complex observations concerning properties and other specific factors may require a greater degree of interpretation, leading to an increase in subjectivity.

The normal procedure is to start with a general description of any aspects and characteristics that can be seen at first glance; looking mainly at lithology and tectonic structure, different zones in the rock mass that are more or less homogeneous are identified. A detailed description and characterisation is then made of the components and properties of these zones. Finally, with these data, geomechanical classification of the rock mass is carried out. Description of each zone must be done separately, and include the study of the intact rock, discontinuities and the rock mass as a whole, with a description of both its intrinsic properties and the external factors conditioning its behaviour.

The stages of the systematic procedure to describe rock mass outcrops can be summarised as follows:

— Identification of the components of the outcrop, and general description.
— Grouping these components into different zones with a general description of each zone.

| Project: | | | |
|---|---|---|---|
| Phase of study: | Element investigated: | | |
| Location and accesses: | | Author: | Date: |
| Observations: | | | |
| PHOTO | | SKETCH | |
| GENERAL GEOLOGICAL DESCRIPTION: | | | |
| BASIC DESCRIPTION OF EACH ZONE: | | | |
| Zone I: | | | |
| Zone II: | | | |
| Zone III: | | | |

*Figure 6.1*    Data sheet for rock mass description and separation into zones.

- Detailed description of each zone.
  - Intact rock.
  - Discontinuities.
- Description of rock mass parameters.
- Global characterisation and rock mass geomechanical classification.

Definitive characterisation of rock masses depends on the appropriate evaluation of each of these aspects, which are dealt with in the following sections.

A **general description of the outcrop** should identify the basic components of the rock mass in terms of their condition and general characteristics, with a description of each of these components: rocks, soils, areas with water, dominant discontinuities, etc.

**Division into zones** that are roughly homogeneous is mainly based on lithological and structural criteria. The number of zones to be established and the area these cover will depend on the degree of heterogeneity of the materials and structures making up the rock mass, the extent of the outcrop, the amount of detail required and the purpose of the investigation. The general characteristics of each zone must be described (*Figure 6.1*).

A **description of each zone** is done separately and in detail. This should be as clear and as objective as possible, and standard terminology should be used so that the same description is available to different observers, thus avoiding differences in the interpretations of observations or measurements from the same zone. During this stage, the physical and mechanical characteristics, of both intact rock and discontinuities, are described together with their properties; aspects of this are set out in *Table 6.1*.

Descriptions are done qualitatively and, wherever possible, numerically. Standard tables, scales, indexes and reference values are available for the purpose of quantifying the different properties and characteristics of the rock mass and its elements. As well as being useful for establishing objective values to work with, the quantifying of parameters is also necessary for the application of rock mass classifications.

Given the large number of parameters to be evaluated, data sheets such as those shown in *Figure 6.2* are of great practical use for systematically gathering data, as they allow observations and measurements to be noted down clearly. Where outcrops are extensive, there should be several measuring points or sites in each zone, with data collected systematically from each. The greater the number of measurements and sites, the more representative the

| Table 6.1 | FIELD DESCRIPTION FOR ROCK MASS CHARACTERIZATION | | |
|---|---|---|---|
| Scope of study | Characteristic or property | Method | Classification |
| Intact rock | Identification | Direct observation or with a magnifying glass | Geological and geotechnical classification |
| | Weathering | Direct observation | Standard indices |
| | Strength | Indices and *in situ* tests | Empirical strength classifications |
| Discontinuities | Orientation | Direct measurement with geological compass | |
| | Spacing | Field measurements | Standard classifications and indices |
| | Persistence | | |
| | Roughness | Observations and field measurements | Comparison with standard profiles |
| | Wall strength | Schmidt hammer Field indices | Empirical strength classifications |
| | Aperture | Observations and field measurements | Standard indices |
| | Infilling | | |
| | Seepage | | |
| Rock mass | Number of sets of discontinuities | Field measurements | Standard indices and classifications |
| | Block size | | |
| | Intensity of jointing | | |
| | Extent of weathering | Field observations | Standard classifications |

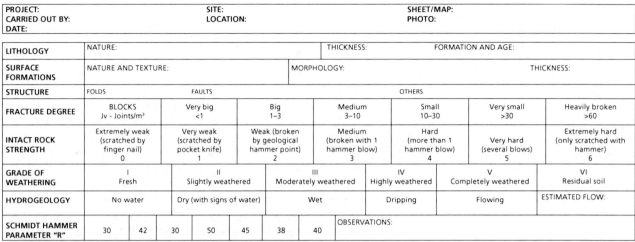

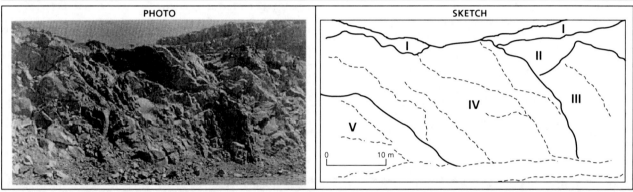

Figure 6.2  Data sheet for gathering geomechanical data in the field.

results will be with respect to a global characterisation of the rock mass.

The **description of rock mass parameters** is based on data collected at each site in the field. The parameters to be established include the number and orientation of discontinuity sets and the typical characteristics of each, including their relative importance, the size and shape of the blocks defined by them and making up the rock mass, together with its fracture degree, along with other factors influencing the rock mass behaviour, such as the degree of weathering and its hydrogeological properties.

The final stage of the descriptive process is **global rock mass characterisation**, which should give the geological and geomechanical conditions of the rock mass as a whole. With these results, **geomechanical classifications** can be applied, providing information on the mechanical quality and strength of the rock mass, together with numerical data for application to different engineering projects, such as tunnels and slopes. This last stage requires greater experience and must integrate knowledge of the site with that of the regional geology of the study area. Results from the geomechanical characterisation of rock outcrops are presented in the form of detailed maps and geological-geotechnical cross sections.

## 6.2 Description and zoning

The first step in the study of rock mass outcrop is to identify its components and describe them so that the mass can be divided into different sectors or zones, which are described separately. Photographs and schematic drawings of the outcrop, showing the basic characteristics of each zone, are of great use at this stage.

Provided the outcrop is not very extensive, the different zones of the rock mass can usually be distinguished at first glance from general differences in appearance and rock type, so an initial separation into zones can be made according to different types of lithology, structural elements, the degree of fracturing and the state of weathering, etc. This makes subsequent description and the systematic application of data-gathering and measuring procedures easier, although difficulties occasionally arise, for example, in cases where outcrops are small or few in number, or where the rock mass to be characterised covers an extensive area.

The following sequence is recommended:

**a) Identification of the outcrop**

This includes location, geographical situation, access, extent, geometric characteristics and prevailing conditions. There should also be an indication as to whether the outcrop occurs naturally or results from excavation.

**b) Photographs and diagrams**

**c) General geological description**

— Formation and geological age.
— Lithology.
— Visible large-scale geological structures.
— General structural features: whether the rock mass is stratified, faulted, jointed or massive.
— Weathered areas and their thickness.
— Presence of water, springs, etc.

It is recommended that all data observed should be included, even data considered to be of secondary importance. Likewise, any uncertainties related to the condition of the outcrop should also be included in the description.

**d) Division into zones and general description of each zone**

Dividing the outcrop into zones based on lithological and structural criteria is carried out to reduce the outcrop to sectors that are more or less homogeneous. Not too many zones should be established, although the number and extent of these will depend on the degree of heterogeneity of the materials and structures forming the rock mass, the extent of the outcrop, the degree of detail required and the object of the investigation. Each zone should be given a short general description, without entering into the detail of the intact rock or its discontinuities; this should include data on lithology, the degree of weathering and fracture, and the presence of water. These qualitative descriptions give an idea of the material to be studied but do not give a quantitative evaluation of the rock properties and its components.

The use of data sheets like those shown in *Figure 6.1* is recommended for dividing the outcrop into zones and giving a general description of each one.

**e) Identification of unique zones**

Unique zones are areas, elements or structures which do not re-occur throughout the rock mass but influence its properties and mechanical behaviour, for example faults, dykes, areas of breccias, cavities, areas with flowing water, and so forth. These zones should be described and dealt with on an individual basis, indicating the specific problems they present and the influence they have on global rock mass behaviour.

## 6.3 Intact rock characterisation

Aspects to be described in the field are:

— Identification.
— Weathering.
— Uniaxial compressive strength.

# Identification

Visual identification of a rock is established from its composition and the texture or geometric relationship of its minerals and grains. If it is possible to deduce the genetic characteristics from mineral paragenesis, chemical composition, the form and structure of the deposit and its temporal and spatial relationships with other rocks, then these aspects should also be added to the description.

The most practical observations are:

— Mineralogical composition.
— Grain or crystal shape and size.
— Colour and transparency.
— Hardness.

In order to observe these properties accurately, the rock has to be cleaned and the weathered surface layer removed. Depending on the type of rock, other aspects may be determined, such as the presence or absence of cleavage.

Lithological classification of the rock can be made from its **mineral composition**. If dimensions permit, the most common rock forming minerals can be identified from samples, using a magnifying glass. Detailed identification of minerals requires a petrographic study, commonly requiring the use of a thin section; this should always be carried out in cases of doubt regarding mineral identification.

Once the minerals have been described, the rock is named and classified. The system most recommended is based on geological classifications devised for geotechnical or engineering purposes. Included in Table 6.2 is the classification proposed by the British Standard Institution (BSI).

Identification of the rock is completed by defining its grain size and colour. Grain size appears as a classification criterion in Table 6.2; special terminology based on this parameter is applied to sedimentary rocks.

**Grain size** refers to the average size of the minerals or fragments of rock making up the intact rock. Estimation of grain size is normally done visually either with a ruler or with the aid of grain size comparators. A hand-held magnifying glass is useful for fine-grained rocks. The size of the mineral particles making up the rock may be homogeneous (equigranular rocks) or show considerable variation (heterogranular rocks).

Universally recognised terms and size intervals are shown in Table 6.3.

The **colour** of a rock depends on the minerals it is composed of. Some minerals have a distinctive colour but this is often modified by certain substances or impurities contained in them. Colour description is similar to that used for soils and comprises a main colour, a secondary colour and their intensity (for example, pale greenish-grey granite). In order to avoid any subjectivity in the description, rock colour charts should be used. Observations must be carried out on fresh rock after the altered surface layer has been removed.

When the rock has not been affected by weathering processes and presents its original characteristic colour, it is defined as fresh rock. Colour variations in the outcropping rock indicate that the rock material has undergone weathering processes. Such variations may affect all the minerals involved or only a few of them, an aspect that should be reflected in the description.

**Hardness** is a property directly related to strength, which depends on mineralogical composition and the degree of weathering of the rock material. Description is qualitative and rock density and strength are generally used as criteria, with grade 1 indicating a less dense and hard rock ($\gamma = 15$ kN/m$^3$ and $\sigma_c = 5$ MPa) and grade 14 for the most dense and hard rock ($\gamma = 27$ kN/m$^3$ and $\sigma_c = 180$ MPa).

The Moh's scale is used to evaluate mineral hardness, with a value of 1 given to the softest material (talc) and a value of 10 to the hardest (diamond).

# Weathering

The degree of weathering of the rock is an important consideration as it has a permanent effect on its mechanical properties. As the weathering process progresses, the porosity, permeability and deformability of the rock material change at the same time as its strength diminishes. Physical and chemical weathering processes affecting rocks are described in Section 3.2 of Chapter 3.

The degree of weathering in rock can be identified systematically from the descriptions in Table 6.4. When describing the degree of weathering, four basic terms are used: fresh, discoloured, disintegrated and decomposed; these terms may be followed by other qualifying terms e.g. "wholly" or "partially".

The principal agent of weathering is climate, so weathered rock will therefore vary in appearance according to different climatic regions. Figure 6.3 shows some examples of different degrees of weathering in granitic rock. Weathering processes affect different types of rock unevenly, as described in Section 3.2 of Chapter 3; the greater the exposure time to atmospheric agents, the more intense these processes will be.

# Strength

Intact rock strength can be estimated at the outcrop from field indexes or from correlations with data obtained from simple *in situ* tests, such as the Point Load Test (PLT) or the Schmidt hammer.

The range of rock strength can be roughly estimated from **field indexes**. The criteria for their identification, which should be applied to the rock once the altered surface layer has been cleaned, are described in Table 3.7 of Chapter 3.

The **point load test** (PLT) fails a rock specimen between two platen points, from which a point load strength index $I_s$

# ROCK MASS DESCRIPTION AND CHARACTERISATION

**Table 6.2  IDENTIFICATION AND CLASSIFICATION OF ROCKS FOR ENGINEERING PURPOSES**

| Grain size mm | Bedded rocks (mostly sedimentary) | | | | | | Metamorphic rocks | | | Igneous rocks: generally massive structure and crystalline texture | | | | |
|---|---|---|---|---|---|---|---|---|---|---|---|---|---|---|
| | Grain size description | | ≥50% of grains are carbonate | ≥50% of grains are of fine-grained volcanic rock | | | Grain size description | Foliated | Massive | Grain size description | | | | |
| 20 | RUDACEOUS | CONGLOMERATE Rounded boulders, cobbles and gravel cemented in a finer matrix | Calcirudite | | Fragments of volcanic ejecta in a finer matrix Rounded grains AGGLOMERATE Angular grains VOLCANIC BRECCIA | SALINE ROCKS HALITE AN-HYDRITE GYPSUM | COARSE | GNEISS Well developed but often widely spaced foliation sometimes with schistose bands Migmatite Irregularly foliated; mixed schists and gneisses | MARBLE QUARTZITE GRANULITE | COARSE | GRANITE[1] | These rocks are sometimes porphyritic and are the described for example, as porphyritic granite | DIORITE[1,2] | GABBRO[1,2] | Piroxenite Peridotite |
| 6 | | BRECCIA Irregular rock fragments in a finer matrix | | | | | | | | | | | | | |
| 2 | ARENACEOUS | SANDSTONE Angular or rounded grains, commonly cemented by clay, calcitic or iron minerals Quartzite Quartz grains and siliceous cement Arkose Many feldspar grains. Greywacke Many rock chips | LIMESTONE AND DOLOMITE Calcarenite | | Cemented volcanic ash TUFF | | MEDIUM | SCHIST Well developed undulate foliation; generally much mica | HORNFELS AMPHIBOLITE SERPENTINE | MEDIUM | MICRO-GRANITE[1] | These rocks are sometimes porphyritic and are the described as porphyries | MICRO-DIORITE[1,2] | DOLERITE[3,4] | |
| 0.6 | | | | | | | | | | | | | | | |
| 0.2 | | | | | | | | | | | | | | | |
| 0.06 | ARGILLACEOUS | MUD-STONE SILTSTONE CLAYSTONE | Calcareous mudstone CHALK Calcisiltite Calcilutite | | Fine-grained TUFF Very fine-grained TUFF | | FINE | PHYLLITE Slightly undulose foliation SLATE Well developed foliation planes | | FINE | RHYOLITE[4,5] | These rocks are sometimes porphyritic and are the described as porphyries | ANDESITE[4,5] | BASALT[5] | |
| 0.002 | | | | | | | | MYLONITE Found in fault zones, mainly in igneous and metamorphic areas | | | | | | | |
| Amorphous or crypto-crystalline | | Flint: occurs as bands of nodules in the chalk Chert: occurs as nodules and beds in limestone and calcareous sandstone | | | | COAL LIGNITE | Amorphous or crypto-crystalline | | | Amorphous or crypto-crystalline | OBSIDIAN[5] | VOLCANIC GLASS | | |
| | | Granular cemented, except amorphous rocks | | | | | | CRYSTALLINE | | | Pale ← | | | → Dark | |
| | | SILICEOUS | CALCAREOUS | | SILICEOUS | CARBONA-CEOUS | | SILICEOUS | Mainly SILICEOUS | | ACID Much quartz | INTER-MEDIATE Some quartz | BASIC Little or no quartz | ULTRA BASIC | |

## SEDIMENTARY ROCKS

Granular cemented rocks vary greatly in strength; some sandstone is stronger than many igneous rocks. Bedding may not show in hand specimens and is best seen in outcrop. Only sedimentary rocks, and some metamorphic rocks derived from them, contain fossils.

Calcareous rocks contain calcite (calcium carbonate) which effervesces with dilute hydrochloric acid.

## METAMORPHIC ROCKS

Generally classified according to fabric and mineralogy rather than grain size.

Most metamorphic rocks are distinguished by foliation which may impart fissility. Foliation in gneiss is best observed in outcrop. Non-foliated metamorphics are difficult to recognise except by association.

Most fresh metamorphic rocks are strong although perhaps fissile.

## IGNEOUS ROCKS

Composed of closely interlocking mineral grains. Commonly strong and not porous when fresh, except in many cases for ignimbrite.

Mode of occurrence:

1. Batholiths; 2. Laccoliths; 3. Sills; 4. Dykes;
5. Lava flows; 6. Veins

(BSI, 1999).

is obtained that can be correlated with uniaxial compressive strength. The **Schmidt hammer test** gives a rebound number resulting from the application of the hammer to the rock of interest, from which the uniaxial strength can be estimated. Both tests and their applications are described in Section 5.5 of Chapter 5; the test apparatus can be seen in Figures 5.71 and 5.72 of the same chapter. For both these tests, it is recommended that a high number of measurements be taken and that a statistical analysis be carried out on them. With the strength values obtained from these methods, intact rock can be classified according to the criteria in *Table 6.5*.

| Table 6.3 | CLASSIFICATION OF GRAIN SIZE IN ROCKS | |
|---|---|---|
| Description | Grain size | Equivalent soil types |
| Coarse-grained | >2 mm | Gravel |
| Medium-grained | 0.06–2 mm | Sands |
| Fine-grained | <0.06 mm | Silts and clays |

| Table 6.4 | DESCRIPTION OF THE DEGREE OF WEATHERING |
|---|---|
| Term | Description |
| Fresh | No visible sign of weathering of the rock material. |
| Discoloured | Change in colour of the original rock material. The degree of change from the original colour should be indicated. If the colour change is confined to particular mineral constituents this should be mentioned. |
| Disintegrated | The rock is weathered to the condition of a soil in which the original fabric is still intact. The rock is friable but the mineral grains are not decomposed. |
| Decomposed | The rock is weathered to the condition of a soil in which the original material fabric is still intact, but some or all of the mineral grains are decomposed. |

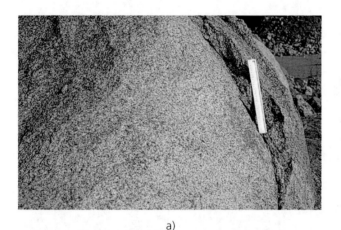

a)

b)

c)

Figure 6.3  Different grades of weathering in a granitic rock. a) slightly discoloured rock with changes in the original colour of the rock; b) rock disintegrated into soil with the original fabric still intact; c) decomposed rock, with decomposed minerals and loss of original texture. The scale shown is 30 cm length in photographs a) and c) and 2 m in b).

# ROCK MASS DESCRIPTION AND CHARACTERISATION

| Table 6.5 | CLASSIFICATION BASED ON ROCK STRENGTH |
|---|---|
| Uniaxial compressive strength (MPa) | Description |
| 1–5 | Very weak |
| 5–25 | Weak |
| 25–50 | Moderately hard |
| 50–100 | Hard |
| 100–250 | Very hard |
| >250 | Extremely hard |

## 6.4 Description of discontinuities

Discontinuities play a decisive role in conditioning the properties of rock masses, and their strength, deformational and hydraulic behaviour. The main factor determining the strength of jointed hard rock masses is the shear strength of discontinuities. To estimate this, it is necessary to define the characteristics and properties of the discontinuity planes.

Section 3.5 of Chapter 3 describes types of discontinuity and the physical and geometric parameters conditioning their properties and mechanical behaviour. Description and measurement of these parameters should be carried out in the field for each set:

— Orientation.
— Spacing.
— Persistence.
— Roughness.
— Wall strength.
— Aperture.
— Filling.
— Seepage.

Some of these parameters, such as roughness, wall strength, aperture and filling, determine the mechanical behaviour and shear strength of the discontinuities.

## Orientation

Systematic discontinuities occur in sets with more or less similar orientation and characteristics. The relative orientation and spacing between different sets in a rock mass define the shape and size of the blocks making up the rock mass.

An unfavourable orientation of engineering works with respect to the orientation of discontinuity planes may permit instability to develop within a rock mass and can induce failure of engineered structures and facilities. Figure 3.68 of Chapter 3 shows examples of how engineering works, such as slopes, dams and tunnels, are influenced by the orientation of planes of weakness.

Orientation of a discontinuity in space is defined by its dip direction (direction of the line of maximum gradient of the discontinuity with respect to north) and dip (gradient of that line with respect to the horizontal). Measurements are carried out with a geological compass or a Clar-type compass.

Dip direction is measured clockwise from north and varies between 0° and 360°. Dip is measured with the clinometer, with values ranging from 0° (horizontal plane) to 90° (vertical plane). The values of dip direction and dip are usually recorded in this order on data sheets, with an

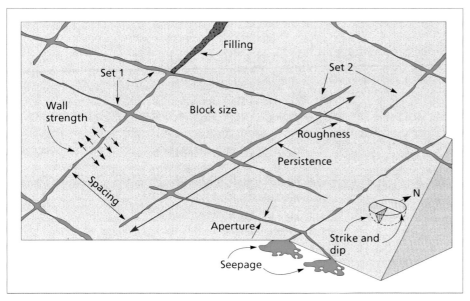

Figure 6.4 Diagram of the geometric properties of discontinuities (Hudson, 1989).

indication of the type of discontinuity they correspond to. For example, a value of $S_0$ 270°/60° indicates a bedding plane with a 60° dip and a dip direction of 270°.

The orientation of a discontinuity plane can also be defined from its strike (the angle formed between a horizontal line drawn on the discontinuity plane and magnetic north) and its dip; dip direction (north, south, east or west) should also be indicated. The strike of the plane and dip direction form an angle of 90° (Figure 6.5). For example: $J_2$ 135°/50° SW indicates a discontinuity plane belonging to a set of joints $J_2$ striking 135° with respect to north and towards the east, with a dip of 50° towards the south-west; the orientation of this plane can also be defined by 315°/50° SW.

In order to define each set adequately it is recommended that a sufficient number of discontinuity orientations are measured. The number of measurements will depend on the size of the area under study, the random distribution of orientations of the planes and the detail of the analysis. If orientations are constant, or there is little dispersion, fewer measurements will be required.

Graphic representation of the orientation of different discontinuity sets can be carried out by:

— Stereographical projection, representing the poles or planes with average values of different sets.
— Rosette diagrams, which enable a large number of numerical measurements of orientations to be represented (Figure 6.6).
— Block diagrams, giving a general view of the sets and their respective orientations, as shown in Figure 3.77 of Chapter 3.
— Symbols on geological maps, indicating average strike values and the dip direction, and value for different types of discontinuity (joints, faults, foliation, etc.).

## Spacing

The spacing between discontinuity planes conditions the block size of intact rock, thus defining the role this will

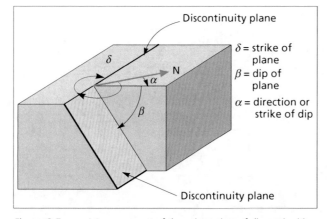

Figure 6.5 Measurement of the orientation of discontinuities.

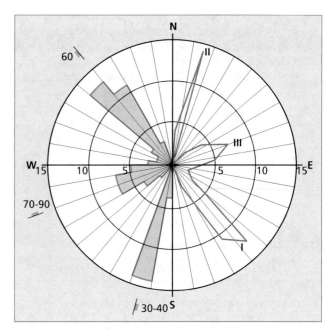

Figure 6.6 Two methods of representing orientation data on a rosette diagram (ISRM, 1981).

play in the mechanical behaviour of the rock mass, and its relevance with respect to the influence of discontinuities. In rock masses with spacing several metres wide, the properties of either the intact rock or the discontinuity planes will govern deformation and failure processes, depending on the scale of the engineering work under consideration and its orientation with respect to discontinuities. With closer spacing, ranging from several decimetres to one or two metres, rock mass behaviour will be influenced by planes of weakness. If spacing is very close, the rock mass will be heavily jointed and may show isotropic behaviour determined by the properties of the sets of more or less uniform blocks as a whole.

Spacing is defined as the distance measured perpendicularly between two discontinuity planes belonging to the same set. Its value normally refers to the average or modal spacing of values measured for discontinuities of the same set.

Measurements of spacing are taken with a tape measure along a length that is sufficiently representative of the frequency of discontinuities (at least three metres). As a rule, measurement length should be ten times higher than spacing. The tape should be placed perpendicular to the planes in order to measure the distance between adjacent discontinuities.

On the exposed surfaces of rock outcrops, it is not generally possible to carry out measurements of spacing perpendicular to the discontinuities; therefore, apparent spacing is measured and corrections must be made to obtain real spacing. Figure 6.7 shows an outcrop face in which only the apparent spacing of three discontinuity sets can be measured. The tape measure is placed perpendicular to the trace

# ROCK MASS DESCRIPTION AND CHARACTERISATION

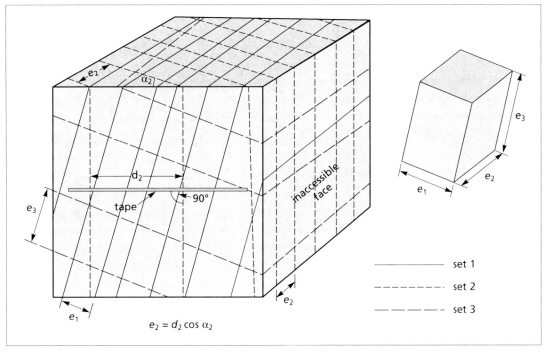

Figure 6.7  Measurement of joint spacing from observation of an exposed rock outcrop (ISRM, 1981).

lines of the planes of each set, and distance $d$ is measured; this is then corrected to obtain real spacing:

$$e = d \cdot \cos\alpha$$

where $e$ is real spacing, $d$ is the average distance measured with the tape and $\alpha$ is the angle between the line of measurement and the strike of the set.

Table 6.6 shows the terms used to describe spacing. Examples of discontinuities with different spacing are shown in Figure 6.8.

## Persistence

The persistence of a discontinuity plane attempts to represent the areal extent or size of a discontinuity within a plane, measured by length along the strike and along the dip of the plane. Although it is a parameter of considerable importance, it is difficult to quantify because what is normally seen in outcrops from simple observation are traces of discontinuity planes along the apparent dip.

Persistence is measured with a measuring tape. If the outcrop permits three-dimensional observation of the discontinuity planes, lengths should be measured along the dip and strike. Discontinuities may or may not terminate against other discontinuities and this should be indicated in the description. The most persistent sets should be highlighted, as these are generally a major factor in rock mass failure. Table 6.7 shows a description of persistence.

Table 6.6  **DESCRIPTION OF SPACING**

| Description | Spacing |
|---|---|
| Extremely close spacing | <20 mm |
| Very close spacing | 20–60 mm |
| Close spacing | 60 mm–0.2 m |
| Moderate spacing | 0.2–0.6 m |
| Wide spacing | 0.6–2 m |
| Very wide spacing | 2–6 m |
| Extremely wide spacing | >6 m |

Table 6.7  **DESCRIPTION OF PERSISTENCE**

| Persistence | Length |
|---|---|
| Very low | <1 m |
| Low | 1–3 m |
| Medium | 3–10 m |
| High | 10–20 m |
| Very high | >20 m |

(ISRM, 1981).

Singular discontinuities, such as faults and dykes, are normally very persistent and represent the largest planes of weakness in rock masses. Special attention should therefore be paid to their characterisation and description.

Figure 6.8  Examples of spacing of discontinuities in outcrops. a) Very close spacing (5 cm) for the main set of discontinuities perpendicular to the ruler; b) Limestone rock mass with two main sets of discontinuities, one vertical with "moderate" persistence and other horizontal with "low" persistence, both with "very close" spacing and forming "very small" blocks; c) Good quality quartzite rock mass with horizontal and vertical discontinuities spaced at around 0.5–1 metres.

## Roughness

The main purpose of describing and measuring roughness is to evaluate the shear strength of a discontinuity, $\tau$. For discontinuities without cohesion, this can be estimated from field data and empirical expressions, such as those described in Section 3.5 of Chapter 3 and in *Box 6.1* at the end of this section. Roughness increases shear strength, which diminishes with an increase in aperture and, in general, with the thickness of the filling.

The term **roughness** is used in a broad sense to refer to both the **waviness** of discontinuity surfaces and small-scale irregularities of the surfaces. These are sometimes defined as first and second order, respectively, and a description of roughness therefore requires two scales (*Figure 6.10*):

— Decimetric and metric scale for surface waviness (large-scale undulation or first order roughness): planar, undulating or stepped surfaces.
— Millimetric or centimetric scale for irregularities or unevenness (small-scale or second order roughness): slickensided, smooth or rough surfaces.

The term "slickensided" should only be used if there is clear evidence of previous shear displacement along the discontinuity (ISRM, 1981).

There are several methods for measuring roughness in the field. The method selected depends on the accuracy and scale of measurement required, and accessibility of the outcrop; they include both qualitative estimations and numerical measurements. The quickest and simplest method is visual comparison of the discontinuity with the **standard roughness profiles** of *Figure 6.11*. In qualitative terms, a discontinuity plane can be described, for example, as undulating-smooth, planar-rough or undulating-rough.

*Figures 6.12* and 3.78 of Chapter 3 show different examples of descriptions of roughness and other parameters for discontinuity surfaces.

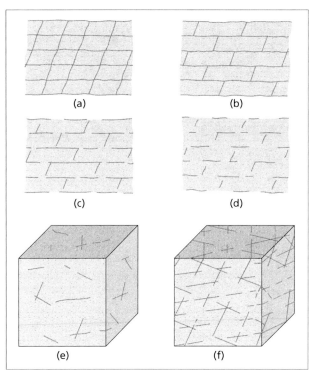

Figure 6.9  Diagrams showing different models of persistence in various sets of discontinuities (ISRM, 1981).

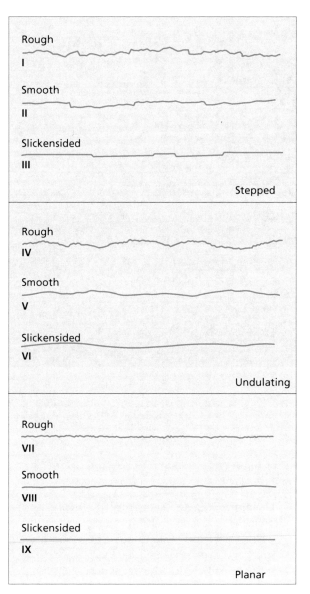

Figure 6.11  Roughness profiles. Profile length ranges from 1 to 10 metres (ISRM, 1981).

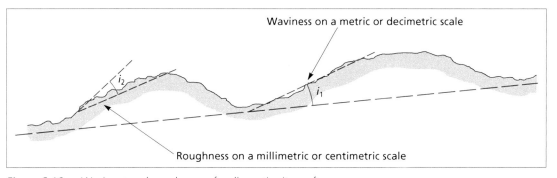

Figure 6.10  Waviness and roughness of a discontinuity surface.

Figure 6.12  a) Highly persistent "smooth, undulating" discontinuity in a volcanic rock mass; b) "Rough, planar" discontinuity in quartzites. The scales of measurement shown are 2 m and 30 cm respectively.

Numerical measurements of waviness and roughness can be carried out using more accurate methods:

— **Linear profiles.** A straight edge is placed on the most prominent asperities usually in either the direction in which shearing is likely to occur or in which shear strength is to be evaluated, and the distance between the edge and the discontinuity surface (considered to be typical of the plane) is measured at regular intervals that are appropriate for giving a detailed record of the x-y values from which the angles of roughness and waviness can be obtained. The distance measured depends on the scale of roughness and ranges from a few decimetres for small-scale roughness to several metres for decimetric or metric scale roughness.

— **Plate method.** This method is used to measure roughness angles of the discontinuity plane in several directions and is of particular use when the direction of potential movement is unknown. The results give local variations in the discontinuity surface with respect to its general dip. A series of flat plates of different diameter (5, 10, 20 or 40 cm, depending on the scale of work) are placed over different areas on the discontinuity and the strike and dip of the plate are measured with a geological compass. When a large plate is used (for example, 40 cm), the roughness angle will be less than when measured with smaller plates, as shown in Figure 6.13. The results can be represented stereographically with respect to possible different directions of sliding or movement on the plane.

Measurements should be carried out on profiles that are representative of the roughness of the planes. In order to establish values of the angles of roughness and waviness, it is recommended that a large number of measurements be taken. If the direction of potential sliding along a discontinuity is known or assumed, it is along this particular direction that roughness should be estimated. If the direction is not known, the roughness for several directions of potential sliding on the discontinuity plane must be characterised.

## Strength of discontinuity wall

The strength of a discontinuity wall influences its shear strength and deformability. It depends on the type of intact rock, the degree of weathering and the presence or absence of filling. In clean, unaltered discontinuities, strength would

# ROCK MASS DESCRIPTION AND CHARACTERISATION

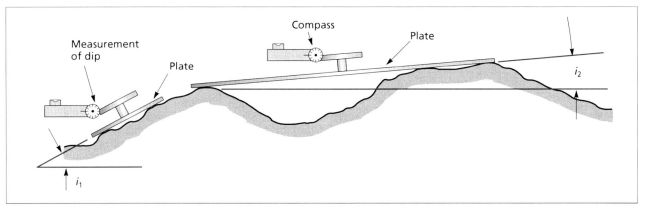

*Figure 6.13* Plates for measuring roughness of discontinuities (ISRM, 1981).

be the same as that of the intact rock, but it is normally less because of weathering of their walls; alteration processes affect discontinuity planes to a much greater degree than they do intact rock (*Figure 6.14*). For this reason, the degree of weathering of the intact rock must be assessed also when the discontinuity wall strength is measured (see *Table 6.4*).

Strength can be estimated in the field by applying the **Schmidt hammer** directly on the discontinuity, using the same procedure for measuring intact rock strength described in Section 5.5 of Chapter 5, or comparing with **field indexes** (see Table 3.7 of Chapter 3), in which rock wall strength generally ranges from $R_0$ to $R_6$.

In both cases, measurements must be carried out on walls that are representative of the state of alteration of the discontinuities, with the most frequent or significant discontinuities in the rock mass also being taken into consideration.

## Aperture

Aperture is the perpendicular distance separating discontinuity walls when there is no filling (*Figure 6.15a*). This parameter may vary considerably in different areas of the same rock mass: while aperture may be high at and near ground level, it decreases with depth and may even close. Aperture has a great influence on discontinuity shear strength, even in tight discontinuities, as its influence on the permeability of discontinuities can modify effective stresses acting on their walls. Displacement or dissolution processes in the discontinuity may give rise to apertures of considerable size.

Measurement is carried out directly with a ruler marked in millimetres. When separation is very small a calliper can be inserted in the aperture. Measurements are taken along at least three metres of the discontinuity to determine whether the aperture shows variations; if this is the case, these should be indicated. *Table 6.8* shows the terminology used for description. Measurements should be carried out for each

*Figure 6.14* Deposition of iron hydroxide and associated alteration by oxidation on the surfaces of discontinuities of a quartzitic rock mass while the intact rock remains fresh.

| Table 6.8 | DESCRIPTION OF APERTURE |
|---|---|
| Aperture | Description |
| <0.1 mm | Very tight |
| 0.1–0.25 mm | Tight |
| 0.25–0.5 mm | Partly open |
| 0.5–2.5 mm | Open |
| 2.5–10 mm | Moderately wide |
| >10 mm | Wide |
| 1–10 cm | Very wide |
| 10–100 cm | Extremely wide |
| >1 m | Cavernous |

(ISRM, 1981).

Figure 6.15   a) Discontinuity in sandstones with a "very wide" aperture and no filling; b) "Rough, planar" discontinuity with a "wide" aperture and dry clay filling; c) "Rough, undulating" discontinuity in limestones with a "very wide" aperture and clay filling. The scale of measurement shown in the photographs are 0.6 m (a) and 30 cm (b and c).

set of discontinuities, with the most representative average values adopted for each one.

## Filling

Discontinuities may be filled with a different material from that of the wall rock. There are many varieties of fill material, with a great diversity of physical and mechanical properties. As the presence of filling determines discontinuity behaviour, it is essential that all aspects related to its properties and state should be recognised and described. It should be borne in mind that weak or altered infilling materials may undergo important variations in their short-term strength properties if there is a change in their water content, or if movement takes place along the joints they fill.

A description of the main characteristics of filling in the outcrop should include its nature, width or thickness, shear strength and permeability (these last two parameters are assessed in an indirect and qualitative way):

— Width is measured directly with a ruler marked in millimetres.
— Description of the filling includes identification of the material, mineralogical description and grain size. If the filling is the result of decomposition and alteration of the discontinuity wall material, the degree of weathering should be evaluated, usually corresponding to the terms decomposed or disintegrated (Table 6.4).
— Strength can be estimated with the field estimation indexes shown in Table 3.7 of Chapter 3 (for soft filling the scale is from $S_1$ to $S_6$), or by use of the Schmidt hammer, described in Section 5.5 of Chapter 5.

## Box 6.1

### Evaluation of discontinuity shear strength from field data

- **Barton and Choubey's equation**

The shear strength of discontinuities with no cohesion can be estimated from field data using Barton and Choubey's equation (1977), as described in Section 3.5 of Chapter 3:

$$\tau_p = \sigma'_n \tan (JRC \log_{10} (JCS/\sigma'_n) + \phi_r)$$

where:

$(JRC \log_{10} (JCS/\sigma'_n) + \phi_r)$ represents the peak friction angle of the discontinuity, $\phi_p$

$\tau_p$ = peak shear strength in rough discontinuities without cohesion.

$\sigma'_n$ = normal effective stress on the discontinuity plane.

JRC = joint roughness coefficient of the discontinuity.

JCS = joint wall compression strength.

$\phi_r$ = residual friction angle of the discontinuity, which can be estimated from the expression:

$$\phi_r = (\phi_b - 20°) + 20° (r/R)$$

where $r$ is the rebound value of the Schmidt hammer on the discontinuity wall, $R$ the rebound value of the Schmidt hammer on the intact rock and $\phi_b$ the basic friction angle of the material. The values of $R$, $r$ and JCS are estimated in the field, as explained in Section 3.5 of Chapter 3; the value of $\sigma'_n$ is calculated according to the vertical (i.e. lithostatic) load on the discontinuity, when specific weight of the rock material and, if present, water pressure are known; and the value of $\phi_b$ can be estimated from tables in the literature (Table 3.13 of Chapter 3). The roughness coefficient value, JRC, is estimated by comparison with the typical profiles that appear in Figure 3.85 of Chapter 3.

- **Tilt test**

The frictional strength of a discontinuity can be easily estimated by carrying out a simple field test, known as the tilt test, described in Section 5.5, Chapter 5. Values obtained from this test can be compared with those calculated with the empirical method above.

*Figure 6.16* Circulation of water along discontinuities in a highly weathered sandstone rock mass.

— The moisture content should be indicated and a qualitative estimation of the permeability of the filling material given.
— If it is recognised that shear displacement has taken place along the filling, this should be indicated, as the properties and mineralogical structure will have undergone changes with respect to their initial conditions.

*Figure 6.15* shows examples of discontinuities with filling.

## Seepage

Water present in a rock mass is generally derived from the flow that circulates through its discontinuities (secondary permeability), although in certain permeable rocks seepage through intact rock (primary permeability) may also be significant. How seepage is observed and described, whether in clean or filled discontinuities, is shown in *Table 6.9*.

## 6.5 Rock mass parameters

In order to carry out a global characterisation of a rock mass based on outcrop data, together with a full description of its components (intact rock and sets of discontinuities), a number of other representative factors must be taken into account:

— Number and orientation of discontinuity sets.
— Block size and fracture degree.
— Degree of weathering.

| Table 6.9 | DESCRIPTION OF SEEPAGE IN DISCONTINUITIES | |
|---|---|---|
| Class | Unfilled discontinuities | Filled discontinuities |
| I | The discontinuity is very tight and dry; water flow along it does not appear possible. | The filling materials are heavily consolidated and dry; significant flow appears unlikely due to very low permeability. |
| II | The discontinuity is dry with no evidence of water flow. | The filling materials are damp but no free water is present. |
| III | The discontinuity is dry but shows evidence of water flow, i.e. iron staining, etc. | The filling materials are wet with occasional drops of water. |
| IV | The discontinuity is damp but no free water is present. | The filling materials show signs of outwash; continuous flow of water (estimate litres/minute). |
| V | The discontinuity shows seepage, occasional drops of water but no continuous flow. | The filling materials are washed out locally; considerable water flow along out-wash channels (estimate litres/minute and describe pressure i.e. low, medium, high). |
| VI | The discontinuity shows a continuous flow of water (estimate litres/minute and describe pressure i.e. low, medium, high). | The filling materials are washed out completely; very high water pressures (estimate litres/minute and describe pressure). |

(ISRM, 1981).

## Number and orientation of discontinuity sets

The mechanical behaviour of a rock mass, its deformability and failure mechanisms are conditioned by the number of discontinuity sets. Orientation of the different sets may also determine the stability of engineering works.

The degree of fracturing or fracture intensity ("blockiness"), and the size and shape of blocks of intact rock result from the number of discontinuity sets and the spacing between them. Each set is characterised by its spatial orientation and the properties and characteristics of the planes.

In field surveys of rock masses, all sets present must be registered and their relative degree of importance evaluated. This can be expressed by assigning to the sets numbers that are correlated to their order of importance; thus, the main set (with most persistence, closer spacing, greater aperture, etc.) would be set Number 1.

The average orientation of a set is evaluated by means of stereographic projection or rosette diagrams from orientation data based on measurements for each discontinuity. With the aid of computer programs, this work can be carried out with speed and accuracy.

The rock mass can be classified by the number of sets as shown in Table 6.10; these vary from massive rock masses, or with a single main set of discontinuities, for example a granitic rock mass, to rock masses with four or more sets, all of which could be significant, such as outcrops of folded and heavily jointed slate. The presence of three main discontinuity sets orthogonal to each other is frequent in sedimentary rock masses, with one of the sets corresponding to the bedding planes.

| Table 6.10 | CLASSIFICATION OF ROCK MASSES BY THE NUMBER OF SETS OF DISCONTINUITIES |
|---|---|
| Type of rock mass | Number of sets |
| I | Massive, occasional random joints |
| II | One joint set |
| III | One joint set plus random |
| IV | Two joint sets |
| V | Two joint sets plus random |
| VI | Three joint sets |
| VII | Three joint sets plus random |
| VIII | Four or more joint sets |
| IX | Crushed rock, earth-like |

(ISRM, 1981).

Discontinuity sets can be represented graphically by means of block diagrams, such as those shown in Figure 6.17 and Figure 3.77 of Chapter 3, which allows spatial visualisation of their relative orientation and the size, and shape, of blocks of intact rock.

## Block size and fracture degree

The role of **block size** is decisive as it conditions both the behaviour of the rock mass and its strength, and deformational

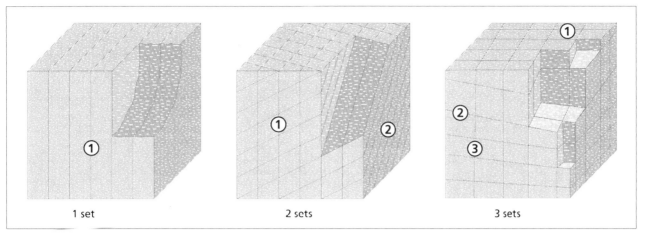

*Figure 6.17*   Block diagrams showing sets of discontinuities.

properties. Block size and shape are defined by the number of discontinuity sets, their orientation, spacing, and persistence. Block size can be described as follows:

— By the block size index $I_b$, which represents the average dimensions of blocks typifying the outcrop. For example, for a sedimentary rock with bedding planes and two discontinuity sets perpendicular to each other, the $I_b$ index would be defined as:

$$I_b = (e_1 + e_2 + e_3)/3$$

where $e_1$, $e_2$ and $e_3$ are the average values of spacing in the three discontinuity sets.

— By the parameter $J_v$ (volumetric joint count) which represents the total number of discontinuities that intercept a unit volume (1 m³) of rock mass. Because three-dimensional observation of an outcrop is difficult, the value of $J_v$ is usually determined for each set by counting the discontinuities that intercept a certain length and measuring perpendicularly to the strike of each set (or, if this is not possible, by correcting the measurement of apparent strike):

$$J_v = \sum \frac{n° \text{ of discontinuities}}{\text{length of measure}}$$

For example, for a rock mass with three discontinuity sets ($J_1$, $J_2$ and $J_3$):

$$J_v = (n° J_1/L_1) + (n° J_2/L_2) + (n° J_3/L_3)$$

The length measured depends on the spacing of each set and usually varies between 5 and 10 metres. A quicker, though less accurate, estimation of the value of $J_v$ can also be done by counting the total number of discontinuities intercepting a length $L$ in any direction of interest (intersecting the greatest number of planes possible); this value will correspond to the discontinuity frequency, $\lambda$:

$$\lambda = \frac{\text{number of discontinuities}}{L \text{ (m)}}$$

or

$$\lambda = \frac{1}{\text{average spacing of discontinuities (m)}}$$

The value of $J_v$ can be related to block size as shown in *Table 6.11*; values greater than 60 correspond to heavily jointed rock masses.

*Table 6.12* shows a rock mass classification based on block size and shape and the fracture degree. *Table 6.13* shows the terminology proposed for rock block dimensions and descriptions.

*Figures 6.18* and *6.19* illustrate examples and give descriptions of different block sizes and of the degree of fracturing in rock masses, which depends on the number of discontinuity sets.

| Table 6.11 | DESCRIPTION OF BLOCK SIZE BASED ON NUMBER OF DISCONTINUITIES |
|---|---|
| Description | $J_v$ (joints/m³) |
| Very large blocks | <1.0 |
| Large blocks | 1–3 |
| Medium-sized blocks | 3–10 |
| Small blocks | 10–30 |
| Very small blocks | >30 |

(ISRM, 1981).

| Table 6.12 | CLASSIFICATION OF ROCK MASSES BASED ON BLOCK SIZE AND SHAPE | |
|---|---|---|
| Class | Type | Description |
| I | Massive | Few joints or very wide spacing |
| II | Blocky | Blocks approximately equidimensional |
| III | Tabular | Blocks with one dimension considerably smaller than the other two |
| IV | Columnar | Blocks with one dimension considerably larger than the other two |
| V | Irregular | Wide variation of block shape and size |
| VI | Crushed | Heavily jointed rock mass |

(ISRM, 1981).

| Table 6.13 | ROCK BLOCK DIMENSIONS AND DESCRIPTION |
|---|---|
| First term | Maximum dimension |
| Very large | >2.0 m |
| Large | 0.6–2 m |
| Medium | 0.2–0.6 m |
| Small | 60–200 mm |
| Very small | <60 mm |
| **Second term** | **Shape of block** |
| Blocky or cubic | Equi-dimensional |
| Tabular | Thickness much less than length or width |
| Columnar | Height much greater than cross section |

(BSI, 2003).

Figure 6.18  a) Small cubic blocks in marly limestone materials formed by sets of discontinuities perpendicular to each other; b) large columnar blocks (approximately 3 m high) in a volcanic rock mass with a greater degree of jointing in the lower part.

Rock mass fracture is defined by the number of discontinuities, their spacing and conditions, whatever their type and origin. The **degree of fracturing** is usually expressed with the **RQD index** (rock quality designation), which is measured from borehole cores as described in Box 5.1 of Chapter 5. Rock mass quality is classified on the basis of this value, as shown in Table 3.11 of Chapter 3.

Although it is useful, this index does not take into account aspects such as orientation, separation, fillings and other conditions present in discontinuities, and it is therefore not suitable for describing the characteristics of fracture in rock masses; additional aspects such as these should be covered by descriptions obtained in the field and from borehole cores. It must be also considered that RQD is a function of the direction in which it is measured.

# ROCK MASS DESCRIPTION AND CHARACTERISATION

Figure 6.19   a) Massive volcanic rock mass divided naturally into very large blocks; b) dolomitic rock mass broken up into very small blocks.

A description of fracture using outcrop data may refer to the number of discontinuity sets and block size, as has already been described. The RQD index can be estimated from outcrop data using empirical correlations, such as that of Palmstrom, 1975 (ISRM, 1981):

RQD = 115 − 3.3 $J_v$     for $J_v > 4.5$
RQD = 100                  for $J_v \leq 4.5$

For example, for a rock mass of acceptable quality with an RQD of 65, the corresponding value of $J_v$ is 15, while for a poor quality rock mass with a RQD of 30, $J_v$ is 26.

The RQD index can also be estimated from the discontinuity frequency, $\lambda$, by means of the following expression that gives the minimum theoretical value of the RQD (Priest and Hudson, 1976) (Figure 6.20):

$$RQD \approx 100\, exp^{-0.1\lambda}\, (0.1\lambda + 1)$$

where $\lambda$ is the inverse of average spacing of the discontinuities.

## Degree of weathering

The description of the weathering is important as most construction on or in a rock mass is undertaken at shallow depth

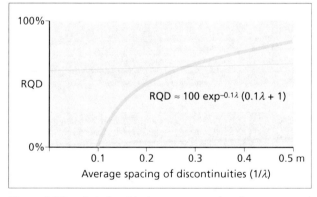

Figure 6.20   Relationship between spacing frequency and RQD index.

within the zone of surface weathering. It should include the degree, extent and nature of weathering so that their influence on the engineering properties of the rock mass can be evaluated (BSI, 1999). The features most commonly to be examined include the following:

— Strength and reduction of strength (from any direct or indirect strength measurements).
— Colour and discoloration.

### Table 6.14 CLASSIFICATION OF THE DEGREE OF WEATHERING IN THE ROCK MASS

| Degree of weathering | Type | Description |
|---|---|---|
| I | Fresh | No visible signs of weathering; perhaps slight discolouration on major discontinuity surfaces |
| II | Slightly weathered | Discolouration indicates weathering of rock material and discontinuity surfaces. All the rock material may be discoloured by weathering and may be somewhat weaker externally than in its fresh condition. |
| III | Moderately weathered | Less than half of the rock material is decomposed and/or disintegrated to a soil. Fresh or discoloured rock is present either as a continuous framework or as corestones. |
| IV | Highly weathered | More than half of the rock material is decomposed and/or disintegrated to a soil. Fresh or discoloured rock is present either as a discontinuous framework or as corestones. |
| V | Completely weathered | All rock material is decomposed and/or disintegrated to soil. The original mass structure is still largely intact. |
| VI | Residual soil | All rock material is converted to soil. The mass structure and material fabric are destroyed. There may have been a large change in volume but the soil has not been significantly transported. |

(ISRM, 1981).

Figure 6.21 Examples of weathering in rock masses. a) Grade II: gneiss with slightly discoloured intact rock and discontinuity surfaces; b) and c) Grade III: moderately weathered calcareous and quartzitic rock masses, with alteration of the discontinuity surfaces and blocks of intact rock; d) Grade IV: highly weathered quartzitic rock mass, with separated weathered blocks of intact rock.

- The nature of weathering products.
- Fracture state and changes therein if attributable to weathering.

The classification of weathering, although may often not be appropriate due to the great variability in lithologies and other rock mass features, is useful for engineering purposes. The degree of weathering in a rock mass is evaluated from direct observation of the outcrop followed by comparison with the standard descriptions shown in *Table 6.14*. A six stage scale is used: fresh, slightly weathered, moderately weathered, highly weathered, completely weathered and residual soil. In order to observe weathering in intact rock, it may occasionally be necessary to break a piece of rock into fragments.

*Figure 6.21* shows examples of rock masses affected by different degrees of weathering.

## 6.6 Rock mass classification and characterisation

Geomechanical evaluation of a rock mass can be carried out using data obtained from the description and measurement of the characteristics and properties of intact rock, discontinuities and rock mass parameters. The application of **geomechanical classifications** based on these data allows the quality and approximate rock strength parameters, in terms of friction and cohesion, to be estimated (as described in Section 8 of Chapter 3). Figure 3.132 of Chapter 3 shows examples of rock mass classification.

In order to complete the global characterisation of the rock mass it is also necessary to evaluate other aspects that have a considerable influence on its mechanical behaviour, such as:

- Strength and deformability.
- Hydrogeological behaviour.
- State of stress.

Although these factors cannot be quantified from outcrop data, the corresponding characteristics of the rock mass can be determined, at least qualitatively, from observation.

Determining the **strength and deformability** of a jointed rock mass can be complex as these factors depend both on the strength properties of the intact rock and on discontinuities; this is further complicated by the fact that several types of discontinuity may coexist in the rock mass, each with its own characteristics. Areas that are tectonised, weathered or wet imply zones of weakness and anisotropy; this is also the case in areas of differing composition, or where certain structures associated with rocks are present, such as folds, faults or dykes; such areas will show different behaviour and characteristics with respect to strength and deformation.

Generally, in a rock mass with hard intact rock, strength is controlled by the different discontinuity sets, with either one set or a combination of several sets predominating, depending on their characteristics and orientation.

Strength values in high quality rock masses are lower than those of the intact rock of which it is constituted, although these may vary greatly depending on the way the discontinuity planes are arranged and oriented. In weak rock masses, the intact rock plays a more important role as the difference between its own strength and that of the discontinuities is less. In these cases, rock mass strength is characterised either by that of the intact rock, or by a combination of intact rock strength and the strength of the discontinuities.

For the estimation of rock mass strength, once the elements that determine it are established (for example, the presence of one or more discontinuity sets, intact rock strength, areas of weakness, prominent discontinuity planes or a combination of these), the corresponding empirical strength criteria, described in Section 3.6 of Chapter 3, can be applied.

In the same way, an approximate evaluation of deformability can be made, using the expressions and empirical criteria described in the same section.

**Hydrogeological factors** that should be considered are: water tables, flow directions, seepage and infiltration. Particular areas or elements of a rock mass which may imply the existence of barriers or preferred flow channels for water, such as faults, open joints and bedding, dykes, cavities and clay fillings, should also be identified.

A further important aspect is the description of the **state of stress** that the rock mass is subjected to. Although quantitative evaluation of stress magnitude is not possible from field data, an indication of stress directions anticipated within the rock mass can be obtained from geological and geomorphological observations. These can be complemented with existing knowledge of the geological and tectonic history of the area (see Section 3.7 of Chapter 3).

## Recommended reading

ISO (2003). Geotechnical investigation and testing—Identification and classification of rock. Part 1: Identification and description. International Standard. ISO 14689-1. Switzerland.

ISRM (1981). Suggested methods for rock characterization, testing and monitoring. ISRM Suggested methods. Ed. E.T. Brown. Pergamon Press.

ISRM (1980). Basic geotechnical description of rock masses. Int. J. Rock Mech. Min. Sci. and Geomech. Abstr. vol. 18, pp. 85–110.

# References

Barton, N. and Choubey, V. (1977). The shear strength of rock joints in theory and practice. Rock Mechanics, vol. 10, no. 1, pp. 1–54.

BSI (1999). BS5930:1999. Code of practice for site investigations. Bristish Standard Institution. London.

BSI (2003). BS EN ISO 14689-1. Geotechnical investigation and testing. Identification and classification of rock. British Standard Institution. London.

Hudson, J.A. (1989). Rock mechanics principles in engineering practice. Butterworths. Ciria, London.

ISRM (1981). Suggested methods for rock characterization, testing and monitoring. ISRM Suggested methods. Ed. E.T. Brown. Pergamon Press.

Priest, S.D. and Hudson, J. (1976). Discontinuity spacing in rock. Int. J. Rock Mech. Min. Sci. and Geomech. Abstr. 13, pp. 135–148.

# 7

# ENGINEERING GEOLOGICAL MAPPING

1. Definition
2. Types of maps
3. Mapping methods
4. Data collection
5. Applications

## 7.1 Definition

Engineering geological maps are used to present geological-geotechnical information for land use and planning, and to plan, construct and maintain engineering infrastructures; they provide data on the characteristics and properties of the soil and subsoil of a specific area to enable its behaviour to be evaluated and to forecast geological and geotechnical problems.

The data included in geological maps (topography, relief, lithology, structure, etc.) allow valuable information to be obtained about material properties, but the geological descriptions are not sufficient by themselves for geological engineering purposes:

— They do not provide quantitative data on the physical and mechanical properties of the materials or on their heterogeneity and anisotropy.
— They do not provide the components of the geological medium with engineering geological significance and their influence on planning and engineering work.
— They do not represent the dynamic nature of the geological medium in relation to engineering.

Engineering geological maps should consider the following general aspects that are relevant to engineering geology:

— Geotechnical descriptions and classifications of soils and rocks.
— Physical and mechanical properties of materials.
— Hydrogeological conditions and distribution of ground water.
— Geomorphological conditions and processes.
— Dynamic processes.

The content, detail and complexity of the map design depend on:

— The scale and geographical area covered.
— The specific purpose for which it is designed.
— The importance of different geological-geotechnical factors to be shown and their relationships.
— The information and data available, and how representative they are.
— The techniques used to represent them.

Engineering geological maps include:

— **Descriptive information** on geological materials and processes.
— **Quantitative data** on the different components of the geological medium and on the physical and mechanical properties of the materials involved.
— **Interpretative information** for their geotechnical or engineering application.

These documents **cannot replace site investigation for a particular project**, but they are an invaluable help to accomplish rational project design, foresee geological-geotechnical problems in a given area, plan the site investigations and interpret the results of field and laboratory tests.

## 7.2 Types of maps

### Classification

Engineering geological maps are prepared on scales appropriate to their purpose, providing geological-geotechnical information which may be basic (e.g., for regional planning) or specific to a particular application (selecting sites, excavations, foundations, etc.). The maps can be classified according to their intended purpose, content and scale, as shown in *Table 7.1*; *Table 7.2* includes a classification of maps according to their scale and content, as well as their methodology and applications.

In simple terms, engineering geological maps can be grouped as follows:

- **Maps for engineering geological ground evaluation**: qualitative maps with general classifications, problem zones, suitability of the ground for different uses, etc. The most common include:

  — Geological map interpretation; usually at scales between 1:50,000 and 1:100,000; and with a geologically based legend; of limited practical use.
  — Geotechnical characteristics of superficial deposits; at scales between 1:25,000 and 1:100,000; data on overburden, soils, alluvium, etc.; qualitative (and sometimes quantitative) description and general overall zoning.

- **Maps for engineering geological characterization**, which may include:

  — General characterization of the ground, on scales between 1:25,000 and 1:50,000, with engineering geological evaluation of the units taken as a whole, with property data and quality indicators.

# ENGINEERING GEOLOGICAL MAPPING

| Table 7.1 | CLASSIFICATION OF ENGINEERING GEOLOGICAL MAPS |
|---|---|
| Criterion | Type |
| Purpose | — **Multipurpose:** providing information covering many aspects of engineering geology for different planning purposes.<br>— **Special or specific:** providing information on one specific aspect or for one specific purpose. |
| Content | — **Analytical or thematic:** providing information and details on an individual component of the geological environment, for a variety of planning and engineering purposes or for specific use:<br>  — engineering geological conditions (according to multipurpose or special classifications)<br>  — engineering geological zoning (evaluation and delimitation of areas according selected criteria)<br>  — engineering geological suitability (delimitation of material distribution in terms of material suitability)<br>— **Comprehensive:** providing information on all components of the geological environment (rocks and soils, groundwater, geomorphic and geodynamic processes…) for a variety of planning and engineering purposes:<br>  — engineering geological conditions (according to multipurpose classifications)<br>  — engineering geological zoning (evaluation and delimitation of areas in terms of their engineering geological conditions)<br>  — engineering geological suitability (delimitation of areas in terms of their suitability)<br>or specific use:<br>  — engineering geological conditions (according to special classifications)<br>  — engineering geological zoning (evaluation and delimitation of special units)<br>  — engineering geological suitability (delimitation of areas in terms of their suitability)<br>— **Auxiliary:** provide information or data on some particular geological or geotechnical aspect (structural contour maps, isopach maps etc)<br>— **Complementary:** maps of basic data (geological, geomorphological, hydrogeological) |
| Scale | — Large scale (local):    >1:10,000<br>— Medium scale:    1:10,000 to 1:100,000<br>— Small scale (regional):    <1:100,000 |

Modified from UNESCO-IAEG (1976) and Hrasna and Vlcko (1994; in Proske et al, 2005).

| Table 7.2 | CLASSIFICATION OF ENGINEERING GEOLOGICAL MAPS ACCORDING THEIR SCALE | | |
|---|---|---|---|
| Type and scale | Content | Mapping method | Applications |
| **Regional**<br><1:10,000 | Geological data, lithological groups, tectonic structures, regional morphological features, large areas affected by processes. General information and interpretations of geotechnical interest. | Aerial photography, previous topographical and geological maps, existing information, and field observations. | Preliminary studies and planning, general information on the region and types of material, and geomorphological processes present. |
| **Local**<br>Preliminary survey phase.<br>1:10,000 to 1:500 | Description and classification of soils and rocks, structures, morphology, hydrogeological conditions, geodynamic processes, location of possible construction materials. | Aerial photography, and ground truth field survey, measurements and other field data. | Planning and viability of works and detailed site reconnaissance. Basic design. |
| **Local**<br>Site investigations phase.<br>1:5,000 to 1:500 | Material properties, geotechnical conditions, and other aspects important for the carrying out of a specific construction project. | All previous data plus data from boreholes and trial pits, geophysical methods, *in situ* and laboratory tests. | Detailed information on sites and geological-geotechnical problems. Detailed design and analyses. |

Modified from Ruíz Vázquez and González Huesca (2000).

— Engineering geological zoning for engineering projects at scales between 1:5,000 and 1:25,000 with quantitative information depending on application (foundations, slopes, excavations, construction materials, etc.).
— Detailed engineering geological mapping, on scales of between 1:100 and 1:2,000, with geotechnical information and data for a specific engineering project.

While small and medium scale engineering geological maps are usually produced by official or research institutions, local maps at scales over 1:10,000 are made by geotechnical or engineering geology specialists or, depending on their content, by structural geologists, geomorphologists and hydrogeologists.

*Figure 7.1* shows an example of a small scale, comprehensive, multipurpose geological engineering map. *Figure 7.2* shows an example of medium scale mapping for an urban planning application with the specific aim of being suitable for a variety of purposes; it corresponds to a part of the geological engineering map of the city of Zaragoza, Spain, on the original scale of 1:25,000. In addition to the content shown on the simplified and reduced legend in *Figure 7.2*, the map also provides data on foundation conditions (allowable loads, type of foundation and complementary investigation); the map is complemented by another one showing geomechanical characteristics and construction conditions.

## Content of engineering geological maps

The following basic information should be included regardless of the type of map produced:

— Topography and toponymy (e.g. Quarry Hill).
— Lithological distribution and description of the geological units.
— Soil, superficial deposits and weathered rocks depths.
— Discontinuities and structural data.
— Geotechnical classification of soils and rocks.
— Properties of soils and rocks.
— Hydrogeological conditions.
— Geomorphological conditions.
— Dynamic processes.
— Existing previous surveys and investigations.
— Geological hazards.

The most important of these factors are detailed below.

### Classification and geotechnical properties of soils and rocks

The delimitation and mapping of **rock or soil units considered to be homogeneous in terms of their physical and mechanical properties** such as strength, deformability, permeability and durability, is carried out based on the **geological properties** most closely related to their geotechnical properties. The mineralogical composition and lithology are directly related to the density and plasticity of the soils. The composition of rocks determines their hardness, strength, alterability and other properties. The mineralogical texture and structure are other aspects which provide information on the mechanical behaviour of the materials in relation to their porosity and density. Their permeability provides information on soil consistency and on the likely weathering conditions for soils and rocks. In the case of rock masses, the frequency, distribution and type of discontinuities, the degree of fracturing and of alteration, or weathering, all provide information on strength, deformability and permeability.

The geotechnical parameters to be represented in engineering geological mapping, depending in each case on the scale and intended purpose of the map and on the information and data available, are:

— Density.
— Porosity.
— Consistency and activity.
— Permeability.
— Uniaxial compressive and tensile strength.
— Strength parameters.
— Deformability.
— Durability.

In addition, special purpose maps, whether thematic or comprehensive, include other geotechnical properties and features, depending on their intended application.

The **geotechnical classification** of soils (USCS unified soil classification system, Casagrande plasticity chart) and rocks (based on different physical and mechanical properties) and the use of **empirical correlations** and field indices allow the geotechnical properties to be evaluated and provide quantitative data. These aspects are dealt with in Chapters 2, 3 and 6 of this book.

The **geotechnical units** and their spatial distribution are generally established from the lithology, origin and geological characteristics of the materials, from field observations and measurements and from photo interpretation and, where possible or necessary, from boreholes, *in situ* and laboratory tests and from sample analysis. Depending on both the mapping scale and the data available, these units are defined with different degrees of homogeneity (see Section 7.3).

The IAEG (1981a) proposed a procedure to follow for the classification and description of soils and rocks for engineering geological mapping, including the following:

— Lithological classification of rocks.
— Geological-geotechnical description and classification of rocks and rock masses (described in Chapter 3, Section 2) in terms of:
  • Intact rock: colour, texture, fabric, porosity, alteration and weathering, strength (Chapter 6, Section 3)

# ENGINEERING GEOLOGICAL MAPPING

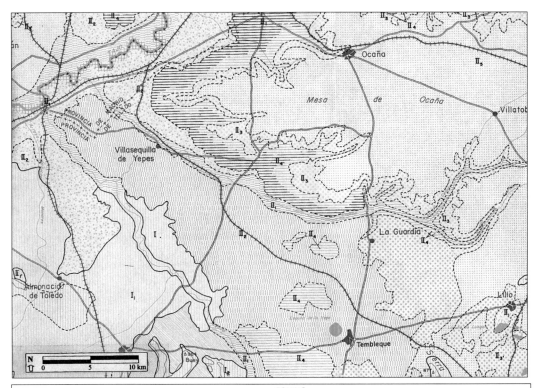

| ZONING | | | |
|---|---|---|---|
| REGION | AREA | | ZONING CRITERIA AND GENERAL CHARACTERISTICS |
| HIGH LANDS | MODERATE RELIEF | $I_1$ | Predominantly granites and gneisses. Flat morphology.<br>Permeability related to presence of sandy layers and fractured zones.<br>Acceptable natural drainage. Water in faults and fractured zones.<br>Very suitable mechanical characteristics except in weathered sandy layers.<br>High allowable loads without settlement. |
| | HIGH RELIEF | $I_2$ | Predominantly shales, quartzites and limestones. Steep morphology with slopes 20%–30% or more.<br>Large areas with surface covered by loose, fractured or disaggregated materials.<br>Impermeable materials. High surface run-off. Good drainage.<br>Very suitable mechanical characteristics. High allowable loads without settlement. |
| LOW LANDS | LOW RELIEF | $II_1$ | Silty clay, gravel and sand. Flat morphology.<br>Variable permeability depending on lithology. Natural drainage. Swampy areas.<br>Extremely variable mechanical characteristics. Low to medium allowable loads. Medium to high settlement. |
| | | $II_{1'}$ | Granular deposits. Flat morphology.<br>Permeable materials over permeable substratum, with poor drainage.<br>Suitable mechanical characteristics. Medium to high load capacity. Very little settlement. |
| | | $II_2$ $II_{2'}$ | Granular deposits with clays. Flat morphology.<br>Medium permeability materials. Good natural drainage. Steady water levels in deep aquifers.<br>Medium mechanical characteristics. Medium load capacity and settlement. |
| | | $II_3$ $II_{3'}$ | Sandy or marly limestone. Flat morphology. Horizontal layers covered with clay deposits.<br>Permeable fractured rock masses. Good drainage along fractures.<br>Suitable mechanical characteristics in fresh rocks. |
| | MODERATE RELIEF | $II_4$ | Clayey marls and gypsum. Sloping relief with sharp changes in angle.<br>Impermeable materials. Good surface drainage. Sulphurous water.<br>Intermediate mechanical characteristics in fresh rock masses.<br>Collapse due to gypsum dissolution processes. |
| | LOW RELIEF | $II_5$ | Sandy clays with gravel layers. Alluvial plain deposits. Flat morphology.<br>Semi-permeable materials over impermeable substratum, poor drainage.<br>Medium mechanical characteristics. Medium load capacity and settlement. |

| LEGEND | | | | |
|---|---|---|---|---|
| GENERAL BUILDING CONDITIONS AND TYPE OF PROBLEMS | | | | |
| SUITABLE | ACCEPTABLE | | UNSUITABLE | VERY UNSUITABLE |
| Geomorphological and geotechnical | Geotechnical | Geomorphological | Geomorph. and geotechnical | Lithological, geomorphological and geotechnical |
| Geotechnical | Geomorph. and geotechnical | Litholog., geomorph. and hydrolog. | Litholol., geomorph. and geotech. | |
| Geomorphological | Lithological and geotechnical | Hydrological and geotechnical | Hydrological | |
| | Geomorph., hydrolog. and geotech. | Lithol., hydrolog. and geotechnical | Geomorph., hydrolog. and geotech. | |
| | Lithological and geomorphological | Lithological and geomorphological | Lithol., hydrolog. and geotech. | |
| | | | Hydrological and geotechnical | |

Figure 7.1  Example of a general geotechnical map of the Toledo area, Spain (original scale 1:200,000; simplified legend) (IGME, 1972).

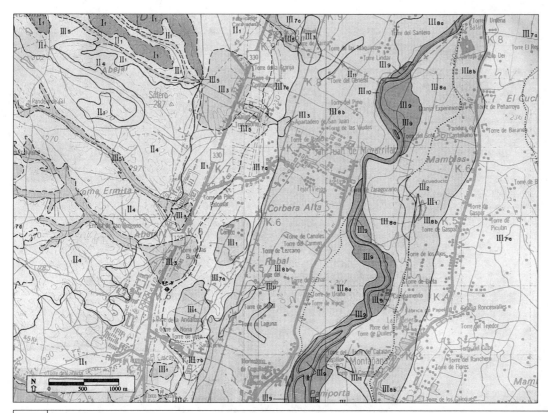

Figure 7.2  Example of a map of geo-mechanical characteristics and geological factors impacting on construction in the Zaragoza area, Spain (simplified legend; original scale 1:25,000) (IGME, 1987).

- Rock mass: structure, number of discontinuity sets, block size and shape, degree and profile of weathering (Chapter 6, Section 5)
- Characteristics of discontinuities (Chapter 3, Section 5; Chapter 6, Section 4)
— Geological-geotechnical classification and description of soils (Chapter 2, Section 2):
  - Name and type: grain size, organic material, plasticity, type of genetic deposit.
  - Description of the material: colour, shape and composition, weathering state, strength.
— Additional geological information: name and age of the geological formations.
— Anthropic fill materials, waste disposal, underground storage.

## Hydrogeological conditions

The presence of water in geological materials affects their properties and mechanical behaviour and should be considered in engineering geological mapping. The hydrogeological aspects are of special importance in geotechnical maps intended for planning and land use, exploitation of water resources and selection of sites for projects directly related to hydrogeological conditions, e.g., sites for urban or other types of waste and reservoirs.

As well as the changes in the conditions of the materials and the variation in their geotechnical properties, surface and ground water cause physical and chemical weathering which may result in changes in the relief and mass movement such as landslides, settlement and subsidence. In turn, the natural flow of water may be affected by engineering works, hydraulic structures, extractions, urbanization, deforestation and by mass movements.

The **hydrogeological data** included in engineering geological maps should allow the forecasting of hydrogeological changes associated with ground engineering and provide the information needed to avoid, minimise or control such changes. The data should include:

— Water distribution and water content of materials.
— Aquifer formations.
— Lakes, rivers and springs.
— Piezometric levels, depth and seasonal or other fluctuations.
— Confined aquifers.
— Flow direction and velocity.
— Infiltration zones and conditions.
— Hydrological parameters: permeability and storage.
— Hydrochemical properties and water quality.

Areas flooded in the past or zones potentially susceptible to flooding should have their limits of flooding mapped indicating the frequency or return period of the floods.

## Geomorphological conditions

The information related to the geomorphological conditions and processes of interest for engineering geological applications includes:

— Topography.
— Elements of relief: valleys, terraces, slopes, scarps, etc.
— Recent history and development of the landscape and relief, and its relationship with geology.
— Origin, evolution and age of the geomorphologic elements.
— Relationship with hydrogeology.
— Relationship with internal and external geodynamic processes.
— Prediction of processes such as erosion, subsidence, slope movements, etc.

These aspects can be shown on special or general purpose maps. The detail depends on the scale. Contour lines should always be shown.

The geomorphological aspects and interpretation of topography are very important to the physical description of an area and provide information on active processes and unstable zones; the site and alignment of many engineering works are conditioned by these factors.

## Geodynamic processes

Engineering geological maps should reflect the dynamic character of the geological medium and its implications for any engineering intended purpose. They should provide spatial information and data on the **internal and external dynamic processes operating**. The information to be included in geotechnical mapping, depending on the scale and the available data, is:

— Location and extension of the processes.
— Limits and associated morphological trends.
— Age.
— Intensity and frequency of occurrence.
— Level of activity, velocity.
— Conditions, causes and conditioning factors.
— Forecasting of potential processes (if possible).

These processes may be represented on all types of maps, and the degree of detail will depend on the scale.

An important aspect related to geodynamic processes are **geological hazards** which may affect populated areas, infrastructures and buildings. Chapters 13, 14 and 15 deal with these maps and include some examples of the mapping of hazards and vulnerability in relation to mass movements.

## 7.3 Mapping methods

For engineering geological mapping the **basic geological information** must be available on a **topographic plan**. If there is no existing topographic base map at the required scale, one should be drawn either by conventional methods or from aerial photography. In the same way, if there is no **geological map** available, at least in outline, or at the required scale, one should be prepared from the information available and from field work; what is usually done, for medium and large scale maps, is to complement the geological information available from maps on scales of 1:50,000 or 1:25,000 with more detailed field observations at a larger scale.

## Geotechnical zoning

Geotechnical zoning is based on the classification of **geotechnically "homogenous" units,** which may include units of different geological ages; these may, in turn, be divided into sub-units. The detail and degree of homogeneity will depend on the scale, the purpose of the map and the data available.

The geotechnical units and their spatial distribution are generally established from the lithology, origin and geological characteristics of the materials, specified from existing information and geological maps, photo interpretation, field observations and measurements. Depending on the scale of the map and the data available, the units are defined with different degrees of homogeneity. Where required for project design, the mapping units are described in more detail using data from *in situ* investigations, laboratory tests and analysis of samples.

The publications of the IAEG (1981a) and of UNESCO–IAEG (1976) propose the following **rock and soil unit classifications for engineering geological mapping** (based on the lithology and origin of the materials), applicable according to the map scale and arranged in descending order of degree of lithological and physical homogeneity:

— **Engineering geological type:** lithologically and physically uniform, characterized by average values of the engineering geological properties determined from *in situ* or laboratory measurements. These are generally shown on large-scale engineering geological maps.
— **Lithological type:** homogeneous in composition, texture and structure, but not usually uniform in physical properties; average mechanical properties cannot be given for the whole unit; usually a range of values is presented. These are used in large-scale and, where possible, medium-scale maps.
— **Lithological complex:** a group of related lithological types, forming at the same time and being under similar tectonic conditions but with variable lithology and physical properties. Data on individual lithological types and on the general behaviour of the unit are provided. They are used in medium-scale and sometimes small-scale mapping.
— **Lithological suite:** various lithological complexes (or engineering formations) developed under similar palaeographic and tectonic conditions; they present certain common lithological characteristics which differentiate them from other lithological suites; only general engineering geological properties are provided. Used in small-scale maps.

Dearman (1991) proposes the terms "Engineering formation" and "Engineering group" as alternatives for the last two units in the previous list.

This classification of units based on geological conditions may be used for specific or comprehensive multipurpose maps. The stratigraphic and structural characteristics for the different units should also be described.

In special purpose maps made for specific applications, zoning should be based on engineering geological parameters or conditions, and on the geological phenomena or processes which affect the intended purpose of the map, e.g., for the feasibility of linear infrastructures. Once the different units or zones have been described, established and drawn on the map, they can be classified in the legend in an **interpretative** way (e.g., the suitability of the territory for the intended purpose: suitable, suitable with restrictions or unsuitable); in general and specific maps, they can be classified in relation to geotechnical, geological, geomorphological or hydrogeological problems which may occur in particular applications.

## Representing data

Various organisations associated with engineering geology have formed committees to propose symbols and procedures for use in engineering geological mapping, and national and international standards have been developed for this purpose (the Geological Society of London, 1972; the Association of Engineering Geologists, AEG; or the International Association of Engineering Geology, IAEG, 1981b). However, due to the complexity of the geological materials and the different applications and purposes to which engineering geological maps are put, there is no standard procedure; there are significant differences between maps, both in the data represented and in the way they are shown.

The information on the characteristics and geotechnical properties of soils and rocks should be shown on the map by:

a) Assigning geological-geotechnical properties to the established lithological suites or units.

b) Dividing the homogeneous units according to some property (strength, density, plasticity, degree of fracturing, degree of weathering, etc.).
c) Zoning geotechnically homogeneous units and assigning them quantitative values.
d) Depicting the distribution of quantitative values by isolines.

*Table 7.3* shows the usual methods of representation. Standard geomorphological and geological graphic symbols are often used, but this is not the case with letters and numbers which are used to define the lithologies and ages of geological formations, since these do not provide information on physical or mechanical properties. In the aforementioned references a key is provided for the map symbols which represent the different types of soils and rocks, the hydrogeological properties, geomorphological and structural features, and also symbols for site investigations, among others. Dearman (1991) also includes a detailed list of symbols.

The **legend** should detail and clarify the information contained in the map, and is often wide ranging and explicit, including classification tables and complementary data. The sheets may also include auxiliary or complementary synthetic maps, on a much smaller scale than the geotechnical map, alongside the legend.

In the **memoir** which accompanies the sheets, the contents of the map and its legend are extended and detailed, presenting the results of the surveys and investigations carried out, the data obtained, the methodologies and criteria used in the preparation of the map, etc. and often including geological-geotechnical classifications of materials, borehole descriptions, test results and photographs of some aspects of engineering geological interest.

Engineering geological maps and their corresponding legends should be independent documents with no need to consult the memoir to interpret them.

Some geotechnical parameters and borehole or test data can be represented in small graphics or diagrams divided into sectors according to the information they contain, arranged on the map over the units to which they refer and explained in the legend (*Figure 7.3*).

| Table 7.3 | CARTOGRAPHIC REPRESENTATION OF BASIC ELEMENTS ON GEOTECHNICAL MAPS | | | | |
|---|---|---|---|---|---|
| Scale | Classification of soils and rocks | Geological-geotechnical properties | Hydrogeological conditions | Geomorphological conditions | Dynamic processes |
| Small | — Colours and ornaments<br>— Letters and numbers | — Colours and ornaments | — Symbols and numerical values | — Contour lines<br>— Specific symbols for geomorphological elements | — Symbols |
| Medium | | | — Outlines and lines<br>— Numerical values | — Contour lines<br>— Detailed morphological boundaries and features | |
| Large | | — Colours and ornaments<br>— Numerical values<br>— Diagrams and graphs | — Isolines<br>— Numerical values | | — Outlines and lines |

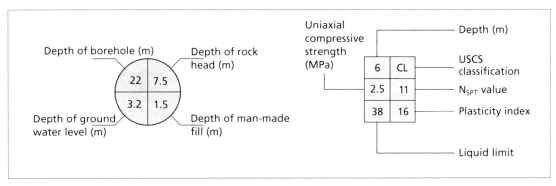

*Figure 7.3* Example of diagrams to represent geotechnical data from boreholes and tests.

## Computer aided mapping

The use of computer software and geographical information systems (GIS) in engineering geological mapping allows:

— Processing and analysis of data.
— Mapping of individual and combined factors or elements.
— Preparation of data bases.
— Ongoing updating of map data and information.
— Creation of 3D models and simulation of on-site action.

Computer aided mapping helps to eliminate errors and subjective interpretations during the characterization of geotechnical units, and when combined with GIS applications for computerized processing and analysis of the information, enables land to be zoned and classified based on specific criteria, depending on the purpose of the maps; this is normally achieved through the superposition of information layers or thematic maps prepared in advance. Only trained professionals with experience in geological engineering should specify the criteria and the weighting given to each of the factors considered, and interpret the results obtained.

## Geotechnical cross-sections

The geotechnical data and the subsoil information of interest can be represented in cross sections along relevant directions; these complement the engineering geological maps and present such information as variations of the material properties with depth, the limits of weathered zones and depth to bedrock. Boreholes and other subsoil investigation techniques are of considerable help when preparing engineering geological cross-sections, allowing different properties measured or estimated indirectly to be correlated.

Profiles and geotechnical cross-sections are essential in projects for foundations, dams or underground projects. In Chapter 10, dealing with tunnels, Figure 10.10 shows an example of a geomechanical zoning profile (cross-section).

The number and direction of the cross sections chosen depend on the geomorphology, type and orientation of tectonic structures and trends traits, lithological variation and complexity of conditions with depth. They should be carried out at appropriate scales for representing the information available; the depth should be equal to that reached in the investigation and subjective interpretations should not be included unless they are supported by sufficient data.

## 7.4 Data collection

The geotechnical characteristics and properties of soils and rocks can be evaluated from direct or indirect measurements or, as indicated in Section 7.2, through the application of geotechnical classifications, indices and empirical correlations which allow quantitative data to be obtained.

*Table 7.4* shows the data collection methods used for producing engineering geological maps; these methods, in general use in geological engineering, are described in detail in Chapter 5. Direct and indirect survey methods (bore holes, tests and geophysical methods) are used depending on the map type and purpose, the complexity of the zone under consideration and the budget and time frame available; the classification and description of the geological-geotechnical units will be made with varying degrees of homogeneity and detail depending on the quantity, quality and representability of the data.

All the available topographical, geological maps and other types (e.g. geomorphological, hydrogeological) should

| Table 7.4 | DATA COLLECTION METHODS FOR GEOTECHNICAL MAPPING |
|---|---|
| **Method** | **Data** |
| Photo-interpretation and remote sensing | — Soil and rock mapping<br>— Geological structures<br>— Stream and drainaje networks<br>— Dynamic processes |
| Surveying and collecting field data | — Geological and geomorphological details<br>— Geological-geotechnical data and measurements |
| Geophysical methods | Electrical resistivity:<br>— Porosity, fracturing, saturation, salinity<br>— Depth of water table<br>— Depth of bedrock<br>Seismic methods:<br>— Density, deformation moduli<br>— Degree and depth of weathered zones<br>— Depth of bedrock. |
| Boreholes, trial pits and sampling | — Provide representative samples<br>— Allow direct inspection of materials<br>— Physical properties and characteristics of the ground<br>— Hydrogeological conditions |
| *In situ* tests | — Stress and strain properties<br>— *In situ* stresses<br>— Permeability, water pressure<br>— Data from borhole tests |
| Laboratory tests | — Physical and mechanical properties of materials |

be collected beforehand as well as any information already existing on the site (reports, projects, publications, etc.). This existing information will help to ensure the process of engineering geological description is rational.

**Aerial photography** is very useful in the initial mapping phases, and for the preliminary and feasibility studies. The main disadvantages of photo-interpretation are that it does not provide information on the subsoil or lithological details, and the information has to be checked in the field. **Field surveys** are always necessary, and are normally carried out in several stages throughout the preparation of the map and in tandem with the completion of the geotechnical description. The methods above can be applied to all types of engineering geological mapping.

**Geophysical surveys,** at ground level and in boreholes, allow the depth of the bedrock and surface formations, the degree of weathering and depth of the weathered zones and the dynamic deformability parameters of the materials to be determined.

The quantitative information included on a map or in its legend must be reliable and representative of the zone it is assigned to, which means that the **data selection, analysis and interpretation,** and its grouping must be carried out rigorously by experienced specialists. If sufficient data on the properties of the materials, field measurements etc., is available, then statistical analysis can be carried out and the results presented in the form of histograms.

## 7.5 Applications

## Land and urban planning

The maps produced for land and urban planning are comprehensive, multipurpose maps which provide information on different geological-geotechnical aspects for different geotechnical or geological engineering purposes, such as regional, local and urban planning. They are normally drawn on a small or medium scale and include information on factors with implications for construction, required for an understanding of the problems related to foundations, excavations, ground stability, natural resources, water reserves and waste disposal. Another important environmental aspects are geological hazards; as well as spatial information, the maps should include information wherever possible on the magnitude and return periods of these processes.

Mapping for urban planning and development contributes to (González de Vallejo, 1977):

— Selecting environmentally favourable or sustainable sites and areas for urban planning and development.
— Reaching agreements on technical, economic and environmental aspects.
— Problem solving during the development, design and construction phases.

*Figure 7.4* shows an example of a medium scale urban planning map, where geological-geotechnical zones have been established based on the lithology, geomorphology and geotechnical classification of materials, and defined on the map by colours and fill patterns. From this information, and from the geotechnical properties of the materials, other maps of the suitability of the area are prepared in relation to the different engineering geological conditions.

*Figure 7.5* shows another example of mapping for urban planning. The engineering geological zones and subzones have been established on lithological criteria, on soil and superficial deposit thickness and on the depth of the bedrock.

One important aspect of land planning is the selection of waste disposal sites (for inert, municipal and special waste, but not radioactive waste). Proske *et al.* (2005) present a complete study of the different aspects to take into consideration when drawing up special purpose maps for the selection of waste disposal sites, describing the geological and engineering geological features and attributes to be considered, such as the criteria selection and the suitability evaluation. *Figure 7.6* shows an example of a land suitability zoning map for the selection of sites.

## Engineering

Engineering geology maps made for specific applications in geotechnical engineering serve two different purposes:

— Preliminary or feasibility studies for site or alignment selection.
— Project and construction information and data.

In both cases, the information on the map should be accompanied by geotechnical cross-sections.

Among the most important applications of engineering geological mapping are feasibility studies and the selection of alternatives for the alignment and construction of **linear infrastructures** such as roads and railways. The geological-geotechnical ground conditions may determine the route design. The maps should provide information on the topographical and morphological conditions, water courses, geological-geotechnical problems, general properties of the soils and rocks and construction materials; the scale of these maps is usually between 1:10,000 and 1:2,000. The maps should include a strip of at least 500 m on either side of the planned route, and wider than this where necessary (e.g., if there are extensive areas affected by landslides or other problematic conditions which could affect the planned route).

Detailed engineering geological mapping for particular projects is produced on larger scales, normally 1:500–1:2,000.

**362** GEOLOGICAL ENGINEERING

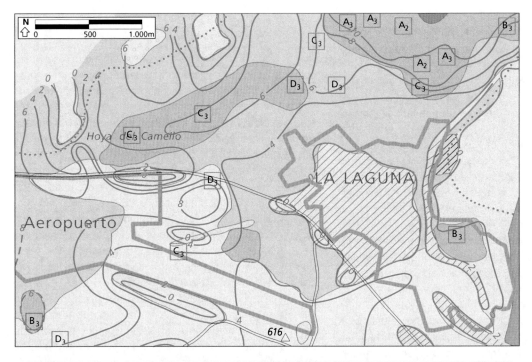
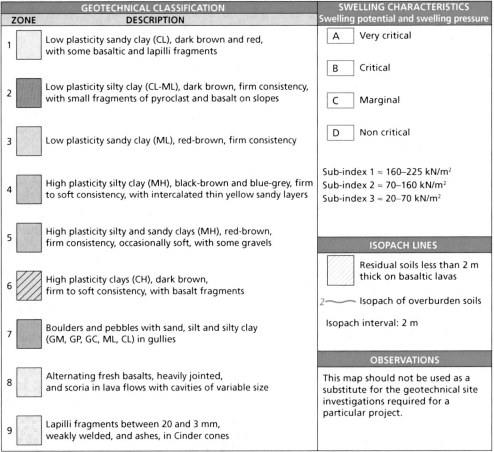

Figure 7.4  General geotechnical map for an area of the island of Tenerife (simplified legend; original scale 1:25,000) (González de Vallejo, 1977).

ENGINEERING GEOLOGICAL MAPPING

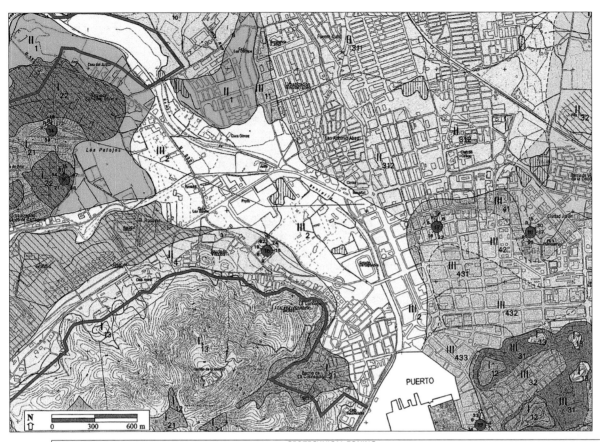

| | | GEOTECHNICAL ZONING | | | |
|---|---|---|---|---|---|
| | ZONES. Lithological description | | SUB-ZONES. Lithological description | Water level | Penetration strength |
| HARD ROCKS $I_1$ | Sub-volcanic, metamorphic and carbonated rocks | $I_1$ | | Not detected | Rock |
| WEAK ROCKS $I_2$ | Sedimentary rocks | $I_{21}$ | Micaceous and calcareous marls and sandstones | Not detected | Rock or fill 0–2 m thick |
| | | $I_{22}$ | Limestones, calcretes, pebbles, sands and conglomerates | Not detected | |
| HARD SOILS II | Chalky boulders with carbonated silt-sandy matrix | $II_1$ | | Water level at 3–10 m | Very dense granular or hard clay soils with $N_{spt} > 50$ |
| | Sandy silts and silty clays with boulders | $II_{21}$ | Polygenic cemented clays | Not detected | Intermediate soils with $N_{spt} = 4$–50; very dense granular soils or hard clay soils with $N_{spt} > 50$ |
| | | $II_{22}$ | Silty-sandy clays with gravel layers | Level to S of zone: 10–20 m Level to N of zone: 30–50 m | |
| | Silts and sandy clays. Levels of pebbles and sand | $II_{31}$ | Sandy clays >6 m thick | Near surface (0.5–5 m) Level to NW of zone: 30–50 m | Fills or granular very dense or hard clay soils with $N_{spt} > 50$ or rock |
| | | $II_{32}$ | Clays and sandy silts <6 m thick | Near surface (2–6 m) | |
| SOFT SOILS AND FILLS III | Sand-silty clays with levels of sand and gravels | $III_{11}$ | Alluvium <8 m thick | Deep water level to the N of the zone (>30 m) Near surface (2–5 m) | Fills or intermediate soils with $N_{spt} = 4$–50 |
| | Silty clays with pebbles, gravel and sands | $III_{12}$ | | Near surface (0–3 m) | |
| | Fill: clay, gravel and wastes on soil and bed rock | $III_{31}$ | 1–3 m thick, on phyllites and schists | Not detected | Fills or intermediate soils with $N_{spt} = 4$–50 |
| | | $III_{32}$ | 1–3 m thick, on phyllites and schists | Near surface (2.7 m) | |
| | Silt materials and fills: wastes, clays, silts, sands, mud and gravel on hard soils or bedrock | $III_{41}$ | Very soft fills and sandy clay on stiff clays, <6 m thick | Near surface (1–5 m) | Fills or loose granular soils or soft clay soils with $N_{spt} < 4$ |
| | | $III_{42}$ | Fills, sandy clay and mud on sandstone and marl, 12–20 m thick | | |
| | | $III_{43}$ | Fills, clays, silts, mud and gravels on schists or marly bedrock, 12–20 m thick | | |
| | Compacted fills and wastes (4–16 m thick) | $III_5$ | | Near surface (1–2.5 m) | Fills |

*Figure 7.5* Fragment of the geotechnical map of the city of Cartagena, Spain (simplified legend; original scale 1:15,000) (IGME-COPOT, 2000).

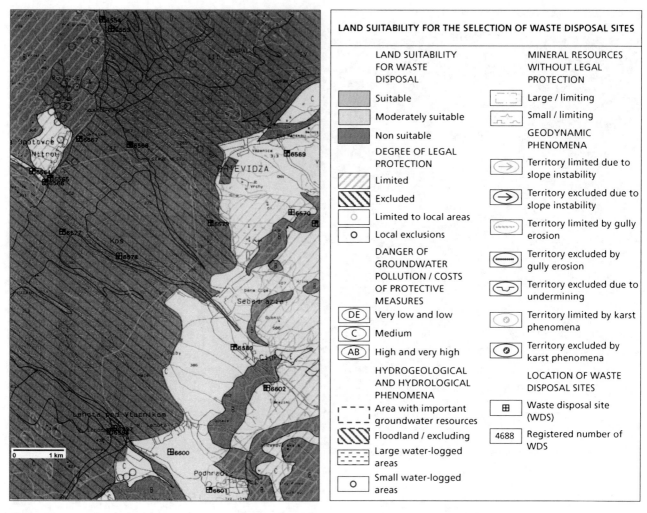

*Figure 7.6* Land suitability map for the selection of waste disposal sites (Holzer *et al.*, 1993, in Proske *et al.*, 2005).

As well as the general geological-geotechnical conditions described in Section 7.2, they should provide specific information on the:

— Location of boreholes and surveys carried out.
— Dynamic processes (landslide and subsidence), unstable areas and geological hazards.
— Geotechnical suitability for foundations and support for earth works.
— Stable slopes to be expected for excavations.
— Drainage conditions.
— Location and quality of construction materials.

Where the maps apply to **tunnels**, they must provide information on the material properties at depth. The working scales used will be similar to those for surface linear infrastructures. To produce the detailed maps and geotechnical profiles, boreholes which reach at least the depth of the tunnel must be drilled; the information to be included is as follows:

— Lithology, discontinuities and faults.
— Strength and deformability of the materials.
— Water flow and drainage.
— *In situ* stresses.
— Excavation methods.
— Support methods.
— Portal zones: slope stability, overburden, weathered materials, presence of water, etc.

Engineering geological maps for **dam** construction should be made both for the foundation area of the dam site and for the area where the reservoir is planned. The usual scales for preliminary studies are 1:10,000–1:2,000; maps for the design of the dam are drawn in great detail, 1:1,000– 1:500, and the usual scale for the reservoir area is between 1:5,000 and 1:2,000.

When mapping the dam site area special attention must be paid to the following aspects:

— Lithology, discontinuities and faults.
— Strength and deformational properties of the foundation rock masses.
— Rock mass stability in the abutment zones.
— Permeability and hydrogeological conditions.
— Seismicity and other natural hazards.
— Superficial formations and depth of weathered zones.
— Slope stability.

Maps of the reservoir area should consider the following:

— Geological hazards.
— Slope stability.
— Superficial deposits.
— Sources of construction materials and quarries.
— Sources of leakage.

# Recommended reading

Dearman, W.R. (1991). Engineering geological mapping. Butterworth-Heinemann. London.

Dearman, W.R. and Fookes, P.G. (1974). Engineering geological mapping for civil engineering practice in the United Kingdom. Quarterly Journal of Engineering Geology, vol. 7, pp. 223–256 (also in: Mapping in Engineering Geology. Compiled by J.S. Griffiths, 2002. The Geological Society of London, Key issues in Earth Sciences. pp. 79–113).

Geological Society of London (1972). The preparation of maps and plans in terms of engineering geology. Quarterly Journal of Engineering Geology, vol. 5, pp. 297–367 (also in: Mapping in Engineering Geology. Compiled by J.S. Griffths, 2002. The Geological Society of London, Key issues in Earth Sciences. pp. 7–79).

IAEG (1981). Rock and soils description and classification for engineering geological mapping. Report by the IAEG Commission on Engineering Geological Mapping. Bull. of the IAEG no. 24, pp. 235–274.

Smith, A. and Ellison, R.A. (1999). Applied geological maps for planning and development. A review of examples from England and Wales, 1983 to 1996. The Quarterly Journal of Engineering Geology, vol. 32, supplement.

UNESCO–IAEG (1976). Engineering geological mapping. A guide to their preparation. Commission on Engineering Geological Maps of the IAEG. Earth Sciences, 15. The Unesco Press.

# References

Dearman, W.R. (1991). Engineering geological mapping. Butterworth-Heinemann. London.

Geological Society of London (1972). The preparation of maps and plans in terms of engineering geology. Report by the Geological Society Engineering Group Working Party. Quarterly Journal of Engineering Geology, vol. 5, pp. 297–367.

González de Vallejo, L. (1977). Engineering geology for urban planning and development with an example from Tenerife (Canary Islands). Bull. of the IAEG, no. 15, pp. 37–43.

IAEG (1981a). Rock and soils description and classification for engineering geological mapping. Report by the IAEG Commission on Engineering Geological Mapping. Bull. of the IAEG, no. 24, pp. 235–274.

IAEG (1981b). Recommended symbols for engineering geological mapping. Report by the IAEG Commission on Engineering Geological Mapping. Bulletin of the IAEG, no. 24, pp. 227–234.

IGME (1972). Mapa geotécnico general de Toledo. Hoja 53. Escala 1:200.000.

IGME (1987). Mapa geotécnico y de riesgos geológicos de la ciudad de Zaragoza. Escala 1:25.000 y 1:5.000.

IGME-COPOT (2000). Mapa geotécnico y mapas de peligrosidad natural de la ciudad de Cartagena. Escala 1:15.000.

Proske, H., Vlcko, J., Rosenbaum, M.S., Dorn, M., Culshaw, M. and Marker, B. (2005). Special purpose mapping for waste disposal sites. Report of IAEG Commission 1: Engineering Geological Maps. Bull. of Engineering Geology and the Environment, vol. 64, no. 1, pp. 1–54.

Ruiz Vázquez, M. y González Huesca, S. (2000). Geología aplicada a la ingeniería civil. Ed. Limusa Noriega. México.

UNESCO–IAEG (1976). Engineering geological mapping. A guide to their preparation. Commission on Engineering Geological Maps of the IAEG. Earth Sciences, 15. The UNESCO Press.

# PART III

APPLICATIONS

# 8
# FOUNDATIONS

1. Introduction
2. Shallow foundations
3. Deep foundations
4. Foundations on rock
5. Foundations in complex geological conditions
6. Site investigation

## 8.1 Introduction

All structures are supported by the ground beneath them thus the ground itself can be considered as one of the materials used in the construction process, although compared with other building materials like concrete or steel, the ground is usually much weaker and more deformable. This means that in most cases the ground cannot stand up to the same stress as the building upon it and so structures have to be built with **foundations** to support them and distribute the stress, transmitting these to the ground as appropriate, depending on how strong and elastic it is.

The shape and dimensions of the foundations will depend on the load and on the type of ground. **Shallow foundations** are normally used where possible to distribute the structural loads on a horizontal supporting plane (Figure 8.1a). These foundations, built just below the ground surface, are also known as **surface foundations**. Where the ground near the surface is either not strong or not rigid enough to provide direct support, **deep foundations** are needed, to transmit the structural loads vertically, either by distributing them or by transmitting them to deeper strata where they can be carried (Figure 8.1b).

## Basic design criteria

Correctly designed foundations must meet the following conditions:

1. They must be **stable**; i.e. have an adequate factor of safety (the ratio between the load they place on the ground and the ultimate shear strength of the ground).
2. Their elasticity or **deformability** must be **acceptable**; i.e. that displacements such as settlement, horizontal movements or rotation, caused by the deformation of the ground due to stresses transmitted by the foundations, will be tolerated by the structure and permit the structure to function as designed.
3. They must not affect **nearby buildings**; i.e. the effect of the foundations on the ground must be tolerable outside the limits of the structure under construction. This means ensuring that they do not have any negative effect on other buildings around them.
4. They must be **durable**; i.e. the above conditions will last throughout the intended life of the structure. This means considering the possible change of the initial conditions due to:

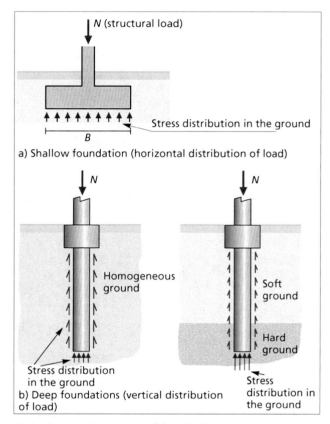

Figure 8.1   Basic types of foundations.

- Changes in volume of the ground due to collapse of poorly compacted landfills, or naturally collapsible soils such as loess or some types of silts.
- Changes in volume of the ground due to changes in the moisture content of potentially expansive soils.
- Undermining in river beds and banks.
- Soil erosion from broken sewers or other water pipes.
- Deterioration of the concrete used for the foundations from contact with aggressive soil or groundwater.
- Fluctuations in the water table, which may cause changes in effective stress or in the strength and deformability of the soil.
- Landslides, if the construction is on an unstable slope.

## Stages in foundation design

There are four basic stages in the design of foundations:

— Collecting data on the type of structure required, on the loads envisaged and on their distribution. In buildings these loads are mainly vertical, but there may be horizontal loads and moments due to wind, lateral ground pressures or seismic forces.
— Obtaining a geotechnical site report (Section 8.6), paying particular attention to ground shear strength and deformability. Strength is expressed as the angle of internal friction and cohesion, or as the undrained shear strength in the case of clay soils. Deformability is expressed as the deformation modulus or compression index, and where long-term settlement is predicted, as the coefficient of consolidation. The non-uniformity or heterogeneity of the ground should also be considered, as differential settlements may occur if a foundation covers ground that has different properties.
— Defining the ultimate shear strength of the soil and applying a factor of safety, to obtain the allowable pressures. Deformation or settlement from the loads applied to the ground have to be estimated and analysed to determine if the structure will tolerate them.
— Designing the foundations by defining their dimensions, their shape and their reinforcement.

## 8.2 Shallow foundations

## Types of shallow foundations

Shallow foundations distribute the structural loads onto a horizontal supporting plane. Common variants of these foundations are called **surface foundations**, where the supporting plane or base is located at a depth equal to, or less than, 5 times the width or minimum dimension of the ground plan.

The two main types of shallow foundations are footings and rafts. **Footings** may be either individual, supporting a single column, or combined, with various columns using the same footing. This type of combined foundations is known as **pad or spread footings**. A special case of combined footings is a **continuous or strip footing** which supports a series of aligned columns or a wall. A **raft foundation** is an extended layer of reinforced concrete used to distribute the load from the structure over the whole area of the structure. A **mat or grid foundation** is a cross between a footing and a raft and consists of a series of continuous footings in two or more directions, linked at their intersections (Figures 8.2 and 8.3).

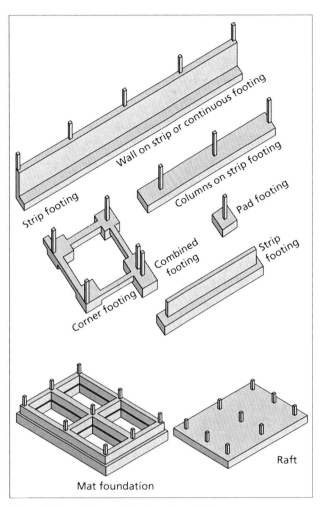

Figure 8.2   Types of shallow foundations.

Figure 8.3   Construction of a large raft foundation for a shopping centre.

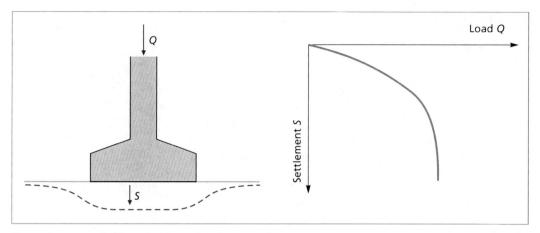

Figure 8.4   Definition of ultimate bearing capacity.

## Ultimate bearing capacity

When a progressively increased load is applied to a shallow foundation, the ground underneath the foundation is deformed, as shown in Figure 8.4. While the loads are small or moderate, settlement increases at a rate approximately proportional to the load applied; the ground resists, performing with a certain degree of elasticity. If the load continues to increase, eventually the ground will not be able to resist any further however, deformation continues, and this is when structural failure occurs.

The pressure at which the ultimate bearing capacity of the ground is reached and failure occurs is a function of its shear strength, the dimensions of the foundations, how deep they are, the unit weight of the soil above this depth, and the position of the water table. How the ground fails under shallow foundations when the ultimate bearing capacity is reached is shown schematically in Figure 8.5. As can be seen, sliding occurs along the surface ABC and can be accompanied by ground uplift.

The following sections describe how to evaluate the ultimate bearing capacity and the safe ultimate bearing capacity for shallow foundations.

## Basic definitions

(See Figure 8.6)

1. **Total gross pressure** ($q_g$): the total vertical pressure exerted on the foundation base (ratio of total load to foundation area); it includes all vertical components: overburden, structural weight, dead weight of the foundations, etc.
2. **Total net pressure** ($q_{net}$): the difference between $q_g$ and the total vertical ground pressure ($q_0$) at the foundation base level ($q_0$ = total weight of the ground above the level of the foundation = bulk or total unit weight of the ground ($\gamma$) × depth of the foundation

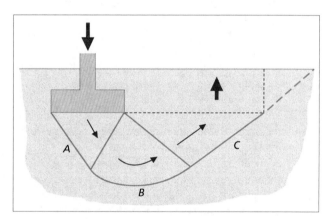

Figure 8.5   Shape of the failure surface.

($D$)); $q_{net}$ is usually the net increase in total vertical stress at that level.
3. **Effective gross pressure** ($q'_g$): the difference between the total gross bearing pressure and the pore pressure ($u$) at foundation level.
4. **Effective net pressure** ($q'_{net}$): the difference between $q'_g$ and the effective vertical pressure ($q'_0$) due to earth overburden at foundation level (NB: $q_{net} = q'_{net}$)

$$q'_{net} = q'_g - q'_0 = (q_g - u) - (q_0 - u) = q_g - q_0 = q_{net}$$

5. **Ultimate bearing capacity** ($q_{ult}$, $q'_{ult}$): the maximum vertical pressure that the soil can bear before failure; this may be expressed in terms of total or effective stress, gross or net.
6. **Safe ultimate bearing capacity** ($q_s$, $q'_s$): the maximum vertical pressure at which there is an adequate factor of safety against failure; it can be expressed in terms of total or effective stress, gross or net. This pressure does not have to be the one finally selected as allowable for the structure, since although it provides enough security against failure, it does not include any special limitations to prevent settlement,

so that there may be excessive structural deformation, even though the structure does not collapse.
7. **Allowable bearing capacity** ($q_a$, $q'_a$): is the allowable vertical pressure for a specific structure, taking its safety against failure and its settlement tolerance into account; obviously it will be equal to or less than $q_s$. It can be expressed in terms of total or effective stress, gross or net. The maximum load that can be applied to a soil will be greatest for $q_{ult}$, smaller for $q_s$ and even smaller for $q_a$.

## Calculating the ultimate bearing capacity

The ultimate bearing capacity of a strip foundation (infinite in the direction perpendicular to the footing width B, Figure 8.7), is normally calculated using the expression proposed by Terzaghi:

$$q_{ult}(\text{gross}) = cN_c + q_0 N_q + \frac{1}{2} B \gamma N_\gamma$$

where:

— $N_c$, $N_q$, $N_\gamma$ are the **bearing capacity factors** and depend exclusively on the effective angle of internal friction of the soil.
— $c$ is the ground cohesion
— $q_0$ is the vertical stress due to the total soil overburden at foundation level.
— $B$ is the width of the footing
— $\gamma$ is the unit weight of the ground underneath the foundations.

The three terms of the equation above show three different contributions to the strength: the cohesion, the effect of the soil overload on the supporting plane and the unit weight of the ground under the foundations. This equation can be expressed in total or effective stress, gross or net.

The equations below can be used to determine the bearing capacity factors directly and reliably:

$$N_q = \frac{1+\sin\phi'}{1-\sin\phi'} e^{\pi \tan\phi'} \quad N_c = (N_q - 1)\cot\phi'$$

$$N_\gamma = 1.5(N_q - 1)\tan\phi'$$

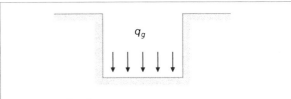

1. Total gross pressure.

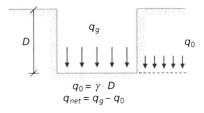

$q_0 = \gamma \cdot D$
$q_{net} = q_g - q_0$

2. Total net pressure.

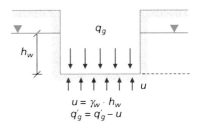

$u = \gamma_w \cdot h_w$
$q'_g = q_g - u$

3. Effective gross pressure.

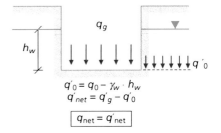

$q'_0 = q_0 - \gamma_w \cdot h_w$
$q'_{net} = q'_g - q'_0$

$\boxed{q_{net} = q'_{net}}$

4. Effective net pressure.

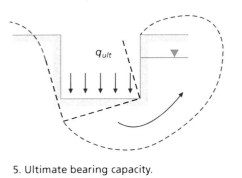

5. Ultimate bearing capacity.

Figure 8.6   Definition of pressures.

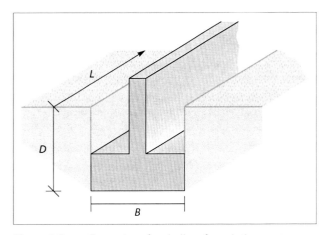

Figure 8.7   Geometry of a shallow foundation.

## Ultimate bearing capacity in undrained conditions

### Strip load

This is the simplest theoretical case, applicable to cohesive (saturated) soils, where the most unfavourable conditions tend to occur immediately after the load is applied i.e. before the excess pore pressure generated by the imposed loads have been dissipated.

The undrained calculation is made for total stress, taking $\phi = 0$, $c = S_u$. For a zero angle of friction, the bearing capacity factors adopt the following values:

— $N_q(\phi = 0) = 1$
— $N_c(\phi = 0) = 5.14$
— $N_\gamma(\phi = 0) = 0$

As a result, the total gross ultimate bearing capacity is:

$$q_{ult}(\text{gross}) = S_u N_c + q_0 = 5.14 S_u + \gamma D$$

where:

— $\gamma$ is the unit weight of the soil above the foundation base
— $D$ is the depth of the foundation base

### Correction factors

#### a) Influence of the depth of the foundation

Terzaghi's equation does not consider the strength of the ground above the footing. As already mentioned, its stabilizing action ($q_0 = \gamma D$) is only produced by the effect of overburden. However, if the foundations fail, then obviously the theoretical shear surface will "lift" the soil near the footing, as shown in *Figure 8.5*, and also cut through it, mobilizing its resistance to shear. This effect, which of course will increase the available bearing capacity, needs to be taken into account and a depth correction factor is normally used. *Figure 8.8* shows Skempton's suggested correction factor ($d_c$) applied to the cohesion value, depending on the depth/width ratio of the foundations.

Applying this correction factor, the total gross bearing capacity in undrained conditions for a continuous strip foundation produces:

$$q_{ult}(\text{gross}) = 5.14 d_c S_u + \gamma D$$

When using this factor in practice, it is important to note that sometimes it may be better not take it into consideration, e.g. when constructing a shallow footing in a highly plastic clay soil, since in dry periods shrinkage may cause cracks to form down to the full depth of the foundation. Here, the theoretical slip surface would cause soil uplift, but not mobilize the shear strength, since there would be open discontinuities with no resistance. In granular soils or those not prone to cracking when dry, applying the correction factor is not a problem.

Another related but more general aspect to be considered is how to use the depth correction factor. From the ultimate bearing capacity equation it can be seen that the load increases in direct ratio to the depth $D$ of the foundations. For shallow foundations the hypotheses made must be guaranteed to remain valid during the working life of the structure, and possible future excavations to install services alongside the foundations must be considered, since these would reduce or even cancel out the soil overburden.

#### b) Influence of foundation shape

The ultimate bearing capacity of either a circular or a square footing is greater than that of a continuous or strip foundation. When a strip footing (infinite in one direction) reaches failure, the failure surface is clearly two-dimensional, but in a square footing of the same width $B$ the failure surface is clearly three-dimensional as the failure "wedge" extends beyond the geometric limits of the footing; thus continuous strip footings, infinite in one direction, mobilize the resistance of a smaller volume of soil per unit of foundation area than foundations which are finite in two directions.

The effect of this increased strength as a function of shape is taken into account through a shape correction factor ($s_c$) that can be taken as:

$s_c = 1.20$ for circular footing
$s_c = 1 + (0.2 \, B/L)$ for a rectangular footing with base dimensions $B \times L$.

The general expression of the total gross ultimate bearing capacity in undrained conditions will then be:

$$q_{ult}(\text{gross}) = 5.14 \, d_c s_c S_u + \gamma D$$

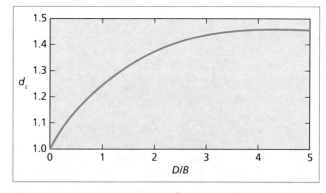

Figure 8.8  Depth correction factor ($d_c$) (Skempton, 1951).

# FOUNDATIONS

## Box 8.1

### Calculating the ultimate bearing capacity

A building is to be founded on a clay stratum. The water table is practically at ground surface level. Laboratory tests show that the unit weight of the soil is $\gamma = 20$ kN/m³ and classify the soil as a firm clay. The uniaxial compressive strength is $q_u = 150$ kN/m².

Supposing that the footings will be square, 2 m wide, and the foundation depth will be 2 m below the ground surface, calculate the ultimate bearing capacity.

**Solution:**

The parameters to be considered in the ultimate bearing capacity equation are:

$$S_u = \frac{q_u}{2} = 75 \text{ kPa}$$

$d_c = 1$ (as it is near the surface)

$$s_c = 1 + 0.2 \frac{2}{2} = 1.2$$

As a result, the pressure required to cause shear failure is:

$$q_{ult}(\text{gross}) = 5.14 \cdot 1.2 \cdot 75 + 20 \cdot 2$$
$$= 502.6 \text{ kPa} \approx 0.5 \text{ MPa}$$

## Ultimate bearing capacity in drained conditions

### Strip load

If the soil is permeable enough to allow any excess pore pressure generated when the foundation loads are applied to be dissipated almost simultaneously, during construction (e.g. as in mainly granular soils) the ultimate bearing capacity is calculated in terms of effective stress. This calculation can also be applied to low permeability soils such as clays, to determine the bearing capacity when pore pressure equilibrium is reached and consolidation processes are complete.

The basic expression of the gross effective ultimate bearing capacity for a strip load (infinite continuous footing) is:

$$q'_{ult}(\text{gross}) = c'N_c + q'_0 N_q + \frac{1}{2} B\gamma N_\gamma$$

when the bearing capacity factors are the same as in the sections above, and the shear strength parameters ($c'$, $\phi'$) are expressed in effective stresses. The unit weight $\gamma$ is taken as needed to calculate the effective stresses under the foundations ($\gamma$ for unsaturated ground or $\gamma' = \gamma_{sat} - \gamma_w$ for soil saturated with hydrostatic water pressure, and is also known as submerged or bouyant unit weight).

### Correction factors

a) **Influence of foundation depth**

The effect of the depth of the foundations has already been analysed for undrained conditions. In drained conditions, the coefficients used for the resistance of the ground above foundation level are:

— $N_c$: $d_c$ (Figure 8.8)
— $N_q$: $d_q = 1$ (normally ignored)
— $N_\gamma$: $d_\gamma = 1$ (no real influence)

b) **Influence of foundation shape**

This effect has also been described for undrained conditions. For drained conditions the following can be used:

— $N_c$: $s_c$, defined above
— $N_q$: $s_q = 1 + 1.5 \tan \phi' \, B/L$
— $N_\gamma$: $s_\gamma = 1 - 0.1 B/L$

Other factors which may influence the ultimate bearing capacity include the inclination of the load, its eccentricity, an existing layer of stiff soil near the surface or foundations near the edge of a slope. The procedure to follow in these cases, often solved with correction factors in the general expression, may be consulted in the recommended reading.

## Factor of safety. Safe bearing capacity

To ensure that the foundations will not fail, a factor of safety must be applied to the calculated ultimate pressure. This factor of safety is usually applied to the net ultimate bearing capacity and so the safe bearing capacity of shallow foundations can be expressed as:

$$q_s(\text{gross}) = \frac{q_{ult}(\text{net})}{F} + q_0 = \frac{q'_{ult}(\text{net})}{F} + q_0$$

## Box 8.2

### Calculating the effective ultimate bearing capacity

The foundation footings for a viaduct are to be sunk into sand and gravel alluvium. According to the borehole data the water table is at a considerable depth, below the foundation area of influence.

The unit weight of the ground is $\gamma = 20$ kN/m$^3$ and the effective angle of internal friction, as deduced from penetrometer tests is $\phi' = 35°$.

Find the effective ultimate bearing capacity for the footings of the viaduct, 4 m wide (B) and 8 m long (L) on the ground plan, if they are sunk 2 m below ground level.

### Solution:

Since this is granular soil, the effective cohesion can be taken as zero ($c' = 0$). Applying the expressions in the previous sections gives:

$$N_q = 33.29 \quad N_\gamma = 33.92 \quad s_q = 1.52$$
$$S_\gamma = 0.95 \quad d_q = 1 \quad d_\gamma = 1$$

The effective vertical pressure at the base of the foundation, $q'_0$ is:

$$q'_0 = \gamma \cdot D = 2 \cdot 20 = 40 \text{ kPa}$$

Finally, the ultimate bearing capacity will be:

$$q'_{ult} \text{(gross)} = q'_0 s_q d_q N_q + \frac{1}{2} B \gamma s_\gamma d_\gamma N_\gamma$$
$$= 40 \cdot 1.52 \cdot 33.29 + 0.5 \cdot 4 \cdot 20 \cdot 0.95 \cdot 33.92$$
$$= 3,313 \text{ kPa}$$

As can be seen, the ultimate bearing capacity obtained is very high (>3 MPa). This effect is common in granular soils, except in loose ones, and means that the allowable bearing capacity is mainly limited by the allowable settlement of the structure rather than from shear failure.

### Notes:

1. For undrained calculations, total stress (gross or net) is used to express the safe bearing capacity.
2. For drained calculations, it is much better to use effective stresses, applying the factor of safety to the effective net ultimate bearing capacity, $q'_{ult}$.
3. In normal practice a factor of safety of 3 is usually adopted for long term conditions. For short term conditions (temporary works construction phase, earthquake shaking, etc.), or for earthwork foundations (embankments, reinforced earth, etc.), in practice smaller factors of safety can be used, with each case considered individually.

## Distribution of pressures under shallow foundations

The design and size of shallow foundations must be taken into account all the forces exerted on them. The forces transmitted by the structure are known beforehand, but there are other more complex ones, such as the distribution of the reaction of the ground under the footing, which depends on the type of ground and the rigidity of the foundations.

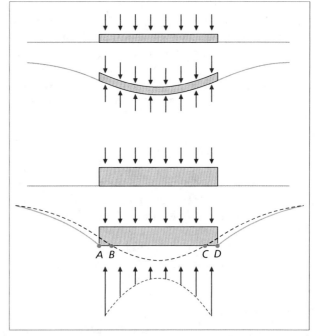

Figure 8.9 Pressure distribution in flexible and rigid foundations.

## Box 8.3

### Calculating the effective ultimate bearing capacity and safe ultimate bearing capacity

a) Calculate the effective net ultimate bearing capacity for the example given in Box 8.2. What would the effective gross safe ultimate bearing capacity be for a factor of safety $F = 3$?

b) In the example given in Box 8.2, if the water table is 2 m below the ground surface level, coinciding with the base of the foundations, by how much would the gross safe ultimate bearing capacity be reduced? (Assume saturation does not alter the apparent unit weight of the ground).

c) What would the reduction be if the water table were to rise to ground level?

**Solution:**

a) In the example in Box 8.2, the result was $q'_{ult}(\text{gross}) = 3{,}313$ kPa so that:

$$q'_{ult}(\text{net}) = q'_{ult}(\text{gross}) - q'_0 = 3{,}313 - 40 = 3{,}273 \text{ kPa}$$

$$q'_s(\text{gross}) = \frac{q'_{ult}(\text{net})}{F} + q'_0 = \frac{3{,}273}{3} + 40$$

$$= 1{,}131 \text{ kPa} \approx 1.1 \text{ MPa}$$

b) In this case the only modified parameter is the unit weight of the ground under the foundation, which should be taken as submerged:

$\gamma' = \gamma_{sat} - \gamma_w \approx 10$ kN/m³. The effective net ultimate bearing capacity will be:

$$q'_{ult}(\text{net}) = q'_0(s_q d_q N_q - 1) + \frac{1}{2} B \gamma' s_\gamma d_\gamma N_\gamma$$

$$q'_{ult}(\text{net}) = 40(1.52 \cdot 33.29 - 1) + 0.5 \cdot 4 \cdot 10 \cdot 0.95 \cdot 33.92$$

$$= 2{,}628.5 \text{ kPa}$$

and the gross safe ultimate bearing capacity will be:

$$q'_s(\text{gross}) = \frac{2{,}628.5}{3} + 40 = 916 \text{ kPa} \approx 0.92 \text{ MPa}$$

c) In this case the effective pressure also varies due to the height of water above the base of the foundation:

$$q'_0 = q_0 - u = 20 \cdot 2 - 10 \cdot 2 = 20 \text{ kPa}$$

$$q'_{ult}(\text{net}) = q'_0(s_q d_q N_q - 1) + \frac{1}{2} B \gamma' s_\gamma d_\gamma N_\gamma$$

$$q'_{ult}(\text{net}) = 20(1.52 \cdot 33.29 - 1) + 0.5 \cdot 4 \cdot 10 \cdot 0.95 \cdot 33.92$$

$$= 1{,}636.5 \text{ kPa}$$

and the gross safe ultimate bearing capacity will be:

$$q'_s(\text{gross}) = \frac{1{,}636.5}{3} + 20 = 565.5 \text{ kPa} \approx 0.56 \text{ MPa}$$

*Note:* The rise in the water table may cause a substantial reduction in the safe ultimate bearing capacity.

---

Suppose that a uniform load is applied to an infinitely flexible footing (*Figure 8.9*) laid directly on a horizontal ground surface. The effect of this load will be settlement of the ground and the footing. The settlement will be greater in the centre than at the edges and will not be restricted to the load bearing area but will also extend to a certain distance on both sides. As the footing is infinitely flexible, it is not able to bear bending moments and so the distribution of pressures from the ground reaction will be exactly the same as the uniform distribution of the pressures exerted on the footing.

Now suppose that an infinitely rigid footing is laid on horizontal ground. When the same uniform load is placed on it, the settlement which occurs is very similar to the average settlement of a flexible footing. However, the rigid footing makes the settlement profile curve obtained for the flexible footing impossible to achieve, so that between points A and B and between points C and D, the settlement will be greater than with the flexible footing; on the other hand, the settlement of the rigid footing between B and C will be less than with the flexible footing. This means that the pressures between A and B and between C and D on the rigid footing, will be greater than those on the flexible footing but, in contrast, the pressures between B and C will be lower than the uniform pressure applied to the footing.

This gives a non-uniform distribution of pressure under the footing, characterized by maximum values at the ends and a minimum value in the centre.

If the ground were elastic and infinitely strong, the stresses under edges A and B would be infinite. As the ground strength is not infinite, these stresses may or may not be very high, but they will have a finite value. In clays (*Figure 8.10a*), the pressure distribution is very similar to the previous theoretical example.

However, finite ground strength produces areas of elasticity at both ends. In sand, since very high pressures cannot occur at the surface edge, the pressure distribution is in the shape of a parabola, as shown in *Figure 8.10b*.

Analytical methods allow an approximate estimate to be made of the real pressure distribution under rigid footings. However, in practice, this is not usually determined. Generally speaking, a trapezoidal or triangular linear pressure distribution under the footing is taken to occur.

*Figure 8.11* shows a footing subjected to a centred vertical load Q and a moment at the base of the column M. This combination is equivalent to the vertical load Q, located at a certain distance from the axis of the column e, its eccentricity ($e = M/Q$).

When the eccentricity is less than one sixth of the width of the footing, the maximum and minimum stresses are expressed as:

$$\sigma_{max} = \frac{Q}{B}\left(1+\frac{6e}{B}\right)$$
$$\sigma_{min} = \frac{Q}{B}\left(1-\frac{6e}{B}\right)$$
$$e \leq \frac{B}{6}$$

Where the eccentricity is greater than one sixth of the footing width, then the pressure distribution is triangular, not trapezoidal, and the value of the maximum stress and of the width of the footing subjected to pressure are expressed as follows (*Figure 8.12*):

$$\sigma_{max} = \frac{4}{3}\frac{Q}{B-2e}; \quad b = \frac{3}{2}(B-2e); \quad e > \frac{B}{6}$$

## Stress distribution under loaded areas

### Fundamentals. Criteria for use

The previous section focused on ground strength and how pressures, up to those that could cause the ground to fail, are distributed directly under shallow foundations. However, this information is not enough to estimate ground settlements. To do this, an understanding is first needed of how the pressures transmitted by the foundations are distributed through the ground at depth. It has been shown that the influence of load over a finite area is limited to a certain area of influence

*Figure 8.10*   Pressure distribution in rigid footings on clays and sands.

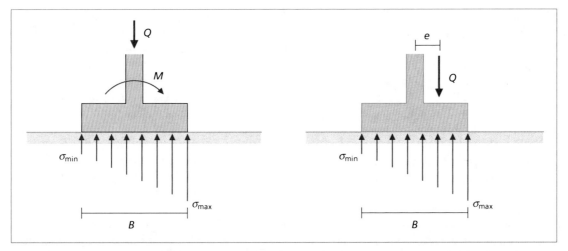

*Figure 8.11*   Eccentric loads and distribution of theoretical pressures underneath the footing (e < B/6).

around it and this is why it is important to analyze how the stresses caused by such loads are distributed through the ground.

The solutions to this problem are based on the theory of elasticity, which provide reasonable parameters for estimating the vertical stress increments caused by foundations.

The following sections include some of the usual solutions applied in practice.

## Point load on an elastic half-space

This was the original problem studied by Boussinesq (Figure 8.13). Other solutions for elastic half-space come from the appropriate integration of other, more complex load hypotheses.

The expression for the vertical stress increment caused by the point load P of Figure 8.13 gives:

$$\Delta\sigma_z = \frac{3P}{2\pi}\frac{z^3}{R^5}$$

where $z$ and $R$ are geometric parameters which define the position of the point or element of the ground where the stress is to be calculated.

## Vertical stresses under the corner of a uniformly loaded rectangle

Newmark's solution to this problem allows the vertical stress increment to be obtained at depth $z$ under the corner of a rectangle loaded uniformly with a distributed load $q$ (Figure 8.14).

If:

$$m = L/z$$
$$n = B/z$$

then

$$\Delta\sigma_z = \frac{q}{4\pi}I_\sigma$$

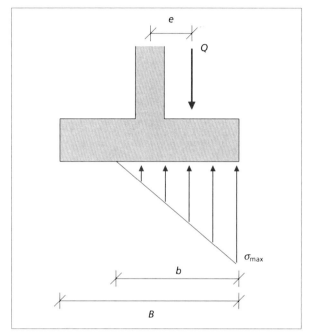

Figure 8.12   Eccentric loads and distribution of theoretical pressures underneath the footing ($e > B/6$).

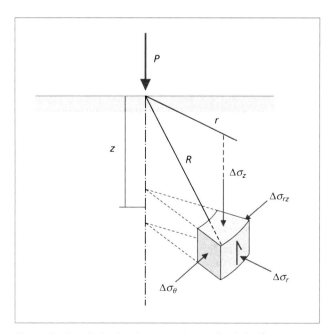

Figure 8.13   Point load stresses in an elastic half-space.

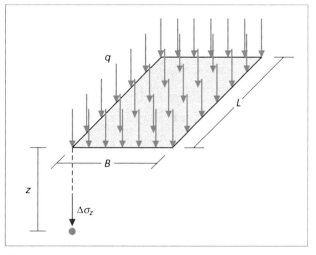

Figure 8.14   Vertical stresses under a uniformly loaded rectangle.

where $I_\sigma$ is an influence coefficient determined from geometric considerations by the following expression:

$$I_\sigma = \frac{2mn(m^2+n^2+2)\sqrt{m^2+n^2+1}}{(m^2+n^2+1+m^2n^2)}$$
$$\times \arcsin\left[\frac{2mn\sqrt{m^2+n^2+1}}{m^2+n^2+1+m^2n^2}\right]$$

The relative complexity of the formula above can be solved simply using direct entry charts.

*Figure 8.15* shows Fadum's chart, which allows the influence coefficient to be obtained directly as a function of the geometric parameters $m$ and $n$. As shown, the divisor $4\pi$ of Newmark's solution is already included in the expression of this coefficient.

It is usually necessary to understand the stresses produced by rectangular loads under their corners, and also at other points on the ground (e.g. under the centre and edges, and even under a point at some distance from the loaded area). The superposition principle of elasticity theory can be used for this, by combining the increased stress produced by a series of appropriately selected rectangles; e.g. the increased stress under the centre of a rectangle with sides $L$ and $B$ would be the sum of the stress increments originating under the corners of 4 sub-rectangles with sides $L/2$ and $B/2$.

## Stresses under a uniformly loaded circular area

Integrating Boussinesq's equation under the centre of the load area gives the following expression (*Figure 8.16*):

$$\Delta\sigma_z = q \cdot \left[1 - \left(\frac{1}{1+\left(\frac{R}{z}\right)^2}\right)^{\frac{3}{2}}\right]$$

*Figure 8.17* can be used at different points below the load area. This represents the vertical stress increments due to

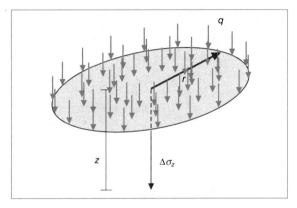

Figure 8.16  Vertical stresses under the centre of a uniformly loaded circular area.

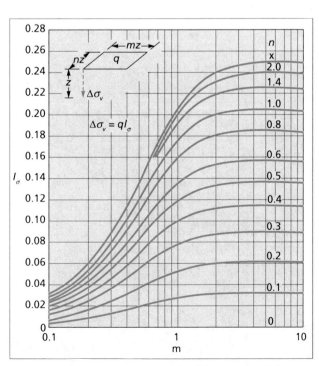

Figure 8.15  Vertical stress under the corner of a uniformly loaded rectangle (Fadum, 1948).

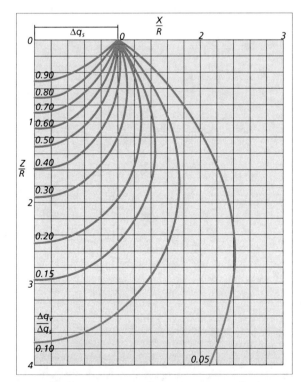

Figure 8.17  Increment in vertical stress under a uniformly loaded circular area.

# FOUNDATIONS

## Box 8.4

### Calculating the ground stress distribution

Define the increase in vertical stress caused by the uniformly loaded rectangle shown in the diagram at depth z under point c, some distance from it.

Since the calculation point required is not found on the vertical of the corner of the loaded rectangle, the superposition principle has to be applied. To do this, a series of sub-rectangles has been drawn in the diagram.

If $\Delta\sigma_z\begin{pmatrix} 1 & 2 \\ 3 & 4 \end{pmatrix}$ is the vertical stress increment originating at depth z under the corner of a rectangle with vertices 1, 2, 3, 4 uniformly loaded with the distributed load q, then:

$$\Delta\sigma_z(c) = \Delta\sigma_z\begin{pmatrix} a & c \\ e & g \end{pmatrix} - \Delta\sigma_z\begin{pmatrix} a & c \\ d & h \end{pmatrix}$$
$$- \Delta\sigma_z\begin{pmatrix} b & c \\ e & f \end{pmatrix} + \Delta\sigma_z\begin{pmatrix} b & c \\ d & i \end{pmatrix}$$

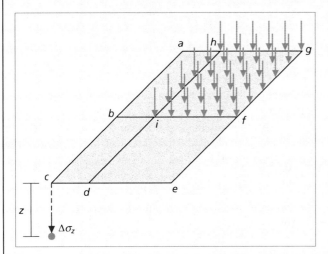

Note that all the sub-rectangles selected have a corner at the vertical of the calculation point required, so that the appropriate addition or subtraction of n effects allows the required stresses to be determined. Then, the stress increments at depth z corresponding to the corner of each rectangle in the expression above has to be determined and substituted.

pressure of intensity $\Delta q_s$, uniformly distributed on the surface of a circle with radius R. This diagram is adimensional and for symmetry only represents half of the half-space, with the y-axis (relative depths z/R) coinciding with the centre of the circle. The x-axis shows the relative distance to the centre of the load circle X/R. Finally, the stress increments are represented by a series of lines of equal incremental stress, relative to the surface stress, $\Delta q_v/\Delta q_s$.

This diagram allows the concept to be easily understood and shows the practical limitations of the effect of a non-infinitely extensive load. For most applications, the problem areas are where the vertical stress increment is up to 10% of the load applied at the surface. Outside these areas, the overstress on the ground is small enough in most cases for its effects to be negligible.

Figure 8.17 shows that the zone defined by 10% of the surface load forms a "bulb" extending down to a depth of approximately 2 diameters (or 2 widths) of the load area. This area of special influence is known as the **stress bulb**.

The corollary of this (Figure 8.18), is that if a shallow foundation with width B is planned, the minimum depth of site investigations such as boreholes, should be approximately

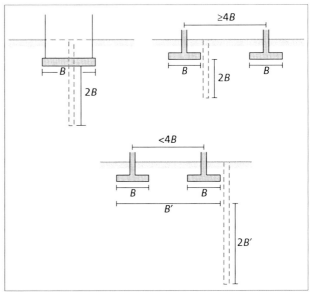

Figure 8.18  Recommended depth of site investigations depending on the proximity and type of foundations (modified from Tomlinson, 2001).

2B below the planned foundation level. On the other hand, if the foundation has footings less than 3 to 4 widths apart, the stress bulbs will overlap at depth, so that the foundations will have to be considered as if they were the whole width of the building, i.e. as a raft foundation.

# Settlement in soils

## General considerations

Knowing how stresses applied to a foundation are distributed through the ground allows the settlement they can cause to be calculated. Calculating settlements is one of the most complex problems in soil mechanics.

There are currently various methods of approach, giving widely differing results, which suggests that they are not really satisfactory. The problem becomes even more complex when differential settlements in a structure have to be predicted, since the interaction between the structure and the ground causes a reorganization of loads and related movements.

The difficulty in predicting settlement is due mainly to the nature of the soil itself. This means that the site investigation must provide a reliable ground profile to obtain the ground deformation parameters.

Sophisticated constitutive models have recently been developed which analyse complex situations using numerical calculation tools such as finite elements. The main problem with these methods is that they usually require the specification and use of a considerable number of soil parameters which are difficult to estimate, both on site and under laboratory conditions, so that it is often not viable to use these advanced analysis systems. In normal practice, it is preferable to concentrate on determining simple, representative ground parameters and to use an equally simple calculation method giving "good enough" results.

The following sections describe a settlement estimation method for shallow foundations, for use in conventional i.e. normal conditions.

## Immediate and consolidation settlement

Settlement due to soil consolidation in one-dimensional conditions is the result of the distribution and dissipation of pore pressures caused by an applied load (see Chapter 2, Section 2.5). This settlement is only part of the total ground settlement. *Figure 8.19* shows the settlement-time ratio after applying an instant finite load to the ground. Some settlement occurs immediately, as soon as the load is applied. This is called **immediate settlement** ($S_i$). If the soil is saturated and its permeability is low, as in the case of saturated clays, then initially the water is not able to escape from the soil pores, and there is no drainage, so that in fact this initial settlement corresponds to soil distortion, and there is no change in volume.

After this immediate settlement, the excess pore pressure generated is gradually dissipated. This is the **consolidation settlement** described in Chapter 2, Section 2.5, which is called **primary consolidation settlement** ($S_c$).

Finally, some soils may continue to settle after all the excess pore pressure has completely dissipated. This settlement, which is produced at constant effective stress and therefore corresponds to a creep phenomenon, is called **secondary consolidation settlement** ($S_s$).

This means that the **total settlement** will be the sum of the three components above: immediate, primary and secondary consolidation settlement:

$$S_t = S_i + S_c + S_s$$

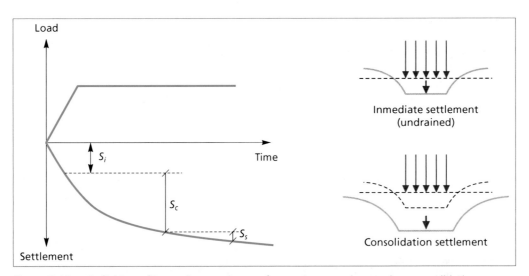

*Figure 8.19*  Definition of immediate settlement from primary and secondary consolidation.

In highly permeable (granular) and partly saturated soils, settlement due to loading occurs almost simultaneously with the load application, so that $S_i$ and $S_c$ are the same. Secondary consolidation can be studied from the consolidation curves obtained in the oedometer test. In most non-organic soils there is little settlement of this type.

## Immediate and primary consolidation settlements in saturated clays

The load conditions in foundations do not normally lead to a one-dimensional state of zero lateral deformation. In practical terms, however, for saturated clay soils, Burland et al. (1977) suggest calculating settlements with the one-dimensional compression method described in Chapter 2, Section 2.5, using the following corrections (without taking creep settlement into account):

— The consolidation settlement is obtained by the one-dimensional or **oedometer method** (Chapter 2, section 2.5); this settlement is called $S_{1 \times D}$.
— If the clay is normally consolidated, it can be assumed that:

$$\left. \begin{array}{l} S_c \approx S_{1 \times D} \\ S_t \approx 1.1 S_c \end{array} \right\} \Rightarrow S_i \approx 0.1 S_c$$

This means that the primary consolidation settlement will be approximately equal to that calculated with the oedometric or one-dimensional method, and the instantaneous settlement will be 10% of this.

— If the clay is over-consolidated, i.e. it has been unloaded in the geological past, it can be assumed that:

$$\left. \begin{array}{l} S_t \approx S_{1 \times D} \\ S_i \approx 0.5 S_t \end{array} \right\} \Rightarrow S_c \approx 0.5 S_t$$

As shown, here the total settlement is approximately equal to the consolidation settlement calculated by the oedometric method; approximately 50% of this settlement is produced instantaneously or immediately and the other 50% will occur over time as primary consolidation settlement.

This means that to calculate the settlement produced by finite loads using the one-dimensional method, the process can be divided into the following steps (see Figure 8.20), without taking the effects of secondary consolidation into account:

1. Divide the compressible stratum into sub-layers of height $H_i$.
2. Calculate the effective initial stress in the centre of each sub-layer ($\sigma'^i_{v0}$).
3. Calculate the total vertical stress increments produced by loading in the centre of the sub-layers ($\Delta\sigma^j_v$).
4. Calculate the oedometric or one-dimensional settlement of each sub-layer when the excess pore pressure has dissipated (and the total stress increment ($\Delta\sigma^j_v$) has been transformed into the effective stress increment ($\Delta\sigma'^j_v$); to do this, different compressibility parameters can be used as described above in Chapter 2, Section 2.5:

$$S^i_\infty = \frac{1}{E_m} \Delta\sigma'_v H_i$$

or

$$S^i_\infty = H_i \frac{C_c}{1+e_0} \log \frac{\sigma'^i_{v0} + \Delta\sigma^j_v}{\sigma'^i_{v0}}$$

(e.g. for the virgin compression curve).

5. Apply the corrections in the previous section taking into account whether the clay is normally or

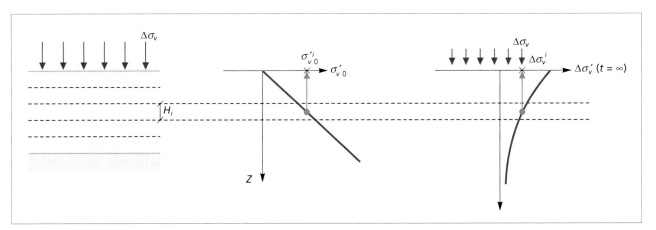

Figure 8.20  Procedure for calculating oedometric or one-dimensional settlement under non-infinite loads.

over-consolidated and determining the immediate and primary consolidation settlements of each sub-layer.
6. Add together the sub-layer settlements to obtain the final total settlement.

## Settlements in granular soils

Settlement in granular soils is usually estimated by empirical methods. Using the methodology proposed by Burland and Burbridge (1985), the most probable settlement of a foundation in sandy soils is related to its resistance to the standard penetration test expressed as:

$$S_i = f_l f_s q' B^{0.7} I_c$$

where:

— $S_i$: average settlement at completion of construction in mm.
— $q'$: effective gross pressure applied to the foundation base (in kN/m²)
— $B$: width of footing or raft (in m.)
— $I_c$: compressibility index, defined from the average value of the SPT test in a specific zone of influence under the footing or raft, $Z_I$ (Figure 8.21).

The coefficients to be applied are as follows:

1. $f_s$: footing shape coefficient, expressed as

$$f_s = \left[\dfrac{1.25\dfrac{L}{B}}{\dfrac{L}{B}+0.25}\right]^2$$

2. $f_l$: correction factor for a rigid layer below the footing at depth $H_s \leq Z_I$ where $Z_I$ is the influence depth under the footing; this is expressed as:

$$f_l = \dfrac{H_s}{Z_I}\left[2 - \dfrac{H_s}{Z_I}\right]$$

3. For over-consolidated soils or for foundations at the base of an excavation located at a depth where the maximum effective vertical stress was $(\sigma'_{v0})$, the $q'$ to insert in the settlement equation is:

$$q' - \dfrac{2}{3}\sigma'_{v0} \quad \text{when} \quad \sigma'_{v0} \leq q'$$

$$\dfrac{q'}{3} \quad \text{when} \quad \sigma'_{v0} > q'$$

4. The compressibility index is determined by the expression:

$$I_c = \dfrac{1.7}{N_{mean}^{1.4}}$$

where $N_{mean}$ is the arithmetical mean of the number of blows $N_{SPT}$ throughout the area of influence.

The following additional conditions should be noted:

— The number of blows $N_{SPT}$ is not corrected by the effect of the depth.
— If the soil is composed of fine sands and silty sands below the water table, Terzaghi's correction can be used:

$$N_{SPT}(\text{corrected}) = 15 + 0.5(N_{SPT}(\text{measured}) - 15)$$

— For gravels and sandy gravels better predictions are obtained using:

$$N_{SPT}(\text{corrected}) = 1.25\, N_{SPT}$$

Sometimes, even in granular soils, delayed settlements have been observed. To take this effect into account the following expression is suggested by Burland and Burbridge (1985):

$$S_t = f_t S_i$$

where:

— $f_t = 1.5$ for static loads over a 30-year period
— $f_t = 2.5$ for cyclical loads over the same time period.

## Settlements in stiff clays

The use of the one-dimensional or oedometric method in stiff clays, which in most cases are over-consolidated, often leads to estimates for settlement that are far greater than in fact take place. The elastic calculations made from deformation

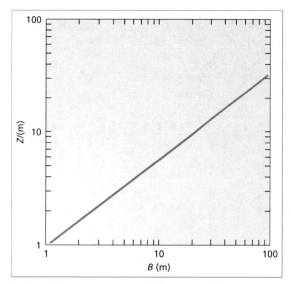

Figure 8.21   Area of influence depending on foundation width B (Burland et al., 1977).

## Box 8.5

### Estimating settlement

a) Estimate the expected settlement after the construction of the viaduct foundations in the example in Box 8.2 if, for reasons of safety from failure, a working load equal to the safe ultimate bearing capacity is applied.
b) Calculate the settlement if the working load is reduced to $q' = 0.5$ MPa.

**Solution:**

a) The gross effective safe ultimate bearing pressure has been calculated in the example in Box 8.3, giving the result $q'_s$(gross) = 1,131 kPa.

The ground has an angle of internal friction of 35%, which according to the correlation of Peck et al. (1967) is equivalent to a medium blowcount $N_{SPT} = 25$.

The correction factors and ratios to apply, taking into account that this is a stratum of undefined depth and a footing with width $B = 4$ m and length $L = 8$ m, are:

$$f_s = \left[ \frac{1.25 \frac{8}{4}}{\frac{8}{4} + 0.25} \right]^2 = 1.23; \quad f_i = 1$$

$$I_c = \frac{1.7}{25^{1.4}} = 0.0187; \quad B^{0.7} = 4^{0.7} = 2.639$$

On the other hand, the foundation is constructed at a depth of 2 m in a soil with a unit weight $\gamma = 20$ kN/m³, so that:

$$\sigma'_{v0} = 20 \cdot 2 = 40 \text{ kPa} \Rightarrow$$
$$\Rightarrow q' - \frac{2}{3}\sigma'_{v0} = 1{,}131 - \frac{2}{3} 40 = 1{,}104.3 \text{ kPa}$$

which replaced in the settlement formula gives:

$$S_i = 1.23 \cdot 1{,}104.3 \cdot 2.639 \cdot 0.0187 = 67 \text{ mm}$$

b) If the gross effective working load is reduced to $q' = 500$ kPa, then:

$$S_i = 1.23 \left( 500 - \frac{2}{3} 40 \right) 2.639 \cdot 0.0187 = 28.7 \text{ mm}$$

---

modulus obtained in triaxial tests also tend to give higher estimates of settlements than really occur. There are many different reasons for these discrepancies, including disturbance of samples, deformability of the test apparatus itself, the trimming procedure of the sample, non-linearity of soil deformation modulus or high rigidity with slight deformations.

A reliable estimate of settlement in these materials requires the use of specialized laboratory or *in situ* testing techniques which are outside the scope of this book.

In any case, since the estimates given by conventional methods tend to be greater than the real settlement, elastic methods from deformation modulus obtained in triaxial test or appropriately verified *in situ* tests can be used.

## 8.3 Deep foundations

When the superficial layers of the ground are not strong enough or are highly compressible, shallow foundations may not guarantee either an adequate factor of safety or settlement within the allowable values for the structure. Even in competent soil, if loads are very high it will be difficult to reach an acceptable factor of safety. In special circumstances, where differential settlements have to be strictly delimited, shallow foundations may not be the appropriate solution. Here the loads have to be supported by transferring them to deeper soil or rock with a higher bearing capacity.

When structurally unstable soils are present, such as swelling clays, collapsible soils such as loess and organic soils such as peat, a possible solution is to transfer the loads to deeper levels instead of founding them on these soils. In all these cases pile foundations should be used.

A **pile** is basically a structural foundation component where length is its maximum dimension. A deep foundation is normally considered to be a pile when its total length is equal to or greater than eight times its width or minimum dimension.

*Figure 8.22* Piles forming a retaining wall.

*Figure 8.23* Installing reinforcement for bored cast-*in-situ* (or cast-in-place) piles.

## Types of pile

There are many types of piles and ways to use them, but they can be classified into two basic groups:

— precast driven piles
— cast-in-situ (or in-place) bored piles.

**Precast piles** are straight pieces of timber, reinforced concrete, pre-stressed concrete or steel which are driven into the ground by blows or vibrations until they reach the required depth. They are also known as **displacement piles**, because of this method of instalation, since when they are driven into the ground they displace the soil occupying that space.

**Bored piles** are sunk by boring holes of the required depth and diameter. A steel cage of reinforcement is installed as required and then the hole is filled with concrete (*Figures 8.23* and *8.24*). As a volume of soil has to be excavated and removed to construct the piles they are also known as **substitution piles.**

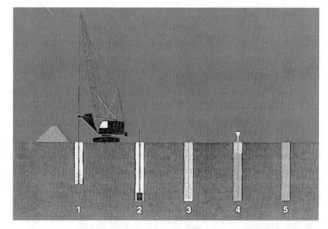

*Figure 8.24* Construction process for bored cast-*in-situ* piles. 1) Drilling with screw or bucket. 2) Clearing the base with bucket, if required. 3) Installing reinforcement. 4) Concreting with tremie tube. 5) Finished pile.

# FOUNDATIONS

Figure 8.25   Driving in pre-cast piles.

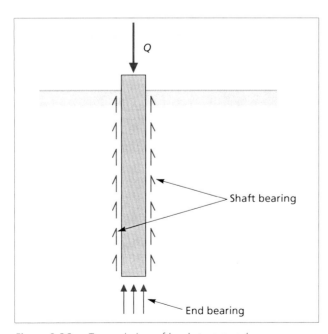

Figure 8.26   Transmission of loads to ground.

Sinking bored piles depends on the soil and water conditions. If the soil is competent enough for the sides of the holes to remain stable, shoring systems will not be needed. Holes are bored at the required depth using spiral or bucket augers attached to rotary drilling machines. When stability cannot be guaranteed, thixotropic slurry or casing is used.

Precast piles are used in soft or loose soils and are popular in areas where contaminated ground exists as no ground has to be excavated and taken from site; they are driven in with a hammer until refusal point is reached (Figure 8.25).

## Single piles

A column is not normally supported by a single pile, and in fact this system should not be used except in the case of a large-diameter column with a high bearing capacity, but the design of piled foundations usually starts with the consideration of the single isolated pile subjected to a vertical load.

The load applied at the top of the pile is transmitted to the ground, partly through friction between the soil and the pile shaft (**shaft bearing**) and partly through the end or tip of the pile (**end bearing**) (Figure 8.26). Depending on the ratio between the load transmitted by the tip and by the shaft, a distinction is made between two basic types of piles: column piles and floating piles.

In a **column or end-bearing pile** all or most of the load received at the top is transmitted integrally to the tip, with only a small part of the load transmitted by the shaft (Figure 8.27a). An example of this would be a pile driven into very soft soil with the tip resting on hard, competent ground, such as rock. The very soft soil is hardly capable of bearing any load, so that all the load is transferred to the tip of the pile.

On the other hand, in a **friction pile** (Figure 8.27b) most of the load is transmitted to the ground through the skin friction of the shaft. In this case, the load reaching the tip may be smaller than that transmitted along the length of the pile shaft, as happens when piles are sunk into homogeneous soil.

The cases described above are not very common. Normal situations fall somewhere between the two—ranging from the column pile in the strictest sense, where the shaft contribution is almost non-existent, to that of the friction pile with a negligible tip bearing, as occurs in the case of very long piles in homogeneous soil.

Whichever the case, the end-bearing and shaft-bearing transmission mechanisms are not completely independent, since mobilizing the strength at the pile tip requires the contribution of the shaft at the deepest point. As shown in Figure 8.28, mobilizing the end bearing resistance creates plastic areas underneath the pile tip, which affect the ground above along a certain length of the pile shaft.

The length of the shaft needed for the complete mobilization of the end bearing capacity is considered as equal to 8 times the pile diameter for sands and 4 times for clays. This

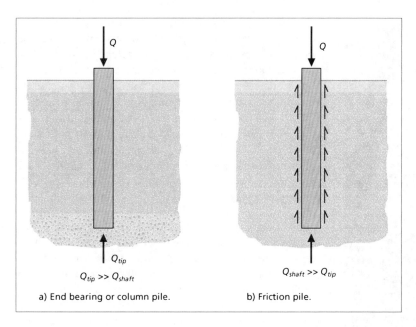

Figure 8.27   Column pile and friction pile.

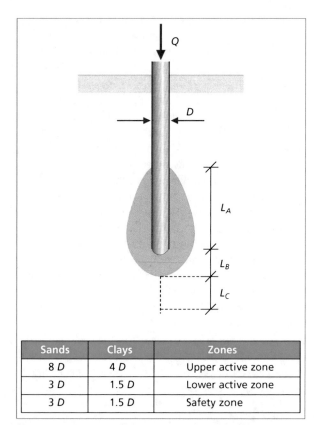

| Sands | Clays | Zones |
|---|---|---|
| 8 D | 4 D | Upper active zone |
| 3 D | 1.5 D | Lower active zone |
| 3 D | 1.5 D | Safety zone |

Figure 8.28   Definition of zones in pile foundations according to Spanish codes.

area is called the **upper active zone**. The length of the plastic bulb zone underneath the pile tip is approximately 3 diameters for sands and 1.5–2 diameters for clays. This area is called the **lower active zone**. For the total ground bearing capacity to be mobilized at the tip of the pile, the ground underneath the plasticized bulb must have either the same or better characteristics than in the upper or lower active zones, for a length of around 3 diameters in sands and 1.5–2 diameters in clays. This is called the **safety zone**.

When load is increased progressively on a pile, some settlement occurs. The load-settlement ratio is approximately linear, then it starts to curve until it reaches the point at which no further increase in load can be sensibly obtained for the further settlement that occurs. This is when the **ultimate load capacity** for the pile is reached. This has two components: the ultimate end bearing load and the ultimate shaft bearing load (Figure 8.29).

The way in which the end and shaft components are mobilized, as well as how the ultimate load values are reached, depends on whether the pile works primarily through the end or the shaft (column or friction).

The **allowable load capacity** ($Q_a$) for a pile is obtained by dividing the **ultimate load capacity** ($Q^{ult}$) by an appropriate factor of safety. Taking into account the differences between the mobilization of the end and shaft resistance, different safety factors are usually applied for each component. A safety factor of 3–4 is normally used for the tip and 1.5–2 for the shaft, so that the allowable load capacity becomes:

$$Q_a = \frac{Q_{tip}^{ult}}{3 \text{ to } 4} + \frac{Q_{shaft}^{ult}}{1.5 \text{ to } 2}$$

# Ultimate load capacity of a pile

The ultimate load capacity of a pile depends on the end or base resistance and the shaft friction.

## Ultimate end load capacity

The ultimate end load capacity at the tip of a pile is equal to the area or cross section of the pile multiplied by the unit pressure at the tip of the pile (Figure 8.30). The general expression of this unit pressure ($q_{tip}$) in terms of effective gross pressure is:

$$q'_{tip} = c' N_{ct} + q'_0 N_{qt} + \frac{1}{2} D \gamma N_{\gamma t}$$

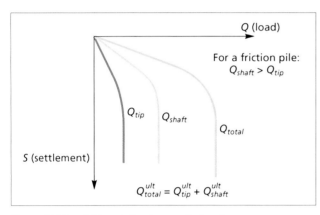

Figure 8.29  Ultimate load capacity in piles.

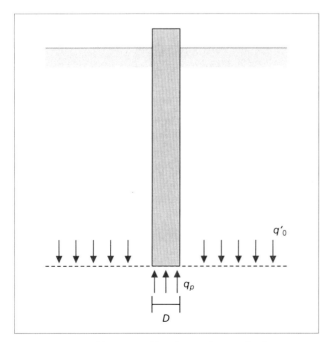

Figure 8.30  Ultimate end load capacity at pile tip.

where:

— $N_{ct}$, $N_{qt}$, $N_{\gamma t}$ are the bearing capacity factors which depend exclusively on the effective angle of internal friction of the ground and can be obtained from the chart in Figure 8.31.
— $c'$ is the effective ground cohesion.
— $q'_0$ is the effective vertical stress from the ground overload at the pile tip.
— $D$ is the diameter of the pile.
— $\gamma$ is the unit weight of the soil.

This is similar to the equation to obtain the ultimate bearing capacity for shallow foundations, although for piles the third term in this expression is normally much smaller than the other two and tends to be ignored.

When piles are sunk in granular soils such as gravels, sand and non-cohesive silts, the effective cohesion is zero, which gives:

$$q'_{tip} = q'_0 N_{qt}$$

From this equation, and for a specific soil with a specified angle of internal friction, the bearing capacity factor $N_{qt}$ is constant and therefore the unit pressure at the tip increases to $q'_0$ in linear proportion to the depth of the tip, although this is only completely true to certain depths. In full scale tests it has been proved that in fact the unit pressure at the tip in granular soils increases in proportion to the length of the pile, but below a certain depth it remains approximately constant. This depth is approximately equal to 10 times the pile diameter for loose sand and 20 times the pile diameter for dense sand and below it no increment at all should be considered.

In clay soils the most unfavourable conditions tend to occur immediately after loading, i.e., before the excess pore pressure generated by loading has been dissipated. The undrained calculation is made under total stress, taking $\phi = 0$, $c = S_u$.

For a zero angle of friction $N_{qt} = 1$, so that discounting the weight of the pile itself, the total net ultimate end load capacity $q_{tip}$ is equal to:

$$q^{net}_{tip} = S_u N_{ct}$$

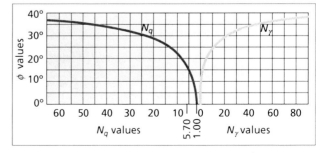

Figure 8.31  Bearing capacity factors (modified from Hansen, 1970).

Thus if $N_{ct} = 9$, the net ultimate end load in clays is equal to 9 times the undrained shear strength, equal to 4.5 times the uniaxial compressive strength.

## Ultimate shaft load capacity

The soil exerts horizontal stress along the full length of the pile shaft and so generates a friction resistance force with an effective pressure $\sigma'_h$ which varies with depth (Figure 8.32).

The general expression of the tangential stress $q_{shaft}$ which produces the failure of the pile shaft is:

$$q_{shaft} = c'_a + \sigma'_h \tan \delta'$$

where:

- $\sigma'_h$ is the effective horizontal stress.
- $\delta'$ is the ground-pile friction angle.
- $c'_a$ is the effective available ground-pile adhesion.

As seen here, the unit strength $q_{shaft}$ is made up of two factors, one due to the ground-pile friction and the other to the adhesion between the pile and the ground. The first is similar to frictional strength and the second to cohesive strength.

The effective horizontal pressure on a vertical plane in the ground is proportional to the effective vertical pressure at that depth, i.e.:

$$\sigma'_h = K \sigma'_v$$

Assuming that sinking the pile does not alter the original ground conditions, this coefficient will be the earth pressure at rest ($K_0$) defined in Chapter 2, Section 2.5.

The total pile shaft strength will therefore be obtained by integrating the shaft stresses needed to produce failure along the whole length and perimeter of the pile, as follows:

$$Q_{shaft}^{ult} = \int_0^L \pi D q_f dz$$

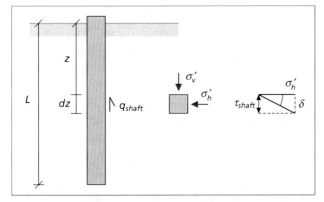

Figure 8.32   Determining tangential stresses on a pile shaft.

If the soil is homogeneous and the water table is near the surface, this gives the following solution:

$$Q_{shaft}^{ult} = \left[ \frac{1}{2} K (\gamma_{sat} - \gamma_\omega) \tan \delta' L^2 + c'_a L \right] \pi D$$

Where the ground is heterogeneous and has different strata, the integration should be done partially for each stratum and the critical loads given for each type of soil added together.

In fact, the pile installation method will always influence the mobilization of shaft strength to some extent. This means that in the case of bored piles, drilling to excavate the hole for the pile can be expected to relieve the horizontal stress and therefore the coefficient $K$ will probably be lower than the earth pressure at rest, $K_0$. On the other hand, displacement piles tend to increase the lateral compression of the surrounding soil, so that the $K$ coefficient will probably be higher than $K_0$.

The value of the effective adhesion, $c'_a$, which can be mobilized will also depend on how the pile is formed; if there is some remoulding of the ground around the pile, as in driven piles, then the adhesion between the pile and the ground will, most probably, be lost.

Finally, a pile shaft is normally fairly rough, so that if critical stress $q_{tip}$ is reached failure will probably occur through the soil around the pile, within the remoulded soil area. This type of failure is probably easier to generate than failure which has to adapt to the irregularities of the pile surface, and suggests that the effective angle of friction, $\delta'$, may be taken as equal to the effective angle of internal friction of the remoulded soil.

Burland et al. (1977) suggest the following expressions for the ultimate shaft load capacity:

— **Normally consolidated clays:**

$$q_{shaft} = (1 - \sin \phi') \sigma'_v \tan \phi'$$

where:

- $\phi'$ is the internal angle of friction of normally consolidated clays
- $K = K_0 = 1 - \sin \phi'$ in the case of an ideal pile (where the installation does not disturb the soil), and a slightly lower value can be taken for cast-in-situ piles and a slightly higher one for driven or displacement piles.

— **Over-consolidated clays:** in this case the fundamental difference is in the coefficient of the initial earth pressure at rest; the following expression can be used to give an approximate determination of this coefficient:

$$K_0 = (1 - \sin \phi') OCR^{\sin \phi} \approx (1 - \sin \phi') \sqrt{OCR}$$

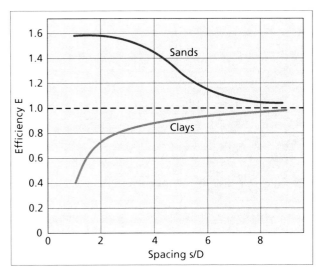

Figure 8.33 Efficiency coefficient for pile groups (Oteo, 1980).

In soft and medium clays, the efficiency factor is generally lower than 1, which means that the strength of the group is proportionally less than the strength of each individual pile. However, in the case of loose and medium sands, sinking the piles, especially if these are pre-cast and driven-in, may produce densification and some improvement in the ground strength and deformability characteristics, so that normally $E > 1$, although in calculations a value of $E$ greater than 1 is not normally adopted.

The efficiency factor obviously depends not only on the type of soil, but also on the distance between the axes of the piles within the group; in other words, it depends on the spacing $s/D$ where $s$ is the distance between the axes of the piles and $D$ is their diameter.

Figure 8.33 shows the empirical values normally used for the efficiency factor of the group. It can be seen that for clays, $E$ ranges between 0.70 and 1 for spacings between $s/D = 2$ and $s/D > 8$.

## Negative friction on piles

If a pile is sunk through soft soils and its tip is embedded in a hard, competent stratum (Figure 8.34a) and if extensive fill or some similar overburden is applied to the ground surface, settlement will occur in the soft soil, and may be significant. The same problem can arise if instead of a surface overburden, the water table in the soft soil falls.

In the former case an increase in vertical effective stress will result and in the latter case the resulting reduction in pore pressure will increase the effective vertical stress; both cause ground settlement. In this case, the pile, well supported at the tip, will prevent the soil immediately around it from settling, and the settlement profile will be as shown in Figure 8.34b.

This settlement of soft soils around the piles produces a downward tangential stress or down-drag on the pile shaft which is called **negative skin friction**.

These stresses may add a very considerable increment to the load on the pile. The tangential stress due to negative friction can be expressed as:

$$f_n = K_s \sigma'_v \tan \delta'$$

Depending on the type of ground, the product $K_s \tan \delta$ may be taken as equal to (Bjerrum, 1973):

| Type of soil | $K_s \tan \delta'$ |
|---|---|
| Silty clays | 0.25 |
| Low plasticity clays | 0.20 |
| Plastic clays | 0.15 |
| High plasticity clays | 0.10 |

where:

- $\phi'$ is the angle of internal friction of normally consolidated soils (eliminating the "peak" effect caused by the remoulding)
- OCR is the over-consolidation ratio. In these conditions, the critical shaft strength is expressed as:

$$q_{shaft} = (1 - \sin \phi') \sqrt{OCR} \, \sigma'_v \tan \phi'$$

— **Granular soils**

$$q_{shaft} = K \sigma'_v \tan \delta'$$

where:

- $\delta' \approx 2/3 \cdot \phi'$.
- $K = 0.5$ for loose sands.
- $K = 1.0$ for dense sands.

## Pile groups

The allowable load capacity for a group of piles is not generally equal to the product of the allowable load for the individual piles by the number of piles in the group. To obtain the ultimate load capacity for the group, the ultimate load capacity of each pile must be multiplied by the number of piles in the group ($n$) and by a specific **efficiency factor** ($E$):

$$Q^{ult}_{group} = E n Q^{ult}_{individual}$$

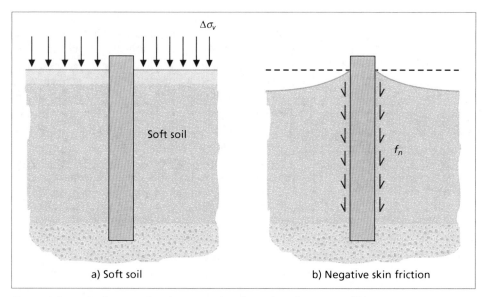

*Figure 8.34   Surface overburden around a pile and settlement profile.*

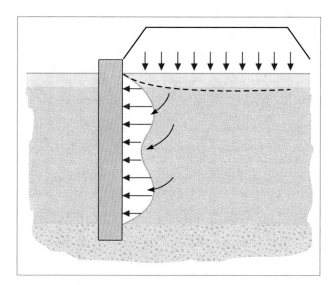

*Figure 8.35   Laterally loaded piles.*

## Laterally loaded piles

As shown in *Figure 8.35*, if piles are driven through cohesive soft soils and an asymmetric overburden is placed on the ground surface, i.e., on only one of its sides, settlement and horizontal ground displacement will occur. This horizontal "flow" produces thrust and flexure on the pile, which in extreme cases may cause pile failure.

This problem has to be taken into account when the acting overburden ($\Delta\sigma_v$) is 3 times greater than the undrained shear strength of the soil:

$$\Delta\sigma_v \geq 3 S_u$$

## 8.4 Foundations on rock

Just as the ultimate bearing capacity for soil foundations can be calculated using the Mohr-Coulomb criteria, the load which produces plastic deformation in a rock mass can be obtained using the analytical method developed by Serrano and Olalla (1996 and 1998). This method can be applied where the rock medium is homogeneous and isotropic and follows the Hoek & Brown failure criteria, i.e. groups I, IV and V shown in *Figure 8.36*.

In simplified cases, the analytical method mentioned is resolved with direct entry charts. The geometric definition used is represented in *Figure 8.37*, where:

— $P_{ult}$: ultimate load capacity.
— $i_2$: load angle.
— $\sigma_1$: vertical stress acting at "contour 1" located beside the foundation; this corresponds in simplified cases to where the soil overburden is above the base of the footing.
— $\alpha$: slope of the ground surface beside the footing (contour 1).

The **ultimate load capacity**, $P_{ult}$, can be estimated by the following expression:

$$P_{ult} = \beta(N_\beta - \zeta)$$

where:

$$\beta = \frac{m\sigma_{ci}}{8} = \frac{m_i\sigma_{ci}}{8}\exp\frac{RMR-100}{28}$$

$$\zeta = \frac{8s}{m^2} = \frac{8}{m_i^2}\exp\frac{RMR-100}{25.2}$$

# FOUNDATIONS

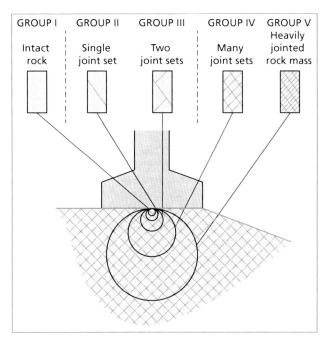

Figure 8.36  Simplified representation of the influence of scale on the behaviour of the rock mass for the design of shallow foundations (modified from Hoek and Brown, 1980).

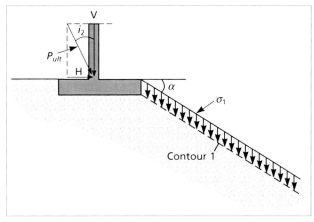

Figure 8.37  Geometric definitions used to determine, by simplified analytical methods, the allowable pressures in rock.

being $m$, $s$ and $m_i$ the Hoek and Brown criterion parameters and $\sigma_{ci}$ the uniaxial compressive strength of the intact rock (Chapter 3, Section 3.6).

The bearing capacity factor $N_\beta$ is a generalization of the Prandtl parameters $N_c$ and $N_q$, and it is a function of the ground slope, of the load angle and of the normalized external overburden acting around the footing, $\sigma_{01}^*$:

$$\sigma_{01}^* = \frac{\sigma_1}{\beta} + \zeta$$

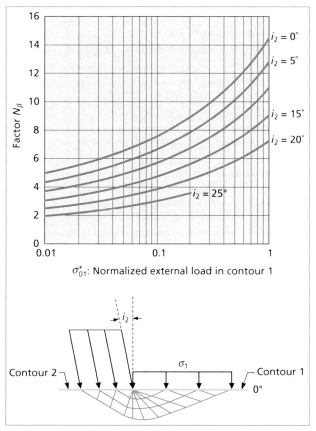

Figure 8.38  Values for the bearing capacity factor $N_\beta$ depending on the normalized external load and the angle of slope of the loads (horizontal ground surface $\alpha = 0°$) (Serrano and Olalla, 1998).

*Figure 8.38* shows the values for $N_\beta$ when the ground is horizontal and the x-axis the external overburden.

The expression for $P_{ult}$ given above does not consider the effect of the dead weight of the ground, which may be very beneficial in large foundations. When jointing and other discontinuitueis make the srength of a rock mass anisotropic their influence can be taken into account using a reduction factor, depending on the direction of the discontinuities and their strength, as described in Serrano and Olalla (1998).

Without going into considerations relating to settlement analysis, which in very specific cases could determine the design of a rock foundation, the **allowable load capacity** is established by dividing the ultimate load capacity, $P_{ult}$, by a factor of safety $F$ which can be expressed as the product of two factors:

$$F = F_p F_m$$

The factor $F_p$ considers the statistical variations of the rock parameters and its value is linked to the probability of

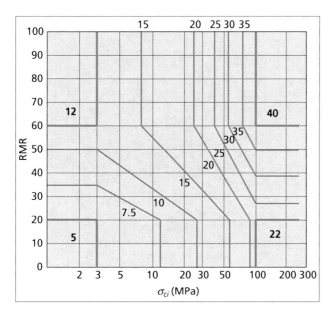

*Figure 8.39* Partial factor of safety coefficients $F_p$ for a failure probability of $<10^{-4}$ (Serrano and Olalla, 1996).

foundation failure. The factor $F_m$ includes the possibility that failure of a part or of the whole of the foundation does not fit Hoek and Brown's model.

*Figure 8.39* shows the partial safety factor $F_p$ to be used depending on the value of the RMR and the uniaxial compressive strength, $\sigma_{ci}$, for a failure probability of $<10^{-4}$.

The partial safety factor $F_m$ depends on the strength uncertainties related to the plastic criterion. In the case of rock foundations in Hoek and Brown's group I (*Figure 8.36*), the brittle behaviour of the rock mass must be taken into account; if $\sigma_{ci} > 100$ MPa, the rock mass has a brittle response and values of 5 or 8 should be used for $F_m$. The failure of the rock mass can be considered as plastic if $\sigma_{ci} < 12.5$ MPa and for this reason it is not necessary to include any related factor of safety ($F_m \cong 1$).

When the rock mass is heavily fractured and jointed, the behaviour of the foundation is not brittle and there is no reason to adopt a special safety factor for $F_m$ (Hoek and Brown's groups IV and V).

## 8.5 Foundations in complex geological conditions

Foundations for any structure require stable geological conditions. First, the site should be **free from geological hazards** or, if these hazards are random or recurrent, the structural design must take the possibility of their presence into account (see Chapter 15). Second, the foundation must be remain stable if engineering geological problems occur, and although these are not as significant as geological hazards, special measures may have to be adopted for the foundations, including ground improvement. These problems tend to occur in the following situations:

a) Lithologic heterogeneity, where there are materials of different strength and deformability within the same foundation area.
b) Deep competent substratum below foundation level, that may require deeper or more expensive foundations to reach them than normal shallow ones.
c) Adverse geo-environmental factors, such as a high water table, significant seepage, steeply sloping ground, water or aggressive materials.
d) Anthropogenic fills and dumps with very high compressibility, contamination and gas generation, and erratic geometry.
e) Solution cavities and soft, expansive, organic, collapsible or soluble soils.

Points a, b and c do not normally require special foundation solutions and are resolved by conventional procedures (piles, rafts, dewatering, etc.). Points d and e, however, require special measures and ground improvement.

Some of the most frequent engineering geological foundation problems are as follows:

— Expansive or swelling clays.
— Collapsible soils.
— Karstic and volcanic cavities.
— Soft and organic soils.
— Anthropogenic fills.

Some of the solutions to these problems are outlined below.

## Expansive soils

As mentioned in Chapter 2, the problem of expansive or swelling soils is linked to the presence of smectite minerals and changes in the moisture content of the soil, which in turn is conditioned by cyclical changes in their water content. In semi-arid areas, the ground loses water through evaporation during periods of drought, which produces volumetric contraction of the soil and fissuring; later, when the drought is over, the rain water penetrates the fissures and saturates the clay, which then swells back towards its original value. The depth over which this occurs is called the **active layer**, and is the depth affected by climatic changes.

These changes in volume arising from contraction and/or expansion are also caused locally if the moisture content below buildings (especially those built in wet or dry seasons or periods) are changed by leaking pipes, or by the presence of vegetation and trees (roots can cause changes in the moisture under buildings). Swelling of soil is just as important as shrinkage, especially for light-weight structures (one- or two-storey houses, platforms for railways or roads) since it is the change in volume that does the damage, not what kind of change it is.

Various remedial measures are used to solve this problem when building foundations (Figure 8.40):

— Sinking shafts or piles through the active layer reaching minimum depths of around 4 m and filled with limed concrete.; the structural footing is located above this, with the structural slab of the lowest floor of the building placed to rest on it, so as to be isolated from the ground by a space or cavity which can be filled, using slabs of expanded polystyrene etc.
— Sinking piles and micro-piles through the active layer—in some countries such as South Africa this may be up to 15 m deep—in such a way that the piles are founded in an area which does not undergo volume changes; the piles must be able to work in tension, since an increase in the volume of earth above tends to lift them; structures built on such piles should also have a resistant slab on the lowest floor.
— Replacing the expansive ground on the surface (in the active layer) by an inert coarse grain material, laying an impermeable barrier between the ground and this material so that there is no water infiltration at depth; this solution is often used to support structures with a large surface area but small load, e.g. a road structure, which is made extra wide to avoid water penetrating under the road platform. The added material can be the swelling clay itself, adequately compacted (with sheep's foot roller on the wet side of optimum moisture content) or else with added lime.

All water pipes and installations must also be carefully considered, with flexible joints to allow movement, and laid in trenches with lightly compacted granular infill (so that they can absorb volume changes that arise from leaks), avoiding hazards near the installations and eliminating vegetation with extensive root systems.

In some road works, where there has been clearing and grading, good results have been obtained by sealing the

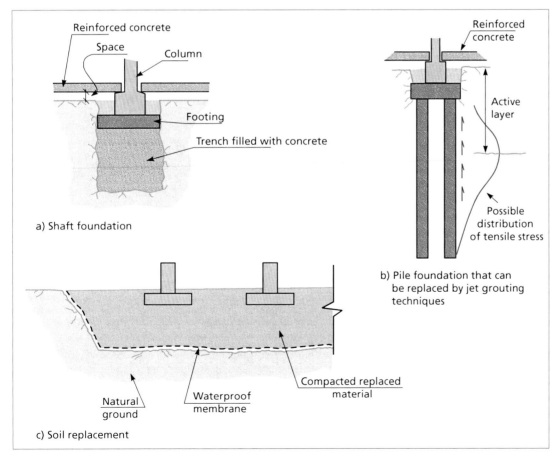

Figure 8.40   Examples of foundation solutions in swelling soils.

base of trenches dug in swelling clays with the same clay, treated with 2–3% lime, making a strong, waterproof seal. This treatment with lime has also been used to build motorway earth embankments.

## Collapsible soils

Collapsible soils are characterized by negative volume change that may be caused by:

— Wetting and elimination of negative pore pressures produced around grain contacts in the menisci formed by the water in partially saturated soils; the saturation reduces the effective stress causing volume reductions that may reach 5–7%, e.g. in Tertiary sands in central Spain.
— Stress concentration at grain contacts producing failure (structural collapse), as may occur after flooding in rockfills and earthfills made from volcanic materials.
— Dissolution of the bridging links between silt particles as a result of flooding, as occurs in gypsum-bearing silt and loess, with considerable reductions in volume.

Remedial measures for collapsible soils include:

— Aggitating and re-compacting sandy soils. Gypsiferous silts can be replaced with the same material. Usual values for dry density may be increased from between 11.5–12.5 $kN/m^3$ in the natural state to 17.0–17.5 $kN/m^3$ when well compacted. Alternatively, other material can be added depending on the type of foundations. In all cases, care has to be taken to ensure that water does not reach the layers of ground that have not been disturbed.
— Compacting an area of the ground, using dynamic compaction. This technique tends to be more effective for sands than for gypsiferous silts.
— Using gravel columns made by aggitating a column of ground with water which leads to the collapse of the ground mass, but at the same time makes the area treated stronger by the addition of the gravel to the hole created this way, and its compaction.
— Strengthening the ground with grout to create a more rigid ground structure than the original one, although this tends to produce an appreciable collapse in the ground during the process.
— Using deep foundations with piles or micro-piles.

## Karstic cavities

Karstic cavities develop in the following materials:

— **Gypsum-bearing and saline materials**, where sinkholes and cavities are produced by rapid dissolution (Chapter 11, Section 11.6). In the case of gypsum-bearing silts, some of the fissures are filled with very soft, wet gypsiferous silts; these phenomena have caused subsidence in agricultural and industrial buildings, and roads, including recorded surface collapse of up to 30 m in diameter in Zaragoza, Spain.
— In **calcareous materials**, water flow causes the gradual formation of sinkholes and cavities, which may be large, but which can be stable for long periods of time due to the rock strength.

In gypsum and saline materials, deep foundations are normally used and all loads are to be supported by the shaft resistance of the piles, in case there are cavities underneath the pile tips. Where cavities are detected, grouting may be advisable (*Figure 8.41*). Another approach is to remove the material and replace it, either with the same material, now compacted, or with alternative material. For karst in hard gypsum, another approach is to clear the soft zones and fill them with weak concrete.

In calcareous karst, different approaches have been used in addition to those mentioned above (*Figure 8.41*):

— Raft or slab foundations to distribute the loads over a larger area and grouting to fill the voids and cavities to a depth similar to the width of the foundation.
— If the cavities are near the surface, the ground can be excavated to a depth of between 0.5 and 4 m and a reinforced concrete slab inserted at the bottom of the excavation, which is then filled in with appropriate material.

## Volcanic cavities

Cavities can often form in volcanic rocks although most of them are small. In basaltic lava flows, cavities can form with very different volumes, horizontal rather than vertical and with an irregular geometry which tends to form bubbles. Caves and volcanic tunnels are less frequent, although they may be very large, like those found on the Canary Islands, the Azores and Hawaii.

These are a hazard for surface foundations, and detailed site investigation is needed to detect them. The most widely-used techniques are borehole drilling and rotary percussion drilling where rates of penetration, and sudden drops of the drilling rods, can be diagnostic of cavities at depth. Surveys using video cameras inside boreholes can be very helpful (see Chapter 13, *Figure 13.41*). Geo-radar and gravimetric geophysical methods have also been used with acceptable results.

Possible solutions for geotechnical-related problems due to these types of cavities in volcanic materials include grouting and filling a cavity with concrete, and transferring

# FOUNDATIONS

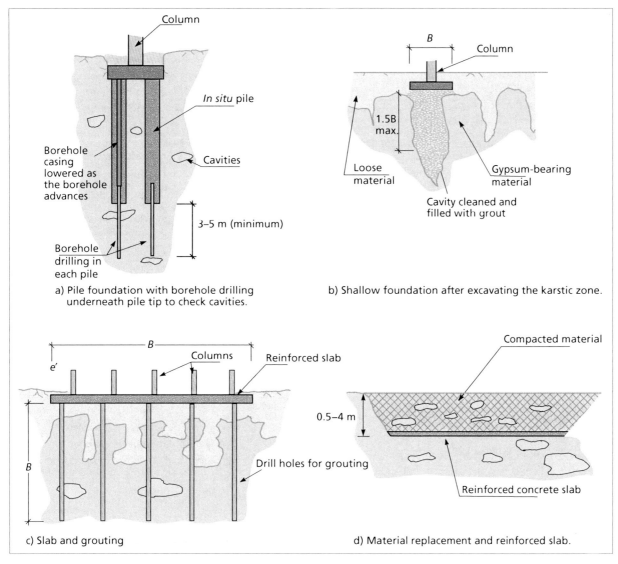

Figure 8.41   Some solutions for foundations in karstic zones.

the structural load to a stronger layer using micro-piles, or piles.

## Soft and organic soils

Highly compressible soft soils are not suitable for shallow foundations. The normal solution is to use piles which transfer the load to a deeper and stronger material, but note; these soils may produce negative skin friction. Depending on the type of structure and its allowable settlement, gravel columns, sand drains and drainage wicks have been used to speed up consolidation; rockfills have also been used. Pre-loading can also be carried out to consolidate the ground prior to construction and can be particularly suitable for earth works foundations.

## Anthropogenic fills

Man–made fills, landfills and waste dumps are very common materials in cities and their surroundings. Given their heterogeneity and very low compaction, these fills are prone to settlement and have a high risk of subsidence or collapse, even under low loads. In general, the most usual solution is to support the structure on piles or shafts, transferring the loads to the resistant sub-stratum below the fill, taking into account the effect of the negative skin friction which these fills can produce. However, if these fills are not very thick, the best solution is to remove them provided that is environmentally acceptable.

Sometimes fills can be treated with conventional grouting, jet-grouting, dynamic compaction, preloading and other techniques including chemical, bio-chemical ground improvement. If collapse occurs in the fills, the remedial measures taken are those described above for collapsible soils.

## 8.6 Site investigation

One of the basic objectives of site investigation for foundations is to obtain geological and geotechnical data to calculate the bearing capacity, settlement and other determining factors of the soil or rock mass behaviour. The safety of the structure depends on the outcome of these site investigations and also the technical and financial suitability of the foundation solution adopted. The need for geotechnical studies has been emphasized throughout the first part of this book. Site investigations are not only fundamental for design and construction purposes, they are also extremely cost effective, as they avoid unforeseen costs and delays which may affect the viability of the project.

The **objectives of site investigation** for foundation design should include the following:

— Nature and arrangement of the soils or rocks that make up the foundation site.
— Strength and deformability properties of the foundation materials.
— Depth of the water table and the hydrogeological conditions.
— Presence of ground anomalies, such as cavities, water collector galleries, remains of old foundations, wells or other excavations.
— Presence of unstable soils such as swelling or expansive clays, collapsible soils, poorly compacted or weak fills and soluble soils.
— Chemical components of groundwater or soils which are potentially aggressive for the construction materials.
— Location of services such as wells, drains, galleries and underground structures, power lines and water pipes.
— Geological hazards, including landslides, active faults, karstification zones, sink holes or subsidence.
— Previous experience from nearby buildings and excavations, retaining walls, drainage installation and local knowledge of suitable types and depths of foundations. It is particularly important to find out about the foundations of neighbouring buildings and their behaviour, as they may be affected by the projected foundations.
— Possible use of the materials excavated to form the foundations for embankment construction, fill or aggregate.

Site investigation methods are described in Chapter 5 and the geomechanical parameters needed to calculate foundations have been dealt with in previous sections, so that with a knowledge of these methods and parameters the type, depth and number of site investigations can be defined. However, the complexity of the geological environment, both because of its heterogeneity and anisotropy, as well as the great variety of possible structures requiring different foundations, loads or foundation depths, means that each site investigation for foundation purposes is different, and must therefore be analysed individually.

The **methodology** to be followed in geotechnical foundation design can be summarized as follows:

1. Structural characteristics of the building or installation to be built.

    — Location and intended floor plan
    — Anticipated level and depth of foundations
    — Stress and load distribution, and geometric excavation data.

2. Site investigation of the foundation site.

    — Literature review, local knowledge and experience
    — Geological and geotechnical surveys
    — *In situ* tests.

3. Geotechnical report

    — Strength and deformability analysis
    — Type of foundation recommended
    — Allowable bearing pressures and foundation depth
    — Building recommendations

4. Foundation design

    — Dimensioning of structural elements
    — Building solutions
    — Cost estimation

5. Geotechnical control during construction

    — Direct observation of the excavations
    — *In situ* testing
    — Checking and comparing design recommendations with ground conditions encountered during excavations and adopting additional measures as required.

### Stages in site investigations

1. Preliminary studies

    — Preliminary engineering geological survey
    — Literature review and local information.
    — Trial pits and borings.
    — Definition of geological model

2. Pre-design investigation stages
   - Borehole drilling, probing, trial pits and trenches
   - Geophysical surveys
   - Laboratory tests.
   - Definition of geotechnical model and outline design
3. Detailed design investigation stages
   - Confirmatory borehole drilling
   - *In situ* test (plate load test, pressiometer, permeability tests, etc.)
   - Confirmatory laboratory tests.
4. Construction stage
   - Plate load test, compaction test and geotechnical instrumentation.
   - Quality control and monitoring

The planning and design of site investigations for building foundations should also consider the following project conditions:

— **Large-scale projects:** housing developments, shopping centres and buildings of exceptional size, height or intended use. In these cases, all the stages mentioned above should be carried out. The investigation should begin with a preliminary study to evaluate the general conditions of the foundation site. In the pre-design stage, the foundation type is evaluated, and geological-geotechnical problems that may require technical solutions are identified and budgeted. At the design stage, the allowable bearing capacity and recommendations for overcoming problems expected during the excavation and construction of foundations are specified. During construction, the foundation design is verified, e.g. as the ground is exposed, or as a pile is driven. Sometimes the number of stages mentioned above can be either reduced, because the geological and building conditions are straightforward, or overlap because time is short.

— **Conventional buildings.** Generally the same phases are followed but simplified and grouped into one or two stages, although carrying out the preliminary study as the first stage is highly recommended. When the geological conditions are simple and the type of structure does not present any particular geotechnical problem, and the work is similar to that done before in the area, the site investigations can be carried out in one stage.

**Cost of site investigations.** As a result of the considerations above, an ideal budget cannot be fixed for site investigations, as these depend on the geological complexity and type of project. Based on experiences in some European Countries, the minimum budget assigned can be estimated as follows:

— Large scale projects and special constructions: 0.5–1% of building cost
— Conventional buildings: 0.3–0.5% of cost
— If the geological conditions are complex (see previous section above) these cost estimates may be much higher.

**Amount of site investigations.** Deciding the amount of imvestigation i.e. the type, number and depth of investigations depends on the professional criteria of either the geotechnical engineer or the engineering geologist, who takes into account the lithological heterogeneity, estimated depth of the bearing layer, the possibility of encountering geological problems and the type of building structures and loads.

**Guidelines for site investigations** can be found in Eurocode 7 (1997) which recommends the following surveys:

— For structures covering a large area, exploration points should be sited at grid nodes; the distance between the points will be 20–40 m. If ground conditions are uniform, the boreholes or trial pits may be partly replaced by probing tests or geophysical surveys.
— For foundations with individual or continuous footings, the depth of the explorations or boreholes under the planned foundation level will normally be 1–3 times the width of the foundation elements; at some points, greater depths should be reached to study settlement and possible problems with ground water.
— For raft foundations, the depth of the *in situ* tests or boreholes should be equal to or greater than the width of the foundation, unless the rock substratum is found at a shallower level.
— For pile foundations, boreholes, probing tests or other *in situ* tests must be carried out to survey the ground conditions below the tip of the pile to guarantee safety, normally to a depth 5 times the diameter of the pile shaft; however, there will be cases where deeper investigations or boreholes are needed. The depth of site investigations must also be greater than the short side of the rectangle containing the pile group which forms the foundation at pile tip level.

The **geotechnical report** for foundation design should contain sufficient data to enable:

— The most appropriate type of foundation and construction methods to be selected.

- Either the depth of the foundations to be specified, or the conditions for establishing these precisely while the work is underway to be given.
- The allowable bearing pressures, in the case of direct foundations, or end and shaft resistance, in the case of pile foundations to be determined.
- Settlement to be calculated.
- Measures to be taken to avoid possible structural damage from concrete attack, or expansive, swelling or collapsible soils.
- Procedures to be established for the excavation and dimensioning of the retaining walls, diaphragm walls or other earth retaining structures.

## Recommended reading

Das, B.M. (1999). Principles of foundation engineering. 4th ed. PWS Publishing.

Eurocode 7 (1997). Geotechnical design (distributed through the National organisations responsible for Standards in their country).

Simons, N.E. and Menzies, B.K. (2000). A short course on foundation engineering. 2nd ed. Thomas Telford, London.

Tomlinson, M.J. (2001). Foundation design and construction. Prentice Hall. 7th ed. Harlow, Essex.

Wyllie, D.C. (1999), Foundations in rock. 2nd ed. E & FN Spon, London.

## References

Bjerrum, L. (1973). Problems of soil mechanics and construction on soft soils. 8th ICSMFE. Moscow. Vol. 2, pp. 27–34.

Burland, J.B., Broms, B. and de Mello, V.F.B. (1977). Behaviour of foundations and structures. State-of-the-art report. Session 2. Proc. 9th ICSMFE. Tokyo. Vol. 2, pp. 495–546.

Burland, J.B. and Burbridge, M.C. (1985). Settlement of foundations on sand and gravel. Proc. Institution of Civil Engineers, 78, pp. 1325–81.

Eurocode 7 (1997). Geotechnical design (distributed through the national organisations responsible for Standards in their country).

Fadum, R.E. (1948). Influence values for estimating stresses in elastic foundations. Proceedings. 2nd Int. Conference on Soil Mechanics and Foundations Engineering. Vol. 1, 77–84.

Hansen, J.B. (1970). A revised extended formula for bearing capacity. Danish Geotechnical Institute Bulletin, no. 28.

Hoek, E. and Brown, E.T. (1980). Underground excavations in rock. Inst. of Mining and Metallurgy. London.

Oteo, C. (1980). Pilotajes. In: Jiménez Salas et al., Geotecnia y cimientos III. Cap. 3. Rueda. Madrid.

Peck, R.B., Hanson, W.E. and Thornburn, T.H. (1967). Foundation engineering, 2nd ed. John Wiley and Sons, New York.

Skempton, A.W. (1951). The bearing capacity of the clay. Proceedings of Building Research Congress. Vol. 1, 180–185.

Serrano, A. and Olalla, C. (1996). Allowable bearing capacity in rock foundations based on a non linear criterion. Int. Jl. Rock Mech. and Min. Sc. Vol. 33, 4, pp. 327–345.

Serrano, A. and Olalla, C. (1998). Ultimate bearing capacity of an anisotropic discontinous rock mass. Part I: Basic modes of failure. Part II: Determination procedure. Int. Jl. Rock Mech. and Min. Sci. Vol. 35, no. 3, pp. 301–348.

Tomlinson, M.J. (2001). Foundation design and construction. Prentice Hall. 7th ed. Harlow, Essex.

Wyllie, D.C. (1999). Foundations in rock. 2nd ed. FN Spon, London.

# 9
# SLOPES

1. Introduction
2. Site investigations
3. Factors influencing slope stability
4. Types of slope failure
5. Stability analysis
6. Stabilization measures
7. Monitoring and control
8. Slope excavation

# 9.1 Introduction

Slope excavation is generally needed when linear infrastructure, such as roads and railways, or other engineering works like canals, pipelines and mining, require the construction of either a level surface on natural slopes or at a pre-determined depth below ground level. Slopes are excavated as steeply as ground strength permits while maintaining acceptable stability conditions.

Slope design is of major importance in geological engineering as it forms part of most construction and mining activities. In civil engineering, maximum slope height is generally 40 or 50 metres (*Figure 9.2*); however, in open pit mines slopes may be several hundred metres high (*Figure 9.3*).

**Permanent slopes** excavated for buildings and infrastructure are designed for **long-term stability**. Where excavations at the required height or angle are impossible for financial or other reasons, additional stabilization measures are needed. Slope design for mining depends on the distribution and depth of the mineral deposits. In the case of non-metallic mineral deposits arranged in horizontal or inclined layers, slopes are generally **temporary** and designed for **short or medium term** stability over several months or years; once the mineral has been extracted, the excavation is either abandoned or infilled. In metal mining operations where the mineral is not found in layers, the geometry of the slopes is modified as excavation progresses, although their gradient is usually maintained. Quarries are a special case where the excavation faces are progressively cut back, and instability is generally related to blocks or groups of blocks that break off along discontinuities in the hard rock masses during quarrying.

**Financial criteria** are fundamental in slope design and excavation for mining, and a certain degree of risk of localized or partial slope failures is often assumed, provided

*Figure 9.2*  Manually excavated slopes in cemented and faulted silty calcarenites as exposed in the Corinth Canal, Greece.

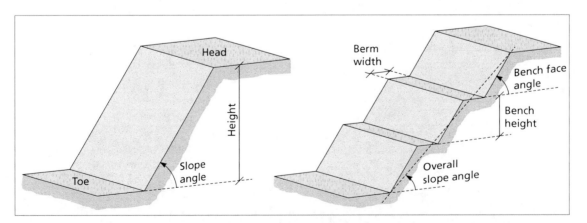

*Figure 9.1*  Uniformly-angled slope and stepped slope with berms and benches.

*Figure 9.3*  Slopes more than 300 m deep at the Corta Atalaya mine, Rio Tinto, S. Spain (courtesy of P. Gumiel).

*Figure 9.4*  Slope excavation for a motorway.

this does not compromise either safety or work schedules; supports and other stabilizing measures are not usually installed on temporary slopes. In civil engineering, however, tolerance of slope failure is strictly limited because of possible effects on nearby structures, and **safety criteria** are of the greatest importance.

The **design of stable slopes**, and the design of **stabilization measures** for unstable slopes, the two main aims of slope stability analysis, are based on geological and geotechnical studies. The design must take into account the specific requirements in each case (short, medium or long-term, the cost-safety ratio, the level of risk assumed).

Slope design is based on **stability analyses**, which include **factor of safety** calculations. These analyses provide the data for the corrective and stabilizing measures to be adopted in cases of real or potential failure. It is essential to understand the geological and geomechanical characteristics of the materials which form the slope, the possible failure models and mechanisms, and the factors conditioning and triggering instability.

The **methodology** followed in slope stability studies is based on knowledge of the geology, hydrogeology and geomechanical behaviour of the rock mass or soil (aspects dealt with in Chapters 2 to 6); this information, together with an analysis of the external factors affecting the ground, will define the behaviour of the materials and their deformation and failure mechanisms.

This chapter does not deal with instability processes in **natural slopes**. These include **landslides, rockfalls** and **surface processes** such as earth or mud flows. These processes are generally conditioned by natural factors, although they may also often be triggered by human activity. A description of these processes can be found in Chapter 13.

## 9.2  Site investigations

The purpose of site investigations for slopes is to provide a geological and geotechnical description of the area of the proposed excavation, to obtain the parameters needed for stability analysis, slope design and excavation of material, and to design stabilization and drainage measures. A description of the different methods of investigation is included in Chapter 5.

As a general rule, the following aspects should be considered for an excavation:

— Intended dimensions (slope length and depth).
— Position of the water table and hydrogeological conditions, and rainfall.
— Rock and soil types, geological structure and seismicity.
— Project requirements (long or short-term slopes, geometric conditions, etc.)

The site investigations should be adapted to these conditions and take place in stages, as described in Chapter 5, Section 5.1, taking financial considerations and the planned time frame into account. Typical procedures include:

— A preliminary geological survey, carried out at the initial stage to plan the site investigations. A geological-geotechnical map is drawn up on a scale between 1/2,000 and 1/500, depending on the type and range of the project. If rock masses are involved, structural data are gathered from field sites on rock outcrops.
— Trial pits in soils and heavily weathered rock to inspect the materials and take undisturbed samples.
— Seismic refraction along the slope profile. As this is a low-cost technique providing the data needed to estimate the rippability and thickness of weathered materials, seismic refraction should be carried out along the whole length of the slope, or at least in representative sections.

— Boreholes drilled along the slope to investigate the slope crest and toe area. The number of boreholes depends on the length and geological complexity of the slope. Core samples should be taken from the boreholes for laboratory tests, and piezometric pipes installed in them to measure water levels.
— Permeability tests in excavations requiring pumping or drainage.

The strength properties of materials, soils or rock masses, are obtained from the appropriate *in situ* and laboratory tests, and the use of empirical criteria and correlations. Typical laboratory tests used in slope design include those for identification and classification, shear strength in soils and direct shear strength on discontinuities (and paleo-shear surfaces if it is the case), and uniaxial compressive strength.

Where excavation is to be carried out on hilly ground, it is essential to investigate the possible presence of **natural landslides**, either active or inactive, as engineering work modifies the initial conditions of the slope (geometry, hydrogeology, state of stress, etc.) and may reactivate movement. The presence of pre-existing natural instabilities may modify the excavation design, and if excavation is found not to be feasible, alternative sites must be found. Aspects to be investigated in these cases include the scale and depth of the natural instability, the failure mechanism, the location of slip surfaces and the position of the water table. Instability processes on natural slopes and methods used to investigate them are described in Chapter 13.

## 9.3 Factors influencing slope stability

Slope stability is determined by **geometric factors** (height and angle), **geological factors** (which dictate the presence of surfaces and areas of weakness and anisotropy on the slope), **hydrogeological factors** (related to the presence of water) and **geomechanical factors** (strength, deformability and permeability).

A combination of these factors may determine failure conditions along one or more surfaces and make the movement of a certain volume of soil or rock mass kinematically possible. The possibility of failure depends mainly on geological and geometric factors, which also determine the mechanisms and models of slope instability.

Geological, hydrogeological and geotechnical factors intrinsic to natural materials are considered to be **conditioning** factors (*Table 9.1*). In soils, the lithology, stratigraphy and hydrogeological conditions determine the strength properties and, in turn, the slope behaviour. In hard rock masses, the main conditioning factor is the geological structure: the distribution and frequency of discontinuities and the degree of fracturing.

| Table 9.1 | FACTORS INFLUENCING SLOPE INSTABILITY |
|---|---|
| **Conditioning factors** | **Triggering factors** |
| — Stratigraphy and lithology.<br>— Geological structure.<br>— Hydrogeological conditions and hydrogeological behaviour of materials.<br>— Physical, strength and deformational properties.<br>— *In situ* stresses and state of stress. | — Static overloads.<br>— Dynamic loads.<br>— Changes in hydrogeological conditions.<br>— Climatic factors.<br>— Changes in slope geometry.<br>— Reduction of strength properties. |

In weak materials, such as shales or mudstones, the lithology and level of weathering can also play an important role.

Apart from conditioning, or "passive", factors affecting slope stability, **triggering** or "active" factors come into play to cause failure if a set of conditions is met. These external factors act on soils and rock masses, modifying their characteristics and properties, and conditions of the slope equilibrium (*Table 9.1*). Awareness of all these factors allows the slope to be correctly analysed to assess its stability and design measures to prevent or stabilize movement where required.

## Stratigraphy and lithology

The kind of material in a slope is closely related to the type of instability which may occur, and different lithologies show different degrees of susceptibility to potential slippage or ultimate failure. The stress-strain behaviour of each material, and therefore its stability, is governed by its strength properties which also depend on the presence of water.

Types of failure and the location of failure surfaces depend on factors such as alternating materials of different lithology, the extent of weathering and the presence of layers of soft material or hard strata. In soils, which are often considered homogeneous materials compared to rock masses, instability may be generated by differences in the degree of compaction, cementation and grain size, which may make certain areas more susceptible to weakness and water flow (*Figure 9.5*). In rock masses, the characterization and analysis of slope behaviour is further complicated by the presence of layers or strata with differing strength and properties, which condition how the rock masses fail.

## Geological structure and discontinuities

Geological structure plays a definitive role in conditioning slope stability in rock masses. Problems that may occur are defined by a combination of structural elements and

*Figure 9.5* Slope excavated in colluvial soils and highly weathered rock with failure caused by the low strength of the materials and seasonal water flow inside the ground.

*Figure 9.6* Slope excavated in jointed and folded shales, with failures controlled by the rock mass structure (area shown: 15 m wide).

geometric slope parameters, such as height, gradient and **orientation** (Figures 9.6 and 9.17).

The structure of the rock mass is defined by the spatial distribution of discontinuities, which provide surfaces that bound relatively hard blocks of intact rock; these blocks are held together by the strength characteristics and properties of the discontinuities. The presence of these surfaces of weakness (bedding surfaces, joints, faults) dipping towards the slope face implies the existence of potential failure planes on which sliding can readily occur. The orientation and spatial distribution of discontinuities will condition the type and mechanism of the instability.

The presence of discontinuities implies an anisotropic rock mass behaviour and mobilization of failures along preferred planes; for instance, a specific system of fracturing will condition both the direction of movement and the size of blocks liable to slide, or the presence of a fault dipping towards the slope will limit the unstable area and condition the failure mechanism. Structural changes and singularities in the rock mass, such as tectonized or shear areas, or abrupt changes in the dip of the strata, indicate heterogeneities from which failure might originate.

The relationship between the dimensions of the slope face and the discontinuity network is significant; depending on this, slope behaviour will be defined by one or more macrodiscontinuities (relative to slope scale), or by several joint sets and networks of other planes of weakness, as these can condition the type and degree of instability.

The influence of the geological structure goes beyond the geometric conditioning of failures since slope stability may be affected by changes to the initial conditions during excavation; for example, the existence of tectonic *in situ* stress associated with compressive or distensive structures, such as folds and faults, may induce destabilizing processes.

# Hydrogeological conditions

Most failures are caused by the effects of water in the ground, including **pore pressures** and erosion of slope materials, both internally and at the surface. Water is generally considered to be the worst enemy of slope stability, together with human action where excavations are carried out without adequate geotechnical care.

The presence of water in a slope reduces stability by decreasing ground strength and increasing forces which favour instability. The main effects of water are:

— Reduction in the shear strength of failure surfaces as effective normal stress, $\sigma'_n$, decreases:

$$\tau = c + (\sigma_n - u) \tan \phi = c + \sigma'_n \tan \phi$$

— Increase in the down slope shear forces as water pressure is exerted in tension cracks.
— Increase in weight of the material due to saturation:

$$\gamma = \gamma_d + Sn\gamma_w$$

where: $\gamma_d$ = dry apparent unit weight; $S$ = degree of saturation; $n$ = porosity and $\gamma_w$ = unit weight of water.
— Softening of soils associated with an increase in their water content.
— Internal erosion (piping) caused by sub-surface or underground flow.
— Weathering and changes in mineralogical composition of soils and rocks.
— Opening up of discontinuities caused by freezing water.

The shape of the **water table** on a slope depends on such factors as the permeability of materials, and the geometry or shape of the slope. In rock masses, the configuration of the water table is greatly influenced by geological structure and the alternation of permeable and impermeable materials, which in turn affect the distribution of pore pressures on any potential slip surface on a slope (*Figure 9.7*).

Changes in the water table may be either seasonal or the result of long periods of rainfall or drought. Figure 13.16, in Chapter 13, shows the distribution of water in a slope. Only part of the rainwater or run-off penetrates the ground, and only a small amount of that reaches the water table. Although any change in the water table generally takes place slowly and over long periods, it may rise fairly quickly after intense rainfall on very permeable materials.

Apart from water inside the ground, the role of surface water (from precipitation and run-off) must also be taken into account. Major problems affecting stability may occur from high pressure caused by water entering cracks and discontinuities; in fact, soil slope failure occurs most frequently when water is readily available as in rainy periods, after heavy storms and during thaws. The phenomena of erosion and leaching in weak or low consistency materials are also associated with the presence of surface water.

The influence of water on the properties of materials depends on their **hydrogeological behaviour**. The greatest effect is produced by the pressure exerted, defined by the **piezometric head**.

The following aspects need to be known to be able to assess the magnitude and distribution of **pore pressures** and the effects of water in a slope:

— Hydrogeological behaviour of the materials.
— Presence of water tables and piezometric heads.
— Water flow in the slope.
— Relevant hydrogeological parameters: permeability coefficient or hydraulic conductivity, hydraulic gradient, transmissivity and storage coefficient.

Pore pressures inside a slope can be measured directly with **piezometers**, as explained in Chapter 5, Section 5.6. These measurements give the value of the pressure exerted by water at a point inside a borehole, or the piezometric head in the layer or formation intercepted by the piezometer (note that if several formations are intercepted, the measured level may not correspond to the formation with the highest piezometric head).

Pressures can also be evaluated indirectly from the slope **flow net**. This method gives an assessment of the water pressure values at different points on the failure surface (*Box 9.1*). The shape of a slope flow net depends on the

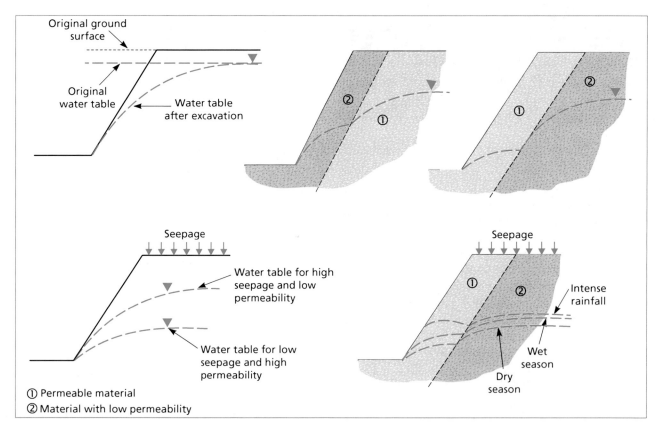

Figure 9.7   Diagrams of the water table in a slope as a function of the distribution of materials and presence of infiltration.

## Box 9.1

### Calculating water pressures in a slope using a flow net

The phreatic surface of an aquifer is defined as the locus of points at which the water pressure is equal to the atmospheric pressure, and it forms the free surface of the aquifer. In unconfined aquifers, the ground is saturated below the water table. From the flow net, the piezometric head or **total head**, $h$, of a point A on the aquifer is defined by the corresponding equipotential line and its depth, with the following expression (Figure A):

$$h_A = z_A + (P_A/\gamma_w)$$

where:

$z_A$ = height of point A with respect to the horizontal plane of reference or datum
$P_A$ = difference between the pressure at point A and atmospheric pressure
$\gamma_w$ = unit weight of water

and the value of **pressure** $P$ is:

$$P_A = (h_A - z_A)\gamma_w$$

Pressure exerted is calculated in the same way on a plane AB across the slope for a discrete number of points on its surface. In the example shown in Figure B, heights $z$ and $h$ (total head with respect to the line of reference) can be measured from the points under consideration:

| Points | 1 | 2 | 3 | 4 | 5 |
|---|---|---|---|---|---|
| Height $z$ (m) | 8.5 | 11 | 14 | 18 | 21 |
| Height $h$ (m) | 10.5 | 14.5 | 18.5 | 22 | 23.5 |

From these values, the values of the pressure acting on each point are calculated:

$$h_1 = z_1 + (P_1/\gamma_w) = 10.5 \Rightarrow P_1 = (10.5 - 8.5)\gamma_w = 20 \text{ kN/m}^2$$

In the same way, the following are obtained:

$$P_2 = 35 \text{ kN/m}^2 \quad P_3 = 45 \text{ kN/m}^2$$
$$P_4 = 40 \text{ kN/m}^2 \quad P_5 = 25 \text{ kN/m}^2$$

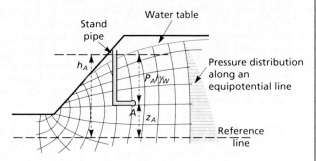

Figure A — Flow net of a slope. Piezometric head corresponding to point A.

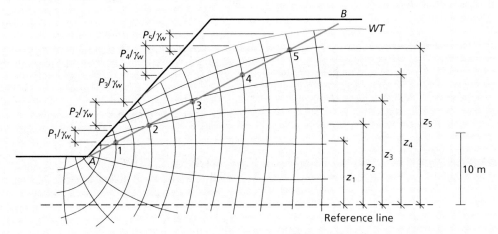

Figure B — Calculation of pressure at points on a plane crossing a slope.

homogeneity and anisotropy of the ground, its permeability in different directions, and on the slope geometry (see Chapter 2, Section 2.3 and Chapter 4, Section 4.5).

Provided that the aquifer is homogeneous and unconfined, when the elements needed to plot the flow net are not known but the position of the water table in the slope is known, the water pressure, $u$, at a given point can be approximately estimated from the weight of the vertical column of water above it:

$$u = z\gamma_w$$

where $z$ is the height of the column of water and $\gamma_w$ the unit weight of the water (this estimate becomes increasingly less realistic as the angle of flow increases from the horizontal and as the ground becomes increasingly inhomogeneous).

Definition of the **pore pressure distribution** model in a slope is difficult and may require assumptions to be made. The usual hypotheses adopted for estimating pressures (flow parallel to the slope surface, hydrostatic conditions, etc.) do not take the parameters controlling the hydraulic regime of the slope (such as non-steady flow) into account, which may lead to error.

In straightforward cases, one way of obtaining an approximate assessment of the total force exerted by water on discontinuity surfaces or tension cracks is to assume triangular distributions of hydrostatic pressure on these surfaces, as shown in *Figure 9.8*. The height of the triangle corresponds to maximum water pressure on the plane. This simplification helps to resolve the equilibrium equations of the slope, with the total force of water acting on the discontinuity given by the area of the resulting pressure triangle considered in two dimensions.

## Geomechanical properties of soil and rock masses

The possible failure of a slope along a surface depends on its shear strength, which basically depends on the strength parameters of the material: cohesion and internal friction.

The influence of geological history (e.g. consolidation, erosion, diagenetic processes, *in situ* stresses, weathering) on the mechanical properties of **soils** means that strength parameters, representative of shear strength, must be determined taking the geological characteristics into account. For example, in marly clay formations found in southern Spain, which are generally classified as high plasticity clays having some carbonates and a uniaxial compression strength of hundreds of kPa, the parameters that agree with slope strength are generally residual ones, representing the strength of discontinuity surfaces in the marly clay masses (either **slickensided** -shear surfaces-, or very fine, slightly silty discontinuity surfaces).

In rock masses, mechanical behaviour is determined by the strength properties of the discontinuities and of the intact rock depending on its degree of fracturing and the nature of the materials and discontinuities within it.

The behaviour of a hard rock mass generally depends on the characteristics of its discontinuities, although the lithology and its geological evolution may also play an important role. The shear strength of surfaces of weakness depends on their nature and origin, persistence, spacing, roughness, type and thickness of infill and the presence of water. Shear strength is the most important aspect determining rock mass stability; Chapter 3, Section 3.5 includes methods for determining shear strength in discontinuities. Failure planes may develop along discontinuities and across "bridges" of intact rock which generally provide strength to the mass as a whole.

## *In situ* stresses

*In situ* stresses have an important role in slope stability. The stress relief from decompression when a slope is excavated may transform its material properties. In rock slopes the weakest areas can be degraded and begin to behave like a soft rock or "granular" soil. This effect has been observed in open-pit coal mines on mudstone or mud-shale slopes subjected to high *in situ* stresses; the rock formation is broken down into a granular material with centimetre-sized fragments several metres thick inside the slope, leading to disintegration and collapse of the slope.

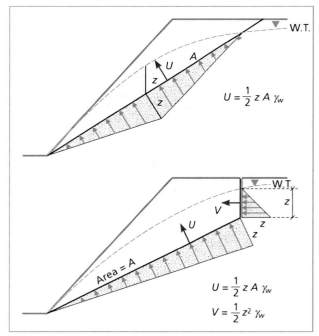

Figure 9.8  a) Water pressure triangle for a single discontinuity plane. b) Pressure triangles where there is a tension crack at the slope head (U and V are forces due to water on the sliding plane and on the tension crack respectively).

The state of stress of a slope depends on its geometric configuration and the pre-excavation state of stress of the mass. *Figure 9.9* shows an example of the distribution of lithostatic stresses after excavation. In deep excavations, high stresses generated at the slope toe or other specific areas may cause unstable conditions or even ductile deformation. Anisotropic state of stress also develops at the slope head, with tensional components that can cause vertical cracks to open.

If a rock mass is subjected to tectonic stress, these will be released and redistributed when the rock is excavated, and this modification of the previous state of stress will mean a decrease of the material strength. Discontinuity planes and areas with compressive structures such as folds may become weak areas when extension or tension stresses appear. Stress relief from excavation tends to the establishment of a new state of equilibrium and may cause displacements in the rock mass to occur with the development of cracks and openings in discontinuity planes; these play an important part in the initial stages of instability processes. This readjustment depends on the type, structure and strength of the rock mass and decreases over time.

The stress-strain state of a rock mass must be considered in stability analyses if it can affect its behaviour and strength properties, especially in excavations more than 50 m deep. The ratio between vertical and horizontal stresses, $K = \sigma_H/\sigma_V$, is an important aspect to consider. Two rock masses subjected to the same vertical force may be able to bear very different horizontal forces depending on their strength. Geological phenomena, such as erosion or neotectonic processes, may contribute to the ratio variation between $\sigma_H$ and $\sigma_V$ in a given area. These aspects are dealt with in Chapter 3, Section 3.7.

## Other factors

**Static and dynamic loads** exerted on slopes modify the forces distribution and may produce instability.

Static loads include the weight of structures or buildings on a slope, or loads derived from fills, waste dumps or heavy vehicles, and when these loads are exerted on the slope head, they create an additional weight that may contribute to the destabilizing forces.

Dynamic loads are mainly due to natural or induced seismicity and to vibrations caused by nearby blasting. These mainly affect jointed rock masses by opening up pre-existing discontinuities, reducing their shear strength, and by displacing rock blocks which can then fall. When there are strong seismic movements, the instantaneous application of forces may lead to general slope failure if existing conditions make it prone to instability. These forces must be included in slope stability analyses in seismic areas or in areas subjected to other types of dynamic forces. For an approximate calculation, dynamic action can be considered as a pseudostatic force, given as a function of maximum horizontal acceleration produced by the earthquake (see Chapter 14, Section 14.6).

**Precipitation and climatic regime** influence slope stability by modifying ground water content. Alternate periods of drought and rainfall produce changes in soil structure that can lead to loss of strength. Slope instability risk criteria can be established based on rainfall measurements. *Figure 9.10* shows Lumb's criterion (1975) for precipitation related movement risk. It shows the relationship between rainfall intensity, over a 15 day period and in the last day previous to the failure, and slope failures in residual soils of granitic origin in Hong Kong. In clay formations in semi-arid climate regions, common of southern Spain, dried-out material is often saturated after rain; *Figure 9.11* compares different cases, showing a failure risk criterion for clay soils in S. Spain, relating monthly rainfall and maximum daily intensity.

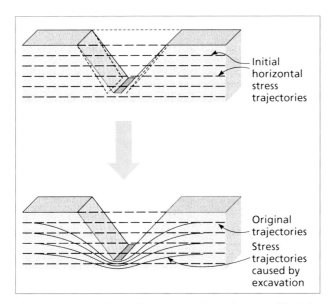

Figure 9.9   Horizontal stress trajectories modified by excavation.

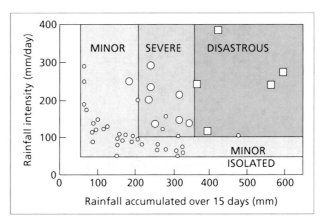

Figure 9.10   Landslide hazard criterion based on accumulated rainfall and daily rainfall intensity in Hong Kong (Lumb, 1975).

**Weathering processes** play a fundamental role in reducing the strength properties of certain types of soils and weak rock masses, causing intense disturbance of their fabric and deterioration of their strength when these materials are exposed to environmental conditions as a result of excavation. This loss of strength may provoke the fall of surface material and result in a general failure if critical areas of the slope, such as the toe, are affected, especially in situations where water is present.

## 9.4 Types of slope failure

### Soil slopes

Soil slopes generally fail along **curved surfaces** with the shape of the failure mainly conditioned by the morphology, stratigraphy and lithology of the slope (*Figure 9.12*):

— When the slope consists either of homogeneous soil or variable strata with homogeneous geotechnical properties, the most usual slip surface is roughly circular with its lower end cutting the slope at its toe (toe slip; *Figure 9.13b*).
— The failure surface may be almost circular, passing under the slope toe (deep slip; *Figure 9.13c*).

In certain conditions, e.g. where there are strata or layers of different strength, slope failure may occur along a **plane surface** or a **polygonal surface** made up of several planes (*Figure 9.13d*).

Soil slope failure that mainly occurs along a single **plane parallel to the slope** is very unusual, although this model may be valid for **natural slopes** on rocks covered with residual soil or debris (*Figure 9.13a*), or on rock slopes where discontinuities daylighting on the slope face may define planar failure surfaces, although these often do not

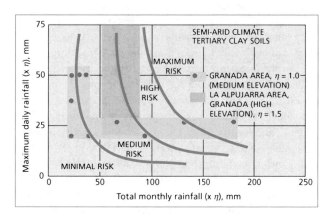

Figure 9.11  Slope instability hazard criterion for clay soils in S. Spain.

Figure 9.12  Curved failure in clays and shales in central Spain, showing the tension crack and vertical displacement of the slope head.

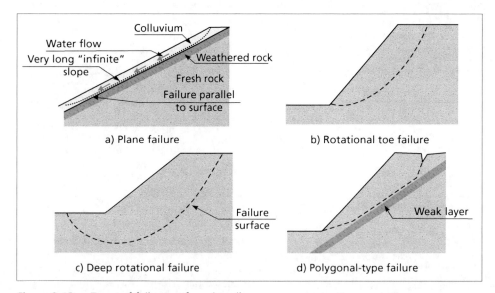

Figure 9.13  Types of failure surfaces in soils.

reach the head of the slope. The **"infinite" slope** model (in which the length up the slope may be considered infinite in relation to the thickness of the failed mass) can be applied to natural slopes where the failure surface is defined by the contact (approximately parallel to the slope) between the surface material, formed by colluvial or residual soil, and the underlying rock (see diagram in *Figure 9.27*).

# Rock slopes

Different types of rock slope failure are conditioned by the type of fractures and the degree of fracturing affecting the rock mass, and by the orientation and distribution of the discontinuities in relation to the slope face. The stability is defined by the strength of the discontinuities and the intact rock. In hard or strong rock masses, the location of the failure planes is determined by the discontinuities. In weak rock masses, intact rock also plays an important part in the development and location of such planes and in the failure mechanism. *Figure 9.14* shows different models of slope failure and the relationship between angles and heights for different types of rock mass.

The most frequent failure types are: plane failure, wedge failure, toppling, buckling and non-planar failure. The structural conditions for some of these are shown in *Figure 9.15*.

## Plane failure

Plane failure takes place along a pre-existing surface, such as a bedding plane, tectonic joint or fault. A precondition for plane failure is the presence of discontinuities dipping

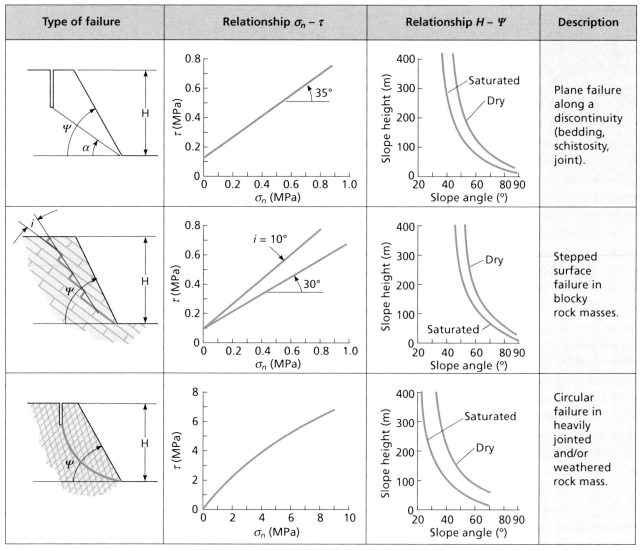

*Figure 9.14* Types of failure in rock slopes, strength curve for the rock mass and relationship between slope height and angle (modified from Hoek and Bray, 1981).

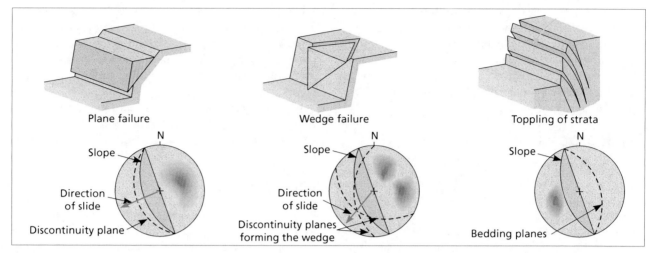

*Figure 9.15* Stereographic representation of discontinuity planes in relation to slope orientation for common types of failure in rock masses (modified from Hoek and Bray, 1981).

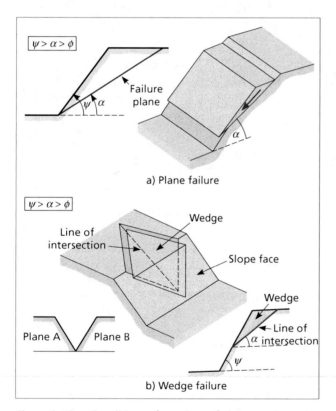

*Figure 9.16* Conditions for plane failure and wedge failure.

*Figure 9.17* Plane failures along bedding planes on slope benches (slope height: 60 m).

generated along the schistosity (*Figure 9.17*). Different types of plane failures depend on the distribution and characteristics of discontinuity sets in the slope. The most frequent are (*Figure 9.18*):

— Failure along a plane outcroping on the face or at the toe of the slope, with or without a tension crack.
— Failure along a plane parallel to the slope face, caused by erosion or loss of strength at the slope toe.

## Wedge failure

This failure type consists of a sliding wedge-shaped block formed by two discontinuity planes whose line of intersection dips towards the slope face (*Figure 9.19*). For this type of failure to occur, the two planes must outcrop on the slope surface and fulfil the same conditions as for plane failure: $\psi > \alpha > \phi$; in this case, where $\alpha$ is the dip of the line of

to the slope with the same strike; the failure plane must be exposed, or "daylight", on the slope face ($\psi > \alpha$), with a dip greater than its friction angle ($\alpha > \phi$) (*Figure 9.16a*). In slopes excavated parallel to the bedding surfaces, plane failure may occur from strata slip; this type of failure is typical of mudstones, schists or shales, where failure planes are

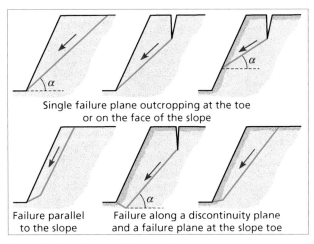

Figure 9.18   Types of plane failure.

Figure 9.19   Planes of a wedge slide on a rocky slope in southern Spain.

intersection (Figure 9.16b). This type of failure usually occurs in rock masses with several sets of discontinuities, and their orientation, spacing and persistence will determine the shape and volume of the wedge.

Figure 9.20 shows a stereographic representation of several sets of discontinuities and the direction of sliding for wedges formed on a slope with a given inclination $\psi$. Comparing the angles of the slope, the line of intersection of the wedge planes and the friction of the planes, determines whether movement is kinematically feasible and if the wedge is stable or unstable.

## Toppling

Toppling occurs on slopes in rock masses where the strata dip steeply away from the slope, striking parallel or sub-parallel to it. In block toppling the strata generally form columns defined by discontinuity sets orthogonal to each other (Figures 9.21

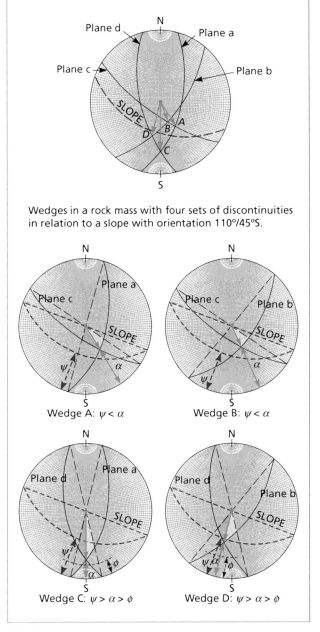

Figure 9.20   Stereographic representation of wedges. Sliding is impossible for wedges A and B as they do not meet the condition $\psi > \alpha$; wedges C and D meet this condition and can move.

and 9.22). The stability of the blocks is not conditioned solely by the shear strength of the discontinuities as they are also subjected to rotational movement. Flexural toppling occurs in thin continuous steeply dipping strata, or columns, that break in flexure as they bend forward.

The toppling process may start by sliding, excavation or erosion of the slope toe, with retrogression back into the rock mass, forming tension cracks. Wyllie and Mah (2004)

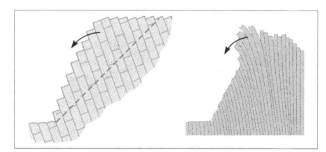

*Figure 9.21* Diagram of slopes with a structure favouring strata toppling: block toppling and flexural toppling.

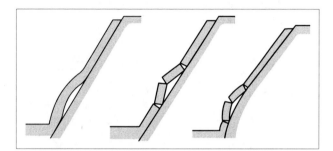

*Figure 9.23* Diagram showing buckling of vertical strata, with bending and fracturing of strata.

*Figure 9.22* Rock blocks on a slope where toppling has occurred.

include detailed descriptions of different types of toppling failure.

## Buckling

Buckling occurs along bedding planes parallel to the slope ($\psi = \alpha$), with a dip greater than the friction angle ($\alpha > \phi$). Failure may occur with or without flexure of the strata; a necessary condition is that the strata are sufficiently thin in relation to slope height to be able to buckle (*Figure 9.23*).

Buckling failure may be caused by:

— Excessive slope height.
— Presence of external forces applied to the strata.
— Unfavourable geometry of the strata.
— Presence of water pressure on the strata.
— Unfavourable stress concentration.

This type of failure usually occurs on the back slopes of open cast mines where excavation is parallel to bedding and strata are closely spaced (*Figure 9.24*).

## Non-planar failure

Non-planar and even circular failure may occur in very weak rock masses or heavily jointed or broken rock masses showing isotropic behaviour, where individual sets of through going discontinuity planes do not control mechanical behaviour (*Figure 9.25*); in such cases, the rock mass behaves like a soil. Nevertheless, the existence of specific areas of weakness and extensive discontinuity planes, such as faults, may condition failure models with other different typologies.

*Figure 9.24* Buckling of strata in mudstone materials with failure of bedding at the base.

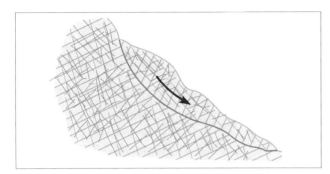

*Figure 9.25* Diagram of a curved failure surface in heavily jointed rock.

## 9.5 Stability analysis

## Introduction

Stability analysis is required to design a slope that is stable and to design stabilization measures when there are problems of instability. An appropriate factor of safety must be selected which will depend on the purpose of the excavation and whether the slope is temporary or permanent; safety aspects and the cost of excavation will be also considered, as well as the risks and impacts of possible failure. As a general rule, the factor of safety adopted for permanent slopes should be equal to or more than 1.5, or even 2.0, depending on the safety level required and the level of confidence in the geotechnical data used for calculations. For temporary slopes, a factor of safety around 1.3 is usual, although lower values may be used in some circumstances.

Carrying out these analyses allows the geometry of the excavation to be defined and, if it is the case, the external forces specified to achieve the required safety factor. In the case of unstable slopes, stability analysis facilitates the design of suitable corrective measures and stabilization procedures to prevent further movements.

**Back-analysis** is carried out once slope failure has occurred and the mechanism, model and geometry of the instability are known. It is especially useful for carrying out a geomechanical characterization of the materials involved, for studying factors influencing failure and for understanding the mechanical behaviour of slope materials, all at slope scale; results can be extrapolated and applied to other slopes with similar characteristics. This type of analysis uses relevant field data (slope geometry, types of material, failure models, hydrostatic pressures, etc.) to determine ground strength parameters; these are usually values $c'$ and $\phi'$ that fulfil in terms of effective stress the condition of limit equilibrium for the slope (i.e. $F = 1.0$) along the failure surface for the real failure conditions. The concepts of factor of safety and limit equilibrium are described below.

**Stability analysis methods** are based on a physical-mathematical approach involving stabilizing and destabilizing forces acting in the slope, which determine slope behaviour and condition the stability; these can be grouped into:

— **Deterministic methods**: these methods indicate whether or not a slope is stable, once its conditions are known or assumed. The appropriate values of the physical and strength parameters determining the behaviour of materials must be selected, and the degree of stability or factor of safety of the slope is defined using these values and the appropriate behaviour criteria. There are two groups: **limit equilibrium** methods and **stress-strain** methods.
— **Probabilistic methods**: these study the probability of slope failure under certain conditions. The distribution functions must be known of different values considered as random variables in the analyses. These methods are more difficult to use because uncertainty about the properties of the materials means a large quantity of data is needed. The factor of safety is calculated from these functions, using iterative processes. The probability density functions and probability distribution of the factor of safety can be generated, and slope stability curves with the factor of safety associated with a specific particular probability of occurrence obtained.

In each case, selection of the most appropriate method of analysis will depend on:

— Geological and geomechanical characteristics of the materials (soils or rock masses).
— Available data on the slope and the surrounding area (geometric, geological, geomechanical, hydrogeological data, etc.).
— The scope and objectives of the study, and how much detail is required.

These factors are all interdependent; detailed analysis cannot be carried out if the necessary data is unavailable or incomplete, just as a case of complex stability cannot be tackled using a simple method because not enough field or laboratory data is available. It is important that both field and laboratory data are obtained in accordance with the stability analysis method to be used and how the data are to be processed. Once the parameters influencing slope stability are known, a method must be selected that represents the particular conditions for each case. Probabilistic methods are not often used as they are difficult to perform and their results are difficult to apply.

## Limit equilibrium methods

Limit equilibrium methods analyse the equilibrium of a potentially unstable mass by comparing the forces tending towards movement along a given failure surface with the forces resisting it. They are based on:

- Selection of a potential failure surface through the slope.
- The Mohr-Coulomb failure criterion.
- Definition of "factor of safety".

Problems of stability are statically indeterminate; resolving them requires a series of initial hypotheses, which differ depending on the method used. The following conditions are assumed:

- The geometry of the failure surface allows sliding, i.e. sliding must be kinematically feasible.
- The distribution of forces acting on the failure surface can be computed using known data (unit weight of the material, water pressure, etc.).
- Strength is mobilized simultaneously along the whole failure surface.

With these conditions, equilibrium equations can be established between forces inducing sliding and those resisting it. The **factor of safety** of the slope, $F$, is obtained for the surface analysed, and when $F = 1.0$ is referred to as the strict or limit equilibrium between acting forces:

$$F = \frac{\text{Forces resisting slide}}{\text{Forces tending to slide}}$$

or, expressed in terms of stress:

$$F = \frac{\text{Resisting shear stress}}{\text{Sliding shear stress}}$$

Once the factor of safety of one hypothetical failure surface has been estimated, other kinematically feasible failure surfaces must be analysed until the one with the smallest safety factor, $F_{min}$, is found; this surface is taken as the potential failure surface of the slope and $F_{min}$ is taken as the factor of safety for the slope in question.

Assuming there are no external forces affecting the slope, the forces acting on a potential failure or sliding surface are those due to the weight of the material, $W$, to cohesion, $c$, and to the friction, $\phi$, of the surface (Figure 9.26). The factor of safety is given by:

$$F = (R_c + R_\phi)/S$$

where:

$R_c$ = cohesive forces = $cA$
$R_\phi$ = frictional forces = $W \cos \alpha \tan \phi$
$S$ = forces tending to slide = $W \sin \alpha$
$A$ = area of failure plane

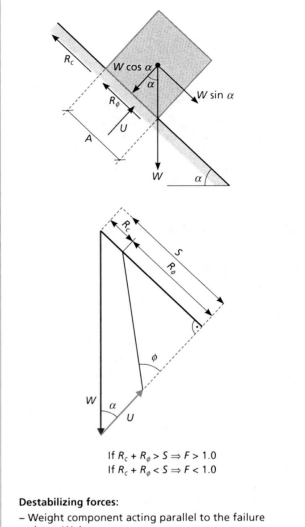

If $R_c + R_\phi > S \Rightarrow F > 1.0$
If $R_c + R_\phi < S \Rightarrow F < 1.0$

**Destabilizing forces:**
- Weight component acting parallel to the failure plane, $W \sin \alpha$.
- Resultant force of the pore water pressure, $U$, acting on the potential failure plane.
- Static and dynamic external loads exerted on the slope acting against stability.

**Resistant forces:**
- Ground shear strength resulting from cohesion and internal friction mobilized along the failure plane.
- Weight component acting normal to the failure plane, $W \cos \alpha$.
- External forces favouring stability.

Figure 9.26   Forces acting on a failure plane in a slope.

in cases where water pressure is present on the failure surface, $U$ being the total force due to water on surface $A$:

$$R_\phi = (W \cos \alpha - U) \tan \phi$$

There are several more or less complex methods of calculating the factor of safety for limit equilibrium which have been devised for applying mainly to **soil-like materials**. Analytical methods, such as those by Taylor or Fellenius, give the factor of safety from the immediate solution of simple equations; numerical methods, such as those by Spencer and Morgenstern-Price, require systems of equations and iterative calculation processes to solve them.

Limit equilibrium methods are classified as:

— Methods analysing the total block or total mass.
— Methods analysing the mass divided into vertical slices.

The first type of method is valid for homogeneous materials, and forces are calculated and compared at only one point on the failure surface. The second type can be used to analyse non-homogeneous materials, and the different methods include their own particular hypotheses in relation to location, position and distribution of forces acting on the slices. Forces acting on each slice of slope are calculated and all the results are finally integrated. The most common classical slice methods are the modified Bishop and Janbu methods. The modified Bishop method is suitable for analysing circular failures; the Janbu method is used for circular, non-circular, polygonal and planar failures.

Methods used for **rock failures** are based in the same way on equilibrium equations between acting forces, established from the specific geometry of each type of failure.

## Soil slopes

Classical methods of analysing soil stability include:

— Analysing planar failures in "infinite" slopes.
— Analysing various blocks of ground interacting with each other, applicable to polygonal-type failure surfaces (multiple wedge method).

— Methods for curved failures that analyse the total equilibrium of a mass sliding on surfaces of circular or logarithmic shape.
— Method of slices.

### Infinite slope

This method is based on the hypothesis that the length of a planar failure surface parallel to the slope may be considered infinite with respect to the thickness of the sliding mass, as shown in the diagram in *Figure 9.27*. It is generally used to analyse the stability of natural slopes under the conditions established in Section 9.4.

It is enough to analyse the forces on a section or slice of slope subjected laterally to forces $E_1$ and $E_2$ and to its own weight, $W$, at its base. The weight produces a tangential sliding force (the component of $W$ parallel to the slope) at the same time as it generates a friction resistance at the base (due to its normal component), as a function of the internal friction of the soil; this resists sliding through its tangential component (parallel to the slope). The possible presence of a resisting force due to cohesion can be added to this strength component when appropriate.

In the simplest case, where cohesion is not present, the factor of safety is given by:

$$F = \frac{\tau}{S} = \frac{\sigma_n \tan \phi'}{S} = \frac{\tan \phi'}{\tan \alpha}$$

with:

$$\sigma_n = \frac{W \cos \alpha}{l} = \gamma H \cos^2 \alpha$$

$$S = \frac{W \sin \alpha}{l} = \gamma H \sin \alpha \cos \alpha$$

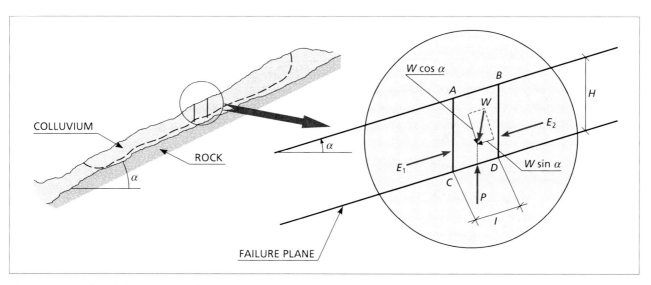

*Figure 9.27*   Plane failure on an "infinite" slope.

where $l$ is the length of the base of the slice, $\phi'$ is the effective friction and $\gamma$ is the apparent unit weight of the soil.

If water pressure, $u$, can be considered constant along the failure plane:

$$F = \frac{(\sigma_n - u)\tan\phi'}{S} = \frac{(\gamma H \cos^2\alpha - u)\tan\phi'}{\gamma H \sin\alpha \cos\alpha} =$$

$$= \left(1 - \frac{r_u}{\cos^2\alpha}\right)\frac{\tan\phi'}{\tan\alpha}$$

where $r_u = u/\gamma H$ is the **pore pressure ratio**.

If there is seepage parallel to the slope surface, the $r_u$ factor is:

$$r_u = \frac{h}{H}\frac{\gamma_w}{\gamma}\cos^2\alpha$$

$h$ being the height of water above the slip surface.

With this notation, the factor of safety can be written as:

$$F = A\frac{\tan\phi'}{\tan\alpha}$$

$A$ being a function of $r_u$ and $\alpha$ (*Figure 9.28*).

If there is cohesion in the ground, the calculation is similar:

$$F = \frac{(\sigma_n - u)\tan\phi' + c'}{S}$$

therefore the final expression can be written as:

$$F = A\frac{\tan\phi'}{\tan\alpha} + B\frac{c'}{\gamma H}$$

$B = 1/(\cos\alpha \sin\alpha)$ being a function of the slope angle, $\alpha$ (*Figure 9.28*).

## Multiple wedge analysis

While the hypothesis of plane failure may be very simplistic for soils, a **polygonal shaped surface** formed by two or more blocks resting on adjacent ones can adequately reproduce in some cases more complex problems such as those involving earth dams, embankments on soft soils, etc., provided that a Mohr-Coulomb failure criterion is applicable to the failure planes.

In this type of analysis (known as the multiple wedge method, although it is not related to wedge-type failure as seen in jointed rock masses), the sliding mass is divided by vertical lines into various blocks, as shown in *Figure 9.29*, and the equilibrium of vertical and horizontal forces for each block is established. The factor of safety is the ratio between the tangential strength available and that required for equilibrium. The method is applied as follows:

— A failure surface is assumed.
— The sliding mass is divided into two or more wedges, in such a way that each straight section of the failure surface affects only one soil type.
— The weight of each wedge is calculated.
— A value is assumed for the safety factor, $F_1$, and values $c_m = c/F_1$ and $\tan\phi_m = \tan\phi/F_1$ are calculated.
— The polygon of forces is drawn for the end wedge (number 2 in *Figure 9.29*) from the value $\phi_m$, assuming

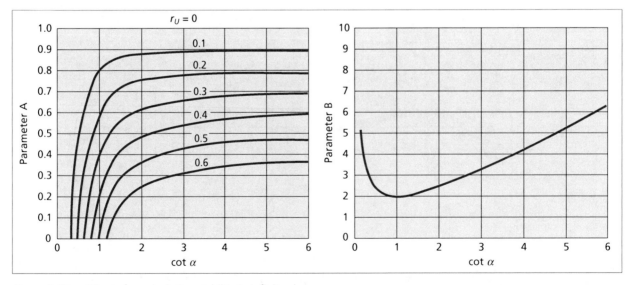

*Figure 9.28*   Charts for calculating stability in infinite slopes.

SLOPES

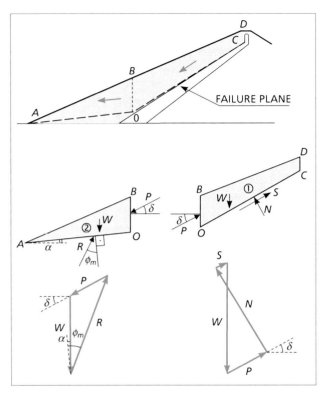

Figure 9.29 Method for analysing polygonal failure.

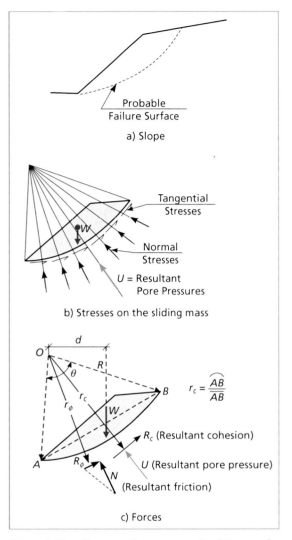

Figure 9.30 Forces acting on a circular failure surface.

a value for angle $\delta$, with which the polygon can be closed.

— From the results obtained, equilibrium is established in the next block, followed by the one after it, until the last block is reached. The force S necessary to close the polygon is calculated and checks made to see if there is equilibrium for the mobilized cohesion and friction values, $c_m$ and $\phi_m$.

— If there is equilibrium, the factor of safety is $F_1$. If not, another value, $F_2$, is assumed and calculation is repeated until the factor of safety for the surface selected is obtained.

— A new polygonal surface is analysed and the process is begun again, until the surface with the lowest value of F is obtained.

It is often assumed that $\delta = 0$, i.e. that forces between blocks are horizontal, although in the case of earth dams a specific value for $\delta$ ($\delta \leq \phi_m$) is usually assumed as this gives a better representation of the kinematics of the problem.

### Total mass methods. Taylor's method

The commonly employed hypothesis of circular failure surfaces in two dimensions represents the problem in slopes where the development of failure surfaces is not clearly defined. Forces acting on the failure surface are shown in diagram form in *Figure 9.30*:

— Weight, W, of the soil mass.
— Pore water pressure distributed along the failure surface, with resultant U.
— A tangential force distributed over the failure surface, with resultant $T(R_c + R_\phi)$.
— Normal force distributed over the same surface, with resultant N.

With the definition of the safety factor, F, already shown, assuming homogeneous ground and using the Mohr-Coulomb failure criterion, the tangential strength acting in order to reach strict equilibrium (F = 1.0) will be:

$$S = \frac{\tau}{F} = \frac{c}{F} + \sigma_n \frac{\tan \phi}{F}$$

therefore the stresses distributed on the failure surface can be substituted by:

— The resultant due to cohesion,

$$R_c = \int_0^\theta (c/F) d\theta$$

acting parallel to the chord $\overline{AB}$.

— The resultant of normal stresses, $\overline{N}$; its magnitude and position are unknown, although it must be normal to the assumed failure surface.

— The tangential resultant due to friction, $R_\phi$; this must be normal to $\overline{N}$ and meet $R_\phi = \overline{N} \tan \phi / F$; however, as the position and magnitude of $\overline{N}$ are unknown, they are also unknown for $R_\phi$.

Thus, there are more unknowns ($F$, magnitude and position of $N$, plus the parameter defining the position of $R_\phi$) than equations available (those of force and moment equilibrium on the failure surface), making the problem statically indeterminate.

If the soil is clayey and undrained failure takes place, strength is expressed by $c = S_u$ and $\phi = 0$, $S_u$ being the undrained shear strength, and the problem is then determinate. The factor of safety can easily be calculated by taking moments of acting forces with respect to the centre of the circle of failure:

$$Wd = R_c r_c \quad \Rightarrow \quad F = \frac{S_u R^2 \theta}{Wd}$$

If $S_u$ is not constant along the circle, it is simply divided into $n$ sections in which it is constant, with amplitude $\theta_i$ for the angle defining them, giving a value of $F$:

$$F = \frac{R^2 \sum_{i=1}^n S_{ui} \theta_i}{Wd}$$

When analysis is performed taking friction into account (i.e. strength is no longer considered as $S_u$, which is some unknown combination of $c$ and $\phi$ but a parameter of known $c$ and $\phi$, i.e. $c \neq 0$, $\phi \neq 0$), due to the nature of the ground or drainage conditions, an additional complementary hypothesis is needed in order to solve the problem. The most widely used is Taylor's hypothesis, which holds that the normal force resultant is concentrated at one single point, producing the so-called **friction circle method** or **Taylor's method** (Taylor, 1948), which requires various graphical or analytical approximations.

Using this method, Taylor analysed the problem dimensionally for homogeneous soils with the aim of devising simple-to-use charts. Figures 9.32 and 9.33 show charts that can be used respectively for soil with $S_u$ only (short term stability in saturated clays, i.e. undrained failure), or for soils with internal friction. In the first case, the presence of a hard layer that limits the depth of failure circles should be considered in the analysis (Figure 9.31). The chart in Figure 9.32 shows the relationship between the parameters $D$ (ratio between

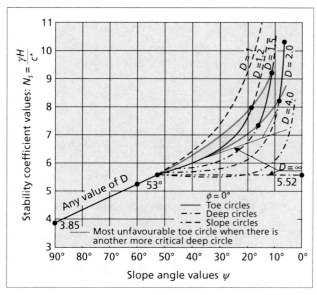

Figure 9.32   Short term stability in saturated clays (Jiménez Salas et al., 1976).

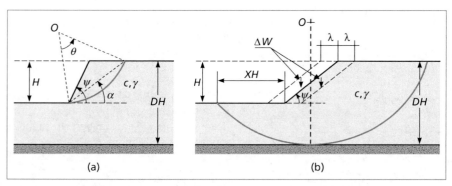

Figure 9.31   Failure surfaces: a) toe circle, b) deep circle (Jiménez Salas et al., 1976).

the depth of the hard layer and slope height), $\psi$ (slope angle) and $N_s$ (stability coefficient). The values of $N_s$ and the safety factor, $F$, can be obtained from the values of $D$ and $\psi$.

$$N_s = \gamma HF/c \implies F = cN_s/\gamma H$$

The most unfavourable failure circles are the following types:

— Toe circles, undercutting the slope toe.
— Deep circles, tangent to the hard layer and centred vertically above the halfway point of the slope.
— Slope circles, cutting the slope face.

The chart in *Figure 9.33* relates $N_e$ (stability number), $\psi$ (slope angle) and $\phi$ for values of $\phi$ between 0° and 25°. An example of its use is shown in *Box 9.2*.

These charts can also be used in cases where a mid-slope water table is present, by using the average unit weight of the soil, as shown in *Figure 9.34*.

## Hoek and Bray charts

The charts developed by Hoek and Bray (1981) are based on Taylor's method. They enable a quick and simple calculation of the slope factor of safety in soils with circular failure at the slope toe, using slope geometric data and soil strength parameters. The following hypotheses are assumed:

— The slope material is homogeneous.
— A tension crack is present.
— Normal stress is concentrated at one point on the failure surface.

Five situations are considered with respect to the position of the water table on the slope, ranging from dry to saturated, with flow parallel to the slope. One of the five calculation charts is selected accordingly. An example of application is given in *Box 9.3*; in cases where the position of the water table is different, the charts shown in Appendix A are applied accordingly.

## Vertical slice method. Bishop method

Taylor's hypothesis assumes that normal stresses on the failure surface are concentrated at one point, which introduces a certain element of error, although generally on the side of safety. Moreover, in Taylor's chart the presence of water may be only studied in cases where the soil is homogeneous and the water table is horizontal (*Figure 9.34*). To avoid this inconvenience, Bishop (1955) used the following hypothesis to develop a **vertical slice method**, known as the **Bishop method** (*Box 9.4*):

— A circular failure surface is assumed.
— The sliding mass is divided into $n$ vertical slices; generally a minimum of five slices.

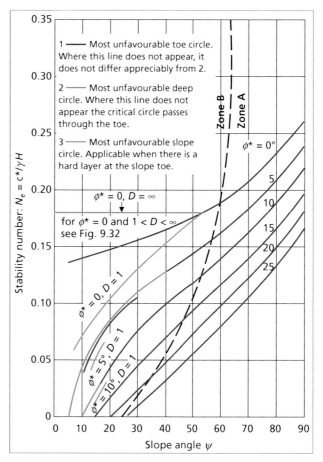

Figure 9.33  Stability of homogeneous slopes in soils with internal friction (Taylor, 1961, in Jiménez Salas et al., 1976). In zone A the critical circle is above the slope toe. In zone B the most unfavourable circle penetrates beneath the slope toe. See Figure 9.31 for the meanings of $D$ and $\psi$.

— Moment equilibrium of the forces acting on each slice is established with respect to the centre of the circle.
— The $N$ forces (those normal to the failure surface) are obtained from the vertical forces for equilibrium for each slice, and are substituted into the resultant moment equilibrium equation.
— Bishop's simplified method (which is the most well-known and commonly used), also assumes that contact forces between each pair of slices are in equilibrium and therefore have no influence.
— The factor of safety expression, $F$, is then obtained for the surface under consideration (*Box 9.4*).

As $F$ does not explicitly appear in this expression, various iterations may be necessary to obtain its value, although convergence is generally very quick.

## Box 9.2

### Example of the application of Taylor's method

Calculate the factor of safety of a 10 m high slope in soil, having a gradient of 30°, and a cohesion of 10 kPa, friction angle of 18° and apparent unit weight of 18 kN/m³.

First, it is assumed that the factor of safety with respect to cohesion, $F_c$, has a certain value, for example 1.3, with which the stability number, $N_e$, can be calculated:

$$N_e = \frac{c^*}{\gamma H} = \frac{c/F_c}{\gamma H} = \frac{10/1.3}{18 \times 10} = 0.042$$

Entering this value and the slope angle (30°) on the chart in Figure 9.33 will give a friction angle for limit equilibrium, $\phi^*$, of around 16.5°. The resulting factor of safety with respect to friction will be:

$$F_\phi = \frac{\tan \phi}{\tan \phi^*} = \frac{\tan 18°}{\tan 16.5°} = 1.09$$

This value is lower than the assumed value of 1.3; therefore, the real factor of safety will be somewhere between the two. If a further rough calculation is done with another $F_c$ value, for example 1.20, the value obtained for $F_\phi$ will be 1.21; the real factor of safety will therefore be equal to 1.2.

To find the angle of a soil slope with a factor of safety $F = 1.5$ and a height of 15 m, with the same parameters as the previous example, the factor of safety of 1.5 is applied to c and to tan $\phi$, obtaining a stability number of:

$$N_e = \frac{10/1.5}{18 \times 15} = 0.0247$$

and the friction value for limit equilibrium:

$$\tan \phi^* = \frac{\tan \phi}{F_\phi} = \frac{\tan 18°}{1.5} = 0.2166 \Rightarrow \phi^* = 12°$$

By inputting the values of $N_e$ and $\phi^*$ on the chart in Figure 9.33, an angle of approximately 18° is obtained for the slope angle, which is the value sought.

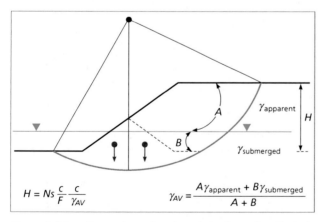

Figure 9.34   Correction of unit weight for using Taylor's chart.

Once the factor of safety F of the surface being studied is obtained, another circular surface is taken and a new F value is determined, and so on, until the minimum value for F is reached. These equations are normally solved by computer programs so that circles with different centres and radii are analysed until one is obtained that gives the minimum value of F (Figure 9.35).

Several more recent slice methods have been devised, aimed at reproducing better the phenomenon of instability; they establish various different hypotheses between forces present in the contacts between slices (something the Bishop method ignores by assuming these do not produce moments). They also consider non-circular surfaces (as in the Jambu method, for example), substituting these with a logarithmic spiral, which is perhaps more consistent with what is observed in the field, or a polygonal surface. In this respect, the methods of Morgenstern-Price and Spencer are worth mentioning as they provide more accurate solutions. No method, however, is without its drawbacks; this means that, in practice, the use of Bishop's simplified method for circular failures is very common, despite problems that may arise for it from high pore pressures and high friction.

Available computer programs which analyse potential failure surfaces in a slope, both circular and non-circular, have the advantage of enabling large numbers of calculations to be carried out in a very short time. Different limit equilibrium methods can be used, either approximate or exact, such as those of Bishop, Jambu, Spencer, Lowe and Karafiath, and analysis can also include external forces and those arising from the presence of water. Such programs

## Box 9.3

### Calculation of the factor of safety of a soil slope using Hoek and Bray charts

The task is to obtain the factor of safety of a partially saturated slope excavated in soil with a height $H = 12$ m and a slope angle $\psi = 35°$. The soil strength parameters are $c' = 15$ kPa, $\phi' = 25°$ and $\gamma = 18$ kN/m³.

The following steps enable a factor of safety to be obtained:

— The appropriate chart is selected according to the position of the water table in the slope; in this case, the water surface resembles most that of Figure A, which corresponds to chart n° 3 (Figure B).

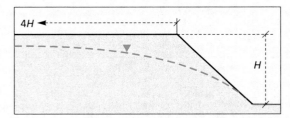

*Figure A*    Hypothesis n° 3 for the location of the water table in a slope: the water table is projected to intersect ground level at a distance 4 times the slope height back from the slope head.

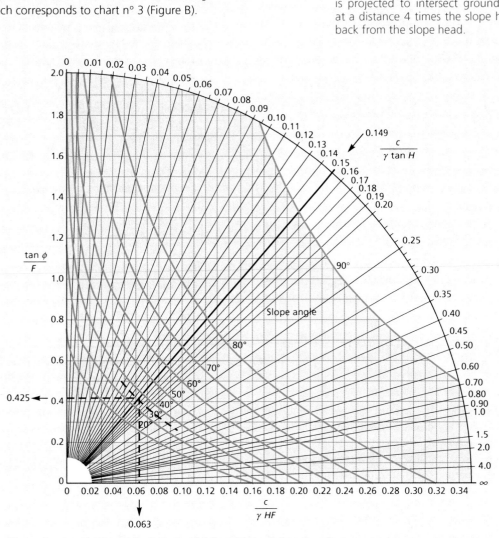

*Figure B*    Hoek and Bray chart n° 3 for circular failure in soils.

> - The value of the expression $c'/(\gamma H \tan \phi')$ is calculated and used in the chart.
> - The values of the expressions $\tan \phi'/F$ and $c'/(\gamma HF)$ can be read on the ordinate and abscissa axes at the intersection of the line corresponding to the previous value with the curve corresponding to the slope angle; from these expressions $F$ can be obtained.
>
> For the data in the example:
>
> $$c'/(\gamma H \tan \phi') = 15/(18 \times 12 \times 0.466) = 0.149$$
> $$\tan \phi'/F = 0.425$$
> $$c'/(\gamma HF) = 0.063$$
>
> from which the value for $F$ obtained is 1.1.
>
> In the same way, values corresponding to $c'$ and $\phi'$ can be obtained from the charts for a particular factor of safety $F$ and slope angle.

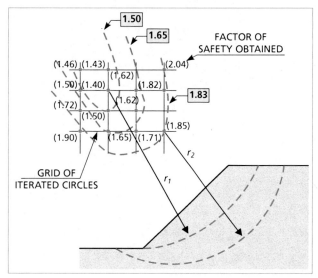

*Figure 9.35* Iterations with various circles with different centres and radii to obtain the minimum factor of safety (F = 1.4) corresponding to the slope.

also have the advantage of providing graphical output, with the results and position of the surfaces analysed (*Figure 9.36*).

## Aspects to be taken into account in stability analysis

— In slope design, the angle of natural stable slopes should be taken into account. *Figure 9.37* includes results from field studies in Southern and Central Spain that enable slope angle to be estimated based on the pre-existing hill slope, according to soil type.
— The above methods of analysis describe mathematical conditions for stresses but these should be adapted to the real drainage conditions of the problem being solved by carrying out analyses with either effective or total stresses as appropriate. In situations where excavation and construction work is to be carried out quickly, analyses may be done with total stresses; for the long-term situation, however, analyses should be done with effective stresses.
— With the exception of the "slice methods", the methods described use single values for the c and $\phi$ parameters; however, these depend on the state of the soil, which may vary along the failure surface. There are normally variations in the dry density value of the soil forming a slope in the order of 10%, and the void ratio may therefore vary 15 to 20%, which in sands can lead to variations of 5°–6° in the friction angle.
— In saturated clays, undrained shear strength, $S_u$, varies according to the overconsolidation ratio (OCR). When strength is being determined, the type of deformation in the soil along the slip surface should also be taken into account so that laboratory tests are representative.
— The methods of analysis themselves partially condition the influence of the parameters. In the case of "infinite slopes", cohesion is normally very low or zero on the slip surface as this is usually an existing discontinuity, usually with water flowing within it. If the value $c = 0$ is taken, then $F = \tan\phi/\tan\alpha$, so for a given geometry $F$ depends linearly on $\tan\phi$, and a variation of 2° will mean a variation in the order of 8 to 12% in the value of $F$. If the factor of safety of a slope is low (around 1.15), $F$ may fall locally to 1.0, generating sliding.
— To calculate pore pressure on the failure surface, the flow net in the slope must first be determined, something that is not always easy to do. It can be obtained from an equivalent static water table or by incorporating the appropriate values of the coefficient $r_u = u/\gamma H$ into the calculations (for each slice, for example).
— There are certain factors influencing slope stability that are not taken into account in these analyses, including environmental factors (evapotranspiration, surface erosion, vegetation, etc.), which may affect aspects such as shear strength parameters and hydrogeological conditions in the ground.
— Correctly applied, the method of slices gives acceptable results. It is advisable to begin by defining the location

## Box 9.4

### Bishop's simplified method

Moment equilibrium can be established from forces acting on each of the slices in the slope under consideration:

$$\sum S \cdot R = \sum W \cdot x = \sum W \cdot R \cdot \sin \alpha$$

Since:

$$F = (cA + N \tan \phi)/S$$

the value of S is:

$$S = (cA + N \tan \phi)/F$$

then:

$$\sum \frac{cA + N \tan \phi}{F} R = \sum (WR \sin \alpha)$$

$$F = \frac{\sum (cA + N \tan \phi)}{\sum (W \sin \alpha)}$$

To isolate the unknown $N$, vertical equilibrium of the slice is established:

$$W + \Delta X = N \cos \alpha + U \cos \alpha + S \sin \alpha$$

substituting $S$ and isolating $N$ gives:

$$N = \frac{W + \Delta X - U \cos \alpha - \left[(cA + N \tan \phi)/F\right] \sin \alpha}{\cos \alpha}$$

from where:

$$N = \frac{W + \Delta X - \left[(cA \sin \alpha / F) + U \cos \alpha\right]}{\cos \alpha + \left[(\tan \phi \sin \alpha)/F\right]}$$

and the factor of safety is (considering $\Delta X = 0$):

$$F = \frac{\sum \left[cA \cos \alpha + (W - U \cos \alpha) \tan \phi\right]\left[1/Mi(\alpha)\right]}{\sum W \sin \alpha}$$

where:

$$Mi(\alpha) = \cos \alpha \left(1 + \frac{\tan \phi \tan \alpha}{F}\right)$$

The chart in Figure B can be used to calculate $Mi(\alpha)$.

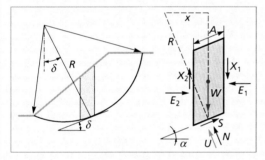

*Figure A*      Forces acting on a slice.

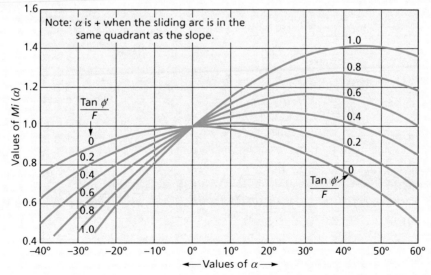

*Figure B*      Chart for obtaining the value of $Mi(\alpha)$ from Bishop's formula.

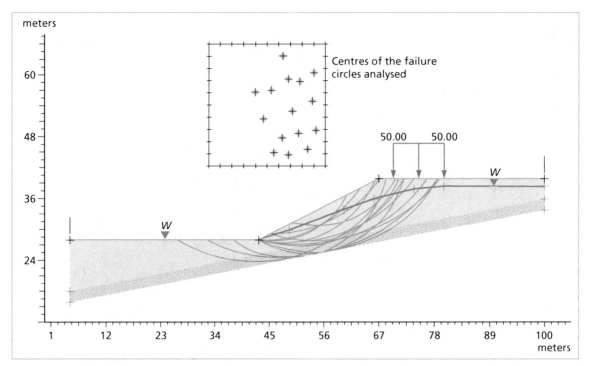

Figure 9.36   Slope stability analysis. Graphical result showing the potential failure planes analysed. SLIDE program.

of possible failure surfaces, and consider similar surfaces for analysis, rather than only inputting data on radii and centres into the computer program without previously noting their position on the slope being analysed.

## Rock slopes

### Plane failure

This is the simplest case for analysis. The factor of safety equation is established from forces acting on the failure surface under consideration (Figures 9.38a and 9.26):

$$F = \frac{cA + (W\cos\alpha - U)\tan\phi}{W\sin\alpha}$$

where:

— $cA$ = force due to cohesion in the sliding plane.
— $(W\cos\alpha - U)\tan\phi$ = force due to friction in the plane.
— $W\cos\alpha$ = stabilizing component of the weight (normal to the slip surface).
— $U$ = total force due to water pressure on the slip surface.
— $W\sin\alpha$ = weight component driving sliding (parallel to the slip surface).

If there is a tension crack full of water (Figure 9.39):

$$F = \frac{cA + (W\cos\alpha - U - V\sin\alpha)\tan\phi}{W\sin\alpha + V\cos\alpha}$$

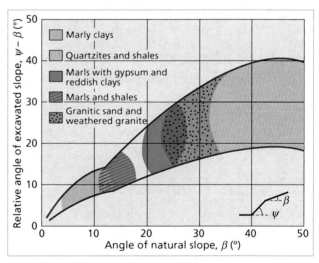

Figure 9.37   Relationship between the angle of the natural and excavated slopes in southern Spain.

$V$ being the force exerted by the water on the tension crack.

The slope weight is calculated from the volume of the sliding block and unit weight of the material. Force exerted by the water can be estimated by (see Figure 9.8):

$$U = 1/2\, \gamma_w z_w A; \quad V = 1/2\, \gamma_w z_w^2$$

$A$ being the length of the slide surface.

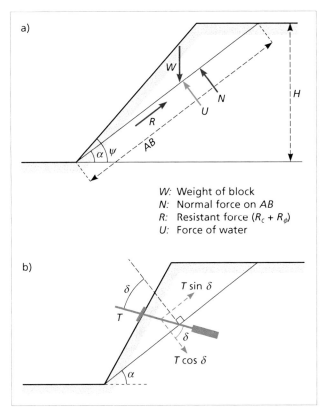

Figure 9.38  a) Forces acting on a planar slide surface. b) Forces exerted by a resistant external element (anchor) applied to the slope.

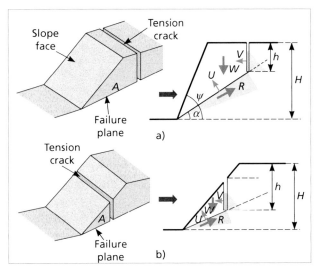

Figure 9.39  Geometry of plane failure in a slope. a) Tension crack at head. b) Tension crack on slope face. (Modified from Hoek and Bray, 1981).

Based on this general formulation and depending on the characteristics and type of plane failure and other factors involved, the different acting forces are introduced into the equations. In cases when a resisting **external load** is applied to the slope (for example, an anchor; see resistant elements in Section 9.5), the factor of safety obtained is (Figure 9.38b):

$$F = \frac{cA + (W\cos\alpha - U + T\cos\delta)\tan\phi}{W\sin\alpha - T\sin\delta}$$

From this equation it is possible to calculate the total anchoring force necessary to obtain a certain factor of safety in a slope. For example, if the value of $F = 1.3$ is required for a potential failure plane of a block weighing 70 tons along a surface dipping at 35°, considering values $c = 0$, $\phi = 32°$ and $U = 22$ tons for the failure surface, and considering an anchor inclined at 30° with respect to the horizontal, that is, $\delta = 25°$:

$$1.3 = \frac{(70\cos 35° - 22 + T\cos 25°)\tan 32°}{70\sin 35° - T\sin 25°}$$

from which the value of $T = 27$ tons is obtained; this force can be applied with either a single resistant element or among several distributed on the slope face. The magnitude of force $T$ varies according to its orientation with respect to the discontinuity plane.

## Wedge failure

Depending on the accuracy required and the object of the analysis, different procedures can be used to analyse wedge stability.

Mathematical expressions of the factor of safety of a wedge using the **analytical method** (Hoek and Bray, 1981) are complicated to resolve. In the simplest case, assuming that only friction exists in the two wedge planes, and that the friction angle for both is the same, the factor of safety is given by (Figure 9.40):

$$F = [(R_A + R_B)\tan\phi] / [W\sin\alpha]$$

where $\alpha$ is the angle of the line of intersection of the two planes forming the wedge with the horizontal.

To obtain $R_A$ and $R_B$ the forces acting parallel and perpendicular to the intersecting line of the wedge planes have to be calculated:

$$R_A \sin(\beta - 1/2\xi) = R_B \sin(\beta + 1/2\xi)$$
$$R_A \cos(\beta - 1/2\xi) - R_B \cos(\beta + 1/2\xi) = W\cos\alpha$$

isolating $R_A$ and $R_B$ and adding:

$$R_A + R_B = [W\cos\alpha \sin\beta]/\sin 1/2\xi$$
$$F = (\sin\beta / \sin 1/2\xi)(\tan\phi / \tan\alpha)$$

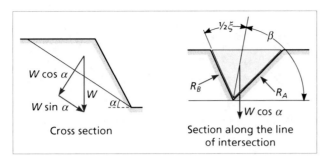

Figure 9.40  Diagram of the forces acting on planes forming the wedge (Hoek and Bray, 1981).

Calculation becomes more complicated if cohesion on the planes and the pressure of water acting on them are considered. A complete explanation of the procedure is included in Hoek and Bray (1981); Simons et al. (2001) give detailed examples of solutions applying Hoek and Bray's method.

Computer programs are available for both deterministic and probabilistic analyses of wedge stability that allow forces due to water pressure, external and seismic forces, etc. to be included; e.g. SWEDGE program, based on Hoek and Bray's analytical method.

In the straightforward case of a wedge formed by two planes with cohesive strength = 0 and no water pressure, the factor of safety can be obtained with the **Hoek and Bray charts** from values relating to the dip, strike and angle of friction of the planes. The charts provide two adimensional parameters, A and B, in the expression:

$$F = A \tan \phi_A + B \tan \phi_B$$

where $\phi_A$ and $\phi_B$ are the angles of friction of the two planes forming the wedge, with plane A being the one with least dip. This quick method is useful for estimating wedge stability in the slope design phase. Box 9.5 shows an example of how it is applied. Charts for analysing different possible cases are included in Appendix A (note that these are only applicable when cohesion = 0).

A complete wedge stability analysis can be carried out using the John method (John, 1968) where the weight of the block must be known. This method is based on **stereographic representation** of the directions of forces acting on the wedge and the planes forming it, and its aim is to find the angles between the different resultant forces to be able to calculate the safety factor (Figure 9.41).

### Block toppling

Block toppling is analysed by studying the equilibrium conditions for each block in the slope and calculating the relationships between them, taking into account the mutual interaction and the geometry of the blocks and slope.

Goodman and Bray (1976) and Hoek and Bray (1981) developed an analytical method for straightforward cases and for slopes with schematic blocks. More complex cases cannot be represented by simple models and cannot be analysed with limit equilibrium methods. Below is a description of the procedure for analysing stability in slopes with the characteristics and conditions needed for block toppling failure.

The slope should be divided into three areas (Figure 9.42a), where the distance between the faces $M_n$ and $L_n$ of each block, in contact with the blocks in front and behind are, respectively:

— Blocks on the slope crest

$$M_n = Y_n - a_2$$
$$L_n = Y_n - a_1$$

— Blocks below the crest

$$M_n = Y_n$$
$$L_n = Y_n - a_1$$

— Blocks above the crest

$$M_n = Y_n - a_2$$
$$L_n = Y_n$$

When certain conditions are met, each of the blocks forming the slope may become unstable, either toppling or sliding, depending on the acting forces and block dimensions (Figure 9.42b):

$$\phi > \alpha \rightarrow \text{no sliding possible}$$
$$\phi < \alpha \rightarrow \text{sliding possible}$$
$$\Delta x/Y_n > \tan \alpha \rightarrow \text{no toppling possible}$$
$$\Delta x/Y_n < \tan \alpha \rightarrow \text{toppling possible}$$

where $\phi$ is the friction angle of the base of the block and $\alpha$ is the angle with the horizontal.

For a block n one of the forces resisting sliding or toppling is $P_{n-1}$, which is transmitted by the block immediately below it. In the case of **toppling**, the equilibrium equation of block n, establishing moments with respect to the point of rotation, is:

$$W_n \sin \alpha \, Y_n/2 + P_n M_n = W_n \cos \alpha \, \Delta x/2 + P_n \tan \phi \, \Delta x + P_{n-1} L_n$$

and the value corresponding to the force $P_{n-1}$, opposed to toppling is:

$$P_{n-1,t} = [1/2 \, W_n (\sin \alpha \, Y_n - (\cos \alpha \, \Delta x)) + P_n (M_n - (\tan \phi \, \Delta x))]/L_n \qquad (1)$$

## Box 9.5

### Calculation of the factor of safety of a wedge using Hoek and Bray charts

A wedge is formed by two planes with the following characteristics:

|         | Dip | Dip direction | Friction angle, $\phi$ |
|---------|-----|---------------|------------------------|
| Plane A | 45° | 115°          | 34.5°                  |
| Plane B | 75° | 225°          | 37°                    |

The procedure for finding the factor of safety is as follows:

— The value of the difference in dip between the two planes is calculated: 75° − 45° = 30°.
— From the different charts available (see Appendix A), the two that correspond to the value of 30° are selected.
— The value of the difference in dip direction between the two planes is calculated: 225° − 115° = 110°.
— On the abscissa axis of Chart a) a line is projected from this value of 110° to intersect the line corresponding to the dip of plane A, giving a reading of the value of parameter A on the ordinate axis of 1.10.
— The same procedure is applied to plane B on Chart b) to obtain 0.45, the value of parameter B.
— Once the values of A and B have been found, the factor of safety can be calculated with the friction angle values of the two planes:

$$F = A \tan \phi_A + B \tan \phi_B =$$
$$= (1.1 \times 0.687) + (0.45 \times 0.753) = 1.09$$

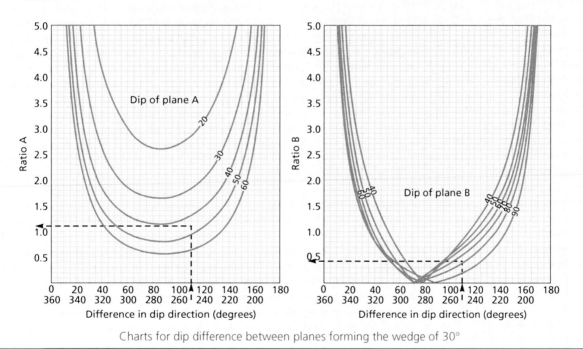

Charts for dip difference between planes forming the wedge of 30°

In the same way, the equilibrium equations for a block $n$ with respect to **sliding** are:

$$S_n = R_n \tan \phi$$
$$W_n \sin \alpha + P_n - P_{n-1} = [W \cos \alpha + (P_n - P_{n-1}) \tan \phi] \tan \phi$$

where $Q_n = P_n \tan \phi$ and $Q_{n-1} = P_{n-1} \tan \phi$.

Isolating the value of force $P_{n-1}$, which is opposed to sliding, gives:

$$P_{n-1,s} = [W_n (\sin \alpha - \cos \alpha \tan \phi)$$
$$+ P_n (1 - \tan^2 \phi)]/[1 - \tan^2 \phi] \quad (2)$$

$$P_{n-1,s} = [W_n (\sin \alpha - \cos \alpha \tan \phi)/(1 - \tan^2 \phi)] + P_n$$

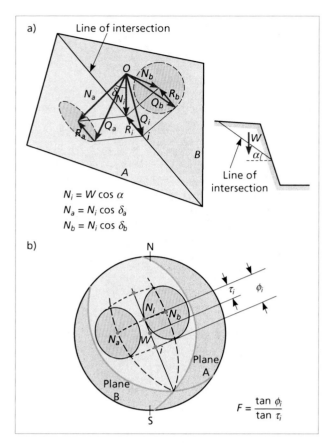

*Figure 9.41*  a) Diagram of the friction cones and forces acting on the planes forming the wedge. b) Stereographic projection.

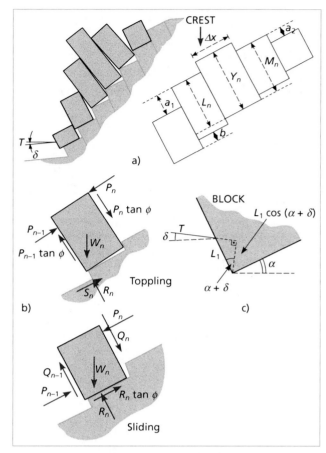

*Figure 9.42*  a) Geometric model for limit equilibrium analysis of block toppling on a slope. b) Limit equilibrium conditions for toppling and for sliding on one of the blocks. c) Anchor force applied to the block at the slope toe. (Modified from Hoek and Bray, 1981).

Slope **stability analysis** is performed in the following steps:

1. The blocks for analysis are defined and starting at the slope top the first block meeting the toppling conditions, $\Delta x/Y_n <  \tan \alpha$, is identified. For this block $n_1$, $P_n$ is taken as $= 0$.
2. The forces needed to prevent toppling and sliding from occurring, $P_{n-1,t}$ and $P_{n-1,s}$, are calculated for block $n_1$ with equations (1) and (2), using geometric data and weight of block, and with an angle $\phi$ assumed to be initially greater than $\alpha$.
3. The greater of the two values obtained is taken and applied to the analysis of the next block (the one immediately below); this will be the value of force $P_n$ of the new block. Calculation of $P_{n-1,t}$ and $P_{n-1,s}$ is repeated for the new block, and the greater of the two values will be the $P_n$ of the following block. If $P_{n-1,s} > P_{n-1,t}$ the block under consideration will be susceptible to sliding; if not, the potential movement will be toppling.
4. Calculations are made this way for all blocks susceptible to toppling. Once a block is reached that meets the condition $\Delta x/Y_n > \tan \alpha$ (where toppling is not possible), analysis will be for sliding only. The analysis is continued until the block at the slope toe is reached.
5. Analysis of the lowest (toe) block on the slope (either for toppling or sliding or for sliding only) will give:

    — $P_{n-1} = 0$: the slope is in limit equilibrium for the assumed value of angle $\phi$.
    — $P_{n-1} < 0$: the calculation is not valid and must be repeated for other values of $\phi$.
    — $P_{n-1} > 0$: the slope is unstable for the assumed value of $\phi$.

Using this method, the **force necessary to stabilize the slope** at its base to prevent toppling and sliding can be calculated. Taking an anchor situated on the toe block with the direction shown in Figure 9.42c, the force $T$ it exerts in order to maintain equilibrium will be equal to $P_{n-1}$, which is the force necessary to prevent the block from toppling or sliding.

For **toppling**, the force that must be transmitted by the anchor is calculated by the formula:

$$T_t = P_{n-1,t} = \frac{P_1(Y_1 - \Delta x \tan \phi) + (W/2)(Y_1 \sin \alpha - \Delta x \cos \alpha)}{L_1 \cos(\alpha + \delta)}$$

and for **sliding**:

$$T_s = P_{n-1,s} = \frac{P_1(1 - \tan^2 \phi) - W(\tan \phi \cos \alpha - \sin \alpha)}{\tan \phi \sin(\alpha + \delta) + \cos(\alpha + \delta)}$$

where the normal and tangential forces exerted on the base of the block are:

$$R_1 = W_1 \cos \alpha + P_1 \tan \phi + T \sin(\alpha + \delta)$$
$$S_1 = W_1 \sin \alpha + P_1 - T \cos(\alpha + \delta)$$

The force that must be applied to the anchor is the highest value obtained for $T_t$ and $T_s$.

## Buckling

Buckling failure is analysed with equations used for beam buckling in analyses for material strength.

The factor of safety for a rock column susceptible to buckling is expressed by:

$$F = P_{cr}/P_D$$

where $P_{cr}$ is the critical buckling load and $P_D$ is the force exerted on the point of the column that undergoes buckling.

Cavers (1981) proposes a simple method for analysing buckling and bending of strata based on the hypotheses that the rock column analysed is elastic and obeys Hooke's law, is weightless, is perfectly straight and has embedded ends.

The following expression gives the maximum or critical bearing load, $P_{cr}$, for the rock column before buckling occurs:

$$P_{cr} = K \pi^2 E I / l_b^2$$

where:

$K$ = constant, factor of the conditions at the ends of the column; for embedded columns $K = 1.0$.

$E$ = elastic modulus of the material.
$I$ = moment of inertia.
$l_b$ = length of the column experiencing bending and buckling.

The length $l_b$ can be estimated as a ratio of the column length. It is generally taken as $l_b/l = 0.5$.

For different conditions at the ends of the column (such as a column with an embedded base and free to buckle in compression), Piteau and Martin (1982) propose the following expression:

$$P_{cr} = \pi^2 E I / 4 l_b^2$$

The value of $P_D$ is defined as:

$$P_D = (W_D \sin \alpha - W_D \cos \alpha \tan \phi - l_D c)$$

where $W_D$ is the weight of the column with length $l_D$, $c$ the cohesion and $\phi$ the friction of the plane (Figure 9.43).

The critical length of the column at which buckling may occur is given by the expression:

$$l = \sqrt{\frac{\pi^2 E d^2}{2.25(\gamma \sin \alpha - \gamma \cos \alpha \tan \phi - c/d)}}$$

where $\gamma$ is the unit weight.

## Circular failure

In low strength heavily weathered or jointed isotropic rock masses, this type of failure can be analysed using methods for circular failure in soils. The most widely used is the simplified **Bishop's method**. The general formulation for this is included in the section on soil slopes.

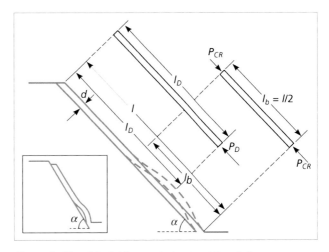

Figure 9.43    Model for analysing buckling failure.

## Stress-strain methods

Stress-strain methods can be used as an alternative to limit equilibrium methods, wherever suitable for specific slope analysis. The main advantage of these methods is that they consider the **stress-strain relationships** of the material during the deformation and failure process; these relationships determine its mechanical behaviour and control its strength.

When subjected to certain loads, ground deforms according to its strength and deformation properties, obeying its behaviour law to the point of critical failure. At the same time, different states of stress are generated at different parts of the slope. Stress-strain methods allow this process to be quantified, from (1) a model representing the geological structure, stratigraphy and hydrogeology of the slope or area under consideration (with specific boundary conditions being applied), (2) the appropriate behaviour law for the material, and (3) the strength and deformation properties of the different lithologies involved. Resolving the elasticity or plasticity equations (or any other behavioural model), by finite elements or alternative mathematical methods, produces values for the displacements, strains and stresses generated throughout the model analysed. This gives the type and magnitude of the displacements in the slope consistent with its state of equilibrium.

Limit equilibrium methods only consider forces acting on one or several points of the failure surface, assuming that failure occurs instantaneously, and that the strength is mobilized at the same time along the whole surface. In contrast, stress-strain methods analyse the deformation process at each point selected in the model. This allows the influence of different parameters on the state of slope stability to be evaluated, such as natural stresses, dynamic forces, and water pressure.

A basic modelling requirement is an understanding of the behavioural laws of the materials involved and a knowledge of the strength and deformation parameters; these are the main limitation of the stress-strain methods.

The currently available computer programs provide graphic output showing the displacements, stresses and strains on slopes, giving the keys for identifying unstable areas and failure mechanisms and models. They also provide the slope factor of safety (which is not exactly the same as in limit equilibrium methods as no specific failure surface is defined). Figures 9.44 and 9.45 show examples of results from stress-strain slope analysis.

A wider application of these methods is slope analysis in soils and in heavily jointed, soft or low strength rock masses (which may be considered as continuous mediums, where failure and deformation are not controlled by pre-existing

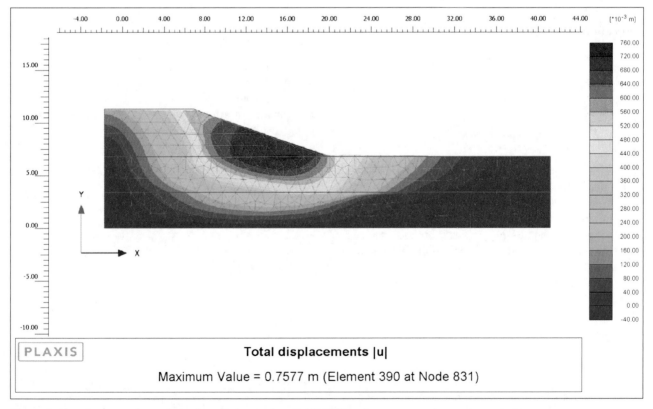

Figure 9.44    Stress-strain analysis of a soil slope using the PLAXIS code.

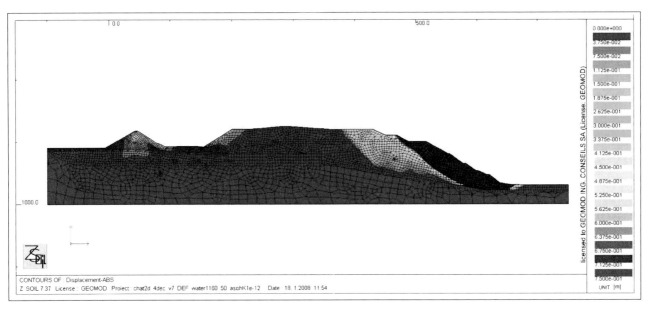

*Figure 9.45* 2D slope stability analysis. Failure mechanism identified by displacement increments color maps. ZSOIL.PC.

discontinuity planes) and in intact rock masses. These methods can also be applied to rock slope analysis at depth, e.g. in open pit mines, where very high stresses and significant plastification and deformation may occur. Other applications, such as the UDEC program, analyse discontinuous rock masses by studying movements between rock blocks along discontinuities.

# Geomechanical slope classification

## Slope mass rating (SMR)

The stability of an excavation can be evaluated by applying an empirical geomechanical classification to its slope. Romana (1993) proposed the SMR classification, based on the RMR classification described in detail in Chapter 3, Section 3.8. This is summarized below.

The slope mass rating (SMR) is obtained from the RMR (Chapter 3, Table 3.26), by adding some adjustment factors to account for the relative orientation of joints and slope and for the method of excavation used; the first of these is the product of sub-factors $F_1$, $F_2$ and $F_3$ (Table 9.2):

— $F_1$ depends on whether the strike of the discontinuities or joints and the slope face are parallel. It varies between 1.0 (when they are parallel) and 0.15 (when the angle between them is greater than 30° as by then the probability of failure is very low). These values conform approximately to the expression:

$$F_1 = [1 - \sin(\alpha_j - \alpha_s)]^2$$

where $\alpha_j$ and $\alpha_s$ are the strike angles of the discontinuity and the slope respectively.

— $F_2$ depends on the dip of the discontinuity or joint. For planar failures, this varies between 1.0 (for discontinuities with a dip > 45°) and 0.15 (for discontinuities with a dip < 20°); this can be adjusted approximately with the expression:

$$F_2 = \tan^2 \beta_j$$

where $\beta_j$ is the dip of the discontinuity. $F_2$ takes value 1.0 for toppling failure.

— $F_3$ reflects the ratio between the discontinuity dip and the slope angle.

The excavation method adjustment factor, $F_4$, is shown in Table 9.2. The final SMR value is obtained as follows:

$$SMR = RMR + (F_1 \times F_2 \times F_3) + F_4$$

The method involves finding the SMR for each set of discontinuities and adopting the lowest value obtained. For weathered and degraded rocks, classification must be done twice, the first time for the initial condition of the fresh rock and the second for the weathered rock. The classification does not include wedge failures.

Depending on the value of the SMR index, five stability classes are obtained, as defined in Table 9.2; the slope support measures to be applied fall into six groups:

— SMR > 65: no support (scaling).
— 70 > SMR > 45: protection measures (toe ditches, slope or toe fences, protection meshes over the slope surface).

| Table 9.2 | SMR SLOPE CLASSIFICATION | | | | | |
|---|---|---|---|---|---|---|
| | Adjustment factors by orientation of discontinuities ($F_1$, $F_2$ and $F_3$) | | | | | |
| | Case | Very favourable | Favourable | Normal | Unfavourable | Very unfavourable |
| P<br>T | $\|\alpha_j - \alpha_s\|$<br>$\|\alpha_j - \alpha_s - 180°\|$ | >30° | 30°–20° | 20°–10° | 10°–5° | <5° |
| P/T | $F_1$ | 0.15 | 0.40 | 0.70 | 0.85 | 1.0 |
| P | $\|\beta_j\|$ | <20° | 20°–30° | 30°–35° | 35°–45° | >45° |
| | $F_2$ | 0.15 | 0.40 | 0.70 | 0.85 | 1.0 |
| T | $F_2$ | 1 | 1 | 1 | 1 | 1 |
| P | $\beta_j - \beta_s$ | >10° | 10°–0° | 0° | 0 – (–10°) | <–10° |
| T | $\beta_j + \beta_s$ | <110° | 110°–120° | >120° | – | – |
| P/T | $F_3$ | 0 | –6 | –25 | –50 | –60 |
| | Adjustment factor by excavation method ($F_4$) | | | | | |
| Method | | Natural slope | Pre-splitting | Smooth blasting | Blasting or mechanical excavation | Deficient blasting |
| $F_4$ | | +15 | +10 | +8 | 0 | –8 |
| | Classes of stability | | | | | |
| Class | | V | IV | III | II | I |
| SMR | | 0–20 | 21–40 | 41–60 | 61–80 | 81–100 |
| Description | | Very poor | Poor | Fair | Good | Very good |
| Stability | | Completely unstable | Unstable | Partially stable | Stable | Completely stable |
| Failures | | Large plane failures or soil-like | Plane or large wedges | Some joints or many wedges | Some blocks | None |
| Support | | Re-excavation | Correction | Systematic | Occasional | None |

P: Plane failure  $\alpha_s$: strike of slope  $\beta_s$: dip of slope
T: Toppling failure  $\alpha_j$: strike of discontinuities  $\beta_j$: dip of discontinuities

(Romana, 1993).

— 75 > SMR > 30: local reinforcement (bolts, anchors).
— 60 > SMR > 20: concreting (shotcrete, concrete, buttresses and/or beams, toe walls).
— 40 > SMR > 10: drainage (surface and deep).
— 30 > SMR > 10: re-excavation.

## 9.6 Stabilization measures

### Introduction

When slope failure occurs, or when deformation creates the risk of instability, stabilization measures have to be adopted. These measures are also required when, for construction, environmental or financial reasons, the angle of the slope to be excavated is greater than that corresponding to the ground strength.

Designing and applying these measures needs an understanding of the:

— Ground properties and their geomechanical behaviour.
— Type of failure and mechanism, including the speed and direction of movement and geometry of the failure surface.
— Geological, hydrogeological and other factors influencing and causing instability. These factors will also define the most appropriate stabilization measures in each case. Data on the position of water tables, water pressure and the permeability of the materials are specially important.

The relevant information is obtained by carrying out appropriate geological and hydrogeological studies. The material properties and geotechnical parameters can be defined from detailed site investigations and complementary laboratory tests. An awareness of environmental conditions and possible anthropogenic influencing factors is also essential.

The following should be taken into account when stabilization measures are being designed:

— Financial and material resources available.
— How urgent the measures are.
— Scale and dimensions of the instability.

The **factor of safety** of a slope can be increased either by reducing the destabilizing forces tending to failure or by increasing the stabilizing forces (Figure 9.46). Any activity carried out on the slope will affect basic factors controlling slope equilibrium: the weight and strength parameters of the materials, and the presence of water. These fundamental concepts should always be borne in mind and applied correctly. For example, removing weight from the head of a slope always helps to improve stability, but if only a very small volume (<5%) is removed, this has practically no effect on the safety factor; when a gabion wall is used to reinforce the slope toe, its foundation must be able to resist the sliding forces; deep ditches or drainage channels should not be excavated at the front toe of a retaining wall as this prevents a passive thrust from developing.

The value of the factor of safety required must be pre-determined when designing corrective measures. While a value of 1.5 may be adequate for a permanent slope (or 1.3 for temporary slopes), the correction of an unstable slope will start from a value lower than 1.0, and it may be enough to arrive at a value 1.2 or 1.25, except in cases of excavations where higher values are required.

Before designing stability measures, a **back analysis** is useful to obtain the approximate values of the "real" geotechnical parameters and deduce the position of the water table when failure began; if geological, geometrical and certain geotechnical data are known, this can be done with a fair degree of certainty. The cohesion and friction values, the geometry of the failure surface, hydrogeological conditions in which failure occurred, etc. should all be tested for a factor of safety of $F \approx 1.0$, using trial and error. In the next step, further analyses are carried out applying possible corrective solutions (changing the slope geometry, lowering the water table, introducing resistant structural elements, etc.) until the required factor of safety is reached.

If there is a likelihood of seismic activity in the area, this should also be taken into account in the analysis.

**Stabilizing measures** may involve:

— Modifying the slope geometry.
— Drainage.
— Increasing ground strength by inserting resistant structural elements into the slope.
— Constructing walls or other retaining elements.

## Stabilization methods

### Modifying the geometry

Modifying the slope geometry redistributes forces due to the weight of the materials to obtain a new, more stable configuration. This is most often done by (Figure 9.47):

— Reducing the slope angle.
— Removing weight at the head of the slope.
— Increasing weight at the slope toe ("heels" or rip-rap).
— Constructing benches and berms (stepping).

**Excavation of the head of the slope** removes weight in this area and reduces destabilizing forces.

It is not always possible to reduce the general angle of the slope or excavate and unload at its head, because of difficult access to the upper part of the slope, or the large volume of material that would have to be removed for measures to be effective, or the environmental problems deriving from the dumping of excavated material, and the land take at the head of the slope, or pre-existing installations in this area. All these prevent such solutions. Therefore, changes in geometry are mainly concerned with **increasing weight at the slope toe**, even though this solution means occupying a large area at the base of the slope where available space is usually scarce. Constructing "heels" to act as counterweights increases normal stress on the failure surface and improves stability. The fill at the slope toe should be drainage material, or else an appropriate drainage system must be installed; if not, any stabilizing effect could be cancelled out by water accumulating at the slope toe. This method is particularly applicable to soil slopes and can be combined with excavations at the head of the slope.

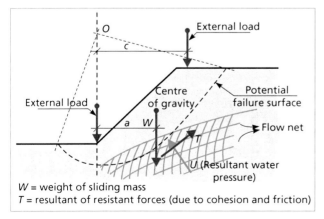

Figure 9.46   Diagram of acting forces on a slope.

If the ground at the slope toe is competent and failure is shallow, a **wall** with proper foundations can be built at the slope toe (in some cases the toe may be cut back slightly). The space between the wall and the slope can then be filled in to add extra weight to the area.

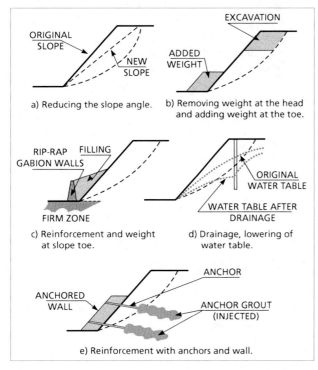

Figure 9.47    Methods for increasing the factor of safety of soil slopes.

*Figure 9.48* shows an example of a slope in clays and sands where additional weight has been loaded by building a gabion drainage wall, founded on competent ground, and spreading a waterproof infill over the slope surface. With these measures, the expansive clays present in the lower half of the slope were prevented from drying out as loss of water content had led to fissuring that was affecting stability.

A quick, straightforward solution often used to increase weight on sliding slopes is to put **rip-rap at the slope toe** (this is in fact replacing the ground); at the same time, this facilitates drainage and increases overall reinforcement. This is why it is often used where there are superficial instabilities in clay materials (*Figures 9.49* and *9.50*).

**Slope stepping** involves the construction of benches and berms that cut into possible failure surfaces and help to prevent failures affecting the whole slope. The application of such measures is normally decided upon before slope excavation is begun. Apart from retaining fallen blocks and the product of local slope failures, berms can also be used for drainage installations and to provide access for repair and control work on the slope.

## Drainage methods

Drainage methods aim to either eliminate or reduce the amount of water present in the slope, along with the pore pressures that are a destabilizing factor on failure surfaces and in tension cracks. In certain types of materials water also reduces their strength. Because water is often the main cause of slope instability, drainage is normally the **most effective**

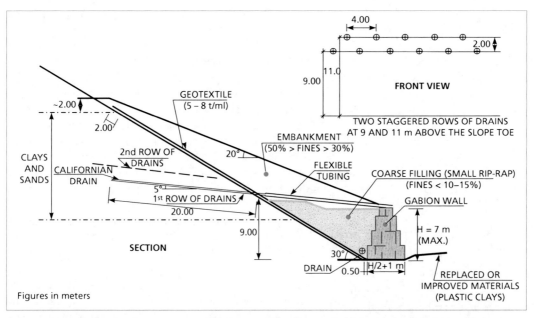

Figure 9.48    Example of road slope stabilization with a foot wall, filling and "californian" drains.

measure. Water can trigger instability by increasing the weight of the potentially unstable mass, raising the water table and pore pressures within it, and creating unfavourably oriented seepage forces, and in some cases, softening the ground or eroding the toe of the slope.

Drainage measures can be at ground level, with drainage ditches and channels, or at depth, using horizontal or "californian" drains, wells or vertical drains, drainage adits and drainage walls.

Draining devices can be sited at specific points (drains and wells) or be continuous (ditches and adits). Figure 9.51 shows some of most common drainage and protection measures used.

**Surface drainage measures** prevent run-off water from infiltrating the slope or penetrating discontinuities and cracks, as this may lead to a higher water table, increased pore pressures and soil saturation. They also prevent the erosive effects of run-off water and leaching through discontinuities

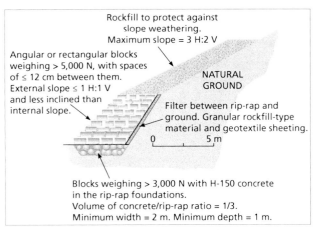

Figure 9.49   Slope stabilization using rip-rap.

Figure 9.50   Methods for stabilizing a road slope: application of rip-rap to reinforce the toe; benching and reshaping of the general slope angle and, in the foreground, a gravel-filled trench to drain the toe area.

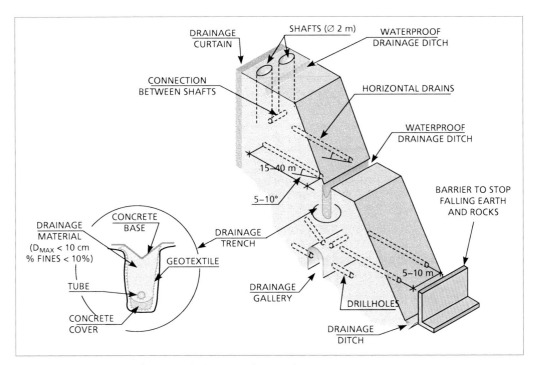

Figure 9.51   Measures for slope drainage and protection.

in rock masses. These are usually designed as preventive rather than stabilizing measures.

Drainage should be designed to prevent water reaching and accumulating on the slope, mainly at the head or on the flat areas of the berms on stepped slopes, where water ponding often occurs after rain.

Run-off water can be channelled and evacuated with trenches or drainage ditches normally made at the head and on the sides of the slope (perimeter trenches). The cross sections and characteristics of these should be calculated according to the volume of water to be drained.

Drainage measures also include channelling and diverting streams and springs in the area around or above the slope, or waterproofing the stream beds.

On large soil slopes with an appropriate gradient, surface drainage systems can be installed using trenches and drainage ditches, running along or across the slope face or in a herring-bone pattern. These collect the water and drain it away from the area of influence of the slope. In this case, the ditches can help to reinforce the ground surface (*Figure 9.52*).

The aim of **deep drainage** is to lower the water table and evacuate water from inside the slope. This solution is often used for rock slopes with instability problems (*Figure 9.53*).

Figure 9.52   Cross-slope drainage.

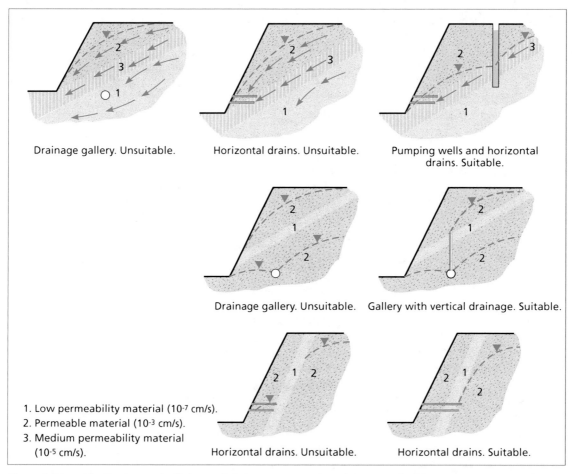

Figure 9.53   Distribution and efficiency of drainage systems in a slope (modified from CANMET, 1977).

The following aspects should be considered when designing deep drainage:

— Permeability and hydrogeological characteristics of the materials, the flow rate to be drained and the radius of action of the drainage element.
— Depth to which drilling go to reach the water level and, where required, to reach the depth the water table is to be lowered to.
— Durability of the drainage system; depending on location and depth, any slope movements may break or disable the drainage devices. This will allow water to reach the failure surface or the unstable mass, and create the opposite effect to that intended.

**"Californian" drains** are sub-horizontal boreholes with a diameter ranging from 100–150 mm and a maximum length of 30–40 m. They are very effective for draining water from the area around the slope toe (Figure 9.54). **Vertical wells**, with a diameter of 30–150 cm or more, discharge the water by internal pumps, actioned when water in the well reaches a certain height, or by gravity through drains connecting the wells with the outside of the slope. Drainage adits excavated in the interior of rock masses are highly efficient but very expensive to build. **Drainage walls** are formed by vertical large diameter shafts (1.5 to 2 m), with interconnecting horizontal drill holes inside them; this technique is used to stabilize both excavated slopes and natural soil or rock slopes (Figure 9.55).

## Resistant structural elements

Resistant elements are inserted in the ground to increase shear strength by:

— Installing elements that improve the shear strength of the failure surface by "replacing" it with unsheared material e.g. piles and micropiles.
— Installing elements that increase shear strength on the failure surface by increasing the frictional forces on it e.g. bolts and anchors (Figure 9.47e).

**Pile walls** are alignments of piles, arranged at intervals to form a more or less continuous structure crossing the sliding mass and embedded in stable ground. The distribution and length of the piles must be studied in great detail, and also their resistance to the forces to which they will be subjected. Diameters vary from 0.65 to 2 m and they are often shored up with beams at the surface.

**Micropile walls** have a similar function; these also cross the sliding zone and are inserted into stable ground (Figures 9.56, 9.57 and 9.58). Micropiles are usually 12–15 cm in diameter and up to 15–20 m long. They are reinforced with a steel tube filled with injected concrete. The system of boring allows rock materials to be penetrated easily.

**Jet-grouting columns** are normally used to stabilize slopes in granular soils and even cohesive ground by cutting into the sliding surface and creating areas with greater shear strength. The ground is penetrated, generally with boreholes

Figure 9.55  Drilling horizontal drains inside a drainage shaft to connect with adjoining shafts (courtesy of P. Berenguer).

Figure 9.54  Drilling horizontal drains in a slope.

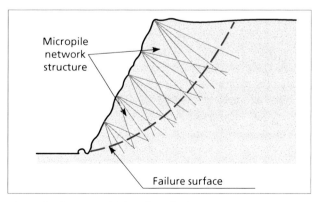

Figure 9.56  A multiple micropile fan.

*Figure 9.57* Anchored micropile wall, N. Italy.

*Figure 9.58* Reinforcement of slope toe with a micropile wall, N. Italy.

0.4–1 m in diameter, and cement is injected at high pressure (at 30–60 MPa) through a high speed rotating drill which penetrates and breaks up the surrounding ground. The result is a high strength column made of the ground itself and the injected material.

**Anchors** are elements consisting of steel cables or bars anchored in stable areas of the rock mass. They work by traction and exert a force opposed to the movement and an increment of normal stress on the failure surface (*Figure 9.59*). Depending on their function, they are classified as **passive** (the anchor is actioned when movement takes place in the rock block or in the ground), **active** (the anchor is tightened after installation until the allowable load is reached) and **mixed** (the anchor is tightened with a load less than its allowable load). Anchors are usually 15–40 m long, with a load capacity normally of 60–120 tons per anchor. The force provided by the anchor is calculated to obtain a satisfactory factor of safety (an example of the calculation of anchor force is given in Section 4, for plane failure in rock). Anchors are frequently used in jointed rock slopes as a very

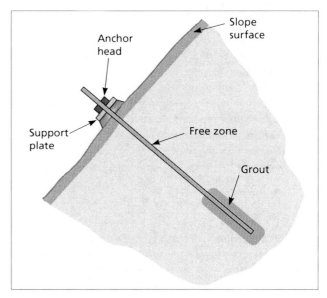

*Figure 9.59* Diagram of anchor parts.

effective measure of stabilizing sliding masses or blocks. The anchor heads may be joined to each other at the surface with concrete beams so that they work together, distributing the stabilizing forces more uniformly over the slope (*Figures 9.60* and *9.61*). In low strength rock masses or soils affected by general instability, a reinforced concrete wall to receive the anchor heads is usually built on the slope, so that the forces transmitted to the ground by these elements are more evenly distributed. In some cases, the wall is replaced with a pile wall (*Figure 9.62*).

**Bolts** are steel bars inserted into the slope and can be considered as low capacity passive anchors. They are usually 3–6 m long and 25 to 40 mm in diameter. They are inserted into holes drilled into the rock, which are then filled with cement or grout; allowable load is generally 5 to 15 ton per bolt (see "Support systems" in Chapter 10, Section 10.8 for more details).

## Walls and retaining elements

**Walls** are built at the toe of a slope to reinforce it and prevent deterioration in this area which is critical for slope stability. These retaining or supporting walls offer resistance and are effective where there is surface instability. The disadvantages of retaining walls are that the slope toe must be excavated before they are built, which in itself favours instability, and that they do not prevent sliding from occurring on failure surfaces above or below the wall. Retaining walls can be built in front of the slope toe, and the space between the wall and the slope subsequently filled in. Revetment walls protect the ground from erosion and provide the slope toe with a stabilizing weight.

*Figure 9.60* Excavation of a large slope stabilized with tied anchors, N. Italy; the area shown at top right is protected with shotcrete.

*Figure 9.61* Installing anchors linked with reinforced concrete beams for an even distribution of force on a slope, N. Spain. (Courtesy of F. Romero)

Different types of walls have different characteristics depending on the stabilization needed, whether flexible or rigid walls are required, or whether the purpose of the wall is to stop movement or provide additional strength. Flexible **gabion walls** consist of fragments of rock, or riprap, enclosed in a steel mesh; they work by gravity and may be constructed in stepped formation facing towards or away from the slope (*Figures 9.63* and *9.64*). One of their advantages is that water from the slope can flow through them. **Diaphragm walls** are made of reinforced concrete sections constructed *in situ* in slots mechanically excavated below the ground surface; their stabilizing action is similar to that of pile walls, although in contrast to these, diaphragm walls are continuous structures.

**Reinforced earth walls** have an outer facing made of prefabricated concrete or metal sheets and a soil infill, reinforced by strips or bands of metal or synthetic material, anchored to the facing and to the slope (*Figure 9.65*). **Anchored walls** are reinforced with anchors, to increase the resistance of the structure to toppling and sliding (*Figure 9.66*).

An important factor to consider when designing and constructing walls is **drainage**, as the ground may become saturated behind the walls, resulting in higher pore pressure and thrusts on the structure. Gabion walls are very permeable structures and drainage takes place naturally, but for concrete walls adequate drainage systems must be designed to ensure that any water which accumulates behind them can drain away; this can be done either with drainage tubes or cross drains running through the wall.

In practice, different stabilization systems are used simultaneously. As water is the principal destabilizing agent in so many slopes, it is normal (and should be a rule) that any stabilization work should include the excavation of ditches to capture run-off water from above the head and at the sides of an unstable slope, and impermeable drainage channels in berms and at the slope toe.

## Surface protection measures

These measures aim to:

— Eliminate rockfall problems.
— Increase slope safety by preventing surface failure.
— Prevent or reduce erosion and weathering on the slope face.
— Prevent infiltration of run-off water.

The most common measures taken are:

— Installing wire meshes.
— Shotcreting the slopes.
— Building revetment walls at the slope toe.

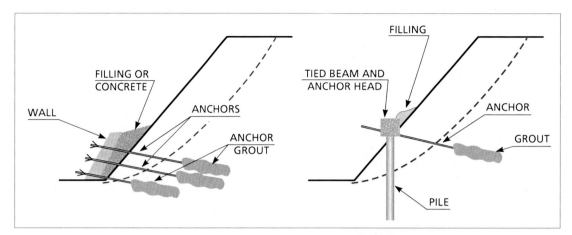

*Figure 9.62* Anchored pile walls.

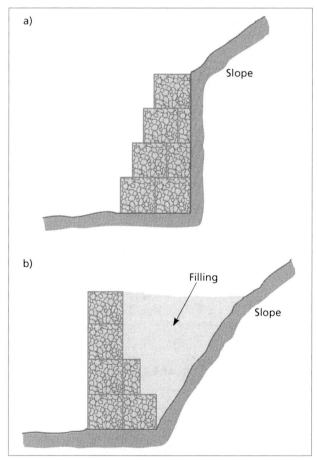

*Figure 9.63* Gabion walls with a stepped external face (a) and with a stepped internal face and filling between the wall and the slope (b).

*Figure 9.64* Gabion wall at the slope toe and drainage ditch.

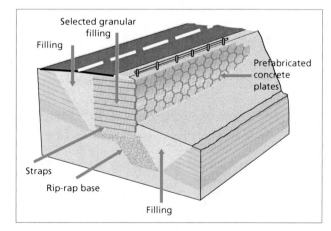

*Figure 9.65* Reinforced earth wall (modified from CANMET, 1977).

— Laying geotextiles.
— Implementing water infiltration control
— Using plant species to help reinforce the ground surface of slopes excavated in soils.

The process of **shotcreting** consists of covering the slope surface with a mixture of shotcrete or gunite (cement, water and aggregate), projected pneumatically with a hose and nozzle. The slope is normally treated with several layers,

*Figure 9.66* Stabilization measures on a road slope: anchored wall at the base and shotcrete and bolts higher up.

*Figure 9.67* Road slope covered with shotcrete, with bolts and anchors.

with a total thickness of 5–8 cm (*Figure 9.67*). Shotcrete can be reinforced by fixing a metal mesh to the slope and spraying the mixture onto it. Holes are drilled through the shotcrete layer to facilitate drainage.

Slopes excavated in jointed rock masses usually present problems of **rock falls** along the discontinuity network. Action taken to stabilize unstable rock blocks includes:

— Installing bolts to fix the blocks; if these are very large, they must be fixed with anchors.
— Using cables and meshes to stabilize heavily jointed areas of the slope. A double or triple twisted metal mesh is laid, with a superimposed cable grid; these

*Figure 9.68* Slope surface protected with metallic mesh. A barrier has been placed at the toe to prevent rocks from falling on to the road.

are anchored to the rock at each end and tightened (*Figure 9.68*).
— Removing blocks with controlled blasting, the use of expansive cement, fragmentation with a jackhammer, manual removal using crowbars, etc. Only relevant blocks should be removed; otherwise, the stability of other blocks that come into contact with them may be affected.

Chapter 13 includes a description of other preventive measures, both active and passive, to protect natural slopes from rock falls.

## 9.7 Monitoring and control

Monitoring is required if signs of instability appear on a slope (tension cracks or failures near the top, bulging or uplift at the toe, etc.), or to control the behaviour of an unstable slope. Information is obtained on the movement characteristics and behaviour, including velocity, displacement rates, location of the failure surfaces and water pressure, etc. to understand the slope behaviour. Recording the velocity of movement allows a behavioural pattern to be established so that decisions on stabilization can be taken; sometimes an approximate forecast can be made of when failure will occur by studying the time-displacement curve and extrapolating to a time frame (*Figures 9.69* and *9.70*).

Monitoring is generally limited to cases where instability affects buildings and infrastructure, mainly because of the cost and equipment required.

Before a slope is monitored, the scale and scope of the required measurements have to be established, to be able to select exactly where to take the measurements and what instruments to use; these must be installed correctly, and great care taken to ensure accurate readings and interpretation of the measurements.

With instrumental monitoring, slope behaviour can be tested and existing models and stability analyses can be verified. Before monitoring is carried out, the characteristics and properties of the materials forming the slope must be defined from preliminary studies based on field observations, data-gathering, laboratory tests, stability analyses, etc.

The subjects normally measured during monitoring are:

— Surface movements.
— Movement inside the ground.
— Movements of tension cracks and/or the distance between blocks.
— Pore pressures and their change.

The techniques and methods for measuring these subjects are described in Chapter 5, Section 6. Different methods, instruments and devices are summarized in *Table 9.3*.

**Measuring displacement** on the surface and in the ground allows movement to be detected either in a specific area or in the slope as a whole, and its velocity and direction determined. Each problem requiring monitoring is different, and so which method is most suitable and where the measuring equipment is located on the slope must be decided by experts on an individual basis in each case. Systems for measuring surface displacement are conditioned by how accurate the results need to be and by the magnitude and speed of the movements expected.

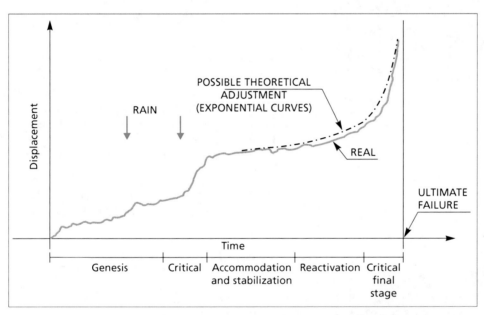

*Figure 9.69*    Stages in the evolution of a landslide and theoretical prediction of ultimate failure.

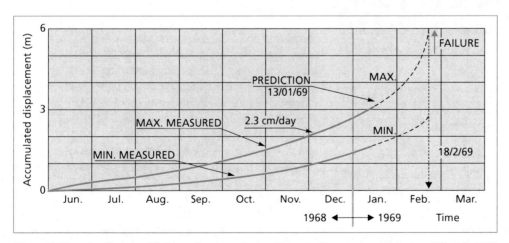

*Figure 9.70*    Prediction of failure of a slope in the Chuquicamata mine, Chile (modified from Hoek and Bray, 1981).

## Table 9.3 GEOTECHNICAL INSTRUMENTATION FOR MEASURING DISPLACEMENTS AND PRESSURES IN SLOPES

| Magnitude measured | Methods | Equipment |
| --- | --- | --- |
| Displacement between points in close proximity | Mechanical reading system. | Convergence tape. Measuring tape. Callipers. Displacement gauges |
| | Electrical reading system. | Potentiometer. LVDT. Vibrating wire. |
| Surface displacements | Geodesic methods, levelling and collimation. | Topographical surveys, DGPS. |
| Displacements inside the slope | Inclinometer. | Vibrating wire and others. |
| | Extensometer. | Wires or rods. Mechanical or electrical reading. |
| Pore water pressures | Piezometer. | Mechanical piezometers. Hydraulic piezometers. Electrical piezometers. |
| Water table | Mechanical measure of distance from ground level. | Slotted stand-pipe. |
| Pressures (anchors, bolts, retaining walls, abutments…) | Pressure cells. | Pneumatic pressure transducer. Hydraulic pressure transducer. Electrical pressure transducer. |

Movements related to **widening cracks** or movement between rock blocks are usually measured mechanically (e.g. with callipers, measuring tape, and wire), or with electrical transducers installed in the slope; for large scale displacement, a tape extensometer is used to measure convergence (see Chapter 5, Figures 5.92 and 5.93).

**Movement inside the ground** is measured with **inclinometers** and **extensometers**. As well as measuring the velocity and direction of movement, these systems allow failure surfaces to be located (Figures 5.94 to 5.98). Other methods for locating failure planes are **cased shafts** (discontinuous to allow for deformation), and the insertion of **markers in cased boreholes** to locate the "blockage" point in the failure zone (Figure 5.99). Readings from radioactive logs can give information on the density and water content of formations, which can help to define failure surfaces. Other geophysical techniques may also be useful to locate areas of weakness and discontinuity inside the slopes.

**Pore pressure** inside the slope is measured with piezometers or standpipes placed in boreholes. Control of **pressure on retaining walls** and of **rock bolt forces** is carried out by installing load cells in contact with these elements and with the slope. Both types of measurement are described in Chapter 5, Section 6 (Figures 5.100 to 5.103).

## 9.8 Slope excavation

Slope excavation is carried out either by mechanical means or by blasting. The method selected will basically depend on the rippability of the materials, the dimensions and geometry of the excavation and how efficient the machinery is.

The principal methods of excavation are:

— **Excavator shovels** and **wheel loaders**, used for direct excavation of granular or poorly consolidated materials and for subsequent loading following fragmentation using other techniques such as blasting.
— **Scrapers**, earthmoving equipment for removing horizontal layers of loose material, also used for layering material and reshaping of excavated slopes.
— **Rippers**, large caterpillar track mounted tractors equipped with scraper or ripper attachments and a shovel to push the fragmented material. The ripper uses a thrusting downwards movement to penetrate the rock and a sideways movements to loosen it. It can also make circular movements and some types of ripper are equipped with a hydraulic impact hammer.
— **Blasting**, which consists of drilling holes in very hard rocks and placing explosive charges inside them. The

## Box 9.6

### Monitoring of movements in an unstable slope

In the case of landslides like the one shown in the figure below, the displacements measured in order to identify movements and their characteristics would be:

— The opening up or separation (horizontal movement) of cracks which form at the head of the slide; measurement can be difficult if vertical displacements typical of tension cracks at the head have occurred.
— The vertical displacement of the slope crest, which gives a good idea of both absolute movement and the velocity of movement.
— Horizontal surface movements on the toe of the slide.
— Horizontal displacement, both at the surface and at depth, in the middle and lower parts of the slide.
— Relative displacements between the sliding mass and the stable area at points along the failure surface.

The first three types of movement can be measured using normal **topographical methods** by fixing reliable references on the slope (concrete blocks or stakes driven into the ground); the first type can also be measured with **tape extensometers**.

The fourth type of measurement is carried out with **inclinometers** (which must be able to reach the stable ground beneath the deepest failure surface). These are special collimated tubes inserted into boreholes along which a torpedo-shaped probe is lowered oriented in a known direction (normally in the direction of slope), which measures the vertical angle of the tube in that direction at intervals (for example every 50 cm); this multiplied by the distance to which the measure is taken will give the horizontal displacements along the borehole, with readings taken from the bottom up. The point where the failure area is crossed is usually defined by changes in horizontal displacements (from which appropriate back-analyses can be carried out); if there are significant displacements, measurement may be hindered by the tube becoming cut.

The fifth type of measurement is more complex and requires special instruments. An example of these is the **sliding micrometer**, which is of more use for rock masses than soils because of its range of measurements.

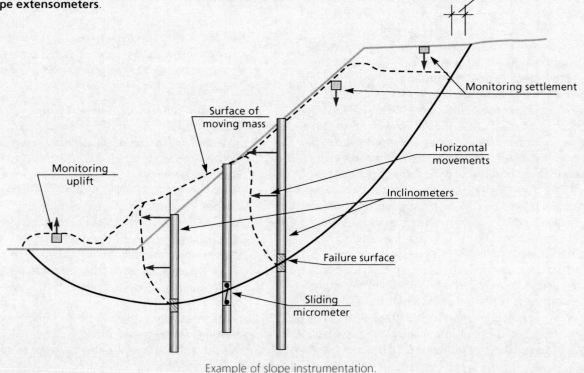

Example of slope instrumentation.

rock will either disintegrate or crack, depending on whether blasting is used for extraction or bulking of the material.

Excavation by conventional mechanical means, such as diggers and earth moving equipment, is only possible with non-cemented soils and heavily weathered rock. Otherwise, the rocks must be extracted either by ripping or blasting. In any linear engineering project involving earthmoving, the techniques applied will have a significant effect on cost, which underlines the importance of selecting the appropriate excavation method in each case.

A rock is said to be rippable when it can be excavated by mechanical means without the use of explosives. **Rippability** depends on the following rock mass parameters:

— Uniaxial compressive strength of the intact rock.
— Tensile strength.
— Seismic wave propagation velocity in the rock mass.
— Degree of fracturing of the rock mass (RQD).
— Characteristics of the discontinuities: spacing, persistence, opening, etc.
— Structure and stratification of the rock, alternating layers of different competence, etc.

It also depends on the specifications of the excavation equipment.

# Rippability criteria

Some of the different methods of estimating rippability or ease of excavation are listed below. These aspects are also dealt with in Chapter 10, Tunnels.

### Seismic wave velocity

The propagation velocity of longitudinal waves in the ground is the most representative parameter for defining its rippability and how deep it can be excavated or ripped before blasting is required. Seismic wave velocity reflects factors which affect rippability, such as how far the materials are compacted, weathered and fractured. Table 9.4 shows a general rippability criterion for materials, based on seismic wave propagation velocity. Although it depends on the power of the machinery used, as a general rule any material is rippable below 2,000 m/s; above 2,500–3,000 m/s ripping becomes extremely difficult and is very costly.

Machinery manufacturers supply ground rippability tables for the type of equipment to be used and the *in-situ* seismic wave propagation velocity measured (Figure 9.71).

### Hadjigeorgiou and Scoble excavation index

The excavation index (EI) is defined by the expression:

$$EI = (I_S + B_S) \, W \, J_S$$

where:

$I_S$ = point load strength index of the rock (PLT)
$B_S$ = block size index
$W$ = weathering index
$J_S$ = relative structural arrangement index.

Table 9.5 shows the values of the parameters that appear in the expression.

### Singh and Denby rippability index

This index is obtained from the sum of the following parameters:

— Tensile strength.
— Degree of weathering.
— Degree of abrasiveness (from the Cerchar index; see Chapter 10, Section 10.7).
— Spacing of discontinuities.

These parameters are shown in Table 9.6 with the values to calculate the index.

| Table 9.4 | GENERAL RIPPABILITY CRITERION BASED ON THE VELOCITY OF SEISMIC WAVES |
|---|---|
| **Seismic velocity (m/s)** | **Excavability** |
| <1,500 | Rocks can be excavated with scrapers, excavators and bulldozers. Blasting not necessary. |
| 1,500–2000 | Easily rippable. Excavation of strata without blasting; quite difficult for excavators or bulldozers with rippers. |
| 2,000–2,500 | Ripping expensive. Light blasting (long lengths of stemming, low specific consumption) helpful. |
| 2,500–3,000 | Light blasting required. Pre-blasting. |
| >3,000 | Significant blasting (closed blast-hole schemes), short lengths of stemming, high specific consumption). |

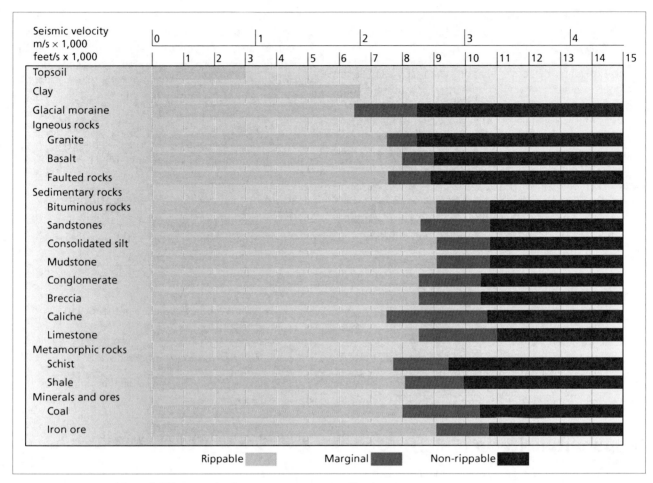

*Figure 9.71* Rippability of different rocks for a D10R-type Caterpillar ripper.

| Table 9.5 | EXCAVABILITY INDEX ACCORDING TO HADJIGEORGIOU AND SCOBLE (1990) | | | | |
|---|---|---|---|---|---|
| Class | 1 | 2 | 3 | 4 | 5 |
| Point load index $I_{s(50)}$ (MPa) <br> Rating ($I_s$) | <0.5 <br> 0 | 0.5–1.5 <br> 10 | 1.5–2.0 <br> 15 | 2.0–3.5 <br> 20 | >3.5 <br> 25 |
| Block size <br> $J_v$ (joints/m³) <br> Rating ($B_s$) | Very small <br> >30 <br> 5 | Small <br> 30–10 <br> 15 | Medium <br> 10–3 <br> 30 | Large <br> 3–1 <br> 45 | Very large <br> 1 <br> 50 |
| Weathering <br> Rating (W) | Completely <br> 0.6 | Highly <br> 0.7 | Moderately <br> 0.8 | Slightly <br> 0.9 | Unweathered <br> 1.0 |
| Relative ground structure <br> Rating ($J_s$) | Very favourable <br> 0.5 | Favourable <br> 0.7 | Slightly unfavourable <br> 1.0 | Unfavourable <br> 1.3 | Very unfavourable <br> 1.5 |
| Excavation index (EI) | <20 | 20–30 | 30–45 | 45–55 | >55 |
| Ease of excavation | Very easy | Easy | Difficult | Very difficult | Blasting |

## Table 9.6 RIPPABILITY INDEX ACCORDING TO SINGH AND DENBY (1989)

| Parameters | Classes of rock mass | | | | |
|---|---|---|---|---|---|
| | 1 | 2 | 3 | 4 | 5 |
| Tensile strength (MPa) | <2 | 2–6 | 6–10 | 10–15 | >15 |
| Rating | 0–4 | 4–8 | 8–12 | 12–16 | 16–20 |
| Weathering degree | Very high | High | Moderate | Light | None |
| Rating | 0–4 | 4–8 | 8–12 | 12–16 | 16–20 |
| Degree of abrasiveness | Very low | Low | Moderate | High | Extreme |
| Rating | 0–4 | 4–8 | 8–12 | 12–16 | 16–20 |
| Spacing of discontinuities (m) | <0.06 | 0.06–0.3 | 0.3–1 | 1–2 | >2 |
| Rating | 0–10 | 10–20 | 20–30 | 30–40 | 40–50 |
| Rippability index Total rating | <22 | 22–44 | 44–66 | 66–88 | >88 |
| Rippability | Easy | Moderate | Difficult | Marginal | Blasting |
| Recommended type of backhoe excavator | Light | Medium | Heavy | Very heavy | None |
| Power (kW) | <150 | 150–250 | 250–350 | >350 | – |
| Weight (ton) | <25 | 25–35 | 35–55 | >55 | – |

# Recommended reading

Hoek, E. and Bray, J.W. (1981). Rock slope engineering. Institution of Mining and Metallurgy. London.

Simons, N., Menzies, B. and Matthews, M. (2001). Soil and rock slope engineering. Thomas Telford. London.

Wyllie, D.C. and Mah, C.W. (2004). Rock slope engineering: civil and mining. 4th ed. Taylor and Francis. U.K.

# References

Bishop, A.W. (1955). The use of the slip circle in the stability analysis of slopes. Geotechnique, vol. 5 (1), pp. 7–17. London.

CANMET (Canada Center for Mineral and Energy Technology, (1977). The pit slope manual. Ministry of Supply and Services, Canada.

Cavers, D.S. (1981). Simple methods to analyze buckling of rock slopes. Rock Mechanics, vol. 2, pp. 87–104.

Goodman, R.E. and Bray, J.W. (1976). Toppling of rock slopes. Proceedings of the Speciality Conference on Rock Engineering for Foundations and Slopes. ASCE, vol. 2. Colorado.

Hadjigeorgiou, J. and Scoble, M. (1990). Ground characterization for assessment of case of excavation. Mine planning and equipment selection. Calgary, Canada.

Hoek, E. and Bray, J.W. (1981). Rock slope engineering. Institution of Mining and Metallurgy. London.

Jiménez Salas, J.A., De Justo, J.L. and Serrano, A.A. (1976). Geotecnia y cimientos, II. Editorial Rueda. Madrid.

John, K. (1968). Graphical stability analysis of slopes in jointed rock. Proceedings ASCE, vol. 94, no. SM2.

Lumb, P. (1975). Slope failures in Hong Kong. Ql. Jr. Engineering Geology, no. 8, pp. 31–65.

Piteau, D.R. and Martin, D.C. (1982). Mechanics of rock slope failure. Third International Conference on Stability in Surface Mining. Brawner C.O. Ed., pp. 113–171. Vancouver.

Romana, M. (1993). A geomechanical classification for slopes: Slope Mass Rating. Comprehensive rock engineering: principles, practice and projects. J.A. Hudson Ed. Volume 3: Rock testing and site characterization. Pergamon Press, Oxford, U.K.

Simons, N., Menzies, B. and Matthews, M. (2001). Soil and rock slope engineering. Thomas Telford. London.

Singh, R. and Denby, B. (1989). Aspects of ground preparation by mechanical methods in surface mining. Symposium on Surface Mining. Future concepts. Nottingham, England.

Taylor, R.D. (1948). Soil mechanics. John Wiley & Sons. New York.

Wyllie, D.C. and Mah, C.W. (2004). Rock slope engineering: civil and mining. 4th ed. Taylor and Francis. U.K.

# 10
## TUNNELS

1. Introduction
2. Site investigation
3. Influence of geological conditions
4. Geomechanical design parameters
5. Rock mass classifications for tunnelling
6. Tunnel support design using rock mass classifications
7. Excavability
8. Tunnel excavation and support methods in rock
9. Tunnel excavation and support methods in soil
10. Geological engineering during tunnel construction

# 10.1 Introduction

The use of underground space provides one of the best solutions for the development of rapid communication routes. Although the cost is greater than for surface alternatives, there are important advantages from an environmental and functional point of view: distances are shortened, safety is enhanced and there is less environmental impact.

Tunnels are usually constructed to overcome natural obstacles in transportation systems, mainly roads and railways, to provide access for underground transport in urban transit systems, to transfer water from one area to another, and to link islands or to cross straits and rivers, in which case they are excavated under water. Underground excavations are also closely related to energy production (hydroelectric and thermal power stations), mineral and rock extraction and underground storage facilities (*Figure 10.1*).

Although there is a wide variety of uses for underground space, most tunnels are excavated to accommodate transport infrastructure and their design is therefore based on **safety** and **economy**. This chapter deals mainly with this type of excavations and in particular with rock tunnelling.

A tunnel is characterized by the longitudinal profile along its central line and by its cross section. These are defined by geometric criteria, such as slope, rail gauge and radius of curvature besides other engineering requirements. For geological engineering purposes, the most relevant data

*Figure 10.1* Excavation of an underground power station.

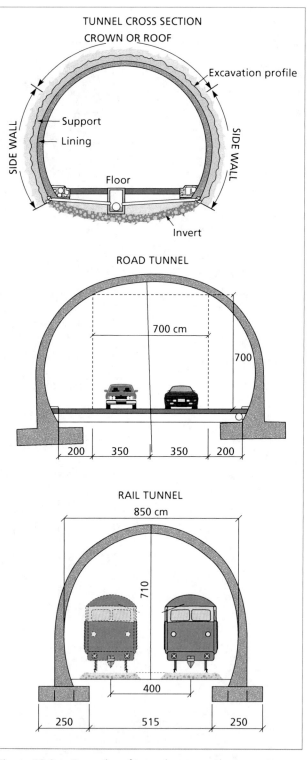

*Figure 10.2* Examples of tunnel cross sections.

refer to the cross section and the longitudinal profile, as well as to particular areas like portals and access adits. *Figure 10.2* shows typical tunnel **cross sections** with the names of the different parts and what they are used for.

Tunnel **support** refers to the structural ground support elements installed immediately after excavation to guarantee stability and safety both during and after construction.

Once the support is installed, the **lining** is applied. This usually consists of a layer of concrete or some other structural element, which provides the tunnel with long-term strength and an even finish. The lining improves aesthetic and functional conditions, as well as aerodynamics, watertightness and luminosity.

Geological and geomechanical surveys are essential where any underground structure is involved, during both the design and construction stages. This chapter describes the **basic methodology** and **objectives** of these studies (see *Figure 10.3*). The main objectives of geological and geomechanical tunnel investigations are to:

— Understand the geological, geomechanical and hydrogeological conditions.
— Identify any peculiarities or areas of greater geological complexity.
— Define geomechanical properties and zonation.
— Establish geomechanical criteria for excavation methods and support design.
— Specify geological and geomechanical conditions for the design of portals and intermediate access.
— Recommend ground treatment for stabilization, reinforcement, drainage or waterproofing.
— Provide monitoring and control during excavation.

## 10.2 Site investigation

In any tunnelling project, a thorough knowledge of the geology of the area is essential for adequate planning and to avoid extra costs, accidents and unforeseen delays. As Terzaghi pointed out, **the difficulty and cost of an underground excavation are determined by the geology, more than by any other factor**.

Geological surveys for tunnels are generally more costly than those for other civil engineering projects. However, the allocation of insufficient resources to this type of investigation may well lead to unforeseen consequences: **ground which has not been studied is risky ground**. Adequate investment in geological engineering investigation may account for as much as three percent of the estimated cost of the construction

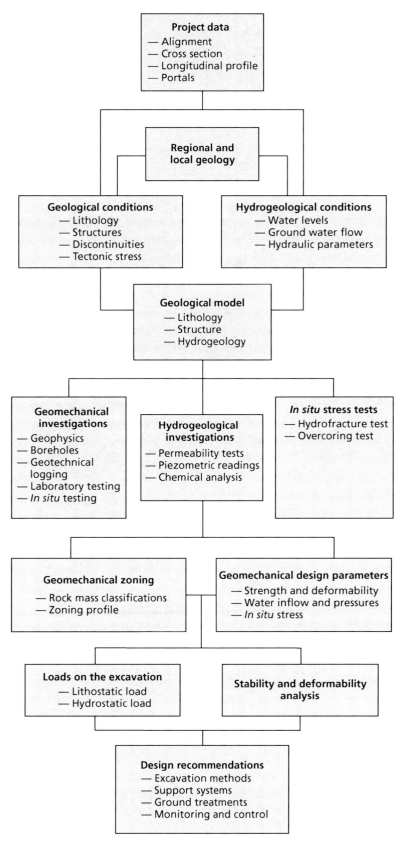

Figure 10.3   A procedure for engineering geological investigations for tunnels.

work and will depend on factors such as geological complexity, tunnel length and overburden thickness. Below this percentage, the number of tunnels where problems occur increases, while above it, unforeseen events are kept to a minimum (Waggoner and Daugharty, 1985).

Because of their technical and economic importance, it is essential that site investigations should be properly planned. **Basic planning criteria** include:

— **Prior information:**
  - Geological description based on previous studies and bibliography.
  - Project information.
  - Access for site investigations.
  - Budget and delivery time.

— **Strategies to follow:**
  - Site investigations should be carried out in successive stages in areas that are significant for:
    - Geological interpretation.
    - Identification of particular problems: complex geological, hydrogeological or geotechnical conditions
    - Portal and access areas.
  - Different site survey techniques should be selected and combined as appropriate, using criteria based on the extent to which they provide a representative picture of the ground using, complementary, cost-effective techniques that suit the logistics of the scheme (accesses, distances and supply routes).
  - Site investigation methods should be in accordance with the aims and scope of the project throughout its different phases (*Table 10.1* and *Box 10.1*). Geological problems that may point to a hazard or complication should be identified at the very early stages as they may condition, or even preclude, tunnel construction.

One of the most important tasks when planning a site investigation is to decide on **boreholes sites**. There are no general rules, although certain criteria should be taken into account:

— Boreholes should cross as many of the geological formations present in the tunnel area as possible and should be strategically located so that their length is kept to a minimum.
— To obtain the maximum amount of information, they should be positioned where geological uncertainties, including hydrogeological problems, are greatest and bored as perpendicular to the geological structures as possible.
— Where feasible, they should extend below the projected invert elevation for the tunnel by a distance at least equal to its diameter.
— Boreholes should be planned to allow different *in situ* tests to be carried out inside them, such as permeability, stress-deformation and geophysical tests.
— There should be a sufficient number of boreholes situated at the portals and intermediate access points to identify and study problems related to stability, seepage and thickness of weathered zones. The number of boreholes will depend on the complexity of the area but as a guideline a minimum of 5 boreholes at each portal zone is recommended (*Box 10.1*).

The most usual **geophysical investigations** include: seismic refraction in areas with a thin overburden, down-hole and cross-hole tests, borehole logging, and seismic reflection for geological structures and in areas where tunnels are deep. The **hydrogeological tests** most often used are Lugeon and Lefranc-type permeability tests, as well as piezometric measurements. **Selection of *in situ* testing** to use depends on the type of material to be investigated and the type of geotechnical problem to be solved. Pressuremetre, dilatometre and hydrofracture tests are some of the most common to be employed.

## 10.3 Influence of geological conditions

When a tunnel is excavated, three types of **geological conditions** may be found that cause loss of strength and **stability problems** in a rock mass (*Figure 10.4*):

— Unfavourable orientation of discontinuities.
— Unfavourable orientation of stress with respect to the tunnel orientation.
— Water flowing inside the excavation along fractures, aquifers or karstified rocks.

These conditions are directly related to geological factors, such as structure, discontinuities, intact rock strength, hydrogeological conditions and the state of stress. In addition, tunnel excavation also generates the following **induced effects**:

— Loss of strength in the rock mass surrounding the excavation as a result of stress relief, opening up of discontinuities, fissuring due to blasting, alterations, and water flowing towards the inside of the excavation.

**Table 10.1  SITE INVESTIGATION PROCEDURES FOR TUNNEL DESIGN**

| Stages and objectives | Tasks | Activities and techniques |
|---|---|---|
| **Preliminary studies**<br><br>• Geological survey of tunnel alignments or corridors<br>• Identification of possible geological hazards for tunnel excavation<br>• Basic geological and geomechanical assessment<br>• Alternative tunnel alignments or corridors<br>• Site investigation planning | Information review | — Topography and geomorphology<br>— Hydrology and hydrogeology<br>— Geological maps<br>— Inventory of tunnels and mines located nearby<br>— Seismicity and other geo-hazards |
| | Photo-interpretation | — Aerial photographs and special techniques for areas covered with vegetation<br>— Remote sensing techniques |
| | Geological survey | — Geomorphology and slope stability<br>— Lithologies<br>— Faults and tectonic structures<br>— Hydrogeological data |
| | Preliminary site investigations | — Boreholes<br>— Surface geophysics |
| | Engineering geological interpretation | — Geological maps and profiles (1:10,000–1:2,000) |
| **Design Stage**<br><br>• Selection of tunnel alignment and portal areas<br>• Detailed geological and geomechanical site investigations<br>• Evaluation of geological and geotechnical construction problems<br>• Geomechanical characteristics of materials<br>• Geomechanical design criteria<br>• Recommendations for support, excavation method and ground improvement treatments | Engineering geological mapping | — Geomorphology<br>— Lithostratigraphy<br>— Structure<br>— Geomechanical characterization<br>— Maps on scale 1:2,000–1:500 |
| | Hydrological and hydrogeological data | — Regional and local scale data collection<br>— Estimation of water flow rates and pressures |
| | Detailed site investigations | — Laboratory tests<br>— Boreholes<br>— Trial pits<br>— Geophysics<br>— *In situ* tests |
| | Geological-geotechnical and hydrogeological interpretation | — Hydrogeological conditions<br>— Geomechanical properties<br>— Geomechanical classifications<br>— Recommendations for support and excavation<br>— Ground treatments |
| **Construction Stage**<br><br>• Geological and geotechnical monitoring and control<br>• Support adjustment to ground conditions<br>• Ground treatment to control instabilitiy and seepage | Monitoring | — Geotechnical instrumentation |
| | Geomechanical control | — Geological-geotechnical mapping of the tunnel front and sidewalls<br>— Boreholes, geophysics and *in situ* tests inside the tunnel excavation. Exploratory adits |
| | Quality control | — *In situ* and laboratory tests |
| | Expert assistance | — Monitoring and control interpretation<br>— On site solutions to reinforcement, seepage control and ground deformation |

## Box 10.1

### Guidelines for planning site investigation in tunnels

**1. PHOTO-INTERPRETATION**

— Scales 1:25,000 to 1:10,000.
— Colour photographs recommended.
— Radar images in areas of dense vegetation.
— Thermal images in faulted areas with water.
— Infra-red images for establishing the limits of geological features that are poorly defined in other types of photos.

**Comments:**

— Low cost for conventional aerial photographs.
— Field verification required.
— Some limitations in areas covered with vegetation.

**2. GEOLOGICAL AND GEOTECHNICAL MAPS**

— Scale for preliminary studies: 1:10,000–1:2,000.
— Design scale: 1:2,000–1:500.
— The geological longitudinal profile along the tunnel axis is essential geological information; sections across the tunnel axis can also be relevant.

**Comments:**

— Essential and most valuable information.
— Very high benefit-cost ratio.
— Subject to uncertainty depending on geological complexity and availability of data.

**3. GEOMECHANICAL FIELD DATA**

— Degree of weathering of the rock mass.
— Structural analysis of discontinuities.
— Jointing characterization.
— Index tests.
— Hydrogeological data.

**Comments:**

— Statistically representative data required.
— Limitations in areas covered by vegetation, soils or heavily weathered rock and inaccessible areas.

**4. TUNNEL INVENTORY**

— The purpose of the Inventory is to understand the behaviour of other tunnels in the same region that have been excavated in similar materials and conditions to those of the designed tunnel.
— Data to be included: geological cross section, geological structure and geomechanical rock mass classification.
— Tunnel cross section, supports installed and construction and maintenance problems related to the prevailing ground conditions.

**Comments:**

— Enables a back-analysis of geomechanical behaviour, stability and support.
— May serve to warn about existing problems as well as provide information for future experiences.
— Information depends on the accessibility and availability of data.

**5. GEOPHYSICAL METHODS**

— **Electrical:** identification of fractures, aquifers and lithological contacts.
— **Seismic refraction, down-hole and cross-hole:** contact between fresh and weathered rock, rippability, deformation moduli, jointing degree.
— **Diagraphs** inside boreholes in order to obtain *in situ* properties such as density, porosity, sound velocity, jointing degree, etc.
— **Seismic reflection:** allows the geological structure to be studied at depth, faults, folds, contacts, etc.

**Observations:**

— The results have to be verified with other field data from boreholes and tests, and require appropriate geological interpretation.
— Seismic reflection can be very costly.

| 6. BOREHOLES | Observations: |
|---|---|
| **Objectives:** <br>— To investigate the geological structure and materials and zones of complex geology. <br>— To obtain data on fracturing of the rock mass and geotechnical core logging. <br>— To obtain samples for test purposes. <br>— To carry out hydrogeological measurements and tests inside boreholes. <br><br>**Appropriate equipment:** <br>— Rotary drills. Minimum diameter NX. <br>— Double or triple core sample tubes; wire-line for deep boreholes. <br>— Equipment for measuring deviation in deep boreholes. <br><br>**Geotechnical logging:** <br>— Standard description of borehole samples. <br><br>**Borehole location:** <br>— In portals, accesses, areas with complex geology and systematically along the tunnel axis. <br><br>**Number:** depends on geological complexity, thickness of the overburden, accesses and costs. Some general guidelines are: <br>— 1 borehole every 50 to 100 metres of tunnel length in areas with complex geology or very variable lithology. <br>— 1 borehole every 100 to 200 metres in more uniform geological areas. <br>— A minimum of 5 boreholes in portal areas. <br>— In tunnels more than 1,000 m long, the total drilling length should be at least 50% of the tunnel length. <br>— In tunnels of less than 1,000 m: 75% of their length. <br>— In tunnels of less than 500 m: 100% of their length. <br><br>**Inclination:** in general, inclined boreholes are better than vertical ones; it is important to measure deviations in deeper boreholes. | — Direct method for obtaining rock samples. <br>— Allows geotechnical and hydrogeological testing to be carried out. <br>— The procedure is slow and costly. <br>— Accesses may be an important conditioning factor to take into account. |
| | **7. *IN SITU* TESTS INSIDE BOREHOLES** |
| | — Observation of discontinuities and cavities: TV cameras, core orientation, callipers, dipmeters, etc. <br>— Deformability: dilatometer and pressuremeter tests, and cross-hole and down-hole tests. <br>— *In situ* stress: hydrofracture tests. <br>— Permeability: Lugeon and Lefranc tests. <br>— Water pressures: piezometers <br>— Water sampling for chemical analysis. <br><br>**Observations:** <br>— Of great use in soft rocks, jointed rock masses or if poor core recovery. <br>— Some of these techniques are sophisticated and costly. |
| | **8. ADITS AND PILOT TUNNELS** |
| | — Used in long tunnels under and complex geological conditions. <br><br>**Observations:** <br>— Allow direct observation; *in situ* tests and drainage measurements before tunnel excavation. <br>— Facilitate monitoring and the implementation of ground improvement treatments both prior to and during excavation. <br>— Slow and very costly. |

— Reorientation of stress fields, creating *in situ* stress changes.
— Other effects including surface subsidence, slope movements and changes in aquifer conditions.

The response of the rock mass to these geological and induced effects determines tunnel stability and consequently the support measures to be adopted. The construction procedure also depends on the excavability of the rock, which is related to its strength, hardness and abrasiveness.

# Geological structure

One of the most influential factors in the stability of an underground excavation is **geological structure**. For folded and stratified rocks, different modes of ground behaviour and thus tunnel stability are conditioned by the strata orientation, including the following factors:

— Dip of the geological structure with respect to the tunnel cross section.

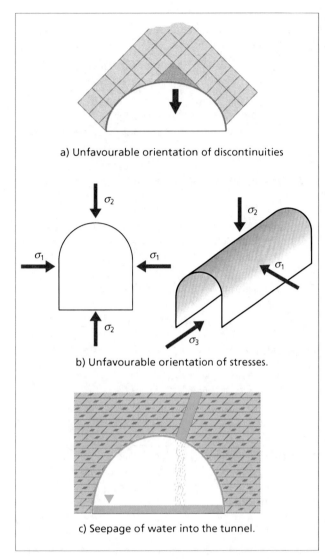

*Figure 10.4* Geological conditions affecting the stability of rock tunnel excavations.

— Strike of the bedding planes with respect to the tunnel axis.
— Folding and types of folds.

*Figure 10.5* shows the influence of geological structure on tunnel stability. In general terms, if the orientation of the geological structure is parallel to the direction of the tunnel axis, stability conditions are unfavourable.

## Discontinuities

Most problems involving stability are due to the tunnel intersecting discontinuity planes (*Figure 10.6*). A distinction is made between systematic and single discontinuities. The surfaces created by joints, bedding planes and schistosity belong to the first group. Systematic discontinuities are present in practically all rock masses, with a greater incidence in shallower zones where processes of weathering are more frequently found, and where the aperture between such discontinuities may also permit the flow of water and contain clay infilling. Because of the effects of confining pressure, apertures in discontinuities are smaller at great depths and may be tightly closed.

With regard to stability one of the most important types of discontinuities are **faults.** These are geological structures that have undergone failure and deformation throughout their geological history and their strength is therefore very low; the presence of faults and mylonitic fills provides preferred flow paths for water circulation. Faults may also accumulate significant tectonic stresses.

Faults are one of the most important geological features to be investigated for tunnelling design purposes. The following subjects should be studied:

— Faulting in relation to regional and local tectonic structure.
— Faulting within the structural geological analysis.
— Identification and classification of faults, based on their origin, age, type and geometry.
— Identification of fault fills, together with their strength and potential to swell
— The influence of faults on groundwater flow; whether they act as barriers to flow or as drains and conduits for flow.
— Influence of faults present on *in situ* stress and seismicity, if applicable.

The study of systematic-type discontinuities should include all aspects dealt with in Chapters 3 and 6 (Section 3.5 and 6.4 respectively). This means that representative structural data must be available in sufficient quantity to carry out statistical analysis. The most commonly used methods of representation and analysis are stereographic projections and rose diagrams, which can be supplemented with block diagrams or cross-sections (*Figure 10.7*).

The **effect of faults** on the stability of an excavation depends on their characteristics, which can be classified as follows:

— Faults that have mechanical contact between different materials and/or have more than one shear surface.
— Faults that contain a zone of variable thickness and/or weak, unstable, plastic or swelling materials.
— Faults that are highly permeable.

The influence of faults on the stability of an excavation also depends on their orientation and their intersection with the tunnel. They are very persistent planes that may cross the

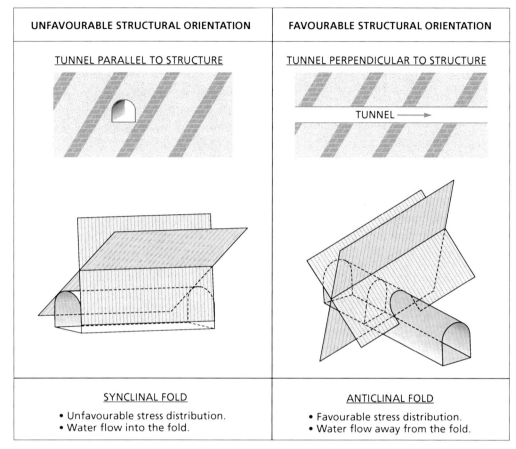

*Figure 10.5*   Influence of geological structure on tunnel stability.

entire tunnel section and intersect systematic-type discontinuities to produce unstable wedges or blocks of considerable size. As their strength is much lower than that of other discontinuities, and as they may be subjected to high pore pressures and/or tectonic stresses they can constitute principal failure planes.

In addition to these considerations, when a fault contains breccia zones or low strength fill material, the filling itself constitutes a zone or plane of failure. Various types of stability-related behaviour have been established which depend on the sort of filling, as shown in *Figure 10.8*.

**Thrust faults** are a particular case of faults characterized by shear surfaces with a very low dip, low strength and great persistence.

When a fault constitutes a preferred flow path for water, different types of behaviour are generated with respect to stability. Depending on the difference in permeability between the materials at each side of the fault and the nature of the filling material, the fault may form either a barrier to water flow or a preferred flow path. In addition to seepage, the presence of weak, karstified or loose materials may cause considerable instability and can give rise to rock falls.

## Intact rock strength

The strength of the intact rock plays a decisive role in the choice of method used for excavation and has a strong bearing on stability.

Three conditions of stability can be distinguished from the **competence factor** $F_c = \sigma_{ci}/\sigma_V$ (where $\sigma_{ci}$ is intact rock strength and $\sigma_V$ is maximum vertical stress):

— $F_c > 10$: the intact rock has a much greater strength than the vertical stress in the rock mass and the excavation is likely to be stable.
— $10 > F_c > 2$: stability is conditioned by time and rock properties; three types of deformation can be identified: elastic, plastic and brittle failure including the risk of rock burst.
— $F_c < 2$: the excavation may be unstable as stresses exceed intact rock strength.

Estimating stability from $F_c$ does not take into account the presence of discontinuities and high horizontal stress. Although not common, this situation may occur in very

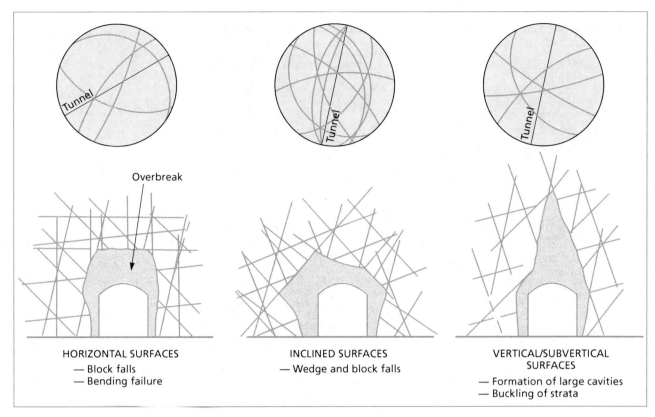

Figure 10.6   Influence of discontinuities on tunnel stability: examples of overbreak.

homogeneous rock masses, e.g. massive crystalline rock and rock salt, or in rock at great depth where the discontinuities are tightly closed.

## Hydrogeological conditions

Excavating a tunnel creates an effect similar to a large drain towards which water will flow from aquifers intercepted during the excavation process. The consequences of this can include:

— Reduction in rock mass strength.
— Increase in pore pressure on the support and lining.
— Swelling and softening in clay materials.
— Rapid formation of cavities in soluble materials such as salt.
— Tunnel construction problems.

Transmissivity in rock masses takes place preferably along fractures, or larger channels in the case of karstified rock. In soils, permeability is closely related to grain size and the presence of particular sedimentary structures, e.g. paleo-channels. The main causes of seepage in rock masses are:

— Faults and fractures.
— Breccia, fault fillings, and highly weathered zones.
— Lithological contacts between rocks with very different permeability.
— Karstic channels and cavities, and volcanic tubes. Seepage in karstic cavities may be an important hazard, and difficult to locate.

In practice, it is not easy to evaluate water pressures and discharge rate along a tunnel, particularly in karstic formations or highly fractured rock; their heterogeneity makes it difficult to assign values for permeability or predict their hydrodynamic behaviour. Mathematical flow models often require data that are not usually available and so simplifications and empirical estimates are used (see *Box 10.3* and Chapter 4).

**Hydrogeological investigation in tunnels** should include the following aspects:

— Water balance in the area of influence of the tunnel, including an inventory of water sources, springs and wells.
— Identification of karstic areas, fractures, faults and highly permeable rocks.
— Piezometric heads in the tunnel area and their seasonal changes.
— Aquifer characterization.
— Hydraulic parameters of the aquifers.

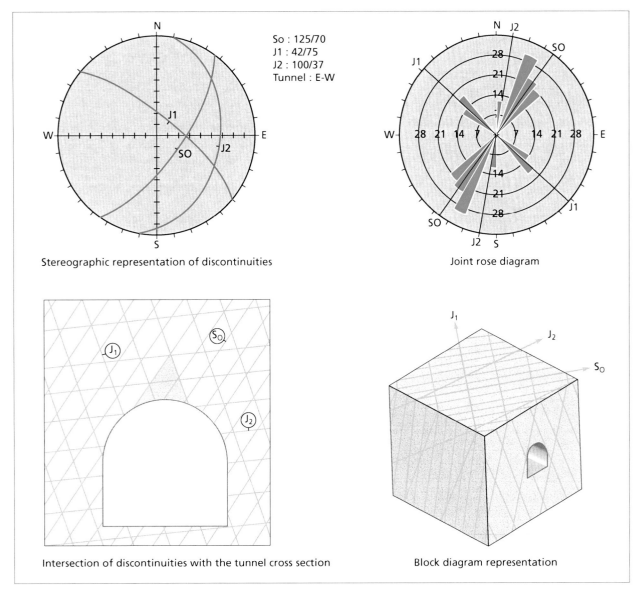

*Figure 10.7* Graphical analysis of discontinuities.

- Flow models.
- Estimation of expected discharges and water pressures inside the tunnel excavation.
- Assessment of high seepage zones.
- The chemical content, quality, temperature and aggressivity of the water.

The results of hydrogeological investigations provide the basic information needed for adopting measures to control water inflow into the excavation and prevent other adverse environmental effects. In particular, the following should be defined:

- Discharges and water pressures throughout the tunnel.
- Drainage and waterproofing measures.
- Influence on changes in aquifers, water chemistry and pollution, and on settlement and subsidence of nearby buildings.

## *In situ* stress

There are two types of *in situ* stress acting on underground excavations: **natural** and **induced**. The first type corresponds to the natural state of stress that results mainly from gravitational and tectonic forces, while the second refers to the redistribution of stress around the excavation as a consequence of the excavation process.

In underground excavation design, the magnitude and direction of *in situ* stresses must be known to assess

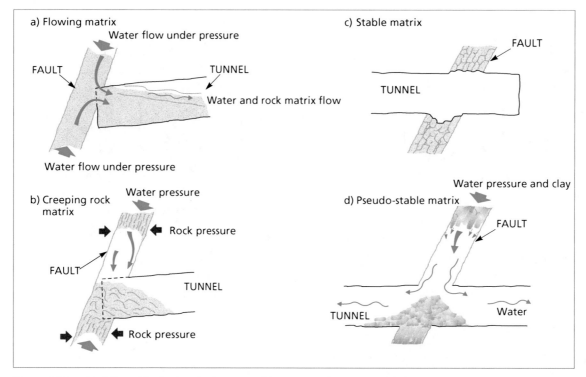

*Figure 10.8* Stability of faulted rocks in a tunnel (Hansen and Martna, 1988).

tunnel stability, tunnel support, cross-section geometry and the construction process. High *in situ* stresses can induce rock burst, rock squeezing and high deformation phenomena. This type of problem can be costly and difficult to resolve. The conceptual aspects of *in situ* stress and the methods used to measure it are described in Chapter 3, Section 3.7.

## Methods of analysis

For tunnel design purposes, the following procedure can be used for *in situ* stress assessment:

1. **Regional tectonic stress:** stress directions can be assessed with respect to the prevailing tectonic regime using published stress direction maps and geological reasoning (Chapter 3, Section 3.7).
2. **Local tectonic structures:** the regional stress field may be locally modified by local geological structures, such as dykes, intrusive bodies, faults, folds, diapirs and volcanoes. Tectonic structures that may accumulate high stress should be identified, as well as other geological anisotropies that may change the direction and magnitude of *in situ* stress.
3. **Stress of gravitational origin:** the stress field is influenced by the effects of erosion and sediment loading. Erosion may generate horizontal stresses higher than the vertical stresses and changes in stress direction. Chapter 3, Section 3.7 shows some of the most typical situations where this occurs. Abrupt changes in relief can also produce significant horizontal stress, although their effects quickly disappear away from its area of influence. These situations are very significant in portal zones and tunnels with a thin overburden.
4. **Estimating the state of stress through geological methods:** geological analyses can provide stress directions, not magnitude. Current stress directions may be indicated by focal mechanisms of earthquakes. *Figure 10.9* shows principal tectonic stress directions in western Europe.
5. **Estimating the state of stress through empirical methods:** using empirical data, a value of maximum vertical stress $\sigma_V = 0.027$ MPa/m can be estimated in areas of smooth topography, where the principal stresses are vertical and horizontal; as a first estimate, maximum horizontal stress $\sigma_H$ can be considered approximately equal to $\sigma_V$ below a depth of 1,000 m, while at shallower depths the value of $\sigma_H$ may exceed that of $\sigma_V$ due to tectonic and gravitational forces.

   The ratio $\sigma_H/\sigma_V$ could be considered equal to $\nu/(1 - \nu)$, ($\nu$ being the Poisson coefficient) under the following circumstances:

   — Lightly deformed sedimentary rocks in which faults, folds and significant tectonic structures are either almost absent or not significant.
   — Rocks of evaporitic, bioclastic and extrusive volcanic origin, except in areas with structures

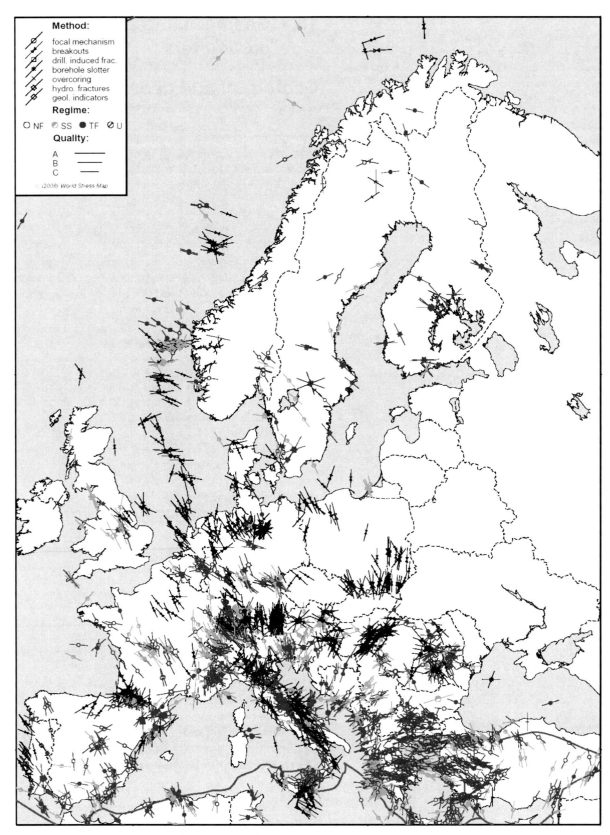

*Figure 10.9* Europe stress map (Heidbach *et al.*, 2008).

or processes that are diapiric, or involve squeezing or intrusion.
— Weak rocks, marls and clays except in areas with stress induced by swelling or squeezing.
— Tunnels with a thin overburden, e.g. less than 50 m.

If the area under study has been subjected to significant tectonic stress, the state of stress in terms of $K$ ($\sigma_H/\sigma_V$) can be estimated from the TSI index described by González de Vallejo and Hijazo (2008). Alternatively, Sheorey's method can be applied if the state of stress is controlled by gravitational forces and depends on rock elasticity. A description of both methods is included in Section 10.4.

6. *In situ* tests: *in situ* tests are carried out in boreholes to measure magnitudes and stress directions as described in Section 3.7 of Chapter 3, although they usually require costly and highly specialized techniques. They are the best methods to determine the *in situ* stress.

## Effects of high stress on tunnelling

Tunnels or mines affected by high *in situ* stress may be at risk of collapse, rock burst or considerable deformation (*Figure 10.35*). Depending on the ratio between the stress magnitude and uniaxial strength of the rock $\sigma_{ci}$, the behaviour of the rock mass in response to stress may be as follows:

— Elastic deformation with brittle-type failure, characteristic of hard rock, generally in deep tunnels, with risk of **rock bursts** and **spalling.**
— Plastic deformation typical of weak rocks, with risk of squeezing and other types of large deformations.

For further reading on tunnels in overstressed rocks it is referred to Hoek and Marinos (2009).

**Rock bursts** may cause accidents. In such cases, special measures of support are needed, so it is important to anticipate these situations. One of the empirical criteria for evaluating stability in deep tunnels with high rock strength was proposed by Hoek and Brown (1980):

$\sigma_V/\sigma_{ci} = 0.1$: stable excavation.
$\sigma_V/\sigma_{ci} = 0.2 - 0.3$: spalling hazard.
$\sigma_V/\sigma_{ci} > 0.5$: rock burst hazard.

**Squeezing** can be assessed from the empirical expression proposed by Sing *et al.* (1992):

$$H > 350\, Q^{1/3}$$

where $H$ is the tunnel depth (m) and $Q$ the quality index, described in Section 10.5.

Ground squeezing in tunnels can also be evaluated using the criteria of Hoek and Marinos (2000). Excessive squeezing is a hazard.

# 10.4 Geomechanical design parameters

## Geological and geomechanical data

Underground excavations require sound geological and geomechanical knowledge to design the support, their excavation method to use and the treatment for suitable ground improvement. The following data are usually needed:

### Project data

— A topographical profile and plan of the tunnel along its alignment.
— A cross-section of the excavation, including portal location, intermediate accesses and any other excavation or construction, either intended or existing, close to the tunnel.

### Geological and geotechnical data

— Geological structure of the area to be crossed, including, lithology, faults and other large discontinuities, systematic joints, geological maps, longitudinal profiles and cross sections.
— Hydrogeological conditions, rock mass permeability and a model of ground water flow.
— Geomechanical properties of the intact rock, discontinuities and rock mass.
— The magnitude and direction of *in situ* stress.
— Rock mass classifications along the tunnel longitudinal profile.

### Geological singularities

— The position of faults and tectonized zones.
— The presence of tectonic structures that are likely to result in highly anisotropic behaviour or high *in situ* stress conditions.
— The existence of weak, expansive, squeezing and swelling materials.
— The location of aggressive and abrasive materials.
— The thickness and location of zones of potentially high seepage rates and large inflows of water.
— The possible presence of gases or high temperatures.

### Portals and tunnel access areas

Portal areas must be investigated in great detail. Low overburden thickness means a portal usually enters a zone of high weathering and high permeability with an increase in water flow. These factors can cause greater deformability and lower strength than in the rest of the tunnel. Landslides and slope instability are the main potential hazards to be considered in any portal area (see Section 10.8).

## Graphical presentation of geological and geomechanical data

The following data should be a presented graphically:

— Geological maps as at ground level and at tunnel elevation.
— Geological cross sections and longitudinal profiles along the tunnel axis.
— Geomechanical zoning along the tunnel axis, including the following data (Figure 10.10).

- Lithology, faults, fractured zones and geological structures.
- Highly permeable materials and potential seepage zones.
- Rock mass classifications.
- Strength and deformability properties.

# Strength and deformability

Direct and empirical methods for calculating strength and deformability in rock masses are described in Chapters 3 and 6. These methods can be summarized as follows:

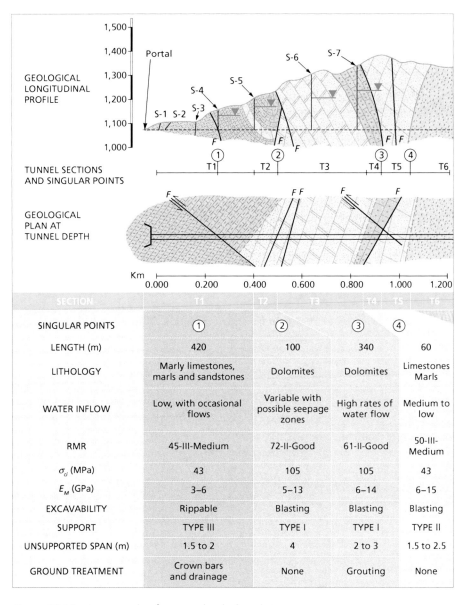

| SECTION | T1 | T2 | T3 | T4 | T5 | T6 |
|---|---|---|---|---|---|---|
| SINGULAR POINTS | ① | ② | | ③ | ④ | |
| LENGTH (m) | 420 | 100 | | 340 | 60 | |
| LITHOLOGY | Marly limestones, marls and sandstones | Dolomites | | Dolomites | Limestones Marls | |
| WATER INFLOW | Low, with occasional flows | Variable with possible seepage zones | | High rates of water flow | Medium to low | |
| RMR | 45-III-Medium | 72-II-Good | | 61-II-Good | 50-III-Medium | |
| $\sigma_{ci}$ (MPa) | 43 | 105 | | 105 | 43 | |
| $E_M$ (GPa) | 3–6 | 5–13 | | 6–14 | 6–15 | |
| EXCAVABILITY | Rippable | Blasting | | Blasting | Blasting | |
| SUPPORT | TYPE III | TYPE I | | TYPE I | TYPE II | |
| UNSUPPORTED SPAN (m) | 1.5 to 2 | 4 | | 2 to 3 | 1.5 to 2.5 | |
| GROUND TREATMENT | Crown bars and drainage | None | | Grouting | None | |

Figure 10.10 An example of geomechanical zoning.

### Intact rock strength

- Uniaxial compressive strength tests, tensile strength tests and triaxial tests.
- Hoek and Brown criterion.

### Shear strength of discontinuities

- Shear strength tests and tilt tests.
- Mohr-Coulomb criterion.
- Barton and Choubey method.

### Rock mass strength

- Hoek and Brown criterion.
- Mohr-Coulomb criterion.

### Rock mass deformability

Rock mass deformability is one of the most complex parameters to evaluate, given the heterogeneity and anisotropy that characterize rock masses. Some methods for evaluating deformability are described in Chapter 3, section 3.6 and Chapter 5, section 5.5. They include the following methods:

- *In situ* tests and geophysical methods.
- Correlations between the rock mass deformability modulus, dynamic modulus and RQD.
- Empirical methods using the RMR, Q and GSI indexes.

*Box 10.2* shows an example of how the strength and deformability of a rock mass are calculated to obtain the geomechanical parameters.

## Magnitude and direction of *in situ* stress

An assessment of the state of stress due to tectonic or gravitational forces is described in Section 10.3. If either the tectonic or gravitational stresses are likely to be high, the following procedures can be used to estimate *in situ* stress:

1. Geological criteria to identify potentially high tectonic or gravitational stresses.
2. Empirical estimations of the value of $K = \sigma_H/\sigma_V$ using the TSI index procedure or Sheorey's method, described below.
3. Estimates of stress direction from applying geological methods (Chapter 3, section 3.7).
4. Hydrofracture tests or other procedures, as described in Chapter 3, section 3.7 and ISMR (2003), to obtain the magnitude and direction of *in situ* stress.

### Estimation of *K* from the TSI index

A procedure for estimating the ratio between *in situ* stresses in rock masses by means of $K$ ($K = \sigma_H/\sigma_V$ where $\sigma_V$ and $\sigma_H$ are the principal stresses), has been proposed by González de Vallejo and Hijazo (2008). The procedure combines the use of the TSI (Tectonic Stress Index) and the probabilistic logic tree method. It provides an assessment of the magnitude of horizontal stress of tectonic origin. The TSI is defined according to the expression:

$$TSI = \log\left(\frac{T}{E \cdot H}\right) NC \cdot SC$$

where:

T = age of the main orogenic period affecting the rock mass (years)
E = elastic modulus of the intact rock (GPa)
H = maximum overburden thickness throughout its geological history (metres)
NC = coefficient of seismotectonic activity
SC = coefficient of topographic influence

The age (T) of the main folding or orogenic period is related to the time elapsed since the main deformation of the rock occurred. This time is one of the factors contributing to the state of stress. It is assumed that the main orogenic period corresponds to peak tectonic stresses and that subsequent to this deformation, tectonic stresses decrease with time. T refers to Hercynian (250–300 M.a.) and Alpine (10–12 M.a.) orogenies, although the Caledonian orogeny (600 M.a.) is also considered. Where a single rock mass is affected by several orogenies, only the oldest of these is considered.

The E parameter is related to the petrophysical properties of the rocks along with processes of diagenesis, compacting, lithification, deformation and recrystallization. High values of E generally indicate high strength rocks able to sustain high stresses before they deform or break. Values of E lower than 25 GPa are not valid for applying the TSI index. If the same rock mass shows several lithologies, the highest E value is considered.

The H parameter indicates the maximum lithostatic load to which the rock has been subjected over its geological history. For sedimentary rocks, H is equivalent to the maximum thickness of sedimentary deposits within a margin of error lower than ±500 m. A rough estimate of H can be made from the difference between the highest elevations in the area and the site elevation. This is applicable to horizontal sedimentary rocks. Moreover, the tectonic structure where the area under study is situated should also be taken into account, since in regions with overthrusts or inverted folds part of the stratigraphic sequence may be repeated, increasing the value of H.

For igneous rocks, H corresponds to the depth at which the rocks were situated and acquired their elastic properties. This depth is estimated within a margin of error of less than ±1,000 m. For extrusive, or volcanic igneous rocks, H is determined as for sedimentary rocks, that is, calculating the current stratigraphic column taking into account possible thicknesses

## Box 10.2

## Calculation of the geomechanical parameters of a rock mass for tunnel design purposes

Calculate strength, deformability, water pressure and *in situ* stress with the object of designing a tunnel support, applying the methods described in Chapter 3.

The tunnel under consideration crosses a Tertiary sandstone formation with a practically horizontal dip.

— Rock mass data: RMR = 80, GSI = 75, RQD = 90, JRC (bedding planes) = 13; $\sigma_v$ is considered = 1 MPa, equivalent to a depth of 38 m.
— Data from laboratory tests on the intact rock: $\gamma$ = 26.7 kN/m³, $\sigma_{ci}$ = 85 MPa, $E_i$ = 40 GPa and $v$ = 0.24; from triaxial tests values of $\phi_i$ = 70° and $c_i$ = 2.9 MPa were obtained.
— Measurements with the Schmidt hammer were (average value): R = 40 (intact rock) and r = 20 (discontinuity wall).
— The rock mass is dry.

### Intact rock strength

- Hoek and Brown's criterion

The compressive strength of the rock is given by:

$$\sigma_1 = \sigma_3 + \sqrt{m_i \sigma_{ci} \sigma_3 + \sigma_{ci}^2}$$

Assuming that $\sigma_3$ = 1 MPa and $m_i$ is taken as equal to 19 (Table 3.14), and $\sigma_{ci}$ = 85 MPa (laboratory): $\sigma_1$ = 95 MPa. If $\sigma_3$ = 0.3 MPa is considered, then $\sigma_1$ = 88 MPa. Tensile strength is:

$$\sigma_t = \frac{1}{2}\sigma_{ci}\left(m_i - \sqrt{m_i^2 + 4}\right) \quad \Rightarrow \quad \sigma_t = 4.5 \text{ MPa}$$

The values of $c_i$ and $\phi_i$ can be obtained from this criterion by adjusting the Mohr-Coulomb $\tau - \sigma_n$ straight line to the Hoek and Brown $\tau - \sigma_n$ curve. If the "Roclab" program is applied (which uses the Hoek and Brown criterion to assess strength; www.rocscience.com), the input data above will give (Figure A):

$$\phi_i = 61°, c_i = 11.2 \text{ MPa}$$

| Hoek-Brown criterion – Intact rock strength |
|---|
| mb = 19, s = 1, a = 0.5 |
| cohesion = 11.27 MPa, friction angle = 60.80 |

Figure A

| Hoek-Brown criterion – Rock mass strength |
|---|
| mb = 7.78, s = 0.0622, a = 0.501 |
| cohesion = 2.39 MPa, friction angle = 62.63 |

Figure B

These values differ from those obtained from laboratory tests; however, they are in accordance with the characteristic ranges of these parameters for the rock under consideration.

- The Mohr-Coulomb criterion

Strength expressed as a function of principal stresses is:

$$\sigma_1 = \frac{2c_i + \sigma_3\left(\sin 2\theta + \tan\phi_i\left(1 - \cos 2\theta\right)\right)}{\sin 2\theta - \tan\phi_i\left(1 + \cos 2\theta\right)}$$

If the most unfavourable theoretical failure plane, corresponding to $\theta = 45 + \phi/2 = 80°$, is considered and it is assumed that $\sigma_3 = 1$ MPa then $\sigma_1 = 65$ MPa.

This value is lower than that obtained in the laboratory for uniaxial compressive strength ($\sigma_{ci} = 85$ MPa), which in theory is not possible.

According to Mohr-Coulomb, uniaxial compressive strength will be:

$$\sigma_{ci} = \frac{2c_i \cos\phi_i}{1 - \sin\phi_i} \Rightarrow \sigma_{ci} = 33 \text{ MPa}$$

which again is much lower than the laboratory value.

Tensile strength is given by:

$$\sigma_t = \frac{2c_i \cos\phi_i}{1 + \sin\phi_i} \Rightarrow \sigma_t = 1.02 \text{ MPa}$$

The Mohr-Coulomb criterion gives strength values well below both those of the laboratory and those obtained with the Hoek and Brown criterion. As pointed out in Section 3.4 of Chapter 3, the Mohr-Coulomb expressions are not representative for assessing intact rock strength.

**Shear strength of discontinuities**

- Barton and Choubey criterion

The frictional strength of a plane is given by (see Section 3.5 of Chapter 3):

$$\phi_p = JRC \log\left(\frac{JCS}{\sigma_n}\right) + \phi_r$$

where:

$$\phi_r = \left(\phi_b - 20\right) + 20\frac{r}{R}$$

The JCS value is obtained from the expression:

$$\log JCS = 0.00088\,\gamma_{rock}\,r + 1.01 \Rightarrow JCS = 29.6 \text{ MPa}$$

where $\gamma$ is in kN/m$^3$ and JCS in MN/m$^2$.

JCS can also be obtained from the Miller chart (Box 5.2), and gives a similar value.

The value of $\phi_r$ is 30°, considering that $\phi_b = 40°$ (Table 3.13), $r = 20$ and $R = 40$.

From the above data, for $JRC = 13$, and assuming $\sigma_n = 1$ MPa, the valued obtained for $\phi_p$ is 49°.

If $JCS/\sigma_n \geq 50$, to calculate $\phi_p$ the expression used would be $\phi_p = \phi_r + 1.7\,JRC$; this ratio would be given for values of $\sigma_n \leq 0.6$ MPa. For horizontal bedding planes, the value $\sigma_n = \sigma_v = 1$ MPa is taken (stress corresponding to a depth of 38 m considering $\gamma = 26.7$ kN/m$^3$), for which $JCS/\sigma_n \approx 30$; the value $\sigma_n \leq 0.6$ would be obtained for discontinuity planes with a dip $\leq 53°$, for the considered value of $\sigma_v = 1$ MPa. In this case, $\phi_p = 52°$ (provided that $JRC = 13$). As can be seen, the value of $\phi_p$ depends on the value of the normal stress exerted on the discontinuity, being not a single value for the plane.

With the $\phi_p$ value, the shear strength of discontinuities can be obtained:

$$\tau = c + \sigma_n \tan\phi_p$$

for $\sigma_n = 1$ MPa and considering $c = 0$, the value of $\tau = 1.15$ MPa is obtained.

If $\sigma_n = 0.6$ MPa and $\phi_p = 52°$ is taken, the result is: $\tau = 0.77$ MPa.

**Rock mass strength**

- Hoek and Brown criterion

Strength is given by:

$$\sigma_1 = \sigma_3 + \sigma_{ci}\sqrt{m\frac{\sigma_3}{\sigma_{ci}} + s}$$

the values of $m$ and $s$ can be obtained from:

a) $m = m_i \exp\dfrac{RMR - 100}{28}$ and $s = \exp\dfrac{RMR - 100}{9}$

for $m_i = 19$ and $RMR = 80 \Rightarrow m = 9.301$ and $s = 0.108$

b) $m = m_i \exp\dfrac{GSI - 100}{28}$ and $s = \exp\dfrac{GSI - 100}{9}$

for $m_i = 19$ and $GSI = 75 \Rightarrow m = 7.78$ and $s = 0.0622$

c) From Table 3.19: $m = 8.78$ and $s = 0.189$.

If the values of $\sigma_3 = 1$ MPa and $\sigma_{ci} = 85$ MPa are taken, the results obtained, as a function of the previous range of values for $m$ and $s$, are:

$$\sigma_{1(a)} = 40.6 \text{ MPa}$$
$$\sigma_{1(b)} = 34.3 \text{ MPa}$$
$$\sigma_{1(c)} = 46.9 \text{ MPa}$$

considering a value of $\sigma_1 \approx 40–45$ MPa as representative.

Uniaxial compressive strength of the rock mass is obtained for $\sigma_3 = 0$: $\sigma_{cm} = \sigma_{ci}\sqrt{s}$; for previous values of $s$ and considering $\sigma_{ci} = 85$ MPa:

$$\sigma_{cm\,(a)} = 27.9 \text{ MPa}$$
$$\sigma_{cm\,(b)} = 21.2 \text{ MPa}$$
$$\sigma_{cm\,(c)} = 36.9 \text{ MPa}$$

with a value of $\sigma_{cm} \approx 30$ MPa being considered representative.

Tensile strength of the rock mass is given by:

$$\sigma_{tm} = \frac{s\sigma_{ci}}{m}$$

and for the previous values of the $m$ and $s$ parameters:

$$\sigma_{tm\,(a)} = 0.98 \text{ MPa}$$
$$\sigma_{tm\,(b)} = 0.68 \text{ MPa}$$
$$\sigma_{tm\,(c)} = 1.83 \text{ MPa}$$

with a value of $\sigma_{tm} \approx 1.4$ MPa being considered representative.

The strength parameters of the rock mass, $c$ and $\phi$, are obtained from the adjustment of a Mohr-Coulomb straight line to the $\tau - \sigma_n$ curve of the Hoek and Brown criterion. Applying the "Roclab" program (www.rocscience.com), considering the values of $m = 7.78$ and $s = 0.0622$, $\sigma_3 = 1$ MPa and GSI = 75, the result obtained is (Figure B):

$$\phi = 62° \text{ and } c = 2.4 \text{ MPa}$$

The cohesion value is very similar to that obtained for intact rock in laboratory tests (2.9 MPa).

## Rock mass deformability

- Correlations with $E_i$

According to laboratory data, intact rock presents a Young's modulus of $E_i = 40$ GPa.

— From the RQD value: $E = E_i\,\mu$; for RQD = 90%, $\mu = 0.5 \Rightarrow E = 20$ GPa.
— Heuze ratio (Table 3.24): $E = E_i/2.5 = 16$ GPa.

- Correlations with RMR and GSI

— $E = 2\,RMR - 100$ (for $RMR > 50$): $E = 60$ GPa.
— The expression

$$E = 10^{\frac{RMR-10}{40}} = 56 \text{ GPa}$$

is not valid for $RMR > 50$.
— The expression

$$E = \sqrt{\frac{\sigma_{ci}}{100}}\,10^{\frac{GSI-10}{40}} = 38.9 \text{ GPa}$$

is applicable to rock masses with $\sigma_{ci} < 100$ MPa. In the case analysed, $\sigma_{ci} = 85$ MPa and GSI = 75.

— Applying the expression proposed by Hoek and Diederich (2006):

$$E = E_i + \left(0.02 + \frac{1 - D/2}{1 + e^{((60+15D-GSI)/11)}}\right) = 32.6 \text{ GPa}$$

A value of 60 MPa is very high and it is not possible for the deformation modulus of the rock mass to be greater than that of the intact rock; the third value is similar to that of the intact rock. Previous expressions overestimate the value of $E$ and the results are not coherent. In view of these uncertainties and taking into account that it is recommended that a range of values, rather than one in particular, should be taken for $E$, the values of 32.6 and 38.9 GPa, reduced by 20–30%, should be acceptable, in which case: $E \approx 25 - 29$ GPa, which would be close to the values obtained from correlation with the intact rock modulus (20 GPa). A representative value of between 25 and 30 GPa can therefore be given for the rock mass deformation modulus.

## Hydrostatic pressure

Not considered.

## *In situ* stress

— $K$ can be estimated by the ratio:

$$K = \nu/(1-\nu) \text{ for } \nu = 0.24 \Rightarrow \sigma_H = 0.316\,\sigma_V$$

— By applying Sheorey's expression:

$$K = 0.25 + 7E_h\left(0.001 + \frac{1}{z}\right)$$

taking $E_h = 25$ GPa and $z = 38$ m, a value of $K > 3$ is obtained, which is considered neither representative nor applicable in this case.

— By the TSI index, $T = 12 \times 10^6$ years, $E_i = 40$ GPa, $H = 1,000$ m, will give:

$$TSI = \log\frac{T}{E_i \cdot H} = 2.48$$

for TSI > 2.46 in alpine rocks, $K < 1.0$ (see Table 10.2), corresponding to low state of stress.

As a conclusion, as the tunnel is at a very shallow depth (38 m), the rock mass is supposed to be decompressed by stress relief and no tectonic stress is expected, the value adopted can be:

$$K = 0.31 \Rightarrow \sigma_H = 0.31\,\sigma_V$$

that have since been eroded. For intrusive, or plutonic, rocks, H is estimated according to the emplacement depth of the intrusive body. This depth may be inferred from several criteria such as texture, grain size, contacts between the wallrock and intrusive rock; three emplacement depth levels can be distinguished:

— Epizone: depth less than 8 km. A vast contrast between the temperature of the magma and surrounding wallrock gives rise to rapid crystallization and small grain sizes. The wallrock is usually affected by contact metamorphism. Contacts between the pluton and wallrock are angular and discordant, and the high temperature difference gives rise to porphyric and aplitic textures.
— Mesozone: In the middle crust at depths of 8–12 km. Contacts are angular and discordant as well as gradational and concordant. Foliation in plutons is scarce and often chemically and mineralogically zoned. The metamorphic contact zone is extensive and the pluton shows flow structures.
— Catazone: Emplaced at depths greater than 12 km. The temperature difference between the magma and wallrock is small and this gives rise to large crystal sizes. The wallrock rock is of high-grade metamorphism. Contacts between the pluton and wallrock are concordant and gradational. Plutons generally show a foliation pattern consistent with that of adjacent metamorphic rocks. Migmatites appear.

For metamorphic rocks, H can be estimated from the mineral composition of the rock as pressure and temperature indicators (geobarometers and geothermometers), and from these variables the depth at which they formed can be estimated. As a simplification, three zones related to the degree of metamorphism can be distinguished:

— Low-grade metamorphism: temperatures between 200–450°C, depths between 5,000–7,000 m. The most common minerals are talc, epidote, chlorite and serpentine. Characteristic rocks: quartzites, slatey granites, mylonitized granites, phylites and slates.
— Intermediate-grade metamorphism: temperatures between 450–650°C, depths between 7,000–12,000 m. The most common minerals are biotite, hornblende, staurolite and andalusite. Characteristic rocks: quartzites, micaceous slates, marbles, amphibolites and eclogites.
— High-grade metamorphism: temperatures between 650°C-fusion, depths between 12,000–20,000 m. The most common minerals are sillimanite, forsterite, wollastonite, garnierite, olivine and garnet. Characteristic rocks: orthogneisses, paragneisses, granulites, graphite and hornblende gneisses.

NC is the coefficient of seismotectonic activity and was fixed at 0.25 for zones close to seismogenetic faults or to plate boundaries; its value is 1 in the absence of these conditions. The SC, or coefficient of topographic influence, was fixed at 0.3 for zones close to steep slopes or deep valleys, and at 1 when these conditions were not met. In cases where both NC and SC are applicable only the 0.25 value must be considered.

Some relationships between $K$ in situ values and TSI have been proposed from empirical data:

— For rocks deformed in the Hercynian orogeny:

$$K_{her} = -1.93 \text{ TSI} + 8.38$$

— For rocks deformed in the Alpine orogeny:

$$K_{alp} = -2.09 \text{ TSI} + 6.15$$

— An assessment of the tectonic stress from ranges of $K$ and TSI values is shown in Table 10.2. For a detailed description of this procedure including the application of probabilistic methods see González de Vallejo and Hijazo (2008).

## Worked examples

### Example 1

Underground excavation at a depth of 400 m in Carboniferous shales, in which the deformation modulus of the intact rock is 48 GPa. Main folding corresponds to the Hercynian orogeny, approximately 300 million years ago. According to regional stratigraphic data, the maximum thickness of sedimentary rocks above the tunnel is in the order of 2,000 m. As the area is not affected by topographic effects or seismicity, NC and SC are not applicable. The value of the TSI index is:

*Table 10.2* **RANGES OF $K$ AND POSSIBLE TECTONIC STRESSES INDICATED BY TSI INDEX**

| Hercynian rocks | | | Alpine rocks | | |
|---|---|---|---|---|---|
| TSI | $K_{her}$ [1] | Stress [2] | TSI | $K_{alp}$ [1] | Stress [2] |
| >3.80 | <1.0 | Low | >2.46 | <1.0 | Low |
| 3.80–3.55 | 1.0–1.49 | Medium | 2.46–2.25 | 1.0–1.49 | Medium |
| 3.54–3.30 | 1.5–2.0 | High | 2.24–1.99 | 1.5–2.0 | High |
| <3.30 | >2.0 | Very high | <1.99 | >2.0 | Very high |

[1] $K_{her}$ and $K_{alp}$ are estimated from the expressions included in Section 10.4. [2] Tectonic stress.

(Gonzalez de Vallejo and Hijazo, 2008).

$$\text{TSI} = \log\left(\frac{300 \cdot 10^6}{48 \cdot 2000}\right) = 3.49$$

which corresponds to a $K$ interval between 1.5 and 2.0: high stress (see Table 10.2).

### Example 2

Underground excavation where rock burst and spalling of material has taken place. The excavation is situated at an average depth of 80 m in a calcareous rock mass folded during the Alpine orogeny. The rocks consist of marbles with low metamorphism and a deformation modulus of 70 GPa. Regional geological data indicate that the age of tectonic deformation was approximately 10 million years ago. The maximum lithostatic load in accordance with the degree of metamorphism is in the order of 3,000 m. NC and NS are not applicable. The TSI is:

$$\text{TSI} = \log\left(\frac{10 \cdot 10^6}{70 \cdot 3000}\right) = 1.67$$

corresponding to a $K$ value higher than 2.0: very high stress (Table 10.2). This result can be consistent with the stress failure phenomena observed.

### Sheorey's method

Sheorey's method (Sheorey, 1994) takes into account membrane stresses and thermal stresses, and those of elastic origin. It does not consider tectonic stress. The following expression is proposed:

$$K = 0.25 + 7\, E_h\, (0.001 + 1/z)$$

where $E_h$ is the average elasticity modulus in the area of the upper crust, measured horizontally in GPa, and $z$ is the depth in metres. An example of its application is shown in Box 10.2.

## Water inflow and pressure

Water inflow inside a tunnel is one of the principal problems and also uncertainties in tunnels that cross either highly fractured or karstified rock masses. In the majority of cases, the most appropriate method for estimating discharge is to use mathematical flow models, but sufficient data is not always available for their application. Approximate results can be obtained from permeability data, piezometric readings and flow nets, provided that they apply to aquifers without significant anisotropy. Alternatively, other analytical methods can be used, such as that of Goodman et al. (1965) shown as an example in Box 10.3.

---

### Box 10.3

### Calculation of the discharge in a tunnel using the Goodman method (Goodman et al., 1965)

**Method**

Discharge in a tunnel is calculated from the following data:

— The permeability coefficient ($k$) of the aquifer or the permeable formation that is crossed.
— The storage coefficient ($S$).
— The length of tunnel that crosses the aquifer formation.
— The pressure head ($H$).
— Average rate of advance of the excavation.

The expression for calculating the discharge is:

$$Q = 2\sqrt{\frac{2 \cdot k \cdot H^2 \cdot S \cdot t}{3 \cdot C}}$$

with

$$t = \frac{3 \cdot Q^2 \cdot C}{8 \cdot k \cdot H^2 \cdot S}$$

**where**

— $Q$: discharge (m³) per metre advanced in time $t$; reference to a one m advance assumes this is the whole section.
— $k$: average permeability coefficient in the formation (m/s).
— $H$: average pressure head (m).
— $S$: approximate values of 0.03 for clays, 0.05 for sandy clays, 0.10 for fine sands, 0.20 for coarse sands and 0.25 for gravels.
— $t$: time interval in which the advance takes place.
— $C$: penetration factor, empirically estimated as:

$$C = 0.12 + 1.24 D$$

in which $D$ is the fraction of aquifer crossed with respect to the total length of the aquifer intercepted by the tunnel:

$$D = 1 - \frac{L_a - L_t}{L_T}$$

where $L_a$ = the total length of the aquifer, $L_t$ = the length of the section crossed and $L_T$ = the total length of the tunnel.

### Example

A tunnel 200 metres long crosses sedimentary materials with the following characteristics along its entire length:

$$k_{average} = 1 \times 10^{-4} \text{ m/s}$$
$$H = \text{from 50 to 80 m}$$
$$S = 0.10$$
$$L_a = 200 \text{ m}$$

The advance is 10 m/day, at 5 day intervals (50 m)

**Calculation of C parameter**

| Section | Days (t) | Length of section $L_t$ (m) | H (m) | D | C |
|---|---|---|---|---|---|
| 1 | 5 | 50 | 50 | 0.25 | 0.43 |
| 2 | 10 | 100 | 50 | 0.50 | 0.74 |
| 3 | 15 | 150 | 80 | 0.75 | 1.05 |
| 4 | 20 | 200 | 50 | 1.00 | 1.36 |

**Section 1**

$t_1 = 5$ days

$$Q_1 = 2 \times \sqrt{\frac{2 \times 10^{-4} \times 50^2 \times 0.1 \times 5 \times 86,400}{3 \times 0.43}}$$

$= 258.8$ m³ per metre advanced in 5 days

$$Q_1 = \frac{258.8 \times 10^3}{5 \times 86,400} = 0.60 \text{ l/s/m}$$

$$Q = \frac{Q_1 + L_1}{L_2} = \frac{258.8 \times 50}{100} = 129.4 \text{ m}^3$$

$t$ corresponding to this $Q$ for $C_2$:

$$t = \frac{3 \times 129.4^2 \times 0.74}{8 \times 10^{-4} \times 50^2 \times 0.1} = 2.1 \text{ days}$$

Section 2

$t_2 = 5$ days $+ 2.1 = 7.1$ days

$$Q_2 = 2 \times \sqrt{\frac{2 \times 10^{-4} \times 50^2 \times 0.1 \times 7.1 \times 86,400}{3 \times 0.74}}$$

$= 235$ m³ per metre advanced in 10 days

$$Q_2 = \frac{235 \times 10^3}{10 \times 86,400} = 0.27 \text{ l/s/m} \qquad Q = \frac{235 \times 100}{150} = 156.6 \text{ m}^3$$

$t$ for $C_3$: $t = \dfrac{3 \times 156.6^2 \times 1.05}{8 \times 10^{-4} \times 50^2 \times 0.1} = 4.4$ days

**Section 3**

$t_3 = 5 + 4.4 = 9.4$ days
$Q_3 = 363.3$ m³ $\approx 0.28$ l/s/m
$Q = 272.4$ m³
$t = 6.8$ days

**Section 4**

$t_4 = 5 + 6.8$ days $= 11.8$ days
$Q_4 = 223.5$ m³ $\approx 0.13$ l/s/m

**Results:** when excavation is complete, for the entire tunnel (200 m) the average discharge will be 26 l/s (93.6 m³/hour), assuming that the tunnel as a whole acts like a drainage hole.

---

The hydraulic load on a tunnel caused by water pressures is evaluated according to the procedures described in Chapter 4; the conditioning factors mentioned in Chapter 3, Section 3.6, which refer to water pressure inside the rock mass, should be taken into account.

## 10.5 Rock mass classifications for tunnelling

Geomechanical classifications are fundamental for the characterization of rock masses as they enable rock quality, strength and deformability parameters of a rock mass to be estimated, as well as tunnel support and excavation to be assessed.

The most frequently used rock mass classifications in tunnels are the RMR and the Q. Although both were developed for the purpose of estimating supports, they have become established as geomechanical indexes for evaluating both rock mass properties and support, and excavation. The RMR Classification System is described in Chapter 3, Section 3.8. The Q System and the SRC classification are described below.

### Q System

The Q System was developed by Barton, Lien and Lunde in 1974 and was based on the study of a large number of tunnels. This system of classification can be used to estimate the geomechanical parameters of a rock mass and the preliminary design of support for tunnels and underground caverns within it. The **Q index** is based on a numerical calculation of six parameters expressed as:

$$Q = \frac{RQD}{Jn} \cdot \frac{Jr}{Ja} \cdot \frac{Jw}{SRF}$$

where:

RQD is the Rock Quality Designation.
Jn is the joint set number.
Jr is the joint roughness number.
Ja is the joint alteration number.
Jw is the joint water reduction factor.
SRF is the stress reduction factor

SRF is the coefficient that takes into account the influence of the rock mass state of stress.

The three factors defining Q index represent:

$\left(\dfrac{RQD}{Jn}\right)$: block size

$\left(\dfrac{Jr}{Ja}\right)$: shear strength between blocks

$\left(\dfrac{Jw}{SRF}\right)$: influence of the state of stress

Table 10.3 shows the criteria for evaluating these parameters. Values of the Q index range from 0.001

| Table 10.3 | VALUES OF CHARACTERISTIC PARAMETERS OF THE Q INDEX | |
|---|---|---|
| **1. Rock Quality Designation** | | **RQD (%)** |
| A | Very poor | 0–25 |
| B | Poor | 25–50 |
| C | Fair | 50–75 |
| D | Good | 75–90 |
| E | Excellent | 90–100 |
| Notes:<br>- Where RQD values are obtained of ≤ 10 (including 0), a nominal value of 10 is used to evaluate Q.<br>- RQD intervals of 5, i.e. 100, 95, 90, etc. are sufficiently accurate. | | |
| **2. Joint set number** | | $J_n$ |
| A | Massive, no or few joints | 0.5–1 |
| B | One joint set | 2 |
| C | One joint set plus random joints | 3 |
| D | Two joint sets | 4 |
| E | Two joint sets plus random joints | 6 |
| F | Three joint set | 9 |
| G | Three joint set plus random joints | 12 |
| H | Four or more joint sets, random, heavily jointed, "sugar-cube", etc. | 15 |
| J | Crushed rock, soil-like | 20 |
| Notes:<br>- For tunnel intersections, use $(3.0 \times J_n)$<br>- For portals use $(2.0 \times J_n)$ | | |
| **3. Joint roughness number** | | $J_r$ |
| a) Rock-wall contact. b) Rock-wall contact before 10 cm shear | | |
| A | Discontinuous plane | 4 |
| B | Rough or irregular, undulating | 3 |
| C | Smooth, undulating | 2 |
| D | Slickensided, undulating | 1.5 |
| E | Rough or irregular, planar | 1.5 |

(continued)

*Table 10.3*  **VALUES OF CHARACTERISTIC PARAMETERS OF THE Q INDEX (CONT.)**

| | | | |
|---|---|---|---|
| F | Smooth, planar | | 1.0 |
| G | Slickensided, planar | | 0.5 |

*Note:*
- Descriptions refer to small-scale features and intermediate scale features, in that order.

| | | | |
|---|---|---|---|
| c) No rock-wall contact when sheared | | | |
| H | Zone containing clay minerals thick enough to prevent rock-wall contact | | 1.0 |
| J | Sandy, gravelly or crushed zone thick enough to prevent rock-wall contact | | 1.0 |

*Notes:*
- Add 1.0 if the mean spacing of the relevant joint set is greater than 3 m.
- $J_r = 0.5$ can be used for planar, slickensided joints with lineations, provided the lineations are oriented for minimum strength.
- $J_r$ and $J_a$ classification is applied to the joint set or discontinuity that is least favourable for stability both from the point of view of orientation and shear resistance, $\tau$ (where $\tau \approx \sigma_n \tan (J_r/J_a)$).

| | | | |
|---|---|---|---|
| **4. Joint alteration number** | | $\phi_r$ | $J_a$ |
| a) Rock-wall contact (no mineral fillings, only coatings) | | | |
| A | Tightly healed, hard, non-softening, impermeable filling, i.e. quartz or epidote. | – | 0.75 |
| B | Unaltered joint walls, surface staining only. | 25–35° | 1.0 |
| C | Slightly altered joint walls. Non-softening mineral coatings, sandy particles, clay-free disintegrated rock, etc. | 25–30° | 2.0 |
| D | Silty- or sandy- clay coatings, small clay fraction (non-softening) | 20–25° | 3.0 |
| E | Softening or low friction clay mineral coatings, i.e. kaolinite or mica. Also chlorite, talc, gypsum, graphite, etc. and small quantities of swelling clays. | 8–16° | 4.0 |
| b) Rock-wall contact before 10 cm shear (thin mineral fillings) | | | |
| F | Sandy particles, clay-free disintegrated rock, etc. | 25–30° | 4.0 |
| G | Strongly over-consolidated non-softening clay mineral fillings (continuous, but <5 mm thickness) | 16–24° | 6.0 |
| H | Medium or low over-consolidation, softening, clay mineral fillings (continuous, but <5 mm thickness) | 12–16° | 8.0 |
| J | Swelling-clay fillings, i.e. montmorillonite (continuous, but <5 mm thickness). Value of $J_a$ depends on percentage of swelling clay-size particles, and access to water, etc. | 6–12° | 8–12 |
| c) No rock-wall contact when sheared (thick mineral fillings) | | | |
| K L M | Zones or bands of disintegrated or crushed rock and clay (see G, H and J for description of clay condition) | 6–24° | 6, 8 or 8–12 |
| N | Zones or bands of silty -or sandy- clay, small clay fraction (non-softening) | – | 5.0 |
| O P R | Thick continuous zones or bands of clay (see G, H and J for description of clay condition. | 6–24° | 10, 13 or 13–20 |
| **5. Joint water reduction factor** | | Water pressure (kp/cm²) | $J_w$ |
| A | Dry excavations or minor inflow, i.e. <5 l/min. locally. | <1 | 1.0 |
| B | Medium inflow or pressure, occasional outwash of joint fillings. | 1–2.5 | 0.66 |
| C | Large inflow or high pressure in competent rock with unfilled joints. | 2.5–10 | 0.5 |
| D | Large inflow or high pressure, considerable outwash of joint fillings. | 2.5–10 | 0.33 |

*(continued)*

## Table 10.3  VALUES OF CHARACTERISTIC PARAMETERS OF THE Q INDEX (CONT.)

| | | | |
|---|---|---|---|
| E | Exceptionally high inflow or water pressure at blasting, decaying with time. | >10 | 0.2–0.1 |
| F | Exceptionally high inflow or water pressure continuing without noticeable decay. | >10 | 0.1–0.05 |

*Notes:*
- Factors C, D, E and F are raw estimates. Increase $J_w$ if drainage measures are installed.
- Special problems caused by ice formation are not considered.

### 6. Stress reduction factor

| a) Weakness zones intersecting excavation, which may cause loosening of rock mass when tunnel is excavated | | **SRF** |
|---|---|---|
| A | Multiple occurrences of weakness zones containing clay or chemically disintegrated rock, very loose surrounding rock (any depth) | 10 |
| B | Single weakness zones containing clay or chemically disintegrated rock (depth of excavation ≤50 m) | 5 |
| C | Single weakness zones containing clay or chemically disintegrated rock (depth of excavation >50 m) | 2.5 |
| D | Multiple shear zones in competent rock (clay-free), loose surrounding rock (any depth) | 7.5 |
| E | Single shear zones in competent rock (clay-free), depth of excavation ≤50 m | 5.0 |
| F | Single shear zones in competent rock (clay-free), depth of excavation >50 m | 2.5 |
| G | Loose open joints, heavily jointed or "sugar-cube", etc. (any depth) | 5.0 |

*Notes:*
- Reduce these values of SRF by 25–50% if the relevant shear zones only influence but do not intersect the excavation.

| b) Competent rock, rock stress problems | | $\sigma_c/\sigma_1$ | $\sigma_\theta/\sigma_c$ | **SRF** |
|---|---|---|---|---|
| H | Low stress, near surface, open joints. | >200 | <0.01 | 2.5 |
| J | Medium stress, favourable stress condition. | 200–10 | 0.01–0–03 | 1 |
| K | High stress, very tight structure. Usually favourable to stability, may be unfavourable for wall stability. | 10–5 | 0.3–0.4 | 0.5–2 |
| L | Moderate slabbing after >1 hour in massive rock. | 5–3 | 0.5–0–65 | 5–50 |
| M | Slabbing and rock burst after a few minutes in massive rock. | 3–2 | 0.65–1 | 50–200 |
| N | Heavy rock burst (strain-burst) and immediate dynamic deformations in massive rock. | <2 | >1 | 200–400 |

*Notes:*
- For strongly anisotropic virgin stress field (if measured): When $5 \leq \sigma_1/\sigma_3 \leq 10$, reduce $\sigma_c$ to $0.75\,\sigma_c$. When $\sigma_1/\sigma_3 > 10$, reduce $\sigma_c$ to $0.5\,\sigma_c$, where $\sigma_c$ = unconfined compression strength, $\sigma_1$ and $\sigma_3$ are the major and minor principal stresses, and $\sigma_\theta$ = the maximum tangential stress (estimated from elastic theory).
- Few case records available where depth of crown below surface is less than span width. Suggest SRF increase from 2.5 to 5 for such cases (see H).

| c) Squeezing rock: plastic flow of incompetent rock under the influence of high rock pressure | | $\sigma_\theta/\sigma_c$ | **SRF** |
|---|---|---|---|
| O | Mild squeezing rock pressure | 1–5 | 5–10 |
| P | Heavy squeezing rock pressure | >5 | 10–20 |

*Notes:*
- Cases of squeezing rock may occur for depth $H > 350\,Q^{1/3}$ (Singh et al., 1992). Rock mass compression strength may be estimated from $q \approx 0.7\gamma\,Q^{1/3}$ (MPa) where $\gamma$ = rock density in kN/m² (Singh, 1993).

| d) Swelling rock: chemical swelling activity depending on presence of water | | **SRF** |
|---|---|---|
| R | Mild swelling rock pressure | 5–10 |
| S | Heavy swelling rock pressure | 10–15 |

(Barton, 2000).

to 1,000, providing the following rock mass quality classification:

| | |
|---|---|
| between 0.001 and 0.01: | exceptionally poor rock |
| 0.01 and 0.1: | extremely poor rock |
| 0.1 and 1: | very poor rock |
| 1 and 4: | poor rock |
| 4 and 10: | fair rock |
| 10 and 40: | good rock |
| 40 and 100: | very good rock |
| 100 and 400: | extremely good rock |
| 400 and 1,000: | exceptionally good rock |

## SRC rock mass classification

The SRC rock mass classification (González de Vallejo, 2003) is based on the RMR classification, but the difference between them is that the SRC includes the rock mass state of stress, as well as some correction factors that take into account the influence of the excavation on rock mass condition and the use of data from outcrops. The following parameters are included in SRC (Table 10.4):

— Intact rock strength.
— Spacing of the discontinuities or RQD.
— Condition of discontinuities.
— Groundwater inflow.
— State of stress

The state of stress is assessed from the following factors:

— Competence factor, $F_c$, described in Section 10.3.
— Tectonic structures: considered when significant faults or tectonic structures are present in the area.
— Stress relief factor: age (in years $\times 10^{-3}$) of the main tectonic orogeny that has affected the region (Alpine or Hercynian orogeny), divided by maximum thickness of the overburden during its geological history (in metres). This factor is estimated from regional geological data (see Section 10.4).
— Seismic activity: considered if the area has a history of significant seismic activity.

Box 10.4 shows an example of the calculation of these factors.

### Correction factors for outcrop data

When data from outcrops is used, the correction factors shown in Table 10.5 are applied.

### Correction factors for construction conditions

Rock mass quality may be affected by the construction process, which includes excavation and support methods and, in particular, the sequence of installation of supports and the time taken to install them. The interaction of adjacent excavations and areas with thin overburden also modifies the state of stress and can change rock mass behaviour. The correction factors to be applied are included in Table 10.6.

### Application of SRC classification

The tunnel is divided along its longitudinal profile into lithological and structurally homogeneous zones and the **SRC** is calculated for each (Table 10.4). These SRC values are then adjusted by the correction factors included in Table 10.5 and the final result is expressed by five classes of rock quality ranging from 1 to 100. If the rock mass is classified for conditions prior to excavation, the **base SRC** is obtained. In order to consider construction conditions, the base SRC index is adjusted in accordance with the correction factors shown in Table 10.6. This new result is the **corrected SRC**, which reflects the influence of the construction process.

For tunnel support assessment as well as to correlate with other rock mass properties, e.g. deformability and strength, Table 10.7 and Table 3.27 in Chapter 3, respectively, are used directly, substituting RMR by SRC, according to RMR criteria.

The following steps are recommended for application of the SRC classification:

1. Divide the longitudinal profile of the tunnel into homogeneous lithological zones.
2. Subdivide these zones in sectors having distinctively different structural or fracturing conditions.
3. Identify singularities along the course of the tunnel. These singularities include any geological or geomechanical features that may affect tunnel excavation.
4. Calculate SRC (Table 10.4).
5. Apply correction factors when data is obtained from outcrops (Table 10.5) and calculate the base SRC.
6. Characterize the rock mass including strength and deformability using the base SRC index in expressions that relate the RMR with the rock mass parameters.
7. Estimate the support using the base SRC index in the same way as RMR (Table 10.7).
8. During tunnel construction, use the corrected SRC (Table 10.6 and 10.7).

Step 8 allows suitable supports to be adjusted to conditions arising during tunnel construction as they take into account the influence of the rock mass once it is affected by excavation. A significant difference between the base SRC and the corrected SRC is an indication that the construction method has affected the conditions of the rock mass.

| Table 10.4 GEOMECHANIC ROCK MASS CLASSIFICATION SRC | | | | | | | |
|---|---|---|---|---|---|---|---|
| **Rock quality indices** | **Range of values** | | | | | | |
| 1. INTACT ROCK STRENGTH Point-load test (MPa) | >8 | 8–4 | 4–2 | 2–1 | Not applicable | | |
| Uniaxial compressive strength (MPa) | >250 | 250–100 | 100–50 | 50–25 | 25–5 | 5–1 | <1 |
| Rating | 20 | 15 | 7 | 4 | 2 | 1 | 0 |
| 2. SPACING OR RQD Spacing (m) RQD (%) | >2 100–90 | 2–0.6 90–75 | 0.6–0–2 75–50 | 0.2–0.06 50–25 | <0.06 <25 | | |
| Rating | 25 | 20 | 15 | 8 | 5 | | |
| 3. CONDITIONS OF DISCONTINUITIES | Very rough surfaces. Not continuous joints. No separation. Hard joint wall | Slightly rough surfaces. Not continuous joints. Separation >1 mm. Hard joint wall | Slightly rough surfaces. Not continuous joints. Separation 1 mm. Soft or weathered joint walls | Slicken-sided surfaces. Continuous joints. Joints open 1–5 mm. Gouge materials | Slicken-sided surfaces. Continuous joints. Joints open <5 mm. Gouge materials | | |
| Rating | 30 | 25 | 20 | 10 | 0 | | |
| 4. GROUNDWATER Inflow per 10 m tunnel length (l/min) General conditions | None Dry | <10 Slightly moist | 10–25 Occasional seepage | 25–125 Frequent seepage | >125 Abundant seepage | | |
| Rating | 15 | 10 | 7 | 4 | 0 | | |
| 5. STATE OF STRESSES Competence factor[1] | >10 | 10–5 | 5–3 | <3 | – | | |
| Rating | 10 | 5 | –5 | –10 | | | |
| Tectonic structures | Zones near thrusts/faults of regional importance | | Compression | Tension | | | |
| Rating | –5 | | –2 | 0 | | | |
| Stress relief factor[2] | >200 | 200–80 | 80–10 | <10 | Slopes 200–80 / <10 | 79–10 / –13 | <10 / –15 |
| Rating | 0 | –5 | –8 | –10 | | | |
| Neotectonic activity | None or unknown | | Low | High | | | |
| Rating | 0 | | –5 | –10 | | | |
| 6. ROCK MASS CLASSES Class number Rock quality | I Very good | II Good | III Fair | IV Poor | V Very poor | | |
| Rating | 100–81 | 80–61 | 60–41 | 40–21 | ≤20 | | |

[1] Uniaxial intact rock strength/vertical stress.
[2] Ratio between the age of the main orogenic deformation affecting the rock mass (in years $\times 10^{-3}$) and maximum overburden thickness (in metres).

## Box 10.4

### Calculation of the state of stress parameter of the SRC classification

#### Example 1

Tunnel excavated in sandstones: average unit weight = 21 kN/m³, average uniaxial compressive strength = 15 MPa. Tunnel overburden = 300 m.

— Competence factor = $\sigma_{ci}/\sigma_V$ = 1,500/(300×2.1) = 2.3 (−10 points)
— Tectonic structure: regional faults in the area where the tunnel is located (−5 points)
— Stress release factor: age of the main folding: Hercynian, approximately 300 million years; maximum thickness of overburden: 500 m, estimated from the actual overburden thickness (300 m) plus the thickness of eroded strata (200 m) according to regional data:

$$\text{Stress Relief Factor} = \frac{300,000,000 \text{ years} \cdot 10^{-3}}{500 \text{ m}} = 600 \text{ (0 points)}$$

— Seismic activity: not applicable (0 points)
— State of stress:

$$-10 - 5 + 0 + 0 = -15 \text{ points}$$

#### Example 2

Tunnel in Hercynian granites: average unit weight = 25 kN/m³; average uniaxial compressive strength = 40 MPa. Tunnel overburden: 300 m.

— Competence factor, $F_c$ = 4,000/(300×2.5) = 5.3 (+5 points)
— Tectonic regime: distensive tectonics (0 points).
— Stress relief factor: age of main folding: Hercynian, approximately 300 million years old; maximum thickness of overburden: according to petrological and geological data, depth at which granites formed is about 5,000 m.

$$\frac{300,000,000 \text{ years} \cdot 10^{-3}}{5,000 \text{ m}} = 60 \text{ (−8 points)}$$

— Seismic activity: not applicable (0 points)
— State of stress:

$$+5 + 0 - 8 + 0 = -3 \text{ points}$$

---

**Table 10.5**    **RATINGS ADJUSTMENT FOR SURFACE DATA FOR SRC ROCK MASS CLASSIFICATION**

**Spacing or RQD.** The maximum score is 25 points

— Compression fractures = 1.3
— Tension fractures = 0.8
— For depths <50 m = 1.0
— Grade of weathering ≥ IV = 0.8
— Grade of weathering III = 0.9
— Grade of weathering I or II = 1.0

**Conditions of discontinuities.** The maximum score is 30 points

— Compression fractures: +5
— Tension fractures: 0
— Not applicable for depths <50 m

**Groundwater**

— Compression fractures: +5
— Tension fractures: 0
— Not applicable for depths <50 m

TUNNELS

## Table 10.6 RATINGS ADJUSTMENT FOR CONSTRUCTION FACTORS FOR SRC ROCK MASS CLASSIFICATION

The total rating from Table 10.4 must be adjusted for the following factors:

**Rock resistance to weathering**[1]

| | |
|---|---|
| Rocks of high durability (low clay content) | 0 |
| Rocks of low durability (high clay content) | −5 |
| Rocks of very low durability (very high clay content) | −10 |

**Discontinuity orientations**[2]

| Strike perpendicular to tunnel axis | | | | | | Dip 0–20° at any direction |
|---|---|---|---|---|---|---|
| Drive with dip | | Drive against dip | | Strike parallel to tunnel axis | | |
| Dip 45–90° | Dip 20–45° | Dip 45–90° | Dip 20–45° | Dip 45–90° | Dip 20–45° | |
| Very favourable | Favourable | Fair | Unfavourable | Very unfavourable | Fair | Unfavourable |
| 0 | −2 | −5 | −10 | −12 | −5 | −10 |

**Excavation method**

| | |
|---|---|
| Tunnel boring machines, continuous miner, cutter machines, roadheaders, etc. | +5 |
| Controlled blasting, pre-splitting, soft blasting, etc. | 0 |
| Poor-quality blasting[3] | −10 |

**Support method**[4]

| | |
|---|---|
| Class I | 0 |
| Class II | |
|     <10 days | 0 |
|     >10 days <20 days | −5 |
|     >20 days | −10 |
| Class III | |
|     <2 days | 0 |
|     >2 days <5 days | −5 |
|     >5 days <10 days | −10 |
|     >10 days | −20 |
| Class IV and V | |
|     <8 hours | 0 |
|     >8 hours <24 hours | −10 |
|     >24 hours | −20 |

**Distance to adjacent excavation**[5]

| | |
|---|---|
| AEF <2.5 | −10 |

**Portals, accesses and areas with small overburden thickness**[6]

| | |
|---|---|
| PF <3 | −10 |

[1] Durability can be assessed by the slake durability test, or indirectly by the clay content.
[2] After Bienawski (1979)
[3] Conventional blasting: 0.
[4] Based on Bienawski's (1979) graphic representation of the stand-up time and the unsupported span, the ratings are applied in relation to the maximum stand-up time in days and hours.
[5] AEF is the adjacent excavation factor, defined as the ratio between the distance to an adjacent excavation (in metres) from the excavation under design and the span of that adjacent excavation (in metres).
[6] PF is the portal factor, defined as the ratio between the thickness of overburden and the span of the excavation, both in metres.

*Table 10.7* **GUIDELINES FOR EXCAVATION AND SUPPORT OF 10 M SPAN ROCK TUNNELS IN ACCORDANCE WITH THE RMR SYSTEM**

| Rock mass class | Excavation | Rock bolts (20 mm diameter, fully grouted) | Shotcrete | Steel sets |
|---|---|---|---|---|
| I Very good rock RMR: 81–100 | Full face, 3 m advance. | Generally, no support required except spot bolting. | – | – |
| II Good rock RMR: 61–80 | Full face, 1–1.5 m advance. Complete support 20 m from face. | Locally, bolts in crown 3 m long, spaced 2.5 m with occasional wire mesh. | 50 mm in crown where required. | None. |
| III Fair rock RMR: 41–60 | Top heading and bench, 1.5–3 m advance in top heading. Commence support after each blast. Complete support 10 m from face. | Systematic bolts 4 m long, spaced 1.5–2 m in crown and walls with wire mesh in crown. | 50–100 mm in crown and 30 mm in sides. | None. |
| IV Poor rock RMR: 21–40 | Top heading and bench, 1.0–1.5 m advance in top heading. Install support concurrently with excavation, 10 m from face. | Systematic bolts 4–5 m long, spaced 1–1.5 m in crown and walls with wire mesh. | 100–150 mm in crown and 100 mm in sides. | Light to medium ribs spaced 1.5 m where required. |
| V Very poor rock RMR: ≤20 | Multiple drifts 0.5–1.5 m advance in top heading. Install support concurrently with excavation. Shotcrete as soon as possible after blasting. | Systematic bolts 5–6 m long, spaced 1–1.5 m in crown and walls with wire mesh. Bolt invert. | 150–200 mm in crown, 150 mm in sides, and 50 mm on face. | Medium to heavy ribs spaced 0.75 m with steel lagging and forepoling if required. Close invert. |

Horseshoe-shaped tunnels, maximum width 10 m, maximum vertical stress 250 kp/cm$^2$.

(Bienawski, 1989).

## Suggested criteria for the application of rock mass classifications

The widespread use of the RMR and Q classifications since the 1970s is evidence of their usefulness. They brought a new concept to rock mass analysis and characterization. Nevertheless, noticeable differences have been found in a considerable number of tunnels between the supports recommended by these classifications and those installed on site, when low or very low quality rock masses are involved. Classifications should therefore not be applied as a matter of routine; their use requires **geological engineering judgement**.

The following results are deduced from a comparative analysis between supports applied on tunnelling excavation and those recommended by rock mass classifications (González de Vallejo, 2003):

— For **good and fair quality** rocks, either the **RMR or Q** systems can be used.

— In **weak rocks under significant *in situ* stress**, the **SRC** classification can predict the rock behaviour during excavation better than the RMR classification.

— Use of a particular classification system should consider both the rock mass type and the parameters involved in the classification, as different classifications are not equivalent. Therefore, **correlations between rock mass classifications are not recommended** for poor and very poor quality rocks.

## 10.6 Tunnel support design using rock mass classifications

Methods to design tunnel support may be analytical, numerical, empirical or observational. **Analytical methods** are based on the theory of elasticity and assume that tunnel behaviour is elastic until a certain critical stress is reached, at which point plasticity occurs. The relationship between this stress and radial deformation of the excavation is expressed by a **ground reaction curve** or characteristic line. The support chosen must be capable of resisting this stress. As the stress deformation law of the support is known, the point where both curves meet defines the situation of equilibrium (see Section 10.8).

**Numerical methods** are based on discretization of the rock mass by means of the finite element method, the discrete element method or the finite difference method. They allow detailed modelling of deformation processes affecting the ground as a consequence of excavation, and analysis of the influence of different factors and parameters related to construction. Appropriate design criteria can then be established and applied to problems that may arise either at the decision-making stage or during excavation, such as instability.

As a detailed explanation of numerical and analytical calculations for tunnel support is beyond the scope of this book, the reader is referred to the recommended bibliography.

**Observational methods** are based on measurements of stresses and deformations produced during tunnel excavation; supports are calculated with the aid of numerical or analytical methods. The most representative observational method is the New Austrian Tunnelling Method (NATM), (Section 10.8).

**Empirical methods** give an approach to tunnel support and are not considered as a design method. However, they can be very useful in fractured rock masses as a means of establishing properties and the support likely to be needed. They are also useful in the preliminary stages of tunnelling design.

Empirical methods for estimating supports are based on the RMR and Q classifications. The following aspects should be taken into account when they are applied:

— The suitability of the chosen geomechanical classification must be analysed; this is based on geological data, the stress-strain behaviour of the rock mass, and the construction process to be used.
— The supports recommended by rock mass classifications represent the average conditions of the section under consideration and may not consider possible specific singularities such as fracture zones, or such features as karstic, evaporitic, swelling or squeezed rocks.

## Tunnel support based on RMR classification

The RMR classification indicates the type of support expected to be used, as shown in *Table 10.7*. The **unsupported roof span** (length of advance without support, *Figure 10.11*), can

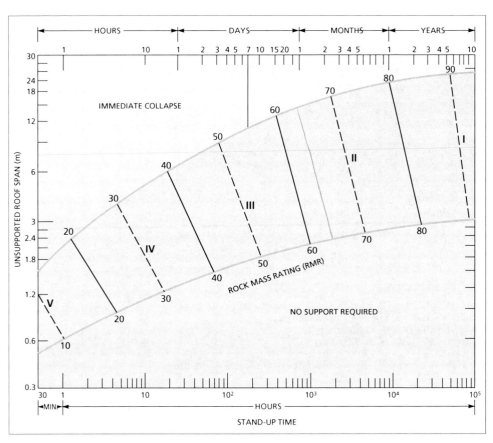

*Figure 10.11* Unsupported roof span and stand-up time for various rock mass classes according to the RMR (Bienawski, 1989).

| Table 10.8 | VALUES OF Q CLASSIFICATION ESR INDEX | |
|---|---|---|
| | **Excavation category** | **ESR** |
| A | Temporary mine openings. | 2–5 |
| B | Permanent mine openings, water tunnels for hydro power (excluding high pressure penstocks), pilot tunnels, drifts and headings for large excavations. | 1.6–2.0 |
| C | Storage rooms, water treatment plants, minor road and railway tunnels, surge chambers, access tunnels. | 1.2–1.3 |
| D | Power stations, major road and railway tunnels, civil defence chambers, portal intersections. | 0.9–1.1 |
| E | Underground nuclear power stations, railway stations, sports and public facilities, factories. | 0.5–0.8 |

(Barton, 2000).

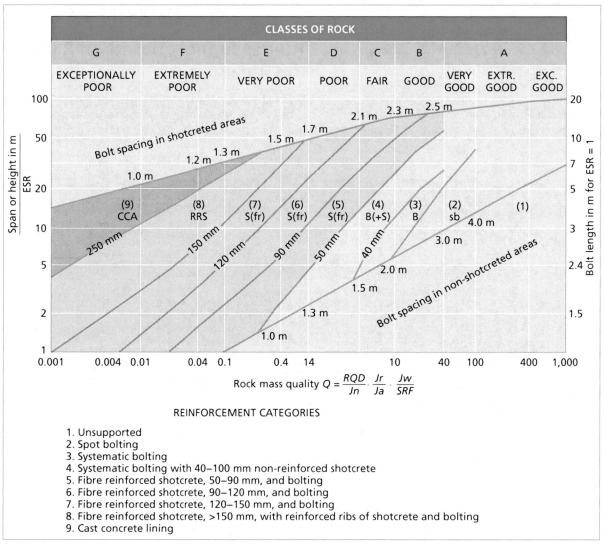

Figure 10.12  Estimated support categories based on the tunnelling quality index Q (Barton, 2000).

be estimated from the RMR. For example, if the RMR is 60, the resulting unsupported roof span is 2 m for a stand-up time of 41.7 days.

The **rock load or pressure** on the support can be estimated from the expression:

$$P = \frac{100 - RMR}{100} \gamma B$$

where $\gamma$ is the unit weight of the rock and $B$ the tunnel width. This empirical expression should be used with caution as results may not be representative.

## Tunnel support based on the Q index

Tunnel support can be estimated from the $Q$ index. The following parameters are defined:

— Equivalent diameter of the tunnel

$$= \frac{\text{width, diameter or height (m)}}{ESR}$$

— ESR (excavation support ratio): the values of this factor are shown in *Table 10.8* and depend on the type of excavation.

Supports are estimated according to the categories in *Figure 10.12*.

The following parameters can also be calculated from the $Q$ index:

— Maximum span (unsupported) = $2\ ESR \cdot Q^{0.4}$ (m)
— Permanent roof support pressure ($P_r$) (kN/m² × 10²)

$$P_r = \frac{2\sqrt{Jn}}{3Jr\sqrt[3]{Q}} \quad \text{(for rock masses with fewer than three sets of discontinuities)}$$

$$P_r = \frac{2}{Jr\sqrt[3]{Q}} \quad \text{(for rock masses with three or more sets of discontinuities)}$$

## 10.7 Excavability

Tunnelling in rock depends on the ease or difficulty of rock removal when using different tunnel excavation methods (Section 10.8).

Excavability depends on the following rock properties:

— Uniaxial compressive strength.
— Hardness and abrasiveness.
— Rock mass fracturing.
— Geomechanical quality indexes.

To assess excavability, some criteria are described in Chapter 9, section 9.8. Examples of those most commonly applied to tunnel excavation are also included in this section.

### Excavability as a function of uniaxial compressive strength

Uniaxial compressive strength and the spacing of discontinuities can be used to differentiate excavation by blasting from excavation by mechanical methods (*Figure 10.13*).

An estimation of excavability for roadheaders, based on the uniaxial compressive strength, $\sigma_{ci}$, and tensile strength, $\sigma_t$ of rock, is as follows:

— Easy to cut: $\sigma_t/\sigma_{ci} < 0.1$
— Difficult to cut: $\sigma_t/\sigma_{ci} > 0.1$

### Abrasiveness: The Schimazek and Cerchar indexes

Abrasiveness and cuttability can be assessed by using the **Schimazek index** (Schimazek and Knatz, 1970) using the formula:

$$F = \frac{Q \times d \times \sigma_T}{100}$$

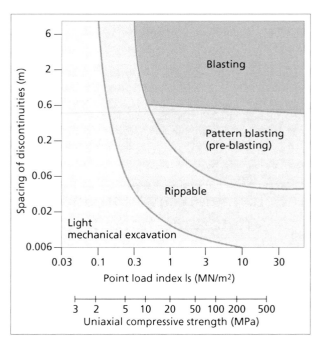

*Figure 10.13* Rock classification for excavability (Franklin, 1974).

| Table 10.9 | ABRASIVENESS OF ROCKS USING THE CERCHAR INDEX | |
|---|---|---|
| **Cerchar index** | **Classification** | **Type of rock** |
| >4.5 | Extremely abrasive | Gneiss, pegmatite, granite |
| 4.25–4.5 | Highly abrasive | Amphibolite, granite |
| 4.0–4.25 | Abrasive | Granite, gneiss, schists, pyroxenite, sandstone |
| 3.5–4.0 | Moderately abrasive | Sandstone |
| 2.5–3.5 | Medium abrasiveness | Gneiss, granite, dolerite |
| 1.2–2.5 | Low abrasiveness | Sandstones |
| <1.2 | Very low abrasiveness | Limestone |

where:

$F$ = abrasiveness factor (N/mm$^{-1}$).
$Q$ = equivalent quartz content in abrasive minerals (%).
$d$ = mean diameter of quartz grains (mm).
$\sigma_T$ = Brazilian tensile strength (MPa).

Mineral examination is carried out using thin sections, taking $SiO_2$ as a reference mineral. According to this index, rock cuttability is as follows:

| Abrasiveness $F$ | Cuttability |
|---|---|
| 0.2–0.3 | Very good |
| 0.3–0.4 | Good |
| 0.4–0.5 | Moderate |
| 0.5–0.6 | Fair |
| 0.6–0.8 | Poor |
| 0.8–1.0 | Very poor |

Abrasiveness can also be determined from the **Cerchar test**. This test involves passing a needle over a sample and measuring the diameter (in tenths of a millimetre) of the scratch it produces. Rocks classified according to this test are shown in *Table 10.9*.

## 10.8 Tunnel excavation and support methods in rock

Excavation and support methods are based on the following basic requirements for tunnels in rock:

— Geological conditions at tunnel depth; overburden thickness; ground water conditions; geomechanical properties and ground behaviour.
— Excavation and support methods are based on geometrical, geological and geomechanical conditions, as well as specific functional and design requirements. Other aspects to be considered include: rate of advance, safety conditions, effects on the surrounding area, e.g. displacements induced by excavation, completion dates, environmental effects, and special conditions in urban tunnels and shallow tunnels with a thin overburden.
— The design of the **primary support** and the **permanent support** depend on the chosen construction system. Prior to installation of the latter, factors such as long term safety, ventilation and type of tunnel use should be considered, bearing in mind that part of the ground load may be absorbed by the primary support.
— Ground monitoring and control must be carried out both during and after excavation to ensure that the ground and the support are behaving properly. Monitoring is also essential for safety purposes.

Tunnelling excavation also involves overcoming a series of problems from a constructional point of view:

1. The tunnel should advance in such a way that the unsupported excavation face should be stable for a certain length of time, during which the primary support is installed, stabilizing the tunnel crown and side walls. This implies that the change in stress induced by excavation has to be compatible with the support and the ground conditions in order to avoid failure. *Figure 10.14a* shows the original state of stress of the ground before excavation. During excavation, the state of stress (points 1 and 6) far from the face will be the same as it was initially, but around the face (points 2, 3, 7 and 8 in *Figure 10.14b*) the state of stress will change. Vertical pressure $\sigma_V$ will decrease with the displacements $u$ due to the relaxation produced by excavation (*Figure 10.14b*). Vertical pressures will stabilize at point 5 (away from the front) when the ground and supports begin to interact and reach equilibrium pressure (point 5). At the same time, horizontal pressures $\sigma_H$ will change and the initial Mohr circle (at points 1 and 6) will now occupy different positions. Before the support is fully active, circle 1 (the tunnel crown)

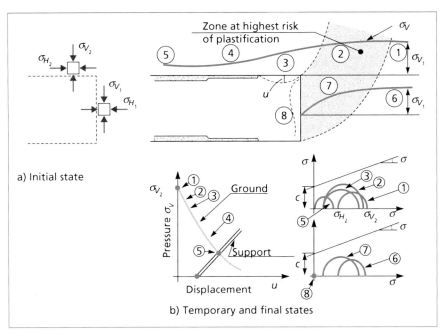

Figure 10.14  Changes of stress at the tunnel front.

moves closer to the line of intrinsic strength (circles 2 and 3, corresponding to points 2 and 3), along which there is risk of failure. However, once the support is active, circle 5 is reached and the situation becomes more stable and safer. A similar phenomenon occurs at point 6, which evolves from circle 6 to 8.

2. Ground excavation always produces stress relief and decompression in the ground, which can be allowed to undergo partial relaxation in such a way that equilibrium pressure is relatively low when the support is installed. Figure 10.15 shows the relationship between ground pressure ($\sigma_v$) and displacements ($u$), usually known as the **ground reaction curve** or **characteristic line**. When tunnel excavation is carried out, pressures drop due to the displacements and stress relief. If the ground is assumed to be elastic, the ground reaction curve is straight. If not, it is a curve, which becomes parallel to the axis $u$ if the cavity is unstable. The support is installed after some displacement, $u_o$, has already occurred and has its own ground reaction curve (in which displacements will increase with the rise in pressure). The point where the two curves meet is the point of excavation equilibrium. Later on, with the passage of time and the deformation of the materials, this point may vary somewhat (Figure 10.15).

3. The **New Austrian Tunnelling Method (NATM)**, described further on, is based on the ground reaction described above, which considers that displacements arising from stress relief mobilize rock strength around the excavation. This, together with the installation of a flexible support, allows the working pressures that have to be balanced by the temporary or preliminary support to be lower than they would have been had no deformation been permitted. Therefore, apart from long-term stress changes, the permanent lining will also have to absorb smaller pressures than would have been the case had no deformation been permitted (Figure 10.15). Nevertheless, given that excavation will affect the surrounding ground and the ground may contain undiscovered characters and singularities, it is not always desirable to allow a substantial deformation of the ground to occur unchecked in an attempt to achieve these benefits. In highly fractured and tectonized rock materials, in some volcanic materials, such as tuffs and pyroclastics, and in soils (where an extrapolation of the NATM may pose a danger because of the loss of bonding in sandy-clay soils and fissured clays), ground deformation may not be admissible (Figure 10.16). In these cases, excavation that results in the least possible deformation to the ground is recommended, with the immediate installation of semi-rigid support.

4. Excavation output should be as efficient as possible, with the aim of completing the tunnel excavation in as few stages as possible, and should be compatible with the need to control deformation: the smaller the excavation, the lower the overall change in ground

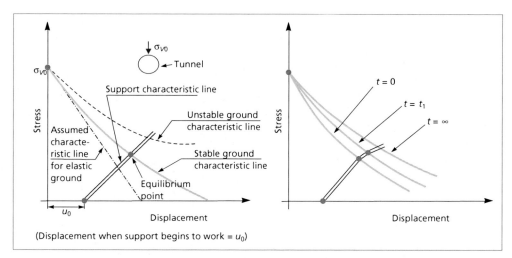

*Figure 10.15*   Ground reaction curves.

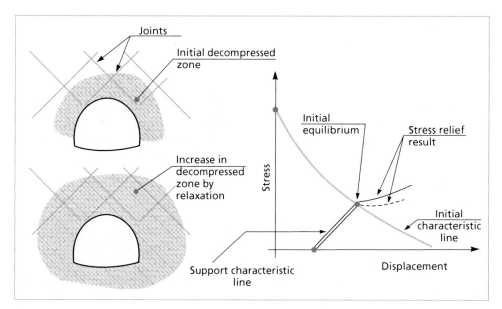

*Figure 10.16*   Decompression around an excavation and its effects on heavily jointed or poorly cemented rocks.

pressures, and so the greater the stability of the system as a whole (*Figure 10.17*).

5. Permanent support and lining will provide appropriate safety conditions on a short- and long-term basis.
6. Safety must be guaranteed in every aspect of tunnel excavation.
7. An economic balance between rate of advance, completion dates and excavation methods should be achieved. The use of highly mechanised construction methods would seem ideal, but this can mean very high costs depending on tunnel size and ground properties, e.g. a tunnel boring machine may cost between 2 and 12 million euros.

The **influence of geological and geotechnical conditions** is basic to tunnel construction; excavation and support are fully dependent on the geomechanical quality of the ground. For these reasons, geological engineering investigations are essential in any tunnelling project. These investigations should focus on the following points:

— Geological and geomechanical characterisation relevant to tunnel design and construction, including slope stability analysis in portal areas and their surroundings.

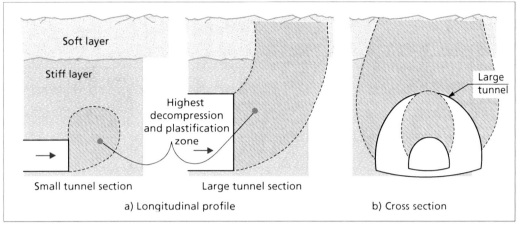

*Figure 10.17   Influence of tunnel size on stability.*

— Identification of potential geological hazards that may affect tunnel construction.
— Recommendations on excavation and support methods.
— Ground displacements induced by tunnel excavations in urban areas and in shallow tunnels.

## Excavation methods

The most commonly used methods for tunnel excavation in rock are **drill and blast,** and **mechanical excavation,** which includes **tunnel boring machines** (TBM).

**Drill and blast** excavation is used in high strength or abrasive rocks, with a seismic velocity in the range of $v_p >$ 2,000–2,500 m/s, where the use of explosives is necessary. This is the most frequently used method in rock tunnelling. The procedure involves drilling holes in the rock face and loading them with explosives, which are then detonated. Drilling is carried out by a drilling carriage or jumbo (*Figure 10.18*). *Figure 10.19* shows the excavation sequence and installation of supports.

One of the main objectives in rock blasting is to avoid excessive breakage of the rock around the excavation. Inadequate blasting gives rise to over breaks and rock falls, which creates additional stability problems. It is therefore essential to carry out **controlled blasting,** which includes techniques such as **pre-splitting** and **smooth blasting,** to minimize structural damage to the rock mass.

**Mechanical excavation** includes the use of road headers or other types of percussive tunnelling machines, such as boom headers and hydraulic hammers.

**Road headers** are machines with an arm that can be extended across the rock face. At the end of this arm there is a rotating head on which the cutting tools or picks are mounted (*Figure 10.20*). Depending on how powerful they are, road headers allow medium or even high strength rocks

*Figure 10.18   Jumbo drilling machine with four arms commonly used in drilling and blasting.*

to be excavated under the repeated blows of these picks. Best results are obtained when the rock has an uniaxial compressive strength of between 20 and 60 MPa. Materials with $v_p$ between 1,900 and 2,500 m/s are seldom rippable and require very heavy machinery. Between 1,900 and 1,600 m/s rippability is average, while below 1,600 m/s they are easily rippable. Together with these criteria, abrasiveness also has to be taken into account.

**Boom headers and hydraulic hammers** are used more selectively. They carry impact tools, or pick hammers, mounted on the end of an articulated arm (*Figure 10.21*). This system is auxiliary to the other methods of excavation described here and is used when either vibration or stability problems preclude the use of explosives, or where the length of tunnel section to be excavated does not warrant changing the construction methods.

**Tunnel boring machines (TBM)** excavate a full circular section by means of a rotating head equipped with discs or cutting picks (*Figure 10.22*). TBMs are generally

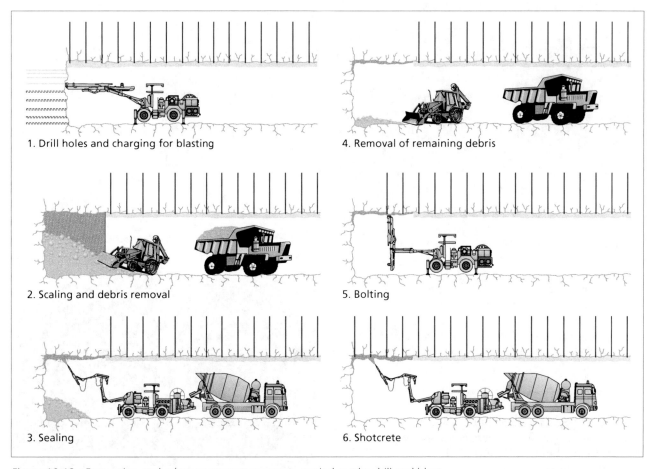

*Figure 10.19* Excavation and advance support sequence carried out by drill and blast.

*Figure 10.20* High-powered roadheader.

*Figure 10.21* Tunnel excavation with a hydraulic hammer. Steel sets and Bernold metal sheets supports.

suitable for a wide range of rocks, from weak to very strong rocks, although in the latter case they are less efficient. Their main advantage is their very high performance as they excavate and support the tunnel in a single continuous process. Certain situations may give rise to limitations, which include: heterogeneous rock masses; fault zones; significant presence of water, squeezing ground or high stress zones. In any case a very thorough knowledge of geological conditions

*Figure 10.22* "Robbins" tunnel boring machine (TBM) for rocks.

*Figure 10.23* Tunnel bench excavation.

in the tunnel is required before a TBM is selected. The choice of excavation method is based on a cost-benefit analysis (once criteria that would exclude the use of a TBM have been ruled out), including factors such as tunnel length, cross section, geological problems, abrasiveness and rock hardness. The following basic aspects should be studied before any decision is made regarding tunnel excavation with a TBM:

— Lithological heterogeneity and lengths of the different types of ground to be excavated.
— Geomechanical conditions of the rock mass.
— Presence of faults, shear zones and highly fracture rocks.
— Seepage zones, squeezing ground and high *in situ* stress.
— Rock hardness and abrasiveness.
— Angle of tunnel with respect to structural anisotropy planes.

Selecting a TBM excavation method is a complex task which must take into account criteria relating to costs, excavation output, geometric characteristics of the tunnel and geological and geomechanical conditions. The latter are fundamental and can be evaluated by applying **specific rock mass classifications for TBM-excavated tunnels**. The $Q_{TBM}$ index (Barton, 2000) and the RME index (Beniawski et al., 2006), allow conditions for excavating tunnels by TBM to be assessed, including TBM performances, average rate of advance, support and design criteria.

## Stages of excavation

When the tunnel cross section exceeds 40 to 60 m², excavation in various stages should be considered especially if ground quality and stability are low. The first stage of excavation is known as **heading or top heading,** and the second stage as **bench excavation** (*Figure 10.23*). Bench material can be excavated either in a single stage or in partial stages, for example, by beginning at the central part of the bench and then by excavating the side benches (*Figure 10.24*).

In a further stage the invert can be excavated forming an **inverted arch** if poor quality ground is encountered. This arch will close the tunnel section and improves the support.

## Support systems

Tunnel stability is provided by the support, which fulfils the following objectives:

— Preventing the ground from losing its strength.
— Preventing rock falling from unstable ground during the excavation process.
— Limiting deformations around the cavity created.
— Controlling seepage and protecting rocks exposed in the tunnel from weathering.
— Providing safety to persons and installations.

In order to attain these objectives, a support system must be installed as early as possible; this is known as **primary support**. Additional support elements, or **secondary support**, may be required to restrain the ground in weak areas. Once excavation is stabilized and supports have been installed, the tunnel is covered with a layer of concrete, which constitutes the **lining.** This has several functions apart from contributing to long-term stability; it improves the tunnel's appearance including luminosity, houses services and reduces air or water friction. The main types of support systems used in rock tunnels are described below; for more detailed description, the reader is referred to the recommended bibliography.

**Shotcrete**, including **gunite** (see Section 9.6) has two main purposes; it seals cracks in the rock surface and prevents decompression and alteration of the rock. The sprayed concrete acquires a strength of 5,000 kN/m² eight hours after application and 28,000 kN/m² after 28 days. It also prevents falls of small blocks and wedges.

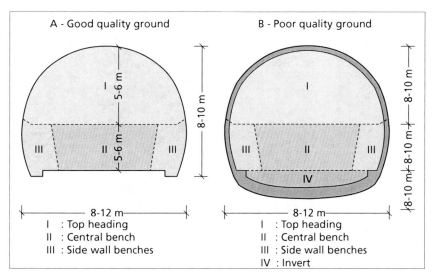

Figure 10.24   Stages in tunnel excavation.

Reinforcing elements are not added to shotcrete used exclusively for sealing the excavation as it is usually less than 5 cm thick. When it is required to function as an element of resistance, shotcrete is reinforced with electro-welded mesh or steel fibres.

**Electro-welded mesh** consists of corrugated rods welded together electrically (Figure 10.25). Because it adapts easily to the shape of the tunnel, it is used for reinforcing shotcrete. The most commonly-used meshes have a rod diameter of 4, 5 or 6 mm, with a mesh size that ranges from 10 to 20 cm. Alternatively **steel fibres** can be used. These are 3 cm long wires with a 0.5 mm diameter that are mixed in with the concrete in a proportion of 40 to 50 kg of fibres to one square metre of concrete; when this system is used the shotcrete should be at least 6 cm thick.

**Bolts** are steel bars from 20 to 40 mm diameter with a length that normally varies between 3 and 6 m; they are installed in appropriately-sized holes that have previously been drilled into the rock mass and will withstand loads ranging from 10 to 25 t. The most commonly-used bolts consist of corrugated bars anchored with grout or resin along their entire length. Other types are **friction bolts**, which are put in place and expanded by water pressure (Figure 10.25), and **polyester resin bolts** reinforced with glass fibre. Bolts may be **active** or **passive**, depending on how they work. The first type functions as an anchor. They have a free shaft and the head (the part that protrudes from the excavation) is held in place with a plate and nut; loads between 5 and 15 t are usually applied. Passive bolts are fixed to the rock along their entire length and loads are not applied to them.

Bolts affect excavation in two ways:

— Keeping discontinuities in the rock mass tightly closed, thus preventing rock mass displacement such as block and wedge rock falls.

Figure 10.25   Installation of "Swellex"-type friction bolts with steel sets, welded wire mesh and shotcrete.

— Providing a confining effect on the rock mass and avoiding stress relief and decompression.

**Steel sets or ribs** are arch-shaped sheet steel profiles that that transfer the load upon them to the ground (Figures 10.21 and 10.25). They need to be firmly supported at their base and positioned so as to make contact with the ground along their entire length. The working load is generally low, that is, in the order of 30–60 kN/m$^2$.

**Forepoling** are bolts installed at an angle of about 40°–45° towards the tunnel heading to avoid rock instability and wedge falls. Forepoling is particularly useful when highly fractured rock is being excavated.

**Crown bar protection** are used in areas of highly fractured or weathered rock, bolts or micropiles can be installed around the crown where rockfall is possible as the

*Figure 10.26* Tunnel reinforcement by crown micropile protection.

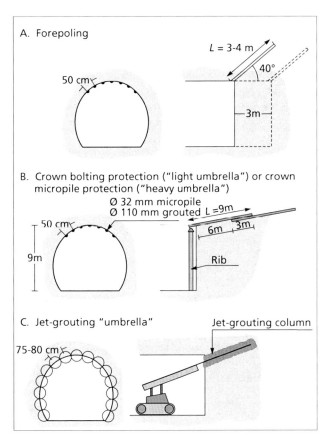

*Figure 10.27* Examples of ground stabilization treatments.

tunnel advances. They are also referred as "umbrellas". The bolts usually have a diameter of 32 mm, whereas the micropiles consist of jet-grouted tubes between 3 and 4 mm thick with diameters ranging from 102 to 150 mm (*Figure 10.26*). When the area to be excavated is large, a succession of bars overlapping each other by 2 or 3 m is installed.

# Ground improvement

When the ground strength is low or has other problems, such as those associated with instability or water inflow inside the excavation, special ground improvement is required. Its application has different purposes: consolidation and reinforcement and waterproofing or drainage (*Figure 10.27*). Ground improvement techniques include:

— **Grouting:** depending on their purpose, grouting techniques may be used for ground consolidation and reducing permeability or waterproofing.
— **Jet-grouting:** can be used in very loose materials (shear and breccia zones, very weathered rock, etc.), the entire perimeter of the tunnel cross section can be reinforced with jet-grouting techniques (see Section 9.6 for a description). Through this reinforcement an arch is formed around the tunnel crown which provides safety and stability to the tunnel front.
— **Drainage** techniques are used to control seepage. Drainage holes and adits are usually made.
— **Freezing** is used only in very special cases. The materials should be saturated or at least have a high water content. This is a very specialised and expensive technique.

# The New Austrian Tunnelling Method

The New Austrian Tunnelling Method (NATM) was developed in Austria in the mid-twentieth century. It is a tunnel construction method based on *in situ* observation of the rheological behaviour of rock masses and the relationship between radial deformation and supports. It is applicable to all types of ground, including tunnels in weak rock or with a thin overburden. Its basic principles can be summarized as follows:

— Immediate application of provisional or primary support, using shotcrete, bolts and, exceptionally, steel arches.
— Monitoring and measurement of *in situ* deformation with time (convergence measurements around the tunnel cross section).
— Additional reinforcement of the support if needed.
— Application of secondary support based on the data obtained from monitoring the behaviour of the excavation.

The NATM is based on the principle of achieving a situation in which the ground around the excavation acts as a load-bearing ring, thus constituting a fundamental element in active support. The following conditions must be met:

1. Weathering and decompression of the rock mass should be avoided as these processes imply a considerable reduction in shear strength. Shotcrete is applied immediately throughout the whole section to the entire excavated surface.
2. Jointed rock masses are very sensitive to uniaxial stresses and so, in these cases, bolts are installed to supplement the shotcrete.
3. Each change or readjustment in the state of stress of the rock mass creates disturbances and failures in the rock around the excavation. For this reason, the whole section should be excavated wherever possible. Partial excavations should be avoided or at least reduced to not more than three stages.
4. Because the "rock ring" around the excavation must remain as strong as possible, an excessive concentration of stresses on the tunnel section must be prevented; corners and marked protuberances should be avoided in favour of rounded sections, especially at the foot of partial excavations.
5. A tunnel construction should not be considered as a static arched structure but as a tube. A tube has much greater load-bearing capacity but only if its section is completely closed. This means that whenever excavation of poor quality ground is involved, the floor of the tunnel (its invert) must be closed by means of an inverted arch or similar structure.
6. Geological and rock engineering assistance is essential throughout the whole excavation process to help interpret movement and thus control deformation.

## Portals

Portals form one of the most critical parts of the tunnel as they are generally sited on slopes with thin overburden. The portal area includes:

— The front and side slopes of the portal excavation; these may be permanent or temporary if the excavation is later filled forming a **cut and cover tunnel.**
— The transitional zone of the front slope and the first few metres inside the tunnel.

The following **geological and geomechanical problems** should be considered in the study of the portal:

— Natural stability: active mass movements and re-activated paleo-landslides.
— Induced instability: the slope excavation may intersect unstable discontinuity surfaces allowing them to daylight onto a free face and permit mass movement.
— Low shear strength zones: highly weathered rocks and surface deposits may provide zones of low shear strength, as well as contacts between colluviums and weathered rocks, or highly weathered rocks lying on fresh rock.
— Weathering: slope excavation can expose surfaces to rapid degradation due to weathering, especially in rocks with clayey composition.
— Stress relief: excavation may lead to the opening up of discontinuities, reducing their strength.
— Seepage: when an excavation intercepts the water table, the flow tends to be towards the excavation which affects the strength and stability of its boundaries.

The location of the portal should be decided taking the geological and geomechanical conditions into account. Portals should be sited on stable slopes and in a rock mass which is strong enough to allow the excavation of the tunnel under acceptable conditions of stability. From the construction point of view, twin portals should be approximately symmetrical and large excavations into the slope should be avoided.

The following **geological and geomechanical criteria** should be applied to portal design:

— Site surveys should include boreholes, seismic refraction geophysics, trial pits, permeability tests, and the installation of piezometers and inclinometers, as described in Section 10.2.
— Areas affected by landslides or other types of slope instability should be avoided where possible and stabilised where this is not possible. Monitoring of the slope may be needed for detailed analysis of these (see Chapter 13).
— Slopes should be designed according to the methods described in Chapter 9, to achieve safety factors equal to or higher than 1.2 for temporary slopes and 1.5 for permanent slopes. Support measures must also be designed where needed to meet the required safety factor (see Chapter 9).
— As well as stabilization measures, slopes must be protected from the following engineering geological problems:

• Rapid alteration and unloading: the slope faces nearest the portals must be treated with shotcrete or bolted, particularly if the rock mass is weathered, highly fractured, or clayey.
• Seepage and its effects: as a preventive measure, run-off water should be channelled or diverted from the tunnel area and the surrounding slopes, and also where the slopes converge towards a portal; drainage holes should be installed where necessary.

*Figure 10.28* Reinforcement of the face of a tunnel portal. Crown micropiling, bolts and welded wire mesh; at a later stage, the slope will be covered with shotcrete.

*Figure 10.29* Portal reinforcement of a tunnel in poor quality rocks with an anchored wall and micropiles.

- Rock falls: slopes should be protected with dynamic barriers or other protection systems, as described in Chapter 13.

— Inside the tunnel excavation, just beyond the portal entrance, the ground should be also reinforced along the cross section or at least along the crown by using forepoling, micropiling, or other means of reinforcement (*Figure 10.28* and *Figure 10.29*).

## 10.9 Tunnel excavation and support methods in soil

### Non-mechanical excavation methods

In the past, tunnels were excavated in small sections using wooden supports. The ground was ripped out with picks and wedges as the section gradually became bigger. Great advances in tunnelling were made in the nineteenth century when major railway tunnels through the Alps and other mountainous areas of Europe and America were excavated. Many of these methods are still in use. Worth mentioning here is the **Austrian method**, which obtained good results by dividing the cross section in two partial sections. The first half section or heading is excavated in the upper half of the tunnel and the second half section or bench is excavated in the lower part. This procedure is called **heading and bench** construction.

The Belgian Method and the Madrid Method are two of many others derived from the Austrian Method. In the **Madrid method** (*Figure 10.30*) the advance in the heading section is around 2.5 m, using metal girders as longitudinal support and wood for the rest of the support. The concrete lining is applied to the crown after 24 hours. The bench and sidewall sections are then excavated 20–25 m. from the heading. Finally the invert is excavated and filled with concrete. Wood and steel sets are used for temporary support and concrete for permanent support. This system has been used successfully in the Madrid Metro.

### Semi-mechanical excavation methods

A significant advance in the excavation of the heading and tunnel face in soils is the use of a **shield.** A series of hydraulic jacks enable the steel shield to advance. This shield can consist of sliding metal blades or lances. Manual or mechanical excavation can take place underneath this cover, due to the reaction provided by the lining, which is erected as soon as excavation has been carried out, with an advance rate of 2.5–3.0 m per period of excavation, or cycle. If the tunnel is large the rest of the section can be excavated following the Madrid Method: sidewalls, bench and invert.

Another method of excavation by partial mechanization is the **Bernold method** (*Figure 10.31*), used for hard soils and poor quality rocks. The advance rate is from 1–3 m per cycle, and steel sets are immediately installed across the entire width of the crown. Perforated metal Bernold sheets are installed resting on the steel sets. Before further advance is made, the space between the metal sheet and the rock is filled in with 15–30 mm thick shotcrete. This may be further reinforced with an external shotcrete layer. The bench is excavated using traditional procedures, supported by extending the steel sets and Bernold metal sheets. The final concrete lining can then be applied.

In hard soils or weak rocks a **mechanical pre-cutting** method can is also be used. This method uses a chainsaw tunnelling machine. A pre-crown is constructed before each

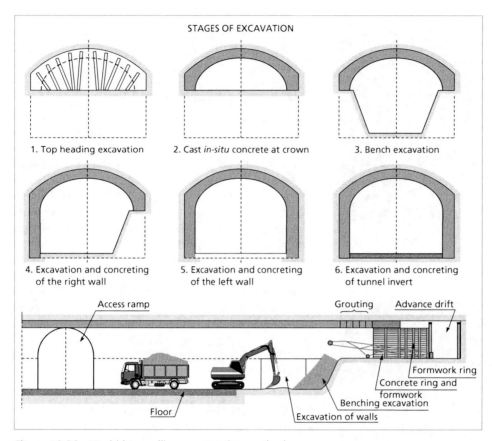

*Figure 10.30* Madrid tunnelling construction method.

advance by a guided saw, filling in the slits cut with concrete and working along the full section (*Figure 10.32*). Theoretically the pre-crown can be 10–30 cm thick and the "tiles" that form the pre-crown are 3–4.5 m long, with 0.5–1 m overlap.

## Tunnel excavation with tunnel boring machines

In both rocks and soils, the complete construction process can be mechanized so that the excavation and support can be carried out systematically, without continually having to make decisions about support, lengths of advance, etc. Tunnel boring machines (TBM) are used for this. TBM can excavate a full section and install a high quality pre-cast lining, which generally consists of reinforced concrete rings made up of 3–7 pre-cast segments normally with a strength of 40–60 MPa. The length of these concrete rings may range from 80 to 170 cm.

TBM were first used in the mid-nineteenth century. They may be open-fronted and excavate with a cutter boom, or they may have a back loaded cutter at the front, armed with picks for soils and cutters for rocks. Excavation advance and reaction is achieved in soils by means of hydraulic jacks resting on and pushing against the pre-cast lining that is being erected inside the metal shield, or in rocks, with grippers or pads that press against the bare rock walls of the tunnel.

During the extension of the Madrid Metro system, six closed-fronted TBM were used, five of them with a diameter of 9.40 m and the other with 7.40 m. The average daily advance rate was between 8 and 18 m, excavating in slightly cemented Pliocene materials and Miocene clays containing gypsum (*Figure 10.33*). These machines were earth pressure balanced (EPB) TBMs. EPBs are equipped with a chamber at the front of the TBM that allows the excavated ground to form a continuous slurry when water and/or foam is added. Pressure can be applied to this slurry simply by removing a larger or smaller quantity of debris from the chamber. In this way, the pressure on the excavation face can be maintained similar to that of the ground as a whole, so increasing stability and reducing deformation in the excavated void at the front of the machine. In soft soils, gravels and loose sands, bentonite or a similar acting fluid additives can be used to help make the excavated soil behave as a controllable slurry.

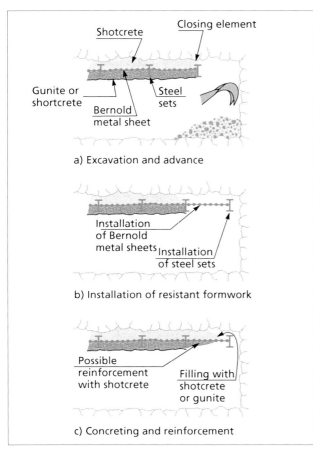

Figure 10.31 Bernold method.

Figure 10.32 Excavation using mechanical pre-cutting.

Figure 10.33 TBM used in the Madrid Metro.

## 10.10 Geological engineering during tunnel construction

Geological and geomechanical problems may arise during tunnel construction. Whether or not they occur depends mainly on how well the geology of the tunnel is understood. To a large extent, the solution to these problems lies in anticipating them and taking the appropriate measures. Unforeseen situations occur in many tunnels, due either to inadequate study of the site or to the inherent uncertainty that is always present in tunnelling, and these may represent an important hazard. It is therefore essential to carry out further site surveys during the tunnel excavation in addition to those carried out before construction. The methods most often used include boreholes, drilled either at the tunnel face or ground surface, and geophysical surveys. In special circumstances, such as very long or very deep tunnels, exploratory adits are also made.

The following engineering geological problems and potential hazards to tunnel excavation should be identified:

— Faults and highly fractured zones intersecting the tunnel cross section.
— Lithological contacts between rocks with significant differences in hydraulic or mechanical properties.
— Large wedges of rocks in weak, loose or highly fractured materials, especially under high water pressures, intersecting the tunnel cross-section.
— Shear and weathered zones across the tunnel section.
— Presence of cavities.
— Highly permeable materials, particularly paleochannels, karstic and evaporite rocks.

- Expansive and swelling materials. Remember that rocks containing anhydrite can induce significative swelling pressures.
- Significant *in situ* stresses that can give rise to high rates of deformation.
- Stress phenomena such as rock burst, spalling, squeezing and large deformation under heavy loads.
- Aggressive rocks, e.g. gypsum, sulphur (e.g. pyrite and marcasite), acidic water, or abrasive rocks with high quartz content.
- Presence of explosive and toxic gases, oxygen deficiency and high temperatures.

*Figure 10.34* shows a tunnel face excavated in shales affected by faulting. *Figure 10.35* shows a sequence of large deformations in weak rocks.

As well as identifying the above listed problems, the following data should be also obtained during tunnel construction:

- Location, depth and extent of the engineering geological problem and its position in the tunnel.
- Lithology of the overburden materials and their hydrogeological conditions.
- Geomechanical properties of the rocks or soil forming the problem zone, including strength, permeability, plasticity, expansivity and aggressivity.
- Water flow rates and pressures where seepage is anticipated.

Geological problems can affect the excavation method and support used including additional reinforcement, consolidation and drainage measurements. Depending on the nature of the problem to be solved some of the following **remedial measurements** can be adopted:

- Immediate installation of the primary support as close as possible to the excavation face.
- Reduction of the excavation section area, even by driving headings and adits from which consolidation grouting, drainage and grouting to reduce permeability can be carried out.

*Figure 10.34* Faults intersecting the tunnel excavation front in RMR Class IV shales.

*Figure 10.35* Large deformations in mudstones and rock salt: a) initial roof failure; b) large failures with high deformations in the side walls; c) excavation is almost closed due to ground deformation and bolt support failure.

- Top heading reinforcement by increasing the length and density of bolts, installing steel sets, and forepoling.
- Invert excavation and floor reinforcement.
- Consolidation of the perimeter of the excavation by means of grouting, jet-grouting, forepoling and micropiling.
- Implementation of drainage and measures to reduce groundwater flow.
- Stabilization of the excavation area by providing central buttresses and sealing the excavation face.

In the case of **tunnels in soil materials and urban environments**, the following aspects should also be considered as they may be affected by additional problems:

- Influence of existing underground excavations (historic tunnels, old military tunnels, metro lines, sewers, underpasses, etc.) and foundations (possibly now abandoned) near the projected tunnel and their effects on the tunnel excavation and its stability.
- Overburden composed of unconsolidated ground, man-made ground, or along former water courses that have been used for waste disposal.
- Buildings and other structures located near the tunnel which may be affected by tunnel construction.
- Weathered or low strength materials, highly permeable zones and contacts between sandy and clayey deposits.
- Aquifer changes induced by tunnel excavation.
- Presence of water levels in sandy layers or "perched" water levels between two impermeable clay layers, which may produce instability in the tunnel excavations.
- Contaminating liquids such as petrol leakage from service stations or kerosene in airport areas. The presence of dangerous liquids floating on the surface of the ground water as well as volatile gases, may make decontamination essential prior to tunnel excavation.

Tunnel construction always involves a degree of uncertainly because of: i) the extrapolation of geological data from a limited number of boreholes and other site investigations; ii) the difficulty in most cases of interpreting geological structures from surface data; iii) the limited number of samples at tunnel depth; iv) the uncertainty of the geomechanical properties assigned to the tunnel materials. It is therefore essential that the excavation must be **monitored and controlled** throughout the entire construction process.

Tunnel monitoring should be carried out to:

- Monitor and control the stability, deformation and seepage problems encountered during construction.
- Detect departures from the ground response anticipated and adjust the support system and the excavation method to the ground conditions encountered during tunnel construction.
- Design remedial measures for slope stability problems in the portal areas.
- Monitor buildings or other nearby structures that may be affected by the tunnel excavation.
- Monitor ground water levels, river waters and soils for potential contamination problems, and the dumping of non-reusable excavation materials.
- Contribute to safety and quality control procedures.

**Tunnel monitoring** is based on the systematic measurement of deformations and stresses that are a consequence of the excavation process and primary support. Measurements are taken in different cross sections at intervals that depend on the geomechanical quality of the ground; these can vary over distances of 20–50 m and may be even smaller in highly variable and poor quality ground. Ideally, monitoring should be continuous and record the excavation of a tunnel as well as the ground's response to it but in practice this does not happen often. Deformation measurements tend to be taken more frequently during the excavation and installation of primary support, generally once a day in the first week. As time goes on, they become less frequent ranging from once a week to once a month.

The **monitoring system** (*Figure 10.36*) is a means to control ground response and tunnel safety, and includes the following activities:

- Displacement measurement in the support and lining, using tape extensometers or optical procedures that allow horizontal movements and relative displacements between the roof and the sidewalls to be monitored.
- Displacement measurements, both at the surface and underground, using single and multiple rod extensometers and inclinometers.
- Settlement and tilting measurement in nearby buildings.
- Ground pressure measurement inside the tunnel excavation, using load cells.
- Ground water and piezometric monitoring of seepage. In some cases seepage can produce a drawdown of the piezometric level in nearby Quaternary materials, which may result in settlement and damage to buildings and installations lying on such materials.

The results of monitoring should be made available on a daily basis and be presented as graphs showing displacement, distances and time from the tunnel front. Interpretation of these measurements can also indicate the level of

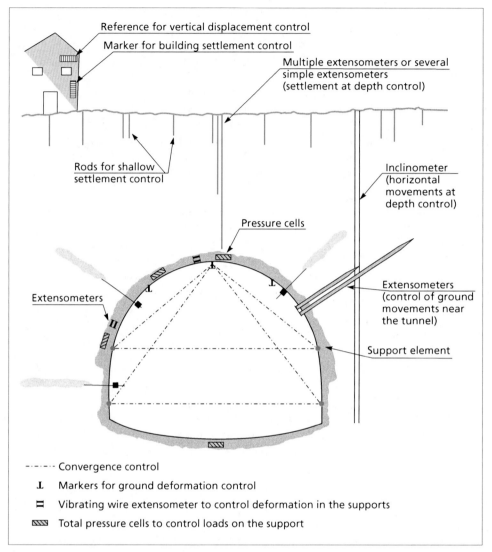

Figure 10.36   Tunnel monitoring systems.

safety conditions and the possible effect of the excavation on nearby buildings. The instrumentation systems are described in Chapter 5, Section 5.6.

**Geological and geomechanical data** obtained during tunnel excavation provide fundamental information for the control of tunnel stability. The following data should be systematically recorded:

— Lithology and rock structure at the tunnel front.
— Faults, fractures and other specific discontinuities.
— Rock mass classification and quality indices.
— Stability problems, swelling, squeezing, rock burst and spalling.
— Seepage, gases and high temperatures.

These data should be analysed in relation to the stability of the excavation and support measurements, and be presented in the form of geological and geomechanical cross sections and plans along the crown and sidewalls. If the geological conditions suggest there is a hazard of swelling, squeezing or very permeable ground, or the possibility of intersecting large tectonic structures, additional boreholes will be required at the excavation face or from ground surface and, where necessary from adits, depending on the scale of the problem.

Geological and geomechanical control during excavation provide basic criteria for determining primary support. This information needs to be supplemented with data obtained from monitoring.

## Recommended reading

Barton, N. (2000). TBM Tunnelling in jointed and faulted rock. Balkema, Rotterdam.
Hoek, E., Kaiser, P.K. and Bowden, W.F. (1995). Support of underground excavations in hard rock. Balkema.
Hoek, E. and Brown, E.T. (1980). Underground excavations in rock. Institution of Mining and Metallurgy. London.
Hudson, J.A. and Harrison, J.P. (1997). Engineering rock mechanics. Pergamon.
Whittaker, B.N. and Frith, R.C. (1990). Tunnelling. Design, stability and construction. Institution of Mining and Metallurgy, London.

## References

Barton, N., Lien, R. and Lunde, J. (1974). Engineering classification of rock masses for the design of tunnel support. Rock Mechanics, 6, 189–239. Springer Verlag.
Barton, N. (2000). TBM Tunnelling in jointed and faulted rock. Balkema, Rotterdam.
Bienawski, Z.T. (1979). The geomechanics classification in rock engineering applications. Proc. 4th Int. Conference on Rock Mechanics. Montreaux, vol. 2, 41–48. Balkema.
Bienawski, Z.T. (1989). Engineering rock mass classifications. John Wiley and Sons.
Bienawski, Z.T., Celada, B., Galera. J.M. and Alvarez, M. (2006). Rock mass excavability indicator. New way to select the optimum tunnel construction method. Tunnelling and Underground Space Technology, 21, 3–4.
Franklin, J. (1974). Rock quality in relation to the quarrying and performance of rock construction materials. 2nd Int. Conf of the IAEG, IV-PC-2. Sao Paulo.
González de Vallejo, L.I. (2003). SRC rock mass classification of tunnels under high tectonic stress excavated in weak rocks. Engineering Geology, 69, 273–285.
González de Vallejo, L.I. and Hijazo, T. (2008). A new method of estimating the ratio between in situ rock stresses and tectonic based on empirical and probabilistic analysis. Engineering Geology, 101, 185–194.
Goodman, R.E., Moye, D.G., Van Schalkwyk, A. and Javandel, I. (1965). Ground water inflow during tunnel driving. Bull. Assoc. Engineering Geologists, vol. 2(1), 39–56.
Hansen, L. and Martna, J. (1988). Influence of faulting on rock excavation. Inter. Symp. on Rock Mechanics and Power Plants. ISRM, Madrid, vol. 1, 317–324. Balkema.
Heidbach, O., Tingay, M., Barth, A., Reinecker, J., Kurfeß, D. and Müller, B. (2008). The release 2008 of the World Stress Map (www.world-stress-map.org).
Hoek, E. and Brown, E.T. (1980). Underground excavations in rock. Institution of Mining and Metallurgy, London.
Hoek, E. and Diederichs, M.S. (2006). Empirical estimation of rock mass modulus. Int. Journal of Rock Mechanics and Mining Sciences, 43, 203–215.
Hoek, E. and Marinos, P. (2000). Predicting tunnel squeezing problems in weak heterogeneous rock masses. Part 1: estimating rock mass strength. Part 2: estimating tunnel squeezing problems. Tunnels and Tunnelling Int. 32/11, 45–51 and 32/12, 33–36.
Hoek, E. and Marinos, P.G. (2009). Tunnelling in overstressed rock. In: Rock Engineering in difficult Ground Conditions. Vrkjan (ed). Taylor and Francis, pp. 44–60.
ISRM (2003). Suggested methods for rock stress estimation. Part 1, 2, 3 and 4. Int. Soc. for Rock Mechanics (www.isrm.net).
Schimazek, J. and Knatz, H. (1970). The influence of rock composition on cutting velocity and chisel wear of tunnelling machines (in German). Glückauf 106: 113–119.
Sheorey, P.R. (1994). A theory for in situ stresses in isotropic and transversely isotropic rock. Int. Jl. Rock Mech. Mining Sci., 31, 23–24.
Singh, B. (1993). Indian experience of squeezing ground and experiences of application of Barton's Q-system. Workshop on Norwegian Method of Tunnelling, CSMRS, New Delhi.
Sing, B., Jethwa, J.C. and Dube, A.K. (1992). Correlation between observed support pressure and rock mass quality. Tunnelling and Underground Space Tech., 7, 59–74.
Waggoner, E.B. and Daugharty, C.W. (1985). Geologic site investigations for tunnels. Underground Space, 9, 109–119.

# 11

# DAMS AND RESERVOIRS

1. Introduction
2. Types of dams and auxiliary structures
3. Site investigation
4. Engineering geological criteria for dam selection
5. Geological materials for dam construction
6. Reservoir water tightness
7. Permeability of dam foundations
8. Reservoir slope stability
9. Engineering geological conditions for dam foundations
10. Seismic actions and induced seismicity

# 11.1 Introduction

Dams are one of the most important development infrastructures in any country and are used for many different purposes including irrigation, water supply, flood control, electricity production (19% of worldwide supply) and storage of mine wastes. Economic growth is directly related to the construction of dams, with more than 45,000 worldwide considered as large dams (i.e. over 15 m high or with a reservoir capacity of over 3 million m$^3$). The shortage of water resources is a major global problem: more than 1,000 million people do not have access to the minimum daily water requirement (50 litres per person/day) while in industrialized countries the consumption is 4–14 times this amount.

Huge hydraulic projects are underway in developing countries such as China, which has more than 22,000 dams (50% of the world total) compared with 22 dams in 1949. In Europe and other industrialized countries, on the other hand, there has been a sharp decrease in the number of dams built since 1990, largely because of public opposition.

**Environmental problems** are among the most hotly debated topics related to dam construction and are the basis for the strongest objections to building more dams. Problems include sedimentation and silting up of reservoirs and soil salinization. Sedimentation affects a large number of dams, with an overall estimated loss of 0.5%–1% of total storage capacity. Soil salinization resulting from higher water tables affects 20% of the areas regulated by dams and makes agricultural land non- productive.

Other environmental factors, such as soil erosion, landslides, induced seismicity, eutrophication, climatic effects and modified river dynamics, along with the social impact (some 40 million people are displaced worldwide because of dams) and the economic impact (many countries incur debt due to the cost of building dams), are also part of the continuing controversy over water demands, sustainable development and environmental impact.

The **safety of dams** is another important factor often debated by their critics, although safety levels are very high and have improved noticeably over recent years, with a 0.5% failure rate recorded since 1950, compared with 2.2% previously. Nevertheless, it is significant that most of these failures have been caused by geological problems. In concrete dams, 21% of failures originate in the foundations, while in embankment dams 31% are due to ground problems such as erosion affecting the foundations.

The relationship between dam safety and geology has been one of the most important research topics in rock mechanics and geological engineering since the 1960s, since many of the most serious accidents have been due to foundation problems. The failures of the Malpasset Dam (France) in 1959, the Vajont Dam (Italy) in 1963, and the Teton Dam (USA) in 1975 are among the best known. Although the problems were different in each case, all three failures were related to the geological conditions.

In **Malpasset** the structure of the gneiss and schist rock mass, dipping downstream, was conducive to the stability of the dam and its foundations as a whole, but the geological surveys for the project did not identify, or at least did not take sufficiently into consideration, a fault transversal to the schistosity on the left slope. The resulting pore pressures when the reservoir was filled created the required conditions for the failure of a large section of the foundation defined by these discontinuities. The foundation failure of the 66.5 m high dam caused the collapse of the dam and immediate emptying of the reservoir. The resulting flooding caused 421 deaths in the town of Fréjus (Section 11.9 deals with this failure in more detail).

The example of **Vajont** is very different. This catastrophe, which caused 2,018 deaths, was the result of a massive landslide of some 300 million m$^3$ into the reservoir, that produced a gigantic 70 m high wave which overtopped the 265 m double curved arch dam without breaching it, and caused flooding and destruction of a large part of the town of Longarone and other villages in the valley of the River Piave.

In this case, the stability studies of the left slope did not forecast the magnitude of possible landslides in spite of considerable evidence of slope movements. The main causes of the landslide were (Semenza and Ghirotti, 2000):

— Pre-existing paleo-landslides.
— Geological structure conducive to failure.
— Presence of a low-strength clay layer on the failure surface.
— An aquifer with a high hydraulic pressure under the clay layer.
— Repeated sudden filling and drawdown of the reservoir.

*Figures 11.1* and *11.2* show views of the dam and reservoir after the disaster.

The **Teton dam** was a 93 m high embankment dam with a rock foundation on highly fractured and permeable

# DAMS AND RESERVOIRS

*Figure 11.1* The Vajont Dam, present-day view from downstream.

*Figure 11.2* The Vajont Dam, present-day view from upstream; note the landslide that has occurred adjacent to the reservoir but the dam itself did not suffer any important damage.

rhyolite tuffs. The central core of the dam was made of low plasticity and highly erodible clay silt. Some months after the dam was filled, leakage was observed on the downstream slope near the right abutment. A few hours later the seepage through the dam caused a large cavity to appear in the face of the dam and the dam rapidly collapsed. Fortunately, many lives were saved in spite of the short time between when water was first seen leaking and the final collapse, and the death toll was only 14. The failure was caused by internal erosion of the materials, as the water flow was not sufficiently controlled. The type of material used in the core and other aspects of the project also played a part in the accident.

At present, great advances in engineering geology and rock mechanics mean that geological surveys for dam projects can detect possible problems in foundations or slope stability, analyse how these could affect the safety of the dams and allow the necessary corrective measures to be taken.

There are many different **information sources** on dams including the International Commission on Large Dams (ICOLD) and the World Commission on Dams (WCD), which provide technical, social and economic data. The International Society for Rock Mechanics (ISRM), International Society for Soil Mechanics and Geotechnical Engineering (ISSMGE) and the International Association for Engineering Geology and the Environment (IAEG) provide specific information on different geological and geotechnical aspects.

## 11.2 Types of dams and auxiliary structures

### Types of dams

Dams can be classified into two groups according to the construction material used: embankment dams and concrete dams.

## Embankment dams

Embankment dams include **earthfill and rockfill dams**. Most geological materials can be used to build dams, except those which may change, dissolve or evolve, so modifying their properties (Section 11.5). The construction method consists of compacting materials arranged in layers, following the procedures described in Chapter 12.

The most characteristic types of embankment dams are:

— **Homogeneous section dams**. All, or almost all, of the dam is built of the same type of material, formed by low permeability compacted soils. To control seepage, different types of drainage measurements can be used as shown in *Figure 11.3a*.
— **Zoned dams with impermeable clay core**. Two or more types of materials are used. The least permeable zone or core acts as the impermeable element. The width of the core and its position within the section may vary considerably (*Figure 11.3b*).
— **Curtain dams**. The impermeable element consists of a relatively slim curtain or membrane. The materials most used include asphaltic or hydraulic concrete, and polymeric or bituminous materials.
— **Rockfill dams**, built with rock fragments of different sizes.

Examples of embankment dams are shown in *Figures 11.4, 11.5* and *11.6*.

## Concrete dams

These dams are made of concrete and their geometry depends mainly on the foundation conditions and the site morphology. The most characteristic types are:

— **Gravity dams**. Their cross-section resists the thrust of the water on its own. Generally they require a greater

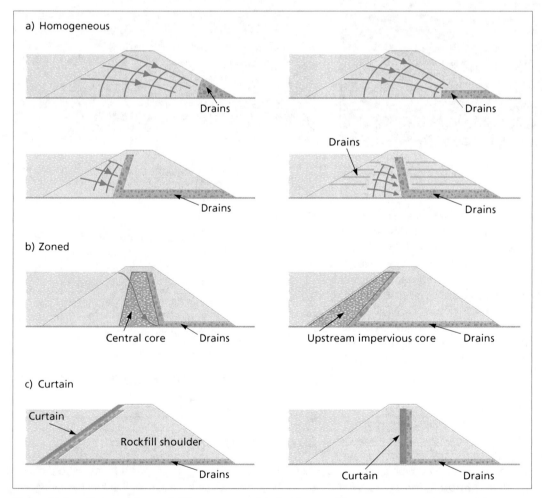

*Figure 11.3* Cross sections of different types of embankment dams.

Figure 11.4   Guadarranque Dam (Spain) on Eocene sandstones and marls; 71 m high earthfill dam with clay core; the spillway can be seen (foreground) and the outlet tower (background).

Figure 11.5   Canales Dam (Spain) viewed from upstream, sited on sandstone and marls; 156 m high embankment dam, with rockfill shoulders and central clay core; note the outlet tower.

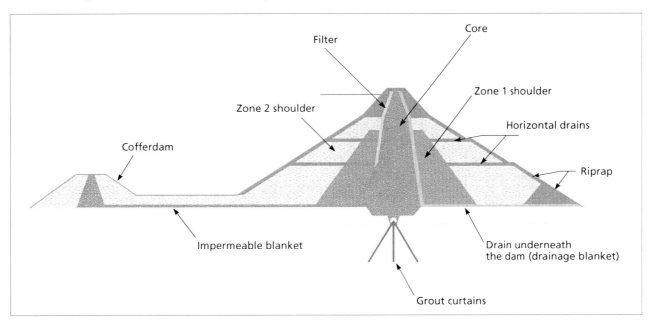

Figure 11.6   Cross section of the Giribaile embankment dam, Spain.

volume of concrete compared with other types of concrete dams. A high strength rock mass near the surface is needed for the foundations. *Figure 11.7* shows an example of a gravity dam.

— **Buttress dams**. This type of concrete dam has structural supports, or buttresses, transversal to the dam section, which reduce the volume of concrete and the uplift pressures. This type of dam needs to be founded on very high strength rocks due to the concentration of loads in the buttresses. They are sited in wide valleys. *Figure 11.8* shows an example of a buttress dam.

— **Arch dams**. To reduce the section of the gravity dams these are designed in an arch shape to transmit part of the load to the abutments or lateral supports for the dam on the valley slopes.

— **Double curved arch dams**. This type of arch dam has a double curvature and is the most complex in design, analysis and construction. These dams are also very slim structures to reduce the volume of concrete required. Their foundations need a very high strength rock mass. The deformability conditions of the rock foundation must be compatible with the estimated

*Figure 11.7*  Sancho gravity dam, 50 m high (Spain).

*Figure 11.9*  Canelles Dam (Spain) on Cretaceous limestones; 151 m high double curved arch dam.

*Figure 11.8*  Aracena Dam (Spain) on Silurian shales; 60 m high buttress dam.

deformability of the dam. Double curved arch dams typically transmit their thrusts to the abutments: they not only require a high bearing capacity rock, but the shear strength of the discontinuities in their abutments must ensure their stability. This type of dam may be very high and they are usually sited in narrow valleys; *Figure 11.9* shows an example.
— **Roller-compacted concrete dams**. Over the last twenty years roller-compacted concrete (RCC) techniques have advanced considerably and are used to build these dams with the equipment and techniques used for embankment dams. The concrete is laid in layers of variable thickness of approximately 30 cm, spread with mechanical shovels and compacted with vibrating rollers. In general terms, RCC dams behave in the same way as gravity dams, although an arch effect may be attributed to them through a geometric curve. *Figure 11.10* shows various typical sections of concrete dams.

## Auxiliary structures

### Cofferdams

These are low dyke-like constructions, which allow the river to be temporarily diverted while the dam is being built. In narrow valleys the river is normally diverted through a tunnel or tunnels excavated through the abutments. These constructions are adapted to use later for emptying or refilling the reservoir, or for a hydroelectric power station. In wide valleys other solutions may be adopted, such as diversion tunnels, canals or galleries which later will be buried under the dam as functional elements for outlet or drainage. In embankment dams the cofferdam is often included in the planned cross-section of the dam (*Figure 11.6*).

### Spillways

These structures allow the reservoir once at its top water level, to discharge further water through the dam itself or through independent structures. They are designed for maximum flows based on studies of real or estimated historical rainfall series in the catchment for the dam site.

The location and design of the spillway depends on the type of dam and on the topographical and geological conditions of the dam site and its surroundings. In

# DAMS AND RESERVOIRS

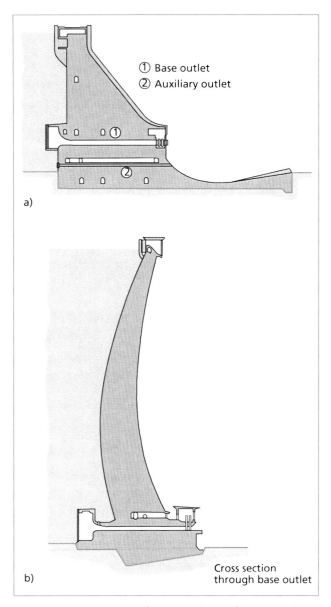

Figure 11.10 Examples of cross sections of concrete dams: a) Gravity dam, b) Double curved arch dam.

embankment dams the spillway is built independently from the body of the dam (*Figure 11.4*).

In embankment dams an insufficient capacity spillway may produce dam failure. Even before the water starts to discharge, the wave action may break the dam crest, causing the throughflow of water and the collapse of the dam.

In concrete dams the spillway can be open or have gates, and can be built into the body of the dam or as a side channel spillway with flip buckets (*Figures 11.7* and *11.8*). These dams may have one or more base outlets. The spillway gates are mechanical installations to control the outflow over the spillway.

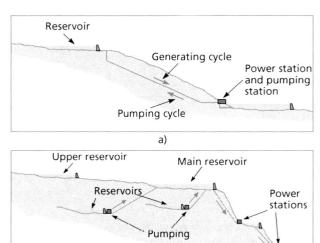

Figure 11.11 Hydroelectric power schemes: a) simple reversible system, b) storage and pumping system with secondary reservoirs situated at intermediate levels.

### Outlets

These structures allow the drawdown of the reservoir at different levels. Their number and capacity depends on the volume stored in the reservoir, on the services which have to be supplied and on the depth of the reservoir (*Figure 11.5*). The base outlets are used to empty the reservoir at levels below the outlets to partially eliminate sediments which may have built up in the reservoir, and to increase the drainage capacity or to carry out inspections of the dam.

### Hydroelectric schemes

*Figure 11.11* shows a hydroelectric power station for generating electricity using off-peak demand periods to pump water to a higher reservoir and then generate power with the downflow of the water at peak demand times. In hydroelectric schemes the power station can be located below or above ground and usually includes several tunnels.

## 11.3 Site investigation

## Planning site investigation

Geological and geotechnical surveys for dams are carried out depending on the different stages of the dam design and construction. *Table 11.1* shows the sequence of these studies which are usually carried out as follows:

| Table 11.1 | SEQUENCE OF ENGINEERING GEOLOGICAL STUDIES FOR DAMS | |
|---|---|---|
| Stage | Type of engineering geological investigations | |
| Feasibility and preliminary studies | Regional geology | — Geological surveys<br>— Reservoir water tightness<br>— Materials<br>— Slope stability<br>— Silting and subsidence |
| | Preliminary investigations of dam sites | — Geological structure<br>— Geophysical surveys<br>— Permeability estimation |
| Design | Dam site and reservoir characterization | — Strength and deformability<br>— Leakage<br>— Auxiliary structures<br>— Materials<br>— Other problems |
| Construction | Geological and geomechanical assistance and control | — Geological mapping<br>— Geophysical surveys<br>— Strength and deformability *in situ* tests<br>— Grouting tests<br>— Ground monitoring and control |
| Operation | Monitoring and control | — Initial reservoir filling<br>— Emergency situations<br>— Safety analysis |

## Preliminary and feasibility studies

The aim of these is to establish the viability of the dam project according to the following geological criteria:

— Absence of significant **geological hazards** for the safety of the dam and the reservoir (e.g. large landslides, intense karstification or active faulting in the dam site).
— Appropriate **geomorphological conditions** of the proposed dam site.

## Selecting the type of dam

At this stage it is necessary to provide geological criteria needed to assess the best type of dam in relation to the alternatives available and the site conditions. Technical, economic and environmental points of view are considered. The engineering geological aspects to be investigated are as follows:

— Availability of construction materials.
— Strength, stability and permeability of the foundations.
— Reservoir slope stability.
— Hydrogeology and watertight reservoir conditions.
— Seismotectonic conditions including induced seismicity.

## Design

The aims of this stage are to provide geological and geotechnical criteria for the design of the dam and its auxiliary structures, the design of ground improvement measures and to provide recommendations for the foundation excavation and dam construction. The aspects to be studied are:

— Detailed geological and geotechnical description of the dam foundations.
— Geological and geotechnical study of the auxiliary structures foundations.
— Seismic and neo-tectonic studies.
— Ground improvement and seepage and drainage control measurements.
— Construction recommendations.

## Construction

Detailed engineering geological surveys are carried out during the dam construction. These include:

— Geotechnical mapping and zoning of the excavations.
— Borehole drilling, geophysical surveys and *in situ* testing.
— Stability analysis of the excavation slopes.
— Monitoring and control of the ground behaviour.
— Seepage control.

- Ground improvement treatments.
- *In situ* and laboratory tests of building materials.

## Operation

During the dam operation monitoring is carried out to ascertain, and where necessary control, the behaviour of the ground and the dam structure while it is in use, especially:

- Geotechnical interpretation of the results from instrumentation and those from monitoring measurements made during the filling of the reservoir and start-up operation of the dam.
- Leakage control, including uplift pressures, slope stability and induced seismicity.

## Site investigation methods

**Site investigations for dams** depend on the geological conditions and on the type of dam, its dimensions and auxiliary structures and should be in accordance with the design specifications. The general criteria for site investigation follow the guidelines given in Chapter 5 referring to the planning of the site investigation carried out in successive phases. This practice is necessary in any important project, and is essential for dams. As has already been mentioned, dam safety is directly related to geological conditions, so that all aspects of the engineering geological studies are essential. These studies must ensure that the geological or geotechnical factors which could affect the safety of the dam are sufficiently investigated and identified.

The cost of site investigations depends on the geological complexity of the site, the type of dam, its dimensions and the auxiliary structures. In general terms, embankment dams require less site investigation than concrete dams, but they need a more detailed study of the auxiliary structures and of the construction materials. Gravity dams require very detailed studies of the foundations. The minimum budget devoted to site investigations varies between 0.5% of the total cost for embankment dams and as much as 2% for concrete dams, percentages which may sometimes double if the geological conditions are complex. *Table 11.2* presents a summary of the most characteristic site investigation methods.

**Table 11.2  SITE INVESTIGATIONS METHODS FREQUENTLY USED FOR DAMS** [1]

| Stage | Site investigation methods | Concrete dams | Embankment dams |
|---|---|---|---|
| Feasibility studies | — Regional geological mapping | Y | Y |
| | — Photo-interpretation and satellite images | Y | Y |
| | — Geological reconnaissance and preliminary geological mapping | Y | Y |
| Preliminary project | — Detailed engineering geological mapping (1:2,000–1:1,000) | Y | Y |
| | — Hydrogeological data and permeability tests | Y | Y |
| | — Geophysical surveys | Y | Y |
| | — Borehole drilling | Y | Y |
| | — *In situ* test inside boreholes | Y | O |
| | — Laboratory tests | Y | Y |
| Project | — Complementary geotechnical mapping | R | O |
| | — Complementary boreholes | R | O |
| | — Exploratory adits | Y | N |
| | — Deformability and strength *in situ* tests | R | N |
| | — Grouting tests<br>— Complementary permeability tests | R | N |
| Construction | — Detailed geological mapping of the excavations and foundations | Y | Y |
| | — *In situ* control tests and ground instrumentation | Y | Y |
| | — Quality control tests | Y | Y |

(Y: necessary; R: recommended; O: optional; N: not necessary).
[1] Remember every dam has to be considered unique and requires whatever investigation is needed for the ground it is on.

In the preliminary studies a geological survey of the dam site, reservoir basin and borrow materials is carried out. It includes an analysis of all the existing information and the study of aerial photographs and satellite images. In areas with few outcrops, open pits and trenches have to be made to observe the lithology, weathering and the bedrock structure; *Figure 11.12* shows a typical case of a granitic valley with slopes covered with vegetation and colluvial deposits, where trenches were excavated at 20 m contour intervals. As well as allowing observations of the rock mass, these trenches were used as access berms for other *in situ* studies (boreholes, geophysical surveys, etc.).

Figure 11.12 Large trenches excavated to survey a dam site in granitic rocks covered with vegetation and colluvial soils.

The geological, hydrogeological and geotechnical conditions of the reservoir area also have to be examined (*Figure 11.13*), especially for the presence of soluble rocks (gypsum, limestone or salts) or very porous rocks (tuffs, volcanic agglomerates or sandstones) and for instability features present on the slopes. The aim of these studies is to establish a preliminary **geomechanical zoning** of the site based on geophysical methods (seismic refraction and electric methods), boreholes and trenches. Some *in situ* tests inside the boreholes are also carried out, e.g. permeability tests.

The number and depth of the site surveys, as well as the types of *in situ* test, depends on geological conditions, the type of the dam and its dimensions. In the case of embankment dams, selecting the type of dam is usually based on the preliminary studies and complemented with additional boreholes at specific points or on the site of the auxiliary structures. For concrete dams, on the other hand, the site investigations are much more detailed and varied, so as to define the deformability parameters and the rock mass strength.

The most common site investigations are the following:

— **Boreholes**. The minimum number of boreholes recommended for dams of less than 50 metres high is 3–4 on each side of the valley and at least 2–3 on the valley floor, preferably inclined and crossed. *Figures 11.14* and *11.15* show various hypothetical layouts of boreholes. How deep these are will depend

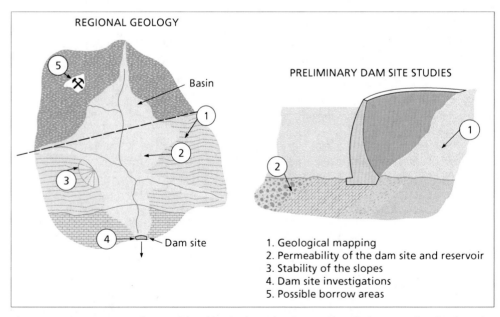

Figure 11.13 Aspects to be considered in site investigations and preliminary studies for dam sites.

## Box 11.1

### Suggested criteria for site investigations of dams

— Site investigations should be carried out in **stages** using methods in the early stages that cover extensive areas and more detailed and sophisticated ones as the project develops.

— *In situ* data should be representative of the investigated rock mass and of sufficient quantity to allow **statistical analysis**.

— Site investigations should be carefully planned and coordinated selecting those methods than can be used for different purposes or can be complementary to each other. Costs and time can be reduced if adequate planning and coordination is carried out.

— The overall interpretation of the geological, hydrogeological and geomechanical parameters should lead to the **geomechanical zoning** of the rock mass foundation (see figure below) that will be used to define the foundation conditions.

— **Specific zones** of the rock mass such us faults, shear zones, dykes, weathered zones, etc., should be considered as particular points and studied on an individual basis.

— *In situ* **large scale tests** are usually carried out in concrete dams. These are limited in number and are located at the critical zones of the rock mass to provide the strength and deformability parameters of the foundations.

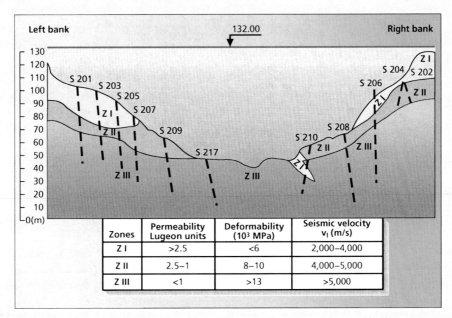

Geomechanical zoning of the foundations of the Cambambe Dam, Angola (Oliveira, 1985).

on the thickness of the weathered rock, the geological structure and on the depth of the bearing rock present in the valley. As a general rule, boreholes should reach a minimum depth of at least half the height of the dam and some of them should be as deep as the height of the dam. However, they should be deeper if there are permeable levels, very weathered materials or faults underneath the foundations. The positioning usually follows alignments along the axis of the dam and transversal to it. Some boreholes should also be sunk upstream and downstream of the dam site and in the reservoir basin. The geological interpretation of these will decide whether further study is needed with more or deeper bores.

— **Geophysical methods**. Seismic refraction profiles and vertical electric soundings are carried out along

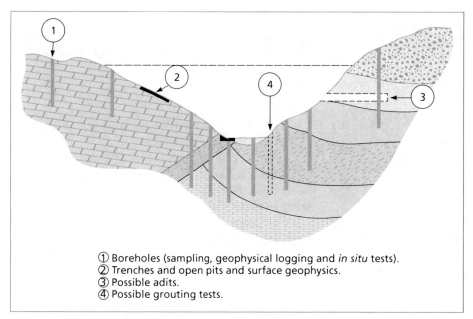

Figure 11.14   Dam site characterization.

① Boreholes (sampling, geophysical logging and *in situ* tests).
② Trenches and open pits and surface geophysics.
③ Possible adits.
④ Possible grouting tests.

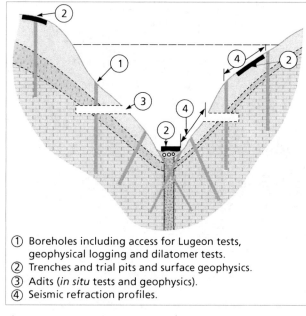

① Boreholes including access for Lugeon tests, geophysical logging and dilatomer tests.
② Trenches and trial pits and surface geophysics.
③ Adits (*in situ* tests and geophysics).
④ Seismic refraction profiles.

Figure 11.15   Site investigations for gravity dams.

the axis of the dam and transversally to it. There should be enough of these to identify the thickness of the weathered zone and alluvial deposits and the location of faults. These geophysical surveys are complemented with downhole and crosshole testing and borehole-logging. Interpretation of the geophysical data is essential as it allows geomechanical zoning, estimation of deformation modulus, identification of faults, weathered zones and dykes, and rock rippability to be assessed.

— **Permeability tests**. These are usual practice in all dam site studies. The most widely used are Lugeon tests on fractured rock and Lefranc tests on highly permeable rocks and soils. They must be carried out according to their prescribed methods.

— **Tracer tests** are used in karstic areas and in general to investigate permeable materials or fractures of high permeability. Chapter 4, Section 4.4 describes these tests.

— **Grouting tests** are carried out to estimate grout intake for consolidation or waterproofing treatments of the ground, and as a complement to permeability tests, especially in areas which have previously been identified as more permeable. They consist of injecting a mixture of cement and water (known as slurry) under pressure and determining the flow intake.

— **Piezometers** must be installed in most of the boreholes. The type will depend on the hydrogeological conditions of the ground and the material properties.

— **Dilatometer tests**. Depending on the type of dam and the rock mass properties, deformation tests with dilatometers are carried out inside the boreholes to obtain the rock deformation modulus.

— **Adits** can provide a direct, large scale observation of the rock mass, needed for concrete dams; they are expensive. They allow tests to be carried out in fault zones and in weak materials. Borings and seismic

profiles can be made from these adits complementing those carried out at surface level. They also give access to work areas during the construction and operation of the dam so they should remain connected to it.
— **Large scale *in situ* tests.** These tend to be used mainly in arch dams, and usually consist of large scale tests on the rock mass or discontinuity planes. The most characteristic tests are direct shear test, flat-jack and plate-load tests. These tests are carried out inside adits or galleries.

## 11.4 Engineering geological criteria for dam selection

### General criteria

The selection of a dam basically depends on the following factors:

— Capacity of the basin (reservoir volume).
— Watertight reservoir conditions.
— Environmental conditions.
— Favourable geomorphologic, geological and geotechnical conditions.
— Value of the flooded land (villages, infrastructure, etc.).
— Availability of construction materials near the dam.
— Favourable conditions for siting spillways, cofferdams and other auxiliary structures.

When it has been decided which of the various alternative sites originally selected is the most suitable for closing off the valley, the type of dam is selected using a cost benefit analysis which provides a detailed analysis of the possible types of dam from different points of view. One of the most important points of view is that for the geological conditions. Factors to be taken into account in this analysis include:

— Height of the dam.
— Geomorphologic conditions of the dam site.
— Geological, hydrogeological and geotechnical conditions of the foundations.
— Availability of construction materials.
— Geological hazards.

When considering the **height of the dam,** it is significant that more than 80% of dams built with a height of under 30 m high are embankment dams; 60% of those 150 m high or more are concrete and 50% of these are arch dams. These statistics reflect the different reasons involved in selecting the type of dam to use, especially geomorphologic ones. Wide valleys only permit low dams and here embankment dams are more economical (where adequate materials are available), while in narrow or very narrow valleys it is usually high dams which are built, with the gravity dam solution the most used.

The **choice of dam type** does not follow fixed rules, as each site has its own characteristics; no two sites present the same geological conditions and therefore construction problems. This means it is essential to take expert engineering geology criteria into account (Kanji, 1994).

To sum up, the following engineering geological problems associated with dams, reservoirs and their auxiliary structures should be considered:

— **Dam site:** strength, deformability and permeability of the foundations and abutments.
— **Reservoir:** stability of slopes, water tightness, silting by sediments and induced seismicity.
— **Quarries and borrow areas:** quality and volume of available construction materials.
— **Spillways:** Stability of the slopes, strength and permeability of foundations and ground erodibility.
— **Canals:** stability of the slopes, erodibility and rippability of materials and foundations.
— **Surface power stations:** stability of slopes, strength and deformability of foundations.
— **Underground power station:** strength, deformability, permeability and *in situ* stress.
— **Tunnels, diversion galleries and base outlets:** excavation stability, strength, deformability, permeability and *in situ* stress.

### Foundation conditions

The **type of dam selected** depends primarily on the **foundation conditions**. Any type of **concrete dam** requires the rock mass deformability to be compatible with that of the concrete; i.e. that certain deformation levels which the dam structure is not able to bear are not exceeded. For this reason concrete dam foundations cannot be made in soils or weak rocks. A concrete dam would not be the appropriate solution either if the depth of the foundations required very deep excavation, because of the cost implications of the presence of highly weathered or tectonized materials in relation to the intended height of the dam; in these cases the cost of the volume of excavation needed would be very high. However, these decisions are complex, as there are various alternatives. In some cases the foundations for concrete dams have been made in low quality rocks in spite of very deep excavations and ground treatment of the rock mass to "consolidate" it i.e. to infill its voids, large and small. In general, when the deformability of the foundation rock is high or its strength is low, the most appropriate solution is the construction of an embankment dam.

## Availability of materials

Another condition for selecting a particular type of dam is the availability of construction materials near the dam site. These are needed both for concrete dams (aggregates) and for embankment dams which depend basically on the material available.

The type of material can affect important aspects of a dam such as seepage conditions and measurements made for their control. For example, two dams built with different types of materials on the same site may generate quite different flownets in the foundations. In *Figure 11.16*, case a) shows the cross section of an homogeneous embankment dam with average permeability $k = 10^{-6}$ cm/s; case b) shows the cross section of a zoned dam of the same height where the core also has a permeability $k = 10^{-6}$ cm/s and where the upstream and downstream rockfill shoulders are very permeable. The water flow in each case is completely different depending on the hydraulic gradients. The flow in case b) is 5 times higher because:

$$v = ki$$

with $i = h/L$, where $h$ is the difference between the downstream water level and the maximum water level of the reservoir.

As a result, in case a) the flow passing through the foundation can be considered low enough not to need treatment to make it watertight, while in case b) the safety and functionality of the dam require appropriate ground treatment, which may be very expensive.

## Siting of auxiliary structures

The **foundation conditions for the auxiliary structures** (spillways, diversion galleries, outlets, etc.) may have a decisive influence on the selection of the dam type; in general, these structures are shorter, simpler and cheaper in concrete dams than in embankment dams. In concrete dams, the spillway is integrated into the dam itself, discharging water by overtopping or through sluice gates or pipes running through the interior of the dam. In embankment dams the spillway is sited independently from the dam itself, and requires excavation of varying width and depth; in some cases where very high water discharges are involved, these excavations require large-scale, expensive work. The site of the spillway may often decide the type of dam.

As well as the cofferdams, spillways and outlets there are other important structures which form part of the hydraulic installations, such as power stations, tunnels and galleries made for different purposes. All of these must be also considered when selecting the type of dam.

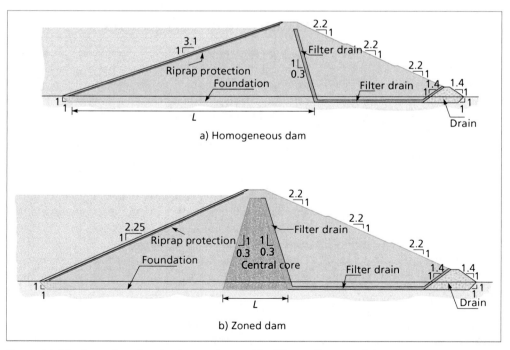

*Figure 11.16* Examples of embankment dams with different flow conditions.

# Conditions for embankment dams

The most important criterion for embankment dams is the availability of materials. The impervious cores require low permeability soils, generally lower than $10^{-5}$ cm/s. These aspects are covered in Section 11.5 and Section 2.7. The geological environment for embankment dam sites may be sedimentary rocks, alluvial soils, over-consolidated clays or weak rocks. Collapsible, erodible, dispersive, organic, soluble or highly plastic soil materials should be rejected as well as highly pervious materials.

Earthfill dams and rockfill dams have different foundation requirements. Earthfill dams are generally flexible and transmit low loads to the foundation. Even under these conditions loose alluvial soils, highly weathered zones, river deposits or, in general, low strength soils are not acceptable as foundation material. Slickensides are often found in hard or over-consolidated soils and can be a potential surface on which sliding could cause the dam to fail. Rockfill dams behave as a more rigid structure and transmit higher loads to the foundations so needing a higher strength foundation.

# Conditions for concrete dams

**Gravity dams** require a high quality rock mass for their foundations to ensure sufficient safety of the dam against failure and deformability. These conditions mean that the dam must be sited on high strength rock mass where the discontinuity planes or weakest surfaces, most critical in terms of stability, have high angles of internal friction. In most cases excavations are needed to remove the weathered and weak zones of the rock mass and the thickness of material to be removed may affect the feasibility of the dam site.

Another important condition for concrete dams is the **control of seepage and uplift pressures** due to pore pressure exerted underneath the dam foundation, aspects which are covered in Section 11.7.

**Buttress dams** present the same foundation problems as gravity dams, but concentrate their loads in the buttress supports, therefore the foundation must be on high strength rocks with low deformability in order to avoid differential settling between the supporting elements. In the same way, seepage in the foundations must be very low, which means that the rock mass must have low permeability.

**Arch dams** are the most demanding of all dam types in terms of their foundation requirements. Their slim structure transmits highly concentrated loads to the foundations and their curved geometry transmits a large part of the hydrostatic thrust to the abutments. This type of dam requires the structural geology of the foundations to be conducive to overall stability under the applied loads and for the rock mass to provide, high strength discontinuity planes and low deformability in the foundation.

# Environmental considerations

Besides the geological and geotechnical conditions, the geoenvironmental effects resulting from dam construction must also considered (Table 11.3), as these effects may have important consequences for the environment. Specific aspects such as the stability of the reservoir slopes and induced seismicity are covered in Sections 11.8 and 11.10 of this Chapter.

| Table 11.3 GEO-ENVIRONMENTAL IMPACTS OF DAM CONSTRUCTION AND OPERATION | | |
|---|---|---|
| **Construction phase** | **Operational phase** | **Emergency situations** |
| — Extraction of materials (quarries and borrow pits)<br>— Waste disposal<br>— Instability of slopes<br>— Noise, traffic, vibrations and dust | **Upstream:**<br>— Landslides and slope stability<br>— Erosion and sedimentation<br>— Silting<br>— Hydrogeological changes<br>— Flooding and loss of mining resources<br>— Induced seismicity<br>— Water quality<br>— Salinization<br><br>**Downstream:**<br>— Erosion and sedimentation<br>— Stability of slopes<br>— Hydrogeological changes<br>— Valley flooding | — Rapid emptying of the reservoir<br>— Flooding downstream<br>— Slope stability<br>— Erosion |

## 11.5 Geological materials for dam construction

The construction of a dam requires a considerable volume of materials. Embankment dams depend fundamentally on the type of material available. The selected materials must combine the following conditions:

— Appropriate volume.
— Adequate quality.
— Operational distance.
— Ease of extraction.
— Suitable environmental conditions.

## Site investigations for dam materials

The materials investigation starts in the initial stages of dam feasibility studies and continues throughout the project and during construction. The geological surveys are based on geological mapping and photo-interpretation. As a preliminary estimate, possible borrow and quarry areas are located. At a later stage, reconnaissance borings, trial pits and geophysical surveys (seismic refraction and geo-electrical profiles) are carried out. Soil samples for classification, index properties and compaction tests are also obtained. In the potential borrow pits and quarries, lithology, structural data, weathering, hydrogeological and geomorphological conditions and slope stability are all considered.

The geomechanical properties of the materials are assessed from laboratory tests on representative samples obtained from the borrow or quarry areas. Once suitable materials have been identified the volumes available can be estimated and extraction conditions evaluated.

## Types of materials

Depending on their use, materials for dams are grouped into the following categories: those suitable for:

— Cores.
— Rockfills and ripraps.
— Filters and drains.
— Aggregates.

*Figure 11.17* shows typical materials for the construction of embankment dams.

Although the majority of geological materials can be used, those which are chemically or physically unstable are not suitable. These include: soluble, organic, collapsible, dispersive, reactive, erodible, expansive, very low density and high plasticity soils. A description of problematic soils is

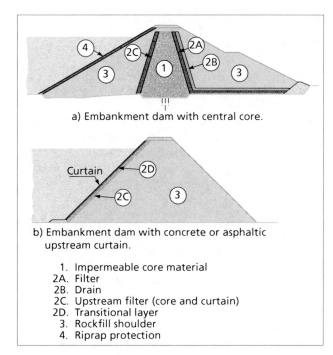

*Figure 11.17* Dam sections showing different types of materials.

carried out in Chapter 2, Sections 2.7 to 2.9. The selection of materials must meet the specific requirements stated in the codes and technical specifications for dam construction.

### Cores

Cores are constructed to prevent water passing through the dam structure. They need low or very low permeability materials, around $10^{-5}$ cm/s or lower. They must be not be collapsible, soluble or contain organic matter and they should be easily compactable. Clays and silts are the most commonly used materials for cores (*Figures 11.18* and *11.21*), although sands with clay content may also be used; their plasticity index should be between 15 and 35. The selection of these materials does not necessarily exclude other low permeability soils which, when suitably compacted and selected, provide adequate impermeability, e.g. marls and argilites. Cores are generally compacted in layers 20–30 cm thick with a sheepsfoot roller or with a vibrating roller (*Figure 11.19*). The properties to be considered in materials to be used in cores are:

— Grain size distribution and plasticity; well graduated soils generally compact better and acquire lower permeability and deformability.
— Mineralogy and its relation to expansivity, soluble salts content, susceptibility and collapsibility (see Chapter 2, Section 2.8).

*Figure 11.18* Compacted clays used for central core.

*Figure 11.20* Rockfills used for shoulders.

*Figure 11.19* Compacting clay core materials with sheepsfoot rollers.

- Dispersibility or susceptibility of the soil to develop erosion; the chemical composition of the clays and the presence of Na may indicate the degree of dispersibility (Chapter 2, Section 2.8). The pinhole test (Chapter 12, Box 12.1) can provide an estimation of the dispersion potential of a soil.
- Strength and deformability as both are fundamental properties for dam stability analysis under the specified design conditions.
- Compactibility, which can be estimated from the Proctor and the modified Proctor tests (Chapter 12, Box 12.1).

## Rockfills and ripraps

The purpose of rockfill is to transmit strength and stability to the dam as well as protecting it from erosion. Rockfill can be made from a variety of rocks in a wide range of sizes (*Figure 11.20*). Ripraps provide protection against erosion. A description of these materials is included in Chapter 12, section 12.3.

Dam shoulders are usually built with rock fill. These materials are laid by extending layers at least 2 m high compacted with vibratory rollers to ensure high strength and low compressibility. The intact rock strength required ranges from 50–150 MPa although rocks with uniaxial compressive strength equal to or greater than 30 MPa are considered acceptable. Ballast and gravel are also used for shoulders, either alone or combined with rockfill; these large size granular materials must not contain fines (less than 5%) and should be well graded.

## Filters and drains

To prevent the core from being eroded and to allow drainage through the dam, filters and drains are used. The volumes required are usually much smaller compared with other materials used although the properties and grading specifications are much more demanding.

The materials for filters and drains should be granular, with very strict grading criteria defined in the specifications, with an absence of fines, with high strength and durability and chemical and physical stability. They may be obtained from natural deposits or by crushing.

## Aggregates

Concrete dams require large quantities of aggregates which must come from quarries or borrow areas located near the dam. The following aspects should be considered for aggregates:

- The material should have a unit weight of at least 25 kN/m$^3$.
- Its uniaxial compressive strength should be greater than that specified for the concrete (24 MPa or above).

- Its water absorption should be lower than 3% and those materials which contain reactive, soluble or weak components should be excluded, e.g. sulphur, gypsum, clays, shale, mica or pyrite.
- Its mineralogical composition should not produce adverse reactions with concrete such as expansivity, chemical alterations and reactions with alkalis.

Site investigations for aggregates are carried out at the different stages of the dam project. In the preliminary stages the following activities are included:

- Geological reconnaissance, photo-interpretation, and geological mapping.
- Geophysical survey (seismic refraction and electrical profiles).
- Boreholes and trial pits.

Aspects to be studied:

- Geological characterisation (lithology, structure and ground water conditions).
- Overburden and its thickness.
- Weathered and highly fractured zones.
- Volume, geometry and homogeneity of the quarry.
- Distance from the dam site and access to it.
- Environmental restrictions.

At a more advanced stage of the dam project, geomechanical characterisation and slope stability analyses are carried out for borrow and quarry areas, as well as laboratory tests of physical and chemical properties. A detailed definition of the quality and quantity of the extractable aggregate is required.

*Figure 11.21* shows an example of embankment dam construction.

*Figure 11.21* Core materials (left) and drain materials (foreground) for an embankment dam.

## 11.6 Reservoir water tightness

One of the basic requirements of a reservoir is that it must be watertight. However, depending on the purpose of the reservoir, some water loss may be acceptable; e.g. a reservoir used for flood control does not have to be as watertight as one used for water supply or irrigation. The water tightness of the reservoir floor and sides should also be analysed in terms of operating requirements as it may be economically viable in some cases to carry out local treatment to reduce leakage in specific zones.

The construction of a dam creates important changes in the hydrology and hydrogeology of the catchment area affected by the reservoir as it involves flooding part of the valley and a rise and periodic fluctuation in water tables. These changes affect aquifer hydrodynamics and may produce either water outflow from the reservoir, or recharge from surrounding aquifers. These conditions are analysed in **hydrogeological investigations** (see Chapter 4).

A preliminary assessment of how watertight the reservoir will be should consider first the direction of the hydraulic gradient then the following factors, which will be crucial when the hydraulic gradient is directed away from the reservoir, as it always is beneath a dam:

- **Massive or slightly fractured rocks** (e.g. igneous, metamorphic and massive sedimentary deposits) are appropriate and generally watertight.
- **Sedimentary rocks with a high clay content** (e.g. marls and mudstones) are very watertight.
- **Fractured sedimentary rocks** are appropriate, as transmissivity takes place through the fractures, which close up with depth, except in highly porous rocks.
- Inflows and outflows of water from the reservoir may be conditioned by **geological structures** such as folds. These structures, and how they are related to any associated faults or discontinuities, must be analysed in depth.
- Large fractures, dykes or heavily weathered areas may form preferential flowpaths This is especially important in hard crystalline and sedimentary rocks, where faults often bring rocks with very different levels of permeability into contact with each other. The hydrogeological control of the structural factors of these fractures is crucial.
- The least appropriate formations, which cause most seepage problems, are **calcareous rocks,** e.g. limestones and dolomites, affected by **karstification processes**.
- Other **porous rocks**, e.g. some types of sandstones, certain volcanic rocks, Quaternary deposits and sedimentary structures such as paleochannels, are generally not watertight.

— Seepage may also be a problem in areas where there are active or derelict **mine workings**; coalfields are particularly problematical as seepage may occur combined with **subsidence.**

From the above it can be seen that the main problems with watertightness are found in soluble or very porous rocks. **Soluble rocks** are formed basically of carbonate and evaporitic rocks. In both types cavities form through dissolution and karstic processes develop. In carbonate rocks the dissolution rate is very slow, but in evaporites it may be very high and therefore significant in terms of engineering geology. Cavities also form in **volcanic rocks** (though not from dissolution), although this happens much less frequently and is due to highly localized phenomena, whereas karstification is a much more generalized and extensive process.

In carbonate rocks the solubility of calcite ranges from 100 to 500 mg/l, and less in dolomites. In evaporitic rocks the solubility of gypsum is 2,400 mg/l, while for halite it is 360,000 mg/l. Under the same hydrodynamic and environmental conditions, the great differences in solubility in evaporites may lead to very rapid dissolution processes, even over short periods of time, as happens in saline rocks. In carbonate rocks, however, the rate of these processes is very slow and consequently has little relevance in terms of engineering geology to the formation of cavities or dissolution phenomena during the lifetime of the structure. The presence of karstic structures is one of the greatest problems in areas rich in calcareous rocks, as is frequently found the Mediterranean region.

As well as cavity formation, dissolution involves changes in the chemical composition of aquifers, with the risk of increasing the soluble solids in the water (e.g. chlorides and sulphates), leading to possible reactions with the concrete and contamination of the water, both in the aquifers and in the reservoir.

## 11.7 Permeability of dam foundations

Problems related to whether reservoirs are watertight have been described in the previous section, but it is also important to consider the permeability of the dam site, not only as a watertight problem but as an engineering geological issue of great significance for the stability and safety of the dam. In general, the dam site should fulfil conditions of low permeability. The effects of seepage, present in most rocks, either through fractures and discontinuities or in weathered zones, are especially critical for the foundations as they lead to the following problems:

— Uplift pressures generated underneath the dam foundation.
— High hydraulic gradients with high seepage rates and risk of erosion.
— Significant loss of water.
— Slope instability at the abutments and in the area downstream from the dam.

## Uplift pressures

The flownet created by a dam may have many different configurations (see Chapter 2, Section 2.3 and Chapter 4, Section 4.5). *Figure 11.22* shows an example where the uplift pressures due to the pore pressures acting beneath the dam are maximum at A and almost negligible at B. To prevent uplift pressures, **cutoffs or diaphragm walls** are provided to modify the flownet, as shown in *Figure 11.23*. The effects of cutoffs are to:

— Decrease the uplift pressures beneath the dam.
— Decrease the flow beneath the dam, its volume and velocity.
— Avoid potential piping and erosion phenomena.

*Figure 11.24* shows a grout curtain acting as an watertight cutoff with drains. The measures to be taken will depend on the flownet conditions and the type of dam.

## Erosion

Erosion is of the greatest importance for the safety of embankment dams, and is the second most common cause of failure of this type of dam (31% from overflow, 15% from

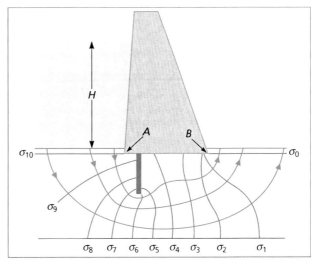

Figure 11.22  Effect of a vertical impermeable curtain (cut-off) on the flownet through the foundation of a gravity dam.

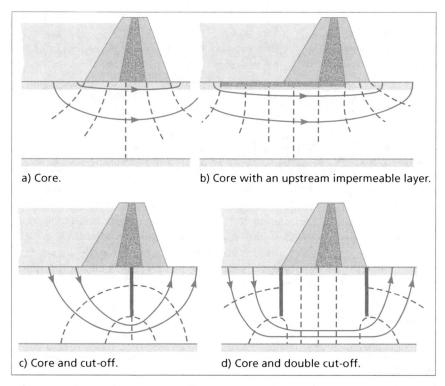

*Figure 11.23* Methods of lengthening flow paths beneath an earth dam (Attewell and Farmer, 1976).

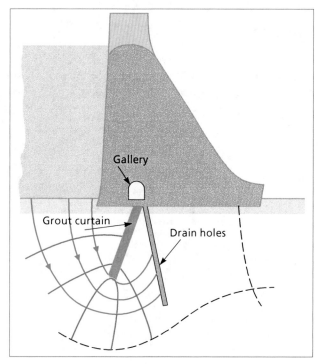

*Figure 11.24* Effects of an impervious grout curtain dipping upstream and a system of drain holes dipping downstream on the flownet through the foundation of a gravity dam.

erosion of the body of the dam and 12% from erosion of the foundations). Erosion consists of the opening up of conduits or paths through the soil caused by high hydraulic gradients (*Figure 11.26c*). Excessive water velocities can produce erosion of the core materials and dam foundations. The materials most susceptible to erosion are fine sands and silts; the pinhole test (Box 12.1 in Chapter 12) can be used to assess how susceptible a soil is to erosion.

Erosion can occur in the following situations:

— In contact areas between the foundation and the dam along fissures in the core and where it is in contact with the foundation.
— In foundations that have not been properly excavated to good ground or have not been sufficiently grouted.
— On discontinuity surfaces in the rock foundations supporting the core that have not been properly cleaned and sealed.
— In contact areas between the core and the concrete.
— In irregular abutments.
— Through the foundations due to the undermining of zones affected by leakage downstream of the dam.
— In materials in the dam structure itself that are inadequately compacted or insufficiently drained.

One of the measurements taken to prevent erosion is the installation of **filters and drains**. Filters prevent the movement and drag of particles towards the foundations out of the structure, which can occur when flows can carry particles through voids between coarser materials; drains are highly permeable elements designed to decrease pore pressures (Figures 11.3, 11.16 and 11.21).

## Leakage control

Leakage may occur through both the foundations and the reservoir basin. Leakage is assessed according to the hydrogeological methods described in Chapter 4 with flownet analysis one of the most commonly used. Whichever method is used, the ground permeability coefficient must be obtained using the tests described in Chapters 4 and 5.

In site investigations for dams the **Lugeon test** is widely used to determine water intake capacity under test conditions. A rock mass is considered to be watertight when water absorption is less than or equal to 1 litre of water per metre of test length, per minute, at 10 bars pressure (1 MPa), i.e. 1 **Lugeon unit**, is equivalent to a permeability of $1.3 \times 10^{-5}$ cm/s. However, the use of the Lugeon test to calculate permeability may not be appropriate as it is very unlikely that Darcy's Law conditions will occur when the test is being carried out on a fractured rock mass; a pressure of 1 MPa will probably produce hydraulic fracture or erosion (Foyo et al., 2005). Taking this into account, the Lugeon test can be used by limiting the pressure up to the threshold of hydraulic fracture, or **critical pressure** of the test. Using this criterion, some correlations between the Lugeon test and permeability have been proposed (Shimizu et al., 1985):

$$k = \frac{LU}{12 \times 10^4} \log \frac{L}{r}$$

in which:

$k$ = coefficient of permeability in cm/s.
LU = Lugeon unit.
$L$ = length of the test section.
$r$ = radius of test borehole.

Once permeability and the flownets underneath the dam have been established the permeability conditions are analysed to determine if they are acceptable, and whether the seepage forces and uplift pressures are compatible with the dam design requirements. If they are not, seepage must be controlled by decreasing the hydraulic gradients and uplift pressures using some means of ground treatment to either reduce permeabilities or head beneath the dam and/or drainage measurements. **Seepage control measurements** include:

— **Grouting** the contact surface between the dam and the foundation (Figure 11.33).
— Construction of **impervious blankets** consisting of compacted clayey materials of low permeability extending from the core to upstream of the dam (Figure 11.6).
— Construction of **impervious cut off curtain or diaphragm walls** to intercept seepage from underneath the dam; these cut offs may be continuous, and can be formed using concrete, clay, bentonite-cements or grout (Figure 11.35).

**Grouting** is the most usual method of leakage control for dam foundations; it consists of inserting fluids into fissures, voids and discontinuities by injecting them via boreholes at a specific pressure. Usual grouting materials include cements, resins and gels. The need for cement injection can be estimated according to the following criteria (ICOLD, 1993):

— For values below 1 Lugeon unit (1 LU), assuming 1 Lugeon is equivalent to $1.3 \times 10^{-5}$ cm/s, grouting is not necessary.
— For 10 LU and above grouting is necessary.
— For 100 LU grouting is necessary for filling open fractures, voids or cavities.

## 11.8 Reservoir slope stability

Filling a reservoir has the following effects:

— The load imposed on the slopes and the bottom of the valley equivalent to the height of the water column at each point in the reservoir.
— The water table in the slopes of the reservoir is raised.
— The hydrogeological conditions in the catchment area change.
— Water tables on the slopes of the reservoir fluctuate as a result of operation and use.

These effects result in stress changes, which may produce instability in the reservoir slopes. Box 11.2 shows the effects of rapid filling or drawdown in the reservoir water levels on slope stability. The most critical situation occurs when there is a sudden drawdown in low permeability materials.

Where instability may occur the following studies should be carried out:

1. Identification of unstable slopes in the reservoir areas.
2. Estimations of types of landslides with their activity, volume, distance from the dam and geometry (see Chapter 13).
3. Stability analysis including rapid reservoir drawdown conditions.
4. If the factor of safety obtained in the stability analysis is near to or lower than 1.0 for a rapid drawdown

## Box 11.2

### Influences of water level fluctuations on the stability of the reservoirs slopes

The operation of a reservoir implies periods of filling and draining. There are times when rapid drawdown may be necessary, for example, to control floods. This is the most unfavourable situation for slope stability. Instability conditions are generated when high pore pressures within the slope do not dissipate at the same speed as the lowering of the water level in the reservoir.

In the attached figures the most characteristic situations are analysed:

a) Initial situation before the reservoir is filled.
b) Situation once the reservoir has been filled slowly and the corresponding water table on its slopes has been established.
c) Situation following a rapid drawdown of the reservoir in materials with low permeability.

As the reservoir fills up the hydraulic conditions of the slopes change; the water table is generally determined by the height of water in the reservoir and the hydrogeological characteristics of the materials.

The ground strength is affected by the generation of pore pressures. However, the water stored in the reservoir exerts a stabilizing force on the slope surface opposite to the destabilizing forces of pore water pressure.

A rapid drawdown of the reservoir level results in the stabilizing force suddenly disappearing, while the pore pressures in the slopes remain high, giving rise to unstable conditions.

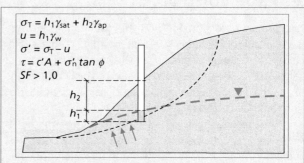

a) Situation prior to the filling of the reservoir: stable conditions.

$\sigma_T = h_1 \gamma_{sat} + h_2 \gamma_{ap}$
$u = h_1 \gamma_w$
$\sigma' = \sigma_T - u$
$\tau = c'A + \sigma'_n \tan \phi$
$SF > 1,0$

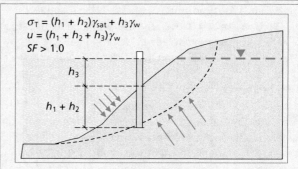

b) New situation after the filling of the reservoir. Water table is established in the slope. The increase in pore pressure is counterbalanced by stabilizing forces generated by the reservoir water on the slope surface: stable condition.

$\sigma_T = (h_1 + h_2)\gamma_{sat} + h_3 \gamma_w$
$u = (h_1 + h_2 + h_3)\gamma_w$
$SF > 1.0$

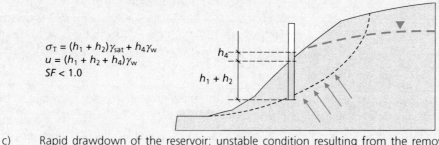

$\sigma_T = (h_1 + h_2)\gamma_{sat} + h_4 \gamma_w$
$u = (h_1 + h_2 + h_4)\gamma_w$
$SF < 1.0$

c) Rapid drawdown of the reservoir: unstable condition resulting from the removal of the hydrostatic load of the reservoir on the slope and the non-dissipation of pore pressures.

situation, remedial measures and slope monitoring systems should be considered. In this case, a **hazard analysis** is also recommended (see Chapters 13 and 15), if there is a large volume of unstable material near the dam.

In the Vajont dam the massive landslide reached an average velocity of 20–40 cm/day, two days before its failure, and 80 cm/day in the hours leading up to the disaster. The rock mass that produced the enormous 70 m high wave moved at a rate of 20–30 m per second. However, the most frequent landslides on the slopes of reservoirs are usually small and take place in weathered and colluvial materials. In most cases their volume and slow movement do not pose any significant risk. Most landslides are usually associated with either geological structures that favour gravitational movements, or geomorphological and climatic conditions that promote large displacements of unstable ground, e.g. mountainous areas affected by glacial or periglacial processes, wetter climate periods, highly active morphotectonic or erosive processes, and areas with important seismic or volcanic activity.

In any case, the effects of landslides on reservoirs should never be underestimated, even when they do not represent any risk for the safety of the dam. The solid material brought down by a landslide and deposited in the reservoir may lead to obstruction of the discharge system and will always decrease the reservoir capacity.

## 11.9 Engineering geological conditions for dam foundations

### General conditions

Dam foundations must fulfil the following conditions to ensure they function safely:

— Adequate strength and stability for both the foundation and abutments.
— Compatible deformability between the dam and the foundations.
— Watertight conditions and means for controlling seepage.
— Stability in relation to uplift pressures and erosive undermining.
— Stability in relation to seismic movements and their effects e.g. liquefaction, densification, collapse and fault displacement.
— Stability in relation to ground movements including landslides, settlement and subsidence phenomena.

### Loads on dam foundations

The construction of a dam substantially modifies the natural stress states in a rock mass and creates the following loads:

— The weight of the dam itself, which produces compressive and shear stresses.
— The hydraulic loads resulting from filling the reservoir that generate compressive, shear and tensile stress.
— Uplift forces exerted by the water retained behind the dam and seepage forces in the foundation and abutments.

— Seismic forces.
— Forces, in concrete dams, from their thermal expansion.

Of these forces, those associated with seepage are the most significant for two important reasons (ICOLD, 1993):

1. Seepage through voids and discontinuities in a rock mass produces internal stresses in the rock which are proportional to the head of water in the reservoir and the hydraulic gradient through its foundations.
2. The hydraulic gradient does not depend on absolute permeability.

The first point draws attention to the fact that water entering a foundation not only has its own hydraulic pressure but that its flow can also impart a **force to the foundations** (a seepage force) and that these can attain the same magnitude as the forces from the weight of the dam and the component of the reservoir it transfers to the foundations. The second cautions that in the same rock mass, the stresses due to water are irrespective of its permeability; (water pressure inside the rock is independent of permeability, it can vary whilst permeability stays largely constant, although it does depend on the flow as the pressure head component of total head can vary along a flow path, depending on its trajectory); the only thing that may vary is the discharge.

Depending on the type of dam and its size, and the combination of loads, the forces exerted on the foundation may vary both in magnitude and direction:

— In embankment dams the stresses on the foundation are the result of the height of the dam at the point under consideration, that is, the stresses are not uniform; the behaviour of this type of dam is flexible and semi-plastic (Figure 11.25a).
— Pressures due to hydraulic loads are hydrostatic and increase linearly with depth.
— The behaviour of concrete dams in relation to foundations is rigid. The loads or stresses on the foundation are not uniform, although deformations may be almost uniform in very strong rock masses (Figure 11.25b and 11.30).
— Due to the effect of filling the reservoir the resulting force from the hydrostatic pressures and the weight of the dam is inclined downstream from the dam (Figure 11.25c and d).
— Forces from hydrostatic loads in arch dams are distributed towards the abutments as their geometry is conditioned by the arch effect (Figure 11.25e).
— In arch dams the system formed by the dam and its foundation is highly **hyperstatic**; that is, the stresses are distributed in accordance with the deformability of the materials.

## Box 11.3

### Failure mechanism in the Aznalcóllar Dam (Spain)

The 28 m high Aznalcóllar tailing dam (see Box 1.3 of Chapter 1) was built with an upstream impermeable curtain. The foundation consists of a thin alluvial layer on underlying blue Miocene marls. The average properties of the materials were: clays 68%, liquid limit 65%, plasticity index 35%, carbonates 21%, unconfined compressive strength 450 kPa, cohesion 31 kPa, residual angle of internal friction 12° and a deformation modulus 100 MPa. The marls contained slickenslides and discontinuities of sedimentary origin.

The failure took place along a surface situated at 14 m underneath the dam on the blue marls dipping 2 degrees downstream. The dam was displaced 53 m after failure (see the attached figure). This deep and progressive failure was due to the following factors:

— Presence of an area subjected to high stresses in the marls because of the weight of the dam and the load from mining wastes.
— Presence of sedimentary-type discontinuity surfaces and shear planes in the marls.
— Development of high pore pressures in the marls that remained high even after the failure.

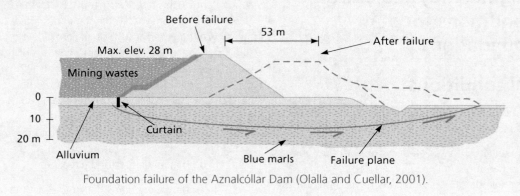

Foundation failure of the Aznalcóllar Dam (Olalla and Cuellar, 2001).

— As a result of the hydrostatic forces the dam will want to slide along the dam-foundation contact and overturn in relation to a point situated downstream from the toe of the dam.

## Dam foundation failure mechanisms

The response of a rock mass foundation to forces exerted by a dam may cause failure. An awareness of this possibility is fundamental for the safety of dams as seen in Section 11.1 and *Box 11.3*.

In **embankment dams** the most characteristic failure mechanisms are caused by the following factors:

— Failure in low strength foundation materials (*Figure 11.26a*).
— Settling caused by the compressibility of soft foundation materials (*Figure 11.26b*).
— Erosion either in the dam or in the foundation (*Figure 11.26c*).

In **concrete dams,** where loads are greater, the concentration of stresses may give rise to significant deformations in softer or less resistant materials. In such cases failure analysis can be carried out using plasticity theory (see Chapter 8), or by limit equilibrium analysis. *Figure 11.27* shows different types of failure in concrete dam foundations. Failures can be due to some of the following geological conditions:

— High shear stresses in weak rocks.
— Low strength discontinuity surfaces dipping downstream.
— Low strength conjugate fractures and joints in hard rocks.
— Discontinuities aligned in the same direction as the thrust forces in the slope abutments.

*Figure 11.28* shows examples of geological conditions which may give rise to different types of failures. However, geological situations that enable and encourage stability can be rendered unstable by water pressure; experience of accidents involving dam foundations has demonstrated the importance of the forces from water as a decisive factor in the mechanisms leading to dam foundation failure.

The failure of the **Malpasset Dam**, already referred to in Section 11.1, was due to the mechanism shown in *Figure 11.29*. The failure of the left abutment was caused by intersecting schistosity planes with a downstream dip and a fault filled with clay material dipping upstream. High

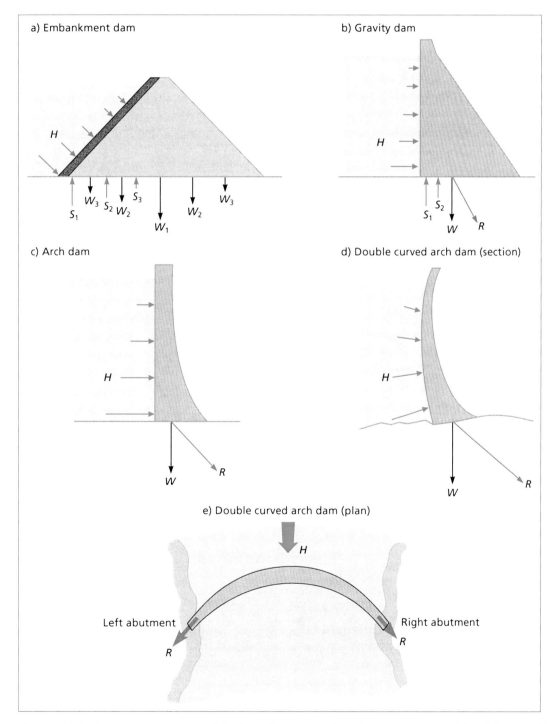

*Figure 11.25* Forces acting on dams. H: hydrostatic pressure; S: uplift pressures; W: weight of the dam; R: resulting force of H and W (full reservoir).

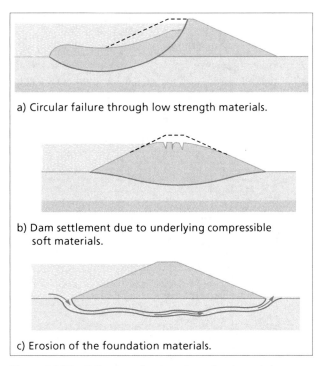

Figure 11.26 Failure mechanisms in embankment dams.

a) Circular failure through low strength materials.

b) Dam settlement due to underlying compressible soft materials.

c) Erosion of the foundation materials.

seepage forces developed across the schistosity planes of the very low permeability rock foundations. These conditions caused the opening up and propagation of a large fracture, allowing the left abutment of the dam to slide along the fault plane.

To sum up, the presence of discontinuities which may act in conjunction to form blocks or wedges that have a kinematic feasibility for failure is an important factor to take into consideration. Sliding failure conditions in rock foundations are possible if the geological structures are aligned in the same direction as the thrust forces from the dam. High water pressures are needed as well as discontinuity surfaces with low shear strength (e.g. fault breccia, clay infilling or weathered zones) and large continuous planes to allow movement in the thrust direction. However, even when geological conditions favour the displacement of large blocks and wedges, **the decisive factor is the contribution made by water pressures**.

Another of the possible mechanisms for sliding foundation failure is the presence of soft sedimentary layers, faults or slickensided surfaces that dip downstream from the dam, or dip horizontally, allowing the displacement of the dam (*Figures 11.27d, 11.27e* and *28f*).

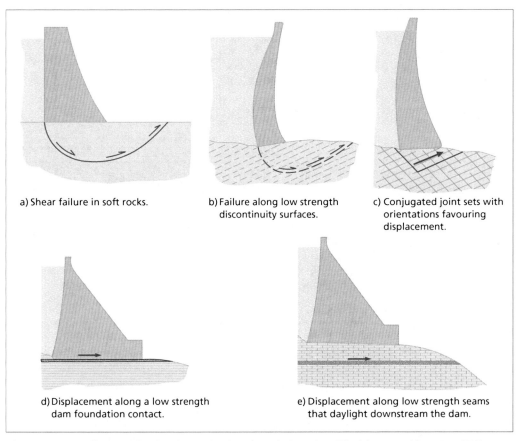

a) Shear failure in soft rocks.

b) Failure along low strength discontinuity surfaces.

c) Conjugated joint sets with orientations favouring displacement.

d) Displacement along a low strength dam foundation contact.

e) Displacement along low strength seams that daylight downstream the dam.

Figure 11.27 Failure mechanism in gravity dam foundations (modified from Wahlstrom, 1974).

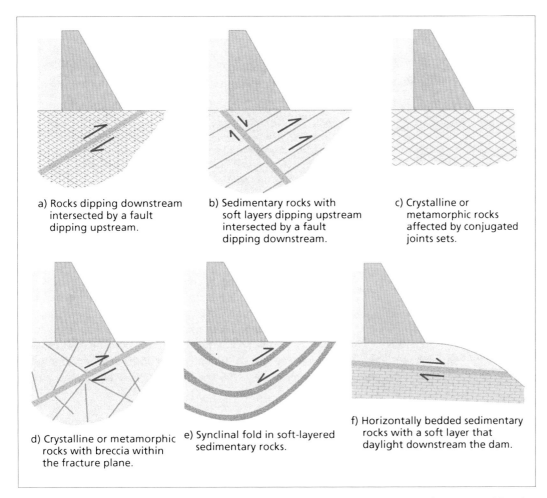

Figure 11.28 Geological conditions in gravity dam foundations that can result in failures (modified from Wahlstrom, 1974).

Box 11.4 shows an example of the stability analysis of a dam along discontinuity planes beneath the foundations, or at the contact between the dam and the foundation.

## Stress distributions in dam foundations

In gravity dams the resulting force of the thrust and the weight of the dam has a downstream direction. This force induces two effects (Wyllie, 1999):

— An overturning moment.
— A non-uniform stress distribution in the foundation.

In practice, the first of these effects can be resolved with the appropriate geometric design of the dam. However, these effects cause compressive stresses in the toe of the foundation and tensile stresses in the heel of the dam (Figure 11.30).

The non-uniform stress distribution implies that deformations in the foundation may be uneven, or not uniform, as a result of the different loads applied. Rocks with a heterogeneous lithology or those affected by significant discontinuities and, in general, anisotropic geological conditions, can affect the strength and the deformation modulus at different points in the dam foundation. The result is a concentration of stresses in the contact areas between the dam and the foundation.

The Funcho Dam (Portugal), a 49 m high double curved arch dam, was built in 1993 on foundation rock consisting of slates and graywackes in a very heterogeneous geological structure across an asymmetrical valley. The right abutment is formed by highly weathered and decompressed shales dipping down towards the river bed; on the left abutment graywackes predominate, dipping towards the interior of the rock mass (Figure 11.31). Geological surveys and in situ tests in galleries allowed differential deformations to be predicted on each side of the abutments as a result of their lithological

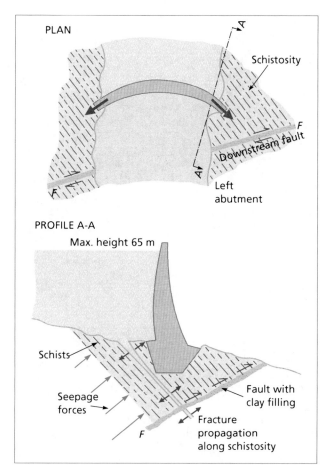

*Figure 11.29* Failure mechanism in Malpasset Dam, France (modified from Wittke, 1990).

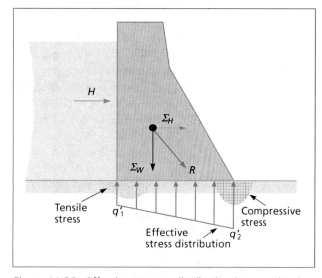

*Figure 11.30* Effective stresses distribution in a gravity dam foundation (modified from Wyllie, 1999).

characteristics. Deformation moduli measured at a depth of a few metres on the right abutment were 25 to 30 times lower than those on the left abutment. A 25 m deep excavation was proposed for the right abutment of the dam, where the ratio between the deformation modulus of the concrete and the rock mass was 10 to 12, and a plinth (or concrete base) was built along its entire length to decrease the contact stresses with the ground. In addition, intensive ground treatment (consolidation grouting) was carried out on the foundation. The reservoir has reached its maximum level, confirming that the dam is performing satisfactorily.

## Foundation improvement measurements

In most dam foundations, especially concrete ones, consolidation treatments are required to improve either the foundation surface supporting the dam or the foundation rock itself.

Excavation of the foundation implies stress changes which require special measurements to be taken in order to avoid the loss of strength of the rock mass and to improve ground properties. Some of these measurements include:

- **Excavation control** (*Figure 11.32*):

  — Blasting control to prevent rock damage (smooth blasting and pre-splitting or similar techniques).
  — Regular shaping of the excavation surface to avoid sharp and irregular surfaces and steep slopes. This is very important for embankment dams.
  — Excavation floor forming steps inclined upstream to increase the sliding resistance of the contact surface between the dam and the rock mass. This is particularly important for gravity dams.

- **Preparation of bearing surfaces** (*Figures 11.33* and *11.34*):

  — Clearing and removal of erodible materials and loose blocks.
  — Sealing of fractures, voids and open joints.
  — Filling of cavities and weathered or fractured zones.

- **Consolidation and leakage control measurements.** If the foundation is not strong enough, has high deformability or seepage problems, consolidation and leakage control measures have to be carried out to increase the rock mass strength, reduce deformability and control seepage.

  These measures include (*Figure 11.35*):

  — Grouting. Deformability can be improved if fractures are grouted because the deformation modulus depends mainly on the aperture of joints. Grouting techniques are generally the most satisfactory methods used.

## Box 11.4

### Sliding failure in gravity dam foundations

Hydrostatic pressures resulting from the filling of the reservoir exerts a force on the dam that must be resisted by both the dam and the foundation rock mass.

**Basic factors:**

— Shear strength along discontinuities and on weak materials.
— Pore pressures exerted on the rock mass and on the sliding surface. These pressures can vary from a maximum in a full reservoir to complete dissipation downstream of the dam.

**Safety factor calculation**

Figure A
a) Sliding surface along the horizontal contact between the dam and the foundation (dam with a vertical upstream wall):

$$SF = \frac{cA + (W-U)\tan\phi}{H}$$

Figure B
b) Sliding surface along the horizontal contact between the dam and the foundation (dam with an inclined upstream wall):

$$SF = \frac{cA + (W + H\cos\psi - U)\tan\phi}{H\sin\psi}$$

Figure C
c) Sliding surface along a layer dipping downstream beneath the foundations:

$$SF = \frac{cA + (W\cos\alpha + W_T\cos\alpha - H\sin\alpha - U)\tan\phi}{W\sin\alpha + W_T\sin\alpha + H\cos\alpha}$$

$c, \phi$: cohesion and angle of friction of surface of displacement.
$A$: area of sliding surface.
$W$: dam weight.
$W_T$: weight of the rock mass wedge sited over the sliding surface.
$U$: uplift pressures on the sliding surface.
$H$: hydrostatic load.

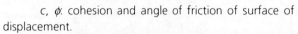

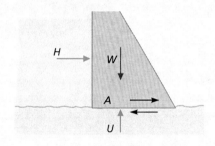

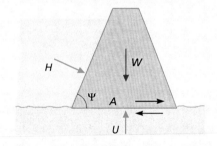

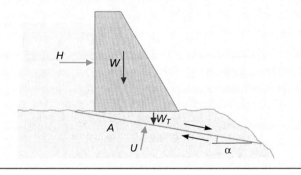

— Curtains and cutoffs. High values of permeability are usually concentrated in open fractures where grout curtains or cutoffs are generally used. Uplift pressures are reduced by means of grout curtains and/or drains, resulting in the increased stability of the dam.

## Dam foundation problems and possible remedial measures

Most foundation problems are related to geological factors and the suitability of the site for the type of dam designed.

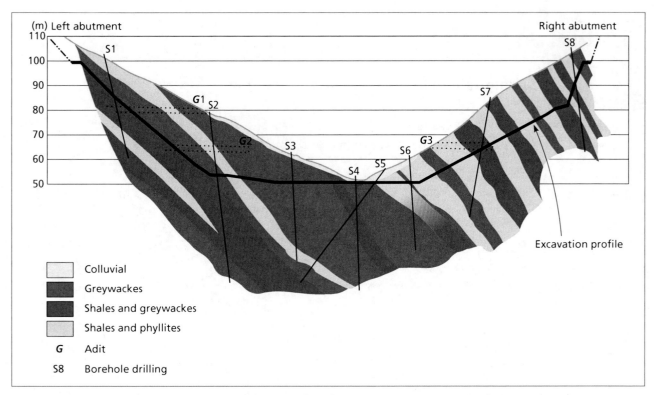

*Figure 11.31* Geological cross section and excavation profile of the Funcho Dam (Portugal).

Some frequent geological problems and possible solutions are described as follows (Kanji, 1994; Oliveira, 1988):

## Weak rocks

— Low strength rocks, underneath the dam (*Figure 11.27a*): In this case a deeper excavation may be required to reach a rock mass with sufficient bearing capacity. If this solution is not possible, it may be necessary to modify the geometry of the dam, to change the type of dam or to improve the rock mass. However, this last option may not the best solution in many cases.
— Loss of strength due to the effect of saturation: most weak rocks lose strength when they become saturated, e.g. cemented and clayey rocks. Under these conditions tests should be carried out on saturated samples.
— Desiccation effects on rocks with a high clay content. Cracking in mudstone or marls can appear on the excavation surfaces which require immediate sealing and concreting.
— Low durability rocks with clayey composition can require immediate sealing and surface protection.
— Rocks with smectite shrink and swell and may either need surface protection or be removed.
— Rock damage due to the effects of blasting can lead to cracking and irregular surfaces which need cleaning and surface protection.

## Hard rocks

The presence of discontinuities or soft layers, either of significant length lying horizontally or dipping downstream at a low angle, represents an important geological problem (*Figure 11.28f*), that is very common in hard rocks. If these discontinuities lie at a shallow depth the solution is to either remove them completely, or partially as the case may be, and to replace them with concrete. This consolidation work can be carried out from adits. Grouting, drainage and sometimes anchoring are also used.

## Sulphur and anhydrite rocks

Rocks containing sulphur compounds like pyrite and marcasite can attack to concrete and steel. The release of sulphuric acid also affects to the ground water composition. Hydration of anhydrite to gypsum produces expansion phenomena that can induce significative swelling pressures.

## Faults and breccia zones

Faults, including breccia filling, and heavily jointed zones are usually not very thick, and the usual treatment consists

Figure 11.32  Abutment excavation for an arch dam.

Figure 11.34  Preparation of rock surfaces for a gravity dam foundation.

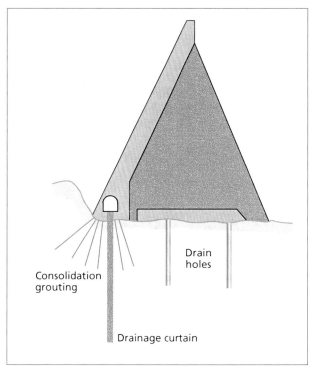

Figure 11.35  Consolidation and curtain grouting in the foundation of a buttress dam (modified from Jaoui et al, 1982; in Wyllie, 1999).

Figure 11.33  Sealing and grouting the foundation surface for an embankment dam.

of excavation to remove the weak material and replacing with concrete or grouting from the surface or from adits. Heavily jointed zones are associated with high permeability, high compressibility and low shear strength. Consolidation treatment and sometimes anchoring are required in these zones.

### Cavities

Caves or cavities are generally filled with concrete. However, the main problem is to identify these caves and determine their size, particularly in karstic areas where it may be unfeasible or uneconomic to seal such cavities.

## 11.10 Seismic actions and induced seismicity

Seismic actions affect the design of a dam and its auxiliary structures, including reservoir slope stability. *Table 11.4* outlines the criteria followed for seismic-resistant dam design. In Chapter 14, the basic concepts of seismic hazards are described.

The possibility of the site being affected by **active faults** also has to be considered. Fault displacement can produce instability and failure both in the dam and its foundations. Section 14.6 of Chapter 14 deals with the hazard of active faulting. A detailed description can be also found in ICOLD (1998).

While there are few known cases of dams sited on active faults, examples exist of sites that have been abandoned after considering the possibility of some faults being active. A controversial case in the 1980s concerned the abandonment of the Auburn Dam in California (USA) after it was suspected that a fault crossing the foundation rock was active. Studies were carried out but opinions of the experts

| *Table 11.4* | SEISMIC CRITERIA FOR DAM DESIGN IN SPAIN | | |
|---|---|---|---|
| **Classification of dams** | **Seismicity** | | |
| | Low | Moderate | High |
| C | NA | DE = 1,000 years RP | |
| B | | DE = 1,000 years RP | DE = 1,000 years RP |
| A | | ME = 3,000 to 5,000 years RP | ME = 10,000 years RP |
| Dam failure or inoperability can cause: A: severe damage including loss of human life. B: serious damage including loss of human life. C: moderate to low damage and occasionally loss of human life. NA: not applicable; DE: design earthquake; ME: maximum earthquake; RP: return period. | | | |

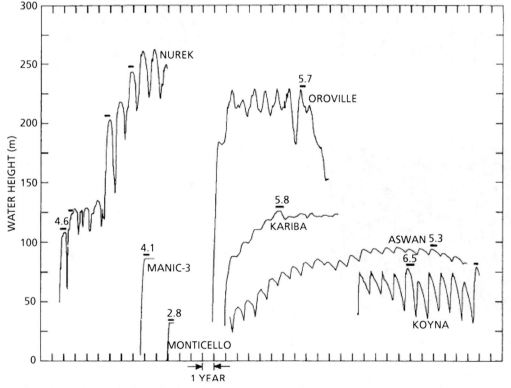

*Figure 11.36* Examples of water levels in reservoirs that triggered seismicity. Vertical scale is absolute water depth at the dam. The horizontal scale is the same for all curves, but the absolute positions have been shifted for clarity. Numbers above the water level curves are magnitudes of the largest earthquakes (modified from Simpson, 1986).

consulted were divided. In the end, the Bureau of Reclamation took the decision to abandon the site and construction of the projected arch dam. In Taiwan, the Shigun gravity dam was affect by a vertical displacement of 7.7 m in the left abutment during the 7.6 magnitude Chi-Chi earthquake of September 21, 1999, which led to failure of the structure (Hesein Juang, 2004).

In large dams and reservoirs, it is especially important to consider seismic action as it affects the filling of the reservoir. In current literature on the subject, it is common to find the expression "induced seismicity" used to define the seismicity associated with the initial period of dam operation. Nevertheless, the term "reservoir triggered seismicity" is also used (Simpson, 1986; Gupta, 2002; McGarr et al., 2002). **Induced seismicity** refers to processes in which the release of energy is directly proportional to the variation of the stresses acting on the ground, as occurs with seismicity associated with the extraction and injection of fluids. The term **reservoir triggered seismicity** more adequately describes the phenomenon by which the action of filling the reservoir triggers seismic activity a fault that is at limit equilibrium. The reservoir load has no significant influence on crustal stresses in comparison with tectonic and gravitational stresses. Therefore, reservoir loading could only promote the occurrence of natural seismicity or in an extreme situation affect the seismogenic fault if this itself is near limit equilibrium.

*Figure 11.36* shows the water levels of some of the most significant cases of reservoir triggered seismicity. This figure also shows the earthquake magnitude reached by the maximum seismic event (Simpson, 1986). Some relevant cases of dams affected by induced seismicity are shown in *Table 11.5*.

The possible mechanisms by which a fault reaches peak strength due to filling the reservoir, can depend on the following factors, jointly or independently:

— Increase in the total stress in the fault zone as a consequence of the reservoir water loading.
— Increase in pore pressures due to the decrease in porosity of the saturated rocks, as a consequence of the increased load.
— Increase in pore pressures due to the head of reservoir water, which produces a change in pore pressures.

When seismicity is associated with filling the reservoir, the recommendation is to control the filling and emptying rate in the first period of the dam operation, to either avoid or minimize this type of seismicity.

## Recommended reading

Fell, R., MacGregor, P., Stapleton, D. and Bell, G. (2005). Geotechnical engineering of dams. Taylor and Francis.
International Commission on Large Dams (ICOLD) (1993). Rock foundations for dams. ICOLD Bull., 88. Paris.
Kanji, M.A. (1994). Engineering geological impacts on the design and construction of dams. Keynote lecture, 7th Int. Conf. of the IAEG, Lisbon.
Wahlstrom, E.E. (1974). Dams, dam foundation and reservoirs. Elsevier, Amsterdam.
Wyllie, D.C. (1999). Foundations on rock. E.F.N. Spon, 2nd ed. New York.

## References

Attewell, P.B. and Farmer I.N. (1976). Principles of engineering geology. Chapman and Hall, London.
Foyo, A. Sánchez, M.A. and Tomillo, C. (2005). A proposal for a secondary permeability index obtained from water pressure test in dam foundations. Engineering Geology, pp. 77, 69–82.
Gupta, H.K. (2002). A review of recent studies of triggered earthquake by artificial water reservoirs with special emphasis on earthquakes in Koyna, India. Earth-Sciences Reviews, 58, pp. 279–310.
Hsein Juang, C. (Ed). (2004). Reconnaissance of the Chi-Chi earthquake, Taiwan. Engineering Geology, Special Issue, 1–2, pp. 1–179.
Kanji, M.A. (1994). Engineering geological impacts on the design and construction of dams. Keynote Lecture, 7th Int. Conf. of the IAEG, Lisbon.
International Commission on Large Dams (ICOLD) (1993). Rock foundations for dams. ICOLD Bull. 88. Paris.
International Commission on Large Dams (ICOLD) (1998). Neotectonics and dams. Guideless and case histories. Bull. 112. Paris.

| Table 11.5 SOME EXAMPLES OF DAMS AFFECTED BY INDUCED SEISMICITY ||| 
|---|---|---|
| Dam | Earthquake magnitude | Dam height (m) |
| Koina (India) | 6.5 | 103 |
| Kremasta (Greece) | 6.3 | 165 |
| Kariba (Zimbawe and Zambia) | 6.2 | 128 |
| Hsinfengkiang (China) | 6.1 | 105 |
| Orovile (USA) | 5.8 | 230 |
| Kurobe (Japan) | 4.9 | 186 |
| Nurek (Tadjikistán) | 4.6 | 317 |
| Volta Grande (Brazil) | 4.2 | 56 |

McGarr, A., Simpson, D. and Seeber, L. (2002). Case histories of induced and triggered seismicity. International Handbook of Earthquake and Engineering Seismology, Part A. Lee, Kanamori, Jennings and Kisslinger Eds. Academic Press, pp. 647–661.

Olalla, C. and Cuellar, V. (2001). Failure mechanism of the Aznalcóllar Dam, Seville, Spain. Geotéchnique, 51, pp. 399–406.

Oliveira, R. (1985). Consolidation of the foundation rock mass of Cambambe Dam (Angola). Proceedings XV ICOLD Congress, vol. IV. Lausanne.

Oliveira, R. (1988). Geotechnical characteristic required for concrete dam rock foundations. Proc. XVI ICOLD Congress, vol. V. San Francisco, USA.

Semenza, E. and Ghirotti, M. (2000). History of the 1963 Vaiont slide; the importance of geological factors. Bull. Engineering Geology and Environment, 59, pp. 87–97.

Shimizu, S., Jojima, S. and Niida, Y. (1985). Design and execution of foundation grouting for multipurpose dams in Japan. XV ICOLD, vol. 3, pp. 429–453. Lausanne.

Simpson, D.W. (1986). Triggered earthquakes. Ann. Rev. Earth Planet. Sci., 14, pp. 21–42.

Wahlstrom, E.E. (1974). Dams, dam foundation and reservoirs. Elsevier, Amsterdam.

Wittke, W. (1990). Rock mechanics. Springer-Verlag, Berlín.

Wyllie, D.C. (1999). Foundations on rock. 2nd ed. E.F.N. Spon, New York.

# 12
# EARTH STRUCTURES

1. Introduction
2. Design methodology
3. Materials
4. Implementation and control
5. Embankments on soft soils
6. Embankments on slopes

## 12.1 Introduction

The basic types of earth structures are:

— Earth structures formed by earthfill or rockfill embankments, built with loose materials compacted until they are semi-continuous and resistant, with low deformability.
— Structures formed by the ground itself, as the result of excavating slopes, trenches and tunnels, with or without the addition of supports or retaining walls.
— Masonry structures, such as bridges and viaducts, using concrete, brick or stone to provide a secure base for the surface platform.

As well as earthfill and rockfill embankments for linear infrastructures (*Figure 12.1*), earth structures are also used for dams, breakwaters and embankments for flood protection.

**Earthfill embankments** are structures made of compacted soil materials, having a particle size generally smaller than 100–150 mm, and with a low fines content; **rockfill embankments** are built of rock fragments.

These structures are layered, with superimposed materials laid down and compacted in succession. **Coarse rockfills** are made of non-compacted rock blocks, in some cases larger than a cubic metre.

The design of earth structures requires the following aspects to be considered:

— A geotechnical understanding of the placed materials (or fill); this is essential.
— The properties of local materials to be used must be considered and also the climatic, hydrological and topographical conditions in which they are to be used and how the project is to be constructed. It is not a question of simply falling back on standard designs and ignoring the topography where the earthwork is to be built.
— Even where water resources are scarce, significant water infiltration may occur occasionally (at 5–10 year intervals) caused by flooding, which may saturate earth structures and the underlying ground. This means that even when stability is guaranteed under dry conditions, if the soil becomes saturated the change in pore pressures can easily reduce the factor of safety with respect to sliding by 30% to 50%. Where the ground under the embankment is considered to be infinite with slope angle $\alpha$, angle of friction $\phi'$ and zero cohesion, the factor of safety with respect to sliding in drained (or dry) conditions will therefore be equal to $\tan\phi'/\tan\alpha$, but if the ground is saturated (producing flow parallel to the slope) the factor of safety will be almost half of its original value. If there is an embankment on this slope, then the combined embankment and slope may fail even before total saturation is reached.
— Erosion may affect or alter the strength conditions of earth structures in the medium or long term.
— The stability of the embankment itself and of the combined embankment and supporting ground must be assured against failure.
— Deformation in the earth structure itself and that it may cause in the underlying ground must be taken into account.

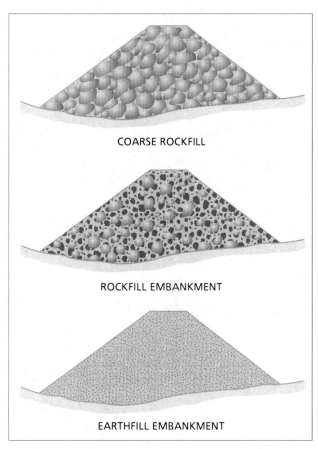

*Figure 12.1* Three types of fill materials.

# EARTH STRUCTURES

All these aspects must be considered if the horizontal surface on top of the earth structure is to meet the short and long term conditions required for its intended function. Different environmental factors can affect the structure, and water is potentially the most damaging.

## 12.2 Design methodology

In general terms, the design of earth structures follows the methodology shown in *Figure 12.2*, with the following requirements:

— An understanding of what types of materials are available locally, where they can be obtained and in what volume. This requires adequate geological, geotechnical and topographical site information.
— A design adapted as far as possible to the available materials and the technical specification requirements.

This could mean that some materials do not meet the technical specifications or relevant standards, as may be the case in areas where clay or saline materials predominate. Zoned embankments may have to be designed where, unlike homogeneous embankments, the foundations, slopes and crest have different characteristics (*Figure 12.3*); cores can be sandwiched between structures which alternate low and high quality materials. Zoning enables the best performance to be obtained from the core and foundations, reducing deformability and permeability, and improving slope strength. Zoning has been used to construct earth and rockfill embankments for highways (*Figure 12.4*), even successfully using materials normally considered to be problematic, such as expansive clays and sands from highly weathered granite.

— The slopes of these embankments may not always be the typical 1.5(H):1(V) (H: horizontal; V: vertical) value generally used for quality materials unless appropriate materials are available. *Figure 12.5* shows the expected values for the shear strength parameters of different types of materials depending on grain size (although other factors may also affect these parameters). *Figure 12.6* shows expected settlement in earthfill and rockfill embankments as a result of their own weight. These criteria can be used to give an initial estimate of the embankment slope stability, e.g. using Taylor's slope stability chart (see Chapter 9, Section 9.5) if they are built on horizontal, high quality ground. For a more detailed design, the geotechnical characteristics have to be taken into account when calculating stability and deformability as a whole, and reinforcing the foundation should be considered where soils are soft or unstable.
— Internal and external drainage systems must be set up as required. The internal systems drain off water which infiltrates the earth structure. The external ones drain off water from around the base of the

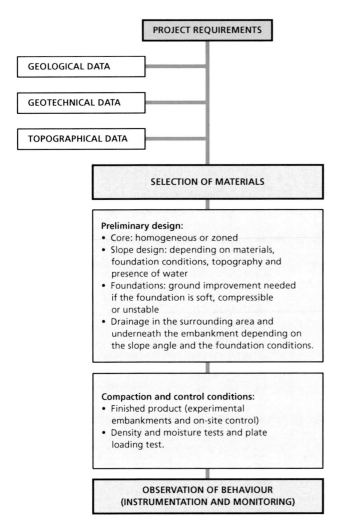

*Figure 12.2*   Earth structures design methodology.

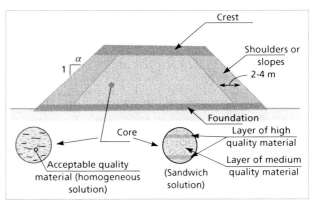

*Figure 12.3*   Example of an earthfill embankment section.

embankment to prevent saturation of the structure and the resulting increased pore pressures inside or underneath it, which may endanger its stability.

— Once the geometric design and type of materials to be used in each area have been decided, the appropriate **compaction** criteria have to be established (i.e. type of machinery and energy to be applied to the materials) to obtain the maximum density for the placed fill and ensure a final product having minimum deformability and acceptable durability. Compaction is used to achieve appropriate quality, not just a pre-established density, although this criterion has been used for many years as a measure of the concentration of solids per unit volume and their response to repetitive loads. For soils, the classic criterion used for judging whether a maximum dry density has been achieved with an optimum moisture content, comes from values defined through compaction tests such as the Standard Proctor or Modified Proctor tests, or else the criterion of determining the apparent deformability of the embankment using the plate load and "footprint" tests. These tests are described in *Box 12.1*.

If the higher energy Modified Proctor test is used, the result obtained is similar to the Standard Proctor test, although a greater optimum density is obtained for a lower moisture content. The "wet side" of the compaction curve turns out to be asymptotic to the line which represents total saturation. The line which joins these optimum points is parallel to the saturation curve (see *figure* in *Box 12.1*).

A compaction curve obtained on site may not coincide with that of the Standard Proctor Test obtained in a laboratory, since nowadays higher compaction energy can be applied on site than in the laboratory and so it will tend to be more like the Modified Proctor curve, although not exactly the same. This is why

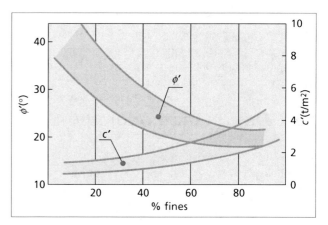

Figure 12.5　Relationship between shear strength and % of fines in earthfill embankments for motorways in Spain.

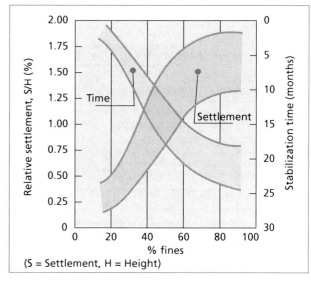

Figure 12.6　Post-construction settlement in earthfill embankments for motorways in Spain.

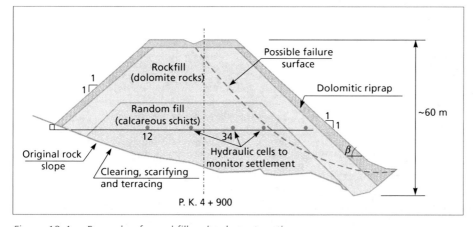

Figure 12.4　Example of a rockfill embankment section.

## Box 12.1

## Testing earth materials

The tests normally used for earth fill embankments are the Standard Proctor Compaction Test, the Modified Proctor Compaction Test, the Harvard Test and the CBR Compaction Tests. The Plate Load Test is carried out for on-site control during construction (see Chapter 5 Section 5.5, deformability tests) as well as the Footprint test and the *in situ* density test (sand and nuclear methods). The Pinhole Test is used to identify dispersive or highly erosionable soils. These tests are described in the different standards (e.g. ASTM and BS). A summary of some of these tests is given below.

### Standard Proctor compaction test

This test is carried out to determine the maximum dry density of a compacted soil and the optimum moisture content at which this density is reached. A litre mould is used, filled in three layers with the material to be tested, each compacted by a specified number of blows with a hammer which always applies the same energy. A number of samples are prepared for the test, each dried and then moistened to a different water content. The dry density of each sample is calculated after the test and; with five or six such values a dry-wet density curve can be drawn, to obtain the maximum dry density value and thus the optimum compaction conditions (maximum dry density and optimum moisture content) as shown in the figure below. This represents the ideal condition to be reached on site since it means the maximum possible concentration of solids and very stable conditions for the compacted material. The 100% saturation curve ($S_r = 100\%$) is practically parallel to the zone with highest water content or wet side. The optimum value corresponds to 85–90% saturation.

### Modified Proctor compaction test

This test is similar to the one described above, but the compaction force used is greater and only material with a particle size less than 20 mm is tested. The result is similar to the Standard Proctor, although greater dry density and lower moisture content are obtained. In the case of clay soils the **miniature Harvard compaction test** is more appropriate, using a different form of energy and type of hammer, by applying the force by steadily increasing pressure and not by percussive impact.

### CBR compaction test

The California Bearing Ratio (CBR) test is carried out to evaluate the bearing capacity of soils in embankments, platforms and base or sub-grade layers of road surfaces. The test consists of compacting samples in standard moulds, submerging the test samples in water and penetrating the sample with a standard piston. The results are represented in dry density-CBR index curves. This index gives the pressure required by the piston for a specified penetration in the sample as a percentage of that required to achieve the same penetration in a standard sample.

### Plate load test

In this test an initial pressure of around 20 kPa is applied followed by two loading/unloading cycles (to a pressure of around 300 kPa) to deduce the deformation modulus of the second cycle $E_2$, and compare it with that of the first cycle $E_1$. A minimum value for $E_2$ is normally required of 30 MPa for embankment core materials and 45 MPa for the crest, and the $E_2/E_1$ ratio should be lower than 2.2 so that the pressure/settlement curves are close and non-recoverable plastic deformations do not occur when successive forces are applied to the soil.

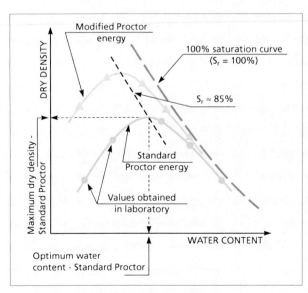

*Curves obtained in compaction test with constant energy*

> **"Footprint" test**
>
> The footprint test is carried out by drawing a 10 m long line on the ground and marking points on it at one metre intervals. These points are levelled and a truck with a pre-determined load is run along beside them. Then the levelling is repeated and the average "settlement" at each of the ten points is obtained. Average settlement of less than 3 mm for the embankment crest and 5 mm for cores is normally required.
>
> **Los Angeles abrasion test**
>
> The extent to which aggregates resist abrasion is determined with the Los Angeles machine, by feeding 5–10 kg of rock sample with a specific particle size into a revolving drum along with steel balls. The sample is subjected to 500 or 1000 revolutions at 33 rpm. Once the revolutions are completed, the material removed from the drum is passed through the ASTM No 12 sieve, to separate the fines produced by the abrasion and breakdown of the aggregate. The result of the test is the difference between the mass of the original sample and the corresponding one for the material retained on the sieve at the end of the test, expressed as a percentage of the original mass.
>
> **Pinhole test**
>
> This test is used to identify dispersive or highly erosionable clay soils. Water is passed through a hole along the length of a prepared sample and the material in suspension within the discharge and the discharge itself are observed.

it may not be appropriate to consider the Standard Proctor optimum density as an absolute reference to validate the product obtained on site, as has the been the case for years, as it does not take the site moisture into account and because the curve on site may be well above that obtained with the Standard Proctor test.
— Where the foundation of an embankment is on a slope, on soft soils or in other inappropriate conditions, the site must be adequately prepared by terracing and draining the slope, and improving and reinforcing the natural soil with drainage ditches and gravel columns.
— If the material properties are unknown or the supporting ground is very soft (in which case settlement at the embankment crest may reach 10–15% of its height), the work in progress will have to be instrumented to monitor its short and long term behaviour (*Figure 12.7*).

## 12.3 Materials

### Earthfill embankments

Ever since the importance of compaction problems in road and earthfill dam embankments was recognised, attempts have been made to establish simple **selection or classification criteria** to determine quickly and cheaply whether or not a material can be used as fill in any given situation. These attempts to establish usable criteria are based on tests to identify the intrinsic properties of the material and do not take into account the condition of the materials when quarried or how they can be laid, since soil may change to an extraordinary extent during extraction operations. This means that the following aspects should be considered when evaluating materials:

— The intrinsic characteristics of the material: grain size and water absorption capacity.

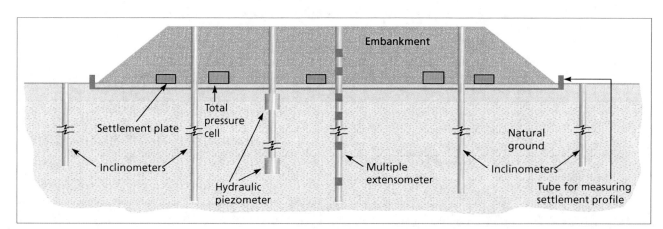

*Figure 12.7* Types of instrumentation for an earthfill embankment.

— Alterations that can be caused by handling operations (these vary depending on the machinery used and the environmental conditions).

Sometimes only the first of these aspects is considered when selecting appropriate materials for use in embankments; this oversimplifies the problem and may lead to conclusions which are not always correct. On the other hand, construction work often has to be carried out with low quality materials because there are no others available within an accessible distance, so the original selection or classification criteria may not be applicable. The operational capacity of the material must always be considered when an acceptable compromise has to be reached with materials which by certain criteria are unsuitable and not recommended (e.g. because they break down).

The **Casagrande soil classification** is the best known, and identifies three types of materials:

a) Highly organic, not recommended for use.
b) Coarse grain (less than 50% passes through ASTM sieve no 200); these are divided into gravels (G) and sands (S).
c) Fine grain (more than 50% passes through ASTM sieve no 200); differentiation based on plasticity.

*Box 12.2* gives a soil classification which includes possible acceptability limits, differentiating five types: **"tolerable", "suitable", "selected", "borderline" and "unsuitable"**. The criteria used to differentiate them are similar to those of Casagrande (grain size, Atterberg limits, organic and material content) but supported by other data related to the work on site (Proctor optimum conditions and CBR index.) In *Figure 12.8* the plasticity index is used to differentiate the soil types.

"Tolerable", "suitable" and "selected" soils must also meet the minimum required conditions indicated in *Box 12.2*. "Unsuitable" soils are those which do not meet the minimum required conditions for "tolerable" soils and as a result their use in embankments is not allowed. "Borderline" soils may be highly plastic and therefore swelling (although under certain conditions, they may be collapsible) and may also contain up to 20% soluble salts. Their use is not allowed without a special study, but in some cases swelling clays have been used in highway embankments, by carefully breaking them down and compacting them (with sheepsfoot rollers and always on the wet side of optimum) and strengthening them with between 1.8% and 2.4% lime, used in varying proportions for the core, slopes and crest of the embankment, as appropriate.

"Tolerable" soils can generally be used in cores, foundations and rockfills, and "selected" soils for the crest (although these can also be used for the rest of the earth structure). "Borderline" soils can be used in all zones if they are adequately studied and improved, although they are used more in cores if improved or treated.

Spanish codes consider neither the conditions nor the energy applied on site when placing fill, leaving the work site managers and contractors considerable freedom to handle the materials and obtain the desired result. The French classification divides materials by grain size, plasticity, salt content and moisture content, and recommends how and when they can be used. *Table 12.1* includes material classifications, with six different main groups: "fine soils" (A), "coarse soils" (B), "intermediate soils" (C), "water sensitive" materials (soils and rocks) (D), "degradeable rocks" (E) and "other materials" (F).

Spanish codes consider two types of materials ("selected" and "suitable") for a fines content of less than 25 and 35%, and the French codes distinguish between the different percentages: 12–35% (B5 and B6), 5–12% (B1 to B4) and less than 5% (D soils). In the group with 12–35%, corresponding to "suitable - selected" soils according to the Spanish codes, the two groups (B5 and B6) are differentiated on their plasticity index, since the permeability of a soil is more closely related to this index than to its liquid limit. In the case of fine soils, which in fact correspond to some "tolerable" ones and also to some "borderline" and "unsuitable" ones, the French codes distinguish four groups (A1 to A4) on the basis of their plasticity index. This classification is more precise in distinguishing between the subgroups, and in theory suggests greater operational capacity, although in practice this is not so clear.

## Rockfill embankments

What is explained in the section above refers to earth embankments, i.e. to earth structures constructed with soils, generally with grain particle sizes smaller than 110–150 mm. Other earth structures may be built with rock materials of much larger grain sizes, for example, breakwaters or rockfill dams, which use blocks of up to 1.0–1.5 m in dimension (weighing 2–3 tons).

*Figure 12.9* shows the standard grain size curves for the three main groups of earth structures made from unbound rock aggregate. The "embankment group" contains under 10% fines and under 30% of particles pass through ASTM sieve no 1, enabling the maximum particle size to be no greater than two thirds of the normal thickness of the compacted layers. In rockfill embankments, materials need to be permeable and very strong, with a high angle of internal friction (40°–50°), which allows rockfills of considerable height (more than 50 m) and steep slopes (1.5H:1V– 1H:1V) as shown in *Figure 12.4*.

Between rockfill and earthfill embankments there are **"random fills"** where coarse soils predominate, but which do not meet the strict conditions for rockfills. Typical grain sizes are shown in *Figure 12.9*.

## Box 12.2

### Soil classification for earth fill embankments based on Spanish Codes

#### Selected soils

These include soils which meet the following conditions:

— Organic material content lower than 0.2% (OM < 0.2%).
— Water soluble salts content, including gypsum, under 0.2% (SS < 0.2%).
— Maximum diameter size not exceeding 100 mm ($D_{max} \leq 100$ mm).
— 15% or less passes through ASTM No 40 sieve (#40 ≤ 15%) or else meets all the conditions below:

- Less than 80% passes through ASTM No 10 sieve (#10 < 80%).
- Less than 75% passes through ASTM No 40 sieve (#40 < 75%).
- Less than 25% passes through ASTM No 200 sieve (#200 < 25%).
- Liquid limit below 30 ($W_L < 30$).
- Plasticity index lower than 10 ($I_p < 10$).

#### Suitable soils

Soils are considered suitable, even if they cannot be classified as selected soils, if they meet the following conditions:

— Organic material content less than 1% (OM < 1%)
— Soluble salts content, including gypsum, lower than 0.2% (SS < 0.2%).
— Maximum size not exceeding 100 mm ($D_{max} \leq 100$ mm).
— Less than 80% passes through ASTM No 10 sieve ( #10 < 80%).
— Less than 35% passes through ASTM No 200 sieve (#200 < 35%).
— Liquid limit below 40 ($W_L < 40$).
— If the liquid limit is higher than 30 ($W_L > 30$) the plasticity index will be higher than 4 ($I_p > 4$).

#### Tolerable soils

Soils considered as tolerable cannot be classified as selected or suitable soils but meet the following conditions:

— Organic matter content lower than 2% (OM < 2%).
— Gypsum content lower than 5% (gypsum < 5%).
— Content in other non gypsum soluble salts lower than 1% (SS < 1%).
— Liquid limit lower than 65 ($W_L < 65$).
— If the liquid limit is higher than 40 ($W_L > 40$), the plasticity index will be greater than 73% of the value which is the result of subtracting 20 from the liquid limit ($I_p > 0.73(W_L - 20)$).
— Settlement in collapse tests under 1%.
— Swelling in expansion test under 3%.

#### Borderline soils

Soils considered as borderline cannot be classified as selected, suitable or tolerable because they do not fulfil all the conditions indicated for these groups, but they do meet the following conditions:

— Organic matter content lower than 5 percent (OM < 5%).
— Swelling in expansion test lower than 5 percent (5%).
— If the liquid limit is higher than 90 ($W_L > 90$) the plasticity index will be lower than 73% of the value resulting from subtracting 20 from the liquid limit ($I_p < 0.73(W_L - 20)$).

#### Unsuitable soils

Soils considered unsuitable are:

— Those not included in the categories above.
— Peat and other soils containing perishable or organic materials.
— Those which may be environmentally unsuitable for use.

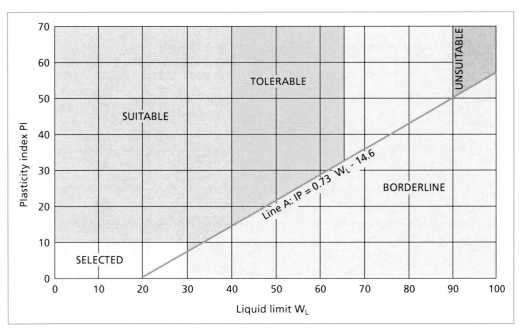

*Figure 12.8* Soil classification for earthfill embankments in Spanish codes.

| Table 12.1 | SOIL CLASSIFICATION FOR EARTH WORKS BASED ON FRENCH CODES | | | | |
|---|---|---|---|---|---|
| A<br>Fine soils | D < 50 mm<br>>35% passes<br>through 80 µm | $I_p < 10$ | | | $A_1$ |
| | | $10 < I_p < 20$ | | | $A_2$ |
| | | $20 < I_p < 50$ | | | $A_3$ |
| | | $I_p > 50$ | | | $A_4$ |
| B<br>Sandy soils and<br>gravels with fines | D < 50 mm<br>5–35% passes<br>through 80 µm | 5–12% passes<br>through 80 µm | Retained by<br>2 mm: <30% | SE > 35 | $B_1$ |
| | | | | SE < 35 | $B_1$ |
| | | | Retained by<br>2 mm: >30% | SE > 25 | $B_2$ |
| | | | | SE < 25 | $B_3$ |
| | | 12–35% passes<br>through 80 µm | $I_p < 10$ | | $B_4$ |
| | | | $I_p > 10$ | | $B_5$ |
| C<br>Soils with fine and<br>coarse fractions | D > 50 mm<br>>5% passes<br>through 80 µm | Small amount passes<br>through 80 µm | Large amount passes through 80 µm | | $C_1$ |
| | | | D < 250 mm | | $C_2$ |
| | | | D > 250 mm | | $C_3$ |
| D<br>Water-sensitive<br>soils and rocks | <5% passes<br>through 80 µm | D > 50 mm | Retained on 2 mm: <30% | | $D_1$ |
| | | | Retained on 2 mm: >30% | | $D_2$ |
| | | 50 mm < D < 250 mm | | | $D_3$ |
| | | D > 250 mm | | | $D_4$ |
| E<br>Degradeable<br>rocks | Fragile, fine structure materials, with no or low clay content, e.g. fine grain sandstones | | | | $E_1$ |
| | Fragile, coarse structure materials, with no or low clay content | | | | $E_2$ |
| | Degradeable clay materials, e.g. marls, clay shales, argilites | | | | $E_3$ |
| F | Materials which are decomposable, combustible, soluble or contaminant, e.g. topsoil, solid waste, peat, certain mining wastes, saline and gypsum bearing soils, some slags, etc. | | | | F |
| SE = Sand equivalent test (test to determine the relative proportions of granular and cohesive soil). | | | | | |

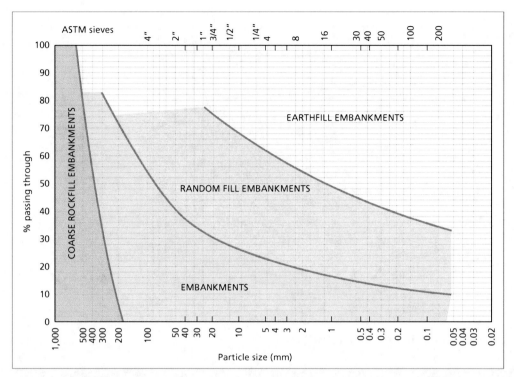

*Figure 12.9* Particle size ranges for different types of earth works.

Various zones, as listed below, can also be differentiated in rockfills and random fills:

a) Base, at least 1 m thick.
b) Core or main central section of the structure.
c) Shoulders or material forming the slopes.
d) Transition adjacent to the core grading between the core and the outer layers of the road, railway or dam crest; this normally has a minimum thickness of two 1 m thick layers.

For rockfill embankments the following are normally used:

— Suitable rocks (with uniaxial compressive strength higher than 50 MPa, with no schistosity or foliation.
— Rocks which need specific analysis, e.g. because they are foliated (slates, schists ), porous (sandstones and volcanic rocks) and alterable (shales and marls) will in all cases have their uniaxial compressive strength higher than 25 MPa.

Unsuitable rocks include those which degrade, disintegrate, dissolve or collapse, and those with a uniaxial compressive strength lower than 25 MPa.

The current trend is to extend the range of possible materials for use, after appropriate testing, and rockfill embankments have even been made with gypsum-bearing materials (Figure 12.10).

*Figure 12.10* Gypsiferous rocks used in road embankments.

The material used must show a certain stability when wet, usually measured by submerging representative samples for 24 hours to test that the material loss, per unit weight, is lower than 2%. Laboratory tests are also carried out using wetting and drying disintegration cycles, and freezing-thawing cycles using magnesium sulphate, which simulates frost action.

The thickness of the layers placed normally varies and depends on the type of rock, on the maximum size obtainable from the quarry and on the machinery available.

# EARTH STRUCTURES

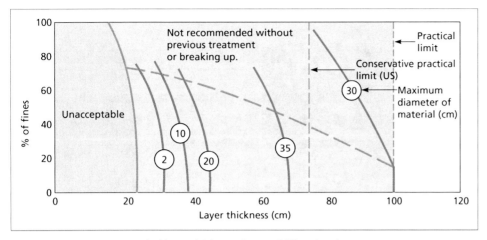

*Figure 12.11* Recommended layer thickness for earthfill embankments.

The maximum thickness may be up to 100 cm although for roads it is normally limited to 60 cm with a maximum aggregate size of around two thirds of the layer thickness (*Figure 12.11*).

## Coarse rockfill

The material used for coarse rockfill has a uniform grain size and the particles are normally large. The rock size for coarse rockfill used for roads should be between the ASTM 20 sieve size (200 kg weight) and the ASTM 8 sieve size (10 kg weight), although larger sizes may be used. The grain size curve for coarse rockfill for roads is as shown in *Figure 12.9*.

For port and hydraulic construction work, the grain size must be uniform, but the volume of each particle has to be much greater, depending on what it is designed to do.

Coarse rockfill is a selected material and its extraction must follow a strict procedure to obtain the appropriate sizes and reduce handling of the material as far as possible. To ensure this, it is extracted from quarries after careful consideration of the type and placing of explosives used, to ensure they give the correct degree of rock breakage to provide the block sizes required.

Coarse rockfill is also used to build roads or railways where the ground is liable to flood and the soils are soft. The rockfill blocks are driven into the ground, partially displacing these soils and reducing settlement.

Coarse rockfills have been used mainly in the construction of dam slopes and are still used in the external layers of breakwaters where large, dense, high quality rocks are required. They are also used as toe weight ballast for stabilizing cut slopes and unstable natural hillsides, placed at the foot of the slope (or replacing the original ground in that area), to provide weight, allow drainage and generally strengthen the whole structure (see Chapter 9, Figures 9.49 and 9.50 and Chapter 12, introductory photo).

## 12.4 Implementation and control

When an earth structure has been designed and its geometry, materials, zoning and other specifications have been defined, the on-site construction process commences; it has various stages:

— Extraction in the quarry, borrow area, nearby cut slopes or tunnels with appropriate machinery (drag-shovel excavators, shovels, etc.).
— Transport of materials to the site without significant changes or alteration in it character (with dumpers, trucks, scrapers, conveyor belts, etc.).
— Dumping and spreading of the material on site (with scrapers, shovels etc); *Figure 12.12* shows different types of machinery used in these pre-compaction phases.
— Compaction of the spread material (*Figure 12.13*); to obtain the greatest possible concentration of solids in a given volume by using appropriate energy. Moisture conditions will allow the particles to bind together to form a semi-saturated structure with the maximum possible stability. Depending on the grain size and petrology of the material, different structures will be obtained (flocculated, oriented, etc.) with different orientation of the grains and different reaction to water; the most sensitive structures are those where silt materials are used.

For **compaction control** on site, data from the laboratory compaction tests mentioned above are used (*Box 12.1*). Once the maximum or optimum dry density ($\gamma_{d,opt}$) has been established using the test considered representative of the situation on site, controls are specified (type of compacting machinery, required depth of layers placed and how many passes of compacting machinery) to determine the apparent

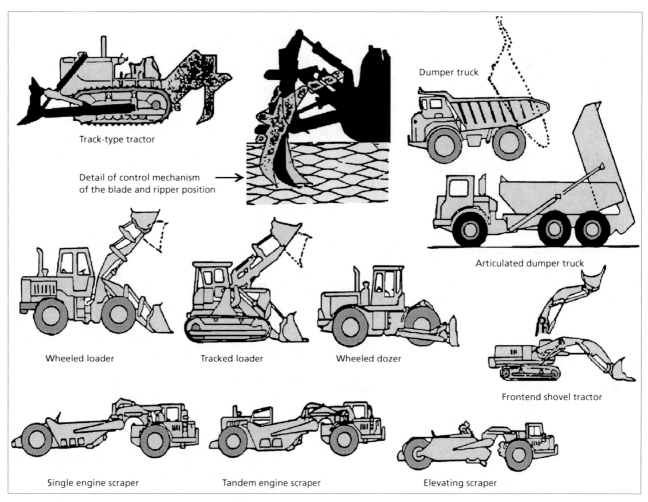

*Figure 12.12* Examples of earth-moving equipment.

*Figure 12.13* Compacting the crest of an embankment with smooth vibratory rollers.

dry density of the soil *in situ* ($\gamma_d$), deduced by defining the apparent density and on-site moisture, and its relationship with the theoretical $\gamma_{d,opt}$ values. Traditionally $\gamma_d$ *in situ* was required to be greater than a specific proportion of the $\gamma_{d,opt}$ value determined by the Standard Proctor test (for roads) or the Modified Proctor test (for earth dams):

$$\gamma_d = \alpha \gamma_{d,opt}$$

This condition, where $\alpha$ may be approx. 0.95–1.0, has two drawbacks:

— It does not consider the influence of moisture content and can be met with a very low moisture level, which could correspond to soils being in a condition where they could collapse when wet.
— The energy on site may be greater than the value from the Standard Proctor; in fact, modern machinery provides far greater energy per surface unit than that used in laboratory tests. This means that although higher values are obtained than even 100% of $\gamma_{d,opt}$, this could be in zones on the dry side or on the wet side of the real curve (with the corresponding "cushioning" of the ground in front of the compacter, without any clear energy absorption).

This is why the criteria used should consider dry and wet density simultaneously, as well as taking into account that geotechnical properties, such as angle of friction and deformation modulus, normally remain constant along a line parallel to the saturation line, as shown in the figure in *Box 12.1*. If the saturation line is considered straight (as it nearly is) then the acceptance criterion will be as shown in *Figure 12.14*. The density is determined on site by the "sand replacement" or "nuclear" method. The acceptance criteria derived from the plate load and footprint tests have to be added to these classic criteria.

The normal requirement for layer thickness is that it must allow the energy transmitted by the compactor to reach the bottom of the layer. The depth is also conditioned by the maximum dimensions of the material, $D_{max}$. *Figure 12.11* reproduces different recommendations for both earthfill and rockfill embankments, set out as a function of the data from different actual construction sites, taking the layer thickness limit for roads as 100 cm, although in dams it may be thicker, around 120–130 cm.

*Figure 12.15* includes practical criteria for the most appropriate compacting machinery, depending on the percentage of fines present in the fill and on its maximum size.

Similar criteria to roads are applied to **embankments for high speed railways**, although the density values required *in situ* are around the optimum obtained in the Modified Proctor test, while for roads the reference values can be lowered to the optimum of the Standard Proctor test.

An example of this is given by the types of the fill used in the Madrid-Zaragoza (Spain) high speed rail line, where the foundation, core, slope and crest zones are different. In the foundations, the same material could be used as in the core up to a height of 2 m, but with a fines content lower than 15% where the foundations were to be in saturation conditions. In the core, a variety of suitable soils were used provided their content of organic material was less than 1% and their sulphates were less than 5% (except in special cases where there is no water infiltration and where, after careful study, up to 15% sulphates were considered acceptable). The core material also has a liquid limit under 50, settlement (measured in an oedometer)

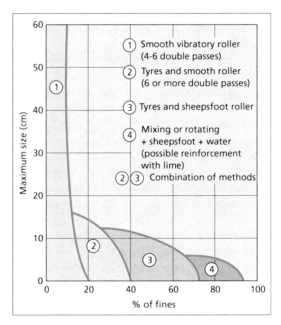

*Figure 12.15* Recommended compaction methods.

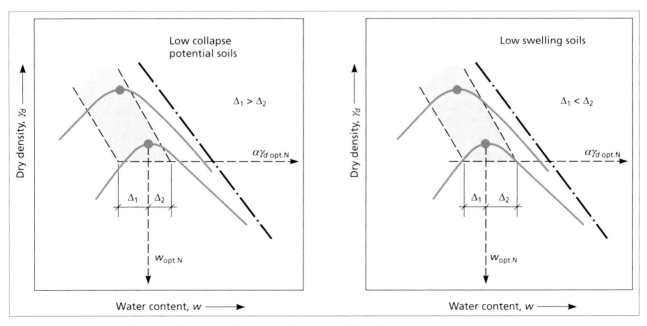

*Figure 12.14* Suggested criteria for compaction according to Spanish codes.

under 1%, optimum density in the Modified Proctor test higher than 17.5 kN/m³, and a CBR index over 5 (with expansion below 1%). At the crest, the allowable liquid limit was lowered to 40, the aggregate size was up to 10 cm and the fines content to 40%. In the slopes, up to 2 m in excess of the basic width was required, depending on anticipated problems with surface erosion, with material having a fines content under 15%.

In the main body of the embankment for this high speed railway, a dry density of over 95% of the Modified Proctor optimum dry density was required. The *in situ* control was mainly based on monitoring dry density and second cycle deformation in plate load tests, with values higher than 45 MPa (in fine soils) or 60 MPa (in granular soils).

In large scale earth moving projects, preliminary **experimental embankments** are generally built, where after various tests and controls with appropriate systems, the compaction process is defined, including the machinery, passes and layer thickness for each material, checking the procedure on-site to allow work to go ahead, rather than using the density and moisture parameters mentioned above. Some tests are carried out from time to time, but the idea is to control the procedure rather than the end product. This means that high volume outputs can be maintained (300,000–500,000 m³/month) with good quality materials.

Plate loading tests can only be used to control the **compaction of rockfill embankments**, when the diameter of the plate is 4–5 times larger than the maximum material size for the rockfill, which means that this test is not often used on the construction site, and then only for special tests. Normally, the **footprint test** is used, with footprints of less than 3 mm required in the transition zone and less than 5 mm in the rest of the embankment. Densities can also be measured *in situ* by digging large pits, weighing the material removed and lining the pit with an impermeable material, and then filling the pit with water. This is considered as a special test.

For random fill structures, compaction control is also carried out, mainly with the footprint test, although *in situ* density and plate load tests can also be used.

## 12.5 Embankments on soft soils

Soft soils characteristically present:

— Low short-term shear strength, which causes embankment stability problems.
— High deformability, normally deformation of 4–10% of the depth of soft soil affected by the embankment.
— A considerable time until deformations have ceased.
— Danger of collapse from flooding or from particle stress concentration in the case of landfills or metastable soils.

**Soft clay soils** are characterized by their low apparent dry density (around 6–14 kN/m³) and high moisture content (40–120%), which can be as high as 400% in peat or peat rich soils. In many cases the organic content of soft soils reaches 10% which considerably influences their deformability. Their resistance to the SPT penetration test may vary from 2 to 10 blows/30 cm, and from 150–800 kPa in the static penetrometer (CPT).

The undrained shear strength, $S_u$, may be very low. Soft soils tend to have an $S_u$ value around 20 kPa although values varying from extremes of 10 kPa to 30 kPa have been recorded. $S_u$ may be around 0.20 $\sigma'_{vz}$, where $\sigma'_{vz}$ is the effective initial vertical stress at depth $z$. The value of $S_u$ is used to find the short-term stability of the embankment on soft soil, although drained shear strength parameters should be used for an analysis in terms of effective stress (*Figure 12.16*).

Instantaneous or initial **settlements** (end-of-construction) should be determined together with any longer term settlement from ground consolidation. In the short term, without drainage, settlement can be estimated by elastic methods, taking a Poisson ratio of 0.5 and a

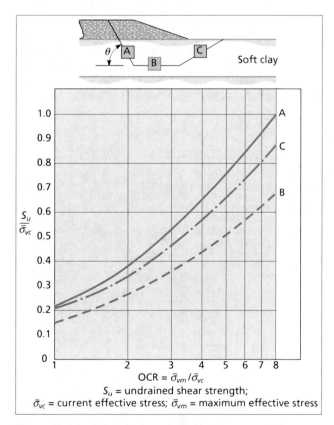

*Figure 12.16* Suggested undrained shear strength parameters in non-cemented plastic clays for total stress conditions (modified from Ladd and Foott, 1977).

deformation modulus $E_u$ from $S_u$: $E_u = \delta S_u$. A $\delta$ value of around 1,000–2,000 will be adequate. In the long term, the value of deformation modulus $E'$ will clearly be lower; it can be estimated empirically from $S_u$: $E' = \delta' S_u$ where the value of $\delta'$ will be around 200–300; i.e. the value of the $\delta/\delta'$ ratio will be around 4–6.

The methods used to define settlement and horizontal movements are those normally used in soil mechanics:

— Oedomer based methods, when the soft soil depth is less than the embankment width.
— Finite element models when the depth is greater than the embankment width.
— Empirical methods.

For stability analysis, as has already been mentioned, analysis of total pressures is usually carried out with failure circles through the soft foundation (generally until they are tangent to a stronger layer at depth if this exists), often using Pilot and Moreau's (1973) charts designed to analyse a homogeneous embankment on a homogeneous soft layer lying on a hard stratum. If the factor of safety against sliding is low and settlement is very considerable, then different solutions involving, for example, ground reinforcement and treatment, are considered:

a) Pre-loading by using the embankment's own load, or by increasing the height of the embankment at a specific point before the construction is completed. The settlement time can be reduced by constructing flexible drains (Figure 12.17). These drains speed up settlement but do not increase ground strength in the short term.
b) Driving in wooden piles particularly in marshy areas (Figure 12.18).
c) Vibroflotation to improve the density of the foundation ground, when this contains loose sands.
d) Gravel columns, usually made with vibroflotation techniques, which reinforce the ground and speed up settling, as they provide a denser and more permeable soil (Figure 12.19).
e) Ground densification with dynamic compaction based on reducing the ground volume by the repeated

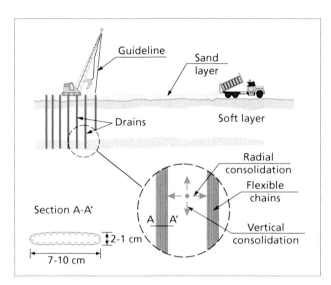

Figure 12.17  Plastic drain construction.

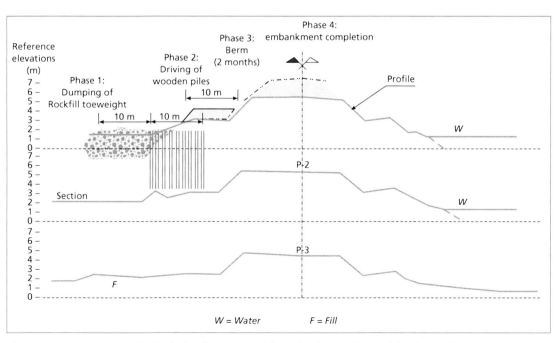

Figure 12.18  Recommended solution for an unstable embankment in marshlands, South Spain.

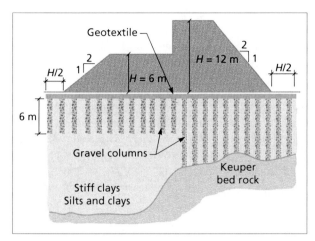

*Figure 12.19* Ground treatment with gravel columns (Oteo and Sopeña, 1994).

application of high force blows on the soil. This is done by letting a heavy weight fall from a height. The blows are regularly spaced on squared grids measuring 4–6 m and the blows are applied in two phases, at the corners and centres of each square.

## 12.6 Embankments on slopes

Most problems in embankments occur after construction, particularly at a point midway on a hillslope, or when the earthfill has been placed on the surface of a natural slope subjected to physical and chemical weathering or even erosion by animals. The hill slope surface is usually composed of the following soil types:

— Transported soils, with clay, silt and coarse grains in a fine matrix having a relatively low permeability and low shear strength.
— Residual soils, altered *in situ* that may be oxidized and fissured (e.g. marls and clay materials) or highly weathered rocks.

These soils are easily saturated through fissures or voids and are influenced by climatic changes; in the case of semi-arid climates, dry seasons cause shrinkage and fissuring in clay materials, allowing the water to penetrate during rainy seasons and saturate the upper and more weathered part of the slope, so that it becomes weakened by moisture and subject to high pore pressures leading to problems of stability. In these cases the fissured and saturated zone has a residual shear strength corresponding to an effective cohesion $c'$ of 0–10 kN/m$^2$ and an effective angle of friction $\phi'$ of 12–15°.

Where this is the case, the following remedial measures have to be taken:

— Draining the embankment base to collect the runoff from the top of the slope.
— Terracing of the natural ground of the slope to provide adequate support for the embankment.
— Reinforcing the foot of the embankment with coarse rockfill materials.

In the rainy season, important problems of stability often appear in embankments, especially if they are constructed in valleys or dry river beds. *Figures 12.20 to 12.23* show some examples of remedial measurements, which include:

— Modifying the road alignment to reduce the height of the embankment.
— Constructing a barrier of large diameter piles to intersect the slipped material and provide appropriate foundations for new additional materials (*Figure 12.21*).
— Creating barriers of micro-piles for embankments less than 10 m high.
— Using angled jet grouting columns to reinforce the ground under the foot of the embankment slope (*Figure 12.23*).

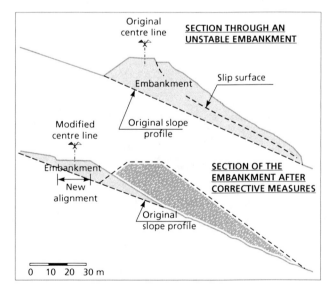

*Figure 12.20* Corrective measures in an unstable embankment.

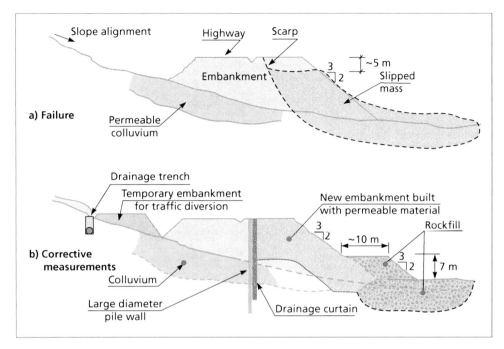

*Figure 12.21* Remedial measurements for unstable motorway embankments (Oteo and Sopeña, 1994).

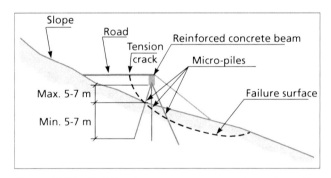

*Figure 12.22* Stabilization measurements using micro-piles in an unstable slope.

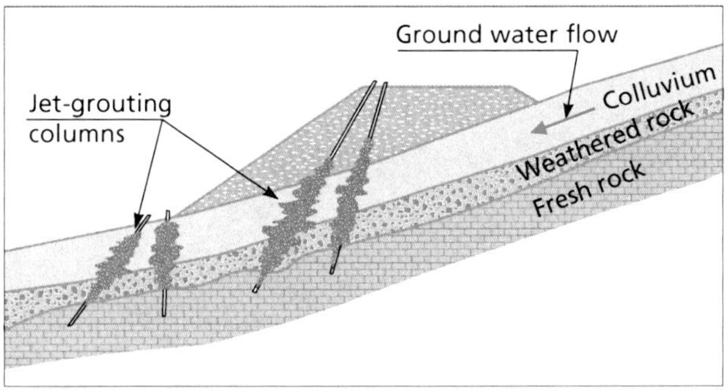

*Figure 12.23* Stabilization measurements using jet-grouting columns on an unstable slope.

# References and recommended reading

Bell, F.G. (2007). Geological materials used in construction. In: Engineering Geology, Chap. 6. Elsevier. 2nd ed.

Ladd, C.C. and Foott, R. (1977). Foundation design of embankments on varved clays. U.S. Department of Transportation. FHWA - Report No. TS-77-214. Washington.

Oteo, C. and Sopeña, L. (1994). Embankment foundations on very soft soils at Medinaceli (Spain). XIV Int. Cong. on Soil Mechanics and Foundation Engineering, New Delhi, Vol. II.

Pilot, G. and Moreau, M. (1973). La stabilité des remblais sur sols mous. Collection du Laboratoire central des ponts et chaussées. Eyrolles, Paris.

U.S. Bureau of Reclamation (1998). Earth Manual. Part 1. 3rd ed. U.S. Department of the Interior, Denver, Colorado.

# PART IV

# GEOLOGICAL HAZARDS

# 13

# LANDSLIDES AND OTHER MASS MOVEMENTS

1. Introduction
2. Slope movements
3. Investigation of landslides
4. Corrective measures
5. Collapse and subsidence
6. Prevention of risks from mass movements

# 13.1 Introduction

Geodynamic processes affecting the earth's surface cause **mass movements** of different types, size and speed. **Slope movement** is the most frequent and widespread type of mass movement, generated by the gravitational down slope displacement of soil and rock masses. **Subsidence** is less common, as it is associated with specific materials and circumstances. Both processes are described in this chapter.

The force of gravity and the progressive weakening of geological materials, mainly due to weathering, together with the action of other natural and environmental phenomena, make mass movements relatively common in the earth's surface.

These processes create potential **geological risks,** as they can cause economic loss and social damage if they affect human activities, buildings and infrastructure. How to avoid these adverse effects is the subject of research into mass movements, including the study of their characteristics, instability mechanisms, controlling factors and causes. To carry out this research it is necessary to know the characteristics and the geological, geotechnical and hydrogeological properties of the soil and rock materials involved, and their mechanical behaviour, as well as the factors that condition and trigger such movements. Studies in this field tend to focus on one or more of the following applications:

— Investigation of particular processes for the design of stabilising measures to either mitigate or reduce damage.
— Analysis of the factors which control and trigger processes at particular locations, in order to prevent possible movements.
— Mapping either unstable or potentially unstable zones, so that the hazardous areas can be delimited and preventive measures can be applied.

The role of geological engineering in preventing the risk caused by landslides and subsidence, is more effective when these processes occur on a geotechnical scale, i.e. with dimensions which allow the processes to be controlled. Mass movements on the much larger geological scale are generally impossible to control, and the only measures possible are prevention and land use restriction.

# 13.2 Slope movements

The geological and climatic processes that affect the earth's surface create topographic relief and define the morphology of slopes, which are modified over time to adjust to new geological or climatic conditions. In general, slopes adopt natural, near-equilibrium angles; when conditions change, the morphology is modified to re-establish equilibrium. In this context, slope movements can be understood as ground adjustments to recover equilibrium under changing conditions.

The areas which are most prone to instability from a global point of view include mountainous regions, areas where the relief is subject to severe erosive and weathering actions, the sides of river valleys, coastal cliffs, areas of soft and loose material, areas of clay, schist and alterable rock mass, seismic zones and areas with heavy rainfall.

The study of **natural slope movements** or mass movements, often included in the general term **landslides**, has much in common with the study of the stability of excavated slopes, discussed in Chapter 9: the geomechanical factors which control failure and instability, the failure mechanisms operating, analysis and modelling methods, stabilisation methods and monitoring.

The study of **excavated slopes** aims to design stable excavations and correct or stabilise failures, which are normally superficial and affect relatively small volumes (although in major mining excavations large scale mass movements may occur). Landslides in natural slopes may be deeper and involve the movement of millions of cubic metres of material; in addition, the failure mechanisms usually are complex, as they are conditioned by factors operating on a geological scale (these include faults, tectonic processes, geomorphological processes and groundwater flow).

*Figure 13.1*  Landslides on the slopes of the Panama Canal delayed work for years.

Large scale movements of tens or hundreds of cubic kilometres are rare, although there is morphological evidence on the earth's surface confirming their occurrence in the past, possibly associated with periods of rainy climate or with high tectonic, seismic and volcanic activity. The largest landslides on earth, with volumes of hundreds of cubic kilometres, occurred in prehistoric times on the flanks of volcanic islands, such as the Canary Islands, Réunion or Cape Verde; the morphologic traces of these mega-landslides can still be clearly seen today.

Slope instability, of both natural and artificial slopes, is due to the imbalance between the internal and external forces acting on the ground, and occurs when destabilising forces tending to cause sliding are greater than stabilising forces tending to resist sliding. This imbalance may be due to a modification of the existing forces or to the application of new static or dynamic external forces.

Although natural instabilities are often complex processes, as mentioned above, the mechanisms and models of ground failure are similar to those of excavated slopes, described in Chapter 9, Section 9.4, and can be grouped in such categories as failures along curved or planar surfaces, failures as wedges or blocks, or others types, depending on whether the materials involved are soil or rock.

Because slope movements are extensive and frequent they are a very important geological risk that affects communication routes, supply pipelines, riverbeds and dams, and occasionally buildings and built up areas.

On the other hand, landslides are perhaps the most predictable of the natural processes and the most responsive to corrective and mitigating measures for preventing the damage they can cause. Slope movement predictions can even be made in cases where they are associated with known factors, such as rainfall.

Slope movements include different types of processes, such as landslides, rock falls, mud, earth and debris flows, and rock avalanches. These different types are described below.

# Types of slope movements

Slope movements are usually classified on the basis of the materials involved, generally distinguishing between rocks, debris and soils, and on their mechanism and type of failure; other aspects are also considered, such as the ground water content and the rate and dimensions of the movement. Some of the most widely used classifications (Varnes, 1984; Hutchison, 1988; EPOCH, 1993; Dikau et al., 1996), each with different criteria and aims, are very useful for a first approach to the study of slope movements and for an understanding of the materials which undergo these processes. Specific classifications have also been set out for some materials, such as clays, or for some specific types of movements, such as flows. Figure 13.2 shows a simplified classification of different types of slope movements, based on failure mechanisms and types of materials involved.

## Landslides

Landslides are movements of masses of soil or rock which slide, moving in relation to the substratum, on one or more failure surfaces, when the shear strength within these zones is exceeded. The mass is generally displaced as a whole, behaving as a unit; the rate of movement may vary considerably, but these processes tend to move rapidly, reaching considerable speeds, and can involve volumes up to several million cubic metres. Sometimes, when the slide material does not reach equilibrium at the toe of the slope (because of its loss of strength, and its momentum), the mass may continue to move for hundreds of metres and reach very high speeds, generally causing a **flow**; rock landslides may also cause **rock avalanches**.

Landslides may occur in debris (e.g. in colluvium), along the contact with the substratum or in very weathered and fractured rocky slopes, along the contact with the competent rock, and generally produce debris flows; these tend to occur in saturated material conditions. The term **debris** refers to loose, unconsolidated material, with a significant proportion of coarse material (Varnes, 1988).

**Rotational landslides** (Figure 13.3) are more common in cohesive, "homogeneous" soils. The failure, which can be superficial or deep-seated, occurs along curved surfaces concave upwards, having the shape of a "spoon". Once instability has begun, the mass begins to rotate, and may divide up into different blocks which slide in relation to each other and give rise to "benches" with surfaces tilted towards the slope and striated tension cracks. The normal dimensions vary from tens to hundreds of metres in both length and width and they may be superficial or deep (the limit can be set at around 10 m). The lower part of the sliding mass accumulates at the toe of the slope forming a lobe type deposit with transversal tension cracks (Figure 13.3). The sliding mass may generate a flow depending on the type of soils involved and their water content.

Figure 13.5 shows different types of curved or rotational landslides. **Successive** landslides occur mainly in stiff fissured clays with gradients similar to their angle of equilibrium and in soft very sensitive clays, where the initial landslide causes an accumulation of remoulded clay which, as it flows, leaves the material higher up the slope without support, so promoting successive failures. These failures are fairly shallow but can have considerable lateral continuity.

Weak rock masses or those with a high degree of fracturing or weathering, where the structural discontinuities do not form preferential surfaces for failure, may also suffer this type of movement.

In **translational slides** the failure takes place along pre-existing planar surfaces or discontinuities (bedding planes,

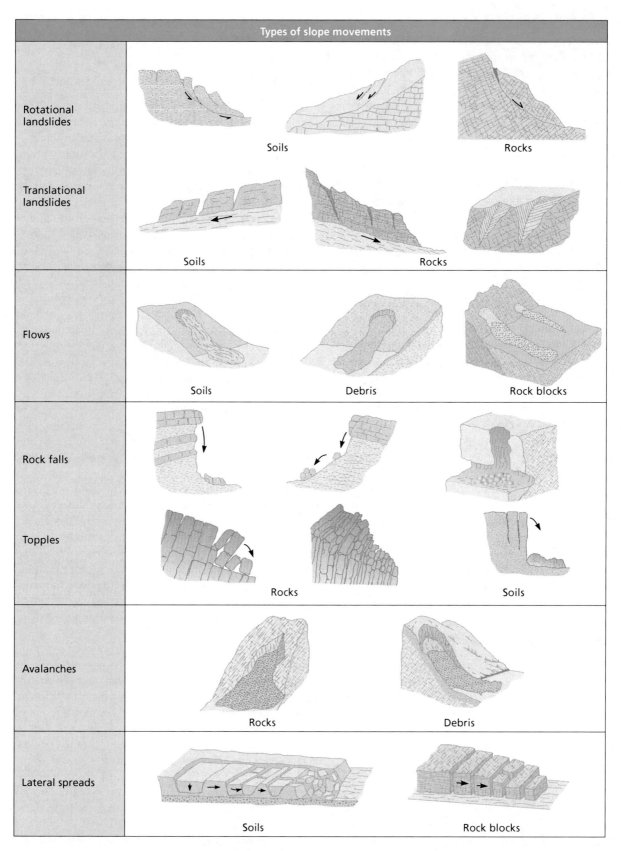

Figure 13.2  General classification of slope movements.

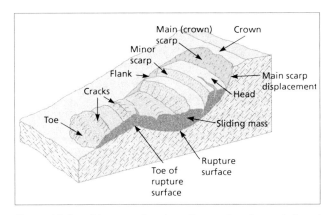

*Figure 13.3* Diagram showing the parts of a rotational landslide (modified from Varnes, 1988).

*Figure 13.4* View of the head of a rotational landslide in clays (eastern England).

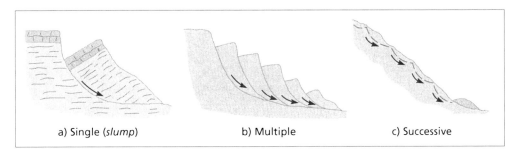

*Figure 13.5* Types of rotational landslides.

*Figure 13.6* Planar displacement surface over a clay layer only a few mm thick, corresponding to the case shown in *Figure 13.17*, NW Italy. Striations on the plane and displaced rock blocks in the background can be seen.

*Figure 13.7* Front of a translational rockslide transporting large masses of strata, NW Italy.

contact between different types of materials, structural surfaces, etc.); sometimes the failure plane is a fine layer of clay material between more competent strata (*Figure 13.6*). They are not normally very thick, although they may be very extensive and reach considerable distances. They may occur in soils and rocks (*Figure 13.7*) and the sliding masses are sometimes rectangular blocks that have become detached from the mass at discontinuities or tension cracks (block landslides). Movements which are either impulse driven or move at different times are common in this type of instability depending on the strength conditions of the slide surfaces, which do not necessarily have to possess a steep gradient. Translational slides generally move faster than rotational ones, because of the simple geometry of their failure mechanism.

## Flows

Flows are mass movements of soil (mud or earth flows), debris (debris flows) or rocks blocks (rock fragment flows) often with a high water content, where the material behaves as a fluid, undergoing continuous deformation, but without having well-defined failure surfaces. Water is the main triggering factor because water decreases the strength of materials having low cohesion. Flows mainly affect sensitive clay soils which show considerable loss of strength when mobilised; these movements are not very deep in relation to their extent; they develop a glacier like morphology and may occur on slopes with low gradients (even less than 10°). The displaced masses may reach several kilometres. *Figures 13.8* and *13.9* show examples of mud and rock block flows.

**Mud or earth flows** occur in predominantly fine and homogeneous materials and can move at a speed of many metres per second; the loss of strength is usually caused by water saturation. They are classified according to the type of material, its strength and its water content (*Figure 13.10*).

Mudflows are generally small scale and slow, but sometimes, especially in saturated conditions, they are extensive and fast, with catastrophic consequences when they reach populated areas. Because of their physical and geomechanical properties, deposits of fine volcanic materials are particularly susceptible to this type of processes.

In loessic soils and dry sands, seismic movements may lead to flows, generally caused by collapse due to the failure

*Figure 13.9* Flow of rock blocks (in the foreground, blocks of several cubic meters) derived from the rock mass which can be seen in the background (southern Spain).

*Figure 13.8* Crown (above) and toe (below) of a mudflow triggered by intense rainfall (northern Spain).

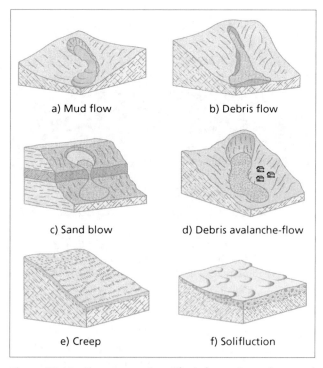

*Figure 13.10* Flow types (modified from Corominas and García Yagüe, 1997).

of the weak bonds between particles; if these materials are saturated or submerged, a non-cohesive mass results which may flow at very high speeds. These abrupt movements caused by the structural collapse of the soil, due to seismic quakes or drying out of the soil, are called **sand and silt blows**, a term which refers more to the cause of the movement.

**Debris flows** are complex movements which include rock fragments, blocks, cobbles and gravel in a fine-grained matrix of sands, silts and clays. They occur on slopes covered with loose or non-consolidated material as in the case of glacial moraine deposits, and especially where there is no vegetation cover. According to data collected by Corominas et al. (1996), the gradients of areas where debris flows originate range from 20° to 45° and in areas where they accumulate range from 5° to 15°, with the flows having speeds of less than 1 m/s to 12–15 m/s.

Flows may be the result of landslides or be triggered by rock falls. Along with landslides they are the most widespread slope movements and affect widely differing types of materials.

In some classifications the flow type movements include different processes with specific characteristics such as **creep** (Figure 13.11), a very slow, almost imperceptible superficial movement (a few decimetres deep), which affects soils and weathered materials, causing continuous deformation that becomes progressively noticeable on slopes over time and seen as fences, walls or posts that are leaning or offset, and trees that are bent. The term "creep" refers to a time-dependent deformation and defines the deformational behaviour of the material, rather than a type of movement. **Solifluction** also affects the saturated surface layer of slopes; this is a slow movement produced by freeze-thaw processes which, because of the daily or seasonal variations in temperature, change the water phase and water content of fine-grained soils in cold regions.

Figure 13.11 Soil creep; notice the leaning trees.

## Rock falls

**Rock falls** are very quick free falls of rock blocks which are dislodged from pre-existing discontinuity planes (tectonic, bedding surfaces, tension cracks). The movement may be by a vertical fall, by a series of bounces or by rolling down the slope face. They are common on steep slopes in mountainous areas, on cliffs and in general on rock walls; the blocks are bounded by different sets of discontinuities, often forming wedge shaped blocks. The factors that cause them include erosion and loss of support for previously separated or loosened blocks in steep slopes, water pressures in discontinuities and tension cracks and seismic shakes. Figure 13.2 shows different types of rock falls.

Although the fallen blocks may be relatively small in terms of volume, rock falls are sudden processes that pose

Figure 13.12 Rock falls often affect villages in mountain areas (northern Spain).

an important risk to communication routes and buildings in mountainous zones and at the foot of steep slopes.

Masses of soil may also fall from vertical natural and excavated slopes, generally due to the existence of tension cracks generated by tensional stresses or shrinkage cracks in ground that has dried.

**Toppling** of strata or blocks of rock may be included in rock falls (the characteristics of this type of failure are described in Chapter 9, Section 9.4). Toppling occurs when the strata dip in the opposite direction to the slope, and also forms naturally inclined blocks which are free to rotate because of failure at the foot of the slope. Toppling tends to occur mainly on rocky slope faces which intersect steeply dipping strata (Figure 13.2).

## Rock avalanches

Rock avalanches, called rock falls or complex movements in some classifications, are rapidly falling masses of rock and debris that detach themselves from steep slopes, sometimes accompanied by ice or snow (Figure 13.13). The rock masses disintegrate during their fall and form deposits of very different block size, with no rounding from abrasion, and a chaotic distribution. Rock avalanche deposits are unstructured and have great porosity.

Avalanches are generally the result of large-scale landslides or rock falls which, because of the steep gradient on which they form and the lack of both structure and cohesion in their materials, travel down steep slopes at great speed. Sometimes, speeds of over 100 km/h are reached, even when the masses are completely dry, because of the reduction in friction caused by the presence of air between the rock fragments. Water from precipitation or thaw, seismic movements and volcanic eruptions may each play an important role in triggering these processes.

**Debris avalanches** are formed from rock material containing a great variety of sizes, and may include large blocks and abundant fines. The material forming moraine deposits is prone to these processes, as well as the loose materials resulting from volcanic eruptions. The main difference with debris flows, apart from water content (which is not necessary in avalanches), is the rate and speed of movement of the avalanchee in areas of steep gradients.

## Lateral displacements

This type of movement (also called lateral spreading in some classifications) refers to the movement of rock blocks or very coherent, cemented soil masses that rest on soft, deformable material. The blocks move slowly down very low gradient slopes. These movements are due to the loss of strength of the underlying material, which either flows or is deformed under the weight of the rigid blocks. Lateral spreading may also be caused by liquefaction of the underlying material or by lateral extrusion of soft, wet clays under the weight of the masses above them (Figure 13.15). These movements occur on gentle slopes and may be very extensive. The coherent upper layers may fragment into blocks, generating cracks, differential displacements, toppling or other types of deformation, and resulting in a chaotic appearance to the zones affected.

Figure 13.14 Chaotic deposits from a large prehistoric rock avalanche on the slopes of the Mombacho volcano in Nicaragua (above) now form a large number of small islands in a lake more than 12 km away from the volcano (below).

Figure 13.13 Rock avalanche from the 1960s (southern Spain) (photo courtesy of J.J. Durán).

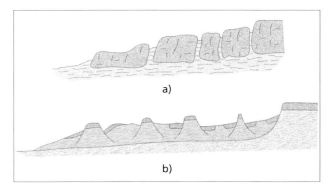

*Figure 13.15* Lateral spread may be caused by: a) flow and extrusion of underlying material, b) liquefaction.

## Causes of slope movements

The factors controlling slope movements are those able to modify the internal and external forces acting on the ground. These factors are described in Chapter 9, Section 9.3 (see Table 9.1). *Table 13.1* shows their effects on the behaviour and properties of materials. **Conditioning** (or "passive") factors depend on the intrinsic nature, structure and morphology of the ground itself and are responsible in general for the types of movement, whereas **triggering** (or "active") factors are external factors which provoke or trigger instability and are responsible in general for the size and speed of the movements.

In relation to their influence on the geomechanical behaviour of soils and rocks, factors which control slope movements can be grouped into those which contribute to reducing the shear strength and those which increase the shear stresses. *Table 13.2* lists these factors.

Various causes contribute to most cases of slope movement, and although these are often attributed to the action of a specific triggering factor (e.g. rain, earthquakes), instability would not have occurred without pre-existing conditions being conducive to the process.

Among the **conditioning factors** are the physical characteristics and strength of the materials (directly related to their lithology and nature) and the morphological and geometrical characteristics of the slope; these factors are fundamental in determining whether a slope is prone to

*Table 13.1* **INFLUENCE OF DIFFERENT FACTORS ON THE CONDITIONS OF MATERIALS AND SLOPES**

| | Factors | Influences and effects |
|---|---|---|
| Conditioning factors | Relief (gradients, geometry) | Distribution of ground weight. |
| | Lithology (composition, texture) | Unit weight, strength. Hydrogeological behaviour. |
| | Geological structure and state of stress | Strength, deformability. Discontinuous and anisotropic behaviour. Weak zones. |
| | Geomechanical properties of materials | Hydrogeological behaviour. Pore pressure generation. |
| | Deforestation | Alteration in hydraulic balance. Erosion. |
| | Weathering | Physical and chemical changes, external and internal erosion, generation of weak zones. |
| Triggering factors | Rainfall and infiltration | Variation in pore pressure and ground weight. |
| | Changes in hydrogeological conditions | Soil saturation. Erosion. |
| | Application of static or dynamic loads | Changes in material weight distribution and state of stress. Increase of pore pressures. |
| | Changes in slope morphology and geometry | Variation of forces due to weight. Change in state of stress. |
| | Erosion or undermining of toe | Changes in slope geometry. Changes in material weight distribution and state of stress. |
| | Climate action (thawing, freezing, drought processes) | Change in ground water content. Generation of tension cracks and weakness planes. Decrease of strength properties. |

## Table 13.2 FACTORS INFLUENCING THE GEOMECHANICAL BEHAVIOUR OF SLOPE MATERIALS

| Reducing shear strength | Increasing shear stresses |
|---|---|
| Initial conditions:<br>— Composition, texture and structure.<br>— Fractures and faults.<br>— Bedding and foliation planes.<br>— Brecciated zones.<br>— Massive rock on plastic materials.<br>— Alternating materials with different permeability.<br><br>Changes in materials from weathering and physical-chemical reactions:<br>— Physical disintegration of rocks.<br>— Hydration of clay materials.<br>— Desiccation of clays and clayey rocks.<br>— Dissolution and leaching of materials.<br>— Increase in plasticity of clays.<br><br>Changes in inter-granular forces due to water and water pressure in pores and fractures:<br>— Rainfall.<br>— Thawing.<br>— Lakes and reservoirs.<br>— Irrigation.<br>— Deforestation.<br><br>Changes in structure:<br>— From fissuring in slates and over-consolidated clays.<br>— From stress relief on rocky slopes in valleys or excavations.<br>— From remoulding of fine soils (sand, loess) and sensitive clays.<br><br>Decrease of strength due to creep processes.<br><br>Actions of tree and shrub roots.<br><br>Excavation of animal burrows. | Action of transitory forces:<br>— Seismic movements.<br>— Vibrations from explosions, machinery and traffic.<br>— Landslides in the vicinity.<br><br>Loss of confinement at the sides and base of slopes:<br>— Erosion of foot of slopes by rivers, streams, waves, tides, etc.<br>— Rainfall.<br>— Piping or underground erosion.<br>— Dissolution and leaching of material.<br>— Mine workings.<br>— Presence of underlying plastic materials.<br><br>Changes in morphology:<br>— Landslides.<br>— Large scale faults.<br><br>Natural overloads:<br>— Weight of rainfall, snow and aquifer water.<br>— Accumulated debris, colluvium or slided masses.<br>— Vegetation.<br><br>Anthropogenic loads:<br>— Fills, wastes, tips.<br>— Buildings and structures.<br>— Cultivation and irrigation.<br><br>Other anthropogenic activities:<br>— Excavations.<br>— Removal of retaining elements.<br>— Constructions of reservoirs and lakes.<br><br>Lateral pressure:<br>— Water in tension cracks and cavities.<br>— Ice in cracks.<br>— Presence of expansive materials.<br>— Mobilization of residual forces.<br><br>Volcanic processes. |

Modified from Varnes, 1988.

instability. Other important factors are geological structure and discontinuities, hydrogeological conditions and state of stress within the slope.

Some aspects of the conditioning factors, which complement the explanations in Chapter 9, Section 9.3, are shown below.

**Relief** plays a fundamental role, as a certain slope angle is necessary for gravitational movements to take place. Mountainous regions are the areas most prone to slope movements. Sometimes, however, and depending on other factors, even a gentle gradient of just a few degrees may be enough for certain types of instability to take place, such as mud or earth flows.

**Geological structure, stratigraphy and lithology** determine the movement potential of different types of rock material and soil, as does the existence of discontinuity planes which can act as failure surfaces. Composition, strength, deformability, the degree of weathering together with fracturing, porosity and permeability, are all factors which determine whether the ground is susceptible to failure and displacement that result from certain triggers. In jointed rock masses and weathered zones, these conditions take precedence over lithology.

The **hydrogeological behaviour** of materials is associated with their lithological and structural characteristics and their degree of weathering, which could be closely related to the climatic conditions of the region. In rainy regions a thick layer of weathered material over the rock strata and high water table, both have a decisive influence on stability conditions. Water reduces the strength of materials in two ways:

— It reduces shear strength by generating pore pressures and it softens certain types of soils.
— It increases shear stresses by increasing ground weight and by generating destabilising forces in tension cracks and discontinuities.

**Strength and geomechanical properties** control the shear strength of the materials and their failure mechanisms. Soils, as they can be considered to be, in general, homogeneous and isotropic, are less complex to characterize, while in rock masses with very different degrees of fracturing, it could be the properties of the intact rock, the discontinuities or the rock mass as a whole which determines their strength and mechanical behaviour.

The most important **triggering factors** (Table 13.1) are rainfall, changes in the hydrogeology of the slope, changes in the shape of the slope, erosion and earthquakes. Some of these, such as changes in water conditions and geometry, are often the result of human activities.

## Rainfall and climatic conditions

The triggering of slope movements by meteorological and climatic factors is basically related to the volume, intensity and distribution of rainfall, and to the climatic regime. Thus, it is important to bear in mind the ground response to intense precipitation lasting hours (storms) or days, as well as seasonal (dry and wet periods throughout the year) and annual fluctuations spread over many years (wet and dry cycles).

Groundwater produces pressures that modify the state of stress and weathering. Pore pressure, increased weight, piping and external erosion all modify the properties and strength of materials, mainly in soils, as do mineralogical changes.

Rainwater infiltration produces sub-surface and underground flows on slopes, increases water content in the non-saturated zone and raises the water table, so recharging the saturated zone (Figure 13.16). The quantity of water which infiltrates the ground depends on the intensity and duration of

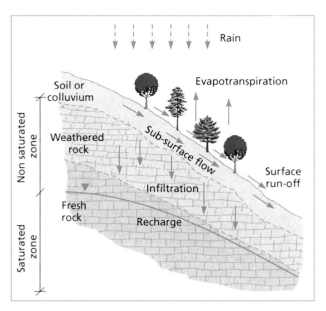

Figure 13.16 Diagram of slope water flow.

the rainfall, the size of the catchment, previous groundwater content (the position of the water table and degree of saturation of the ground), permeability and transmissivity, as well as the topography and other slope characteristics, such as vegetation. Depending on these factors, new states of disequilibrium may be generated, causing slope instability.

**Intense precipitation** lasting hours or days may trigger **surface movements** such as landslides and mud or debris flows, affecting weathered materials and soils, and cause blocks previously detached in rock slopes to fall, as well as reactivate old landslides. In 1982, intense rainfall over 32 hours in the San Francisco Bay area triggered more than 18,000 superficial landslides; in November 1994 intense rainfall over two days (in many areas exceeding 400 mm) caused more than 1,000 translational landslides of large rock blocks, superficial landslides and catastrophic flows in the Piamonte region in the north of Italy (Figure 13.17). During Hurricane Mitch in Nicaragua in 1998 rains caused a huge debris flow at the Casitas Volcano, which stretched for many kilometres, burying a village and killing 2,000 people.

The rapid infiltration of rainwater, saturating the ground surface layer and increasing pore pressures, triggers surface movements. The lack of vegetation on the slopes, the presence of loose materials and pre-existing instability processes can all increase infiltration capacity and contribute to the mobilisation of materials.

For more details on rainwater's effect on the ground and how it relates to triggering slope movements, see Gostelow (1991).

Many authors have established thresholds for the intensity and duration of rainfall which trigger slope movements in

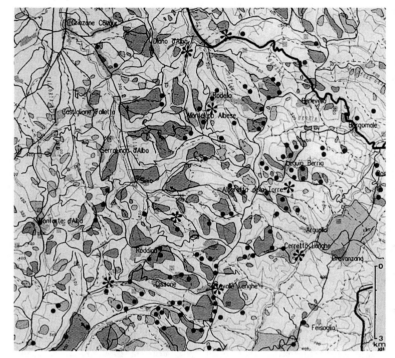

*Figure 13.17* Part of the inventory map of landslides in the Langhe region (NW Italy), showing the location of movements triggered by rainfall in November 1994: these are mainly translational rock block landslides (red dots) and mud flows (the asterisks show the areas with the highest concentration of flows). The base map, prepared some years before these movements, shows the areas with inactive block landslides (yellow) and active (green) over the last 30–40 years (Casale & Margottini, 1995). The photos show examples of the types of movements mentioned above; several buildings destroyed by the surface flows can be seen.

different places in the world. *Table 13.3* shows some of them. The main disadvantage of these thresholds for initiating preventive measures is that they cannot be extrapolated from the area where they were established to other sites, as a wide variety of local factors affect the occurrence of instability processes.

During prolonged **periods of rainfall** (weeks or months), there is normally an important increase in slope movements. The rains during the period 1995–96 in southern Spain re-activated some older deep landslides, and caused many new superficial landslides which mainly affected transportation routes (*Figure 13.18*).

*Table 13.4* shows rainfall values established from the analysis of more than twenty cases of different types of slope movements in Spain. In the first two columns the rainfall in the year before the movement is compared with the annual average; the next three columns show the percentage of rainfall in the 3–4 months before the movement compared with the total rainfall in the previous year and with the annual average rainfall. As can be seen, in some cases the rainfall in the months immediately preceding the movements was 60%–80% of the rainfall of the previous year (for landslides and earth and mud flows) and overall exceeds the mean annual rainfall values (obtained from the meteorological records corresponding to each area investigated). This data indicates the association between movement and unusually rainy periods during the previous year or during the 3–4 months immediately before movement.

The triggering of new **deep or large scale landslides** is not related to seasonal meteorological phenomena but depends instead on long term **climatic conditions**, with rainfall regimes and moisture conditions able to substantially modify the water tables and ground water content. The principal mechanism contributing to instability is the raising of the water table due to net infiltration, with the increase in the ground unit weight as a secondary factor. Generally speaking, the less permeable the materials, the lower the influence of short term precipitation and the greater the influence of the climatic regime and the variations of annual or seasonal conditions.

**Floods** magnify the destabilising effect of rain on the sides of river valleys (above all in meanders and curved or narrow stretches of river beds). The erosive role of the water current at the base of a slope is combined with saturation of the ground by rain, generating flows and landslides or re-activating ancient movements. Landslides in river valleys may clog the riverbed and create lakes or reservoirs upstream which pose new risks: flooding of the valley upstream from the blocked area and flooding downstream from the violent failure of the dam formed by a landslide (*Figure 13.19*).

| Table 13.3 | RAINFALL THRESHOLDS FOR TRIGGERING SLOPE MOVEMENTS | | | |
|---|---|---|---|---|
| Type of movement and country | | Intensity per hour ($I_h$) | Intensity per day ($I_d$) | Accumulated rainfall |
| First time failures | | | | |
| Mud and earth flows | Hong Kong | >40 mm | >50 mm >100 mm | >50 mm (15 days) |
| | Japan | | >125 mm | >182 mm (2 days) |
| | Italy (Tuscany) | | 86 mm | 260 mm (15 days) 325 mm (30 days) |
| | Brazil | 60 mm | >125 mm | >0.4 $P_{annual}$ (annual precipitation) >675 mm (3 days) |
| Debris flows | USA (California) | | | >250 mm (3 days) |
| | Spain | 213 mm | | 52 mm during the event |
| Other types of movements | Japan | 20 mm | | 10–150 mm during the event |
| | USA (California) | | | >180 mm during the event |
| | Italy (Tuscany) | | 143–153 mm | 290–400 mm (15 days) 360–450 mm (30 days) |
| | France | | | >300 mm (60 days) |
| | Spain | | 205 mm | >500 mm (3 days) 476 mm (2 days) |
| | Spain | >60 mm (Atlantic climate) | >150 mm (Atlantic climate) >180 mm (Mediterranean climate) | >300 mm (Mediterranean climate) |
| Reactivation or acceleration of movements | | | | |
| Different types of movements | Italy | | | >520 mm (60 days) small landslides >900 mm (100 days) large landslides |
| | France | | | 300 mm (90 days) |
| | Spain | | | >250 mm (90 days) |
| | Spain | | | 320 mm (15 days) |

Modified from Ferrer and Ayala, 1997; data from various authors.

Other kinds of climate-related actions are seasonal **freeze-thaw processes**, which cause superficial movements (solifluction) in soil slopes in cold regions and rock falls in hard rock masses, where ice causes material weathering and fracture. Quick thawing allows the water content of loose materials to increase rapidly.

## Changes in water level

The elevation of the water level on slopes, as a result of prolonged periods of rain or the filling of reservoirs or lakes, increases pore pressures and may trigger or accelerate landslides. An example of this was the huge landslide in Vaiont, Italy, in 1963 (see Figures 11.1 and 11.2 in Chapter 11).

The most unfavourable scenario for the slope stability of reservoirs and lakes is a **rapid draw-down in the water level**, which generates disequilibrium conditions as the slope materials continue to have high pore pressures which do not dissipate at the same rate as the fall in the water level. This case is shown in Chapter 11, Box 11.2. These circumstances may occur on the slopes of reservoirs designed to control floods in steep sided valleys, subject to seasonal changes in water levels that may exceed tens of metres.

## Erosion

Erosion or undercutting at the foot of slopes, scarps and cliffs, from rivers or other causes, gives rise to loss of strength and modification of the state of stress which, together with unsupported overlying material, may cause instability and generate landslides or rock falls.

A secondary effect of landsliding in river valleys is the obstruction of the river by the slide mass, which may cause flash floods, as mentioned above (*Figure 13.19*).

Coastal slopes exposed to the action of waves and tides are hazardous zones for instability (*Figure 13.20*). The erosion processes in rocky cliffs, causing them to recede, are worth noting. This action is related to maritime storms, especially if these coincide with high tides.

Erosion can also be internal, due to various factors, and affect slope stability. In karstic areas, the formation and collapse of cavities associated with the presence of carbonates and gypsum may trigger instability, above all in gypsum, where the materials are softer and more easily dissolved.

## Earthquakes

Earthquakes can trigger all kinds of slope movements, depending on ground conditions, magnitude and distance from the epicentre. Rock falls, landslides, flows and rock avalanches can occur during seismic shaking (see Chapter 14, Figure 14.20). The seismic forces may also reactivate old landslides whose condition approaches limit equilibrium. In addition, in fine, loose materials, such as sands and silts, **liquefaction** may occur, which also affects old landslides in loose, saturated and non-cohesive materials. These

*Figure 13.18* Landslide on a clay slope caused by the high water content of the material (southern Spain).

*Figure 13.19* Toe of a large rotational landslide on a river valley slope, damming the river and forming a natural reservoir upstream (eastern Spain).

*Figure 13.20* Upper part of a large scale rock landslide on the north coast of Majorca, Spain (photo courtesy of R. Mateos).

| Table 13.4 | RELATIONSHIP BETWEEN PRECIPITATION AND TRIGGERING OF SLOPE MOVEMENTS IN SPAIN | | | | |
|---|---|---|---|---|---|
| | Annual precipitation (mm) | | Precipitation in 3–4 previous months (mm) | | |
| Type of movement | Total in previous year | Annual mean of series (*) | Total in previous months | % total P of previous year | % average annual P of series |
| Landslides | 500–1,000 | 500–800 | 300–500 | 50–60% | ≤30% |
| Earth flows | 500–800 | 600–700 | 300–400 | 50–80% | 50–60% |
| Debris flows | ≥1,300 | 1,100–1,200 | 350–650 | 30–50% | 50–120% |
| Rock falls | 250–700 | 220–450 | 100–250 | ≤30% | 50–130% |

(*) Series analysed between 30 and 70 years. P = rainfall.

aspects are dealt with under seismic hazards in Chapter 14, Section 14.6.

The 1976 Guatemala earthquake (M-7.6) triggered more than 10,000 rock falls and landslides in loose materials. The 1989 Loma Prieta earthquake in California set off more than 4,000 rock falls, flows and landslides, as well as numerous liquefaction processes (Schuster, 1996a). Sometimes deaths attributed to earthquakes are in fact caused by the resulting landslides. In May 1970, a magnitude 7.8 earthquake caused a glacier and rock fall at the summit of the Huascarán mountain in Peru; an avalanche of ice, mud and rocks flowed down the mountainside at a speed of more than 200 km/h, reaching the town of Yungai and burying 18,000 of its 20,000 inhabitants in the space of just a few minutes. The majority of the victims of the 2001 El Salvador earthquake were killed by a sudden landslide of volcanic materials that instantaneously lost their strength (Figure 13.21).

## Volcanism

Volcanic eruptions can cause landslides and rock and debris avalanches of considerable volume and speed on the slopes of volcanic cones, as happened on Mount Saint Helens (USA) in 1980; Schuster (1996b) states that the Mount Saint Helens blast triggered the world's largest historic landslide. Depending on the geotechnical characteristics, on the slope gradient and on the material water content, these landslides and avalanches can flow great distances.

Ash and pyroclastic materials lying on slopes form deposits prone to landslide and flow processes when the materials are saturated by rainfall. In high, snow-covered volcanic areas, the thaw resulting from volcanic activity can produce quick-flows, as happened on the Nevado del Ruiz, Colombia, in 1985, where the flow produced by the thaw of millions of tons of snow on the peak of the volcano buried the town of Armero, killing 25,000 people.

## Human actions

Human impact is one of the most important factors which can modify the conditions and forces acting on natural slopes. Excavations, the construction of dams and reservoirs, the load from buildings, structures, embankments, fills or waste heaps on slopes, and nearby blasting activities, can all modify the state of stress of the ground and its geotechnical properties, so generating instability.

Figure 13.21  Landslide caused by the El Salvador earthquake on 13 January 2001, on a slope above "Las Colinas" in the town of Santa Tecla, Nuevo San Salvador (photo: EFE). The slope is formed by a tuff substratum overlaid with layers of volcanic ash and lapilli. The estimated peak ground acceleration in the area was around 0.5 g, triggering the landslide and causing a very rapid flow which buried part of the town. The crown of the landslide shows a circular failure surface, 6–8 m deep, while the rest shows a flow mechanism. The volume of the displaced mass was approximately 90,000 m$^3$.

The main causes of instability are **changes in geometry** and gradient, changes in **hydrogeological conditions** and the application of external static loads. To a lesser extent, dynamic loads and underground excavations beneath slopes may also affect them.

**Surface excavations** for transportation routes, tunnel portals, mining and other works may modify the equilibrium profiles of slopes and can trigger movements, depending on other conditioning factors such as the geological structure, the strength of the materials or the ground water. Excavation through slopes containing old natural failures, either active or no longer active, can often provokes the re-activation or acceleration of movements; the effects of excavation can also initiate failure in slopes close to limiting equilibrium. The most damaging excavations are those carried out at the foot of a slope (as this area supports the greatest stresses), a common situation when transportation routes are constructed in valleys or on the lower parts of the natural slopes.

Excavations also modify the surface drainage system and affect the hydrogeological behaviour of the slope, causing the water table and flows to vary, or cause an accumulation of water in specific areas.

On natural slopes in **urban areas**, filtration and water leaks from tanks and the supply and sewage networks, may provoke instability, as shown in the case of *Figure 13.22*. Watering gardens and building artificial lakes without taking appropriate measures to avoid water infiltration into the ground can also cause landslides.

## 13.3 Investigation of landslides

The investigation of instability in slopes requires processes to be identified, the causes and factors which control them to be studied, and their movements to be analysed.

*Table 13.5* details the most common investigation methods and techniques depending on whether it is unstable areas or particular movements that have to be analysed. The site investigation methods are described in Chapter 5.

The different phases or steps are developed depending on the scope of the studies. Geological surveys are needed to identify locations susceptible to slope movements. The results of site investigations allow remedial or preventive measures to be planned and stability analysis to be carried out focused on the design of corrective measures to mitigate risks.

The results of the investigations are shown on maps of unstable zones (inventory, susceptibility and hazard maps; Figures 13.23 and 13.24) or in detailed maps, cross sections and models when specific instability problems are to be examined (see *Box 13.1*).

## General field surveys

**Slope surveys** on a regional scale include identifying or evaluating the following aspects:

— Relief; geomorphology and gradients.
— Lithology and stratigraphy.
— Structure of rock materials, including orientation of discontinuity planes.
— Types of soil and thickness, including weathered materials and surface deposits.
— Hydrogeological aspects and natural water courses, drainage systems and springs.
— Existing vegetation on slopes and land use.
— Active natural processes (erosive, seismic, tectonic, etc.)
— Changes in conditions due to natural and human processes.
— Recognition of present and old slope movements: landslides, flows, rock falls, etc.

*Figure 13.22* Rock fall of a large block on a vertical calcarenite slope, caused by leaks in the town water supply network (southern Spain).

| Table 13.5 | INVESTIGATION OF LANDSLIDES | | |
|---|---|---|---|
| Scope | Phases | Methods and techniques | Aim |
| ← — Investigating unstable areas — → <br> ← — Investigating specific landslides — → | Preliminary studies | Review of existing information and existing maps. | Identify processes and type of movements. <br> Identify conditioning factors. <br> General evaluation of stability of the area. |
| | | Interpretation of aerial photos and remote sensing. | |
| | General surveys | Field observations. <br> Processes mapping. <br> Factors mapping. | |
| | Study of processes and causal factors | Field surveys. | Describe and classify processes and materials. <br> Susceptibility analysis based on the existing processes and concurrence of conditioning factors. |
| | | Preliminary underground investigation: geophysical methods. | |
| | Detailed investigation | Observations and measurements at outcrops. | Describe and classify movements. <br> Collect morphological, geological hydrogeological and geomechanical data. |
| | | Boreholes, geophysical methods, *in situ* tests, sampling. | |
| | | Laboratory tests. | |
| | Monitoring | Inclinometers, extensometers, tiltometers, piezometers. | Collect data on speed, direction, situation of failure planes, water pressures. |
| | Stability analysis | Limit equilibrium methods. <br> Stress-strain numerical models. | Define failure models and failure mechanisms. <br> Evaluate stability. <br> Design corrective measures. |

The methods and techniques used **to identify active or old slope movements** and to recognise unstable zones consist basically of identifying the characteristic features of these processes, evidence of past movements and other related signs, such as:

— Erosive and accumulative land forms (anomalies in slope gradients)
— Deposits of transported material.
— Cracks and scarps.
— Damage to buildings or structures, pipelines, linear infrastructures, etc.
— Vegetation types and characteristics.
— Drainage, flood zones, springs.
— Riverbed diversions by landslides, and landslide deposits on flood plains, etc.

These tasks are carried out in the field and also from existing maps (**geomorphological**, topographical, geological or geotechnical maps), **aerial photographs** (*Figure 13.23*) and multi-spectral remote sensing images, with much higher resolution than conventional images, depending on the scale and the aim of the study (Chapter 5, Section 5.2). Relevant information obtained from aerial images can come from morphology, vegetation, drainage conditions and structural alignments.

*Table 13.6* shows some characteristic and predominant features which may help to identify and classify different types of slope movement.

Also important is the identification of human activities and environmental factors which modify natural slope conditions, including excavations, dams, mining, waste disposal, changes in aquifer usage and areas of deforestation or erosion.

Depending on the level of detail of the surveys and on the information available, data can be obtained on the location of the processes and their nature, type, age, and the extent to which they affect a slope; these data can be shown on **inventory maps** of the current or previous processes in the study area (see Chapter 15, Section 15.5). As well as locating unstable zones, these maps also provide

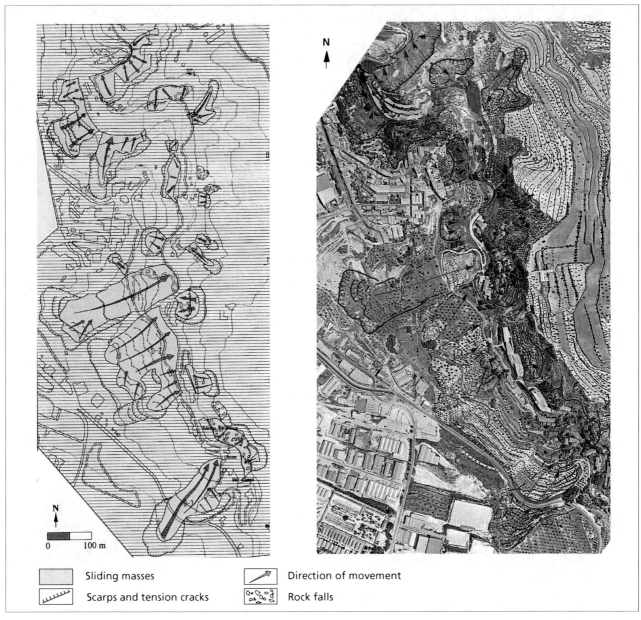

Figure 13.23  Landslide inventory map in the Alcoy area in eastern Spain, prepared from aerial photography with field based support; original scale 1:5,000 (courtesy of IGME).

indirect information on susceptible or potentially hazardous areas. However, as they only show the movements which have actually occurred, not all potentially unstable areas are reflected. *Figure 13.24* provides examples of inventory and susceptibility maps with the location and type of the slope movement shown.

An important aspect to evaluate is the activity of the processes (*Table 13.7*). Cases where landslides have been stabilised by human intervention should be considered, since possible changes in geological, hydrogeological or other conditions could reactivate these old movements.

A preliminary evaluation of the **stability conditions** of an area and its degree of susceptibility, can be carried out by combining factors such as the presence of masses that have moved (e.g., the percentage of the total area studied that is occupied by landslides) and other conditioning factors, such as susceptible lithologies and slope gradients. *Table 13.8* includes a stability classification using these parameters.

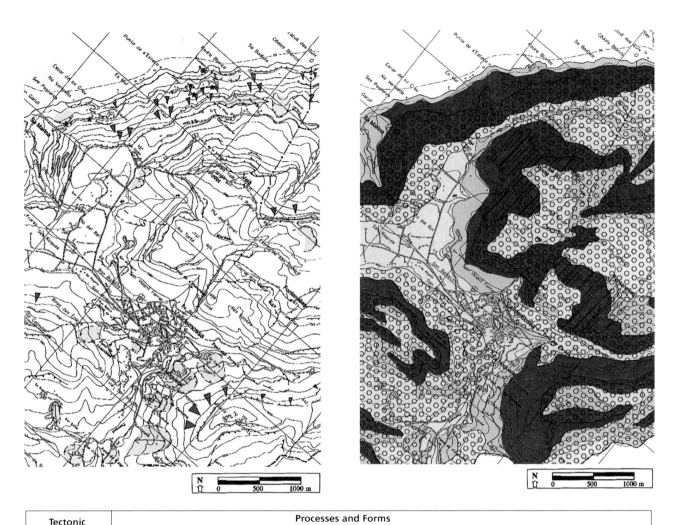

Figure 13.24 Geomorphological and inventory map of slope movements and susceptibility map, Majorca, Spain. Original scale 1:25,000 (courtesy of R. Mateos).

| Table 13.6 | CHARACTERISTIC FACTORS FOR IDENTIFYING SLOPE MOVEMENTS | | |
|---|---|---|---|
| Type of movement | Head and upper part of slope | Lower part of slope | Geometry |
| Rock falls | Irregular and rocky steep slopes with loose material or debris in the upper part. Rock blocks separated by discontinuities or fractures. Tension cracks at the head of the slope. Scarce vegetation. | Accumulation of rock blocks and fragments. | Steep gradients >50°. |
| Rotational landslides | Curved tension cracks concave to the slope. Curved scarps with striation which may be vertical higher up. Tilted ponding surfaces. Contrasting vegetation. Poor drainage conditions and ponding depressions. | Convex, lobed deposits. River beds diverted. | Gradients between 20° and 40°. $D/L < 0.3–0.1$. |
| Translational landslides in rocks or soils | Vertical tension cracks parallel to slope. Shallow vertical scarps. Material broken in blocks with large cracks between them. No ponding at head. Disordered or no drainage. | River beds diverted. Occasional lobe-shaped accumulations of material. | Uniform gradients. $D/L < 0.1$. |
| Lateral spreads | Blocks displaced and tilted in different directions. Smooth or very smooth slopes. Large cracks separating the blocks. Blocks with irregular shape controlled by fractures. Interrupted drainage systems, obstructions in river beds, asymmetric valleys. | — | Gentle gradients, even <10°. |
| Mud flows | Shallow concave head (source area). Few cracks. Vegetation contrasting with stable zones. Ponding. No significant irregularities in drainage. | Lobes. Irregular undulating morphology. | Gradients 15–25%. $D/L = 0.05–0.01$. |
| Earth and debris flows | Concavities and lobules in the source area. Several scarps. Deposits shaped as if by flow in valleys. Absence of vegetation. Irregular and perturbed drainage in the flowed mass. | Lobes, convex deposits. Irregular morphology. | Gradients >25°. $D/L$ very small. |

$D/L$ = depth/length of the displaced mass. Data from Soeters and Van Westen, 1996, and other authors.

## Analysis of the processes

The next step in investigating slope instability processes includes:

— Identifying failure mechanisms, models and types.
— Studying the different factors that control and condition the process.
— Evaluating the extent, frequency and recurrence of movements.

These studies are based mainly on field observations. They attempt to qualitatively analyse the movement processes and the causes which contribute to them and rank their importance, without going into a detailed analysis of the degree of stability of specific slopes.

| Table 13.7 | CLASSIFICATION OF LANDSLIDES BY ACTIVITY | |
|---|---|---|
| **Classification and estimated age** | **Description** | **Characteristic traits** |
| Active<br><100 years | Currently moving.<br>Possible reactivation. | Irregular, lobed topography.<br>Well-defined main scarp without vegetation.<br>Ponding depressions.<br>Streams on flanks.<br>Deposits at toe of slope.<br>Diversion of river beds. |
| Inactive<br>100–5,000 years<br>(Upper Holocene) | No current movement. | Lobed relief, scarps with vegetation.<br>Drained or non-drained depressions.<br>Different vegetation in adjacent zones.<br>Slope toe may be cut by present watercourses. |
| Fossil or ancient<br>5,000–10,000 years<br>(Lower Holocene) | Inactive for thousands of years.<br>Recognizable by relief features. | Gentle relief and scarps. With vegetation.<br>Modified drainage.<br>Different vegetation in adjacent zones.<br>River terraces covered by slid mass.<br>Current river beds on slid mass.<br>Wider flood plains upstream. |
| Relict<br>>10,000 years<br>(Upper Pleistocene) | Inactive for thousands of years.<br>May not be recognizable from the relief. | Gentle undulating topography. Without scarps.<br>Terraces excavated in the slid mass.<br>Uniform current flood plain. |

Modified from Keaton & De Graff, 1996, and other authors.

| Table 13.8 | PRELIMINARY CLASSIFICATION OF TERRITORY IN THE CONTEXT OF POTENTIAL SLOPE MOVEMENTS | | | |
|---|---|---|---|---|
| **Presence of movements and types of materials** | | **Gradient** | | |
| | | **Very low (<10%)** | **Medium–low (10°–20°)** | **Medium–high (>20°)** |
| Without active or old movements | Stable substratum | Stable | Stable | Moderately stable |
| | Non susceptible superficial deposits | | Generally stable | Moderately stable |
| With old movements | Stable substratum | Stable | Moderately stable | |
| | Susceptible substratum | Generally stable | Moderately unstable | |
| | Susceptible superficial deposits | | Moderately unstable | Unstable |
| With current or active movements | | Unstable | Unstable | |

Based on Nilsen, 1979; in Varnes, 1984.

Movement types are described in Section 13.2 of this chapter, and instability and failure mechanisms are described in Chapter 9, Section 9.4.

A general description of slope movements should include (Cruden & Varnes, 1996):

— Type of movement (indicating if complex).
— Material.
— Dimensions.
— Activity (see Table 13.7).
— Distribution of movements within the mass.
— Speed (see Table 13.9).
— Water content (dry, wet, very wet, saturated).

The ground water content is an important aspect which also helps to classify movements; it can be described simply in terms of:

— No signs of water on the slope.
— Water present, but not free; the behaviour of the material may be plastic, but it does not flow.
— Material has enough water to behave partly as a liquid.

| Table 13.9 | VELOCITY SCALE FOR SLOPE MOVEMENTS | | | |
|---|---|---|---|---|
| Class | Description | Speed (mm/s) | Typical values | Probable damage |
| 7 | Extremely rapid | $5 \times 10^3$ | 5 m/s | Violent and catastrophic. Destruction of buildings from impact of sliding mass. Large number of deaths. |
| 6 | Very rapid | $5 \times 10^1$ | 5 m/min | Difficult to escape. Some deaths. Destruction of buildings and structures. |
| 5 | Rapid | $5 \times 10^{-1}$ | 1.8 m/h | Possible to escape. Destruction of buildings and structures. |
| 4 | Moderate | $5 \times 10^{-3}$ | 13 m/month | Some structures may remain temporarily in place. |
| 3 | Slow | $5 \times 10^{-5}$ | 1.6 m/yr | Corrective measures can be applied. Structures and buildings may remain in place. |
| 2 | Very slow | $5 \times 10^{-7}$ | 16 mm/yr | Permanent structures not generally damaged. |
| 1 | Extremely slow | | | Imperceptible if not monitored. Construction is possible with precautions. |

Cruden and Varnes, 1996.

*Figure 13.25* shows a diagram for the description of landslide characteristics and dimensions (see also *Figure 13.3*). For more details, refer to Dikau *et al.* (1996).

The **study of conditioning and triggering factors** which cause instability should pay special attention to the factors which most influence the occurrence of slope movements, i.e.:

— Lithologies and degree of weathering or alteration
— Geological structure, distribution and orientation of discontinuity planes (in the case of rock slopes)
— Geomechanical behaviour of materials and their physical and strength properties
— Hydrogeological parameters, piezometric head and their variation
— Meteorological conditions
— Other possible natural and human triggering factors.

Information on the factors present in a zone must be obtained rigorously, quantifying it wherever possible.

There is a close relationship between these factors and the types of movement: the conditioning factors define the failure mechanisms and movement types, while the triggering factors generally influence the scale of instability. The triggering factors may be variable (such as ground water content) or "transitory" (such as seismic quakes).

Estimating the **frequency** of movement is carried out using historical and statistical studies, analysis of precipitation records (analysing the recurrence or return periods, regimes and historic maximum rainfall) and relationships between slope movements and earthquakes, floods, and volcanic activity, depending on the case in question.

# Detailed investigations

Detailed studies are needed to ascertain the failure mechanism within a slope and carry out an analysis of its stability. The aims are to:

— Define the geomechanical and hydrogeological properties of the ground.
— Determine the movement's characteristics, essentially its speed and the location of failure surfaces.
— Define the stability model that is appropriate and the data needed for stability analysis.

These studies are also applicable to slopes that have not yet failed but are potentially unstable, and to slopes with old landslides or active movements, whenever the speed of the movement is slow enough to allow ground investigation and surveys.

The **geomechanical and hydrogeological investigations** are completed through field work, with appropriate observations and tests made at exposures, boreholes and trial pits; geophysical methods provide information on specific physical and mechanical properties. Laboratory tests complete the description, providing values for the physical and geotechnical parameters of the materials. *Table 13.10* includes the usual site investigation methods, described in Chapter 5.

In addition to large-scale geological maps (1:500–1:2,000, depending on the extent of the area under study), detailed geomorphological mapping, which shows tension cracks, scarps, steps, areas with differential movements, location of investigation and measurement sites, and other features, is also necessary.

# LANDSLIDES AND OTHER MASS MOVEMENTS

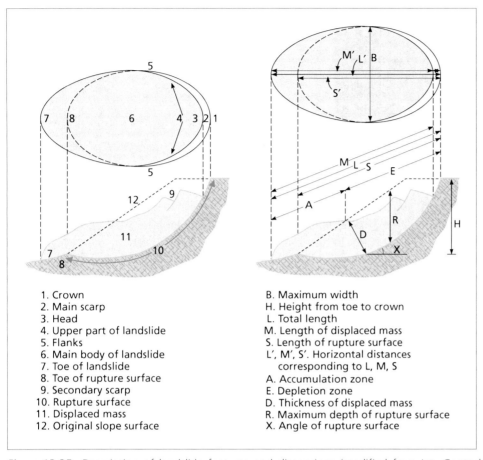

1. Crown
2. Main scarp
3. Head
4. Upper part of landslide
5. Flanks
6. Main body of landslide
7. Toe of landslide
8. Toe of rupture surface
9. Secondary scarp
10. Rupture surface
11. Displaced mass
12. Original slope surface

B. Maximum width
H. Height from toe to crown
L. Total length
M. Length of displaced mass
S. Length of rupture surface
L', M', S'. Horizontal distances corresponding to L, M, S
A. Accumulation zone
E. Depletion zone
D. Thickness of displaced mass
R. Maximum depth of rupture surface
X. Angle of rupture surface

*Figure 13.25* Description of landslide features and dimensions (modified from Int. Geotech. Soc. UNESCO, 1993; in Dikau *et al.*, 1996).

| Table 13.10 DETAILED INVESTIGATION OF LANDSLIDES | | |
|---|---|---|
| **Material properties** | **Failure surfaces** | **Movement speed** |
| Geophysical methods (resistivity, seismic refraction). Trial pits and trenches. Boreholes. Laboratory tests. Field tests. Back analysis. | Field observations. Seismic and electric methods. Excavations and trial pits. Measurements inside boreholes. Detailed geotechnical logging in boreholes. Observations in large diameter wells. Inclinometers and extensometers. Back analysis. | Measurements in tension cracks and scarps. Installation of markers and stakes. Conventional topographic methods. Inclinometers and extensometers. Electronic and GPS measurements. Analysis of satellite images (interferometry). |

To carry out a stability analyses based on limiting equilibrium, or to verify the mathematical/numerical models, if done, **the location of failure surface** (which may be one or more) must be known; with this information, geometric models of the unstable mass can also be generated. Failure surfaces can also be determined from:

— **Surface observations and measurements**: a) at the head and flanks (scarps and steps) and at the toe of the landslide (in some cases the failure surface may outcrop, depending on the behaviour of the unstable mass, on the type of movement and on its evolutionary phase; *Figure 13.26*); b) in rotational landslides the failure surface may be inferred from the geometry of the displaced mass and by measuring the inclination or tilting at the top of the slope; c) in translational landslides, the depth of tension cracks at the top can be measured, if these reach the failure plane.

*Figure 13.26* Main failure surface of a landslide in mudstones showing displacement of the sliding mass over the underlying stable zone (northern Spain).

— **Observations in excavations and trial pits**, carried out at the edges of the landslide or on the slide mass if the depth of failure allows machinery access.
— **Observations and measurements in boreholes** (inclinometers, TV cameras, etc.)
— **Monitoring of displacements** in shafts lined with concrete rings which are moving differentially as a consequence of ground movement (see Chapter 5, Figure 5.99)
— **Geophysical methods**, such as seismic refraction and electric resistivity; the slip surfaces are deduced from property differences between the displaced mass and the stable slope substratum; these do not normally give satisfactory results in translational landslides along planes parallel to the stratification.

Detailed geological surveys sometimes allow at least an approximate estimate of the possible position of slide surfaces from geological, stratigraphical or structural criteria: contacts between different lithologies, the presence of strata or levels of different hardness, contacts between soil and rock substratum or between weathered rocks and fresh material, the presence of faults, etc.; this information offers an important starting point for the subsequent investigation.

Observation and description of **drill cores** can detect soft clay levels, striated surfaces, weathered, brecciated or wet zones and planes with signs of water. Failure may occur along either one or many surfaces or in zones of a finite thickness.

Different observations and measurements can be made **inside boreholes** or drilled wells, if these are deeper than the estimated failure surface of the landslide:

— By inserting a pipe to be cut by the failure surface, then measuring the depth to the cut; the disadvantage of this method is that it does not detect deeper failure surfaces.
— By inserting control markers (a piece of metal tubing, heavy enough to slide easily down inside the borehole) and raising it at regular time intervals until the failure produced by the landslide prevents it from moving upwards, so indicating the depth of the shear surfaces (Chapter 5, Figure 5.99).
— By lowering TV cameras to observe directly inside the boreholes.
— By installing inclinometers and extensometers, taking into account that if there are rapid movements, this may cut the inclinometer pipes.

**Inclinometers** should be extended to a number of meters below the failure surface, so that to install them it is necessary to have an approximate idea of the location of this surface; these instruments measure the deviation (inclination) of the borehole in two directions at right angles to each other, giving displacement curves whose inflection indicates where the failure surfaces are located (see Chapter 5, Section 5.6 and Figure 5.96). **Extensometers** measure relative movements between the mouth of the borehole and one or more points located along their length.

Other systems have been developed to detect failure surfaces, based on inserting different devices or elements into the boreholes or slopes. For more details, see Hutchison (1983).

If the failure surfaces cannot be detected directly, they can be deduced from a comparison between the properties and characteristics of the sliding material and that of the ground not affected by the movement. In some types of materials (soils and soft rocks), the displaced masses tend to be weathered, disorganised, fissured, etc., and in the case of clay soils, their fabric can contrast with that of the ground at the site.

Through petrological and mineralogical analyses of clay materials, the orientation of minerals in the direction of the movement can be identified, and the presence of the shear surfaces deduced.

Sometimes samples can be taken of the slide surface and laboratory tests carried out to obtain the residual values of the strength parameters.

The age of the movement can be determined using **absolute dating** and **dendrochronology** techniques; **micro-palaeontological analysis** can be used to detect discordant contacts between different micro-palaeontological levels.

As mentioned above, the methods described can detect the position of slide or failure surfaces in slow moving active movements and are also valid for **inactive or old movements,** except those methods based on displacement measures.

# Box 13.1

## Benamejí landslide, south Spain

Aerial photograph of the landslide.

Landslide head and main scarp.

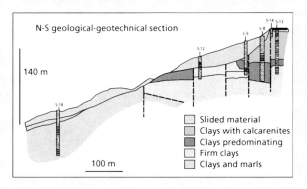

N-S geological-geotechnical section

- Slided material
- Clays with calcarenites
- Clays predominating
- Firm clays
- Clays and marls

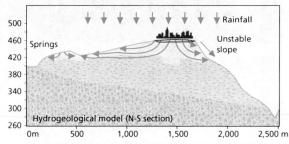

Hydrogeological model (N-S section)

### LEGEND

- Sands and gravels. Quaternary.
- Calcrete, calcarenites, clays and carbonate crusts. Pleistocene–Holocene.
- Calcarenites and clays with conglomerates, microbreccias and calcarenites. Oligocene–Lower Miocene.
- Reddish and greenish clays with calcarenite layers. Pliocene–Eocene.
- Sandy marls, limestone marls and clay marls. Cretaceous–Paleocene.
- Clays and marls with layers of sandstone and gypsum. Keuper facies.
- Landslide materials
- Landslide materials–modified by excavation
- Head tension crack
- Old landslide materials
- Landslide areas with conglomerate blocks, limestone pebbles, cherts and micro-conglomerates.
- Flooded areas and streams
- Dip of beds
- Faults
- Postulated faults

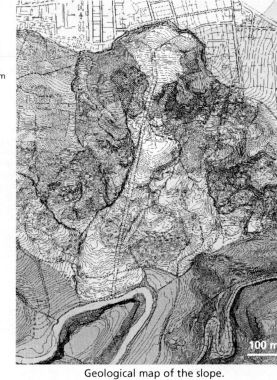

Geological map of the slope.

The number and location of site investigations will depend on the particular conditions of each case, but in general terms the following should be carried out:

— Boreholes at the head, mid-slope and toe of the landslide.
— Geophysical profiles, longitudinal and transversal to the landslide.
— Measurements of superficial movements along the length of the displaced mass and in the main cracks (head and flanks).
— Hydrogeological tests at the head and in the main body of the landslide.
— Inclinometer measurements.

The boreholes and geophysical measurements must be deep enough to reach the stable zones under the moving masses (as suggested above, this limit may sometimes be determined approximately from geological, stratigraphical or structural criteria). It is advisable to divide the investigation program into phases to be able to modify and complement the investigations where necessary.

Other areas that should be investigated are the boundary zones of unstable slopes, to detect other instabilities, and to examine the geological characteristics and hydrogeological behaviour around a slide.

*Box 13.1* presents an example of a landslide in Cordoba, in the south of Spain, showing the different study phases and their results; the aim here was to design stabilisation measures. The landslide, approximately 1 km long, on the slope between the town of Benamejí and the River Genil, had shown periodical re-activations in centuries past, mainly due to intense rainfall periods, which had affected buildings at the southern end of the town. *Box 13.1* shows the preliminary work, field surveys and site investigation, detailed slope mapping and the models developed based on the results obtained. *Figure 13.27* shows a diagram of the actions carried out to stabilise the landslide; *Figure 13.28* shows the slope after stabilisation work was completed.

## Stability analysis

When the geological, hydrogeological and geometric data have been collected, of both the moved mass and of the slope, and the geomechanical properties of the materials are known, then **geological, hydrogeological and**

*Figure 13.28* View of the upper part of the Benamejí slope after stabilising works.

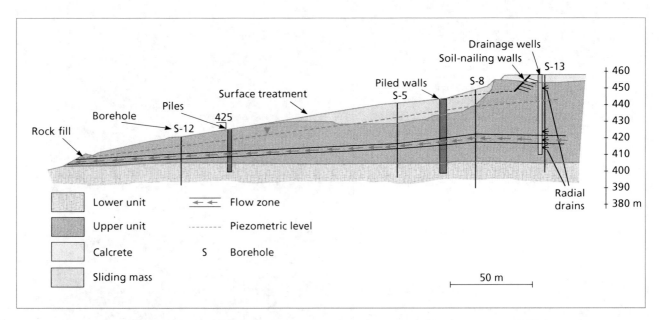

*Figure 13.27* Diagram of stabilizing measures for the Benamejí landslide (southern Spain) (courtesy CEDEX, Ministerio de Fomento).

**geotechnical models** can be devised to carry out a slope **stability back analysis**.

The quantitative determination of stability either in determinist terms (by calculating either the **factor of safety** or by calculating the stress-strain relationships and corresponding displacements), or in terms of **probability**, is a fundamental requisite in engineering projects requiring action on unstable slopes. In Chapter 9, Section 9.5, stability analysis methods are described.

Back analysis by **limit equilibrium methods** gives:

— The factor of safety for the slope, for a specified failure surface and material properties.
— The strength parameters, $c$ and $\phi$, for a specified failure surface and factor of safety value in the model (with analysis in unstable or near equilibrium situations taking $F = 1.00$), which allows the results to be compared with laboratory test results, and parametric or sensitivity analyses to be carried out to obtain values for the most representative strength parameters.

Modelling and analysis by **stress-strain methods** allows:

— The stress-strain behaviour model for the whole slope to be defined from the material properties, and to be compared with real observed behaviour.
— The strength and deformability parameters of the slope materials to be determined, from the model reproducing the features and behaviour observed in the field, and to be compared with parameters obtained in laboratory or *in situ* tests.

The results of both methods should be compared. It should be noted that results of laboratory tests, and even *in situ* tests, tend not to be representative of real scale parameters, above all in rock masses, and higher values are often obtained than those resulting from back analysis.

The use of computer software programs allows detailed modelling and analysis of failure and slope behaviour in soils and rocks. Programs such as FLAC, UDEC, ZSOIL, PLAXIS or PHASE2 allow the analysis of complex cases and of a wide variety of hydrogeological or stress conditions, as well as the effects of stabilisation measures.

## Monitoring

The monitoring of landslides is like listening to their geotechnical pulse and is the most delicate phase of the detailed investigation; its aim is to detect, monitor and predict slope movement, as well as obtain data on the movement process. This monitoring should basically aim at investigating:

— The location of the failure surface or surfaces.
— The distribution of displacements in the slope, their rate and their variation.
— The position of the water table and pore pressures.

The monitoring, observation and measurement time frame depends on various factors, but it should last at least one annual meteorological cycle or longer if the long term influence of climatic conditions is required (Sowers and Royster, 1988). If the investigation is done during a dry period, then the corrective or stabilising measures designed may not be effective when conditions change and there is a period of rainfall.

The movement characteristics and speed depend on factors such as the type of movement, the slope and the water content of the ground. Data on the rate of movement are required to design corrective or mitigating measures. Displacements and rate of movement can be measured by surface and in-depth monitoring (**inclinometers**). The velocity values, i.e., the displacement in terms of time, give an insight into the evolution of movement and in some cases allow the outcome of the failure to be predicted (see Chapter 9, Figure 9.69).

Recently, advanced radar remote sensing techniques (ground base radar) have been used to monitor specific sites were slope instability occurs; by the application of advanced differential interferometry techniques, displacement maps for the study area can be obtained, with a precision of a few millimetres, and the development of displacement with time for every detected ground target can be calculated (*Figure 13.29*).

**Piezometers** give the piezometric head and thus the water pressure at the levels where they are installed. To measure the pressure acting on the failure surface, they should be installed either in the failure surface or immediately above it.

Figure 13.29 Cumulative displacement map for a landslide, retrieved with ground based radar (IBIS-L) developed in Galahad UE project (courtesy of G. Herrera).

Monitoring and related aspects are discussed in Chapter 9, Section 9.7, and the methods used are described in Chapter 5, Section 5.6.

### Alarm systems

These consist of different systems or instruments installed on the slope surface and/or at depth, and designed to detect movements or measure specific parameters related to movements. The most usual are:

— Installing inclinometers and piezometers on landslides or slopes where instability poses important risks (e.g. on slopes of reservoirs or in built up areas).
— Installing fences or wire meshes on rock slopes where there is a hazard of rock falls.

In the first case threshold values have to be set (of displacement for inclinometers or water head for piezometers) above which there is considered to be a risk of either movement or acceleration. It is very important to carry out the data collection correctly, preferably automatically, as well as the interpretation of the measurements obtained, and any decisions taken as a result must be based on expert judgement. The frequency of measurement depends on the characteristics of the landslide and the phase or state of the instability. Care must taken to ensure that the systems work correctly and are well maintained.

Measurements of piezometric head allow correlations to be made with precipitation, and so help define alarm levels or thresholds in relation to maximum hourly or daily rainfall or accumulated precipitation over days or weeks.

Fences or mesh that detect block falls by electrical means or other devices (when the blocks hit them) are generally installed on rock slopes where railway lines or roads exist at a lower level. The device can be connected to a warning system that signals imminent danger.

## 13.4 Corrective measures

The aim of corrective or stabilisation measures is to prevent instability and mitigate damage. Their design and application depends mainly on the type, dimensions and rate of the movement, and they can be carried out either before (in the case of potentially unstable slopes) or during movement, whenever the velocity of movement allows. Large-scale landslides or flows, even moving slowly, are very difficult and often impossible to halt. Stabilisation measures are described in Chapter 9, Section 9.6.

Once the movement is underway, the action to be taken depends on the:

— Volume of the unstable mass.
— Characteristics of the movement, speed of the process and behaviour of the unstable mass.
— Depth of the failure surfaces.
— Slope angle and height.
— Type of materials affected.
— Slope accessibility.

The most effective measures, often also the cheapest, are those which act directly on the factors triggering instability. In the case of landslides, above all when these are of a manageable size or depth, **drainage works and modification of slope shape and height** are measures recommended, because they act on two of the main causes of failure: pore pressures and distribution of forces due to the weight of the ground.

Drainage and other measures to avoid water infiltration are always beneficial for slope stability, but they must be designed according to the hydrogeological characteristics of the slope and based on detailed studies of conditions, especially for deep drainage.

In the case of slow moving and shallow landslides and flow type movements, surface drainage is recommended by excavating perimeter ditches to prevent water from reaching the unstable mass (Figure 13.30). Surface drainage on the slide mass should be carried out once it has stabilised (Figure 13.31).

Other measures, like installation of anchors and bolts, are recommended in potentially unstable rock masses to avoid landslides and rock falls, but are not effective in soil landslides (unless they are installed on walls or beams distributing the forces uniformly); in this case, resistant elements such as piles or diaphragm walls are more effective. Figure 13.32 shows a diagram of the stabilisation of a large-scale superficial landslide in clayey silt residual soils.

Figure 13.30 Excavation of a perimeter drainage ditch to stabilise a landslide.

# Stabilisation and protection against rock falls

Section 9.6, in Chapter 9, includes a description of surface protection measures for excavated slopes, also applicable to natural slopes. In the same section, **active measures** or stabilising procedures to prevent potential rock falls are also described, which consist of:

- Installing bolts and anchors to tie the blocks to more stable ground.
- Installing systems of metallic cables and wire meshes, fixed or anchored to the slopes to stabilise heavily jointed areas; this consists of installing a double or triple torsion metallic mesh, overlaid with a series of cables forming a grid, anchored to the rock at the edges and tightened (Figure 13.33).

Other types of action, known as **passive measures**, are aimed at avoiding the damage rock falls may cause to buildings, structures and communication routes. These consist of:

- Metallic meshes to hold small loose rock fragments.
- Ditches or collection areas to trap the fallen blocks.
- Walls and earth ridges.
- Static barriers to brake and contain the blocks.
- Dynamic barriers for the same purpose.
- Artificial tunnels on roads and railway lines.

**Control mesh**, made of steel wire, is hung from the head of the slope, covering the whole surface to the foot. It is used to "guide" the rock blocks as they fall, preventing them from bouncing and rebounding outwards, so that they pile up at the base where they can be removed. This is effective for blocks smaller than approximately 0.5 m$^3$. Hexagonal, triple torsion, galvanised steel mesh is the strongest.

**Ditches or collection areas** are dug at the foot of the slope to catch rock blocks. Their width and depth depends on the expected volume of the blocks. They are not effective if

Figure 13.31 System of "fishbone" drainage ditches on an unstable slope in northern Italy.

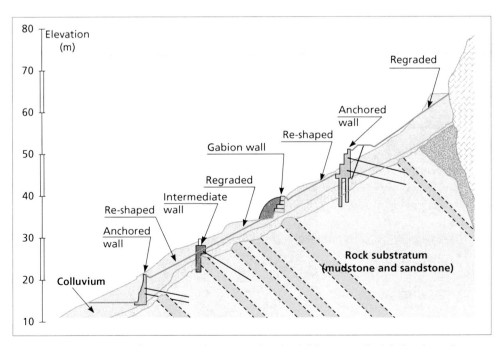

Figure 13.32 Diagram of measures taken to retain a landslide on a colluvial silty clay soil over a substratum of mudstone and sandstone.

the rocks bounce away from the slope surface as they fall, as happens in areas of uneven gradient and hard materials. To ensure the blocks are stopped a layer of gravel or earth is spread in the bottom of the ditch. A catch fence or mesh barrier can also be used to prevent the blocks bouncing clear of the ditch.

**Walls** made of concrete or gabions, stop loose blocks advancing and are generally built at the foot of slopes. The disadvantage of rigid walls is that they break more easily on impact. Earth ridges can also be built in a trapezoidal form.

**Static barriers** are used at the base or on the surface of a slope to intercept and stop blocks. They are normally made of metal posts driven or built into the slope and fitted with resistant wire mesh (*Figure 13.34*). They can also be made with metallic beams, placed close enough together to prevent falling blocks passing between them. Like the rigid walls, they may be damaged by heavier blocks than they were designed to withstand.

Flexible or **dynamic barriers** are able to absorb the impact energy of the blocks through elastic and plastic deformations of the metallic mesh and other components of the barrier. They consist of twined steel cable mesh, supported by steel posts founded and anchored to the slope, and linked to each other by cables (*Figure 13.35*). The system is based on the deformation capacity of the mesh and cables and on the built in braking systems which come into play on the impact of the blocks, with an energy absorption capacity of up to 5,000 kJ. They are normally installed leaning away from the slope and 3 m high, although they may be higher. These barriers are designed on the impact energy of the blocks and their size, speed and trajectory.

These barriers can also be installed above a portal of tunnels to brake the free fall of large blocks (*Figure 13.36*).

**Artificial tunnels** also protect roads and railway lines against rock falls. They are built of concrete and a layer of granular material is laid on their roof to cushion the impact of blocks and prevent them from bouncing (*Figure 13.37*). If the volume of the rock blocks is relatively small, then lighter structures or dynamic galleries can be used instead of artificial tunnels.

The measures mentioned above are designed on the quantity, volume and weight of the blocks, the gradient and trajectory of the blocks, their impact energy and

Figure 13.33   Stabilisation for rock falls with flexible cable network (courtesy of Geobrugg).

Figure 13.34   Static barriers made of posts and meshes for protection against rock falls, Liguria, Italy.

Figure 13.35   Dynamic barrier to retain rock falls, NE Spain.

# LANDSLIDES AND OTHER MASS MOVEMENTS

*Figure 13.36* Horizontal dynamic barrier (shelter) at a tunnel portal to retain rock falls, NE Spain.

*Figure 13.37* Artificial tunnel (rock shelter) for protection from rock falls on a road in the north of Majorca, Spain.

distances they can reach, the level of potential risk of rock falls, accessibility of the slope and the space available to install them.

Field observations and measurements provide the most reliable data for the choice and design of the most appropriate solution, and a combination of several is often used. Estimating the parameters mentioned above is very important, especially when installing barriers and walls, to ensure that they are not damaged significantly and that the blocks do not go over the top of them if they bounce on the slope, or in the zone between the foot of the slope and the catch barrier.

There is specific computer software (e.g. ROCKFALL, ROTOMAP) which can be used to simulate the trajectory of blocks from a slope having a given geometry, source areas of given location and blocks of given size and shape for the materials forming the slope, and other variables. The results from the simulation will define the position and height of the barrier or wall. When the height of the barrier and the distance to the foot of the slope where it has to be installed, to ensure it brakes any possible rock fall, are not acceptable (because it is too high or there is not enough space) then these two parameters (position and height) are decided from probabilistic analysis, assuming an acceptable level of risk of potential damage. Intermediate mid-slope barriers can also be installed depending on the gradient of the slope, the location of the source areas and access.

## 13.5 Collapse and subsidence

## Types of movements and their causes

These processes refer to vertical movements of two sorts, in general terms sudden movements or collapse, and slower movements or subsidence. The following types can be differentiated:

— Collapse of underground cavities in rock, which may or may not be reflected at ground level.
— Shallow collapse in rock or soil that reaches ground level.
— Subsidence or slow, gradual lowering of the ground level.

In the first case, movements tend to occur as a result of the **collapse** of the ceiling of underground cavities, when the strength of the overlying rocks is not sufficient to bear the stresses upon them. The behaviour of the materials is brittle and their failure is violent. Whether the collapse affects ground level or not depends on the strength and geomechanical characteristics of the overlying materials.

## Collapse

**Underground cavities** may be natural or manmade. Their collapse depends on the following factors:

— Volume and shape of the cavities.
— Thickness of the overburden (or depth of the cavity itself).
— Strength and mechanical behaviour of the overlying materials.

Natural cavities or caves are associated with **karstic or soluble materials**, such as carbonate and evaporite rocks, where dissolution processes create caves; when certain dimensions are reached, these generate a disequilibrium, or instability, causing the failure of the roof or ceiling of the cavity; if this is not very thick or the strength is low, then the ground surface will collapse (*Figure 13.38*). Water level variations in karstic materials may produce stress readjustments which cause collapse. Caves are also formed in volcanic materials, as described below.

The surface results of karstic collapse are **sinkholes**, although these may also be generated by the gradual solution of surface rocks or by subsidence in the soft soils covering karstic materials.

The mechanisms which generate karstic cavities and collapse are not described here as they are widely dealt with in the literature on karstic geomorphology. **Evaporitic materials** (salts and gypsums), are much softer than the carbonate materials, and show a greater capacity for dissolution (see Chapter 11, Section 11.6); the re-adjustment of evaporitic materials surrounding cavities is continuous and gradual, compared with the sudden collapse in carbonates. The cavities in gypsum and salt are not normally more than 200 m deep.

**Volcanic lava flows** can have cavities and tunnels which form naturally within them due to the differential cooling of these materials; they are generally tubular. Although collapse is not common because of the considerable strength of these materials, they may be a risk in the context of the loads transmitted by foundations or other engineering work (see Chapter 8, Section 8.5).

The **human activities** which may cause collapse are **underground mining** (excavation of bedded minerals such as coal, or mining by dissolution in saline materials, etc.; *Figures 13.39* and *13.40*) or other infrastructure excavations, such as tunnels for roads, underground railways and storage facilities.

One case with particular repercussions in built-up areas is the risk from excavating an underground system of tunnels, where thin overburden is the main cause for the collapse or subsidence that can occur above the excavation (see Chapter 10, Section 10.10).

*Figure 13.39* Collapse from underground old mining cavities in karst (photo courtesy of J.J. Durán).

*Figure 13.38* Collapse in alluvial materials overlying limestone (photo courtesy of J.J. Durán).

*Figure 13.40* Collapse from salt dissolution, now a salt lake (northern Spain).

## Subsidence

Subsidence is generally a very slow process although it is often accelerated by human activity.

It can affect all ground types, generally soils, and is due to ground stress changes caused by the following:

— Lowering of the water table.
— Underground mining and tunnels.
— Extraction of petroleum or gas.
— Intensive aquifer exploitation.
— Slow processes of material dissolution.
— Morphotectonic and sedimentation processes.
— Consolidation in soft and organic soils.

A **drawdown of the water table**, in periods of drought or because of pumping from aquifers, affects unconsolidated materials, which undergo changes in their state of stress as a result of decreasing water pressures (effective stresses increase as pore pressure decreases), reducing their volume, with a fall in ground level. These are slow processes, but they may affect wide areas. Subsidence in clayey soils is common if water is extracted from alternating sandy aquifers. The level of subsidence depends on the type and thickness of sediments, and the extent of the drawdown in water table. The rate of subsidence may be several centimetres per year. Ground deformations are sometimes partially recoverable if the hydrological conditions are restored, and if the ground has not exceeded its yield point.

Subsidence is a risk when it occurs in populated areas, as it damages and cracks buildings and affects their foundations. Other consequences are flooding in coastal areas or changes in natural drainage systems. An example of subsidence from a fall in the water table in a built-up area is shown in *Box 13.2*.

There are some materials which are particularly susceptible to subsidence processes, such as organic soils or peat, and un-compacted fills. Permafrost type soils in cold regions (see Chapter 2, Section 2.9), where the sub-surface remains permanently frozen, may experience subsidence if for some reason the temperature of the sub-soil rises and the frozen water thaws.

**Gas and petroleum extraction** causes deformation of the overlying ground depending on the depth and volume of extraction. When extractions are very deep, where rocks have low compressibility, the strength of the mineral grains of the rocks containing the fluids plays an important role.

Spectacular examples of subsidence from a combination of petroleum and water extraction have occurred in Long Beach, Los Angeles, and in the San Joaquin Valley, California (USA), reaching over 9 m. In the San Joaquin Valley, the water table in some areas has fallen by 150 m and the subsidence basin is the largest in the world, with an area of 9,900 km$^2$ (Walthan, 1989).

**Mining subsidence** occurs as the result of ground deformation caused by mining. The magnitude and extent of subsidence are related to the type of ground, to the volume excavated and to the working methods used. The most common subsidence is linked to working bedded, sub-horizontal, massive deposits (coal, evaporites) and vertical or steeply inclined metallic deposits (dykes, veins). Subsidence is difficult to predict and may appear years after the mine has been abandoned; where the beds are sub-horizontal, subsidence occurs usually on a smaller scale and it is easier to predict, although it may affect to a large area. Models based on instrumental measurements and numerical analysis allow mining subsidence to be evaluated; empirical relationships also can be used between the maximum subsidence expected and thickness and depth of the beds, and extension of the workings. Subsidence processes from mining may affect soils and rock masses as well as slopes. Subsidence and collapse may appear simultaneously; *Figure 13.40* illustrates a local collapse within an area affected by overall subsidence.

**Tunnel excavations** may also cause subsidence, mostly in areas where the overburden is not very thick and when low strength soils are present.

Subsidence may occur in **karstic materials** from dissolution even in clay filled sinkholes or in zones with a relatively thick cover of soils.

Subsidence may also be related to **tectonic processes** or to the **consolidation of clay sediments**, as in the case of deltas, where the rate of material deposited means an increasing load on the substratum, producing vertical readjustments. Peat (an un-consolidated deposit of decomposed vegetable matter) often is affected by subsidence linked to the reduction of its water content or to external loads, given its high compressibility.

The city of Venice is a classic example of subsidence, aggravated since the 1940s by the pumping of ground water, with the average values of natural subsidence increased several times, from under 0.5 mm/year to 1–2 cm/year by the 1960s. The subsidence slowed when intensive pumping was stopped (Murk *et al.*, 1996; Walthan, 1989).

Mexico City is another example of regional subsidence, with maximum values of over 8 m during the last 250 years, due mainly to the consolidation of the lacustrine clays lying beneath the city (see figures in Tables 1.1 and 1.4 in Chapter 1).

## Investigation of the processes

Surveys to evaluate the likelihood of movement in a given zone should be done to identify:

— Lithologies liable to suffer settlement or collapse from natural processes: carbonate rocks, gypsum, salts, etc.

## Box 13.2

### Example of subsidence caused by water table drawdown in the city of Murcia, eastern Spain

During the period 1994–1996 the city of Murcia suffered the effects of general subsidence caused by a water table drawdown of as much as 8 m, resulting in settlement and cracks in around 150 buildings, the older buildings with shallow foundations being the most affected. The direct losses were estimated at over 36 million euros.

The decisive factors in the process were:

— The location of the city on flood plain deposits, basically clays and silts.
— The drought affecting the area since 1992.
— The extraction of groundwater.

The reduction of the pore pressures in the sub-soil caused consolidation of the superficial soft ground layers. The estimated average settlement was 2–3 cm, with maximum theoretical values of 8 cm, obtained by numerical modelling.

The Figure below shows a map of the estimated settlement in the built-up area.

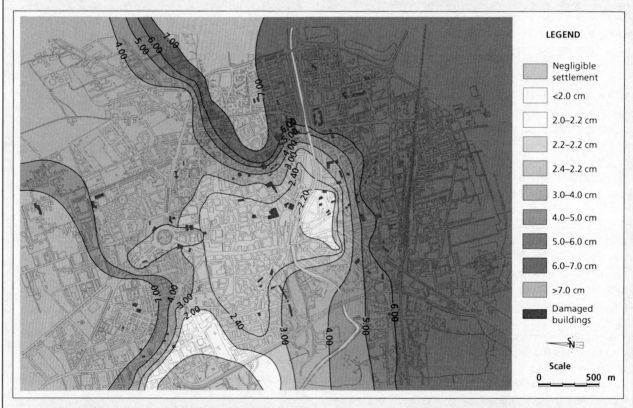

Map of estimated settlements in the city of Murcia from subsidence caused by water table draw down (courtesy of IGME and COPOT, Región de Murcia)

(karstic and saline materials in general) and areas with underground cavities and collapses that reach ground level.
— Soft and deformable lithologies.
— Areas with natural or manmade processes which may trigger subsidence.

In the first case, ground relief and morphology tend to be indicators of the dissolution processes acting and how far these processes have evolved; the lithologies can be recognised immediately from detailed geological maps. In the second case, there are various susceptible lithologies but they are generally fine grained, deformable, soft materials with low strength (flood plain sediments, lake fills, muds, peat bogs, etc.). The third case generally includes areas with underground mining or areas subjected to significant variations in the water table (due to both seasonal changes and those that occur over a number of years, and to aquifer exploitation).

Combining susceptible lithologies and triggering factors allows the potentially hazardous areas to be defined.

**Detecting underground cavities** may be difficult depending on their size and depth. Geophysical methods and boreholes are the most effective means of detection. Geophysical methods can be applied, with limitations, in areas where the existence of cavities is suspected, the most effective methods being gravimetric and micro-gravimetric (for cavities at shallow depths) and geo-radar. **Boreholes** are the most reliable method, but the disadvantage is that if the possible location of the cavities is not known or suspected, then a large number of boreholes will be needed. They allow the depth of cavities to be determined but not the volume. As well as conventional boreholes, percussion drilling, which is cheaper, can be used. TV cameras can be introduced inside boreholes or percussion drilled wells to observe the characteristics and size of the cavities encountered (Figure 13.41).

Even when **subsidence** develops very slowly, it can be measured with appropriate monitoring (extensometers or tachymetric measurements) to estimate the rate of the process and forecast its evolution and future consequences. The analysis of radar images, by advanced differential interferometry techniques, allows the preparation of detailed displacement maps and models, and the study of the evolution of the subsidence process, with a precision of a few millimetres, as well as the calibration and validation of forecasting models (Figure 13.42).

Processes caused by the extraction of water, gas or petroleum may be predicted in advance, depending on the volumes to be extracted.

## Corrective measures

As it is impossible to avoid large scale collapse and subsidence, measures to mitigate their effects should be based on their prevention.

Cavities up to a certain volume should be filled once their volume and depth is known, so ensuring that subsidence is no longer active. Sometimes, if the process is still active, infilling must be repeated periodically (see Chapter 8, Section 8.5).

Subsidence may be prevented and controlled by acting on the processes which cause it. Ground can recover and return to the initial equilibrium conditions if it has not exceeded its elastic deformation, as in the case of ground affected by drawdown of the water table. Subsidence due to underground excavations can be countered by jet grouting and ground consolidation treatment beforehand (see Chapter 10, Section 10.10).

## 13.6 Prevention of risks from mass movements

The damage caused by landslides, collapse and subsidence depends mainly on the speed and size of the processes. The greatest risks are from rapid slope movements which may threaten lives, while slow movements and subsidence are potentially less damaging. Subsidence and collapse cause damage when they affect the surface, so risk is linked to the effect they have on the surface, rather than to the process itself. The highest risk processes are often small scale, such as rock falls and sudden collapses. For large-scale movements, **prevention** is the most effective action to avoid risks.

Slope movements can involve any type of material, relief and climatic zone, anywhere on the earth's surface, whilst subsidence and collapse are limited to areas with certain lithological or other specific characteristics, as described above.

Losses from landslides in Spain have been estimated at around 150 million euros per year. The reduction in these losses as the result of applying preventive and mitigation measures has been estimated at around 90%. In other countries, such as Japan and Italy, estimated losses are more than 2,000 million dollars per year, and 1,300 million dollars in the USA (Schuster, 1996a). According to the UNESCO-IAEG, 200 to 300 deaths per year can be directly attributed to landslides worldwide, excluding landslides caused by earthquakes.

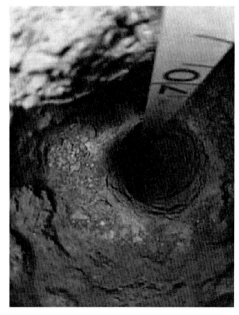

Figure 13.41 Image from a bore hole TV camera inside a borehole formed by percussion drilling, where a cavity can be seen in basaltic lava flows.

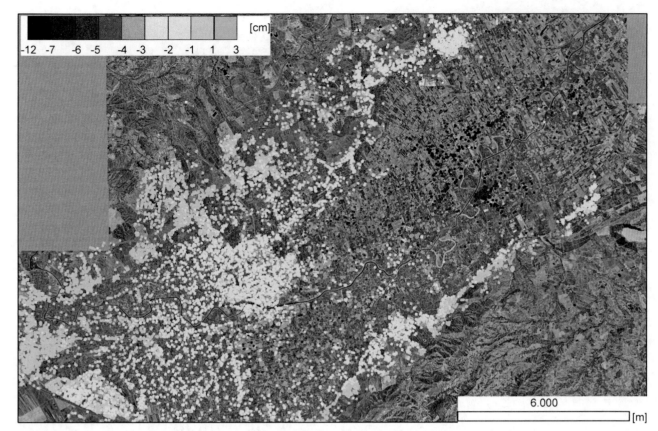

*Figure 13.42* Map of cumulative surface subsidence as retrieved with the A-DInSAR technique SPN (Stable Point Network) within the Terrafirma UE project (courtesy of G. Herrera).

In spite of advances in surveying, prevention and emergency procedures, the damage caused by slope movements continues to increase worldwide. According to Schuster (1996a) the reasons for this are:

— Increased building and development in landslide prone areas.
— Deforestation in potential landslide areas.
— Increased rainfall in certain regions as a result of change in climatic conditions.

When landslides happen as the result of earthquakes or storms, i.e. at very irregular intervals, their occurrence cannot be predicted in time to avoid damage, but what can be identified and mapped are the geological factors and conditions which make an area more prone to landslide, and thus showing hazardous areas in earthquake or storm conditions. These maps help planners, land owners and insurers to evaluate and prevent potential risk.

Large scale landslides or flows of hundreds of thousands of tons, which may reach speeds of 100 km/hour, are not frequent processes; in general, the larger the landslide, the lower the frequency of occurrence. These phenomena normally leave scars and other indicative signs on the landscape, which allow the affected areas to be identified for many years, or even centuries after the event. In some cases, large natural ground movements can be prevented as a result of studying evidence of older or even prehistoric landslides and from hazard studies. The Vajont disaster in Italy (1963) could have been avoided if the many signs of landsliding in the valley had been taken into account, along with the evidence of existing geological conditions conducive to the generation of massive rock failures.

In spite of the speed with which they occur, **collapses** are unlikely to cause death or significant damage, except in urban areas or on transportation routes. In karstic areas, collapses are often caused by structural or building loads. **Subsidence** linked to human activity may cause considerable economic loss when widespread and affecting built-up areas. Subsidence associated with natural geological processes develops at such slow rates that it does not have any short or medium term effect. The damage caused by these processes tends to be very localised and generally consists of:

— Destruction and damage from differential ground settlements and cracks.

- Flooding in low lying areas near rivers, lakes or the sea.
- Leaks and filtration from reservoirs.
- Changes in water flow in canals, drains, sewers, etc.
- Collapse in water and petroleum wells and pipes.
- Groundwater contamination along the cracks produced by subsidence.

The annual losses in the USA from this type of damage are calculated at 500 million dollars.

In general, slope movements, subsidence and collapse can be prevented; this, along with the fact that prevention generally costs less than correction once the processes have been triggered, means that the economic and social damage they cause are directly related to negligence and ignorance of the processes operating in the ground and their interaction with human projects and activities.

Prevention is based on the identification and study of instability and the processes and factors which control them, as these enable the preparation of preventive maps for purposes of land use and planning. Risks can be mitigated by passive measures (land use restrictions, development of movement control systems, etc). **Preventive and mitigation measures** for mass movements are described in Chapter 15, Section 15.4.

## Susceptibility and hazard maps

**Mass movements maps** are prepared to ensure adequate land use and so prevent risks, and also to provide a basis for detailed studies for the design and construction of engineering projects. Mapping may include:

- Location of processes and affected zones (process density, distribution and activity level)
- Representation of the factors controlling the processes.
- Representation of susceptible areas.
- Hazard zoning.

These maps, their contents and general methodologies are described in Chapter 15, Section 15.5. Identification of instability, by either direct or indirect means (i.e. their effects or damage) and of their governing factors does not present any real problem, except for obtaining representative and quantitative data for the characterization of the materials present and processes operating, especially at depth. However, the wide variety of geological, geomorphological, geomechanical or meteorological factors which can control mass movements means that mapping procedures to delimit potential instability zones and possibility of occurrence, are complex, even more so if an attempt is made to quantify the influence of such factors and to carry out a deterministic or a probabilistic analysis of the data. The mapping is based on field studies, backed up by remote sensing techniques, including aerial photography and satellite images, and on geographical information systems (GIS) applications (see *Figure 13.24* and Chapter 15, Figure 15.3).

**Hazard maps** show the possibility or probability of a process occurring in a specific area during a given period of time, zoning the territory into different hazard classes. To do this, all the natural and human triggering factors have to be considered, in addition to the conditioning factors. The general methodology for preparing hazard maps is shown in Chapter 15, Figure 15.4.

## Slope movement maps

**Susceptibility maps** show where landslides may occur; maps can be prepared by different methods:

- "Expert" or **direct evaluation** from field surveys and **geomorphological mapping**; the quality will depend on the professional experience and criteria used when evaluating the potential activity of the study area, based on the geological materials and slopes characteristics, and on existing landslides, which are mapped in the greatest detail the working scale allows. As this method is based on field studies, it is more common in local studies although it can also be applied on a regional scale.
- **Superimposing thematic maps** of governing factors (gradients, lithologies, deposits of loose material, etc.) onto maps of existing unstable processes, help define the susceptible zones and level of qualitative susceptibility by revealing the presence of coincidental factors. The main parameters affecting the distribution of landslides in the study area can be ranked (e.g. steep gradients, soft and weathered materials, water table near the surface, type of vegetation on the slopes, etc.). These maps are applicable on medium scales.
- **Combining factors**, dividing each one into different classes and assigning a weighting to each class (in quantitative terms), depending on how much it contributes to landslide occurrence. This is evaluated from criteria based on field surveys and from the distribution and density of the instabilities present in the area. The study area is normally divided into cells or polygons and mathematical functions or matrices are defined to obtain the final score for each cell, with different combinations of factors, deciding which are the most representative depending on how they match with the process inventory map. This method is valid in areas with homogeneous conditions on medium scales. The use of **geographical information systems** allows automatic data processing and numerous analyses to be carried out with different combinations.

- **Probabilistic methods**, using the same methodology as above, but based on a statistical definition of how far each factor and combination of factors contributes to the occurrence of landslides (by analysing the processes in the study area).
- **Deterministic methods** based on calculating the stability of slopes in a given area; this is applicable only on a large scale (1:5,000 or above).

In the case of **hazard maps**, the temporal prediction of when movements will occur is based on (Alonso, 1987):

- Direct correlation with precipitation.
- Movement measurements.
- Pore pressure measurements.

The first method is based on the relationship between rainfall and slope movements, since rain is a triggering factor; the applicability is higher for areas prone to superficial movements of loose, soft materials and may be used for long-term forecasting. The methods based on observations of movements require control and monitoring of unstable slopes in order to measure displacement, obtain the time-displacement relationship and predict the moment of failure, based on recorded movements (see Chapter 9, Figure 9.69); these methods are applicable on a large scale and in the short term. This is also the case with the third method, which is based on the relationship existing, in some cases, between the position of the water table and landslide occurrence.

The main difficulty with prediction of slope movements with time is that it requires quantitative data on instability triggering factors in a specific area (e.g. detailed historical records, instrumental data). The predictions may be valid for specific landslides or in limited areas, but cannot be applied far beyond these areas on account of their complexity and the high cost of monitoring. Also the data obtained for a specific area usually cannot be extrapolated to others because of variations in conditions.

Correlations can also be established between earthquakes and slope movement occurrence in seismic zones, from observations and historical data.

## Collapse and subsidence maps

The main difference between these maps and maps of slope movements is that both the governing and triggering factors are more limited. In collapse and subsidence mapping two different cases may be considered:

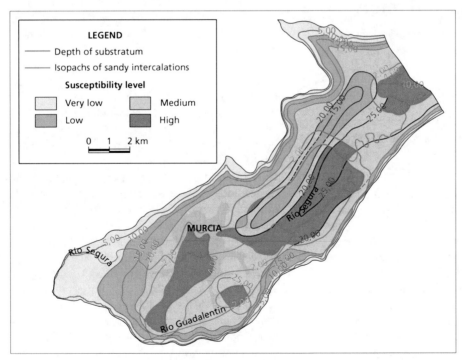

*Figure 13.43* Subsidence susceptibility map of an area of Murcia region (eastern Spain), after water table draw down over a two year drought period; estimations were based on the depth of the firm substratum and the depth and thickness of sandy levels intercalated in the soft materials which form the sub-soil (courtesy IGME and COPOT, Región de Murcia).

— Areas where the process depends on the presence of specific susceptible lithologies, such as karstic zones or soft soils (independently of natural or human triggering factors).
— Areas where human action (normally underground excavation) may trigger processes in materials which initially are not prone to collapse and settlement.

In the first case, **susceptibility maps** are based on mapping existing processes and conditioning factors, just as for slope movements. This means that when the collapse potential of karstic zones is evaluated, the factors to be considered are lithology and degree of karstification, structural alignments (the dissolution processes develop along them), geomechanical properties of the materials, hydrogeological and climatic factors, etc.; seismic and human factors must also be taken into account where relevant. Gypsum and saline karst and karst covered by superficial material are the most prone to movements.

In the case of susceptibility to subsidence, the presence of soft, compressible materials should be considered, as well as possible variations in the water table and other natural and human factors that may have a bearing on the movements. *Figure 13.43* gives an example of a medium scale susceptibility map.

When the causes are human, susceptibility and hazard maps are drawn based on the mining or other excavation sites which may be the cause of instability; the mechanical characteristics of the materials and the influence of variations in the water table on their behaviour must also be considered.

In both cases quantitative information can be included in the maps whenever data is available or where the subsidence values in a given area can be evaluated.

# Recommended reading

Fell, R., Corominas, J., Bonnard, C., Cascini, L., Leroi, E. and Savage, W.Z. (2008). On behalf of the JTC-1 Joint Technical Committee on Landslides and Engineered Slopes. Landslide susceptibility, hazard and risk zoning for land use planning. Engineering Geology, vol. 102, 3–4, pp. 85–98. (a full version of JTC-1 Guidelines for landslide susceptibility, hazard and risk zoning for land use planning is available in www.webforum.com/jtc1).

Dikau, R., Brundsen, D., Schrott, L. and Ibsen, M.L. Eds. (1996). Landslide recognition. Identification, movement and causes. John Wiley & Sons.

Schuster, R.L. and Krizek, R.J. Eds. (1988). Landslides: Analysis and control. Transportation Research Board Special Report 176. 5th printing. National Academy of Sciences. U.S.A.

Turner, A.K. and Schuster, R.L. Eds. (1996). Landslides: Investigation and mitigation. Transportation Research Board Special Report 247. National Academy of Sciences. U.S.A.

Varnes, D.J. (1984). Landslide hazard zonation: a review of principles and practice. Natural Hazards 3, UNESCO.

# References

Alonso, E. (1987). Riesgos geológicos asociados a las avenidas y su previsión. In: Avenidas. Sistemas de previsión y alarma. Berga and Dolz (Eds). Colegio de Ingenieros de Caminos. Madrid, pp. 37–58.

Casale, R. and Margottini, C., Eds. (1995). Meteorological events and natural disaster. Report on a CEC field mission in Piedmont, 4–6 November 1994. Enea-Civita Consorcio. Italy.

Corominas, J. and García-Yagüe, A. (1997). Terminología de los movimientos de ladera. IV Simposio Nacional sobre Taludes y Laderas Inestables. Granada, Vol. III, 1051–1072.

Corominas, J., Remondo, J., Farias, P., Estevao, M. and Zézere, J. *et al.* (1996). Debris Flow. In: Landslide recognition. Identification, movement and causes. Dikau, Brunsden, Schrott and Ibsen (Eds). John Wiley & Sons, pp. 161–180.

Cruden, D.M. and Varnes, D.J. (1996). Landslide types and processes. In: Landslides. Investigation and mitigation. Turner and Schuster (Eds). Special Report 247. Transportation Research Board. National Academy Press. U.S.A. Chapter 3.

Dikau, R., Brundsen, D., Schrott, L. and Ibsen, M.L. (1996). Introduction. In: Landslide recognition. Identification, movement and causes. Dikau, Brundsen, Schrott and Ibsen (Eds). John Wiley & Sons, pp. 1–12.

EPOCH (1993). Flageolet, J.C. (Ed). Temporal occurrence and forecasting of landslides in the European Community, vol.1. Contract no. 90.0025.

Ferrer, M. and Ayala, F. (1997). Relaciones entre desencadenamiento de movimientos y condiciones meteorológicas para algunos deslizamientos en España. IV Simposio Nacional sobre Taludes y Laderas Inestables. Granada. Vol. I, 185–197.

Gostelow, T.P. (1991). Rainfall and landslides. In: Prevention and control of landslides and other mass movements. Almeida-Teixeira, Fantechi, Oliveira and Gomes Coelho (Eds), CEC Report EUR 12918 EN, pp. 139–161.

Hutchinson, J.N. (1983). Methods of locating slip surfaces in landslides. Bulletin Association Engineering Geologists. Vol. XX, no. 3, pp. 235–252.

Hutchinson, J.N. (1988). General Report: Morphological and geotechnical parameters of landslides in relation to geology. Proc. 5th Int. Symposium on Landslides, Lausanne. Vol. 1, pp. 3–36.

Keaton, J.R. and De Graff, J.V. (1996). Surface observation and geologic mapping. In: Landslides. Investigation and mitigation. Turner and Schuster (Eds). Special Report 247. Transportation Research Board. National Academy Press. U.S.A. Chapter 9.

Murk, B.W., Skinner, B.J. and Porter, S.C. (1996). Environmental geology. John Wiley & Sons.

Schuster, R.L. (1996a). Socio-economic significance of landslides. In: Landslides. Investigation and mitigation. Turner and Schuster (Eds). Special Report 247. Transportation Research Board. National Academy Press. U.S.A. Chapter 2.

Schuster, R.L. (1996b). The 25 most catastrophic landslides of the 20th century. In: Landslides. Proceedings of 8th International Conference and Field trip on Landslides. Chacon, Irigaray and Fernández (Eds). pp. 53–62.

Soeters, R. and Van Westen, C.J. (1996). Slope instability recognition, analysis and zonation. In: Landslides. Investigation and mitigation. Turner and Schuster (Eds). Special Report 247. Transportation Research Board. National Academy Press. U.S.A. Chapter 8.

Sowers, G.F. and Royster, D.L. (1988). Field Investigation. In: Landslides. Analysis and control. Schuster and Krizek (Eds). Transportation Research Board. Special Report 176. 5th printing. National Academy of Sciences. U.S.A. Chapter 4.

Varnes, D.J. (1984). Landslide hazard zonation: a review of principles and practice. Natural Hazards 3, UNESCO.

Varnes, D.J. (1988). Slope movement types and processes. In: Landslides. Analysis and control. Schuster and Krizek (Eds). Transportation Research Board. Special Report 176. 5th printing. National Academy of Sciences. U.S.A. Chapter 2.

Walthan, A.C. (1989). Ground subsidence. Blackie. Chapman & Hall.

# 14

# SEISMIC HAZARD

1. Introduction
2. Faults and earthquakes
3. Seismicity studies
4. Seismic hazard analysis
5. Seismic site response
6. Ground effects induced by earthquakes
7. Applications to geological engineering

## 14.1 Introduction

When a large earthquake occurs, the resulting devastation has incalculable social consequences throughout the affected area. At present, the only effective way of avoiding such consequences is by taking measures aimed at either preventing or mitigating their effects. Geological engineering contributes to the study of many of the fundamental aspects of seismic risk evaluation:

— Evaluation of seismic and geological conditions of sites for engineering works.
— Assessment of seismic hazard and effects induced by earthquakes.
— Calculation of the dynamic properties of the ground and its seismic response.
— Geological and seismic criteria for seismic resistant design.
— Preparation of microzonation maps for urban planning.
— Vulnerability analysis of buildings and infrastructures.
— Preventive measures, civil defence and the provision of aid in the event of a disaster.

This chapter deals with the following topics related to the above:

— Influence of faults on seismic hazard.
— Seismic hazard evaluation.
— Influence of local conditions on seismic response.
— Liquefaction, landslides caused by earthquakes and surface rupture through faulting.
— Applications to site selection, microzonation and vulnerability assessment.

## 14.2 Faults and earthquakes

### Faults as the source of earthquakes

Characterization of earthquake sources is one of the basic aspects of the study and evaluation of seismic hazard. The distribution of seismicity on a global scale can be explained by plate tectonics, whereby seismically active areas coinciding with the limits of lithospheric plates are distinguished from relatively stable areas lying inside the plates. In seismic areas, the actual sources of shallow earthquakes are geologically defined structures: faults (*Figure 14.1*). Tectonic activity on faults is responsible for the release of energy during the earthquake.

Efforts to explain the 1906 San Francisco earthquake, caused by an abrupt slip on the San Andreas Fault, led to the formulation of the **elastic-rebound** model, which states that blocks of rock separated by an **active fault** (see Section 14.4) tend to undergo displacement relative to each other, although the fault remains locked until a critical threshold of strength is exceeded, resulting then in sudden slip along the fault plane (*Figure 14.2*).

A relationship between faults and shallow earthquakes has been observed when greater precision in the instrumental location of earthquake epicentres has been available. Alignments of epicentres can be correlated with

*Figure 14.1*   Normal fault at Cape Kidnappers, New Zealand.

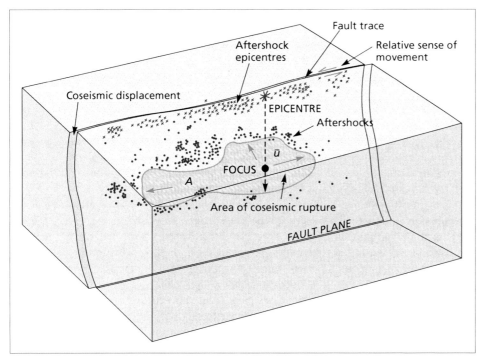

Figure 14.2   Diagram of a seismic fault during and after an earthquake. The strike-slip fault is displaced first at the focus after which the rupture zone spreads rapidly at velocity $\bar{u}$ until it affects an area $A$ that is a fraction of the total surface of the fault plane. The epicentre is the projection of the focus above the earth's surface. Earthquake magnitude is proportional to the area of rupture and coseismic displacement. After the earthquake, readjustments take place for a while in the area around the rupture, giving rise to aftershocks with epicentres aligned parallel to the fault trace.

Quaternary movements along fault lines. Furthermore, increased accuracy in locating the aftershock epicentres has allowed seismologists to identify fault zones responsible for earthquakes (Figure 14.2).

Also in continental areas within the plates, considered to be stable, earthquakes of surprisingly large magnitude have taken place. Examples include the earthquakes at Cutch, India (1819), Marryat Creek (1986) and Tennant Creek (1988), Australia, Killari, India (1993) and Sichuan, China (2008). These all occurred in areas with no known seismic significance and may have been related to the reactivation of fault movements (Crone et al., 1992). Therefore, in addition to seismic data, characterization of potentially seismogenic faults present in the area is necessary for a satisfactory quantitative hazard assessment. The development of more accurate models for carrying out this characterization has given rise to new branches of geology, such as neotectonics, active tectonics and paleoseismology, and the use of innovative methodologies.

## Stick-slip regimes and the seismic cycle

In seismic hazard studies, a model of how seismic faults work is essential for understanding which source parameters are significant. The effects of a regional stress field on an active fault will cause two blocks to move relative to each other. This displacement leads to the deformation and fragmentation of rocks in the fault zone and the generation of fault material.

Seismic faults, which cause earthquakes, differ from aseismic faults by their mechanical behaviour, deformation mechanisms and the displacement regime. Seismic fault material behaviour is brittle and deformation mechanisms comprise **brittle fracturing, brecciation** and **crushing**, giving rise to fault material composed of angular fragments of the original rock, with a cohesive matrix at depth that becomes non-cohesive towards the surface. Materials generated, in decreasing order of grain size, are: **fault breccia, fault gouge** and **fault flour** (non-cohesive matrix), and **cataclasite** and **ultracataclasite** (cohesive matrix).

Motion regime in seismic faults follows a pattern of relatively long periods of locking alternating with short events in which sudden movement, or **coseismic displacement**, occurs. This type of regime is known as a **stick-slip regime**. During the periods of locking between two seismic events (the pre-seismic period), energy accumulates in the form of elastic deformation, which is then partially released as elastic waves during an earthquake. Friction slows coseismic displacement of the fault and it eventually becomes locked once more to begin a new period of energy accumulation. This cycle of gradual accumulation of energy and its sudden release during short events is known as the **seismic cycle**.

The increase in temperature with depth in the earth's crust causes ductile behaviour in materials, and rocks below a certain depth tend to become deformed in a process known as "ductile creep". Minerals are affected by intracrystalline deformation mechanisms and certain intercrystalline processes, such as superplasticity, causing a ductile flow at relatively high temperatures, or cataclastic flow at lower ones.

The behaviour of different minerals can be determined from so-called "creep maps", which relate deformation velocity to constant levels of temperature and tectonic stress (secondary creep). The rheology of the upper part of the earth's crust is controlled by the behaviour of quartz, a mineral present in the granitic rocks that predominate in this part of the crust. Rock behaviour here is brittle, but below the depth at which 300° C is reached, where the behaviour of quartz is already ductile, the crust also becomes ductile. Earthquakes therefore take place in the brittle upper part of the crust, known as the schizosphere (or broken part of the crust); thus, a fault behaves seismically in the upper part of the crust but is aseismic in the lower crust, known as the plastosphere (*Figure 14.3*). This explains the frequency distribution of focal depths in shallow earthquakes, and also why the depth of large earthquakes is usually in the 10 to 15 kilometres range.

## The seismic fault model

Energy released by a fault as seismic waves represents only about 1 to 10 per cent of the energy involved in the rupture and deformation of fault material, and the displacement generated during the process of either sliding or slip in the stick-slip regime. The rupture phenomenon has been studied in detail in a series of earthquakes, and models have been made in an attempt to understand how rupture, once initiated, is propagated, how displacements are distributed along the ruptured zone of the fault, and how much energy is released.

The initial point of failure is the earthquake focus. From this point, rupturing and displacement take place in a lateral direction, with a variable velocity that can be measured in metres per second (*Figure 14.2*). Around the fault rupture zone, deformation builds up and is released in form of aftershocks. The energy accumulated during the seismic cycle determines the **size of the rupture zone** and the **amount of displacement.** These two parameters are proportional to the seismic moment and the earthquake magnitude.

In major earthquakes, with magnitudes larger than 6, the rupture area may reach a length of several hundred kilometres, and coseismic slips may be of several metres and be visible at the surface. At depth, rupture may have involved the entire brittle crust.

Using data from field observation, a considerable number of **empirical relationships** have been established

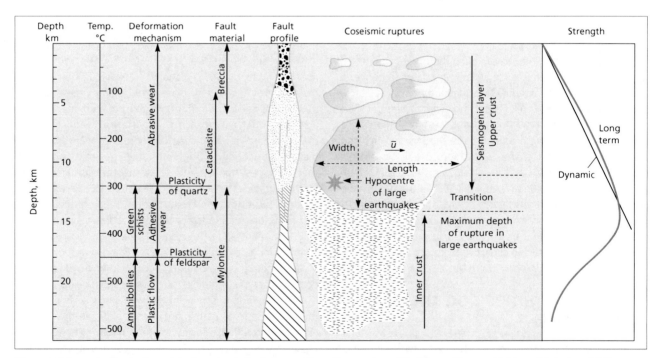

*Figure 14.3* Mechanical model of a seismic fault in continental areas (modified from Sibson, 1983 and Scholz, 1990).

between earthquake magnitude and fault rupture parameters. Although there is a good correlation between rupture area and magnitude, rupture length-magnitude and displacement-magnitude relationships are very variable and of low reliability if extrapolated to different areas.

Some of the most commonly used relationships are those of Wells and Coppersmith (1994):

- **Rupture area - moment magnitude**

$$M_w = 4.07 + 0.98 \log (RA), s = 0.24$$
$$\log (RA) = -3.49 + 0.91 M_w, s = 0.24$$

where $M_w$ is moment magnitude, $RA$ the rupture area in km$^2$ and $s$ the standard deviation of the estimate.

- **Surface rupture length - moment magnitude**

$$M_w = 5.08 + 1.16 \log (SRL), s = 0.28$$
$$\log (SRL) = -3.22 + 0.69 M_w, s = 0.22$$

where $M_w$ is moment magnitude, $SRL$ the surface rupture length in km and $s$ the standard deviation of the estimate.

- **Maximum displacement - moment magnitude**

$$M_w = 6.69 + 0.74 \log (MD), s = 0.40$$
$$\log (MD) = -5.46 + 0.82 M_w, s = 0.42$$

where $M_w$ is moment magnitude, $MD$ maximum displacement at the surface in metres and $s$ the standard deviation of the estimate.

The relationships proposed by Stirling et al. (2002) may be more suitable for estimating moment magnitude and average displacement from the surface rupture length of prehistoric earthquakes:

$$M_w = 5.89 + 0.79 \log (L), R.s.d. = 0.21$$
$$\log (D) = -0.09 + 0.35 \log (L), R.s.d. = 0.33$$

where $L$ is surface rupture length in km, $D$ is average surface displacement in m, and $R.s.d.$ the residual standard deviation of $M_w$.

## Slip rates and recurrence periods

The magnitude and recurrence period of large earthquakes are related to the average velocity of fault movements (Figure 14.4). Faults with a high slip rate, for example in the order of 10 mm/year, accumulate large amounts of elastic energy over a short time, which means they have a short seismic cycle. This gives rise to earthquakes with large magnitudes of 6 or 7 and relatively short recurrence periods of around 200 years. In contrast, slow faults, with velocities of 0.1 to 0.01 mm/year, generate earthquakes of the same magnitude over much longer periods, ranging from 45,000 to 500,000 years.

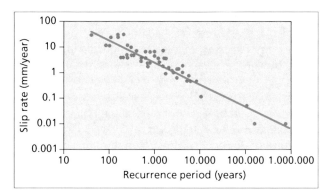

Figure 14.4   Relationship between slip rate and earthquake recurrence period (Villamor and Berryman, 1999).

A common initial premise is that coseismic fault rupture occurs at regular intervals, with constant increments in slip, and that future fault behaviour will follow a similar pattern. However, the way in which this takes place in the fault as a whole, over periods of time, may follow very different models. In each event, rupture and displacement may not be constant along the entire fault; different sectors may move with a particular increment of displacement for each quake. Several models have been proposed, such as the variable displacement model, the uniform displacement model, the characteristic earthquake model (see Section 14.5), the overlap model and the coupled slip model (Berryman and Beanland, 1991). For a particular point on the fault, with a constant slip rate, the model may consider either variable displacement or constant displacement. In the case of the former, it can be assumed that the time taken for a new rupture to occur depends on the extent of the previous slip (earthquake predictions in terms of time), or that displacement in the new rupture will be proportional to the time that has elapsed since the previous event (earthquake predictions in terms of slip or magnitude). If slip is constant the model will be either the characteristic earthquake type or the coupled type, with recurrence time also constant.

A significant feature of these fault movement models is that they consider fault rupture takes place over the whole fault length in every seismic event, and that it may do so along **defined segments** which maintain their individual behaviour throughout the evolution of the fault, and therefore throughout the accumulation of events that lead to sudden slip or displacement. There has been much discussion about this possibility, because in the case of a fault acting as segments of defined length, earthquake magnitude would be limited by simple relationships, and knowledge of former earthquakes would not be required to evaluate the magnitude. The existence of fault segments remains a controversial issue.

**Seismic segments** or **rupture segments** have been identified in the San Andreas fault from field surveying and monitoring of the rupture area of historic earthquakes. In order to define these segments, it is necessary to demonstrate that surface rupture in former earthquakes has been limited to one portion of the fault in at least two or three seismic events. For the majority of faults this is not possible and it is necessary to resort to geological criteria. The premise is that there are barriers along the fault plane that put a limit on the extension of the rupture, thereby dividing the trace into **fault segments.** Geological features considered as possible segment limits are static criteria with very diverse characteristics (offsetting of the fault trace, pull-apart or push-up structures, branches on the fault or transversal structures).

## Geological recording of fault activity

Accumulated slip in active faults and coseismic displacement can interfere with geomorphological and stratigraphical processes that take place at ground surface, and can also induce post-sedimentary changes in the structure and composition of sediments that give rise to a geological record which can be used to evaluate the seismogenic potential of a fault.

The landscape of a particular area evolves in accordance with the so-called Davis' erosion cycle. According to this model of evolution, which assumes a time-scale of several million years, a tectonic movement is followed by a longish period in which gradients and relief are gradually worn down by geomorphological processes. This can be divided into three periods: youth, maturity and old age, which tend towards the eventual development of a surface of low relief and gentle outlines known as a peneplain. The occurrence of another tectonic impulse will interrupt the cycle and bring about rejuvenation of the relief. The variables controlling this theoretical cycle (tectonic, climatic and geomorphological processes) vary over much shorter periods of time, in the order of thousands of years.

Tectonic movements on faults produce scarps and changes in the ground elevation, due to both accumulated slip and coseismic events, interfering with the evolution of the relief and giving rise to forms that are studied by analysing the **tectonic geomorphology.** Sedimentary processes are also influenced by tectonic movements which deform and disarrange previous formations, conditioning the location, thickness and facies of sediments during and after coseismic events.

McCalpin (1996) distinguishes three levels of organization for systematizing geological features related to paleoseismic activity: genesis, location and time of occurrence. He differentiates primary features, formed by tectonic deformation, from secondary features resulting from the ground shaking as seismic waves pass through it. At the same time, features generated along the fault itself from those situated outside it can be distinguished, and those produced instantaneously during the seismic event from those that are post-seismic.

Relief features generated directly by active faults take on characteristic forms. The most significant is the **fault scarp** (*Figure 14.5*), which degrades over time through a progressive reduction of slope angle and height by the agents of erosion, generating an accumulation of fallen material and sediment at its base. If the velocity of this evolution process is appropriately assessed, it is possible to identify and evaluate the time that has elapsed following a displacement or slip event on a fault.

Depending on the type of fault, other geomorphological and structural elements are also found which serve as criteria for identifying active faults. Typical of strike-slip faults (*Figure 14.6*) are linear valleys, the displacement and offsetting (diversion) of river courses, shutter ridges (*Figure 14.7*), sag ponds, springs, pressure ridges and benches. All the features of strike-slip faults, such as pull-aparts, are found together with these structures. In normal faults, scarps show typical features such as triangular facets or elongated depressions along small grabens bounded by the main fault. Folds associated with coseismic displacement are characteristic of inverse faults, with anticlinal elevation in the raised block and synclinal depression in the sunken one.

In the case of sediments, it is the faulting and folding of sedimentary strata that is most visible as well as different types of fractures.

Away from the fault, coseismic deformation of the ground gives rise to tilting of the surface, uplift and sinking along coastlines, and, eventually, deposits from tsunamis. After the event, other features occur, such as tectonic alluvial terraces and erosive discordances brought about by uplift, sinking or tilting of the ground, colluvial wedges, fills in fissures and angular discordances.

## The study of seismic faults

The study of seismic faults include:

— Neotectonic framework.
— Paleoseismic analysis of faults.
— Evaluation of seismic parameters in present-day earthquakes.

### Neotectonic framework

The neotectonic framework for a region is defined by a number of related issues, such as its relationship to tectonic plates, and its present-day tectonic framework; this includes tectonic stress fields and fault systems, in which a distinction can be made between inactive faults and active faults, both new and reactivated.

A geodynamic framework is established by ascertaining the location of lithospheric plates using information from magnetic-anomaly bands in the oceans, studies of

# SEISMIC HAZARD

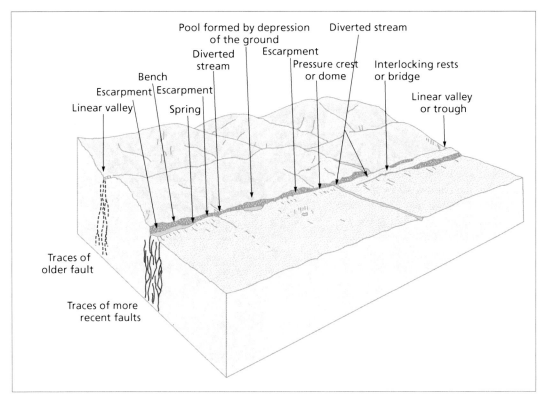

*Figure 14.5*  Geomorphological features associated with strike-slip faults (Keller and Pinter, 1996).

*Figure 14.6*  Interlocking spurs associated with the Mohanka fault (Glendale, New Zealand), a dextral shear fault displacing streams and spurs of land.

*Figure 14.7*  Change in level of a Holocene terrace caused by an escarpment on the Poukawa fault; Argyll, New Zealand.

paleomagnetism (polar drift) and the results of studies of absolute plate movements from long baseline interferometry or laser monitoring on artificial satellites.

The stress field is obtained from studies of fault population analysis and analyses of the focal mechanisms of earthquakes. Deformation currently present in the plates is assessed from long baseline interferometry studies and repeated GPS measurements using networks of fixed field stations. Geological observations are made in conjunction with these new methodologies to detect either the raising or the lowering of coastlines and islands areas, the diversion of drainage networks and the distribution of thermal springs.

Active faults and their distribution can be detected using remote sensor studies (mainly satellite imagery and aerial photography), reinterpretations of previous studies and geological maps, analyses of digital models of terrain and quantitative geomorphological analysis (geomorphological indexes). These indexes measure such things as changes in gradient of valley slopes, shape of drainage networks, evolution of meanders and asymmetry of valleys. Neotectonic fault activity can also be established using criteria such as the correlation between fault traces and the location of earthquake epicentres, and the link between focal mechanisms and motion in the fault.

## Paleoseismic analysis of faults

The aim of this analysis is to determine the average velocity of fault movement, assess the magnitude of paleoearthquakes and estimate the date of the last paleoseismic event. The slip rate is obtained from studies of faults exposed in specially dug trenches, usually perpendicular to the fault. An analysis is made of the sedimentary levels or edaphological evidence in the soil, and of tectonic deformations due to the movement of the fault and of the geomorphological evolution of the scarps (Figure 14.8). The width of the excavation normally varies between 5 and 8 metres, its length is as required and its depth is usually 2 or 3 metres. A reference grid made from strings is fixed on the trench walls, reference levels and significant structures are marked on the grid and a detailed geometrical survey is then carried out. Evidence of coseismic events can be determined from any quantifiable deformation of the reference levels. Samples are then taken of deformed layers that can be dated, for example, from volcanic ash or soil deposits with carbonaceous materials.

Dating methods include those outlined in Table 14.1, depending on the type of material collected. The average displacement rate is obtained from dated displaced layers. Paleoearthquake magnitudes can be assessed from the empirical relationships between magnitude and slip or displacement in fault scarps, and from the appearance of different types

Figure 14.8  Paleoseismic study of faults revealed in trenches, near Rotorua, New Zealand.

| Table 14.1 | METHODS OF DATING TECTONIC AND PALEOSEISMIC ACTIVITY | | | | |
|---|---|---|---|---|---|
| Absolute methods | | Relative methods | | | Correlations |
| Nearest year | Radiometrics | Radiological | Simple processes | Complex processes | Correlation |
| Historical records | Carbon 14 | Traces of uranium | Amino acid racemization | Soil development | Stratigraphy |
| Dendrochronology | Uranium series | Thermoluminescence and electron spin resonance | Obsidian hydration | Weathering of rocks and minerals | Tephrachronology |
| Sedimentary varves | Potassium-argon | | Tephra hydration | Progressive modification of the relief | Paleomagnetism |
| | Fission track | Cosmogenic isotopes | Lichenometry | Sedimentation rate | Fossils |
| | | | | Geomorphological position and incision rate | Stable isotopes |
| | | | | Deformation rate | Tectites and microtektites |

(Pierce, 1986).

# SEISMIC HAZARD

## Box 14.1

### Sedimentary structures and other ground effects originated by earthquakes

**Seismites** are commonly referred to those sedimentary structures originating in soft sediments and granular soils from the cyclic shaking of the ground as seismic waves pass through them. Most are a result of **liquefaction** phenomena (see Section 14.6 of this chapter). For seismites to form, the material must be granular, not very densely packed, non-cohesive and saturated. As the seismic waves pass, cyclic shear stresses overcome the strength of the granular skeleton, making it more densely packed. The velocity and cyclicity of the phenomenon prevent the expulsion of water and pore pressure increases almost instantaneously, leading to the deformation and flow of sediment with water, that is, fluidization or liquefaction. This flow gives rise to upward injections of sandy material into overlying beds and its eventual outflow at the surface. During this process, some structures such as sand volcanoes or blows, clastic dykes and clastic sills are formed.

In underwater environments, for example in lacustrine areas, other types of structures form, such as convoluted laminations, pillars and lens-like structures. An understanding of the phenomenon of liquefaction and structures resulting from it allows a quantitative evaluation to be made of the size of paleoearthquakes.

For liquefaction phenomena to take place there is a threshold of minimum earthquake intensity. In highly susceptible soils this intensity threshold is VI on the Modified Mercalli Scale, or a magnitude greater than 5.5 (see figure).

For further reading on seismic liquefaction see Obermeier (1996 and 2005).

The paleoseismicity of an area can also be studied by identifying and analysing other earthquake-induced phenomena, such as **landslides** and effects on **tree growth (dendroseismology)**.

The identification and analysis of recent landslides associated with major earthquakes can be easily carried out. However, problems arise when working with older landslides which have undergone modification of their geometry and features over time. Different types of slope instability may occur during an earthquake, the most common being landslides and rockfalls. If a series of induced landslides of the same age and regional distribution are identified, it is possible to assess the magnitude of the earthquake. The minimum magnitude required to induce falls of blocks of rock or small landslides is 4.0, whereas magnitudes of 6.0 to 6.5 (or intensities ≥ VIII) are needed to generate rock and soil avalanches (see Section 14.6).

The effects of earthquakes on trees may cause breakage or induce a slowing down of the growth rate due to rupture of its branches and reduced photosynthesis, resulting in thinner rings.

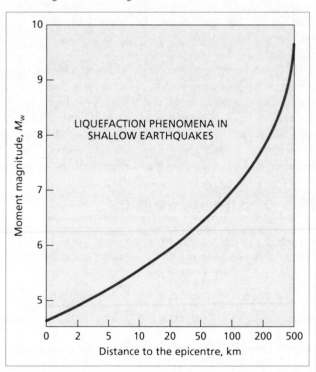

Maximum distance at which liquefaction structures have been observed in shallow earthquakes in relation to magnitude (Obermeier, 1996).

of sedimentary structures originated by earthquake ground effects (*Box 14.1*). In the characteristic earthquake model, this magnitude is an approximation of the one corresponding to the fault or fault segment being studied. An estimation of the mean recurrence period can be obtained from the slip rate.

The dating of seismic events is based on the age of displaced layers. Greater slip accumulates in older layers than in the newer ones, which may pass upwards into undeformed beds.

Apart from the excavation of trenches, other geological methods can be used to study the age of deformations associated with faults, especially when estimating the date of the last coseismic slip. In this respect, it is common practice to date fault scarps using quantitative geomorphological methods. Analysis of the evolution of scarps resulting from prehistoric earthquakes of a known date, or artificial scarps (slopes, quarries, etc.), allows the approximate age of other fault scarps under investigation to be calculated. As a rule, the gradients of scarps lessen with age, which enables different scarp systems and associated facets to be dated and used to calculate the age of the most recent events; in favourable cases, slide rates between successive uplift events can be determined.

## Evaluation of seismic parameters in present-day earthquakes

Seismic data such as epicentre location, focal depth, magnitude and focal mechanism are used to identify the fault responsible for an earthquake and calculate the energy released. In addition, studies on ground deformation associated with the seismic event are carried out. Surface ruptures produced by earthquakes can be studied by mapping scarps and ground breakage such as cracks, ground tilting displacements and seismite distribution.

Increasing use is being made of **radar interferometry** techniques for measuring surface deformation by means of satellite radar imagery of the ground surface, taken before and after the earthquake. The small radar wavelength makes it possible to detect differences in altitude measured in millimetres. Interferograms appear as maps with concentric coloured bands interpreted as areas with the same vertical ground displacement.

# 14.3 Seismicity studies

Seismotectonic studies are based on analysis of the relationship between tectonics and seismicity (or frequency of earthquakes per unit area). The following seismic information is required:

— A seismic catalogue of seismicity showing earthquake location, date, time, magnitude, intensity, focal depth, etc.
— Maps of historic and instrumental epicentres.
— Knowledge of focal mechanisms and stress distribution, with analysis and identification of stress fields.
— Observations of intensities, effects on the ground and on buildings; isoseismal maps.
— Records of strong-ground motion.

Analysis of seismicity requires the following:

— A review of historic seismicity.
— An evaluation of seismic parameters.
— The determination of ground motion attenuation relationships.
— Knowledge of earthquake distribution and recurrence, and
— Seismic regionalization, and
— Seismotectonic relationships.

The following aspects should also be taken into consideration when carrying out these studies:

— Data selection systems.
— Seismic catalogue completeness and standardization.
— The acquisition of ground motion attenuation relationships
— Epicentral and focal error.
— Criteria for seismic regionalization.

## Common relationships used in seismicity studies

### Relationships between different scales for measuring earthquake size

The size of an earthquake is measured by two different scales: intensity and magnitude. Intensity measures the effects of the earthquake on people, buildings and the surrounding environment. It is represented in roman numerals on a progressive scale. In Europe, the most common scales are the Mercalli-Cancani-Sieberg (MCS), used mainly in Italy, and the more widely used Medvedev-Sponheuer-Karnik (MSK) scale; both are on a scale of I to XII. In recent years, an interest has been shown at a European level in standardising and updating the criteria for different measurements of intensity, resulting in the European Macroseismic Scale (EMS) (Grünthal, 1998).

Unlike intensity, magnitude is an instrumental measurement. The first scale of magnitude was devised in 1935 by Richter, who defined it as the base 10 logarithm of the maximum amplitude registered in micrometers on a Wood-Anderson seismograph situated 100 km from the earthquake epicentre. This is known as local Richter magnitude ($M_L$). Since then more than ten different scales of magnitude have been defined, depending on the type of waves and periods measured on the seismogram. The most commonly used are surface wave magnitude ($M_S$) and body wave magnitude ($m_b$). $M_S$ is measured on the amplitude of Rayleigh surface waves with a 20 s period, and it is used to measure large shallow focus earthquakes (depth h < 70 km) at long distances (length d > 1000 km). The body wave magnitude $m_b$ is determined from the maximum amplitude of P waves with a period of around 1 s, and gives a better estimation of magnitude for deep and/or small to medium-sized earthquakes. The main problem with these magnitude scales is that they saturate at around magnitudes 7.5 and 6.5 respectively. In order to solve this problem, the moment magnitude ($M_W$) is used (Hanks and Kanamori, 1979):

$$M_W = 2/3 \log M_o - 10.7$$

where $M_o$ is the seismic moment in units of dynes cm.

In seismic hazard studies (Section 14.4) it is essential to have an earthquake catalogue that uses the same scale of magnitude. There is plenty of literature on empirical conversion ratios for the different types of magnitude discussed above (e.g., Nuttli, 1985; Ambraseys, 1990). However, the main problem with standardizing a seismic catalogue lies in converting data on intensity to that of magnitude. The effects felt from an earthquake are not only related to the quantity of seismic energy released but to other factors, such as the amplification of movement due to the effect of sediments, the presence or absence of people to perceive the earthquake and the type and quality of construction. In other words, earthquakes of different magnitudes may produce the same levels of intensity. Several methods are used to convert data from intensity to magnitude; the most common is a linear adjustment using magnitude - epicentral intensity ($I_o$) relationships. This type of relationship is generally only valid for the region in which the data were obtained. By way of example, the relationship proposed for the Iberian Peninsula by Rueda and Mezcua (2001) is:

$$M_W = 0.575 \, I_{max} + 1.150 \pm 0.56$$

— **Relationship between earthquake size and frequency**

One of the requirements in seismic hazard studies (Section 14.4) is an estimation of the frequency of earthquake occurrence based on size. The relationship generally used for this was put forward by Gutenberg and Richter (1954):

$$\log N = a - bM$$

where $N$ is the number of earthquakes with a magnitude greater than or equal to $M$, and $a$ and $b$ are fitting parameters. Parameter $b$ represents the relationship between large and small earthquakes, and parameter $a$ the level of seismicity in the region. If fitting is carried out normalizing $N$ with respect to the temporal size of the sample, the accumulated annual rate of occurrence is obtained. It should be noted that the Gutenberg-Richert relationship is also used in terms of intensity.

— **Relationship between strong-ground motion and earthquake size**

The perception and recording of strong-ground motion basically depend on earthquake magnitude, distance from the source of released energy and geotechnical conditions of the ground. Attenuation relationships are empirical expressions that can be obtained by means of statistical studies of the variation in ground motion parameters with respect to magnitude and distance. Such relationships are very useful as they make it possible to predict the value of a particular motion parameter, generally that of acceleration, depending on earthquake magnitude and distance from the epicentre or focus. As will be seen later, they are fundamental in calculating seismic hazard.

An attenuation relationship widely used in European countries is that of Ambraseys et al. (1996):

$$\log(a) = -1.48 + 0.266 \, M_s - 0.922 \log(r) \\ + 0.117 \, S_A + 0.124 \, S_s + 0.25 \, P \\ r = (d^2 + 3.5^2)^{1/2}$$

where $a$ is acceleration in units $g$, $M_s$ the shear wave magnitude, $d$ the minimum distance in km to the surface projection of the fault, and $S_A$ and $S_S$ soil coefficients with values:

$S_A = 0$ and $S_S = 0$ for rock ($V_s > 750$ m/s)
$S_A = 1$ and $S_S = 0$ for stiff soil ($360 < V_s \leq 750$ m/s)
$S_A = 0$ and $S_S = 1$ for soft soil ($180 < V_s \leq 360$ m/s)

and $P$ is a constant that takes a value of 0 to obtain the value of $\log(a)$ corresponding to the mean, and 1 to obtain the 84 percentile (0.25 being the standard deviation of $\log(a)$). The limits of application of this relationship are imposed by the characteristics of the database used, specifically earthquakes of $M_S \geq 4.0$ with a focal depth of less than 30 km. Other attenuation relationships frequently used in Europe are those of Sabetta and Pugliese (1987 and 1996), obtained from recordings of Italian earthquakes. A compilation and discussion on attenuation relationships is to be found in Douglas (2003).

In countries with low or moderate levels of seismic activity, where there are no significant records of large magnitude earthquakes (the ones that hold most interest from an engineering perspective), ground motion attenuation is often considered in terms of intensity. Such relationships are built using information provided by isoseismal maps, which represent the spatial distribution of the levels of intensity following an earthquake. An attenuation relationship of epicentral intensity with distance can be obtained from studying statistically representative data. Because local parameters, such as geotechnical conditions, types of construction, population distribution, and so forth, exert a strong influence on the intensity value, the use of this type of relationship is only valid for the region where the data has been gathered. Moreover, one of the drawbacks of this type of relationship in engineering applications is that the intensity value has to be converted into a strong-ground motion parameter. In order to overcome this limitation, numerous equations have been devised to convert intensity into acceleration (e.g. Trifunac and Brady, 1975; Schenk et al., 1990; Wald et al., 1999). These relationships show strong dispersion, both with respect to the sample under consideration and their comparison with each other.

### Sources of seismic information

The main institutions providing on-line seismic information are the International Seismological Centre (www.isc.ac.uk), the National Earthquake Information Center (http://neic.usgs.gov) and, in the Mediterranean region, the European Mediterranean Seismological Centre (www.emsc-csem.org). More seismic information and related software for downloading can be found in the ORFEUS website (Observatories and Research Facilities for European Seismology) (www.orfeus-eu.org). The United States Geological Survey database (http://earthquake.usgs.gov/regional/sca/) and, for Europe, the European Strong-Motion Database (www.isesd.cv.ic.ac.uk) are very useful for obtaining data on strong-ground motion. Seismic hazard maps covering the whole world can be found on the website of the international Global Seismic Hazard Assessment Project (www.seismo.ethz.ch/gshap). With regard to practices used to assess seismic hazard for legislation on earthquake resistance in different European countries, the reader is referred to García-Mayordomo et al. (2004).

## 14.4 Seismic hazard analysis

The **object of seismic hazard analysis** is to determine the maximum strong-ground motion that might affect an installation during its operational lifetime, or the strongest earthquake that might occur at a site or in a region over a given period of time.

The first methods used in seismic hazard assessment were deterministic, that is, they were based on historical records of large earthquakes. However, these were soon brought into question and replaced by **probabilistic** methods, which give a better cost-risk balance. Neither of these methods is entirely satisfactory, as essential elements of the models on which they are based are unknown, and in general there is a lack of available data. In spite of this, from a practical point of view, the need to find acceptable answers means they are the best option available at the present time. For a more detailed account of different methods of assessing hazard see Reiter (1990) and González de Vallejo (1994).

Figure 14.9 shows the methodology followed for assessing seismic hazard and its application to geological engineering.

## Deterministic method

This method assumes that future seismicity will be the same as in the past, and assessment of the strongest earthquake expected is based on the strongest ever recorded.

The method consists of five stages (Figure 14.10):

### 1. Characterization of seismogenic sources

A seismogenic (or seismic) source is a general term that includes any seismotectonic source. According to the USNRC (1997a and 1997b), the following categories can be distinguished:

— **Seismogenic source.** This is a part of the crust considered to have uniform seismicity, which can be anything from a well-defined fault to a region covering an extensive area (**seismotectonic province**).
— **Capable tectonic source.** This is a tectonic structure that may generate earthquakes or surface deformations, that is, an **active fault**. The concept of active fault is defined in relation to the age of the last displacement. From a neotectonic perspective, it is a fault showing deformations that have taken place during the present tectonic regime. From the perspective of geological engineering applied to projects involving dams and other large structures, active faults are those which have shown movement during the last 10,000 years (Holocene); where radioactive installations are involved, this period is extended to the last 500,000 years.
— **Seismotectonic province.** This can be defined as a region of geological, geophysical and seismological similarity and can therefore be assumed to have uniform potential seismicity. Earthquakes may occur in any part of the province, even when seismic data indicates the existence of preferred clusterings or locations.

### 2. Selection of controlling earthquake

The controlling earthquake is the largest earthquake expected at a seismogenic source; for each source a maximum potential earthquake is assigned, based on historical or paleoseismicity data.

### 3. Shifting of controlling earthquake

The controlling earthquake is situated at the nearest distance to the site in each seismogenic source. In cases where the site is inside a seismotectonic province, the controlling earthquake is located at a distance of 15 km from the site

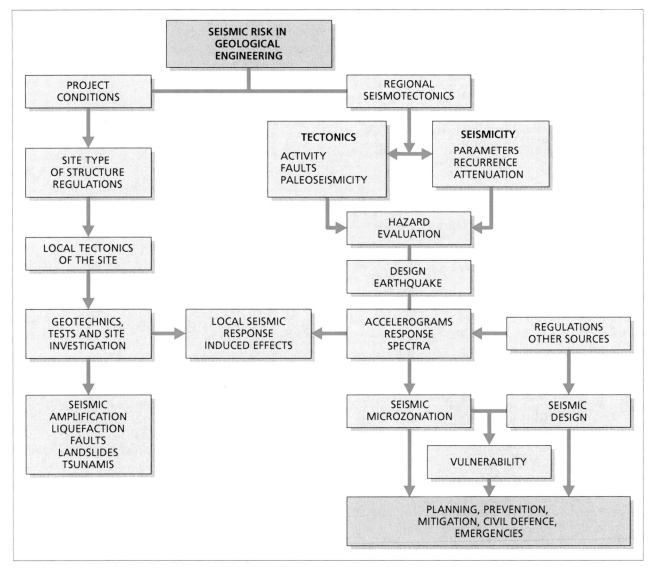

*Figure 14.9* Seismic risk studies in geological engineering.

according to USNRC (1997a). This procedure means that the controlling earthquake may occur at any point in the seismogenic source.

### 4. Determination of maximum ground motion at the site

The controlling earthquake selected is the one producing the greatest intensity at the site. Effects of the controlling earthquake at the site are assessed using an appropriate ground motion or intensity attenuation relationship. Attenuation relationships should be based on earthquakes that have taken place in the region or are representative of it; in other words, they must reflect the influence of wave propagation over distance in the area under study. Results of the effects of an earthquake at the site are expressed in intensity or acceleration.

### 5. Determination of seismic hazard at the site

Results obtained from the previous stage are expressed according to intensity, acceleration or any other measure of seismic motion at site.

This method considers hazard to be defined by the greatest value of ground motion generated by different controlling earthquakes. The results of the deterministic method can therefore give rise to extremely conservative conclusions, except where installations are concerned in which safety would be seriously compromised by the unacceptable consequences of failure due to seismic activity.

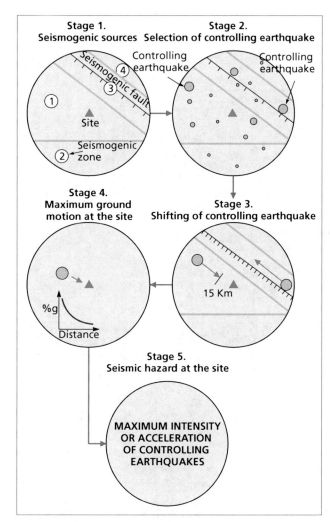

*Figure 14.10*   Stages in the deterministic method.

# Probabilistic methods

Probabilistic methods make use of intensity or magnitude-frequency relationships obtained from recorded seismicity to give probability curves for different levels of motion.

Of the probabilistic methods proposed, the one developed by Cornell (1968) has been the most widely accepted. It consists of the following stages (*Figure 14.11*):

### 1. Definition of seismogenic sources

The same approach is followed as used in the deterministic method, except that sources are explicitly defined as uniform with regard to seismic potential; in other words, the possibility of an specific size earthquake occurring inside the source is the same, independently of its location.

The size of the source has a great influence as it modifies seismicity recurrence relationships. In intraplate areas the size of the sources is usually large. USNRC regulations stipulate using radii of 320 km around sites for regional studies, and of 40 km for active fault characterization.

### 2. Source activity

Seismic parameters characterizing each source are defined by the frequency distribution of earthquakes according to Gutenberg and Richter's expression referred to previously (Section 14.3), in which the number of earthquakes in a region diminishes exponentially with magnitude. To assess the $a$ and $b$ parameters for each source, seismic data should be first revised and standardized, and catalogue completeness verified.

### 3. Earthquake effects at a site

Ground motion attenuation relationships are used, as with the deterministic model, but in this case a ground motion attenuation relationship is considered for each characteristic earthquake of each source. In this way, a family of ground motion attenuation relationships is obtained, relating epicentral intensity to intensity or acceleration felt at the site.

### 4. Hazard assessment

The Poisson model is normally used when probabilistic methods are considered in calculating hazard. In this model, each earthquake occurs randomly and independent of time, and each event is independent from others and does not influence them or condition their distribution. Results from a probabilistic analysis are expressed as the **annual exceedance probability** of different levels of intensity or ground acceleration during a given period of time, which generally coincides with the operational lifetime of buildings or infrastructures (*Figure 14.12*).

However, with respect to seismic hazard, the term "return period" ($T$) is more frequently used. This is defined as the inverse of the mean annual occurrence rate ($\lambda_a$) that exceeds a certain level of ground motion ($a$):

$$T = \lambda_a^{-1}$$

For low (<0.1) annual exceedance probabilities ($P[A > a]_{t=1}$), it can be assumed that:

$$P[A > a]_{t=1} \approx \lambda_a$$

If $T$ is known, the exceedance probability can be calculated for any period of time ($t$):

$$P[A > a]_t = 1 - [1 - 1/T]^t$$

The levels of risk usually observed in analyses of seismic hazard assume an exceedance probability of 10% during the operational lifetime of the infrastructure (see *Table 14.2*). *Figure 14.13* shows an example of a seismic hazard map.

# SEISMIC HAZARD

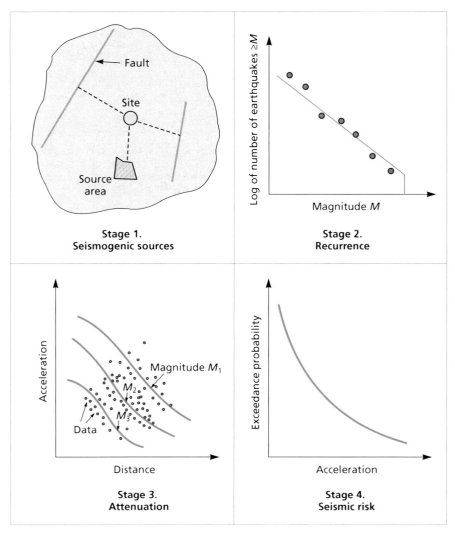

Figure 14.11  Stages in the probabilistic method.

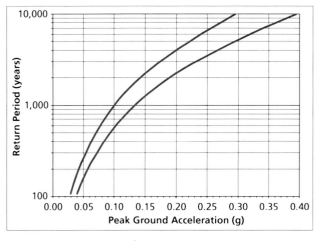

Figure 14.12  Example of hazard curves obtained at a same site in the Canary Islands using two different strong-ground motion attenuation relationships (González de Vallejo et al., 2006).

| Table 14.2 | COMMONLY USED EQUIVALENCES OF RETURN PERIODS HAVING A 10% PROBABILITY OF BEING EXCEEDED DURING THE OPERATIONAL LIFETIME OF STRUCTURES |
|---|---|
| **Operational lifetime (years)** | **Return period (years)** |
| 50 (ordinary structures) | 475 ($\approx$ 500) |
| 100 (important structures) | 950 ($\approx$ 1,000) |
| 1,000 (critical structures) | 9,492 ($\approx$ 10,000) |

## 14.5 Seismic site response

Regional seismic hazard studies generally assess strong-ground motion in rock conditions. Nevertheless, local conditions specific to each site (nature of the soil, topography, depth of water table, etc.) may imply a different

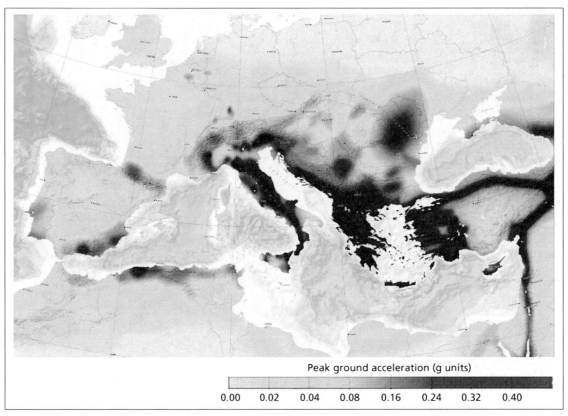

Figure 14.13   Seismic hazard map of the Mediterranean area, expressed as the peak horizontal acceleration for a 475 year return period (equivalent to a 10% probability of exceedance in 50 years) for stiff soil conditions (Giardini et al., 2003).

seismic response within the geographical area under consideration. These effects, also known as **local** or **site effects**, are of great importance in seismic-resistant planning and design.

## Design earthquake

Different values for ground motion (intensity, acceleration, etc.) for various return periods (100, 500, 1,000 years, for example) are obtained from hazard analysis. The **design earthquake** corresponds to the selected return period. For example, when a dam is studied, it is usual practice to consider return periods of 1,000 years; assuming that an intensity of VIII corresponds to this period, the probability of such an earthquake occurring within the structure's operational lifetime of 100 years would be 9.5% (Figure 15.1, Chapter 15). For building purposes a return period of 500 years is normally adopted. For relevant infrastructures, such as dams and bridges, this is 1,000 years, and for high security installations (nuclear power stations and the storage of radioactive waste) it is in the order of 10,000 years. These aspects are dealt with in Section 15.2 of Chapter 15.

## Seismic parameters of ground motion

The movement of the ground due to an earthquake (ground motion or strong motion) is expressed by a series of physical parameters that form the basis of seismic design. If accelerometer records are available for the site, parameters such as acceleration, velocity, displacement, fundamental period and duration can be obtained. A more detailed explanation of these parameters can be found in Kramer (1996) and Bommer and Boore (2004).

The **earthquake accelerogram** (graphic representation of the variation in acceleration with time recorded by an accelerometer) allows the calculation of maximum acceleration for a specific damping and dominant period, through numeric integration. Representation of these maximum accelerations with relation to different vibration periods of a single-degree-of-freedom oscillator constitutes the so-called ground **response spectrum** (Figure 14.14), which shows amplification of ground motion with respect to acceleration, velocity or displacement. The response spectrum is used in seismic-resistant structural design, in which it is essential that the ground motion response spectrum does not exceed that of structure design.

# SEISMIC HAZARD

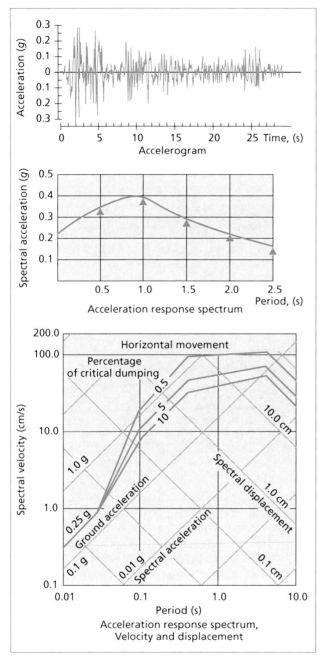

Figure 14.14 Examples of response spectra of ground motion caused by an earthquake.

If representative accelerograms of local earthquakes are unavailable, international databases can be consulted for records of events of similar size, with distances and geotechnical conditions comparable to those at the site. Response spectra can also be constructed directly from the results of seismic hazard analysis if a spectral strong-ground motion attenuation relationship is used in the calculations (Ambraseys et al., 1996; Sabetta and Pugliese, 1996). If the probabilistic method has been used the resulting response spectrum can be considered uniformly probable.

In conventional construction work, regulations on seismic resistance established for each country, or by Eurocode-8 in Europe, must be adhered to. These regulations commonly define a standard spectral shape based on a given acceleration value anchored at a vibration period of zero seconds, which is generally associated with an exceedance probability of 10% in 50 years (equivalent to a return period of 475 years).

# Modification of ground motion by local conditions

The seismic characteristics of a particular earthquake (design earthquake), characterized by its accelerogram, may be modified by local conditions (type of soil, topography, etc.). This can lead to an amplified seismic response with respect to that defined by the design earthquake.

Factors having most influence on the seismic response at a site are:

— Type and lithological composition of the materials, especially superficial deposits with a geotechnical behaviour corresponding to that of soils.
— Thickness of sediments and the depth of substratum.
— Dynamic soil properties.
— Depth of water table.
— Surface and substratum morphology.
— Presence of faults, their situation and characteristics.

The effects of local conditions are of great importance in that they determine the chances of surface rupture, soil liquefaction and landslides arising from faults, as well as seismic signal amplification. Figures 14.15 and 14.16 show examples of these effects. In the case of Figure 14.16 the effects would be:

— The greater the soil thickness, the greater the amplification of acceleration (thickness of Basin 1 > Basin 2).
— Soil properties influence amplification: soils in Basin 2 ($N = 10$) are softer than in Basin 1 ($N = 20$), with greater amplifications in Basin 2.
— The greater the extension, the less the edge effect in the subsoil in the response spectrum (Basin 1 is more extensive than Basin 2).
— The greater substratum depth the higher vibration period (Basin 1 is deeper than Basin 2).
— The presence of a high water table and soft soils (Basin 2) may be considered sufficient to generate a liquefaction hazard.
— The proximity of an active fault may amplify acceleration and induce surface rupture (point C).
— Topography may amplify acceleration (accelerations at point B greater than point A).

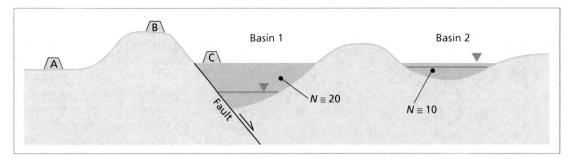

*Figure 14.15* Example of the influence of local conditions on seismic response (modified from Dowrick, 2000).

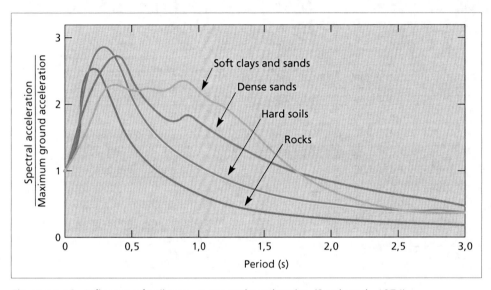

*Figure 14.16* Influence of soil type on spectral acceleration (Seed et al., 1974).

Investigation of the **geotechnical properties** of the ground is carried out by borehole surveys, *in situ* tests, geophysical methods and laboratory testing. The most usual *in situ* tests are the SPT and the CPT. Geophysical techniques include downhole and crosshole tests and surface waves. Laboratory tests generally used to determine the dynamic properties of soils are:

— The **resonant column test**, from which soil deformation values and the damping coefficient can be obtained.
— The **cyclic triaxial test**, to determine the dynamic properties of a sample through the application of a confining pressure equivalent to existing stress in the ground, and inducing dynamic forces as a result of cyclic vertical stress.
— The **cyclic shear test**, used to assess the shear strength of a sample subjected to dynamic forces of a cyclic nature and to analyse liquefaction phenomena.

If accelerograms typical of local site conditions are not available, it is possible to assess local seismic response by using **indirect methods**. One of these methods determines the "type columns" of different soils present in the area, including thickness, apparent density, grain-size distribution, SPT, tangential deformation modulus, shear wave velocity, and depth of water table. A reference accelerogram is then selected which, if not from the same region, should at least resemble the conditions being analysed (see previous section); this accelerogram is used to simulate ground response for each of the "type columns" defined above. Another method is to use the soil amplification factors provided by the corresponding national seismic code, or Eurocode-8 in Europe.

*Figure 14.17* shows an example of local soil amplification in Cartagena (southeast Spain). The SHAKE program (Schnabel *et al.*, 1972) was used in the calculations. The greatest amplifications are found in soft deposits, with peak acceleration values up to 4 times higher than the reference accelerogram.

Another factor that modifies local seismic response is the ground **morphology**, both at the surface and the substratum. Some effects are:

— Greater amplifications are produced in areas of high relief than in smooth relief or flat areas.
— Earthquake duration is increased in high relief areas.

# SEISMIC HAZARD

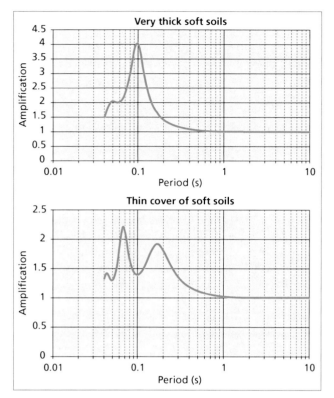

*Figure 14.17* Example of seismic amplification in Cartagena soils, southeast Spain.

— Differential displacements may occur on slopes.
— The horizontal motion component is increased more than the vertical in areas with scarps or on slope edges.

For more details on these topics see Dorwick (2000), AFPS (1990) and Eurocode-8.

## 14.6 Ground effects induced by earthquakes

As well as the characteristic vibratory motion, earthquakes may induce a series of effects that give rise to considerable deformation and rupture of the ground:

— Soil liquefaction.
— Landslides and rockfalls.
— Surface rupture due to tectonic faults.
— Tsunamis.

With the exception of tsunamis, these induced effects are directly related to the geological and geotechnical behaviour of the ground. Their main characteristics and methods of analysis are described below.

## Liquefaction potential

Liquefaction occurs when certain types of soils affected by earthquakes undergo rapid rises in pore pressure (undrained), resulting in a reduction in shear strength and failure of the soil, which behaves as if it were a liquid (*Box 14.1*). This phenomenon leads to foundation and slope failure and landslides. The types of soil susceptible to losing much of their strength through dynamic stresses are loose and fine sands and poorly graded sands and silts. Another condition necessary for liquefaction to take place is a high water table, near the surface, and a low degree of compaction equivalent to SPT $N$ values of less than 20 blows.

In areas affected by liquefaction the following relationships have been observed:

— Liquefaction is associated with earthquakes of magnitude 5.5 or more, with ground accelerations equal to or above 0.2 g.
— Liquefaction does not take place below a depth of 15 m.
— In most cases where liquefaction has been observed the water table was at a depth of less than 3 m; at below 5 m, liquefaction potential is very low.

The characteristic properties of liquefiable soils are:

— 100% saturation.
— Average diameter $D_{50}$ from 0.05 to 1.0 mm.
— Uniformity coefficient $Cu = D_{60}/D_{10} < 15$.
— Fines content less than 10%.
— Low degree of compaction, that is, $N < 10$ for depths <10 m, and $N < 20$ for depths >10 m.

**Liquefaction potential** can be assessed by several methods (Youd and Idriss, 2001); the **Seed and Idriss method** (Seed and Idriss, 1971) is one of the most commonly used.

According to this method, soil will liquefy if the cyclic shear stress ratio (CSR) produced by an earthquake is greater than the shear strength of the soil:

$$\text{CSR} = \frac{\tau_{cm}}{\sigma'_v} = 0.65 \times \frac{\sigma_v}{\sigma'_v} \frac{a_{max}}{g} r_d$$

where

$\tau_{cm}$ = average cyclic shear stress
$\sigma_v$ = total stress
$\sigma'_v$ = effective stress
$a_{max}$ = maximum horizontal acceleration
$g$ = gravity acceleration
$r_d$ = reduction factor with depth ($r_d = 1 - 0.015z$, where $z$ is the depth)

A soil's liquefaction potential can be assessed from empirical data. *Figure 14.18* shows the possibility of liquefaction occurring in earthquakes of several different sizes calculated from the CSR value according to the previous expression, for a value of $(N_1)_{60}$.

$(N_1)_{60}$ is the SPT value, normalized for an approximate pressure of 10 t/m² and an effective energy of 60%. The following expression is applied to obtain this value:

$$(N_1)_{60} = NC_N C_E C_B C_R C_S$$

where

- $N$ = the number of SPT blows.
- $C_N = (P_a/\sigma'_v)^{0.5}$, ($C_N$ should not exceed 1.7).
- $P_a$ = atmospheric pressure (10 t/m² ~ 100 kPa).
- $C_E$ varies according to the energy delivered to the SPT sampler; for the "doughnut" type of drop hammer $0.5 < C_E < 1.2$; for the "automatic" type $0.8 < C_E < 1.3$; and for the "safety" type $0.7 < C_E < 1.7$.
- $C_B$ indicates the influence of the diameter of the borehole, being equal to 1.0 for: 65 mm < Ø < 115 mm.
- $C_R$ varies according to the length of the rods ($L$); for $L < 3$ m, $C_R = 0.75$; for $4$ m $< L < 6$ m, $C_R = 0.85$ and for $10$ m $< L < 30$ m, $C_R = 1.0$.
- $C_S = 1.0$ for standard samplers.

The Seed and Idriss method has been revised by Youd and Idriss (2001). To assess liquefaction potential it is necessary to calculate the cyclic shear stress ratio (CSR) exerted by the earthquake on the ground, and the soil's capacity to resist this stress (CRR). CSR is defined in the same way as in the method of Seed and Idriss, with the value of $r_d$ modified as follows:

$$r_d = 1.0 - 0.00765z \text{ for } z \leq 9.15 \text{ m}$$
$$r_d = 1.174 - 0.0267z \text{ for } 9.15 < z \leq 23 \text{ m}$$

CRR can be calculated either from the cone penetration test (CPT) (Section 5.5 of Chapter 5), from the SPT, or from shear wave velocity $V_s$, the CPT method giving the best results.

Figure 14.19 shows the relationship between CRR or CSR and the value of $(N_1)_{60}$, indicating the limit between soils susceptible to liquefaction and non-susceptible soils for an earthquake of magnitude 7.5, and different fines content.

The ordinate axis in Figure 14.19 shows CSR resulting from a generic earthquake of magnitude ($M_s$) 7.5 for sandy soils with different $(N_1)_{60}$. The curves represent the soil's CRR in relation to its fines content. CSR values above the CRR curves will represent cases of soils susceptible to liquefaction.

If magnitudes other than 7.5 are being studied, corrections are made by multiplying the ordinate axis by the magnitude scaling factor (MSF). This takes into account the variation in the number of equivalent load cycles in relation to earthquake magnitude (Table 14.3).

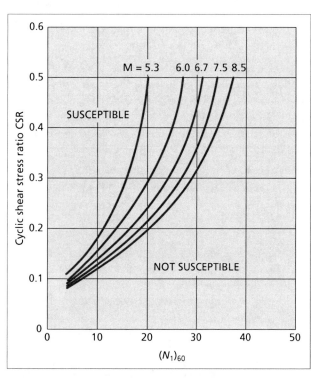

Figure 14.18  Liquefaction potential of a soil based on $(N_1)_{60}$ and the cyclic shear stress ratio, CSR (Ho and Kavazanjian, 1986).

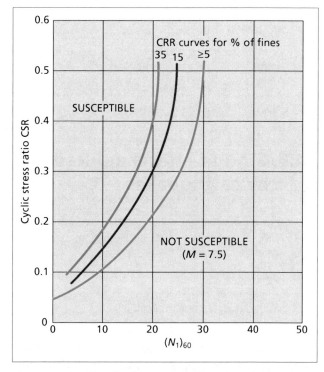

Figure 14.19  Liquefaction potential based on the SPT $(N_1)_{60}$ for an earthquake of magnitude 7.5 (modified from Youd and Idriss, 2001).

# SEISMIC HAZARD

## Box 14.2

### Calculation of liquefaction potential: worked example

Determine the liquefaction potential of sands with low fines contents lying at a depth of 3 m with the water table at 2 m, with $N = 12$, for a design earthquake of magnitude ($M_s$) 7.5 situated at 18 km from the site. In accordance with these data and using the ground motion attenuation relationship of Ambraseys et al. (1996) for soft soils, the acceleration predicted at the site would be around 0.3 g.

- If the method of Seed and Idriss is applied:

$\sigma_v = 52$ kN/m$^2$
$\sigma'_v = 42$ kN/m$^2$
$a_{max} = 0.3$ g
$r_d = 0.95$

$$CSR = 0.65 \times \frac{52}{42} \times 0.3 \times 0.95 = 0.23$$

In accordance with Figure 14.18, for a magnitude of 7.5 and $(N_1)_{60} = 15$, liquefaction is possible.

- If the modification made by Youd and Idriss (2001) is applied, then $r_d = 0.977$ and CSR = 0.23. Introducing the values CSR = 0.23 and $(N_1)_{60} = 15$ in Figure 4.19, and considering the curve for ≥5% of fines, it can be seen that liquefaction is possible.

Note that in order to apply the Seed and Idriss method it is necessary to know the acceleration and therefore the magnitude and distance of the earthquake, as well as the SPT (or CPT or $V_s$) value at the depth specified.

*Table 14.3* **VALUE OF MSF (MAGNITUDE SCALING FACTOR) FOR DIFFERENT MAGNITUDES (EUROCODE-8)**

| Magnitude ($M_s$) | MSF |
|---|---|
| 5.5 | 2.86 |
| 6 | 2.20 |
| 6.5 | 1.69 |
| 7 | 1.30 |
| 7.5 | 1.00 |
| 8 | 0.67 |

(Ambraseys, 1988).

## Landslides induced by earthquakes

Of all the causes of damage associated with earthquakes, landslides are among the most frequent (Chapter 13), though intensity has to be high for these to take place (Figure 14.20 and Box 14.1). According to empirical data, no landslides of any significance have been experienced for intensities <VIII. Factors to be considered in assessing the potential for earthquake-induced landslides include the following:

— Presence of unstable slopes or slopes in a precarious state of stability prior to the earthquake.
— Steep slopes.
— Soils with low strength or a metastable structure (quick clays, collapsible soils, etc.)
— Rocky scarps with potential rockfall hazard.

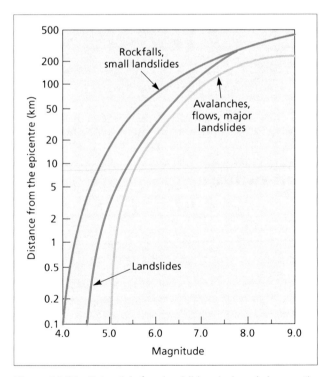

Figure 14.20 Potential for landslides induced by earthquakes (Hays, 1990).

Analysis of susceptibility to earthquake-induced landslides can be carried out either by qualitative methods which consider the various relevant factors, or by analytical methods, such as the pseudo-static method, in which seismic action can be assigned to a static horizontal force incorporated into the limit equilibrium equation. This pseudo-static force is the product of a seismic coefficient $k_h$ and of the weight $W$ of the soil mass analysed:

$$p = k_h W = (a_h/g) W$$

being

$$k_h = a_h/g$$

where $a_h$ is maximum horizontal acceleration and $g$ is gravity acceleration.

The above expression gives conservative results; for a more accurate calculation of $k_h$, refer to Eurocode-8.

For limit equilibrium conditions the factor of safety would be (Figure 14.21):

$$F = LSR/(WE + k_h WD)$$

where:

L = length of failure surface (ab)
S = shear strength per unit of area
R = radius of circle of failure or distance with respect to resistant moment.
E and D = distances with respect to slide moment due to weight and seismic forces.
W = weight per unit of length perpendicular to the plane of the section under analysis.

This calculation refers to rotational landslides and it cannot be applied to other forms and shapes of failure.

## Fault rupture

One effect of earthquakes is rupture of the ground surface caused by the displacement of active faults. Earthquake-induced dislocation along a fault may manifest itself at the surface in the form of a scarp, cracking or landslides, along the rupture area.

Ground surface displacements due to active faults are associated with earthquakes of magnitude 6.0 or greater; their effects depend on numerous factors, for example:

— The nature of the ground and its dynamic properties.
— The overburden thickness above the substratum or hard layer.
— Earthquake characteristics: magnitude, focal depth and distance to the epicentre.
— The type of fault and associated faults.

The amplifying effect of accelerations along faults has been demonstrated in large earthquakes, with maximum horizontal acceleration values peaking at more than double the acceleration recorded in areas close to the epicentre. This amplifying effect, together with the ground motion produced along the fault, is potentially very destructive and constitutes a determinant factor in seismic hazard analysis and its applications in geological engineering (site assessment for infrastructures, urban and regional planning, etc.). It is therefore essential to locate possible seismogenic faults, including so-called "hidden" faults (faults covered with sediments that are not visible).

Observations of surface faulting include (Wang and Law, 1994):

— Surface rupture may be induced by creep-type movement; this is less likely if overburden thickness is more than 5 m.
— Dislocation produced in a fault "hidden" by overburden may give rise to faulting at ground surface.
— Tectonic dislocation at the surface is very likely to take place for earthquakes of magnitude $M \geq 6$ at a shallow depth (less than 30 km).
— When the thickness of overburden is more than 30 m the possibilities of surface rupture are low.
— The width of the surface rupture zone is usually in the order of several metres or tens of metres, forming a narrow corridor, although fracture length may extend over hundreds of metres.

Overburden thickness has a significant influence on surface faulting, given its capacity to absorb energy; surface rupture may therefore depend on this thickness. However, the type of fault is a determining factor in the types of ground displacement which occur. Translational faults produce surface ground displacements independent of sediment thickness, but normal or reverse faults need considerable overburden thicknesses to absorb deformation.

Section 14.2 includes various relationships between the ground displacement, length and area of the fault, and earthquake magnitude.

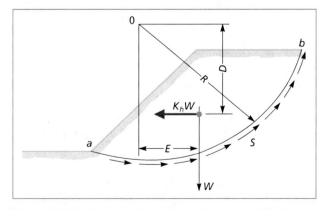

Figure 14.21 Pseudostatic analysis of slope stability (Wang and Law, 1994).

## 14.7 Applications to geological engineering

Seismic hazard studies involve numerous disciplines and activities, such as geology, seismology, engineering, architecture, planning and civil defence; the aim they all have in common is to either avoid or reduce damage from earthquakes. In geological engineering, the principal applications of these studies are the following:

— **Site assessment:** the use of seismic, geological and geotechnical criteria when considering sites for dams, bridges, industrial plants, power stations, radioactive installations, etc.
— **Urban and regional planning:** regional seismic zonation and urban seismic microzonation (microzonation maps) for the purposes of planning, urban codes, seismic resistant design and land use.
— **Seismic prevention and mitigation measures:** to provide criteria for assessing the extent of damage or losses that may be caused by earthquakes expected in the area. Typically, these studies are concerned with the vulnerability of structures, buildings, installations, etc., estimations of possible losses and proposals for mitigating measures and preventing damage.

## Seismic hazard studies applied to site assessment

Dams, special buildings, high security installations and any type of major infrastructure in general all require sites that offer adequate conditions of geological, seismic or geotechnical stability. The following sequence of outcomes related to studies of seismic hazard should be therefore carried out:

1. **Seismic hazard assessment** at a regional level, in accordance with the criteria described above, in order to obtain the seismic parameters of earthquakes corresponding to a specific return period.
2. **Design earthquake**, according to the type of structure and degree of safety required at the site. *Table 15.2* of Chapter 15 shows useful lifetime periods of different installations, and Section 15.3 the return periods used for selecting design earthquakes.
3. **Effects induced** by the earthquake in relation to geological and geotechnical conditions at site. The main hazards to consider are:

    — Liquefaction.
    — Landslides.
    — Surface faulting.
    — Tsunamis.

4. **Local seismic response**, taking into account soil properties, morphology and induced effects, soil amplification and response spectrum.
5. **Site assessment.** In principle, processes or hazards such as liquefaction, surface rupture due to faults, landslides, etc. are considered to be an **exclusion factors** as far as structural supports or foundations are concerned. However, it is essential to take into account the probability of hazard occurrence and the level of safety required for both site and structures. In addition to identifying the presence of active faults, a detailed analysis is needed of seismic hazard in the region, together with an assessment of the probability of the occurrence of an earthquake capable of inducing the phenomenon.

In areas of high seismicity, the presence of some of the hazards mentioned are considered to **preclude** for any type of significant structure, building or installation. In areas of moderate seismicity the risk factors and safety levels required for each structure must be carefully evaluated and assessed. It is also necessary to determine whether there is a reasonable probability of ground surface rupture taking place during a structure's lifetime.

A further aspect to consider in site selection is seismic response amplification, which is fundamental to the design of structures and foundations. This may be a **conditioning feature** for certain types of installation, especially in areas of high seismicity.

Local seismic response should be analysed in relation to regulations for seismic resistance regulations and the type of structure and its function. It should also be borne in mind that some infrastructures have **lifeline** functions, that is, not only must they be structurally stable, they must also be able to remain operative in the event of an earthquake (hospitals, power stations, supply networks, communications, etc.).

## Seismic microzonation

Seismic microzonation is of great importance in the study of seismic hazard for urban areas. It involves the identification and characterization of lithological units, generally soils, which have similar dynamic responses to earthquakes, the analysis of induced effects, such as liquefaction and faults, and the assessment of their hazards. The resulting **microzonation maps** are presented with a topographical base valid for building and urban planning purposes. Scales used usually range from 1/15,000 to 1/5,000, although these may vary depending on information available and the degree of detail required.

The most significant aspects to be considered in microzonation studies are:

— Geotechnical properties of soils, their thickness and density, shear wave velocity, stiffness modulus, strength, SPT, CPT, depth of water table, etc.

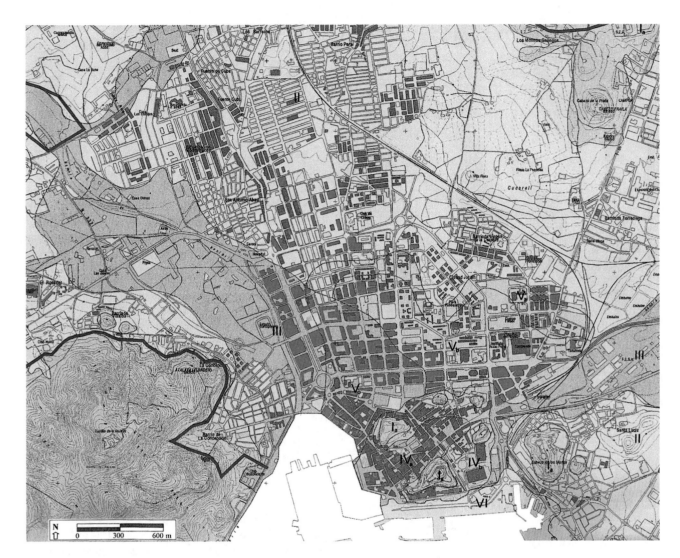

| Seismic response and vulnerability of building structures |||||||
|---|---|---|---|---|---|---|
| Zone | Ground vibration || Percentage of buildings ||| Type of vulnerable buildings | Seismic response |
| | Maximum amplification (*) | Period (s) | Type 1 | Type 2 | Type 3 | | |
| Ia | Very low | – | 100 | – | – | None | Very low |
| Ib | Very low | 0.04 | 95 | 5 | – | None | Very low |
| II | Medium | 0.043 | 78 | 17 | 5 | None | Low |
| III | Medium | 0.25 | 79 | 3 | 18 | 1 | Medium-high |
| IVa | Medium | 0.37 | 20 | 50 | 30 | 2-1-3 | Medium-high |
| IVb | Very low | – | 40 | 50 | 10 | None | Very low |
| Va | Very high | 0.095 | 27 | 51 | 22 | 1 | Medium |
| Vb | Medium | 0.07 | 13 | 48 | 39 | 1-2 | Medium |
| Vc | High | 0.3 | 19 | 34 | 47 | 2-1-3 | Medium-high |
| VI | Low | 0.45 | 90 | 10 | 0 | 2-3 | Low |

| Controlling earthquake: | Type of building | Nº of storeys | Natural vibration period (s) |
|---|---|---|---|
| Richter magnitude: 4.7 Probability of occurrence in 100 years: 9.5% (*) for the controlling earthquake | 1 | 1–2 | 0.1 to 0.3 |
| | 2 | 3–5 | 0.3 to 0.6 |
| | 3 | >5 | >0.6 |

*Figure 14.22* Seismic response map for the city of Cartagena, southeast Spain; original scale 1:15,000 (courtesy of Geological Survey of Spain, IGME).

- Liquefaction potential and landslides and surface rupture susceptibility due to faults.
- Topographical conditions which may amplify seismic response.
- Tsunamis hazard in coastal areas.

There are several methods of analysing seismic response for microzonation purposes. The so-called direct methods analyse the seismic signal recorded on an accelerogram, either from an earthquake that has taken place in the area or from a large artificially-generated vibration. When such information is unavailable, the seismic response can be simulated for each type of soil present in the area with the accelerogram characteristic of the design earthquake (Section 14.5). The results of microzonation studies are presented on maps showing isovalue lines or cartographic units, in which seismic response is similar for a specific return period. An example is shown in *Figure 14.22*.

## Seismic vulnerability assessment

To carry out seismic risk studies applied to urban planning and emergency management, it is essential to have information about the possible damage and losses a major earthquake could cause; this is fundamental in matters concerning prevention and the adoption of mitigation measures.

Methods used for assessing the vulnerability of structural elements in the event of an earthquake are based on **damage probability matrices** and **vulnerability functions**. In the case of the former, it is necessary to be familiar with the type of construction, the possible behaviour of each type of buildings in the event of an earthquake of certain magnitude or intensity, and the structural damage expected at different levels of intensity. Vulnerability functions are graphic relationships between structural vulnerability (or degree of damage in different types of building) and intensity, or any other significant earthquake parameter (Figure 15.2, Chapter 15).

*Figure 14.23* shows an example of the relationship between two response spectra and the dominant periods for different types of building.

The vulnerability studies allow assessments to be carried out of the degree of losses or damage affecting a city or particular structure in a given "seismic scenario". Such studies can identify the most vulnerable structures or areas in a city and give an estimation of the extent of possible damage/losses; they can even indicate which installations may be very seriously affected or put out of service entirely. Loss of human life can also de estimated. These data are of great importance in devising plans aimed at prevention of loss and seismic mitigation; in addition to the technical measures involved, such plans also include social aspects

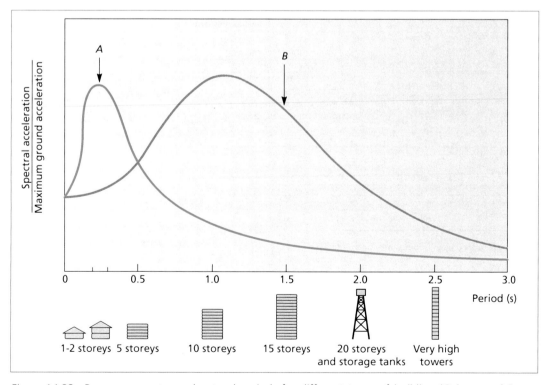

*Figure 14.23* Response spectra and natural periods for different types of building (Coburn and Spence, 1992). The dominant period for lower buildings coincides with the spectrum A period, while that of the higher buildings (10 to 15 storeys) coincides with the spectrum B period. Spectrum A could be characteristic either of an earthquake in the vicinity of the site or hard soils, while spectrum B would correspond to a distant earthquake, or soft soils.

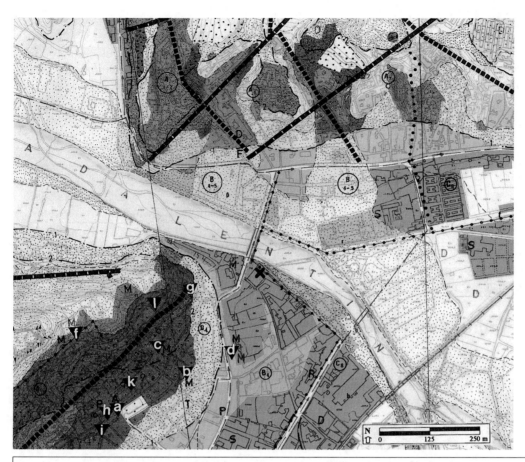

*Figure 14.24* Seismic hazard and vulnerability map for the town of Lorca, southeast Spain; original scale 1:5,000 (courtesy of Geological Survey of Spain, IGME).

# SEISMIC HAZARD

## Box 14.3

### The Kocaeli (Turkey) earthquake of August 17th, 1999

At 03:02 hours local time a large earthquake of magnitude 7.4 shook part of northwest Turkey following rupture of one of the faults of the 1,500 km long North Anatolia Fault system. The length of the displaced fault-segment was 150 km. As a consequence of the earthquake, there were 17,127 deaths, 43,963 people injured and more than 250,000 made homeless. The economic loss may have been as high as 3% of Turkey's GNP.

Although buildings in the affected areas were made from reinforced concrete, more than 20,000 buildings collapsed due to the low quality of construction and disregard for the legislation in force. Nevertheless, much of the damage was due to ground failure brought on by local conditions. In the fault area strong accelerations of up to 0.42 g were recorded, which caused total destruction along a corridor from 5 to 22 m wide, corresponding to surface rupture of the fault. Maximum lateral displacement along

Figure A  Lateral displacement of 445 cm caused by surface fault.

Figure B  Lateral displacement of 290 cm caused by surface fault.

Figure C  Tectonic subsidence caused by normal faulting during the Kocaeli earthquake, resulting in flooding of the first floor of the building.

Figure D  The same building as in Figure C destroyed by seismic aftershocks a month later.

*Figure E*    Ground failure caused by liquefaction in a park bordering the sea, in Gölcük.

*Figure F*    Displacement of overhead decks in a viaduct on the Istanbul-Ankara motorway.

the length of the fault was 5 m (*Figures A* and *B*). In addition to those ruptures in the ground, other ruptures were recorded caused by collapse, subsidence and particularly soil liquefaction. The coastal areas of towns situated on alluvial deposits and artificial fills suffered the worst effects (*Figures C, D* and *E*).

The Trans-European Motorway between Ankara and Istanbul was severely damaged, despite its recent construction in compliance with the California code on seismic resistance. One of its main viaducts, with 3.4 km long, was put out of service although its supporting pillars were undamaged. Huge accelerations and twisting movements caused displacements in the overhead decks which reduced its stability (*Figure F*). The viaduct was situated very close to the seismogenic fault.

Experience from this earthquake highlights the following:

— The earthquake took place in a region with a long seismic history where similar earthquakes were expected.
— Failure to observe regulations on seismic resistance and low quality construction were the cause of failure for most of the buildings.
— During urban planning liquefaction hazard was not taken into account; neither was surface rupture due to faults and other local seismic effects.
— The unsuitable location of the viaduct near the fault was a decisive factor given the seismic amplification, which rendered the viaduct unusable as a consequence.

concerned with information, education and organization, all fundamental for seismic hazard mitigation. For a detailed description of different aspects of seismic vulnerability, see Coburn and Spence (1992).

An example of a seismic hazard and vulnerability map is shown in *Figure 14.24*.

## Recommended reading

Bommer, J.J. and Boore, D.M. (2004). Engineering seismology. In: Encyclopaedia of Geology. Elsevier.

Coburn, A.W. and Spence, R.J. (1992). Earthquake protection. John Wiley & Sons, New York.

Dowrick, D.J. (2000). Earthquake resistant design. John Wiley & Sons, London, 2nd Ed.

Kramer, S.L. (1996). Geotechnical earthquake engineering. Prentice Hall.

Reiter, L. (1990). Earthquake hazard analysis. Columbia University Press. New York.

Yeats, R.S., Sieh, K. and Allen, C.R. (1997). The geology of earthquakes. Oxford University Press.

## References

AFPS (1990). Recommandations pour la rédaction de règles relatives aux ouvrages et installations à réaliser dans les régions sujettes aux séismes. École Nationale des Ponts et Chaussées, Paris.

Ambraseys, N.N. (1988). Engineering seismology. Earthquake Engineering and Structural Dynamics, 17, 1–105.

Ambraseys, N.N. (1990). Uniform magnitude re-evaluation of European earthquakes associated with strong-ground motion records. Earthquake Engineering and Structural Dynamics, 19, 1–20.

Ambraseys, N.N., Simpson, K.A. and Bommer, J.J. (1996). Prediction of horizontal response spectra in Europe. Earthquake Engineering and Structural Dynamics, 25, 371–400.

Berryman, K. and Beanland, S. (1991). Variation in fault behaviour in different tectonic provinces of New Zealand. Journal of Structural Geology, 13 (2), 177–189.

Bommer, J.J. and Boore, D.M. (2004). Engineering seismology. In: Encyclopaedia of Geology. Elsevier.

Coburn, A.W. and Spence, R.J. (1992). Earthquake protection. John Wiley & Sons, New York.

Cornell, C.A. (1968). Engineering seismic risk analysis. Bull. Seism.. Soc. America, vol. 58, 5 pp. 1533–1606.

Crone, A.J., Machette, M.N. and Bowman, J.R. (1992). Geologic investigations of the 1988 Tennant Creek, Australia, earthquakes. Implications for paleoseismicity in stable continental regions. U. S. Geol. Sur. Bull., 2032-A, 1–51.

Douglas, J. (2003). Earthquake ground motion estimation using strong-motion records: a review of equations for the estimation of peak ground acceleration and response spectral ordinates. Earth Science Reviews, 61, 43–104.

Dowrick, D.J. (2000). Earthquake resistant design. John Wiley & Sons, London, 2nd Ed.

Eurocode-8 (2001). Design of structures for earthquake resistance. Part 5: Foundations, Retaining structures and Geotechnical Aspects.

García-Mayordomo, J., Faccioli, E. and Paolucci, R. (2004). Comparative study of the seismic hazard assessments in European National Seismic Codes. Bulletin of Earthquake Engineering, 2, 51–73.

Giardini, D., Jiménez, M.J. and Grünthal, G. (Eds) (2003). The ESC-SESAME European-Mediterranean Seismic Hazard Map, scale 1:5,000,000. ETH-CSIC-GFZ, Zürich.

González de Vallejo, L.I. (1994). Seismotectonic hazard for engineering projects in moderate seismic regions. Keynote Lecture. 7th Inter. IAEG Congress, vol. III, pp. XIX–XXXVIII. Lisbon. Balkema.

González de Vallejo, L.I., García-Mayordomo, J. and Insua, J.M. (2006). Probabilistic seismic hazard assessment of the Canary Islands. Bulletin of the Seismological Society of America, 96(6), 2040–2049.

Grünthal, G. (Ed.) (1998). European macroseismic scale. European Seismological Commission. Cahiers du Centre Européen de Géodynamique et de Séismologie, Luxembourg, vol. 15, 99 pp.

Gutenberg, B. and Richter, C.F. (1954). Seismicity of the Earth and associated phenomena. Princeton University Press, 2nd Ed.

Hanks, T.C. and Kanamori, H. (1979). A moment magnitude scale. Journal of Geophysical Research, 84(B5), 2348–2350.

Hays, W. (1990). Earthquake vulnerability. Cooperative Project for Seismic Risk Reduction in the Mediterranean Region. UNDP/OPS/UNDRO, Trieste.

Ho, C.L. and Kavazanjian, E. (1986). Probabilistic study of SPT liquefaction analysis. Proc. In Situ '86, ASCE Conference on Use of In-Situ Tests in Geotechnical Engineering, Blacksburg, Virginia, ASCE Geotechnical Special Pub. No. 6, 602–616.

Keller, E.A. and Pinter, N. (1996). Active tectonics. Earthquakes, uplift and landscape. Prentice Hall.

Kramer, S.L. (1996). Geotechnical earthquake engineering. Prentice Hall.

McCalpin (Ed.) (1996). Paleoseismology. Academic Press, 588 pp.

Nuttli, O. (1985). Average seismic source parameters relations for plate margin earthquakes. Tectonophysics, 118, 161–174.

Obermeier, S.F. (1996). Use of liquefaction-induced features for paleoseismic analysis. Engineering Geology, 44, 1–76.

Obermeier, S.F. (2005). Paleoliquefaction and appraisal of earthquake hazards. In: Obermeier (Ed). Special Issue. Engineering Geology 76, 3–4.

Pierce, K.L. (1986). Dating methods. In: Active Tectonics. Studies in Geophysics. National Academic Press, Washington, D.C., 195–214.

Reiter, L. (1990). Earthquake hazard analysis. Columbia University Press. New York.

Rueda, J. and Mezcua, J. (2001). Seismicity, seismotectonic and seismic hazard in Galicia (in Spanish). Publicación Técnica, 35. Instituto Geográfico Nacional. Madrid. 64 pp.

Sabetta, F. and Pugliese, A. (1987). Attenuation of peak horizontal acceleration and velocity from Italian strong-ground motion records. Bulletin of the Seismological Society of America, 77(5), 1491–1511.

Sabetta, F. and Pugliese, A. (1996). Estimation of response spectra and simulation of nonstationary earthquake ground motions. Bulletin of the Seismological Society of America, 86(2), 337–352.

Schenk, V., Mantlik, F., Zhizhin, M.N. and Tumarkin, A.G. (1990). Relation between macroseismic intensity and instrumental parameters of strong motions - a statistical approach. Natural Hazards, 3, 2, 111–124.

Schnabel, P.B., Lysmer, J. and Seed, H.B. (1972). A computer program for earthquake response analysis of horizontally layered sites. Rep. EERC 72–12. University of California at Berkeley.

Scholz, C.H. (1990). The mechanics of earthquakes and faulting. Cambridge University Press.

Seed, H.B. and Idriss, I.M. (1971). Simplified procedure for evaluating soil liquefaction potential. Jl. of Soil Mech. and F. Div. ASCE, vol. 97 SM 7.

Seed, H.B., Ugas, C. and Lysmer, J. (1974). Site-dependent spectra for earthquake-resistant design. Rep. EERC 74-12. University of California at Berkeley.

Sibson, R.H. (1983). Continental fault structure and shallow earthquake source. Jl. Geol. Soc. London, 140: 747-767.

Stirling, M., Rhoades, D. and Berryman, K. (2002). Comparison of earthquake scaling relations derived from data of the Instrumental and preinstrumental era. Bulletin of the Seismological Society of America, 92(2), 812-830.

Trifunac, M.D. and Brady, A.G. (1975). A study on the duration of strong earthquake ground motion. Bulletin of the Seismological Society of America, vol. 65, 3, 581-626.

USNRC (1997a). Seismic and geologic siting criteria for nuclear power plants. Appendix A. 10 C.F.R.E. Part 100.

USNRC (1997b). Identification and characterization of seismic sources, deterministic source earthquakes and ground motions. Regulatory Guide 1165.

Villamor, P. and Berryman, K.R. (1999). La tasa de desplazamiento de una falla como aproximación de primer orden en las estimaciones de peligrosidad sísmica. 1er Congreso Nac. Ingeniería Sísmica, Murcia, Spain, 153-163. Spanish Assoc. Earthquake Engineering, Madrid.

Wang, J.G.Z.Q. and Law, K.T. (1994). Siting in earthquake zones. Balkema.

Wald, D., Quitoriano, V., Heaton, T. and Kanamori, H. (1999). Relationships between peak ground acceleration, peak ground velocity and Modified Mercalli intensity in California. Earthquake Spectra 15(3), 557-564.

Wells, D.L. and Coppersmith, K.J. (1994). New empirical relationships among magnitude, rupture length, rupture width, rupture area and surface displacement. Bulletin of the Seismological Society of America, 84(4), 974-1002.

Youd, T.L. and Idriss, I.M. (2001). Liquefaction resistance of soils. Summary report from the 1996 NCEER and 1998 NCEER/NSF Workshops on Evaluation of Liquefaction Resistance of Soils. ASCE. Jl. of Geotech. and Geoenvironmental Engineering, vol. 127, 4, 297-313.

# 15

## PREVENTION OF GEOLOGICAL HAZARDS

1. Geological hazards
2. Hazard, risk and vulnerability
3. Safety criteria in geological engineering
4. Prevention and mitigation of geological hazards
5. Hazard and risk maps

## 15.1 Geological hazards

Geological hazards are natural processes that may constitute damaging events. The geodynamic processes that affect the earth's surface produce **hazards** of varying characteristics, magnitude, frequency, extent and speed that may become **geological risks** if they affect populated areas or human activities directly or indirectly (Table 15.1).

Geological hazards can be classified according to their origin in internal geodynamic processes, such as earthquakes or volcanic activity, and external geodynamic processes, such as landslides and collapses. Each type of hazard is characterised by its location, intensity and frequency.

Different kinds of natural phenomena such as erosion, seismic shaking, volcanic eruptions and heavy rain may cause landslides, rockfalls, earth and debris flows on slopes, collapse, subsidence, etc. These ground movements reflect the dynamic nature of the geological environment and the natural evolution of topography, but they may also be caused or triggered by human interference that modifies natural conditions in an area.

Some ground movements, such as a large landslide or a powerful earthquake, are potentially dangerous events that may cause high loss of life or injury, property damage, social and economic disruption or environmental degradation.

Natural disasters, including floods and cyclones, affected more than 800 million people and killed an estimated 3 million worldwide in the twenty years from 1980 to 2000; according to the World Bank, natural disasters caused losses in excess of 40 billion dollars from 1990 to 1996 (Murck et al., 1996); more recently, it has been estimated that natural hazards cost the global economy over 50 billion dollars per year, two thirds of this sum corresponding to damages, and the remainder representing the cost of predicting, preventing and mitigating (Bell, 2003).

From 1975 to present the number of reported natural disasters in the world has increased tenfold. In recent years the economic losses and number of people killed by natural disasters has increased dramatically; e.g. in 2005 reported losses were over 400 billion dollars, and in 2008 almost 280 billion dollars (ISDR, 2009).

One of the main aims of **engineering geology**, as a science applied to the study and solution of problems produced by the interaction of the geological environment and human activity, is the **evaluation, prevention and mitigation** of geological risks, i.e. of the damage caused by geodynamic processes.

The problems arising from the interaction between human activities and the geological environment make appropriate actions to balance natural conditions and land use, with geological hazard prevention and mitigation methods essential at the planning stage. These actions should have as their starting point an understanding of geodynamic processes and of the geomechanical behaviour of the ground.

The **damage** related to a specific geological process depends on:

— The **speed, magnitude and extent** of the process; geological hazards may be violent and catastrophic (earthquake, sudden large-scale landslide, collapse) or slow (flows and other slope movements, subsidence, etc.).
— The chances for **prevention or prediction** and the **warning time** available; some processes, such as earthquakes or flash floods, cannot be forecast, and they give very little warning time or none at all.
— Whether **actions can be taken to control the process** or protect elements exposed to its effects.

The effects of ground movements may be direct or indirect, short or long term or permanent. Some tectonic or isostatic processes develop on a geological time scale, what means that their effects cannot be considered on a human scale.

Only certain processes, when they occur on an "engineering" or "geotechnical" scale, can be **controlled** by human action, such as landslide or rockfalls, erosive processes, subsidence and floods. Others, such as earthquakes, tsunamis,

| Table 15.1 | GEOLOGICAL AND METEOROLOGICAL PROCESSES WHICH MAY CAUSE RISKS |
|---|---|
| Generated at or close to ground level | — Landslides and rockfalls<br>— Collapse and subsidence<br>— Erosion<br>— Expansive and collapsibile soils |
| Generated below ground level | — Earthquakes and tsunamis<br>— Volcanic activity<br>— Diapirism |
| Meteorological processes | — Torrential rain and heavy precipitation<br>— Flooding and flash floods<br>— Gully erosion processes<br>— Hurricanes<br>— Tornados |

volcanic eruptions and large scale landslides or avalanches of millions of cubic metres in mountainous areas, are outside the scope of human control.

The following sections deal with aspects related to the evaluation and prevention of geological hazards and their influence on engineering projects. Here the importance of considering the influence of natural dynamic processes on the design and safety of engineering works and installations must be emphasised, as well as the evaluation of the geotechnical safety. This means that engineering geological studies should include:

— Geotechnical safety criteria for the case of ground failure.
— Geological safety criteria in relation to geo-hazards.

Chapters 13 and 14 dealt with slope movements and seismic hazard, as these are the processes that are most directly related to the characteristics and geotechnical behaviour of the ground.

Movements caused by expansive clays and sensitive or collapsible soils are considered more as geotechnical problems, and are dealt with in Chapters 2 and 8.

There are other kinds of hazards related to geological materials and processes that do not involve ground movements. These can be classified as geochemical hazards which include water contamination, naturally occurring toxic or explosive gases and radioactive minerals, among others. These hazards are not dealt with in this chapter.

## 15.2 Hazard, risk and vulnerability

To avoid or reduce geological risks and consider their influence in land planning and use, hazard and risk have to be evaluated.

In hazard studies special terminology is used to define hazard, risk and vulnerability. The term "hazard" refers to any more or less violent process which may affect people or property; it is often taken to be synonymous with "risk", although the two concepts are not the same. **Hazard** refers to the **geological process**, **risk** to the **losses** and **vulnerability** to **damage**. These concepts will now be defined, according to how they are generally used.

**Hazard**, $H$, refers to the frequency with which a process occurs and its location. It is defined as the probability of occurrence of a potentially damaging phenomenon at a specified level of intensity or severity for a given time within a specific area (Varnes, 1984). To evaluate hazard, the following information is needed:

— Where and when the processes occurred in the past.
— Their intensity and magnitude.
— The areas where future processes may occur.
— The frequency of the occurrence.

This last point can only be estimated if the process timeframe is known (e.g. the return period for earthquakes or floods, from historical or instrumental data series), or for the triggering factors (e.g. the return period for rainfall that triggers landslides in a certain area).

Hazard, as it has been explained, can be defined as the probability of occurrence of a phenomenon of specific intensity within a given period, but can also be expressed using the **return period** $T$ (years elapsing between two events or processes of similar characteristics), which is the inverse of the annual exceedance probability, $P_{(a)}$:

$$T = 1/P_{(a)}$$

The probability $p$ that a specific intensity value (e.g. an acceleration value in the case of earthquakes) corresponding to an average return period $T$ (years) will be exceeded during a specific time period $t$ is expressed as:

$$p = 1 - \left(1 - \frac{1}{T}\right)^t$$

The time $t$ (years) can be the service life of a dam or building, that is, the expected exposure time or useful life of the structure.

Table 15.2 shows the service life of different installations; Figure 15.1 gives the probability of exceedance curves as a function of this parameter and of the return period $T$.

The concept of **risk**, $R$, includes socio-economic considerations and is defined as the **potential losses** due to a specific natural phenomenon (human lives, direct and indirect economic losses, damage to buildings or structures, etc.).

At the present time, the risk of earthquakes is the most widely developed of such studies. **Seismic risk** is defined as the expected losses structures will suffer during the period they are exposed to seismic activity; this time period is known as the **exposure time** or **service life** of the structure, as has been mentioned above.

*Table 15.2* **SERVICE LIFE OF DIFFERENT INSTALLATIONS ($t$)**

| Structure or installation | $t$ (years) |
|---|---|
| Storage of radioactive waste | 10,000 |
| Nuclear power stations | 40–80 |
| Dams | 100–150 |
| Bridges, tunnels and major infrastructure works | 100 |
| Storage of toxic waste | 250 |
| Conventional buildings and structures | 50–70 |

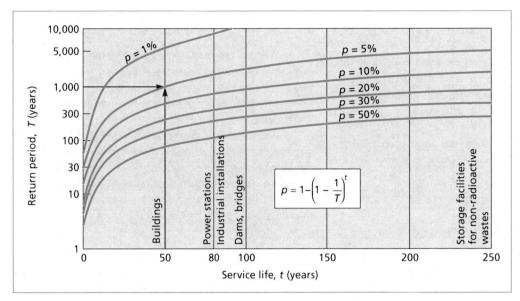

*Figure 15.1* Probability of exceedence (*p*) of an event of known return period occuring in the service life of a structure. Example: What is the probability of a building experiencing a magnitude 6 earthquake if it has a service life or exposure period of 50 years and the return period of the earthquake is 1,000 years? Answer: The probability of exceedence of the earthquake during the service life of the structure is 5%.

Evaluating geological risk is complex, and so is the evaluation of the terms which define it. Risk is evaluated starting from the hazard corresponding to a particular process (cause) and the effects of this on the elements exposed to the hazard (consequences). These effects on the exposed elements (buildings, infrastructures, people, etc.) may be expressed by different parameters: vulnerability, losses, costs, exposure, etc. The risk and the hazard refer to a specified time period, and may be evaluated in either deterministic or probabilistic terms.

The risk can be calculated from the expression:

$$R = H \times V \times C$$

where $H$ is the hazard of the process in question, $V$ is the vulnerability of the elements exposed to the process (elements at risk) and $C$ is the cost or value of these elements. As described above, the risk is expressed in losses (human or economic); in the expression above, these "units" correspond to $C$, while $H$ is a probability and $V$ an adimensional parameter, as is explained below. The value of $C$ can be expressed in either deterministic or probabilistic terms; if the latter, the risk will also be obtained in terms of probability.

If any of the factors is zero, the risk will be zero; this means that in a high hazard zone, the risk will be zero if there are no elements exposed, or if the vulnerability of these is nil. People may increase the risk by occupying hazardous zones, affecting the intensity of the processes or triggering them and by constructing vulnerable buildings or structures. The risk can be reduced by reducing the hazard (acting on the process control factors where this is possible) or the vulnerability (acting on the elements exposed to the risk).

According to Smith (2001) **risk** can be defined as the probability that a hazard will occur and cause losses, and is evaluated from the expression:

$$R = P \times Le$$

where $P$ is the occurrence probability of the process, or hazard, and $Le$ the expected losses.

According to other authors (Varnes, 1984) the product $H \times V$ is known as **specific risk** and is defined as the **level of losses** expected during a given time period resulting from the occurrence of a specific process, expressed in terms of probability. In this case, a quantitative evaluation of losses cannot be made. According to the UNESCO definitions, the risk can be evaluated as follows:

$$R = H \times V \times E$$

where $E$ is the exposure of the elements at risk. Because of the difficulty of quantifying the variable $E$ and considering that for some authors exposure is included in vulnerability (an element is not vulnerable if it is not exposed to risk), the expressions above are more appropriate, when the cost of either the exposed elements, $C$, or the expected losses, $Le$, are considered directly for a specific occurrence.

**Vulnerability**, V, is the expected degree of damage or loss in an element or group of elements at risk resulting from the occurrence of a hazard of specific intensity or magnitude. It depends on the characteristics of the element considered (not on its economic value) and on the intensity of the phenomenon; it is usually evaluated on a scale from 0 (no damage) to 1 (total loss or destruction of the element) and from 0 to 100% damage.

In the case of seismic risk, the vulnerability of a structure or group of structures, or of a whole urban area, is defined as its intrinsic predisposition to sustain damage if a seismic movement of a specific intensity occurs. This will depend on the structural design characteristics and on the intensity of the earthquake; it means that the vulnerability of a masonry building is higher than that of a concrete building during an earthquake.

This parameter is usually defined through vulnerability functions (Figure 15.2) that can be established from the damage or losses such processes have caused in the past and/or from the hypothetical potential damage these phenomena would cause were they to occur. In both cases, present-day measures to reduce or mitigate the potential damage have to be taken into account, as these reduce the vulnerability of the exposed elements.

**Social vulnerability** depends on population density, condition of the buildings and structures, warning and alert systems and emergency and evacuation plans (Table 15.3). Under-developed countries, as has frequently been demonstrated, are particularly vulnerable because of deficiencies in their constructions, high population density in urban areas, etc. This can be evaluated in terms of the percentage of population affected by a specific process.

The **exposed elements**, or elements at risk, may be people, assets, property, infrastructure, services, economic

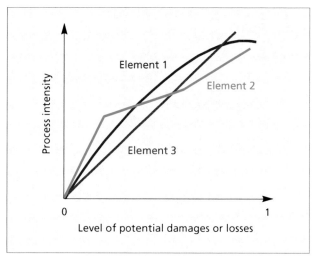

Figure 15.2  Examples of vulnerability functions. A single element, or a group of elements, will be more vulnerable to phenomena of greater intensity. On the other hand, the vulnerability of the individual elements is different for an event of specific intensity.

| Table 15.3 | ELEMENTS TO BE CONSIDERED WHEN EVALUATING VULNERABILITY | |
|---|---|---|
| **Vulnerability** | **Damages or losses** | |
| Social | — Dead and missing<br>— Wounded and disabled<br>— People left homeless<br>— People left jobless<br>— Epidemics and diseases | Social vulnerability depends on:<br>— Intensity and speed of phenomenon<br>— Population density<br>— Structural vulnerability<br>— Warning time<br>— Emergency and response systems |
| Structural | — Damage to buildings and structures<br>— Damage to contents<br>— Loss of profits<br>— Effects on persons | Structural vulnerability depends on:<br>— Intensity and speed of phenomenon<br>— Building type and characteristics<br>— Concentration in populated areas |
| Economic | Direct damage:<br>— Costs of replacement, repair or maintenance of structures, installations or properties, communications systems, power etc.<br>Indirect damage:<br>— Reduction in value of assets<br>— Interruption to transport systems | — Lost productivity of agricultural or industrial ground<br>— Lost income from taxation<br>— Lost human productivity<br>— Lost commercial profit<br>— Lost taxation collection<br>— Cost of preventive or mitigation measures<br>— Reduced water quality and contamination |

## Box 15.1

## Examples of risk evaluation

**Example 1. Probabilistic risk calculation**

This example is based on the preparation of hazard and vulnerability maps for an area where the potential risk these hazards present to the buildings, structures and services of the area cause them to be grouped into four categories.

Once the **hazard** has been estimated, that is, the probability of occurrence of a specific process of intensity $i$ for a return period $T$, and the **vulnerability** of the relevant structures has been evaluated, the potential **risk level** is obtained from the product of these. Figure a) shows the area studied with three hazard levels; in Figure b) the area has been classified in terms of the vulnerability of the elements exposed (four groups of buildings or constructions), evaluating the level of damage which would be caused by the occurrence of the hazard defined in the map in Figure a). The potential risk level, in probabilistic terms, is shown in the map in Figure c).

**Example 2. Probabilistic risk analysis**
(after Smith, 2001)

Assuming that the processes which occurr can be evaluated from historical data, then it can be known that a specific process $E_j$ has an occurrence probability $P_j$ and will cause loss $L_j$ (e.g. evaluated in euros or in number of victims), where $j = 1, n$ and $P_1 + P_2 + \cdots + P_n = 1$. Ordering the $n$ processes by losses, from lower to higher ($L_1 < L_2 < \ldots < L_n$), the accumulated probability for a particular process can be calculated as:

$$P_{j(A)} = P_j + \cdots + P_n$$

indicating the probability of occurrence of an event with losses equal to or greater than $L_j$.

If all the possible processes can be categorized in terms of economic losses, then a risk analysis can be carried out:

| Losses L (euros) | Probability P | Accumulated probability |
|---|---|---|
| 0 | 0.950 | 1.000 |
| 100,000 | 0.030 | 0.050 |
| 500,000 | 0.015 | 0.020 |
| 1,000,000 | 0.005 | 0.005 |

The table indicates there is a 95% probability that there will be no losses, and only a 2% probability of losses of 500,000 euros or more. The risk is defined by the total probable losses and can be evaluated as follows:

$$R = P_1 L_1 + \cdots + P_n L_n$$

giving a resulting value in this case of 15,500 euros.

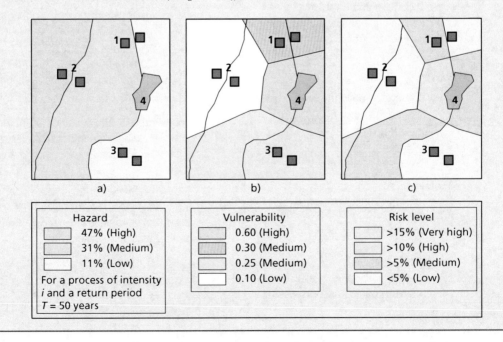

activities, land resources, environmental qualities, etc. which may suffer the direct or indirect consequences of a geological process in a specific area, The **cost** or **value** of these elements can be expressed using different criteria: construction cost of buildings or structures, cost of repairing the damage caused, insured value, etc.; the costs derived from the interruption of communication routes, economic activities, services, etc., could also be considered.

Box 15.1 gives examples of probabilistic risk evaluation using two different approaches: the first uses the hazard level and vulnerability level of the different elements exposed in an area and the second uses the probability of occurrence of the processes and the potential losses caused by them.

## 15.3 Safety criteria in geological engineering

In geological engineering it is normal practice to estimate safety under geological and geotechnical conditions by using a factor of safety, as a deterministic indicator of the relationship between the stabilizing and destabilizing forces in a limit equilibrium situation ($F = 1.00$). The **factor of safety**, $F$, can be defined as the coefficient by which the ground shear strength must be reduced for a slope, excavation, foundation etc. to reach a state of limit equilibrium (Morgenstern, 1991). The value chosen for this factor depends on **how much is known** about the ground strength parameters, hydrostatic pressures, potential shear surfaces and the magnitude of the external forces which act or may act on the ground (Hoek, 1991).

Engineering construction works causes several actions in the ground, as summarized in Table 15.4. A satisfactory solution to the geological and geotechnical problems which may arise from interactions between the ground and the structures depends on the correct selection of geomechanical parameters, the application of the appropriate analytical tools and the choice of reasonable safety and acceptability criteria. Tables 15.5, 15.6 and 15.7 show the different **acceptability criteria** for various types of structures.

| Table 15.4 | HOW ENGINEERING WORKS AFFECT THE GROUND | | | |
|---|---|---|---|---|
| **Engineering works** | **Actions on ground** | **Ground response** | **Induced effects** | **Environmental impact** |
| **Foundations** | Static and dynamic loads. Excavations. Water table changes. | Stresses and deformations. Flow path changes. | Failures. Settlements. | Vibrations. Settlement in nearby structures. |
| **Excavations** Slopes and quarries | Excavations. Water table changes. | Stress relief. Deformations. Flow path changes. | Failures. Instabilities. Alterations. | Settlement induced by water table downdraw. Waste dumps. |
| Tunnels and underground mines | Excavations. Changes in state of stress. Water table changes. | | Stress relief. Instabilities. Seepages. | Waste dumps. Changes in aquifers. Subsidence. Water contamination. |
| **Hydraulic and maritime works** Dams Canals Maritime works | Static and dynamic loads. Hydrostatic pressures. Uplift water pressures. | Stresses and deformations. Flow path changes. | Seepage. Internal erosion. Failures, settlements. Instabilities. | Changes in coastal dynamics (maritime works). Silting. |
| **Reservoirs** Dam reservoirs | Hydraulic loads. Water table changes. | Stresses and deformations. Flow path changes. | Seepage. Instability of slopes. Sedimentation. Dissolution. | Silting. Changes in aquifers. Salinization of aquifers. Induced seismicity. |
| Storage and waste reservoirs | Excavations. Static and hydraulic loads. Percolation of chemical substances, fluids etc. Water table changes. | | Instability. Seepage. Chemical-physical changes in soils and water. | Contamination of aquifers and soils. Gas emissions. Thermal gradients. |
| **Earth works** | Loads. Water table changes. | | Instability. | Modifications to drainage network. |

# GEOLOGICAL ENGINEERING

**Table 15.5  TYPICAL PROBLEMS, CRITICAL PARAMETERS, METHODS OF ANALYSIS AND ACCEPTABILITY CRITERIA FOR SLOPES**

| Structure | Typical problems | Critical parameters | Analysis methods | Acceptability criteria |
|---|---|---|---|---|
| Landslides. | Complex failure along a circular or near circular failure surface involving sliding on faults and other structural features as well as failure of intact materials. | • Presence of regional faults.<br>• Shear strength of materials along failure surface.<br>• Groundwater distribution in slope, particularly in response to rainfall or to submergence of slope toe.<br>• Potential earthquake loading. | Limit equilibrium methods which allow for non-circular failure surfaces can be used to estimate changes in factor of safety as a result of drainage or slope profile changes. Numerical methods such as finite element or discrete element analysis can be used to investigate failure mechanisms and history of slope displacement. | Absolute value of factor of safety has little meaning but rate of change of factor of safety can be used to judge effectiveness of remedial measures. Long term monitoring of surface and subsurface displacements in slope is the only practical means of evaluating slope behaviour and effectiveness of remedial action. |
| Soil or heavily jointed rock slopes. | Circular failure along a spoon-shaped surface through soil or heavily jointed rock masses. | • Height and angle of slope face.<br>• Shear strength of materials along failure surface.<br>• Groundwater distribution in slope.<br>• Potential surcharge or earthquake loading. | Two-dimensional limit equilibrium methods which include automatic searching for the critical failure surface are used for parametric studies of factor of safety. Probability analyses, three-dimensional limit equilibrium analyses or numerical stress analyses are occasionally used to investigate unusual slope problems. | Factor of safety >1.3 for "temporary" slopes with minimal risk of damage. Factor of safety >1.5 for "permanent" slopes with significant risk of damage. Where displacements are critical, numerical analyses of slope deformation may be required and higher factors of safety will generally apply in these cases. |
| Jointed rock slopes. | Planar or wedge sliding on one structural feature or along the line of intersection of two structural features. | • Slope height, angle and orientation.<br>• Dip and strike of structural features.<br>• Groundwater distribution in slope.<br>• Potential earthquake loading.<br>• Sequence of excavation and support installation. | Limit equilibrium analyses which determine three-dimensional sliding modes are used for parametric studies on factor of safety. Failure probability analyses, based upon distribution of structural orientations and shear strengths, are useful for some applications. | Factor of safety >1.3 for "temporary" slopes with minimal risk of damage. Factor of safety >1.5 for "permanent" slopes with significant risk of damage. Probability of failure of 10 to 15% may be acceptable for open pit mine slopes where cost of clean up is less than cost of stabilization. |
| Vertically jointed rock slopes. | Toppling of columns separated from the rock mass by steeply dipping structural features which are parallel or nearly parallel to the slope face. | • Slope height, angle and orientation.<br>• Dip and strike of structural features.<br>• Groundwater distribution in slope.<br>• Potential earthquake loading. | Crude limit equilibrium analyses of simplified block models are useful for estimating potential for toppling and sliding. Discrete element models of simplified slope geometry can be used for exploring toppling failure mechanisms. | No generally acceptable criterion for toppling failure is available although potential for toppling is usually obvious. Monitoring of slope displacements is the only practical means of determining slope behaviour and effectiveness of remedial measures. |

*(continued)*

### Table 15.5 TYPICAL PROBLEMS, CRITICAL PARAMETERS, METHODS OF ANALYSIS AND ACCEPTABILITY CRITERIA FOR SLOPES (CONT.)

| Structure | Typical problems | Critical parameters | Analysis methods | Acceptability criteria |
|---|---|---|---|---|
| Loose boulders on rock slopes. | Sliding, rolling, falling and bouncing of loose rocks and boulders on the slope. | • Geometry of slope.<br>• Presence of loose boulders.<br>• Coefficients of restitution of materials forming slope.<br>• Presence of structures to arrest falling and bouncing rocks. | Calculation of trajectories of falling or bouncing rocks based upon velocity changes at each impact is generally adequate. Monte Carlo analyses of many trajectories based upon variation of slope geometry and surface properties give useful information on distribution of fallen rocks. | Location of fallen rock or distribution of a large number of fallen rocks will give an indication of the magnitude of the potential rockfall problem and of the effectiveness of remedial measures such as draped mesh, catch fences and ditches at the toe of the slope. |

(Hoek, 1991).

### Table 15.6 TYPICAL PROBLEMS, CRITICAL PARAMETERS, METHODS OF ANALYSIS AND ACCEPTABILITY CRITERIA FOR DAMS AND FOUNDATIONS

| Structure | Typical problems | Critical parameters | Analysis methods | Acceptability criteria |
|---|---|---|---|---|
| Zoned fill dams. | Circular or near-circular failure of dam, particularly during rapid drawdown. Foundation failure on weak seams. Piping and erosion of core. | • Presence of weak or permeable zones in foundation.<br>• Shear strength, durability, gradation and placement of dam construction materials, particularly filters.<br>• Effectiveness of grout curtain and drainage system.<br>• Stability of reservoir slopes. | Seepage analyses are required to determine water pressure and velocity distribution through dam and abutments. Limit equilibrium methods should be used for parametric studies of stability. Numerical methods can be used to investigate dynamic response of dam during earthquakes. | Safety factor >1.5 for full pool with steady state seepage; >1.3 for end of construction with no reservoir loading and undissipated foundation porewater pressures; >1.2 for probable maximum flood with steady state seepage and >1.0 for full pool with steady state seepage and maximum credible horizontal pseudo-static seismic loading. |
| Gravity dams. | Shear failure of interface between concrete and rock or of foundation rock. Tension crack formation at heel of dam. Leakage through foundation and abutments. | • Presence of weak or permeable zones in rock mass.<br>• Shear strength of interface between concrete and rock.<br>• Shear strength of rock mass.<br>• Effectiveness of grout curtain and drainage system.<br>• Stability of reservoir slopes. | Parametric studies using limit equilibrium methods should be used to investigate sliding on the interface between concrete and rock and sliding on weak seams in the foundation. A large number of trial failure surfaces are required unless a non-circular failure analysis with automatic detection of critical failure surfaces is available. | Safety factor against foundation failure should exceed 1.5 for normal full pool operating conditions provided that conservative shear strength values are used ($c' \approx 0$). Safety factor >1.3 for probable maximum flood (PMF). Safety factor >1 for extreme loading – maximum credible earthquake and PMF. |

(continued)

Table 15.6 TYPICAL PROBLEMS, CRITICAL PARAMETERS, METHODS OF ANALYSIS AND ACCEPTABILITY CRITERIA FOR DAMS AND FOUNDATIONS (CONT.)

| Structure | Typical problems | Critical parameters | Analysis methods | Acceptability criteria |
|---|---|---|---|---|
| Arch dams. | Shear failure in foundation or abutments. Cracking of arch due to differential settlements of foundation. Leakage through foundations or abutments. | • Presence of weak, deformable or permeable zones in rock mass.<br>• Orientation, inclination and shear strength of structural features.<br>• Effectiveness of grout curtain and drainage system.<br>• Stability of reservoir slopes. | Limit equilibrium methods are used for parametric studies of three-dimensional sliding modes in the foundation and abutments, including the influence of water pressures and reinforcement. Three-dimensional numerical analyses are required to determine stresses and displacements in the concrete arch. | Safety factor against foundation failure >1.5 for normal full poof operating conditions and >1.3 for probable maximum flood conditions provided that conservative shear strength values are used ($c' \approx 0$). Stresses and deformations in concrete arch should be within allowable working levels defined in concrete specifications. |
| Foundations on rock slopes. | Slope failure resulting from excessive foundation loading. Differential settlement due to anisotropic deformation properties of foundation rocks. | • Orientation, inclination and shear strength of structural features in rock mass forming foundation.<br>• Presence of Inclined layers with significantly different deformation properties.<br>• Groundwater distribution in slope. | Limit equilibrium analyses of potential planar or wedge failures in the foundation or in adjacent slopes are used for parametric studies of factor of safety. Numerical analyses can be used to determine foundation deformation, particularly for anisotropic rock masses. | Factor of safety against sliding of any potential foundation wedges or blocks should exceed 1.5 for normal operating conditions. Differential settlement should be within limits specified by structural engineers. |
| Foundations on soft rock or soil. | Bearing capacity failure resulting from shear failure of soils or weak rocks underlying foundation slab. | • Shear strength of soil or jointed rock materials.<br>• Groundwater distribution in soil or rock foundation.<br>• Foundation loading conditions and potential for earthquake loading. | Limit equilibrium analyses using inclined slices and non-circular failure surfaces are used for parametric studies of factor of safety. Numerical analyses may be required to determine deformations, particularly for anisotropic foundation materials. | Bearing capacity failure should not be permitted for normal loading conditions. Differential settlement should be within limits specified by structural engineers. |

(Hoek, 1991).

## PREVENTION OF GEOLOGICAL HAZARDS

**Table 15.7** TYPICAL PROBLEMS, CRITICAL PARAMETERS, METHODS OF ANALYSIS AND ACCEPTABILITY CRITERIA FOR UNDERGROUND CIVIL ENGINEERING EXCAVATIONS

| Structure | Typical problems | Critical parameters | Analysis methods | Acceptability criteria |
|---|---|---|---|---|
| Pressure tunnels in hydro-power projects. | Excessive leakage from unlined or concrete lined tunnels. Rupture or buckling of steel lining due to rock deformation or external pressure. | • Ratio of maximum hydraulic pressure in tunnel to minimum principal stress in the surrounding rock.<br>• Length of steel lining and effectiveness of grouting.<br>• Groundwater levels in the rock mass. | Determination of minimum cover depths along pressure tunnel route from accurate topographic maps. Stress analyses of sections along and across tunnel axis. Comparison between minimum principal stresses and maximum dynamic hydraulic pressure to determine steel lining lengths. | Steel lining is required where the minimum principal stress in the rock is less than 1.3 times the maximum static head for typical hydroelectric operations or 1.15 for operations with very low dynamic pressures. Hydraulic pressure testing in boreholes at the calculated ends of the steel lining is essential to check the design assumptions. |
| Soft rock tunnels. | Rock failure where strength is exceeded by induced stresses. Swelling, squeezing or excessive closure if support is inadequate. | • Strength of rock mass and of individual structural features.<br>• Swelling potential, particularly of sedimentary rocks.<br>• Excavation method and sequence.<br>• Capacity and installation sequence of support systems. | Stress analyses using numerical methods to determine extent of failure zones and probable displacements in the rock mass. Rock-support interaction analyses using closed-form or numerical methods to determine capacity and installation sequence for support and to estimate displacements in the rock mass. | Capacity of installed support should be sufficient to stabilize the rock mass and to limit closure to an acceptable level. Tunnelling machines and internal structures must be designed for closure of the tunnel as a result of swelling or time-dependent deformation. Monitoring of deformations is an important aspect of construction control. |
| Shallow tunnels in jointed rock. | Gravity driven falling or sliding wedges or blocks defined by intersecting structural features. Unravelling of inadequately supported surface material. | • Orientation, inclination and shear strength of structural features in the rock mass.<br>• Shape and orientation of excavation.<br>• Quality of drilling and blasting during excavation.<br>• Capacity and installation sequence of support systems. | Spherical projection techniques or analytical methods are used for the determination and visualization of all potential wedges in the rock mass surrounding the tunnel. Limit equilibrium analyses of critical wedges are used for parametric studies on the mode of failure, factor of safety and support requirements. | Factor of safety, including the effects of reinforcement, should exceed 1.5 for sliding and 2.0 for falling wedges and blocks. Support installation sequence is critical and wedges or blocks should be identified and supported before they are fully exposed by excavation. Displacement monitoring is of little value. |

(continued)

Table 15.7 **TYPICAL PROBLEMS, CRITICAL PARAMETERS, METHODS OF ANALYSIS AND ACCEPTABILITY CRITERIA FOR UNDERGROUND CIVIL ENGINEERING EXCAVATIONS (CONT.)**

| Structure | Typical problems | Critical parameters | Analysis methods | Acceptability criteria |
|---|---|---|---|---|
| Large caverns in jointed rock. | Gravity driven falling or sliding wedges or tensile and shear failure of rock mass, depending upon spacing of structural features and magnitude of in situ stresses. | • Shape and orientation of cavern in relation to orientation, inclination and shear strength of structural features in the rock mass.<br>• In situ stresses in the rock mass.<br>• Excavation and support sequence and quality of drilling and blasting. | Spherical projection techniques or analytical methods are used for the determination and visualization of all potential wedges in the rock mass. Stresses and displacements induced by each stage of cavern excavation are determined by numerical analyses and are used to estimate support requirements for the cavern roof and walls. | An acceptable design is achieved when numerical models indicate that the extent of failure has been controlled by installed support, that the support is not overstressed and that the displacements in the rock mass stabilize. Monitoring of displacements is essential to confirm design predictions. |
| Underground nuclear waste disposal. | Stress and/or thermally induced spalling of the rock surrounding the excavations resulting in increased permeability and higher probability of radioactive leakage. | • Orientation, inclination, permeability and shear strength of structural features in the rock mass.<br>• In situ and thermal stresses in the rock surrounding the excavations.<br>• Groundwater distribution in the rock mass. | Numerical analyses are used to calculate stresses and displacements induced by excavation and by thermal loading from waste canisters. Groundwater flow patterns and velocities, particularly through blast damaged zones, fissures in the rock and shaft seals are calculated using numerical methods. | An acceptable design requires extremely low rates of groundwater movement through the waste canister containment area in order to limit transport of radioactive material. Shafts, tunnels and canister holes must remain stable for approximately 50 years to permit retrieval of waste if necessary. |

(Hoek, 1991).

When geological processes may occur with potentially damaging results, these processes must be included when calculating the stability and safety of installations. Once the process has been identified (earthquake, flood, landslide, etc.) and the level of severity has been defined (using parameters such as seismic acceleration, water height, speed and scope of the process, etc.), these parameters are integrated into the factor of safety calculation.

To consider the influence of certain geological hazards such as flash floods or earthquakes, **return periods** are used; obviously the highest return periods correspond to processes of greatest intensity.

There are standards or regulations that specify factors of safety, return periods and other safety and acceptability criteria that must be used depending on the type of structure. However, quite often this is left to the judgement of the project designer or the person responsible for the study and in this case the following criteria are suggested:

### 1. For geotechnical failure

— Short-term engineering works with no structures involved (opencast mining, temporary slopes, etc., which do not form a supporting part of foundations or structures): $1.2 \leq F < 1.5$.
— Long-term engineering work with no structures involved: $F \geq 1.5$.
— Foundations and excavations involving structures: $1.5 \leq F \leq 3.0$.
— In all cases, the acceptability criteria given in Tables 15.5, 15.6 and 15.7 should be taken into consideration.

### 2. For geological hazards

What has to be taken into account above all is the incidence of possible hazards which may affect the safety of engineering structures. The severity or intensity of the process is estimated for the following return periods $T$:

# PREVENTION OF GEOLOGICAL HAZARDS

## Box 15.2

### Example of geological safety analysis

The photo below shows an eleven storey building resting on alluvium consisting mainly of coarse sands and gravels, and a sub-vertical slope with water visible at the toe. A fault cuts through the alluvium. The region is subject to high seismicity, with an estimated return period of 150 years for a magnitude M = 7 earthquake. Analyze the geological safety conditions.

1 **Safety in relation to a geotechnical failure**
The building foundation consists on footings calculated for a safety factor of 2.0, in relation to the bearing capacity and load of the building. The effect of the proximity of the slope has been taken into account, discounting a possible failure (see Chapters 8 and 9 for coverage of these points).

2 **Seismic hazard**
The probability of exceedance of a magnitude 7 earthquake during the service life of the building is 28%, so that the structure must be designed to resist the seismic actions of this earthquake, considering site amplification factors (see Chapter 14, Section 14.4).

3 **Susceptibility to liquefaction**
Given the type of ground, this process can be ruled out (see Chapter 14, Section 14.6).

4 **Susceptibility to earthquake induced landslides**
The distance between the building and the edge of the slope, considering the strength of the ground, rules out this possibility (see Chapter 14, Section 14.6).

5 **Susceptibility to ground failure by faulting**
The presence of an active fault and the seismicity of the region indicate that the ground may failure along the existing fault, with a probability that can be assumed equivalent to that of the earthquake considered, so that the site is not acceptable in terms of this criterion.

Photo: W. Hays

### Conclusions

— The safety is acceptable within the context of the geotechnical ground failure, liquefaction and landslide, evaluated deterministically.
— The seismic hazard, evaluated on probabilistic criteria, requires a seismic resistant design to withstand the probable seismic actions.
— The hazard of ground failure from the existing fault means that the site is not acceptable.

— Minor or conventional buildings and structures: $100 \leq T \leq 500$ years.
— Major structures, dams, bridges, landmark buildings, etc.: $T = 1,000$ years.
— Critical facilities: $1,000 \leq T \leq 10,000$ years or the equivalent of recorded maximum level of intensity.

When the geological process causing a potential hazard has been identified, with a specified intensity and return period, then the probability $p$ of this hazard being exceeded during the service life of the structure is calculated, using the following criteria:

— Major structures: $p \leq 10\%$.
— Critical facilities: $p \leq 5\%$.

Excluded from this analysis are some exceptional or extreme geological phenomena (e.g. major tsunamis or large landslides, maximum potential earthquake according to geological data, etc.) with very low probability.

In Chapter 14, Section 14.1, these criteria are analysed in relation to the seismic hazard.

Box 15.2 shows an example of application safety criteria for geological hazards.

## 15.4 Prevention and mitigation of geological hazards

**Preventing** geological hazards involves knowing beforehand where and if possible when a phenomenon will occur, so as to:

— Avoid the process.
— Control or slow down the process.
— Warn, prepare and protect.

The type of action to be taken will depend on the characteristics, speed and magnitude of the process.

Prevention is based on a knowledge of the process characteristics and laws, on the analysis of past data, on scientific observation (detailed research into the process) and on the monitoring and detection of anomalies, changes in physical parameters, and precursor phenomena.

**Prediction**, i.e. announcing what is going to happen, is sometimes used in the same way, although it does not mean the same.

Phenomena such as hurricanes and floods can be predicted in the short term as far as intensity and place are concerned; volcanic eruptions tend to be preceded by previous phenomena, both medium and short term; attempts have been made to establish long term earthquake predictions in terms of probability, e.g. that an earthquake of intensity higher than VII will occur within 30 years in a certain zone.

The zones and places where geological processes are active and will act again can be recognised, for example in seismic and volcanic zones and in landslide-prone areas. However, as explained above, some processes cannot be forecast in time, nor avoided or controlled (earthquakes or large scale landslides) so that where these can cause damage, all that can be done is to protect against them and mitigate their effects.

**Mitigation** consists of moderating or reducing losses and damage through control of the processes (wherever this possible) and/or protection of exposed elements, thus reducing their vulnerability.

Table 15.8 lists the different means of mitigating hazards and possible action to be taken in each case, depending on the characteristics of the process (velocity, magnitude or intensity, extension, etc.) and on whether it can be prevented. These actions are usually known as **preventive measures**, although this concept also includes actions designed to avoid the geological processes and their effects, as well as to mitigate the effects of a hazard that cannot be avoided.

*Table 15.8* **PREVENTION AND MITIGATION OF GEOLOGICAL RISKS**

| Processes | Prevention and prediction | Risks mitigation | Mitigating actions | |
|---|---|---|---|---|
| | | | Structural | Non-structural |
| Landslides and rockfalls | Where and when[1] | Process control[2] Protection Evacuation | Correction and stabilization measures and protective works | Prohibiting or restricting occupation of high hazard zones |
| Subsidence and sinking | Where and when [1] | Process control[2] Protection Evacuation | Consolidation and in-fill measures | Use land planning

Standards and regulations

Alert and warning systems

Emergency planning

Information and education for the general public |
| Earthquakes and tsunamis | Where | Protection Evacuation[3] | Earthquake-resistant design | |
| Volcanic eruptions | Where and when – short term | Evacuation Protection | Diverting and containing lava currents and flows | |
| Flooding and flash floods | Where and when | Process control[2] Protection Evacuation | Diverting, containing and regulating works Works and drainage systems | |

[1] Preventing *when* only if the recurrence of triggering factors is known.
[2] Only when the processes are "geotechnical" in size or scale.
[3] In the case of tsunamis, when there is enough time, or in continuous seismic crises.

The most effective, and generally cheapest measures, are **non-structural** based on land use planning. These are especially effective in newly or recently developed areas, where there are no pre-existing conditions on land use. However, they have the following limitations:

— Knowledge of the potential processes and hazards that may affect an area is essential.
— It is difficult, or perhaps even impossible, to apply such measures to areas that are already developed.
— The high cost of preparing the detailed inventories and maps of the different factors involved in land planning.
— Political or economic interests, opposed or reactionary to the adoption of restrictive measures.

There are zones which could be affected by a process of considerable intensity or magnitude and thus **should not be occupied under any circumstances** (e.g. active faults, dry river beds or valleys, coastal cliffs, etc). To identify these areas, or those which could be used under certain restrictions or conditions, maps of **susceptibility and/or hazard** must be produced (see section 15.5), dividing up the land according to potential hazard levels. Geological surveys are essential before land is allocated for use or the building of infrastructure commences.

Mapping also allows **structural measures** to be designed to protect people and assets and to mitigate damage (Table 15.8). These measures are essential in the case of occupation or use of hazardous zones where there is a probability of ground movements. These measures include work or actions to control the processes (drainage systems or retaining walls to stabilize landslides, hydraulic works to avoid flooding, etc.) and appropriately designed engineering works to avoid damage (earthquake-resistant buildings and structures, dams, bridges and drainage works that are appropriate for the maximum forecast flow etc.). Chapters 13 and 14, respectively, deal with these aspects with reference to landslides and earthquakes.

Other important aspects of prevention and mitigation of geological hazards are to **inform the public and raise awareness** and to **introduce administrative and legislative measures, including the control and inspection of how far these measures are adhered to.**

## 15.5 Hazard and risk maps

These maps are the most effective method of presenting information on hazards and risks in an area or region, and should be used by planners, architects, engineers, scientists and technical staff responsible for taking emergency measures. This mapping is intended to divide the territory into zones or units each having a different level of potential hazard or risk.

Table 15.9 lists the different types of maps. Each of them is produced from the information contained in the previous one plus the analysis of additional data, so that the susceptibility maps are a prerequisite for preparing the hazard maps, and these in turn are required for the risk maps and so on (see Figures 15.3 and 15.4).

**Inventory maps** include the spatial location of the processes and/or affected zones as well as their characteristics. With reference to earthquakes, a map of this type will include the epicentres of earthquakes that have already occurred and the isoseismal magnitudes; for landslides, it will represent the points or zones of current and previous processes and areas

### Table 15.9 TYPES AND CONTENT OF HAZARD MAPS

| Type of map | Content | Methodology |
| --- | --- | --- |
| Inventory | Location and extent of current and past processes and/or areas affected. Process characteristics (type, size, speed, intensity, etc.). | Data collection (documents, maps, aerial photos, field). Survey of process types and characteristics. |
| Susceptibility | Areas with different levels of susceptibility to the occurrence of a process type. | Process analysis. Analysis of conditioning factors. Superimposing factors. |
| Hazard | Areas with different hazard levels. | Analysis of triggering factors. "When" and "where" prediction of process occurrence. |
| Vulnerability | Locating elements or areas with different vulnerability levels. | Identifying elements exposed to a hazard. Evaluating their vulnerability. |
| Risk | Territorial zoning based on risk or level of risk. | Evaluating losses due to a specific process. |
| Multi-risk | Zoning based on risk or level of risk. | Evaluating total losses caused by different processes. |

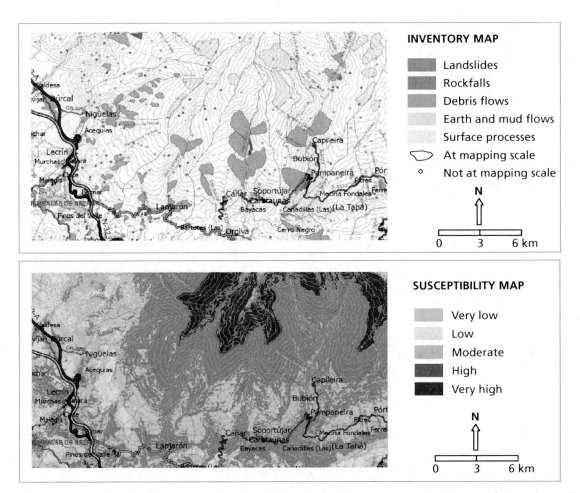

*Figure 15.3* Example of inventory and susceptibility maps for slope movements in the area of Granada, south Spain (courtesy of IGME).

affected and, depending on the detail required, may include the movement type, age, activity level, etc. It is essential that maps of external geodynamic processes contain topographic and geomorphologic information.

**Susceptibility** may be defined as the possibility an area has of being affected by a specific process, expressed at different qualitative and relative levels. This depends on the factors that control or condition the process occurrence, and which may be intrinsic or external to the geological materials.

**Susceptibility maps** can be drawn based on:

— Inventory maps: areas that experience or have experienced processes which may do so again.
— Factor maps: areas where specific factors come together which condition the processes in a specific zone or region; because even though these have not occurred up to the present they may occur in the future.

In this latter case, the basic methodology is to prepare thematic maps of the conditioning factors and, by layering them, to establish the susceptibility level as a function of the weighting given to each factor. These maps are usually prepared with GIS (Geographic Information Systems) techniques which allow automatic data analysis and the building of associated databases (*Figure 15.3*).

Inventory maps are drawn on regional or small scales: (1:100,000 or less) although for certain types of processes, such as landslides, subsidence or collapse, inventory maps are drawn on larger scales to show the features and characteristics of the movements. Susceptibility maps are usually medium scale (1:25,000 to 1:100,000) depending on the type of process, number of conditioning factors and their complexity, the data available, etc.

The general methodology for preparing **risk maps** is as follows (*Figure 15.4*):

— Estimate the hazard of the geological process under consideration, for a selected intensity or magnitude and a given time period (or return period); for this, the "where and when" of the process occurrence must be estimated.

# PREVENTION OF GEOLOGICAL HAZARDS

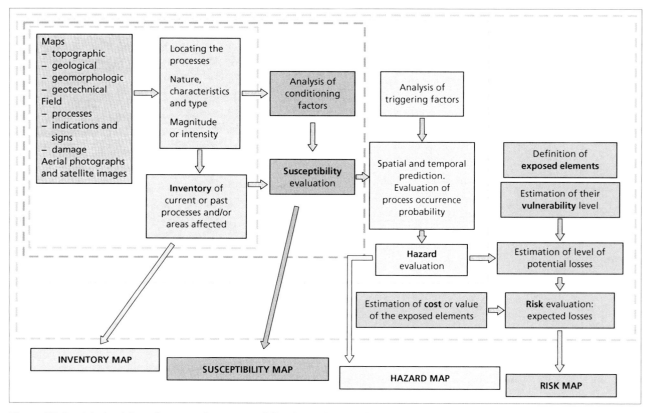

*Figure 15.4* Methodology for preparing susceptibility, hazard and risk maps.

— Identify and evaluate the social, structural and economic elements (and environmental and cultural elements if appropriate) which may be affected.
— Evaluate the social, structural and economic vulnerability (and environmental and cultural vulnerability if appropriate) of the elements exposed.
— Estimate the risk from the hazard and the vulnerability and cost or value of the elements, sets of elements or zones under consideration.

As well as the social and structural elements and economic activities, cultural elements can also be considered, such as historic monuments and buildings, and also environmental elements such as parks or protected areas. In both cases, estimating the value or "cost" is a difficult and complex task.

The data collected in the phases described can be represented on different maps (hazards, vulnerability and risk) or on an integrated map which reflects all the aspects. Hazard maps are generally produced on a medium scale (1:25,000) and risk maps on a more detailed scale.

Chapters 13 and 14 include some examples of susceptibility and hazard maps. The level of detail and the information contained in the map and its legend depend on the data available, the level of analysis and the scale of the map.

Because it is so difficult to make time-bound forecasts to estimate hazards and define probability, hazard is often expressed either qualitatively (high, medium or low hazard) by referring only to the spatial localization of the processes; thus many so-called hazard maps, and even risk maps, are in fact susceptibility maps.

## Recommended reading

Bell, F.G. (2003). Geological hazards. Their assessment, avoidance and mitigation. E & FN Spon, London.

Bryant, E. (2005). Natural hazards. 2nd ed. Cambridge Univ. Press.

Hutchinson, J.N. (2001). Reading the ground: morphology and geology in site appraisal. The 4th Glossop Lecture. Quarterly Journal of Engineering Geology and Hydrogeology, 34, 7–50.

Maund, J.G. and Eddleston, M. (Eds) (1998). Geohazards in engineering geology. The Geological Society, London. Engineering Geology Special Publications 15.

Smith, K. (2004). Environmental hazards. Assessing risk and reducing disaster. 4th ed. Routledge, London.

# References

Bell, F.G. (2003). Geological hazards. Their assessment, avoidance and mitigation. E & FN Spon, London.

Hoek, E. (1991). When is a design in rock engineering acceptable?. Proc. 7th Int. Conf. on Rock Mechanics. ISRM. Aachen, Germany. Vol. 3, pp. 1485–1497.

ISDR (2009). International Strategy for Disaster Reduction. www.unisdr.org.

Morgenstern, N.R. (1991). Limitations of stability analysis in geotechnical practice. Geotecnia, 61, pp. 5–19.

Murck, B.W., Skinner, B.J. and Porter, S.C. (1996). Environmental Geology. John Wiley and Sons.

Smith, K. (2001). Environmental hazards. Assessing risk and reducing disaster. 3rd Ed. Routledge, London.

Varnes, D.J. (1984). Landslide hazard zonation: a review of principles and practice. UNESCO.

# APPENDIX A

## CHARTS FOR CIRCULAR AND WEDGE FAILURE ANALYSIS

### Circular failure charts (Hoek and Bray, 1981)

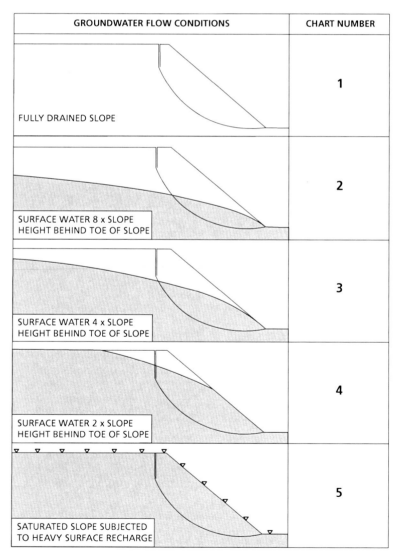

Groundwater flow conditions.

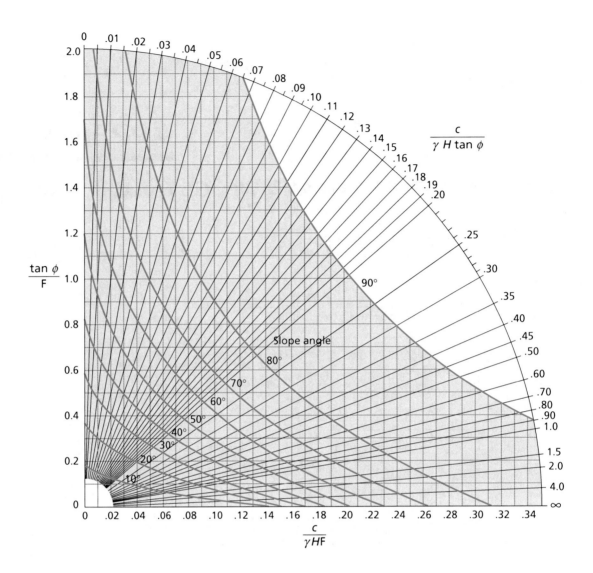

Circular failure chart no 1

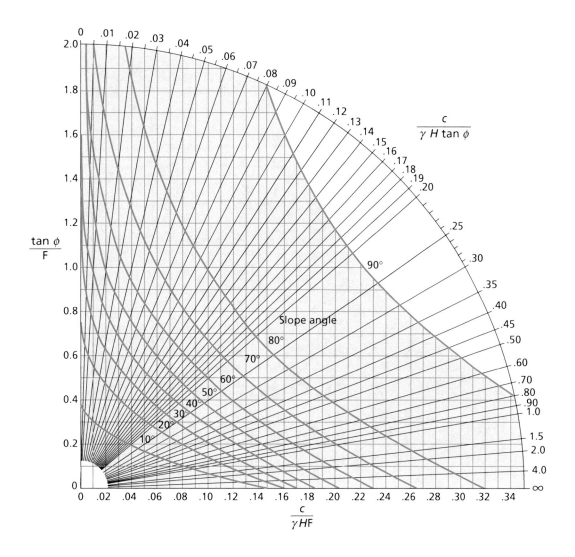

Circular failure chart no 2

# APPENDIX A

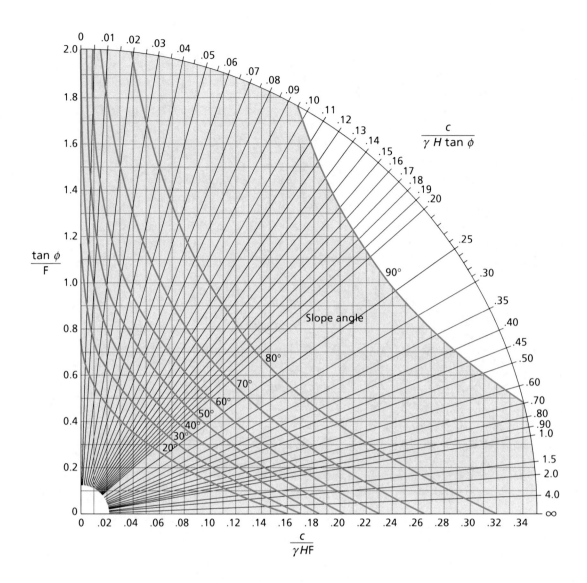

Circular failure chart no 3

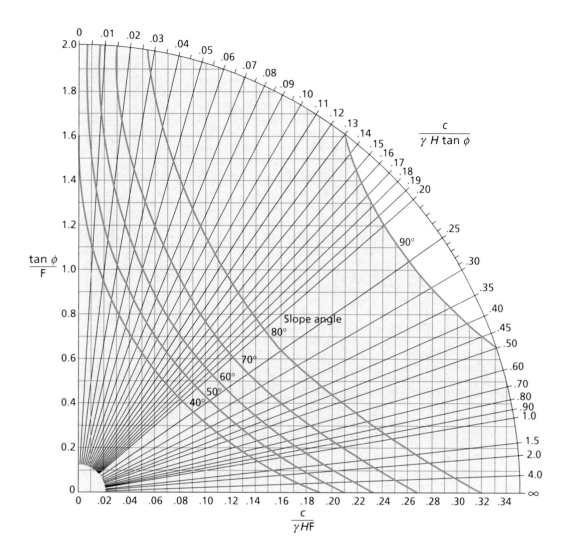

Circular failure chart no 4

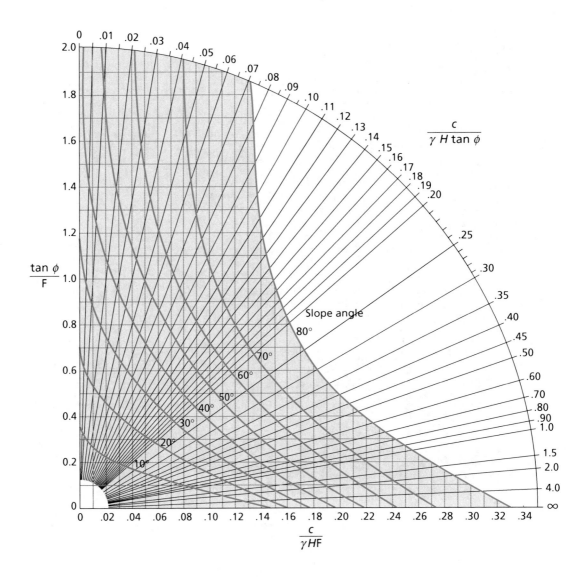

Circular failure chart no 5

# Wedge stability charts for friction only (Hoek and Bray, 1981)

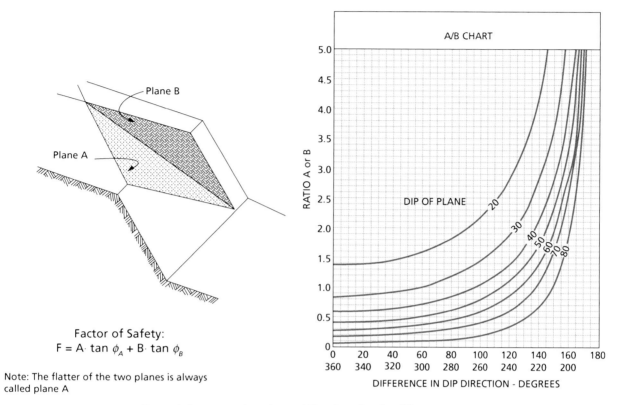

Factor of Safety:
$$F = A \cdot \tan \phi_A + B \cdot \tan \phi_B$$

Note: The flatter of the two planes is always called plane A

General diagram and wedge stability chart for dip difference = 0°

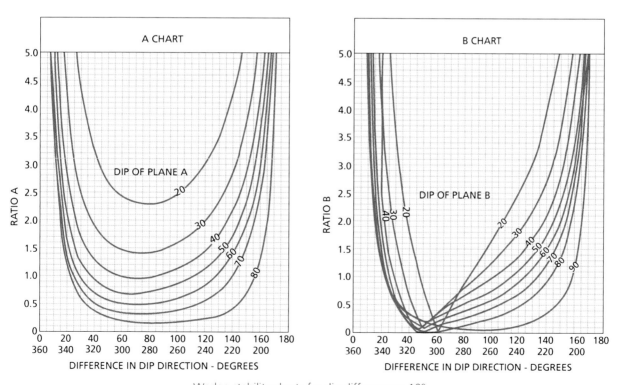

Wedge stability charts for dip difference = 10°

# APPENDIX A

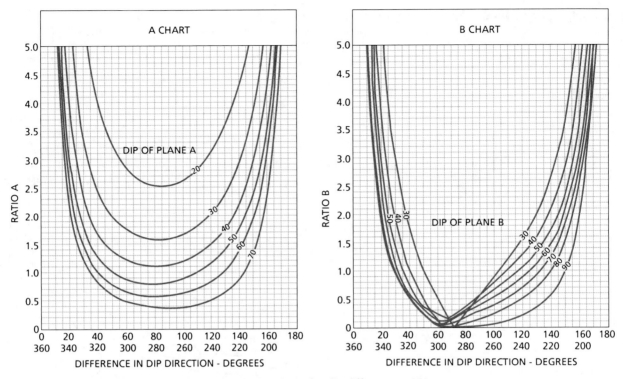

Wedge stability charts for dip difference = 20°

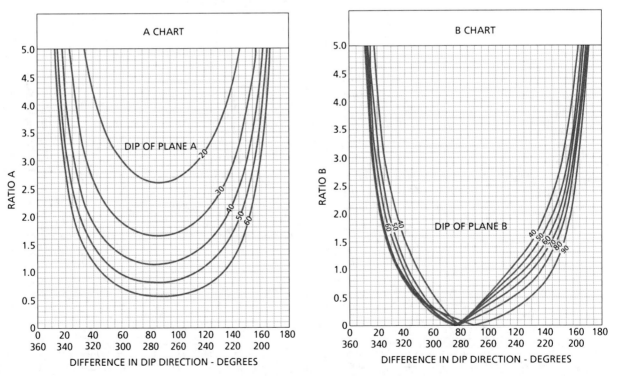

Wedge stability charts for dip difference = 30°

APPENDIX A

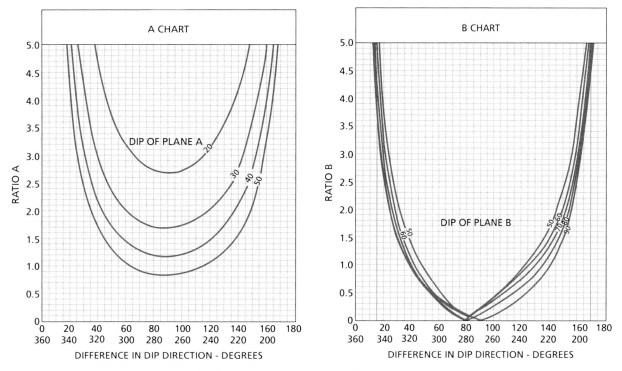

Wedge stability charts for dip difference = 40°

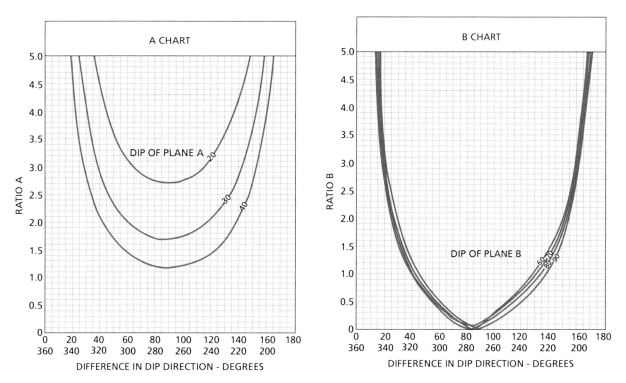

Wedge stability charts for dip difference = 50°

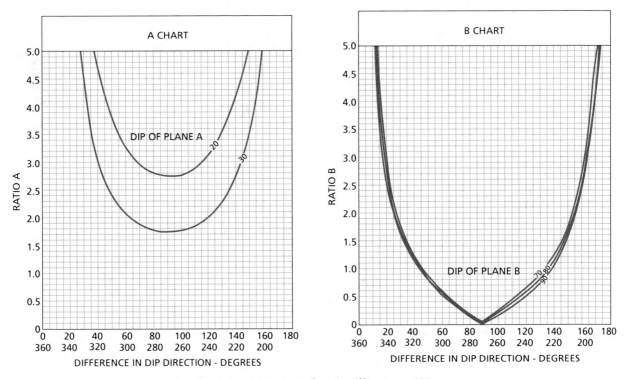

Wedge stability charts for dip difference = 60°

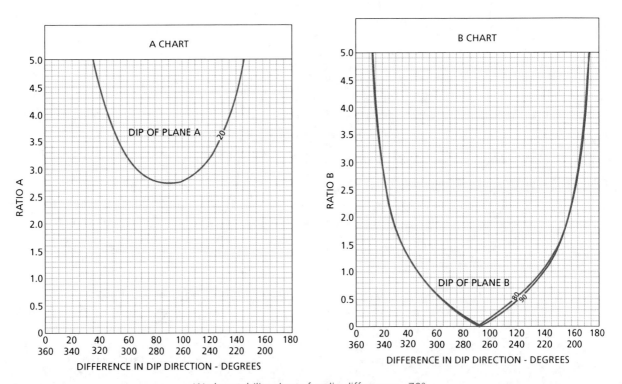

Wedge stability charts for dip difference = 70°

# APPENDIX B

## PRESSURE UNITS CONVERSION CHART

| | | | | | | |
|---|---|---|---|---|---|---|
| **kp/cm²** | 1.0 | 10 | $9.81 \times 10^4$ | 98.1 | $9.8 \times 10^{-2}$ | $9.81 \times 10^5$ |
| **t/m²** | 0.1 | 1.0 | $9.81 \times 10^3$ | 9.81 | $9.8 \times 10^{-3}$ | $9.81 \times 10^4$ |
| **N/m²** | $1.02 \times 10^{-5}$ | $1.02 \times 10^{-4}$ | 1.0 | $10^{-3}$ | $10^{-6}$ | 10 |
| **kN/m²** | $10^{-2}$ | 0.102 | $10^3$ | 1.0 | $10^{-3}$ | $10^4$ |
| **MN/m²** | 10.2 | $1.02 \times 10^2$ | $10^6$ | $10^3$ | 1.0 | $10^7$ |
| **dyne/cm²** | $10^{-6}$ | $1.02 \times 10^{-5}$ | 0.1 | $10^{-4}$ | $10^{-7}$ | 1.0 |
| **Pa** | $1.02 \times 10^{-5}$ | $1.02 \times 10^{-4}$ | 1.0 | $10^{-3}$ | $10^{-6}$ | 10 |
| **MPa** | 10.2 | $1.02 \times 10^2$ | $10^6$ | $10^3$ | 1.0 | $10^7$ |
| **GPa** | $1.02 \times 10^4$ | $1.02 \times 10^5$ | $10^9$ | $10^6$ | $10^3$ | $10^{10}$ |
| **bar** | 1.02 | 10.2 | $10^5$ | $10^2$ | 0.1 | $10^6$ |
| **kbar** | $1.02 \times 10^3$ | $1.02 \times 10^4$ | $10^8$ | $10^5$ | $10^2$ | $10^9$ |
| **atm** | 1.033 | 10.33 | $1.013 \times 10^5$ | $1.013 \times 10^2$ | 0.1013 | $1.013 \times 10^6$ |
| **psi** | $7 \times 10^{-2}$ | 0.7 | $6.89 \times 10^3$ | 6.89 | $6.89 \times 10^{-3}$ | $6.89 \times 10^4$ |
| | kp/cm² | t/m² | N/m² | kN/m² | MN/m² | dyne/cm² |

| | | | | | | |
|---|---|---|---|---|---|---|
| $9.81 \times 10^4$ | $9.8 \times 10^{-2}$ | $9.8 \times 10^{-5}$ | 0.981 | $9.8 \times 10^{-4}$ | 0.968 | 14.2 |
| $9.81 \times 10^3$ | $9.81 \times 10^{-3}$ | $9.8 \times 10^{-6}$ | $9.81 \times 10^{-2}$ | $9.8 \times 10^{-5}$ | $9.68 \times 10^{-2}$ | 1.42 |
| 1.0 | $10^{-6}$ | $10^{-9}$ | $10^{-5}$ | $10^{-8}$ | $9.9 \times 10^{-6}$ | $1.45 \times 10^{-4}$ |
| $10^3$ | $10^{-3}$ | $10^{-6}$ | $10^{-2}$ | $10^{-5}$ | $9.87 \times 10^{-3}$ | 0.145 |
| $10^6$ | 1.0 | $10^{-3}$ | 10 | $10^{-2}$ | 9.87 | 145.2 |
| 0.1 | $10^{-7}$ | $10^{-10}$ | $10^{-6}$ | $10^{-9}$ | $9.9 \times 10^{-7}$ | $1.45 \times 10^{-5}$ |
| 1.0 | $10^{-6}$ | $10^{-9}$ | $10^{-5}$ | $10^{-8}$ | $9.9 \times 10^{-6}$ | $1.45 \times 10^{-4}$ |
| $10^6$ | 1.0 | $10^{-3}$ | 10 | $10^{-2}$ | 9.87 | 145.2 |
| $10^9$ | $10^3$ | 1.0 | $10^4$ | 10 | $9.87 \times 10^3$ | $1.451 \times 10^5$ |
| $10^5$ | 0.1 | $10^{-4}$ | 1.0 | $10^{-3}$ | 0.987 | 14.52 |
| $10^8$ | $10^2$ | 0.1 | $10^3$ | 1.0 | $9.87 \times 10^2$ | $1.452 \times 10^4$ |
| $1.012 \times 10^5$ | 0.1012 | $1.012 \times 10^{-4}$ | 1.013 | $1.013 \times 10^{-3}$ | 1.0 | 14.66 |
| $6.89 \times 10^3$ | $6.89 \times 10^{-3}$ | $6.9 \times 10^{-6}$ | $6.9 \times 10^{-2}$ | $6.9 \times 10^{-5}$ | $6.8 \times 10^{-2}$ | 1.0 |
| **Pa** | **MPa** | **GPa** | **bar** | **kbar** | **atm** | **psi** |

# APPENDIX C

## SYMBOLS AND ACRONYMS

## Symbols

When the symbols in the following list have different meaning than that described here, this is indicated in the text. The symbols not included in the list are defined in the text where they appear.

| | |
|---|---|
| $A$ | Area |
| | Pore pressure parameter (Skempton coefficient) |
| | Constant depending on the value of $m$ (H-B criterion) |
| $a$ | Acceleration |
| | Seismic parameter (Gutemberg-Richter relationship) |
| | Aperture of a discontinuity |
| $B$ | Pore pressure parameter (Skempton coefficient) |
| | Constant depending on the value of $m$ (H-B criterion) |
| | Volumetric elasticity modulus of water |
| | Leakage factor |
| | Width |
| $b$ | Aquifer thickness |
| | Seismic parameter (Gutemberg-Richter relationship) |
| | Spacing of discontinuities |
| $C$ | Shape factor (Lefranc test) |
| | Penetration factor (calculation of discharge in tunnels) |
| | Cost of the exposed elements (risk analysis) |
| $C_j$ | % of material that passes through a sieve with diameter $D_j$ |
| $C_u$ | Coefficient of uniformity of soil particles |
| $C, C'$ | Cohesion (total, effective) |
| $c_c$ | Compression index |
| $C_p, C_p'$ | Peak cohesion (total, effective) |
| $C_r, C_r'$ | Residual cohesion (total, effective) |
| $c_s$ | Swelling index |
| $c_v$ | Coefficient of consolidation |
| $D$ | Depth |
| | Distance |
| $D_e$ | Equivalent diameter |
| | Cone diameter |
| $D_j$ | Sieve diameter |
| $D_r$ | Relative density |
| $d$ | Drawdown in a well |
| | Distance |
| | Diameter |
| $d_c, d_q, d_\gamma$ | Depth correction factors |
| $E$ | Modulus of deformation, elastic modulus or Young's modulus |
| | Efficiency factor |
| | Energy |
| | Exposure of the elements at risk (risk analysis) |
| $E_D$ | Dilatometric deformation modulus |
| $E_d$ | Dynamic deformation modulus |
| $E_h, E_k, E_p$ | Potential, kinetic and pressure energy |
| $E_i$ | Modulus of deformation of the intact rock |
| $E_m$ | Oedometric modulus |
| $E_P$ | Presiometric deformation modulus |
| $e$ | Void ratio |
| | Spacing of discontinuities |
| | Eccentricity |
| $F$ | Force |
| | Factor of safety |
| | Abrasiveness factor |
| $F_c$ | Competence factor of the intact rock (SRC classification) |
| $F_t$ | Tensile force |
| $FS$ | Factor of safety |
| $f_l, f_t$ | Correction factor (footing on granular soils) |
| $f_n$ | Tangential stress due to negative friction (piles) |

| | | | |
|---|---|---|---|
| $f_s$ | Lateral friction (CPT) | $M_w$ | Moment magnitude |
| | Shape coefficient (foundations on granular soils) | $m$ | Mass |
| | | | Rock mass constant (H-B criterion) |
| $G$ | Specific gravity of solid soil particles | $m_i$ | Intact rock mass constant (H-B criterion) |
| $g$ | Gravity acceleration | $m_v$ | Coefficient of volumetric compressibility |
| $H$ | Height | $N$ | Normal force on a plane |
| | Total hydraulic head | | Number of blows (SPT) |
| | Maximum overburden thickness (TSI index) | $N_c, N_q, N_\gamma$ | Bearing capacity factors |
| $h$ | Piezometric level or head | $N_d$ | Number of hydraulic potential drops in a flow net |
| $I$ | Intensity of a earthquake | $N_e, N_s$ | Stability number (Taylor's method) |
| $I_b$ | Block size index | $N_f$ | Number of flow channels in a flow net |
| $I_c$ | Compressibility index | $n$ | Porosity |
| $I_D$ | Density index | $n_e$ | Effective porosity |
| | Durability index | $P$ | Load or force |
| $I_{dis}$ | Dispersion index | | Pressure |
| $I_L$ | Liquidity index | | Probability |
| $I_s$ | Point load strength index (PLT) | | Annual exceedance probability |
| $i$ | Hydraulic gradient or piezometric gradient | $P_{(a)}$ | Annual exceedance probability |
| | Roughness angle in discontinuities | $P_{cr}$ | Critical bearing load (buckling) |
| $i_C$ | Critical hydraulic gradient | $P_F$ | Yield pressure (pressuremeter test) |
| $J_a$ | Joint alteration number (Q system) | $P_f$ | Fracture initiation pressure or breakdown pressure (hydraulic fracturing test) |
| $J_n$ | Joint set number (Q system) | | |
| $J_r$ | Joint roughness number (Q system) | $P_r$ | Fracture reopening pressure (hydraulic fracturing test) |
| $J_v$ | Volumetric joint count | | |
| $J_w$ | Joint water reduction number (Q system) | $P_s$ | Shut-in pressure (hydraulic fracturing test) |
| $K$ | Intrinsic permeability | $P_{ult}$ | Ultimate load capacity (foundation on rock) |
| | Stress ratio ($\sigma_H/\sigma_V$) | $PI$ | Plasticity index |
| | Stiffness | $PL$ | Plastic limit |
| $K_s$ | Ballast coefficient | $p$ | Probability |
| $K_0$ | Coefficient of earth pressure at rest | $Q$ | Discharge |
| $k$ | Coefficient of permeability, effective permeability, hydraulic conductivity or Darcy permeability | | Volume of water |
| | | | Load |
| | Stiffness of discontinuities | | Rock mass quality index (Q system) |
| | | | Quartz content (abrasiveness index) |
| $L$ | Length | $Q_a$ | Allowable bearing or load capacity (piles) |
| $LI$ | Liquidity index | $Q_{ult}$ | Ultimate load capacity (piles) |
| $LL$ | Liquid limit | $q_a, q'_a$ | Allowable bearing or load capacity (total, effective) |
| $l$ | Length | $q_c$ | Cone resistance (CPT) |
| $l, m, n$ | Direction cosines | $q_g, q'_g$ | Gross pressure (total, effective) |
| $M$ | Moment | $q_{net}, q'_{net}$ | Net pressure (total, effective) |
| | Stiffness | | |
| | Magnitude of an earthquake | $q_s$ | Safe bearing capacity |
| $M_L$ | Local Richter magnitude | $q_{shaft}$ | Ultimate shaft load capacity (piles) |
| $M_S$ | Shear wave magnitude | | |

| | | | |
|---|---|---|---|
| $q_{tip}$ | Ultimate end load capacity at the tip (piles) | $V_L$ | Velocity of the longitudinal elastic waves measured in laboratory |
| $q_u$ | Uniaxial compressive strength (soils) | $V_p$ | Velocity of the longitudinal or compression elastic waves |
| $q_{ult}$ | Ultimate bearing capacity (shallow foundations) | | |
| $q_0$ | Total weight of the ground above the foundation | $V_{Rayleigh}$ | Rayleigh waves velocity |
| $R$ | Rebound (Schmidt hammer, intact rock) | $V_s$ | Velocity of the transversal or shear elastic waves |
| | Rebound (SPT) | $v$ | Velocity |
| | Radius of influence (pumping tests) | $W$ | Weight |
| | Distance | | Total dry weight of the soil sample |
| $R_c$ | Resultant cohesive force on a plane | | Strain energy (uniaxial compression test) |
| $R_\phi$ | Resultant frictional force on a plane | $W_{(u)}$ | Well function |
| $r$ | Radius | $w$ | Water content |
| | Rebound (Schmidt hammer, discontinuities) | $Z$ | Position head (Bernoulli's theorem) |
| $r_u$ | Pore pressure ratio | | Depth, height |
| $S$ | Area | $z$ | Depth, height |
| | Settlement | $\alpha$ | Angle between the failure plane and the horizontal |
| | Storage coefficient | | Dip angle of a discontinuity |
| | Saturation | | Dimensionless constant (H-B criterion) |
| | Tangential or shear force on a plane | $\beta$ | Angle between the plane in question and the direction of the major principal stress $\sigma_1$ ($\beta = 90° - \theta$) |
| $S_c$ | Primary consolidation settlement | | |
| $S_i$ | Immediate settlement | | |
| $S_r$ | Degree of saturation | | Compressibility of water |
| $S_s$ | Secondary consolidation settlement | $\gamma$ | Unit weight or specific weight |
| $S_t$ | Total settlement | $\gamma_{ap}$ | Bulk unit weight |
| $S_u$ | Undrained shear strength | $\gamma_d$ | Dry unit weight |
| $St$ | Sensitivity | $\gamma_{max}$ | Maximum dry density |
| $s$ | Rock mass constant (H-B criterion) | $\gamma_{min}$ | Minimum dry density |
| $s_c, s_q, s_\gamma$ | Shape correction factors (shallow foundations) | $\gamma_s$ | Unit weight of solids |
| $T$ | Transmissivity | $\gamma_{sat}$ | Saturated unit weight |
| | Torque moment | $\gamma_w$ | Unit weight of water |
| | External force applied to a slope | $\delta'$ | Effective ground-pile friction angle |
| | Anchor force | $\varepsilon$ | Deformation, strain |
| | Age of the main orogeny (TSI index) | $\varepsilon_l$ | Longitudinal strain |
| | Return period | $\varepsilon_{ax}$ | Axial strain |
| $T_v$ | Time factor (consolidation) | $\varepsilon_t, \varepsilon_r$ | Radial, transversal strain |
| $t$ | Time | $\varepsilon_v$ | Volumetric deformation |
| $U$ | Degree of consolidation | | Vertical deformation |
| | Force due to water pressure on a plane | $\eta$ | Kinematic viscosity coefficient of water |
| $u$ | Pore water pressure | $\theta$ | Angle between the normal to the plane in question and the direction of the major principal stress $\sigma_1$ ($\theta = 90° - \beta$) |
| | Displacement | | |
| $V$ | Volume | | |
| | Force due to water pressure on the tension crack | $\lambda$ | Discontinuity frequency |
| $V_F$ | Velocity of the longitudinal elastic waves measured in the field | $\mu$ | Dynamic viscosity coefficient of water |

| | | |
|---|---|---|
| $\nu$ | Poisson ratio | |
| $\nu_d$ | Dynamic Poisson ratio | |
| $\rho$ | Density | |
| $\rho_{dry}$ | Dry density | |
| $\rho$ | Resistivity | |
| $\rho_a$ | Apparent resistivity | |
| $\sigma, \sigma'$ | Stress (total, effective) | |
| $\sigma_c$ | Uniaxial compressive strength | |
| $\sigma_{ci}$ | Uniaxial compressive strength of intact rock | |
| $\sigma_H$ | Horizontal principal mayor stress | |
| $\sigma_h$ | Horizontal stress / Horizontal principal minor stress | |
| $\sigma_{max}$ | Maximum stress | |
| $\sigma_{min}$ | Minimum stress | |
| $\sigma_n$ | Normal stress | |
| $\sigma_p$ | Peak strength | |
| $\sigma'_p$ | Pre consolidation pressure | |
| $\sigma_r$ | Residual strength | |
| $\sigma_t$ | Tensile strength / Tangential stress | |
| $\sigma_V, \sigma'_V, \sigma_v, \sigma'_v$ | Vertical stress (total, effective) | |
| $\sigma_x, \sigma_y, \sigma_z$ | Stresses along axis x, y, z | |
| $\sigma_y$ | Yield strength | |
| $\sigma_z$ | Vertical stress at depth z | |
| $\sigma_1, \sigma_2, \sigma_3$ | Principal stresses, mayor, intermediate and minor | |
| $\tau$ | Shear strength / Tangential or shear stress | |
| $\tau_{cm}$ | Average cyclic shear stress | |
| $\tau_f$ | Failure tangential stress | |
| $\tau_{max}$ | Maximum tangential stress | |
| $\tau_p$ | Peak shear strength | |
| $\phi, \phi'$ | Angle of internal friction, angle of friction, angle of shearing resistance (total, effective) | |
| $\phi_b$ | Basic angle of friction | |
| $\phi_m$ | Mobilized angle of friction | |
| $\phi_p$ | Peak angle of friction | |
| $\phi_r$ | Residual angle of friction | |
| $\psi$ | Angle of the slope with the horizontal | |

# Acronyms

| | |
|---|---|
| AEG | Association of Engineering Geologists |
| ASCE | American Association of Civil Engineers |
| ASTM | American Society for Testing and Materials |
| BS | British Standards |
| CBR | Californian Bearing Ratio |
| CH | Clays of High plasticity |
| CL | Clays of Low plasticity |
| CPT | Cone Penetration Test |
| CPTU | Cone Penetration Test Undrained |
| CRR | Cyclic Resistance Ratio |
| CSR | Cyclic Shear stress Ratio |
| EI | Excavation Index |
| EM | Electromagnetic |
| ESR | Excavation Support Ratio (Q system) |
| FS | Factor of Safety |
| GIS | Geographical Information System |
| GPR | Ground Penetration Radar |
| GSI | Geological Strength Index |
| IAEA | International Atomic Energy Agency |
| IAEG | International Association of Engineering Geology and Environment |
| ICOLD | International Commission of Large Dams |
| IGME | Instituto Geológico y Minero de España |
| ISRM | International Society of Rock Mechanics |
| ISSMGE | International Society of Soil Mechanics and Geotechnical Engineering |
| JCS | Joint wall Compressive Strength |
| JRC | Joint Roughness Coefficient |
| LL | Liquid Limit |
| LU | Lugeon Unit |
| MH | Silt of High plasticity |
| ML | Silt of Low plasticity |
| NATM | New Austrian Tunnelling Method |
| NC | Coefficient of seismotectonic activity (TSI index) |
| OCR | Overconsolidation Ratio |
| OH | Organic clay or silt of High plasticity |
| OL | Organic clay or silt of Low plasticity |

| | | | |
|---|---|---|---|
| PI | Plasticity Index | SPT | Standard Penetration Test |
| PL | Plastic Limit | SRC | Surface Rock mass Classification |
| PLT | Point Load Test | SRF | Stress Reduction Factor (Q system) Stress Relief Factor (SRC classification) |
| ReMi | Refraction Microtremor | TBM | Tunnel Boring Machine |
| REV | Representative Elemental Volume | TSI | Tectonic Stress Index |
| RMR | Rock Mass Rating | UCS | Uniaxial Compressive Strength |
| RQD | Rock Quality Designation | USCS | Unified Soil Classification System |
| SAR | Synthetic Aperture Radar | USNRC | United States Nuclear Regulatory Commission |
| SASW | Spectral Analysis of Surface Waves | VES | Vertical Electric Sounding |
| SC | Coefficient of topographic influence (TSI index) | VLF | Very Low Frequency |
| SDT | Slake Durability Test | WCD | World Commission of Dams |
| SMR | Slope Mass Rating | WT | Water Table |
| SP | Spontaneous potential | | |

# APPENDIX D

## LIST OF BOXES

**Box 1.1** Geological engineering: education and professional practice — 5
**Box 1.2** El Berrinche landslide, Tegucigalpa (Honduras) — 7
**Box 1.3** The failure of Aznalcóllar Dam: An example of underestimation of the geological and geotechnical conditions with serious environmental consequences — 14
**Box 2.1** Using open pipe piezometers — 30
**Box 2.2** Calculating pore water pressure. Worked example — 32
**Box 2.3** Flow net in an anisotropic medium. Worked example — 37
**Box 2.4** Permeability calculation. Worked example — 39
**Box 2.5** Shear stress and Terzaghi's principle — 42
**Box 2.6** Stress in a homogeneous soil layer. Worked example — 43
**Box 2.7** Stress in stratified soils. Worked example — 45
**Box 2.8** Piping conditions. Worked example — 49
**Box 2.9** Stress distribution. Worked example — 54
**Box 2.10** Vertical and volumetric strain in one-dimensional conditions — 58
**Box 2.11** Calculating the degree of overconsolidation. Worked example — 59
**Box 2.12** Calculating settlement. Worked example — 63
**Box 2.13** Calculating the coefficient of earth pressure at rest ($K_0$) and horizontal stress. Worked example — 64
**Box 2.14** Calculating consolidation time. Worked example — 69
**Box 2.15** Settlement-time curves. Worked example — 70
**Box 2.16** Calculating shear and principal stresses. Worked example — 73
**Box 3.1** Rock to soil transition — 114
**Box 3.2** Intact rock, discontinuities and rock mass — 115
**Box 3.3** Physical and mechanical properties of rocks — 117
**Box 3.4** Principal stresses — 135
**Box 3.5** Graphical and analytical methods for calculating normal and shear stresses acting on a plane — 137
**Box 3.6** Models for stress-strain behaviour in rocks — 145
**Box 3.7** Calculating the elastic constants for the rock: Young's modulus, $E$, and Poisson's ratio, $v$ — 157
**Box 3.8** Example of the calculation of strength parameters $c$ and $\phi$ from triaxial tests — 163
**Box 3.9** Calculating the strength parameters $c$ and $\phi$ for discontinuities — 178
**Box 3.10** Variation in the relationship $\sigma_H/\sigma_V$ due to erosion — 206
**Box 3.11** Determining stress direction through stress relief methods in outcrops — 208
**Box 3.12** Example of a hydraulic fracturing test in a deep borehole — 214
**Box 5.1** RQD calculation — 298
**Box 5.2** Uniaxial strength assessed using the Schmidt hammer — 307
**Box 5.3** Uniaxial strength calculated using the Point Load Test (PLT) — 309
**Box 6.1** Evaluation of discontinuity shear strength from field data — 343
**Box 8.1** Calculating the ultimate bearing capacity — 375
**Box 8.2** Calculating the effective ultimate bearing capacity — 376
**Box 8.3** Calculating the effective ultimate bearing capacity and safe ultimate bearing capacity — 377
**Box 8.4** Calculating the ground stress distribution — 381
**Box 8.5** Estimating settlement — 385
**Box 9.1** Calculating water pressures in a slope using a flow net — 407
**Box 9.2** Example of the application of Taylor's method — 422
**Box 9.3** Calculation of the factor of safety of a soil slope using Hoek and Bray charts — 423
**Box 9.4** Bishop's simplified method — 425
**Box 9.5** Calculation of the factor of safety of a wedge using Hoek and Bray charts — 429
**Box 9.6** Monitoring of movements in an unstable slope — 446
**Box 10.1** Guidelines for planning site investigation in tunnels — 456

**Box 10.2** Calculation of the geomechanical parameters of a rock mass for tunnel design purposes ... 467
**Box 10.3** Calculation of the discharge in a tunnel using the Goodman method (Goodman et al., 1965) ... 471
**Box 10.4** Calculation of the state of stress parameter of the SRC classification ... 478
**Box 11.1** Suggested criteria for site investigations of dams ... 511
**Box 11.2** Influences of water level fluctuations on the stability of the reservoirs slopes ... 522
**Box 11.3** Failure mechanism in the Aznalcóllar Dam (Spain) ... 524
**Box 11.4** Sliding failure in gravity dam foundations ... 529
**Box 12.1** Testing earth materials ... 539
**Box 12.2** Soil classification for earth fill embankments based on Spanish Codes ... 542
**Box 13.1** Benamejí landslide, south Spain ... 579
**Box 13.2** Example of subsidence caused by water table drawdown in the city of Murcia, eastern Spain ... 588
**Box 14.1** Sedimentary structures and other ground effects originated by earthquakes ... 603
**Box 14.2** Calculation of liquefaction potential: worked example ... 615
**Box 14.3** The Kocaeli (Turkey) earthquake of August 17th, 1999 ... 621
**Box 15.1** Examples of risk evaluation ... 630
**Box 15.2** Example of geological safety analysis ... 637

# APPENDIX E

# PERMISSIONS TO REPRODUCE FIGURES AND TABLES

## Chapter 2

**Figure 2.24 and Figure 2.37** Lancellotta, R. (1995). Geotechnical Engineering. Taylor & Francis. The Netherlands. Reproduced by permission and courtesy from Taylor & Francis Group.

**Figure 2.32** Skempton, A.W. (1970). The consolidation of clays by gravitational compaction. Quart. J. Geol. Soc., vol. 125, part. 3, no. 499, pp. 373–412, London. Reproduced by permission and courtesy from the Geological Society of London.

**Figure 2.63** Day, R.W. (1999). Geotechnical and foundation engineering: Design and Construction. McGraw-Hill. Reproduced by permission from McGraw-Hill.

**Figure 2.64** Tsige, M., González de Vallejo, L.I., Doval, M. and Oteo, C. (1995). Microfabric of Guadalquivir blue marls and its engineering significance. Proc. 7th Int. Congress of Eng. Geol. IAEG. Lisbon. Balkema. Vol. II., pp. 655–704. Reproduced by permission and courtesy from A.A. Balkema.

**Figure 2.75** Bennet, R.H. and Hulbert, M.H. (1986). Clay microstructure. Int. Human Resources Dep. Co. Boston, Houston, London. Reproduced by permission and courtesy from IHRDC Publishers.

**Figure 2.96** Sherard, J.L., Dunnigan, L.P., Decker, R.S. and Steele. E.F. (1976). Pinhole test for identifying dispersive soils. Journal of The Geotechnical Eng. Division, ASCE, vol. 102 (GT1), pp. 69–85. Reproduced by permission from the American Society of Civil Engineers.

## Chapter 3

**Figure 3.9** Attewell, P.B. and Farmer, I.W. (1976). Principles of engineering geology. Chapman and Hall, London. Reproduced by permission and courtesy from Springer.

**Figure 3.11** Embleton, C. and Thornes, J.B. (1979). Process in geomorphology. Arnold, London. Reproduced by permission and courtesy from Hodder and Stoughton Ltd.

**Figure 3.18, Figure 3.112 and Figure 3.114** Hudson, J.A. and Harrison, J.P. (2000). Engineering rock mechanics. An introduction to the principles. Pergamon Press. Reproduced by permission and courtesy from Elsevier.

**Figure 3.56, Figure 3.61, Figure 3.84 and Figure 3.96** Brady, B.H.G. and Brown, E.T. (1993). Rock mechanics for underground mining. 2nd ed. Kluwer Academic Publishers. Reproduced by permission and courtesy from Springer.

**Figure 3.57** Wawerssick, W.R. and Fairhurst, C. (1970). A study of brittle rock failure in laboratory compression experiments. Int. J. of Rock Mech. and Min. Sci., vol. 7, no. 5. pp. 561–575. Reproduced by permission and courtesy from Elsevier.

**Figure 3.59, Figure 3.95, Figure 3.106 and Figure 3.115** Hoek E. and Brown E.T. (1980). Underground excavation in Rock. The Institution of Mining and Metallurgy, London. Reproduced by permission from the Institute of Materials, Minerals and Mining.

**Figure 3.66 and Figure 3.126** Johnson, R.B. and De Graff, J.V. (1988). Principles of engineering geology. John Wiley & Sons Ed. Reproduced by permission from John Wiley & Sons Ltd.

**Figure 3.70 and Figure 3.74** Blyth, E. and de Freitas, M. (1984). Geology for engineers. Ed. Edward Arnold, London. Reproduced by permission and courtesy from Hodder and Stoughton Ltd.

**Figure 3.77** ISRM (1981). Rock characterization. Testing and monitoring. Int. Soc. for Rock Mechanics. Suggested methods. Brown, E.T. (Ed.). Commission on testing and monitoring, ISRM. Pergamon Press. Reproduced by permission from Elsevier.

**Figure 3.83, Figure 3.117 and Figure in Box 3.10** Goodman, R.E. (1989). Introduction to rock mechanics. 2nd ed. John Wiley & Sons Ed. Reproduced by permission and courtesy from John Wiley & Sons Ltd.

**Figure 3.86** Bandis, S.C., Lumsden, A.C. and Barton, N. (1981). Experimental studies of scale effects on the shear behaviour of rock joints. Int. J. of Rock Mech. and Min. Sci., Abstracts, vol. 18, pp. 1–21. Reproduced by permission and courtesy from Elsevier.

**Figure 3.87 and Figure 3.90** Hoek, E. and Bray, J.W. (1981). Rock slope engineering. 3rd ed. The Institution of Mining and Metallurgy, London. Reproduced by permission from the Institute of Materials, Minerals and Mining.

**Figure 3.89** ISRM (1981). Rock characterization. Testing and monitoring. Int. Soc. for Rock Mechanics. Suggested methods. Brown, E. T. (Ed.). Commission on testing and monitoring, ISRM. Pergamon Press. Reproduced by permission and courtesy from Elsevier.

**Figure 3.92 and Figure 3.100** Hoek, E and Brown, E.T. (1997). Practical estimates of rock mass strength. Int. J. of Rock Mech. and Min. Sci., vol. 34, no. 8, pp. 1165-1186. Reproduced by permission and courtesy from Elsevier.

**Figure 3.94 and Table 3.14** Hoek, E. and Marinos, P. (2000). Predicting tunnel squeezing (Problems in weak heterogeneous rock masses). Tunnels and Tunnelling Int. Part 1 Estimating rock mass strength. V. 32:11, pp. 45–51. Reproduced by permission and courtesy from Tunnels & Tunnelling.

**Figure 3.97** Zhang, L. and Einstein, H.H. (2004). Using RQD to estimate the deformation modulus of rock masses. Int. J. of Rock Mech. and Min. Sci. 41, pp. 337–341. Reproduced by permission from Elsevier.

**Figure 3.98** Coon, R.F. and Merritt, A.H. (1970). Predicting in situ modulus of deformation using rock quality indexes. Am. Soc. Test. Mater. (ASTM), Spec. Tech. Publ. 477, pp. 154–173. Reproduced by permission and courtesy from ASTM International.

**Figure 3.99 and Figure 3.105** Bieniawski, Z.T. (1984). Rock mechanic design in mining and tunnelling. Balkema. Reproduced by permission and courtesy from A.A. Balkema.

**Figure 3.101** Barton, N. (1995). The influence of joints properties in modelling jointed rock masses. Keynote lecture. Proc. 8th ISRM Congress. Fuji, T. (Ed.). pp. 1023–1032. Balkema. Reproduced by permission and courtesy from A.A. Balkema.

**Figure 3.102** Hoek, E. and Diederichs, M.S. (2006). Empirical estimation of rock mass modulus. Int. J. of Rock Mech. and Min. Sci., 43, pp. 203–215. Reproduced by permission and courtesy from Elsevier.

**Figure 3.103 and Figure 3.107** Cunha, A.P. (1990). Scale effects in rock mechanics. In: Scale effects in rock masses. Cunha, A.P. (Ed.). Balkema. Reproduced by permission and courtesy from A.A. Balkema.

**Figure 3.108 and 3.109** Barton, N. (1990). Scale effects or sampling bias?. In: Scale effects in rock mechanics. Cunha, A.P. (Ed.). Balkema. Reproduced by permission and courtesy from A.A. Balkema.

**Figure 3.110** Cunha, A.P. and Muralha, J. (1990). About LNEC experience on scale effects in the deformability of rock masses. In: Scale effects in rock masses. Cunha, A.P. (Ed.). Balkema. Reproduced by permission and courtesy from A.A. Balkema.

**Figure 3.111** Natau, O. (1990). Scale effects in the determination of the deformability and strength of rock masses. In: Scale effects in rock masses. Cunha, A.P. (Ed.). Balkema. Reproduced by permission and courtesy from A.A. Balkema.

**Figure 3.113** Haimson, B.C. (1990). Scale effects in rock stress measurements. In: Scale effects in rock masses. Cunha. A.P. (Ed.). Balkema. Reproduced by permission and courtesy from A.A. Balkema.

**Figure 3.119** Selmer-Olsen, R. and Broch, E. (1977). General design procedure for underground openings in Norway. Rockstorage 77. 1st Int. Symp. on Storage in Excavated Rock Caverns, Sweden. Vol. 2 (11–22). Reproduced by permission and courtesy from Elsevier.

**Figure 3.123** Herget, G. (1988). Stresses in rock. Balkema. Reproduced by permission and courtesy from A.A. Balkema.

**Figure 3.127 and Figure 3.128** Kim, K. and Franklin, J.A. (1987). Suggested methods for rock stress determination. Int. J. of Rock Mech. and Min. Sci. Geomechanical abstracts. 24-1, pp. 53–73. Reproduced by permission and courtesy from Elsevier.

## Chapter 5

**Figure 5.2** Fookes, P.G. (1997). Geology for engineers: the logical model; prediction and performance. The First Glossop Lecture. Geological Society of London. Ql. Jl. Engineering Geology, vol. 30, n° 4, pp. 293–424. Reproduced by permission and courtesy from the Geological Society of London.

**Figure 5.5** Photogrammetric engineering and remote sensing (2000). Amer. Soc. of Photogrammetric and Remote Sensing, vol. 66, n° 4. Reproduced by permission and courtesy from Amer. Soc. of Photogrammetric and Remote Sensing.

**Figure 5.8** Landsat Data Users Notes (1993). EOSAT Ed. Vol. 8, n°2. Reproduced by permission and courtesy from Landsat Data Users Notes. Landsat.org, Global Observatory for Ecosystem Services, Michigan State University (http://landsat.org).

# APPENDIX E

**Figure 5.32, Figure 5.33 and Figure in Box 5.1**  Clayton, C.R.I., Matthews, M.C. and N.E. Simons (1995). Site investigation. Blackwell Science. Reproduced by permission and courtesy from Blackwell Science Ltd.

**Figure 5.63**  Thornburn, S. (1963). Tentative correction chart for the standard penetration test in non cohesive soils. Civ. Eng. and Publics Works 58; 683: 752–753. Reproduced by permission and American Society of Civil Engineers.

**Figure 5.64**  de Mello, V.F.B. (1971). Standard penetration test. 4th Pan-American Congress on Soil Mechanics and Foundation Engineering. Vol. I, pp. 1–86. Puerto Rico. Reproduced by permission from American Society of Civil Engineers.

**Figure 5.69**  Robertson, P.K. and Campanella, R.G. (1983). Interpretation of cone penetration test. Part I. Sand. Canadian Geotechnical Journal, 20, 4, pp. 718–733. Reproduced by permission and courtesy from NRC Research Press. © 2008 NRC Canada.

**Figure 5.74**  ISRM (1981). Rock characterization. Testing and monitoring. Int. Soc. for Rock Mechanics. Suggested methods. Brown, E.T. (Ed.). Commission on testing and monitoring, ISRM. Pergamon Press. Reproduced by permission from Elsevier.

**Figure 5.76**  Barton, N. (1981). Shear strength investigations for surface mining. 3rd Int. Conference on Stability in Surface Mining. Vancouver. Pp. 171–196. June 1981. Reproduced by permission and courtesy from the Society of Mining Engineers.

**Figure 5.85-A and Figure 5.85-C**  Kim, K. and Franklin, J.A. (1987). Suggested methods for rock stress determination. Int. J. of Rock Mech. and Min. Sci. Geomechanical abstracts. 24–1, pp. 53–73. Reproduced by permission and courtesy from Elsevier.

**Figure 5.85-B**  Brady, B.H.G. and Brown, E.T. (1993). Rock mechanics for underground mining. 2nd ed. Kluwer Academic Publishers. Reproduced by permission and courtesy from Springer.

## Chapter 6

**Figure 6.4**  Hudson, J.A. (1989). Rock mechanics principles in engineering practice. Butterworths. CIRIA, London. Reproduced by permission from CIRIA.

**Figure 6.6, Figure 6.7, Figure 6.9 and Figure 6.13**  ISRM (1981). Rock characterization. Testing and monitoring. Int. Soc. for Rock Mechanics. Suggested methods. Brown, E.T. (Ed.). Commission on testing and monitoring, ISRM. Pergamon Press. Reproduced by permission from Elsevier.

**Table 6.2**  BSI (1999). BS5930: 1999. Code of practice for site investigations. British Standard Institution. London. Reproduced by permission and courtesy from British Standard Institution.

**Table 6.13**  BSI (2003). BS IN ISO 14689-1. Geotechnical investigation and testing. Identification and classification of rock. British Standard Institution. London. Reproduced by permission and courtesy from British Standard Institution.

## Chapter 7

**Figure 7.6**  Proske, H., Vicko, J., Rosenbaum, M.S., Dorn, M., Culshaw, M. and Marker, B. (2005). Special purpose mapping for waste disposal sites. Report of IAEG Commission 1: Engineering Geological Maps. Bull. of Engineering Geology and the Environment vol. 64, no 1, pp. 1–54. Reproduced by permission and courtesy from Springer.

## Chapter 8

**Figure 8.18**  Tomlinson, M.J. (2001). Foundation design and construction. Prentice Hall. 7th ed. Harlow. Essex. Reproduced by permission and courtesy from A.A. Balkema.

**Figure 8.21**  Burland, J.B., Broms, B. and de Mello, V.F.B. (1977). Behaviour of foundations and structures. State of the art report. Session 2. Proc. 9th ICSMFE. Tokyo. vol. 2, pp. 495–546. Reproduced by permission and courtesy from Institution of Civil Engineers.

**Figure 8.31**  Hansen J.B. (1970). A revised extended formula for bearing capacity. Danish Geotechnical Institute Bulleting, n° 28. Reproduced by permission and courtesy from Danish Geotechnical Institute.

**Figure 8.38 and Figure 8.39**  Serrano, A. and Olalla, C. (1996). Allowable bearing capacity in rock foundations based on a non linear criterion. Int. Jl. Rock Mech. and Min. Sci. vol. 33, 4, pp. 327–345. Reproduced by permission and courtesy from Elsevier

## Chapter 9

**Figure 9.10**  Lumb, P. (1975). Slope failures in Hong Kong. Ql. Jr. Engineering Geology, n° 8, pp. 31–65. Reproduced by permission from the Geological Society of London.

**Figure 9.14, Figure 9.15. Figure 9.39, Figure 9.40, Figure 9.42, Figure 9.70, Figure in Box 9.3 and Figure in Box 9.5**  Hoek, E. and Bray, J.W. (1981). Rock slope engineering. 3rd ed. The Institution of Mining and Metallurgy, London. Reproduced by permission from The Institute of Materials, Minerals and Mining.

**Figure 9.31, Figure 9.32 and Figure 9.33** Jiménez Salas, J.A., De Justo, J.L. and Serrano. A.A. (1976). Geotecnia y cimientos, II. Editorial Rueda. Madrid. Reproduced by permission and courtesy from Editorial Rueda.

**Figure 9.53 and Figure 9.65** CANMET (Canada Center for Mineral and Energy Technology). (1977). The pit slope manual. Ministry of Supply and Services, Canada. Reproduced by permission and courtesy from Canada Center for Mineral and Energy Technology.

## Chapter 10

**Figure 10.8** Hansen, L. and Martna, J. (1988). Influence of faulting on rock excavation. Inter. Symp. on Rock Mechanics and Power Plants. ISRM, Madrid, vol. 1, 317–324. Balkema. Reproduced by permission and courtesy from A.A. Balkema.

**Figure 10.9** Heidbach, O., Tingay, M., Barth, A., Reinecker, J., Kurfeβ, D. and Müller, B. (2008). The release 2008 of the World Stress Map (www.world-stress-map.org). Reproduced by permission and courtesy from the World Stress Map Project.

**Figure 10.11** Bieniawski, Z.T. (1989). Engineering Rock Mass Classifications. John Wiley & Sons Ed. Reproduced by permission and courtesy from John Wiley & Sons Ltd.

**Figure 10.12** Barton, N. (2000). TBM tunnelling in jointed and faulted rock. Balkema, Rotterdam. Reproduced by permission and courtesy from A.A. Balkema.

## Chapter 11

**Figure 11.23** Attewell, P.B. and Farmer, I.N. (1976). Principles of engineering geology. Chapman and Hall, London. Reproduced by permission and courtesy from Kluwer Academic Publishers.

**Figure 11.27 and Figure 11.28** Wahlstrom, E.E. (1974). Dams, dam foundation and reservoirs. Elsevier, Amsterdam. Reproduced by permission and courtesy from Elsevier.

**Figure 11.29** Wittke, W. (1990). Rock Mechanics. Theory and Applications with Case Histories. Reproduced by permission and courtesy from Springer.

**Figure 11.30 and Figure 11.35** Wyllie, D.C. (1999). Foundations on rock. 2nd ed. E.F.N. Spon. New York. Reproduced by permission and courtesy from Taylor and Francis Group.

**Figure 11.36** Simpson, D.W. (1986). Triggered earthquakes. Ann. Rev. Earth Planet. Sci., 14, pp. 21–42. Reproduced by permission from Annual Review.

**Figure in Box. 11.13** Olalla, C. and Cuellar, V. (2001). Failure Mechanism of the Aznalcóllar Dam, Seville, Spain. Geotechnique, vol. 51, no 5, pp. 399–406. Reproduced by permission and courtesy from the Institution of Civil Engineers.

## Chapter 12

**Figure 12.16** Ladd, C.C. and Foott, R. (1977). Foundation design of embankments on varved clays. U.S. Department of Transportation. FHWA - Report No. TS-77-214. Washington. Reproduced by permission and courtesy from U.S. Department of Transportation.

## Chapter 13

**Figure 13.3** Varnes, D.J. (1988). Slope movement types and processes. In: Landslides. Analysis and control. Schuster and Krizek (Eds). Transportation Research Board. Special Report 176. 5th printing. National Academy of Sciences. U.S.A. Chapter 2. Reproduced by permission and courtesy from National Academy of Sciences

**Figure 13.25** Dikau, R., Brundsen, D., Schrott, L. and Ibsen, M.L. (1996). Introduction. In: Landslide recognition. Identification, movement and causes. Pp. 1–12. Dikau, Brundsen, Schrott and Ibsen (Eds). Reproduced by permission and courtesy from John Wiley & Sons Ltd.

## Chapter 14

**Figure 14.3** Sibson, R.H. (1983). Continental fault structure and shallow earthquake source. Jl. Geol. Soc. London, 140: 747–767. Reproduced by permission and courtesy from the Geological Society of London. Scholz, C.H. (1990). The mechanics of earthquakes and faulting. Cambridge University Press. Reproduced by permission and courtesy from Cambridge University Press.

**Figure 14.5** Keller, E.A. and Pinter, N. (1996). Active Tectonics. Earthquakes, uplift and landscape. Prentice Hall. Reproduced by permission and courtesy from A.A. Balkema.

**Figure 14.13** Giardini, D., Jiménez, M.J. and Grünthal, G. (Eds). (2003). The ESC-SESAME European-Mediterranean Seismic Hazard Map, scale 1:5,000,000. ETH-CSIC-GFZ, Zurich. Reproduced by permission and courtesy from European Seismological Commission.

**Figure 14.15** Dowrick, D.J. (2000). Earthquake resistant design. London, 2nd ed. John Wiley & Sons Ed. Reproduced by permission and courtesy from John Wiley & Sons Ltd.

**Figure 14.16** Seed, H.B., Ugas, C. and Lysmer, J. (1974). Site dependent spectra for earthquake resistant design. Rep. EERC 74-12. University of California at Berkeley. Reproduced by permission and courtesy from University of California at Berkeley.

**Figure 14.18** Ho, C.L. and Kavazanjian, E. (1986). Probabilistic study of SPT liquefaction analysis. Proc. In Situ'86, ASCE Conference on Use of In-Situ Test in Geotechnical Engineering, Blacksburg, Virginia, ASCE Geotechnical Special Pub. No. 6, 602–616. Reproduced by permission from American Society of Civil Engineers.

**Figure 14.19** Youd, T.L. and Idriss, I.M. (2001). Liquefaction resistance of soils: summary report from the 1996 NCEER and 1988 NCEER/NSF Workshops on Evaluation of Liquefaction Resistance of Soils. Vol. 127, n° 4, pp. 297–313, April 2001. ASCE. Reproduced by permission from American Society of Civil Engineers.

**Figure 14.20** Hays, W. (1990). Earthquake vulnerability. Cooperative Project for Seismic Risk Reduction in the Mediterranean region. UNDP/OPS/UNDRO, Triestre. Reproduced by permission and courtesy from United Nations Development Programme.

**Figure 14.21** Wang, J.G.Z.Q. and Law, K.I. (1994). Sitting in earthquake zones. Balkema. Reproduced by permission and courtesy from A.A. Balkema.

**Figure 14.23** Coburn, A.W., and Spence, R.J. (1992). Earthquake protection. New York. John Wiley & Sons Ed. Reproduced by permission and courtesy from John Wiley & Sons Ltd.

**Figure in Box. 14.1** Obermeier S.F. (1996). Use of liquefaction induced features for paleoseismic analysis. Engineering Geology, 44, 1–76. Reproduced by permission and courtesy from Elsevier.

## Chapter 15

**Table 15.5, Table 15.6 and Table 15.7** Hoek, E. (1991). When is a design in rock engineering acceptable?. Proc. 7th Int. Conf. on Rock Mechanics. ISRM. Aachen, Germany. Vol. 3, pp. 1485–1497. Reproduced by permission and courtesy from A.A. Balkema.

## Appendix A

**Charts for Circular and Wedge Failure Analysis** Hoek, E. and Bray, J.W. (1981). Rock slope engineering. 3rd ed. The Institution of Mining and Metallurgy, London. Reproduced by permission from The Institute of Materials, Minerals and Mining.

# INDEX

## A
accelerogram 611
acceptability criteria 631
active fault 532, 596, 606
activity 26, 89
adits 512
aerial photo interpretation 269
aerial photographs 571
aggregates 517
aggressive soils 101
allowable load capacity 393
alluvial deposits 95
amplifying effect 616
anchored walls 441
anchors 440
angle of friction 77, 78, 148, 310
angle of internal friction 147
anhydrite 103, 530
aperture 341
apparent resistivity 276
aquicludes 224
aquifers 224, 225
aquifuges 224
aquitards 224
arch dams 505, 515
artificial tunnels 584
attenuation relationships 605
auger drilling 291
Aznalcóllar Dam 14, 524

## B
back analysis 415, 435
ballast coefficient 312
Barton and Choubey's criterion 310
basaltic rock 129
bearing capacity 372
bearing capacity factors 373
Benamejí landslide 579
bench excavation 489

Bernold's method 493
Bernoulli's Theorem 29, 227
biological contaminants 257
Bishop's method 421, 425
block samples 296
block size 344
block size index 345
block toppling 428
bolts 440, 490
boom headers 487
bored piles 386
borehole geophysics 286, 297
boreholes 510
Borros test 303
brittle behaviour 142, 145
brittle failure 140
buckling 414, 431
bulk density 27
buttress dams 505, 515

## C
californian drains 439
caliper logging 287, 319
capillarity 125
carbonate rocks 127
Casagrande cup 24
Casagrande plasticity chart 25
Casagrande soil classification 541
cation exchange capacity 88
caves 396
cavities 396, 531, 586
CBR compaction test 539
Cerchar test 484
chemical contaminants 257
chemical processes 256
chemical quality 255
circular failure 410, 431
clays 23, 78, 86
coarse rockfill 536, 545

coarse-grained soils 24
coefficient of consolidation 71
coefficient of earth pressure at rest 62, 64
coefficient of permeability 30, 31, 33, 129, 233, 248
coefficient of uniformity 24
coefficient of volume compressibility 68
cofferdams 506
cohesion 147, 148
cohesive soils 85
collapse 141, 585, 590
collapse potential 105
collapse tests 101, 105
collapsible soils 101, 105, 396
collimation 320
colluvial deposits 95
compaction 538
compaction control 545
competence factor 459
compressibility 56
compressibility index 384
compression failure 141
compression index 61
concrete dams 313, 504, 513, 524
conductivity sonde 287
cone penetration test 303
confined aquifers 226
confining pressure 161
consistency 24, 28, 275
consolidation 50, 56
consolidation settlement 382
consolidation time 69
constant head test 316
contaminant 256
contamination indices 257
contamination of groundwater 257
controlling earthquake 606
core samplers 294
cores 516
creep 143, 561
critical hydraulic gradient 48
cross-hole 288
crown bars 490
curtain dams 504
curtains 529
cutoffs 519
cyclic shear test 612
cyclic triaxial test 612

# D

dam 364
dam materials 516

Darcy's law 31, 33, 233
dating methods 602
debris avalanches 562
debris flows 560
deformability 150, 187
degree of consolidation 71
degree of saturation 28
degree of weathering 123
density 275
density index 28
depth correction factor 374
design earthquake 610
deterministic methods 415, 592, 606
deviator stress 82
diaphragm walls 441, 519
dilatometer test 312, 512
dilatometric deformation modulus 312
dilution processes 257
direct shear test 72, 175
direct tensile test 162
direct tension failure 141
direction cosines 133
discharge 233, 251, 253
discharge velocity 31
discontinuities 115, 165, 335
discontinuity frequency 345
discontinuity strength 170, 175, 340
dispersion 91
dispersion index 104
dispersive soils 101, 103
displacement piles 386
disturbed samples 294
double curved arch dams 505
down-hole 288
drainage 436, 491
drainage walls 439
drained conditions 52
drains 517, 520
drill and blast 487
drilling records 297
dry density 27
ductile behaviour 142, 143, 145
ductile failure 140
durability 120
dynamic barriers 584
dynamic deformation moduli 316
dynamic loads 409

# E

earthquake accelerogram 610
earthquake sources 596

earthquakes 568
eccentricity 378
effective permeability 232, 233
effective porosity 119, 235, 250
effective stress 42, 52
efficiency factor 391
El Berrinche landslide 7
elastic behaviour 143
elastic constants 153, 156, 157
electric charge 88
electric sounding 277
electrical methods 276
electrical resistivity 287
electrical tomography 277
electromagnetic methods 282
electro-welded mesh 490
elevation head 29
embankment dams 504, 516, 524
engineering geological maps 352
engineering geology 4
equipotential lines 34, 229, 252
erosion 519, 567
evaporitic materials 586
excavability 483
excavation support ratio 483
expansive soils 103, 394
experimental embankments 548
extensometers 320, 578

## F

factor of safety 416, 426, 435, 634
failure criteria 72, 140, 149, 181, 186
failure mechanisms 140
failure surfaces 410
fall cone test 25
fault breccia 597
fault rupture 600, 616
faults 458, 530
fill material 342
filters 517, 520
fine soils 85
fine-grained soils 24, 85
flat jack test 189, 210, 313
fleximeter 319
flexure failure 140
flocculation 91
floods 566
flow lines 34, 230, 252
flow net 34–37, 251, 252, 407
flow velocity 235
flows 560

footings 371
footprint test 540, 548
forepoling 490
fracture 140
free swell test 103
freeze-thaw processes 567
friction pile 387
frictional soils 85
Funcho Dam 527

## G

gabion walls 441
gamma-gamma sonde 287
geodesic methods 319
geo-engineering 5
geoenvironmental effects 515
geographical information systems 360, 591
geological engineering 4, 5
geological hazards 6, 626
geological maps 352
geological model 15
geological risks 626
geological structure 457
geomechanical classifications 472
geomechanical model 15
geomechanical parameters 467
geomechanical zoning 510
geomorphological conditions 357
geomorphological mapping 591
geophysical methods 188, 275, 511, 578
georadar 284, 396
geotechnical classification 354
geotechnical engineering 4
geotechnical instrumentation 319
geotechnical parameters 354
geotechnical report 399
geotechnical units 354
geotechnical zoning 358
glacial deposits 96
granular soils 76, 85
gravel columns 396
gravels 23
gravimetric geophysical methods 285, 396
gravity dams 504, 515
ground improvement 491
ground penetration radar 284
ground reaction curve 485
ground response spectrum 610
ground retention capacity 257
grouting 491, 521, 528
grouting tests 512

GSI 184
gunite 489
gypsum 396, 530

## H

halloysite 86
hardness 332
Harvard compaction test 539
hazard 627
hazard maps 591, 592
Hoek and Bray charts 423
Hoek's cell 176, 310
Hoek-Brown's criterion 150, 181, 468
hurricane Mitch 7
hydration 125
hydraulic conductivity 232, 233
hydraulic fracturing test 210, 214
hydraulic gradient 31
hydraulic hammers 487
hydraulic head 29, 227, 228
hydroelectric schemes 507
hydrofracture tests 210, 466
hydrogeological data 357
hydrogeological parameters 238
hydrolysis 126
hydrostatic conditions 29
hydrostatic pressure 29, 30
hydrostatic stress 139

## I

igneous rocks 127, 466
illites 86
immediate settlement 382
impervious blankets 521
*in situ* stress 201, 205, 213, 316, 408, 461, 466, 469
inclinometers 320, 446, 578, 581
indirect tensile or Brazil test 163
induced seismicity 532, 533
infinite slope 417
influence coefficient 380
injection tests 248
intact rock 115, 118, 147
intact rock deformability 150
intact rock strength 147, 459, 467
interlaminar layer 88
internal erosion 46
intrinsic permeability 233
inventory maps 571, 639
inverted arch 489
ionic content 255

ionic interchange 256
isomorphic substitution 88

## J

Jacob's method 244, 245
jet-grouting 439, 491
joint roughness coefficient 310

## K

kaolinite 86
karstic 225
karstic cavities 396
karstic materials 225, 586, 587
Kocaeli earthquake 621

## L

lacustrine deposits 95
Lambe's test 103
landfills 397
landslide hazard 409
landslide or landslides 7
landslides 556, 557, 615
lateral spreading 562
laterites 99
leakage 521, 528
Lefranc test 316
levelling 319
limit equilibrium methods 415, 581
lining 453, 489
liquefaction 106, 568, 603, 613–615
liquefaction potential 613
liquid limit 24
liquid urban waste 257
liquidity index 28
load cells 324
loess 106
Los Angeles abrasion test 540
Lugeon test 249, 317, 521
Lugeon unit 318, 521

## M

magnetic methods 285
Malpasset Dam 502, 525
man-made fills 105
mapping scales 8
marcasite 530
mass movements maps 591
mechanical excavation 487
mechanical pre-cutting 493
metamorphic rocks 129, 470

microcracks 162
microfabric 89, 92
micropile walls 439
micro-piles 395
microzonation maps 617
mining subsidence 587
mitigation 638
Mohr's circle 135
Mohr-Coulomb criterion 72, 146, 149, 185
molecular adsorption 256
moment magnitude 599, 605
monitoring 443, 497, 578, 581
montmorillonite 87
moraine debris 97

## N

natural gamma sonde 287
negative friction 391
neutron-gamma sonde 287
neutron-neutron sonde 287
New Austrian Tunnelling Method 485, 491
normal stress 134, 137
normally consolidated clays 390

## O

oedometer method 383
oedometer test 65
oedometric modulus 68
outlets 507
over-consolidated clays 390
overconsolidation 58
overconsolidation ratio 58
oxidation 126
oxidation-reduction 256

## P

*packer test* 317
paleoseismic analysis 602
partially saturated soil 40
particle size distribution 23
peak strength 140, 146
penetration resistance 301
percussion drilling 292
permafrost 101, 106
permeability 39, 119, 129, 130, 177, 193
permeability tests 316, 512
phreatic surface 229
physical processes 256
piezocone 303
piezometer 30, 48, *322*, 512, 581

piezometric level or head 29, 31, 227–229
pile 385
pile walls 439
pinhole test 540
piping 44, 46, 48
piston samplers 295
plane failure 411, 426
plastic limit 24
plasticity criteria 146
plasticity index 25
plate loading test 311–314, 548
point load test 306, 309, 332
Poisson's ratio 143, 152, 157
pore pressure coefficient 84
pore water pressure 42, 136, 149, 161, 322, 406
porosity 26, 119, 230
portals 464, 492
position head 227
precast piles 386
preconsolidation pressure 58
pressiometric deformation modulus 311
pressure cells 324
pressure head 29, 227
pressuremeter test 311
prevention 589, 626
primary permeability 129
primary support 489
principal stresses 135
probabilistic methods 415, 592, 606, 608
probability 627
probing tests 302
Proctor compaction test 538, 539, 546
pseudostatic analysis 616
pumping tests 238
pyrite 530

## Q

Q index 483
Q System 472, 473
quick clays 93, 106

## R

radar interferometry 604
radioactive contaminants 257
radioactive sondes 287
radioactive tracers 249
raft foundation 371
reduction 126
relative density 28, 77, 302
remote sensing 271
residual soils 20, 114, 125, 550

residual strength 79, 140, 305
resistivity 276
resonant column test 612
return period 608, 627
Reynolds number 234
rippability 447
ripraps 517
risk 628
risk maps 640
RMR classification 216–219, 481
road headers 487
rock 112
rock avalanches 557, 562
rock burst 464, 496
rock classification 122
rock cuttability 484
rock deformability 150
rock falls 443, 561, 583
rock load 483
rock mass 110, 113, 115
rock mass classification 124, 215, 480
rock mass deformability 187, 192, 469
rock mass strength 179, 186, 468
rock slopes 426
rock strength 147, 308
rockfill dams 504
rockfill embankments 536, 541
rockfills 517
roller-compacted concrete dams 506
rotary drilling 289
roughness 171–175, 338
RQD 124, 190, 216, 298, 346

## S

saline materials 98, 104, 396
sands 23
saprolites 127
saturated soils 40, 52
scale effects 174, 195
Schimazek index 483
Schmidt hammer test 305, 307, 334
sclerometer 306
secondary permeability 129
secondary support 489
sedimentary rocks 466
sedimentation process 57
Seed and Idriss method 613
seepage 47, 252, 343, 515
seepage control 521
seepage forces 44, 46
seepage velocity 33

seismic cycle 598
seismic hazard 605
seismic hazard analysis 606
seismic hazard map 610
seismic information 606
seismic prevention 617
seismic refraction 277, 278, 281
seismic risk 607, 627
seismic tomography 288
seismogenic sources 606
seismotectonic province 606
sensitive soils 101, 106
sensitivity 93
settlement 382, 548
shape correction factor 374
shear strength 72, 147, 408
shear strength of discontinuities 170, 175, 468
shear strength of soils 79
shear strength test 72, 308
shear stress 134, 137
shear stress failure 140
Shelby tubes 295
Sheorey's method 464, 471
shield 493
shotcrete 442, 489
shrinkage 102
shrinkage limit 24
silts 23
sinkholes 586
sinking piles 395
slake durability test 120
slip rates 599
slope mass rating 433
SMR 433
soft soils 397, 548
soil 21, 112
Soil Classification System 26
soil compression 41
soil sensitivity 305
soil structure 41
soil swelling 41
solid urban waste 258
solifluction 561
soluble materials 249, 519, 586
sonic velocity 152, 164
spacing 336
specific gravity 27
specific risk 628
specific surface area 88
spillways 506
spontaneous potential 287
SPT 77, 613